Introduction to Molecular Biology, Genomics and Proteomics for Biomedical Engineers

Biomedical Engineering Series

Edited by Michael R. Neuman

Published Titles

Electromagnetic Analysis and Design in Magnetic Resonance Imaging, Jianming Jin

Endogenous and Exogenous Regulation and Control of Physiological Systems, Robert B. Northrop

Artificial Neural Networks in Cancer Diagnosis, Prognosis, and Treatment, Raouf N.G. Naguib and Gajanan V. Sherbet

Medical Image Registration, Joseph V. Hajnal, Derek Hill, and David J. Hawkes

Introduction to Dynamic Modeling of Neuro-Sensory Systems, Robert B. Northrop

Noninvasive Instrumentation and Measurement in Medical Diagnosis, Robert B. Northrop

Handbook of Neuroprosthetic Methods, Warren E. Finn and Peter G. LoPresti

Signals and Systems Analysis in Biomedical Engineering, Robert B. Northrop

Angiography and Plaque Imaging: Advanced Segmentation Techniques, Jasjit S. Suri and Swamy Laxminarayan

Analysis and Application of Analog Electronic Circuits to Biomedical Instrumentation, Robert B. Northrop

Biomedical Image Analysis, Rangaraj M. Rangayyan

An Introduction to Biomaterials, Scott A. Guelcher and Jeffrey O. Hollinger

Foot and Ankle Motion Analysis: Clinical Treatment and Technology, Gerald F. Harris, Peter A. Smith, Richard M. Marks

Introduction to Molecular Biology, Genomics and Proteomics for Biomedical Engineers, Robert B. Northrop and Anne N. Connor

Introduction to Molecular Biology, Genomics and Proteomics for Biomedical Engineers

Robert B. Northrop and Anne N. Connor

CRC Press is an imprint of the
Taylor & Francis Group, an informa business

CRC Press
Taylor & Francis Group
6000 Broken Sound Parkway NW, Suite 300
Boca Raton, FL 33487-2742

First issued in paperback 2019

ISBN-13: 978-1-4200-6119-2 (hbk)
ISBN-13: 978-0-367-38655-9 (pbk)

Library of Congress Cataloging-in-Publication Data

Northrop, Robert B.
Introduction to molecular biology, genomics and proteomics for biomedical engineers / authors, Robert B. Northrop, Anne N. Connor.
p. ; cm. -- (Biomedical engineering series)
"A CRC title."
Includes bibliographical references and index.
ISBN 978-1-4200-6119-2 (hardcover : alk. paper)
1. Molecular biology. 2. Genomics. 3. Proteomics. 4. Biomedical engineers. I. Connor, Anne N. II. Title. III. Series: Biomedical engineering series (Boca Raton, Fla.)
[DNLM: 1. Biomedical Engineering. 2. Genomics. 3. Molecular Biology. 4. Proteomics. QT 36 N877i 2009]

QH506.N665 2009
572.8--dc22 2008038134

Visit the Taylor & Francis Web site at
http://www.taylorandfrancis.com

and the CRC Press Web site at
http://www.crcpress.com

Dedication

We dedicate this text to our spouses, Adelaide and Michael.

Contents

Preface xv
Authors xix

Chapter 1 Introduction 1

1.1 Scope of the Text 1
1.2 Some Definitions 2
  1.2.1 Molecular Biology 2
  1.2.2 Genomics 2
  1.2.3 Proteomics 3
  1.2.4 Bioinformatics 3
  1.2.5 Genetic Engineering and Biotechnology 4
  1.2.6 Systems Biology 5
1.3 Complex Systems 5
  1.3.1 Introduction 5
  1.3.2 Properties of Nonlinear Systems 7
  1.3.3 Parametric Control 8
  1.3.4 Modeling Complex Systems 9
1.4 Optimum Use of Reference Resources 9
1.5 Summary 10
Chapter 1 Home Problems 10

Chapter 2 Life and Death 13

2.1 What Is Life? 13
  2.1.1 Introduction 13
  2.1.2 Properties of Life 14
2.2 The Domains and Kingdoms of Life: Their Origins 16
  2.2.1 Introduction 16
  2.2.2 The Domains of Life 19
2.3 Nonliving, Self-Replicating Genomic Machines 20
  2.3.1 Introduction 20
  2.3.2 Viruses 20
  2.3.3 Viroids 23
  2.3.4 Plasmids 24
  2.3.5 Satellites and Virusoids 24
2.4 The Origins of Life 25
  2.4.1 Introduction 25
    2.4.1.1 Carbon Dating 26
  2.4.2 The First Life 29
  2.4.3 Discussion 34
2.5 Cell and Animal Death 34
  2.5.1 Introduction 34
  2.5.2 Cell Death by Apoptosis (Programmed Cell Death) 35
  2.5.3 Cell Necroptosis 40
  2.5.4 Metazoan Death: Failure of Critical Organ Systems 41

2.6 Summary ... 43
Chapter 2 Home Problems ... 44

Chapter 3 Review of Basic Cell Anatomy and Physiology ... 45

3.1 Introduction ... 45
3.2 The Cell Membrane ... 45
3.3 Inside the Cell Membrane ... 50
3.4 Prokaryotic Cells ... 50
3.4.1 Introduction ... 50
3.4.2 Archaeans ... 51
3.4.3 Eubacteria ... 51
3.5 Eukaryotic Cells ... 53
3.5.1 Introduction ... 53
3.5.2 Single-Cell Eukaryotes (Protists) ... 54
3.5.3 Multicelled Eukaryotes ... 56
3.6 Anatomical Components of Eukaryotic Cells ... 56
3.6.1 The Cytosol ... 56
3.6.2 The Nucleus and Nucleolus ... 57
3.6.3 Ribosomes ... 59
3.6.4 Endoplasmic Reticulum ... 59
3.6.5 Golgi Apparatus ... 60
3.6.6 Lysosomes ... 61
3.6.7 Peroxisomes ... 61
3.6.8 Mitochondria ... 61
3.7 The Roles of Transmembrane Proteins ... 63
3.7.1 Cell Adhesion Molecules ... 63
3.7.2 Transport of Substances Across the Cell Membrane ... 65
3.7.2.1 Diffusion and Osmosis ... 65
3.7.2.2 Aquaporins ... 66
3.7.2.3 Ion and Molecular Pumps ... 68
3.7.2.4 Regulated Ion Channels in Membranes ... 69
3.7.2.5 The Basis for the DC Transmembrane Potential ... 73
3.7.2.6 Cell Signaling by Changes in Transmembrane Potential ... 76
3.7.3 Electrical Synapses: Gap Junctions ... 78
3.7.4 Chemical Synapses ... 80
3.7.4.1 Introduction ... 80
3.7.4.2 Neural Transmembrane Chemical Signal Receptors ... 83
3.7.4.3 G-Protein-Coupled Receptors ... 83
3.7.4.4 Summary ... 84
3.8 Review of Cell Reproduction ... 84
3.8.1 The Cell Cycle: Mitosis ... 84
3.8.2 Meiosis ... 86
3.9 Summary ... 90
Chapter 3 Home Problems ... 92

Chapter 4 Introduction to Physical Biochemistry and Biochemical Systems Modeling ... 95

4.1 Introduction to Chemical Reactions ... 95
4.1.1 Introduction ... 95
4.1.2 Chemical Bond Energy ... 95

4.1.3 Types of Chemical Bonds 97
4.1.4 Chemical Kinetics and Mass Action 98
4.2 Important Coupled Reactions 101
4.2.1 Introduction 101
4.2.2 Glycolysis 102
4.2.3 The Citric Acid Cycle 104
4.3 Other Metabolic Pathways 106
4.3.1 The Chemistry of Organic Redox Reactions 106
4.3.2 Review of Carbohydrate Metabolism 109
4.3.3 Review of Lipid Metabolism 112
4.3.4 Review of Lipid Catabolism 114
4.3.5 Review of Amino Acid Metabolism 115
4.3.6 Review of Amino Acid Catabolism 116
4.4 Photosynthesis 118
4.4.1 Introduction 118
4.4.2 The Pigments of Photosynthesis 118
4.4.3 Structure of the Chloroplast 121
4.4.4 Chemical Reactions and Energy in Photosynthesis 123
4.4.5 The Calvin Cycle 124
4.4.6 The Light Reactions 124
4.5 The Challenge of Modeling Complex Biochemical Systems 126
4.5.1 Introduction 126
4.5.2 Generalized Approaches to Modeling Biochemical Systems 127
4.5.3 Models for Biochemical Oscillators 128
4.5.4 The Hodgkin–Huxley Model (1952) for Action Potential Generation 136
4.6 Some Simulation Languages for Biochemical Systems 143
4.7 Summary 150
Chapter 4 Home Problems 150

Chapter 5 The Basis of Genetic Inheritance 155

5.1 Introduction 155
5.2 Mendelian Inheritance 156
5.2.1 Gene Swapping during Reproduction and Genetic Variation 157
5.3 Epigenetic Inheritance 158
5.4 Genetic Imprinting 159
5.5 Nuclear and Mitochondrial DNA 159
5.6 Genes and the Genome: The Central Dogma 160
5.6.1 Introns and Exons 161
5.6.2 Protein Synthesis 164
5.6.2.1 Transcription 165
5.6.2.2 Translation 169
5.7 The Control of Development 170
5.8 Summary 172
Chapter 5 Home Problems 173

Chapter 6 Nucleic Acids and Their Functions 177

6.1 Introduction 177
6.1.1 DNA: The Command and Control Nucleic Acid 177
6.1.2 RNA: The Executive Nucleic Acid 179

6.2 RNAs Coded by DNA 181
6.2.1 Introduction 181
6.2.2 Messenger RNAs 182
6.2.3 Transfer RNAs 182
6.2.4 Ribosomal RNAs 183
6.2.5 Small Nuclear RNAs 184
6.2.6 Small Nucleolar RNAs 185
6.2.7 Short Interfering RNAs (siRNAs) 185
6.2.8 Micro-RNAs (miRNAs) 186
6.2.9 Ribozymes 187
6.2.10 XIST RNA 188
6.2.11 Discussion 188
6.3 DNA Repair 189
6.3.1 Introduction 189
6.3.2 DNA Repair Mechanisms 189
6.3.3 Discussion 191
6.4 Gene Regulation 191
6.4.1 Introduction 191
6.4.2 Gene Regulation in Prokaryotes 192
6.4.3 Some Molecular Factors in Gene Expression in Eukaryotes 193
6.4.4 Epigenetic Gene Regulation 194
6.5 Summary 195
Chapter 6 Home Problems 195

Chapter 7 A Review of Proteins 197

7.1 Introduction 197
7.1.1 Amino Acids and the Peptide Linkage 197
7.1.2 Primary, Secondary, Tertiary, and Quaternary Structure of Proteins 199
7.2 Examples of Protein Functions 202
7.2.1 Introduction 202
7.2.2 Enzymes 203
7.2.2.1 The Physical Chemistry of Enzymes 203
7.2.2.2 The Regulation of Protein Enzymes 203
7.2.3 Protein Hormones and Intercellular Command, Communication, and Control (C^3) 204
7.2.3.1 Introduction 204
7.2.3.2 Protein Hormones 204
7.2.4 Structural Proteins 206
7.2.4.1 Introduction 206
7.2.4.2 Microfilaments 206
7.2.4.3 Intermediate Filaments 206
7.2.4.4 Microtubules 206
7.2.4.5 Cilia and Flagella 206
7.2.4.6 Muscle Fibers 208
7.2.4.7 Histone Proteins 215
7.2.5 Cell Membrane Receptors and Ion Pumps 216
7.2.5.1 Second Messenger Systems 216
7.2.5.2 Ion Pumps 217
7.2.5.3 Photon-Driven Proton Pumps 218

7.2.6 Proteins in the Immune System 219
7.2.6.1 Introduction 219
7.2.6.2 A Summary of the Cells of the Immune System 219
7.2.6.3 Complement 228
7.2.6.4 Antigen Presentation 229
7.2.6.5 Autacoids: Immunocytokines, Proteins, and Glycoproteins Secreted by Immune System Cells 232
7.2.6.6 Discussion 240
7.2.7 Lectins 241
7.2.8 Glycoproteins 242
7.2.9 Blood Proteins 243
7.2.9.1 Introduction 243
7.2.9.2 Hemoglobin 243
7.2.9.3 Serum Proteins and Antibodies 245
7.2.9.4 Hemocyanins in Invertebrate Circulatory Systems 247
7.2.10 Myoglobin 248
7.3 Errors in Protein Structure 248
7.3.1 Introduction 248
7.3.2 Nongenetic Causes of Protein Errors 249
7.3.3 Heat Shock (Stress) Proteins 251
7.3.4 When Good Proteins Go Bad: Prions 252
7.4 Posttranscription Regulation of Gene Expression 257
7.4.1 Introduction 257
7.4.2 Inteins and Protein Engineering 257
7.5 Protein Destruction and AA Recycling 259
7.5.1 Proteolytic Enzymes 259
7.5.2 The Proteosome 259
7.5.3 Proteolysis in Apoptosis and Cell Necrosis 261
7.5.4 Autophagy 261
7.6 Summary 265
Chapter 7 Home Problems 265

Chapter 8 The Genetic Basis for Certain Inheritable Diseases: Genomic Medicine 267

8.1 Introduction 267
8.2 Examples of Genetic Diseases 268
8.3 Genomic Medicine 270
8.3.1 Introduction 270
8.3.2 Gene Defects and Disease 273
8.3.3 Gene Therapy 274
8.3.4 GM Viruses Used to Fight Cancers 275
8.3.5 Recombinant Vaccines 276
8.4 Epigenetic Therapy for Cancer 278
8.4.1 Introduction 278
8.4.2 Epigenetics and Cancer 278
8.4.3 Epigenetic Therapies for Cancer 279
8.4.4 Some Other Diseases with Epigenetic Etiologies 279
8.5 Summary 281
Chapter 8 Home Problems 281

Chapter 9 Some Instrumental Methods Used in Genomics, Proteomics, and Forensic Science ... 283

9.1 Introduction ... 283
9.2 Biotin, Avidin, and Streptavidin: Applications in Genomics and Proteomics ... 283
9.2.1 Introduction ... 283
9.2.2 Biotin ... 283
9.2.3 Avidin ... 283
9.2.4 Streptavidin ... 284
9.2.5 Immobilization in Microarrays: Applications of Avidin–Biotin Systems ... 284
9.2.5.1 Nonspecific Immobilization ... 285
9.2.5.2 Specific Immobilization ... 285
9.3 The Polymerase Chain Reaction (PCR) and Its Applications in Genomics and Forensic Science ... 286
9.3.1 Introduction ... 286
9.3.2 The PCR Process ... 286
9.3.3 Direct DNA Tests for Genetic Diseases ... 289
9.3.4 The "454" System ... 290
9.4 Fluorescent Molecular Tags ... 291
9.4.1 Introduction to Fluorescence ... 291
9.4.2 Fluorescent In Situ Hybridization (FISH) ... 293
9.4.3 Fluorescence Resonance Energy Transfer (FRET) ... 294
9.4.4 Phase Fluorometry and FLIM-FRET ... 295
9.5 Introduction to Forensic Applications of DNA Restriction Enzymes and STRs ... 298
9.5.1 Introduction ... 298
9.5.2 Restriction Fragment Length Polymorphism (RLFP) Analysis ... 298
9.5.3 Short Tandem Repeat Markers ... 299
9.5.4 Validity of STR DNA Evidence ... 301
9.6 Rolling-Circle Replication of ssDNA Circles ... 302
9.6.1 Introduction ... 302
9.6.2 RCA Protocols ... 302
9.6.3 The JGI RCA Protocol ... 303
9.6.4 Padlock Probes in RCA ... 304
9.7 Methods for the Analysis of Protein Structure ... 306
9.7.1 Introduction ... 306
9.7.2 Electrophoretic Separation of Proteins ... 306
9.7.3 Sequencing Protein AAs ... 307
9.7.4 ELISA Tests for Specific Proteins and Antibodies ... 307
9.7.5 Protein Microarrays ... 310
9.7.5.1 Introduction ... 310
9.7.5.2 The First Large-Scale Protein Microarrays ... 311
9.7.5.3 Antibody Microarrays ... 311
9.7.5.4 Improvements in Fluorescent Microarrays ... 312
9.7.6 The Immuno-PCR Method ... 314
9.7.7 Bio–Bar Codes ... 316
9.7.8 Structural Analysis of Protein "Crystals" Using X-Ray Crystallography ... 317
9.7.8.1 Introduction ... 317
9.7.8.2 Protein Crystals ... 318
9.7.8.3 X-Ray Crystallography ... 320
9.7.9 Cryoelectron Microscopy ... 324

9.8 Summary 326
Chapter 9 Home Problems 327

Chapter 10 Applications of Genomics 329

10.1 Introduction 329
10.2 Genetically Modified Organisms 329
10.2.1 Introduction 329
10.2.2 Examples of Transgenic Plants 330
10.2.2.1 The FlavrSavr™ Tomato 330
10.2.2.2 Golden Rice 332
10.2.2.3 Transgenic Maize 332
10.2.2.4 Genetically Modified Soybeans 333
10.2.2.5 Genetically Modified Cotton 333
10.2.2.6 Transgenic Arabidopsis for Detection of Explosives in the Soil 334
10.2.3 Transgenic Animals and Animal Cells 335
10.2.3.1 Introduction 335
10.2.3.2 Uses of TGAs and Cells 335
10.2.4 Recombinant DNA Technology 337
10.2.4.1 Introduction 337
10.2.4.2 rDNA Insertion Methods 337
10.2.4.3 Preparation of a Plasmid Vector 337
10.2.4.4 The Use of Phages as cDNA Vectors 338
10.2.4.5 Other Viral Vectors in Gene Therapy 339
10.2.4.6 The Gene Gun 339
10.2.4.7 Electroporation and the Effects of Nanosecond Pulsed Electric Fields on Cells 341
10.2.4.8 Ultrasonic Poration 342
10.2.5 Problems with GMOs 342
10.3 Animal Reproductive Cloning 345
10.3.1 Introduction 345
10.3.2 Altered Nuclear Transfer 345
10.3.3 The Roslin Technique 346
10.3.4 The Honolulu Technique 346
10.3.5 Problems in Cloning Mammals 347
10.3.6 Therapeutic Cloning 347
10.4 Stem Cells 347
10.4.1 Introduction 347
10.4.2 Types of Stem Cells 348
10.4.3 Sources of Stem Cells 349
10.4.4 Stem Cell Applications in Regenerative Medicine 353
10.4.5 Cancer Stem Cells 354
10.4.5.1 Introduction 354
10.4.5.2 New Pathways for Cancer Therapy 355
10.4.5.3 Summary 359
10.5 Parthenogenesis 359
10.5.1 Introduction 359
10.5.2 Artificial Parthenogenesis 359
10.5.3 Human Parthenotes? 360
10.6 Chimeras 362
10.6.1 Introduction 362

10.6.2 Cellular Chimeras 362
10.6.3 Somatic Chimeras 363
10.7 Summary 364
Chapter 10 Home Problems 365

Chapter 11 Ethical Issues in Genetic Engineering 367

11.1 Introduction 367
11.2 Framework for Ethical Decision Making 367
11.2.1 Defining Ethics 367
11.2.2 Ethical Frameworks 368
11.2.3 Other Resources 369
11.3 Genetically Modified Food Crops 369
11.3.1 Rogue Herbicide Resistance 369
11.3.2 Antibiotic-Resistant Bacteria 370
11.4 Chimeras 370
11.4.1 Transgenic Organ Farming 370
11.4.2 How Human Is Too Human? 371
11.4.3 Religious Concerns 371
11.4.4 Retrovirus Transfer 371
11.5 Human Embryonic Stem Cells 372
11.5.1 Religious Concerns 372
11.5.1.1 Alternative Sources of Stem Cells 373
11.5.2 Embryonic versus Adult Stem Cells 373
11.6 Summary 374
Chapter 11 Home Problems 375

Appendix 377

Bibliography and References 383

Glossary 407

Index 435

Preface

This introductory textbook is intended for use in a one-semester classroom course that has the purpose of acquainting biomedical engineers and interested scientists with molecular biology. Biomedical engineers in particular will find it valuable. It has been written to provide an introduction to the broad and complex fields of molecular biology, genomics, and proteomics for engineers and scientists who have not formally studied these disciplines, and who are working with, or planning to work with, living systems. It assumes that such readers already have had courses found in engineering curricula on college algebra, basic calculus, ordinary differential equations, and perhaps linear algebra.

Efficient extraction of information from this text and the understanding of its material also requires that the reader has had introductory college courses in chemistry, cell biology, or human physiology and anatomy, and is thus familiar with what is meant by such basic terms as amino acid, cell, cell membrane, chromosome, DNA, egg, gene, germ cell, meiosis, mitochondria, mitosis, nucleus, nucleolus, protein, ribosome, RNA, sperm, zygote, etc. This textbook elaborates on the molecular biological basis for these terms and illustrates man's progress in manipulating the genome at a molecular level. Knowledge of genomics and molecular biology will also shed light on evolutionary theories and future trends in genetic medicine and stem cell research. The reader will come to appreciate the incredible complexity of the biochemical systems required to sustain life in its many forms.

RATIONALE

It may be argued that the 20th century was the age of electronics. However, the 21st century has emerged as the age of molecular biology. Researchers have identified and manipulated the structures of, and reactions between, numerous biomolecules. Their motivation has been driven largely by the search for cures for diseases and medical conditions, and to allow us to feed the world's ever-growing human population. Molecular biology is a very different discipline from the "hard sciences" of physics and chemistry and certain engineering specialties, in that it is multidisciplinary, involving not only physics, chemistry, mathematics, and engineering, but a plethora of information about hundreds of thousands of protein molecules, nucleic acids, genes, and the regulatory mechanisms making cellular biochemical reaction pathways stable, and life possible. Molecular biologists must embrace complexity.

Engineers and scientists in the 21st century very often find themselves working on the interdisciplinary interface between living and inanimate systems. Whether their work involves modeling global warming, developing new medical imaging systems, devising instruments to sense tumors early in their growth, inventing new ways to introduce manmade genetic material into cells, or designing new systems to quickly identify bacterial and viral pathogens, it is important that engineers and scientists not initially trained in molecular biology, biochemistry, genomics, and proteomics have a working background in these areas to be able to communicate effectively and work with bioscientists.

For example, engineers trained in systems analysis and mathematical modeling have cooperated with molecular biologists to model the dynamics of biochemical pathways. However, this type of interdisciplinary interface is not without turbulence. Clearly, the more biology and biochemistry the engineers know, and the more mathematics the cell biologists know, the better will be their communication and the more fruitful their endeavors.

The very complex biochemical relations that permit life to exist are described and illustrated in this text. It is largely about the specialized organic molecules in living organisms and how they interact and react. However, it is a primer, not intended as an academic shortcut or a replacement for formal, in-depth course work in biochemistry, molecular biology, genomics, and cell biology. It is intended to broaden the background and pique the interest of undergraduate students in all branches of engineering; in particular, biomedical, electrical, and chemical engineers, some of whom may choose to enter the field of genetic engineering or become involved in the design of analytical instruments for use in molecular biology, genomics, proteomics, and even ecological systems. It will also help students in chemical engineering, electrical and computer engineering, and biomedical engineering appreciate the instrumental methods used to diagnose genetically based diseases.

The reader will gain knowledge about the awesomely complex chemistry of life and how it may be modeled and manipulated in genomic-based medicine and genetic engineering. Controversial topics such as stem cell research, cloning, parthenogenesis, and chimeras are described, and ethical problems in molecular biology are covered.

DESCRIPTION OF THE CHAPTERS

In Chapter 1, Introduction, key terms in biology are defined and discussed. Also considered are the properties of linear, nonlinear, and complex nonlinear systems and the challenge of modeling them. Living systems and biochemical systems in living organisms are seen to use feedback to effect self-regulation (homeostasis), and the dominant feedback mechanism in molecular biological systems is observed to be parametric control.

In Chapter 2, Life and Death, definitions and properties of life are considered. An overview of the domains and kingdoms used to classify living organisms is given. Viruses are considered to be nonliving, self-replicating genomic "machines." Also covered are theories for the origin of life on Earth. Cell death is described; the molecular biological systems for apoptosis and necroptosis are outlined. Death in metazoans (including human brain death) is viewed from the point of the failure of critical (modular) organ-systems.

In Chapter 3, Review of Basic Cell Anatomy and Physiology, we consider the cell as the basic unit of life. The distinction between Prokaryotes (bacteria) and Eukaryotes (single-celled, metazoan and plants) is presented. Subsystems in eukaryotic cells are described (structure and function), including cell membranes, transmembrane proteins and their various roles, the cytosol, and the roles of the diverse intracellular components of eukaryotes (e.g., nucleus, nucleolus, ribosomes, endoplasmic reticulum, Golgi apparatus, lysozomes peroxisomes, mitochondria, etc.). Also covered are the process of cell division and the cell cycle. The differences and similarities between mitosis and meiosis are described. Finally, intercellular signaling by changes in transmembrane potential are covered.

In Chapter 4, Introduction to Physical Biochemistry and Biochemical Systems Modeling, we review the basic physical chemistry of biochemical reactions. Types of chemical bonds, the role of catalysts and enzymes, and chemical kinetics based on mass action are described. Two ubiquitous, coupled reactions (glycolysis and the citric acid cycle) and their stoichiometry and energy balances are described in detail, followed by a section detailing other metabolic pathways (for carbohydrates, lipids, and amino acids). Photosynthesis is described. Finally, the challenge of modeling biochemical system dynamics is treated, and a number of software programs that can be used in quantitative molecular biology are listed.

In Chapter 5, The Basis of Genetic Inheritance, classical Mendelian inheritance is described. Exceptions to Mendel's rules, epigenetic inheritance, and genetic imprinting are introduced. The role of a cell's genome is presented, including exons and introns, and the enigma of the sequential biochemical regulation of development (gene expression, protein synthesis) is reviewed. The genetic code and the central dogma of protein synthesis (the codon, transcription, and translation) are treated in detail.

In Chapter 6, Nucleic Acids and Their Functions, the roles of DNA and RNA in a cell's command, control, and communication (C^3) structure are given. The many types of RNAs used by a cell, together with their functions, are also covered. DNA repair mechanisms, and gene regulation are introduced.

In Chapter 7, A Review of Proteins, proteins are the tail that wags the molecular biological dog. Whereas there are about 23,000 human genes, there may be over 230,000 proteins possible from the human genome. This large chapter begins with a description of the 20 AAs and the peptide linkages used to form the primary (linear) structure of proteins. How proteins achieve their tertiary structures is discussed, along with the role of chaperone proteins. Numerous examples of protein functions are given, ranging from structural applications and hormones to ion pumps. The misfolded prion protein, the cause of "mad cow disease," is examined in detail as an example of the consequences of an acquired error in protein structure. Finally, we describe the processes of protein destruction and AA recycling, including the role of the proteosome.

In Chapter 8, The Genetic Basis for Certain Inheritable Diseases: Genomic Medicine, the linkages between certain diseases and defects in certain genes are described. The list is large and growing yearly. We see that many diseases are only correlated with gene defects, others are caused by them. This chapter also introduces gene therapy where genetically modified (GM) viruses have been used experimentally to fight certain cancers.

In Chapter 9, Some Instrumental Methods Used in Genomics, Proteomics, and Forensic Science, we describe some of the physical techniques used to characterize and measure nucleic acids and proteins. The polymerase chain reaction (PCR), rolling circle amplification (RCA), and padlock probes are described as means of amplifying and characterizing DNA and RNA oligos. The use of fluorescent molecular tags is described. How DNA is manipulated in forensic science is clearly presented. Section 9.7 details a number of techniques used to characterize large protein molecules, including the use of microarrays, ELISA tests, and x-ray crystallography.

In Chapter 10, Applications of Genomics, we describe the controversial introduction of genetically modified organisms (GMOs) into the human and farm animal food chain. Examples of GMOs are given, illustrating their pros and cons. Recombinant DNA technology is described, including the physical means by which new genetic material is introduced into plants and animals. This chapter also explains animal reproductive cloning. The sources of stem cells, and how they can benefit regenerative medicine, is treated in detail. There is also a unique section on the role of cancer stem cells in the propagation of tumors. The controversial topics of parthenogenesis and chimeras are also described.

In Chapter 11, Ethical Issues in Genetic Engineering, we consider the trichotomy between short-term financial profits aiding humanity by producing more (GM) food and the potential harm that can befall the ecosystems where certain GM foods are grown and also the consumers of GM foods. Also considered are the ethical problems caused by harvesting embryonic stem cells from IVF embryos and the creation of animal chimeras.

FEATURES

Some of the unique features of this text are:

- The first three chapters review basic biology and cell biology; the components of prokaryotic and eukaryotic cells are described. The origin of life and the significance of cellular and organismic death are considered.
- Chapter 4 introduces the reader to physical molecular biology, including types of biochemical bonds, reaction energetics, catalysts, and enzymes. Mass-action kinetics are introduced and used to model biochemical oscillators ("clocks") and the generation of the nerve action potential. For those interested in quantitative biology, there is an extensive listing of available biochemical systems simulation software.

- Chapter 7 covers the ubiquitous roles of *proteins* in all life processes and underscores their importance in all living systems, as well as in viruses.
- Chapter 8, on genetic diseases, introduces the role of our genomes and epigenomes in determining our health.
- Chapter 9 describes in detail many of the important instrumental techniques used in characterizing and quantifying nucleic acids, proteins, and other biomolecules.
- Chapter 10 describes the role of stem cells in regenerative medicine, their future potential in effecting cures, and the sources of stem cells. The control of stem cell differentiation is seen to be a major problem.
- Chapter 10 also has a unique section (10.4.5) on the role of adult stem cells in the growth and metastasis of cancer.
- Chapter 11 treats ethical concerns inherent in the production of GMOs and in the harvesting of embryonic stem cells.

This text has a large, comprehensive glossary with detailed definitions to aid the nonbiologist in accumulating the vocabulary of the biological sciences. The bibliography and references will aid the reader in pursuing topics in detail. (Entries are from textbooks, current journal articles, and, of course, the Internet.) Over 30 color diagrams aid in the interpretation of molecular structures and biochemical reactions.

Robert B. Northrop
Chaplin, Connecticut

Anne N. Connor
San Antonio, Texas

Authors

Robert B. Northrop, Ph.D. majored in electrical engineering at Massachusetts Institute of Technology (MIT), graduating with a bachelor's degree in 1956. At the University of Connecticut (UCONN), he held a graduate assistantship while he studied for a master's degree in systems engineering, which he earned in 1958. In 1958, as the result of a long-standing interest in living systems, he entered the Ph.D. program there in physiology as a research fellow in cell biology. He did research on the neuromuscular physiology of molluscan catch muscles and earned his Ph.D. in 1964.

In 1963, Northrop rejoined the electrical engineering department of UCONN as a lecturer and was appointed assistant professor in 1964. In collaboration with his Ph.D. advisor, Dr. Edward G. Boettiger, he secured a 5-year training grant in 1965 from the National Institute of General Medical Science (NIGMS) of the National Institutes of Health (NIH) and started one of the first interdisciplinary biomedical engineering graduate training programs in New England. UCONN currently awards M.S. and Ph.D. degrees in this field of study and has a robust undergraduate specialization in biomedical engineering, based in the Electrical and Computer Engineering Department.

Throughout his career, Dr. Northrop's areas of research have been broad and interdisciplinary and centered on biomedical engineering. He has conducted sponsored research on the neurophysiology of insect and frog vision and devised theoretical models for visual neural signal processing. He also conducted sponsored research on electrofishing, and he developed, in collaboration with Northeast Utilities Service Co., effective working systems for fish guidance and control in hydroelectric plant waterways on the Connecticut River using underwater electric fields.

Another area of Dr. Northrop's sponsored research has been in the design and simulation of nonlinear, adaptive, digital controllers to regulate in vivo drug concentrations or physiological parameters, such as pain, blood pressure, and blood glucose in diabetics. An outgrowth of this research led to his development of mathematical models for the dynamics of the human immune system that were used to investigate theoretical therapies for autoimmune diseases, cancer, and HIV infection.

Biomedical instrumentation has also been an active research area: An NIH grant supported studies on the use of the ocular pulse to detect obstructions in carotid arteries. Minute pulsations of the cornea from arterial circulation in the eyeball were sensed using a no-touch, phase-locked, ultrasound technique. Ocular pulse waveforms were shown to be related to cerebral blood flow in rabbits and humans.

Recently, he addressed the problem of noninvasive blood glucose measurement for diabetics. Starting with a Phase I SBIR grant, Dr. Northrop developed a means of estimating blood glucose by reflecting a beam of polarized light off the front surface of the lens of the eye and measuring the very small optical rotation resulting from glucose in the aqueous humor, which, in turn, is proportional to blood glucose. As an offshoot of the instrumental techniques developed in micropolarimetry, he developed a magnetic sample chamber for glucose measurement in biotechnology applications. The water solvent was used as the Faraday optical medium. His current research interest lies in complex systems.

Dr. Northrop has written seven textbooks, with topics including analog electronic circuits, instrumentation and measurements, physiological control systems, neural modeling, signals and systems analysis in biomedical engineering, instrumentation and measurements in noninvasive medical diagnosis, and analysis and application of analog electronic circuits in biomedical instrumentation.

Dr. Northrop was on the electrical and systems engineering faculty at UCONN until his retirement in June 1997. Throughout this time, he was director of the biomedical engineering graduate program. As emeritus professor, he still teaches courses in biomedical engineering, writes texts, sails, and travels. He lives in Chaplin, Connecticut, with his wife.

Anne N. Connor, M.A. is a writer, researcher, and analyst for Methodist Healthcare Ministries, a medical nonprofit organization in San Antonio, Texas. Her educational background includes a bachelor's degree from Dartmouth College, where she was a teaching assistant in the English Department. She received honor citations in chemistry and sociology. Her master's degree in communications is from the University of New Mexico at Albuquerque. She considers herself an autodidact and has immersed herself heavily in the field of genomics for the past 4 years. She is a graduate of the Leadership Texas Class of 2006 and has received numerous awards for her work, most recently a humanitarian award from the health care community in San Antonio. She is a member of Phi Beta Kappa, Phi Kappa Phi, and the writers' organization Gemini Ink.

1 Introduction

1.1 SCOPE OF THE TEXT

In this chapter, we describe our motivation for writing this text, who will profit by reading it, and what it contains.

Modern 21st century biology is a complex discipline. Its descriptions involve the interactions of hundreds of thousands of different molecules (proteins, nucleic acids, hormones, cytokines, transcription factors, and other control substances). Its complexity became more apparent toward the end of the 20th century as instrumental techniques became available to efficiently separate and analyze proteins, sequence genes, and examine the roles of RNAs and enzymes in the regulation of gene expression.

Back in the middle of the 20th century, linear algebra and matrices began being used by engineers to analyze linear electronic circuits and linear control systems [Guillemin 1949]. Laplace transform and z-transform methods were also used to extend linear dynamic analysis into the frequency domain and to predict system transient behavior and the response to noise inputs. With the rapid evolution of digital computers, these mathematical techniques led to the development of computer software suitable for modeling linear systems [Northrop 2003].

Biomedical engineers soon began to appreciate that these same analytical techniques for linear systems could be applied to modeling simple, linearized physiological systems [Stark 1968]. At the same time, systems physiologists recognized that, although physiological systems are nonlinear, they can be described by systems block diagrams with both nonlinearities and linear dynamics [Guyton 1991]. The interest in modeling physiological systems was facilitated by the early development of computer programs by IBM for nonlinear systems modeling such as IBM's *Continuous System Modeling Program* (CSMP®) software series. Gradually, programs for modeling nonlinear systems evolved more sophistication, became user-friendly, and developed better graphics user interfaces (GUIs). Appearing in the late 1980s and early 1990s, GUI programs that run on personal computers made the modeling of simple nonlinear biochemical systems easier. More recently, specialized software applications have been developed for the simulation of complex biochemical systems (see Chapter 4, Section 4.5 of this text).

Biomedical engineers, electrical engineers, chemical engineers, physicists, and biophysicists, by virtue of their mathematical training, are well suited to tackle problems in *systems biology*. We predict that there will be a huge, international, interdisciplinary research effort in the first few decades of the 21st century, motivated by (1) the need to understand the genomic and biochemical command, communication, and control (C^3) networks used to regulate stem cells in embryonic development; (2) the need to find an inexpensive, noncontroversial source of pluripotent stem cells for *regenerative medicine*; (3) the need to find and destroy cancers, in particular, cancer stem cells; and (4) the need to stimulate the immune response to fight deadly viruses such as *Ebola, HIV,* and *H5N1*.

Biomedical engineers, other engineers, physicists, and chemists who wish to participate in these quests need to have a mastery of a broad range of academic disciplines including, but not limited to, cell biology, biochemistry, molecular biology, and the state of the art in instrumentation applied to molecular biological systems. In writing this text, we have assumed that the reader has at least had an introductory college course in cell biology and wishes to increase his or her knowledge in molecular biology and related subjects.

At the time of this writing, the only engineering university that has a general education requirement for introductory cell biology is the Massachusetts Institute of Technology (MIT). However, undergraduate students following curricula in biomedical engineering generally take cell biology and/or human physiology and anatomy as part of their programs. Biophysics majors and sanitary engineers (in civil engineering) generally take cell biology or a bacteriology service course.

The following chapters of this text are not meant to replace formal courses in molecular and cell biology, biochemistry, and physiology but to give the interested reader a perspective on the awesome complexity and depth of these fields, and how they are interrelated, and to show how application of engineering systems analysis can improve our understanding of living systems.

The chapter titles following this one are: Chapter 2, Life and Death; Chapter 3, Review of Basic Cell Anatomy and Physiology; Chapter 4, Introduction to Physical Biochemistry and Biochemical Systems Modeling; Chapter 5, The Basis of Genetic Inheritance; Chapter 6, Nucleic Acids and Their Functions; Chapter 7, A Review of Proteins; Chapter 8, The Genetic Basis for Certain Inheritable Diseases: Genomic Medicine; Chapter 9, Some Instrumental Methods Used in Genomics, Proteomics and Forensic Science; Chapter 10, Applications of Genomics; Chapter 11, Ethical Issues in Genetic Engineering. In addition, you will find that this text has robust Glossary, Index, Bibliography and References sections, and an Appendix.

The following sections in this introductory chapter elaborate on terms used in this text to describe areas of endeavor in modern biological science.

1.2 SOME DEFINITIONS

1.2.1 Molecular Biology

Simply stated, *molecular biology* is the study of all types of biological systems at the molecular level. It is all about the molecules inside cells (and also outside them) that make life possible. Needless to say, there are many, many kinds of molecules, so molecular biology is a broad science of considerable complexity. Topics include how DNA, RNA, amino acid, protein, lipid, carbohydrate, steroid, etc., molecules are synthesized, regulated, broken down, metabolized, etc. Molecular biology certainly encompasses knowledge from the more specialized disciplines of *biochemistry, physical chemistry, biophysics, genomics, proteomics, genetics,* and *bioengineering.*

The term *molecular biology* was first coined in 1938 by Warren Weaver at the Rockefeller Institute (now Rockefeller University) [Stanford University 2005]. Because all of life involves a plethora of interacting molecules, molecular biology can be considered an umbrella discipline for the specialized life sciences described in the following subsections.

1.2.2 Genomics

Genomics is the study of an organism's *genome* and its use of its genes. An organism's *genes* code the primary structures of proteins and various RNA molecules used in regulating gene expression and posttranscriptional editing of protein structures. However, the genome includes more than just genes; it consists of all of the nucleotide (Nt) sequences in nuclear *and* mitochondrial DNA. The field of genomics covers structural genes, regulatory sequences including RNA coding, and sequences that apparently have no known function (the enigmatic *introns*).

The term *genomics* was introduced by Thomas Roderick in 1986; it has now evolved into *functional genomics* and *structural genomics*. Functional genomics seeks to elaborate gene function—in particular, control of gene expression in all phases of an organism's development. Structural genomics seeks to map the nucleotide sequences in the genome in detail and to study how evolution has affected genes in closely related species, as well as variable features, such as *short tandem repeat*

(STR) *sequences* used in forensic DNA identification. Also, by analyzing the Nt sequences in the "normal" genome and comparing them with the sequences of individuals with certain "genetic" diseases, structural genomics has indicated possible therapies and cures for patients with those genetic diseases (see Chapter 8 in this text).

The first genome to be completely sequenced was that of the bacteriophage virus, Φ-X174, in 1980. It has 5.368 Megabases (Mb) and only 10 genes. The genome of the bacterium *Haemophilus influenzae* was sequenced in 1995 (1.830 Mb DNA, and 1,739 genes) and that of *E. coli* K-12 in 1997 (4,639,221 bases, arranged in a double-stranded DNA [dsDNA] ring or *plasmid,* and 4,290 genes encoding proteins and 87 encoding RNAs). Much of the human genome has been sequenced; it is estimated that about 3.1×10^9 *base pairs* (bps) are in it. As you will see, only about 1% of these base pairs are parts of genes (i.e., they actually code the primary [linear] structure of proteins).

1.2.3 Proteomics

Proteomics is the comprehensive study of the structures and functions of proteins [Genetics Home Reference 2007; Tyers & Mann 2003]. It is a much tougher and more challenging discipline than genomics because a protein's biochemical function depends on its 3-D, tertiary molecular structure, not just the linear sequence of amino acids (AAs) initially coded by its gene. Determining a protein's 3-D structure is challenging because it relies on the ability to crystallize the purified protein and use high-resolution x-ray diffraction techniques or nuclear magnetic resonance (NMR) on the crystal. Proteins can also be disassembled chemically and their smaller subunits analyzed. Fluorescent and isotope-tagged antibodies can be used to identify certain AA sequences and structures.

Tertiary structure is critical for the molecular *specificity* of protein enzymes, cell surface receptors, gates, and ion pumps, as well as antibodies. A surprising fact is that there may be as many as from 200,000 to a half million different proteins in the human *proteome,* whereas only about 22,000 genes code proteins in humans [ENCODE 2007]. This large amplification in the number of proteins is due to *alternative splicing* and *posttranslational modification* of protein structures by other protein enzymes and special molecules. The broad uncertainty in the number of proteins is due to the fact that proteins often have short half-lives, and that certain proteins are not expressed all at the same time, or at all in certain types of cells. There are also technical problems in isolating and characterizing proteins.

Proteomics depends on the following subdisciplines: (1) *protein separation*; (2) *protein identification*; (3) *protein quantification*; (4) *protein sequence analysis*; (5) *structural proteomics*; (6) *interactional proteomics*; (7) *posttranslational modification,* including *phosphorylation* and *glycosylation*; and (8) *cellular proteomics.* This includes the control of protein synthesis; and where and when the proteins end up in the cell. Are they exported? How are old or damaged proteins recycled in the cell?

1.2.4 Bioinformatics

The terms *bioinformatics* and *computational biology* are often used interchangeably [Nilges & Linge 2002]. They are disciplines in biology that tend to attract engineers, physicists, and mathematicians. They involve the application of quantitative techniques from *engineering systems analysis, information theory, statistics,* and *computer science* to solve problems in molecular biology and physiological systems. Some research efforts in the field include, but are not limited to:

1. *Dynamic modeling of complex biological systems,* ranging from immune system function to the spread of infectious disease in a population to the control of gene expression.
2. *Protein 3-D structure prediction* from its known AA sequence, including protein structure alignment. Prion protein misfolding is associated with BSE (mad cow disease), variant

Cruetzfeld–Jakob disease in humans, and chronic wasting disease in deer and elk (see Chapter 7, Section 7.4.3).

3. *Prediction of protein–protein interactions.*
4. *Gene finding:* A protein has been isolated; where is its gene located in the organism's genome?
5. *Gene comparison:* Finding mutations that may be etiological for a disease.
6. *Modeling evolution* by means of comparison of the base pair sequences of critical genes.
7. *Analysis of point mutations* in many genes of cancer patients. Hidden Markov models are being used to extract "signals" from noisy data.

A recent text edited by Etheridge et al. [2006] adds the disciplines of thermodynamics and nonequilibrium statistical mechanics to the solution of problems in computational biology. Stochastic kinetic modeling applied to systems biology, including chemical kinetics, is described in a text by Wilkinson [2006]. An overview of the broad field of computational molecular biology is found in the *Handbook of Computational Molecular Biology* [Sahni 2006].

1.2.5 Genetic Engineering and Biotechnology

Genetic engineering is closely related to *biotechnology.* In both disciplines, one or more of an organism's genes may be silenced, modified, or turned on permanently, or a completely new gene or genes may be added to the organism's genome. Genetic engineering is the discipline in which *genetically modified organisms* (GMOs) are designed and made. (GMOs are discussed in detail in Chapter 10, Section 10.2 and Chapter 11, Section 11.3.) In 1982, one of the first successful examples of a benign application of genetic engineering was the insertion of the gene for human insulin into the dsDNA plasmid genome of the bacteria *E. coli.* This genetically modified (GM) *E. coli* makes *human* insulin molecules, which are harvested and purified. Before the development of GM *E. coli,* insulin used to treat human diabetes was extracted from the pancreases of human cadavers or from swine and cattle pancreases. Patients were often allergic to nonhuman insulin.

Controversy has surrounded many aspects of genetic engineering. Criticism of work in this area has largely come from three sources: environmentalists, ecologists, and religious fundamentalists. Chapter 11 of this text discusses the moral and ethical issues surrounding genetic engineering. Many of the environmental and ecological issues are addressed in Chapter 10. The results of the *Law of Unintended Consequences* are evident in the long-term effects of many genetically engineered crops. However, not all GM plants have high or questionable environmental impacts; for example, a GM form of thale cress (*Arabidopsis sp.*) is being developed to detect trace NO_2 gas emitted by buried land mines. The NO_2 changes the leaf color from green to red. The plant is used as a sensor; it is not a food crop [Report 2006]! Many of the current issues with GMOs may be resolved with technological advances, a potentially lucrative challenge for any student considering this field.

The term *biotechnology* was coined in 1919 by Karl Erecky, a Hungarian agricultural engineer. Biotechnology has been defined in several ways [Washington State University 2007]. One definition of biotechnology is: "… a technology based on biology, especially when used in agriculture, food science and medicine." The United Nations Convention on Biological Diversity stated: "Biotechnology means any technological application that uses biological systems, living organisms, or derivatives thereof, to make or modify products or processes for specific use."

Biotechnology involves inserting genetic material into an organism's genome to alter its growth, cause it to express a foreign protein, or cause it to develop beneficial properties. In agriculture, agronomy, and plant science, applications of biotechnology are called *green. Red biotechnology* is applied to medicine, for example, causing bacteria to produce insulin or antibiotics, or using gene manipulation to fight cancer or diseases with genetic etiologies. (See Chapter 8 in this text for a comprehensive discussion of applications of red biotechnology.) We will see a rise in *white biotechnology* in the decades to come. White biotechnology has industrial applications, such as using genetically engineering microorganisms to break down organic wastes in order to more efficiently ferment corn, or using sugar cane to make ethanol for fuel, or producing methane or hydrogen fuel from waste.

1.2.6 Systems Biology

Systems biology is a field of endeavor that seeks to describe quantitatively how biological systems function at the systems level (i.e., physiologically). Rather than break a system down into its component parts and examine each component separately, systems biology uses a holistic approach to study how each biological subsystem is involved in the overall network of command, communication, and control (C^3) [Alon 2006]. All levels of interaction are considered: cell signaling circuits, metabolic pathways, organelles, membrane receptors, cells, physiological systems, and organisms. One experimental approach in systems biology is the use of massively parallel protein and nucleic acid analytical microarrays (see Chapter 9, Section 9.2.5) to quantify changes in mRNAs and proteins during cellular processes, such as the embryonic differentiation of stem cells. The recent development of high-throughput array technologies has allowed systems biologists to examine many more system states simultaneously and avoid the errors inherent in trying to characterize a complex, nonlinear, biological system using reductionism. With the collection of massive amounts of quantitative data from arrays, systems biologists can formulate more robust mathematical models to describe a system's behavior in response to changes in its environment and human-generated inputs. For example, what molecular inputs are required to make an adult stem cell revert to embryonic, pluripotent status, then differentiate into a desired tissue type? The elaboration of this process is the "holy grail" of *regenerative medicine.*

1.3 COMPLEX SYSTEMS

1.3.1 Introduction

Before addressing the concept of *complex systems,* we need to establish a working, scientific/engineering definition of *system.* We hear the word *system* almost every day: There are many kinds of systems: control systems, communication systems, weather systems, economic systems, weapons systems, ecosystems, systems of oceanic currents, and physiological systems, including, but not limited to: immune, digestive, circulatory, respiratory, central nervous, autonomic nervous, reproductive systems, etc. In this text we will describe genomic, biochemical, metabolic, immune, photosynthetic systems, etc.

The word *system* has various shades of meaning that are context dependent. One broad, acceptable dictionary [Morris 1973] definition is: "A group of interacting, interrelated, or interdependent elements forming or regarded as forming a collective entity." The interactions occur in the form of feed-forward and feedback relations between the *elements.* In a scientific or engineering context, the *elements* are characterized by quantifiable parameters known as *state variables* (SVs). The *state* of a *linear system* is the smallest number of state variables such that a knowledge of these variables at $t = t_0$, taken together with the system's inputs for $t \geq t_0$, completely determines the system's behavior (i.e., the values of its state variables) for $t \geq t_0$ [Ogata 1970].

Systems can be subdivided into two broad categories: *linear* and *nonlinear.* Linear systems are generally analyzed and simulated by engineers using computationally direct, state-variable, linear algebra. Biological and biochemical systems are generally *nonlinear* and in most cases are *complex,* as well. A *complex system* is any system featuring a *large number* of interacting components (elements, state variables, etc.) whose overall activity is nonlinear and whose behavior may exhibit self-organization under selective pressure and stable, bounded behavior in response to disturbance inputs. Complexity not only depends on the large number of states, but in the number ways they can interact (feed-forward and feedback pathways).

So, what is a *large number* of components (states)? Largeness is in the eyes of the person describing and analyzing the system. An example of a *very large,* complex system is the genomic control system for cellular differentiation in an embryo. The state variables can be the concentrations of the cell's protein enzymes, messenger RNAs, etc. Inputs can be chemical energy, hormones, or other signaling molecules. The important thing is that there are many, many proteins, many genes, and many control molecules in play at any time. We personally consider *large* as greater than 10 states.

An example of a large, nonbiological, nonlinear, complex system is a meteorological system such as a hurricane. In this example, the number of types of state variables is relatively small; however, their total number is not. Some of the variables used are atmospheric pressure, temperature, humidity, energy inputs (solar IR, thermal), and (vector) wind velocity at different altitudes and locations. These variables are sampled at time intervals in and around the storm. These 4-D sampled data are entered into a dynamic computer model that attempts to predict the storm's path and peak winds using thermodynamics and fluid dynamics. There are several computer models used by NOAA and the National Weather Service to predict hurricane paths, often not too well.

An example of a *simple,* nonlinear, nonliving system is a two-gimbal mechanical gyroscope. Its mechanical behavior can be well modeled by a set of three ordinary nonlinear differential equations. The nonlinear ordinary differential equations (ODEs) that describe the gyro's behavior in response to input rotations contain certain trigonometric functions (sines and cosines). These equations are easily linearized to satisfy the engineering mind [Northrop 2005].

The human immune system (IS) is an example of a *complex, nonlinear, biological system.* There are many types of immune system cells with different roles in fighting infection, as well as many molecules made by IS cells such as *interleukins, immunoglobulins, prostaglandins, complement,* and *antibodies.* The whole IS is coordinated and communicates by various cytokines secreted by IS cells; some are activators and others are suppressors of IS functions. Thus, there is a humoral, molecular, C^3 structure tying together and regulating the IS. There are also exogenous hormonal inputs regulating the level of IS responses. The author has simulated the immune system's responses to HIV infection, pathogens, and AIDS therapy drugs using the nonlinear ODE solving software, Simnon®. Thirty-two state variables were used in one nonlinear model [Northrop 2000]. By adding more interleukins, another model had 42 state variables. See Chapter 7, Section 7.2.6 of this text for a short description of the human IS.

Many complex biological systems can be considered to be *time variable* or *nonstationary.* That is, the causal linkages between state variables have values that vary with time. A *linear* time-variable (LTV) system's dynamic model written in state-variable format has the form shown in Equation 1.1:

$$\mathbf{x} = \mathbf{A}(t)\,\mathbf{x} + \mathbf{B}(t)\,\mathbf{u} \tag{1.1}$$

where $\dot{\mathbf{x}}$ is the column vector of first (time) derivatives of the system's state variables (SVs), $\mathbf{x}$ is the column vector of the system's SVs, $\mathbf{u}$ is the column vector of inputs to the system, and $\mathbf{A}(t)$ and $\mathbf{B}(t)$ are matrices with one or more time-variable elements.

Why are biological systems complex? Evolution apparently has led to greater complexity. To investigate arguments supporting this premise, see *The Evolution of Complexity* [Bonner 1988] and *Hidden Order: How Adaptation Builds Complexity* [Holland 1995]. Human cells contain very large numbers of different molecules (e.g., about 22,000 genes and over 220,000 proteins) that coexist inside cells and on cell surfaces in a regulated manner. Molecular biologists are just now beginning to understand how they work together. In all higher plants and animals, many smaller, biochemical subsystems operate *inside cells,* stabilized by negative feedback pathways, and many cells communicate with others of their kind and with other organ systems using signaling molecules (hormones, cytokines, etc.). Feedback also exists to stabilize hormonal systems.

Modeling biological systems is a challenge because they are immense, interconnected, nonlinear, and in most cases time variable, and they use *parametric feedback* [Carpenter 2004; Northrop 2000]. But most important of all, the quantitative dynamic relationships between the many molecules are poorly known or unknown. This is especially true when considering embryonic development.

1.3.2 Properties of Nonlinear Systems

In addition to nonstationarity, *nonlinearity* is an important property in biological systems modeling. *All* biochemical and physiological systems are nonlinear, and most have so many interdependent states that they are certainly classifiable as complex systems. The chief property of nonlinear systems is that they *do not* obey *superposition.* Unlike *linear systems,* nonlinear systems exhibit dynamic behavior that depends on their initial conditions as well as the amplitude (and sometimes frequency or rate of change) of their inputs. An excellent, simple example of this property is seen in the behavior of the venerable 1952 Hodgkin–Huxley (H–H) model for nerve spike generation, described in Chapter 4, Section 4.5.4.

Another property of nonlinear systems is that they can exhibit bounded or stable *limit-cycle oscillations* in their SV concentrations. The H–H model acts as a current-to-frequency converter oscillator, given a constant, depolarizing current input [Northrop 2001]. A second example of oscillations in a small biochemical system having eight state equations is given in Chapter 4, Section 4.5.3. (This system is a hypothetical model for a biochemical "clock" oscillator.)

Another perplexing property of *nonlinear complex systems* (NCSs) is that they can exhibit counterintuitive behavior. Here, one thinks he or she understands the model of an NCS and predicts that *increasing* a certain input will lead to the *decrease* of a certain SV. Instead, it increases! Evidently, our understanding was not complete enough! In a mathematical model of an NCS, counterintuitive behavior may indicate that our model was not accurate (complete) enough or it may be a real property of the system.

What are the sources of nonlinearity in biological systems? There are two main sources of nonlinearity, which can best be described in terms of the mathematical models used to simulate biological systems. The first is *static* or amplitude nonlinearity. Picture a "black box" with a single input, x, and output, y. If the box represents a *linear, static* relation between x and y, then y is given by the *linear* algebraic equation, $y = mx + b$, where m and b are real constants. However, if $y = \mathrm{f}(x)$ (other than the linear relation), the box contains a *static nonlinearity.* When a nonlinearity has a "memory" such as a hysteresis loop, it acquires dynamic properties. A nonlinearity without memory is said to be *single valued* (i.e., there is only one y [output] value for every x [input] value). Note that in *all* biochemical systems, where y represents a molecular concentration or density, y must be nonnegative. Clearly, there are no negative concentrations or densities in nature.

Another source of static nonlinearity in biological systems is *amplitude saturation.* There is a finite number of cell membrane receptors for a given signal molecule, X. Thus, once all the receptor molecules on the cell's surface have bound to the X molecules, further increase in the concentration, $\boldsymbol{x}$, will not produce any additional effect inside the cell. All cells have a finite number of membrane receptors; good examples are hormonal systems and immune system cells. This type of saturation effect is said to be *soft*. That is, $\boldsymbol{y}$ gradually approaches $\boldsymbol{y}_{\mathbf{max}}$ as $\boldsymbol{x}$ increases.

Dynamic nonlinearities are the second source of nonlinearity in biological systems. They generally arise as the result of nonlinear algebraic terms in the set of ODEs describing the kinetics of a biochemical system. That is, the ODEs can contain terms of the form shown in Equation 1.2, written for the kth SV:

$$\dot{\boldsymbol{x}}_k = k_{kj}\boldsymbol{x}_k\boldsymbol{x}_j + k_{kp}\boldsymbol{x}_p^n + k_{kq}\frac{\boldsymbol{x}_q}{1+k_q\boldsymbol{x}_q} + k_{krp}\boldsymbol{x}_k\boldsymbol{x}_r\boldsymbol{x}_s - k_{kL}\boldsymbol{x}_k \tag{1.2}$$

Here, $\boldsymbol{x}_k$ is the kth concentration or density, and we see that the terms involve double and triple products of nonnegative SV values, an exponential term ($n \neq 1$), and a soft saturation for the $\boldsymbol{x}_q$ state. All the terms in Equation 1.2 are nonlinear (except the right-hand term) and can be the result of the use of chemical kinetics to model a system of biochemical reactions. In some cases, dynamic nonlinearities can lead to bizarre and unexpected results in a system simulation. These can include bounded limit cycle oscillations; input amplitude-sensitive behavior, where the system is stable for an input below a certain value; and large inputs that induce it to oscillate or saturate unexpectedly.

The human IS offers a significant example of a complex nonlinear system's behavior. The initial introduction of a foreign protein into the body can prime the IS for a *hypersensitivity reaction* on the next exposure to the protein. However, if minute quantities of the protein are periodically introduced into the body (e.g., allergy injections), the IS can be gradually desensitized to the protein, eliminating hypersensitivity. This type of behavior is a challenge to model.

1.3.3 Parametric Control

Finally, we address a ubiquitous property of biochemical and physiological systems using feedback for self-regulation. This is the principle of *parametric control* (PC) [Northrop 2000]. Broadly stated, PC occurs in a biochemical system when a chemical reaction rate constant or diffusion constant is modulated by the level of an SV in the system introducing feedback that leads to the decrease (or an increase) in the level of the SV. This modulation can be from an alteration in the configuration of an enzyme that catalyzes a chemical reaction step or from a decrease in the concentration of the enzyme, slowing the reaction. An exogenous input, such as a hormone, can also alter SV levels in a biochemical system by PC.

An example of PC is the function of the hormone *insulin,* which increases the permeability to glucose in an insulin-sensitive cell's membrane glucose receptors. Insulin also increases the rate of synthesis of glycogen in liver cells, lowering blood glucose concentration.

Another example of a physiological system with PC is the system that regulates intraocular pressure (IOP) in the eyeball. Excess IOP is called glaucoma; it can lead to blindness by restricting blood flow through the retina. In one model, excess IOP causes the outflow hydraulic resistance to aqueous humor (AH) to *decrease*, expediting its loss through the Canal of Schlemm. Excess IOP may also decrease the production of AH in the eye as well, limiting the maximum IOP [Northrop 2000].

A mathematical example of a simple system with parametric control is as follows. The input rate of active transport of a molecule X into a *compartment* is determined by the concentration of the transported molecule inside the compartment $\boldsymbol{x}_i$ according to the rule:

$$\dot{\boldsymbol{x}}_i = [\mathrm{K}_{\mathrm{Tmax}} - \rho\, \boldsymbol{x}_i]\, \boldsymbol{x}_{\mathrm{o}} - \mathrm{K}_{\mathrm{L}}\, \boldsymbol{x}_i, \quad \mathrm{K}_{\mathrm{Tmax}} \geq \rho\, \boldsymbol{x}_{i\mathrm{max}} \tag{1.3}$$

where $\boldsymbol{x}_i$ is the concentration of the substance X *inside* the compartment, $\boldsymbol{x}_{\mathrm{o}}$ is its concentration *outside* the compartment, K_{L} is the loss rate constant of X from inside the compartment, and the input transport rate is a function of the interior concentration: $\mathrm{K}_{\mathrm{T}}(\boldsymbol{x}_i) = [\mathrm{K}_{\mathrm{Tmax}} - \rho\, \boldsymbol{x}_i]$. Assume $\boldsymbol{x}_{\mathrm{o}}$ is constant; thus, in the steady state (as $\boldsymbol{t} \to \infty$), $dx_i/dt = 0$, and we can easily solve for $\boldsymbol{x}_{i\mathrm{SS}}$. Simple algebra yields

$$x_{iSS} = \frac{x_o \mathrm{K}_{\mathrm{Tmax}}}{\mathrm{K}_{\mathrm{L}} + x_o \rho} \tag{1.4}$$

Thus, the parametric feedback limits the internal concentration of the compartment to a steady-state maximum of $\boldsymbol{x}_{i\mathrm{SSmax}} = \mathrm{K}_{\mathrm{Tmax}}/\rho$ concentration units for large $\boldsymbol{x}_{\mathrm{o}}$.

1.3.4 Modeling Complex Systems

So, how *do* we model nonlinear complex systems (NCSs)? If we attempt to subdivide them into smaller, more manageable subsystems (this is called *reductionism*), we can destroy their observed intrinsic behavior, which can be a property of the whole system's network. Attempts to linearize a nonlinear complex system model can have the same chilling effect on system behavior, which, after all, is a result of its nonlinearity as well as its size. The NCS might be described by as few as 10 nonlinear state equations to as many as several thousand SVs. Modern personal computers with clock speeds in excess of 3 GHz and multiprocessor designs can easily run simulations of NCSs, provided we know all the functional pathways and rate constants affecting each and every SV. In many NCSs, we can only estimate or guess at rate constants. In order to validate a mathematical model, it has to emulate observed, in vivo NCS behavior given certain known inputs. The ultimate value of a nonlinear state model of an NCS is its ability to predict behavior for novel inputs or conditions. For example, if we are modeling the immune system versus cancer, we are interested in using the model to find putative therapies that will cause the IS to destroy the cancer cells, before resorting to animal models and drugs.

In Chapter 4, Section 4.6 of this text we describe a number of software packages written specifically to model biochemical systems and even whole-cell biochemical systems.

1.4 OPTIMUM USE OF REFERENCE RESOURCES

One goal of the authors in writing this text is to stimulate readers to pursue the topics described herein in greater depth, to take formal, advanced courses in molecular and cell biology, etc., and to do further reading on selected topics. To aid this latter pursuit, this text has an extensive bibliography and references section. About 90% of the entries were accessed by the authors on the Internet. In spite of a deplorable trend for some journal publishing houses to have a "pay-for-view" policy, most do not, and many academic (`*.edu and *.ac`) Web sites are rich sources of current information on the rapidly moving field of molecular biology and all related disciplines. This information is generally in the form of refereed journal articles, refereed review articles, college lecture notes, and online texts by respected scholars. Note that certain Web references can be ephemeral. That is, a Web site is frequently modified, and certain older references are deleted to make space for new ones, or revisions. This high-information "vapor pressure" is one price we pay for the sit-at-your-desk convenience of using the Internet as a scholarly resource.

We have referenced certain textbooks because they are "classics"; they teach and summarize well-established principles and facts about biological systems. Interested readers are encouraged to overcome the "activation energy" required to visit their college or university library to examine these text references.

Finally, we wish to warn the reader not to put faith in the validity of Wikipedia® as an inerrant reference source. Some Wiki definitions are timely, up to date, and correct. They are written by experts in the areas and include bibliographies with refereed journal and text citations. Other Wiki definitions may lack the accuracy of refereed information and lack academic provenance.

We have observed that molecular biology and its subdisciplines are the most rapidly growing areas of science today. You, the reader, will find that almost weekly there is a new discovery or development in molecular biology described in the media (and in the "fast" journals such as *Nature* and *Science*) that has the likelihood of affecting the human condition. For example, at the time of this writing (June 2007), we have just seen that researchers have developed techniques to allow differentiated, adult mouse epithelial cells to be transformed back into pluripotent stem cells, eliminating the need for embryos or ova (see Chapter 10, Section 10.4.3 in this text). Although this research was done on mice, its results may soon be extended to humans.

1.5 SUMMARY

The subject matter of this text is characterized by the rapidity with which information is being added to the fields of molecular biology, genomics, and proteomics. This is especially true in the fields contributing to regenerative medicine (i.e., stem cell research), in genetic medicine, and in cancer research. Although specific developments subsequent to this writing do not appear in the text, we have endeavored to address the major trends in current research in molecular biology. Our approach to the following material in this text has been influenced by our backgrounds (i.e., systems engineering, animal physiology, biomedical engineering, and natural science).

Do not be surprised to see equations in material in this text dealing with computational biology, physical chemistry, systems physiology, and biophysics. Some are chemical reaction (balance) equations, others are algebraic in form, and there are a plethora of ordinary differential equations describing biochemical system dynamics. Know that equations are a useful way of describing and summarizing dynamic relations in systems and modeling their behavior.

One thing we tend to do as humans when faced with making a decision involving a complex system (e.g., economics, weather, and molecular biology) is to try to oversimplify it and reduce it to a single-input, single-output relationship. The *single-cause mentality* is built into the thinking of *Homo sapiens*. We have to fight it when considering complex systems. Many molecular biological systems are causally interrelated, both intra- and extracellularly.

CHAPTER 1 HOME PROBLEMS

1.1 Define *homeostasis*. Contrast cellular versus organismic homeostasis.

1.2 What role does *parametric control* play in homeostasis at the cellular level?

1.3 Compare the study of *animal physiology* with *systems biology*.

1.4 Illustrate how *end-product regulation* is a form of parametric negative feedback in a biosynthetic reaction.

1.5 A hormone H in the extracellular fluid (ECF) at concentration $\boldsymbol{h}$ controls parametrically the passive diffusion of molecule M from the ECF into a cellular compartment with volume V_C. Let $\boldsymbol{m}_e$ be the extracellular concentration of M and $\boldsymbol{m}_i$ be the intracellular concentration. The intracellular M is lost at a rate $K_L \boldsymbol{m}_i$. The transmembrane diffusion constant is modeled by the function, $K_D = K_{Do} + \beta \boldsymbol{h}$.

- **A.** Write the system's state equation for $\boldsymbol{m}_i$. (Note that in a diffusion system, the rate of mass flow (μg/min) is proportional to the concentration differences.)
- **B.** Give an expression for the steady-state $\boldsymbol{m}_i$, given a constant $\boldsymbol{m}_o$.
- **C.** Give an expression for the system's time constant, $\tau = f(\boldsymbol{h}, K_L, K_{Do}, \beta, V_C)$.

1.6 A hormone H in the extracellular fluid (ECF) at concentration $\boldsymbol{h}$ controls the passive diffusion of molecule M from the ECF into a cellular compartment with volume V_C. Let $\boldsymbol{m}_e$ be the extracellular concentration of M and $\boldsymbol{m}_i$ be the intracellular concentration. The intracellular M is lost at a rate $K_L \boldsymbol{m}_i$. The transmembrane diffusion constant is modeled by the function $K_D = K_{Do} + \alpha \boldsymbol{h}^2$. (See Figure HP1.6.)

- **A.** Write the system's state equation for $\boldsymbol{m}_i$. (Note that in a diffusion system, the rate of mass flow (μg/min) is proportional to the concentration differences.)
- **B.** Find an expression for the steady-state $\boldsymbol{m}_i$ versus $\boldsymbol{h}$.
- **C.** Plot and dimension the system's natural frequency $\omega_o = \tau^{-1}$ as a function of $\boldsymbol{h}$ and system parameters.

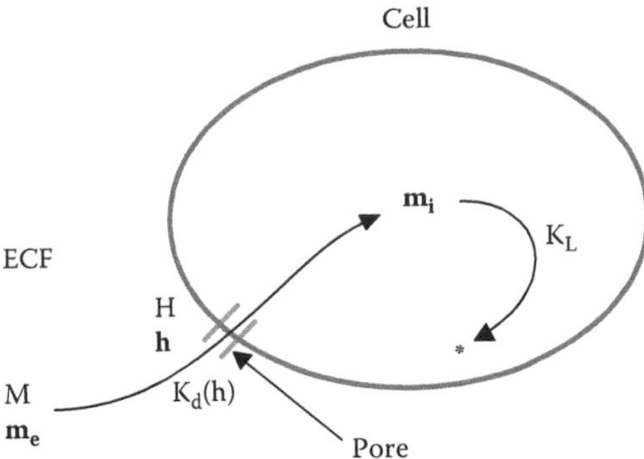

FIGURE HP1.6

1.7 *Parametric control* is used to charge a capacitor from 0 V to $v_c = V_D$. A simple series circuit is used, shown in Figure P1.3. As v_c approaches V_D, the voltage-controlled conductance decreases linearly to 0 S at $v_c = V_D$. That is, $G(v_c) = (G_o - k\, v_c)$ for $0 \leq v_c \leq V_D$, and 0 for $v_c > V_D$. Also, $k = G_o/V_D$. (See Figure HP1.7.)

Simulate the behavior of the circuit as the capacitor charges. Let $V_S = 10$ V, $V_D = 1$ V, $G_o = 10^{-3}$ S, $C = 10^{-3}$ F, and $v_c(0) = 0$. Plot $v_c(t)$, $i_c(t)$ mA, and $\tau(t) = C/G$ s. Use *Simnon* or *Matlab/Simulink*.

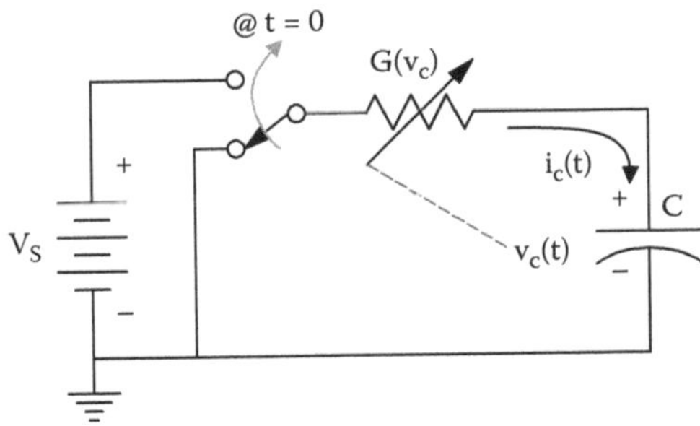

FIGURE HP1.7

1.8 Repeat Problem 1.3 using $G(v_c) = G_o \exp(-v_c/2V_D)$ for $0 \leq v_c \leq V_D$. When $v_c > V_D$, set G = 0. Use the parameter values in Problem 1.3.

1.9 *Graph theory* is used as the descriptive basis for many complex molecular biological systems. Summarize graph theory, and give a list of key terms used in describing graphs and their functions. (Note that the *signal flow graphs* used by systems engineers are a special type of graph used to describe linear systems. Biochemical systems are nonlinear.)

1.10 Describe in detail the human glucoregulatory system. Show how it uses parametric control.

2 Life and Death

2.1 WHAT IS LIFE?

2.1.1 Introduction

Ever since primitive humans became aware that there was a difference between alive and recently dead organisms, they contemplated the origin of life and what happens to the life experienced by humans and animals when they die. Our purpose here is not to discuss the spiritual, faith-based, and metaphysical attributes of life and the soul but to focus instead on the science of life and death, particularly when viewed through the lenses of molecular biology and physiology.

In order to examine the biochemistry of life, we need to agree on a definition of life. Definitions of life have evolved through the ages: The Greek philosophers Thales (640–546 BC), Anaximander (611–547 BC), Empedocles (490–444 BC), and Aristotle (384–322 BC) wrote essays on life, and in particular, on the *spontaneous generation* of life from inert matter. Anaximander claimed (erroneously) that everything living originates in sea ooze and goes through a succession of stages in its development [Oparin 1953]. (The irony here is that recent scientific data obtained by Italian researchers suggest that about 65% of the DNA present in the world's oceans is found in deep sea sediments, and 90% of that DNA is extracellular, forming available substrates of nucleotides and phosphorus for the proliferation of living marine microorganisms [Paterlini 2005]).

Roman and early Christian philosophers such as Plotinus, Basilius (315–379 AD), and St. Augustine (354–430 AD) taught that the "spontaneous" appearance of grasshoppers, flies, mollusks, mice, etc., was indeed God's will. Even in the 15th and 16th centuries, great minds such as Descartes (1596–1659) and Newton (1643–1727) still cleaved to the theory that the spontaneous generation of living creatures from lifeless matter occurred. Indeed, this view was challenged by the Tuscan physician Francesco Redi (1626–1697), who demonstrated experimentally that maggots in rotting meat were the product of eggs laid by flies. Unfortunately, he still held that certain worms (intestinal, wood) still might be produced by spontaneous generation.

The theory of the autogeneration of life persisted until the time of the great French microbiologist, Louis Pasteur. In 1862, Pasteur received a prize from the French Academy of Sciences for experimentally demonstrating, once and for all, microbes grew in various media because they "dropped in" from nonsterile air or from contact with nonsterile surfaces. He completely refuted the hypothesis that decaying infusions generate microorganisms. He demonstrated that, on the contrary, the decay of the infusion was due to the activity of microorganisms introduced along with the medium. He discovered that heat-resistant bacterial spores were generally responsible for it. Given adequate high-temperature sterilization, bacteria did not grow and the infusion did not decay. Indeed, at the present time, we still rely on "Pasteurization" to preserve milk and certain fermented beverages. Pasteur's work necessarily dealt with bacteria, yeast cells, and spores visible under the light microscopes of the era. Not until the development of the transmission electron microscope (TEM) and the scanning electron microscope (SEM) in the second half of the 20th century was the structure of viruses realized. We will return to the consideration of bacteria and viruses later in this text. For more discussion on spontaneous generation, see the classic text by Oparin [1953], *The Origin of Life*.

2.1.2 Properties of Life

In the 20th century, theologians, philosophers, physicists, biologists, and a Nobel laureate biochemist have written on the topic of "what is life." Needless to say, definitions have varied widely because the topic has been viewed from different perspectives, such as scientific and faith based, believer and skeptic. What most definers of life have done is to list a set of necessary properties of living creatures. However, it appears that every time someone comes up with a list of the fundamental attributes of life, others find exceptions to the list.

One reason it is important to have a robust definition for life is that we may encounter "life" in our solar system during future manned or robotic planetary explorations. We want to recognize it when we find it. We are familiar with life on Earth, but will we recognize it or its fossils, for example, on Jupiter's ice-moon *Europa,* on Saturn's moon *Titan,* or on Mars? Will its molecular biology be similar to that of Earth's life?

What we will do in this section is try to distill a consistent and comprehensive definition of life (on Earth) from the definitions of others and our knowledge of contemporary cell biology. First, we list some basic, necessary (required by obligation) attributes of living cells:

1. *The basic unit of life is the cell*; all life on Earth exists either as unicellular organisms (*Archaea, Bacteria, Protists, Algae*) or as multicellular (metazoan) animals and plants formed from aggregates of eukaryotic cells.
2. *All cells have cell membranes* which separate their interiors from their external surroundings. Cell walls differ in their biochemical structures over the domains and kingdoms of life. The purpose of a cell membrane is to separate the biochemically regulated internal environment from the unregulated exterior environment.
3. *Living cells metabolize;* that is, unless they are dormant, they constantly expend chemical energy to maintain an ordered, regulated, internal environment. They obtain chemical energy from external nutrient molecules that are admitted through their cell membranes (e.g., glucose and O_2) or from photons and external molecules (e.g., photons and CO_2). In some cases, these nutrient molecules diffuse passively inward; in other cases, specialized protein molecules in the cell membrane facilitate their entrance into the cell. These molecules may include hydrogen (H_2), oxygen (O_2), sugars, iron, sulfur (S), amino acids (AAs), proteins, carbohydrates, etc. Photon (electromagnetic) energy (in plants and certain protists and algae) also drives metabolisms based on photosynthesis. Metabolic chemical reactions produce "high-energy" molecules such as *adenosine triphosphate* (ATP); the energy released by its hydrolysis is used to drive certain necessary, coupled, internal chemical reactions in the living cell. The waste products of metabolism generally include heat, H_2O, CO_2 and, in the case of *methanogen bacteria,* methane gas (CH_4). During photosynthesis, plants take up CO_2 gas and release molecular O_2 gas.
4. In response to physical and chemical changes in their external environment, *cells actively maintain their interior systems in a steady state* through active chemical processes using their chemical energy inputs. This group of stabilizing chemical reactions is called *homeostasis.* For example, living cells attempt to maintain a nearly constant difference in osmotic pressure between the inside and outside of the cell. They also autoregulate their metabolisms. The internal biochemical reactions of cells generally are regulated by a complex network of feedback and feed-forward pathways.
5. *Cells die naturally* through the processes of *apoptosis* (programmed cell death), *necrosis, necroptosis,* or misadventure (including physical damage; damage by heat, ionizing radiation, caustic chemicals, starvation; infection by viruses or bacteria, etc.). At cell death, all metabolism ceases, the cell wall ruptures, and the orderly structures within the cell disintegrate and their molecules reach chemical and thermal equilibrium with the extracellular environment. (This may be thought of as *thermodynamic death.*)

6. *All living cells contain the cell's DNA genome.* The DNA contains the information required for the cell's duplication, metabolism, homeostasis, growth, reproduction, differentiation, and even the cell's programmed death. One exception to this seemingly universal property is the mature mammalian red blood cell (erythrocyte or RBC). Mature mammalian RBCs have lost their nuclei and DNA in the process of differentiation; they also have no *endoplasmic reticulum* or *mitochondria.* Thus, they cannot reproduce themselves from their mature form. However, they do practice metabolism (they make ATP via glycolysis) and do homeostasis. We consider mature RBCs to be "alive." They live about 120 days before being destroyed [Guyton 1991]. In comparison, the erythrocytes of nearly all other vertebrates (except one genus of salamander) have nuclei. The human platelet, essential for blood coagulation, is another anucleate blood cell. It is formed in the bone marrow by *megakaryocytes* and has a half-life in the blood of about 10 days. A platelet has a cell membrane, mitochondria, contractile proteins, and remnants of the Golgi apparatus, and endoplasmic reticulum. The normal concentration of platelets in the blood is from 1.5 to 3 $\times 10^5$ per microliter [Guyton 1991].
7. *During adaptation and evolution,* the DNA of all life-forms is subject to errors in its replication. This damage may come from environmental factors such as excess thermal energy, oxidizing chemicals, ionizing electromagnetic radiation (e.g., UV radiation, γ-rays, x-rays, TeraHertz waves, and microwaves), and energetic ionizing particles (β, α, and neutrons). DNA damage may also be the result of errors in the enzymes used to reproduce it in cell *mitosis* and *meiosis.*

   Genetic errors lead to *mutations.* As you will see, eukaryotic cells contain enzyme systems that can repair breaks and defects in DNA. However, these repairs are not always perfect. Nucleotide base substitutions in genes cause errors in protein primary structure. Errors in protein structure can range from no effect, to fatal, to beneficial for the mutated cell (or organism containing the mutated cells). Not all mutations affect protein structure; however, some affect the structure of RNAs and the regulation of gene expression. Mutations affect an animal's ability to survive in a given environment; hence, *natural selection* operates on the mutated life-form. "Survival of the fittest" is a statistical paradigm.

   These seven properties of life are nearly universal. The attributes listed in the following text have exceptions as noted:
8. *Certain cells can self-replicate, but others cannot.* In metazoan animals, this includes embryonic development, in which undifferentiated stem cells differentiate into adult forms. The self-replication is guided and controlled by the cell's DNA genome. As noted earlier, mature mammalian RBCs and blood platelets have lost their nuclei and DNA and cannot replicate.

   Central nervous system (CNS) neurons in birds and in most mammals cannot reproduce. Certain marsupials, however, do have the ability to regenerate CNS cells. The gray short-tailed possum, *Monodelphis domestica,* can regenerate a crushed or completely severed spinal cord up to the age of 11 days [Wintzer et al. 2004]. The genome of this possum was recently sequenced by a team at the Southwest Foundation for Biomedical Research in San Antonio, Texas [Mikkelsen et al. 2007].
9. *Mature metazoan animals can reproduce,* either sexually or asexually (i.e., by *parthenogenesis).* For an exception, consider the mule: a live mule has grown from an embryo; its cells have differentiated and reproduced; but because it is a genetic *hybrid* (jackass + horse mare), it is *sterile* and cannot reproduce sexually.
10. *Most single cells are capable of gross motion of some sort in response to external stimuli* by deforming their cell membranes to ingest external objects (*pinocytosis*), extending and contracting pseudopods (amoeba), or propulsion by moving their cilia (certain bacteria and protists). Certain motile cells may obey *chemotaxis*; that is, they can follow a chemical gradient toward its source.

In metazoans, most cells are "cemented" in place to their neighboring cells' membranes. Muscle cells can contract in response to motor neuron stimulation, and ciliated epithelial cells wave their cilia to move mucus, for example. Other cells such as mature liver cells just exist as biochemical factories. Their cell membranes apparently do not make gross motions.

In addition, we note that s*ome life-forms can become dormant.* Dormant life includes bacterial and fungal spores and seeds. Dormant life is essentially life without water. Cell metabolism, homeostasis, growth, and reproduction are all suspended until the spore or seed absorbs enough water to restart its biochemical machinery. A strange, eight-legged invertebrate called a *tardigrade* goes dormant (a condition called cryptobiosis) when deprived of water, and it can exist in the dormant state for tens of years. Dormancy is not hibernation.

2.2 THE DOMAINS AND KINGDOMS OF LIFE: THEIR ORIGINS

2.2.1 Introduction

Taxonomists have established robust scientific evidence supporting theories regarding the precursors of the existing life on Earth. This evidence has come from the fossil and geological records, isotopic dating, the science of comparative anatomy, and the science of comparative genomics. From fossils in dated rocks, it has also been possible to estimate when certain species may have first appeared.

Using the resources cited, biologists have been able to construct *phylogenetic trees* that show the evolutionary interrelationships (similarities and differences) between various contemporary species believed to have common ancient ancestors. Extinct and fossil life-forms can also be "treed," generally on the basis of comparative anatomy and habitat. In a phylogenetic tree, each *node* with descendants represents the most recent common ancestor of the descendants, and branch lengths correspond to time estimates. Each node in a phylogenetic tree is called a *taxonomic unit* (TU). Internal nodes are called *hypothetical taxonomic units* (HTUs) because they cannot be observed directly [Kimball 2006h; Joyce 2002]. Figure 2.1 illustrates a phylogenetic tree based on the *three-domain system* of classifying life.

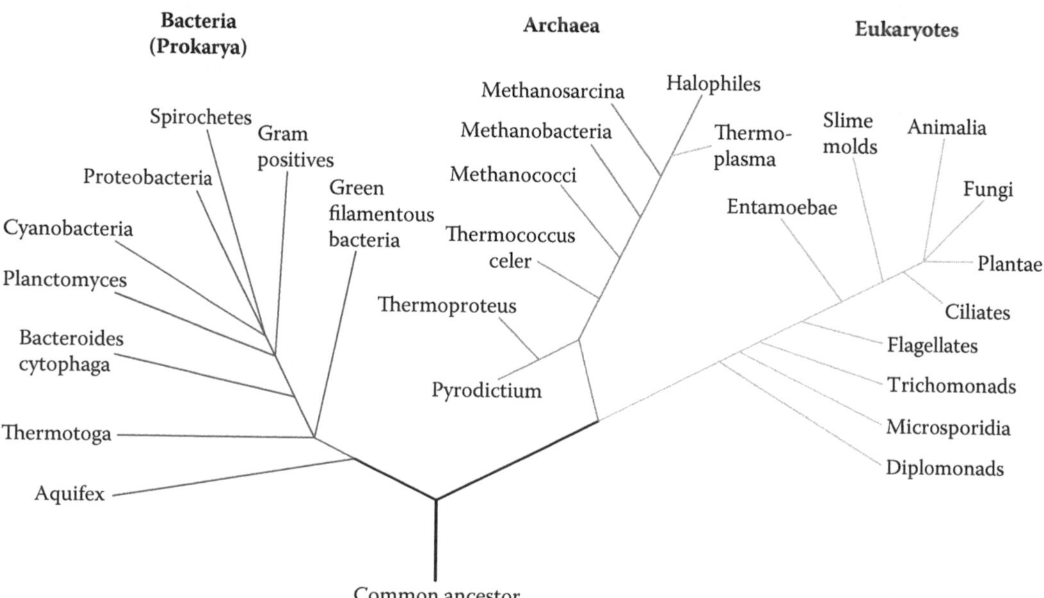

FIGURE 2.1 A phylogenetic tree classifying members of the three domains of life. The structure is based on genomic similarities and differences, morphology, as well as dates estimated from the fossil record.

The Earth itself probably was formed between 5 and 4.54 billion years ago. The oldest Moon rocks from the Apollo missions were dated by two different methods and found to be 4.4–4.5 billion years old. Seventy different meteorites found on Earth were dated by five independent radiometric dating methods and are from 4.4 to 4.6 billion years in age. The oldest rocks in western Greenland, dated by four independent radiometric methods, have an age of 3.7–3.8 billion years [USGS 2004].

Rock dating is approximate using radioisotope decay methods. The reason that isotopic dating is approximate is that the process of radioactive decay itself is a random process, and there are instrumental errors in mass spectrometers used to measure isotope ratios. Errors can also occur if the mineral has not "closed." The loss or gain of parent and daughter isotopes upsets the isotope mass ratios used to estimate age. Closure occurs below a critical temperature; after closure, no parent/daughter isotope can be lost from the rock, and its interior becomes a sealed compartment [USGS 2004]. Scientists who date rocks are aware of the sources of error in their measurements and compensate for them [Meert 2004].

It is quite probable that the first unicellular life appeared on Earth about 3.8 billion years ago. It is not clear whether this life was Prokaryote- or Archaea-type cells, or whether the genomic information stored in these primitive cells was in the form of RNA or DNA. (We shall not comment on the source of this unicellular life at this point.) Geochemical fossil evidence from 2.7-billion-year-old Australian rocks suggests that eukaryotic life existed at that time [Armstrong 1997]. We do know now that *all* present life uses DNA genomes, and that RNA is relegated to regulatory and communication roles.

Prokaryotic life includes all the bacteria (Kingdom: *Monera*). A prokaryotic cell does not have a membrane-enclosed nucleus; instead, it has a closed ring of double-stranded DNA (dsDNA) that carries its genome and genetic control sequences. Bacterial prokaryotes have no mitochondria or chloroplasts, and their ribosomes differ in structure from those of eukaryotic cells. Rigid bacterial cell walls are generally made of *peptidoglycan*; the cell's plasma membrane is a phospholipid bilayer but contains no cholesterol or other steroids. Prokaryotes reproduce mostly asexually. There is no *meiosis* in their sexual reproduction.

The Archaea (Archaebacteria) are microscopic prokaryotes that are generally classified as lying in a separate domain of life on the basis of their biochemistry. They generally live under harsh environmental conditions and may have lived on Earth as long as 3.5–2.7 billion years ago [Armstrong 1997]. They are smaller than most bacterial prokaryotes; usually less than 1 μm long. Some are spherical (coccus), others rod-shaped (bacillus), and some others have strange triangular or flat shapes. The genome of Archaea is also a dsDNA loop (plasmid). There are several features of the Archaean transfer RNA (tRNA) that are more similar to eukaryotic tRNA than that of bacteria. The same is true of Archaean ribosomes.

The cell walls of Archaeans are unique. They must be very robust because Archaeans are found in extreme environments (temperature, pressure, chemical). The basic unit of an Archaean cell wall is composed of a pair of branched, *isoprenoid chains* attached to an L-glycerol phosphate molecule with *ether linkages*. In the cell wall, there is an inner and an outer layer of isoprenoid-L-glycerol phosphate molecules, forming a bilayer in which the center is the facing isoprenoid molecules. The isoprenoid chains are a type of 20-carbon atom *terpene* molecule. Additional wall strength comes from chemical bonds between isoprene side chains and the neighboring isoprene molecules. (See Section 3.2 for a more detailed description of the structure of cell membranes.)

The Archaeans pose a puzzle to evolutionary theorists. On one hand, they are specialized to live in harsh environments, such as might be found on early Earth, and on the other hand, they are genomically closer to the more complex eukaryotic cells found in higher life-forms. If primitive life is found on Mars, or a moon, perhaps it will be closer to the Archaea in their anatomy and biochemistry.

Eukaryotes are structurally and biochemically complex cells. There is geochemical evidence that eukaryotes may have lived on Earth as long as 2.7 billion years ago. They have bilipid cell

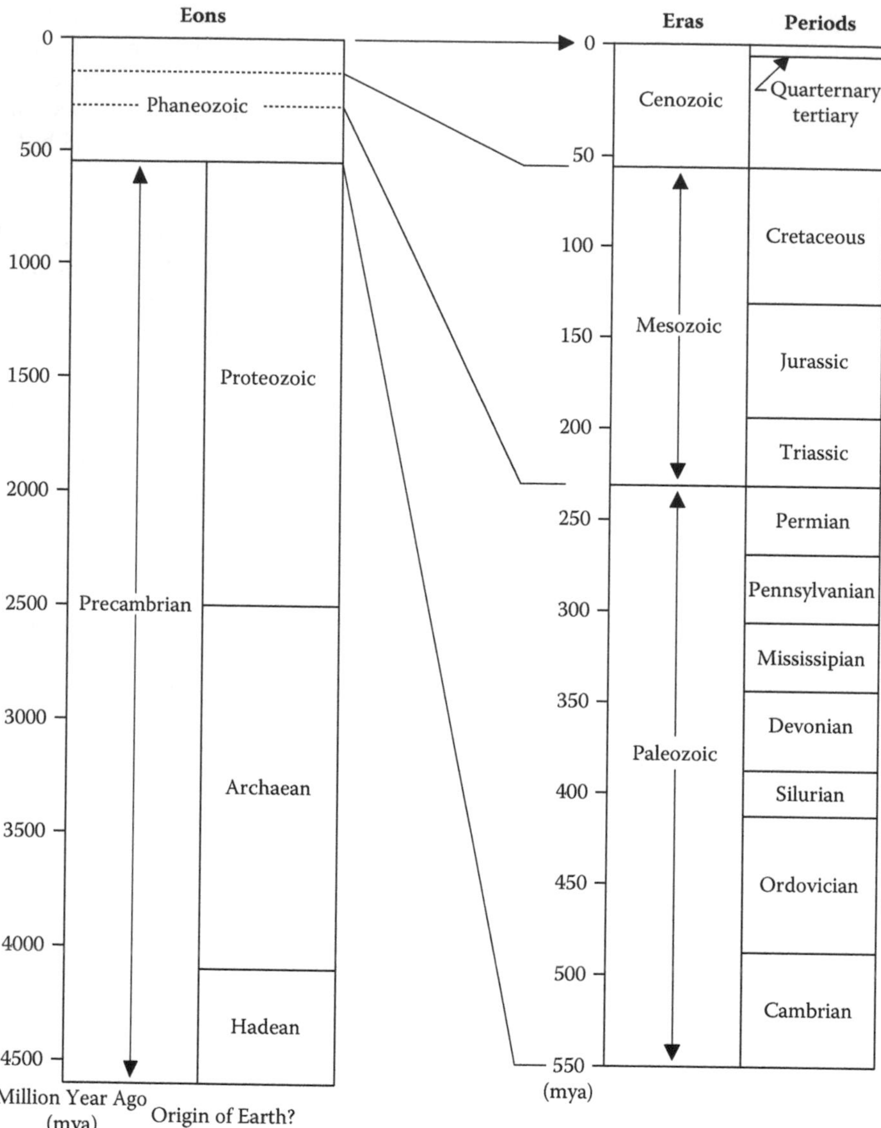

FIGURE 2.2 The eons, eras, and periods over which life has evolved.

membranes through which are embedded complex proteins that serve as receptors, ingestors, ion pumps, etc. Eukaryotes have a membrane-enclosed nucleus, an internal membrane-derived Golgi apparatus, membrane-enclosed mitochondria and ribosomes, and have generally complicated metabolisms and reproductive cycles (mitosis and meiosis).

Figure 2.2 illustrates the generally agreed-on nomenclature for *times and eras* over which life appeared. *Precambrian time* stretches from the first appearance of prokaryotic life about 4.5 billion to 500 million years ago. *Phanerozoic time* stretches from 500 million years ago to the present. Eukaryotic life is dominant in Phanerozoic time (Phanerozoic time is also called eukaryotic time). The Precambrian time can be subdivided into two eras: the *Archaean era* or era of Archaebacteria (4.5 to about 2.8 billion years ago) and the *Proterozoic era,* during which Cyanobacteria capable of photosynthesis and O_2 production appeared. The Proterozoic era is from about 2.8 billion to 0.5 billion years ago.

2.2.2 The Domains of Life

Early approaches to classification relied on an organism's embryological and adult gross morphology. Then ultrastructural features as revealed by electron microscopy (e.g., cell membrane structure and cell organelle structures) were used. Finally, in the last decade of the 20th century, genomic analysis allowed classifiers to compare the structure of homologous genes in species and also examine an organism's entire genome in detail and genomically trace evolutionary branch points.

One approach to classifying life on Earth is the *domain* or *superkingdom* approach (See Figure 2.2).

Three Domains (Superkingdoms) of Living Organisms

I. **Bacteria:** Most of the known Prokaryotes:
   Kingdom *Monera:*
   Division (Phylum) *Proteobacteria:* Nitrogen-fixing bacteria
   Division (Phylum) *Cyanobacteria:* Blue-green bacteria (have photon-capture pigments)
   Division (Phylum) *Eubacteria:* True Gram-positive bacteria
   Division (Phylum) *Spirochetes:* Spiral bacteria (e.g., Borrelia and syphilis)
   Division (Phylum) *Chamydiae:* Intracellular parasites
II. **Archaea (aka. Archaebacteria):** Prokaryotes living in extreme environments
   Kingdom *Crenarchaeota:* Extreme thermophiles
   Kingdom *Euryarchaeota:* Methanogens, halophiles, and thermoacidophiles
   Kingdom *Korarchaeota:* Some hot springs microbes
III. **Eukarya:** Organisms with eukaryotic cells
   Kingdom *Protista:* [about 2.5×10^5 species]: Unicellular protozoans and unicellular and multicellular (macroscopic) algae with 9 + 2 flagellae, and cilia
   Kingdom *Fungi:* [about 10^5 species]: Haploid and binucleate cells, multicellular, generally heterotrophic, without cilia and eukaryotic (9 + 2) flagellae (undulipodia)
   Kingdom *Plantae:* [about 2.5×10^5 species]: Haplo–diploid life cycles, mostly autotrophic, retaining embryo within female sex organ on parent plant
   Kingdom *Animalia:* [about 10^6 species]: Multicellular animals. Animalia can be subdivided into Phylum Invertebrata and Chordata (including Vertebrata).

As an example of classification, consider humans: Their Domain is Eukarya, kingdom Animalia, phylum Chordata, subphylum Vertebrata, class Mammalia, order Primates, family Hominidae, genus *Homo*, and species *sapiens*.

For another example, consider the small marginate conch, *Strombus,* found in the oceans around the Philippines. *Strombus* is a gastropod mollusk about 5 cm in length. It is in the kingdom Animalia, phylum Mollusca, class Gastropoda, order Prosobranchia, family Strombidae, genus *Strombus,* species *marginatus,* and subspecies *septimus,* and the first person to classify it was Linnaeus. This animal is formally described as *Strombus marginatus septimus* Linnaeus. It has a pretty shell.

Animal classification has traditionally been based on anatomy, habitat, embryonic development, and more recently, protein structures (key enzymes, etc.) and the sequencing of its genome and RNAs. An animal's genetic code can be used to establish a phylogenetic tree by comparison with the genomes of other animals. By examining the genes for highly conserved proteins such as *cytochrome* **c** and *ubiquitin* and comparing them with the genes for more rapidly changing protein structures, it is possible to construct a tree that provides supporting evidence for evolutionary relationships.

2.3 NONLIVING, SELF-REPLICATING GENOMIC MACHINES

2.3.1 Introduction

We take the position in this text that *viruses, viroids, plasmids, satellites*, and *virusoids* are not life. In our opinion, they are self-replicating, *genomic machines* (containing DNA or RNA) that require a living host cell to reproduce their genome and accessory molecules. In this respect, they are *molecular obligate endoparasites.*

Outside their host cells, they have no metabolism, homeostasis, or organelles found in cells necessary to support metabolism, homeostasis, protein synthesis, or genomic reproduction. Nor do they have even the remotest capability of doing these things. The genomes of viruses can mutate once inside their host cell, which makes them difficult to fight as disease pathogens. Not all cell and molecular biologists agree with our position.

2.3.2 Viruses

Anyone who has suffered from the flu may find it hard to accept that the influenza virus is not living. They certainly can reproduce rapidly, which is one cause of our misery. Viruses can destroy host cells; this destruction contributes to inflammatory reactions mediated by the immune system. A virion is biochemically inert outside of the host cell, but it destroys the host cell in order to reproduce. It has none of the attributes of life described in Section 2.1.2. Viruses can infect bacteria, plants, and animals and in some cases one another (see Section 2.3.5 in the following text). Some of the human diseases caused by viruses are listed in the following text; they range from benign to annoying to invariably fatal. The deleterious effect on the host animal comes from the viral proteins manufactured inside the infected cells, the eclosion of the new viruses destroying host cells, and the immune system's attempts to destroy infected cells. For a complete listing of diseases with viral etiologies, see Sander [2005]. A partial list of human viral diseases follows:

AIDS (acquired immunodeficiency syndrome)
Bird flu (birds)
Borna disease
Bornholm disease
Cervical cancer
Chicken pox
Cold, common
Cold sores
Colorado tick fever
Cow pox
Dengue fever
Distemper (dogs and cats)
Eastern equine encephalitis
Ebola hemorrhagic fever
Encephalomyelitis
Fifth disease
Foot and mouth disease
Genital HSV
Hepatitis (A,B,C,D,E)
Herpes (various)
Influenza (various)
Lassa hemorrhagic fever
Measles
Meningitis, viral

Mononucleosis
Mumps
Pharyngitis, viral
Pneumonia, viral
Poliomyelitis
Rabies
Roseola
Rubella
SARS
Shingles
Smallpox
Viral encephalitis
Viral pneumonia
Warts
West Nile disease
Western equine encephalitis
Yellow fever
Zoster

Outside of their host cells, free viruses or virions have fairly simple structures compared to living cells. An outer cover, the *capsid,* is made of proteins. The capsid generally has several proteins associated with it that have an affinity for specific cell-surface proteins of the targeted host cells (these targets may be on certain immune system cells, lung cells, blood cells, gastrointestinal epithelial cells, etc.). The capsid protects the viral genome, which can be either DNA or RNA (never both). *DNA viruses* have either dsDNA or ssDNA. *RNA viruses* can be classified into four groups:

1. Those with *single-stranded, (−) sense RNA* (e.g., measles, Ebola, mumps, and rabies).
2. Those with *ss, (+) sense RNA*, which can act as mRNA (e.g., poliovirus, yellow fever, avian infectious bronchitis aka bird flu (H5N1), and hepatitis A).
3. Those with *several pieces of* dsRNA (e.g., *reoviruses*).
4. Those with ssRNA that is copied into the host's DNA by their *reverse transcriptase.* These are called *retroviruses* (e.g., HIV-1).

Figure 2.3 schematically illustrates a human HIV retrovirus virion, showing its components.

There are few viral genes (3–100), depending on the species and viral complexity. The viral genes encode the proteins required for viral reproduction inside the host cell. These are proteins that the host cell generally does not supply directly. Once the viral genome has entered the host cell, it commandeers the cell's complex biochemical machinery to reproduce many copies of its genome, then its necessary proteins. The viral capsid proteins enclose each viral genome. The host cell is then caused to lyse, releasing the daughter virions to infect new host cells. When host cells lyse, their disintegrating parts attract immune system cells, which leads to inflammation and misery.

Bacteriophages are viruses that use prokaryote cells (bacteria) for hosts. Phages range in size from 24 to 200 nm in length. The T4 bacteriophage is relatively large: about 200 nm long and 80 to 100 nm wide (see Figure 2.4). They prey on the ubiquitous *E. coli.* Strains of phages are used in laboratory cultures to identify bacterial genera because of their infective specificity. Certain phages have also been used to fight infections by attacking specific bacteria [Mayer 2004; Kutter 1997, 2000; Weber-Dubrowska et al. 2000]. Fortunately, they do not attack human cells.

An important class of viruses discovered in the 1960s is the *Parvoviruses.* The canine parvovirus is a highly contagious virus generally affecting puppies, and "parvo" immunizations are standard veterinary practice. As their name suggests, the virions are small icosahedrons, about 22–28

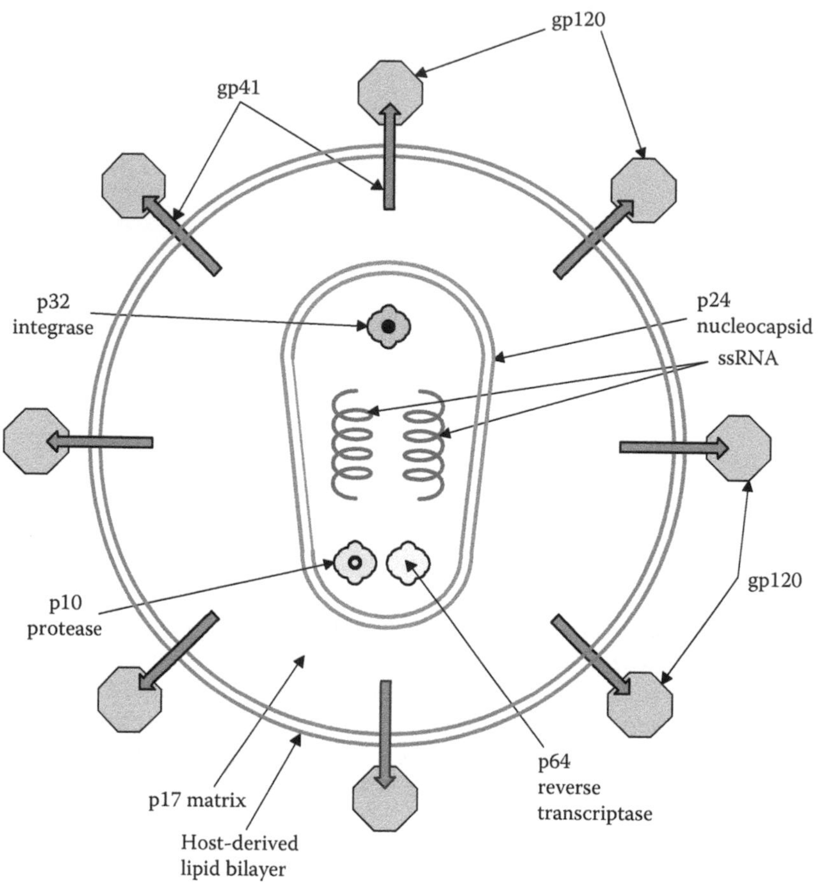

FIGURE 2.3 (See color insert following page 44.) Schematic cross section of the human HIV retrovirus virion. (Adapted from *Endogenous and Exogenous Regulation and Control of Physiological Systems*. Chapman and Hall/CRC Press, Boca Raton, FL. With permission from Taylor & Francis.)

nm in diameter. They are known to infect insects and mammals. There are two subfamilies of the Parvoviridae: the Parvovirinae infect mammals, and the Densovirinae infect insects.

In 1975, the *parvovirus B19* was discovered to infect humans. The B19 parvovirus is packaged into an icosahedral protein shell of about 28 nm in diameter. The capsid consists of 60 structural subunits, about 95% of which are the major viral protein, VP2. The other structural protein, VP1, differs from VP2 only in an N-terminal "unique region" composed of 227 AAs that are mostly located outside the virion and serve as attachment sites to the host cell [Kaufmann et al. 2004]. The B19 genome is ssDNA, with a total of 5596 Nts. An internal coding sequence of 4830 Nts has the genes for three proteins (VP1, VP2, and a single nonstructural protein, NS1). It is flanked by two terminal repeat sequences of 383 Nts each.

B19 has been associated with a variety of generally nonfatal human diseases. These include *"Fifth disease,"* where the patient, usually a child, exhibits a rash on his or her extremities or face. It also causes arthritis (arthralgia), hemolytic anemia, and *hydrops fetalis*. Because it is extremely contagious (spread in saliva and nasal secretions), about 60% of adults are seropositive to this virus and are probably immune to it [Siegal 1998]. It has been shown that the B19 receptor is the human P antigen found on various erythroid progenitor cells and megakaryocytes, endothelial cells, and fetal myocytes [Heegaard & Brown 2002].

In closing this section, we remark that often deadly viruses that quickly kill their host animal survive because of secondary viral hosts in which they are not deadly. An example is the *La Crosse*

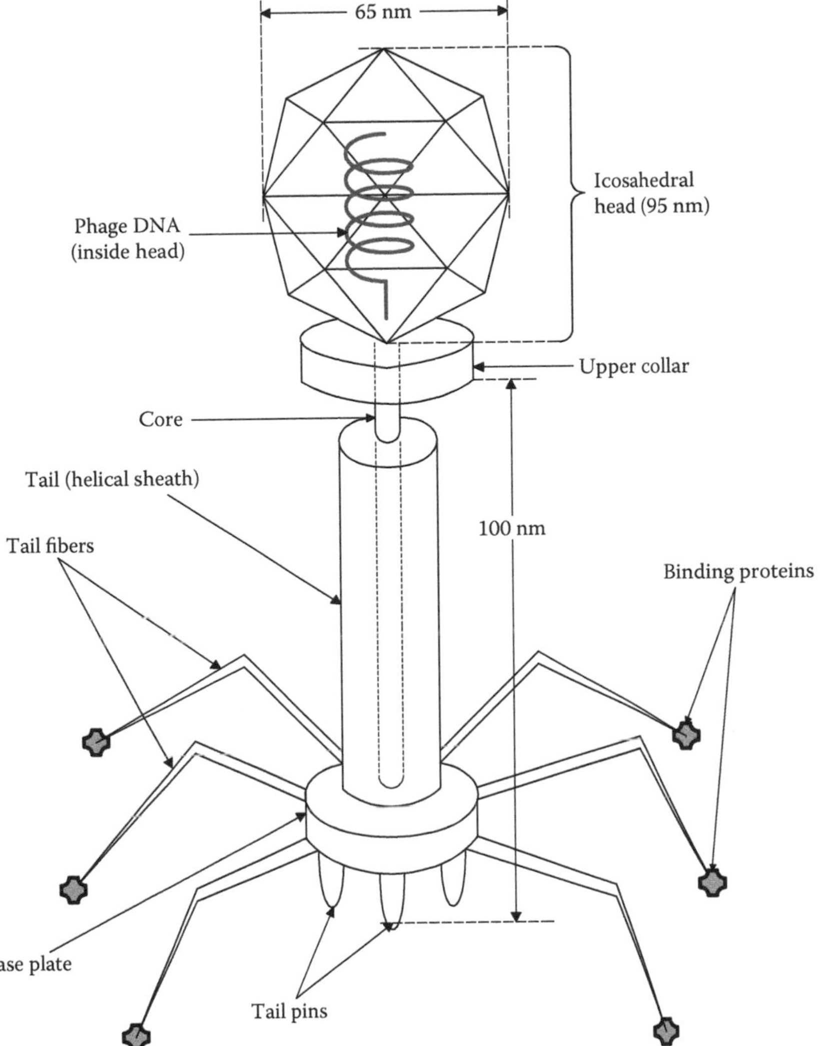

FIGURE 2.4 Schematic figure of a T4 bacteriophage virus. Notice the structural complexity compared to the HIV virion.

virus which causes pediatric encephalitis. The La Crosse virus has an ecological reservoir in the midgut epithelium of the mosquito vector, *Aedes triseratus* [Domingo et al. 1998].

2.3.3 Viroids

Viroids are small, circular ssRNA molecules that lack the protein capsid of viruses. They cause several infectious diseases of higher plants (such as tomatoes, potatoes, and cucumbers). They enter the host cell by pinocytosis and then insert themselves into the cell's nucleus. Their RNA genome has from 240 to 400 nucleotides; it replicates in the host's nucleus by symmetric or asymmetric *rolling circle*. Viroids are classified into two major groups: the *Potato Spindle Tuber Virus* (PSTV) group and the *Avocado Sunblotch Viroid* (ASBV) group. The genomes of most viroids discovered resemble that of the PSTV.

PSTV-like viroids use the host cell's nuclear *RNA Polymerase II* to reproduce their genome. The reproduction method of ASBV-like viroids is not well understood; it may involve *RNA Polymerase I* in the nucleus or RNA polymerases in chloroplasts. The genome of a given viroid carries species specificity (how?) and a specific host pathogenic effect (how?) and the ability to move from cell to cell in the host plant. Compared to the genomes of conventional encapsulated viruses, viroid genomes are simple. Some other key questions about viroids remain to be answered: (1) What are the enzymes and cofactors used in their replication? (2) How is the switch from (–) to (+) RNA synthesis achieved? (3) Which proteins protect the circular viroid and facilitate its movement through the host? (4) How does the viroid escape the host's innate defenses, particularly against invading nucleic acids? (5) What is the relationship of the viroid's genome to epigenetic modifications of the host's genome? (See Tabler & Tsagris [2004].)

2.3.4 Plasmids

A natural *plasmid* is a small, autonomously replicating, circular molecule of dsDNA, devoid of protein, and not essential for the survival of its host cell. Plasmids have a great range in size: from about 4,350 to 240,000 bps. Plasmids are not life-forms. Similar to viruses and viroids, they have no metabolisms and require living cells to reproduce. Some important classes of plasmid are (1) R plasmids, (2) F plasmids, (3) Colicin plasmids, (4) virulence plasmids, (5) metabolic plasmids, and (6) tumor-causing plasmids [Baez 2005].

Synthetic plasmids are used in *genetic engineering.* In genetic engineering, they are called *cosmids* and *phasmids.* Cosmids are synthetic circular loops of DNA containing plasmid DNA together with an arbitrary sequence of up to 45,000 bps of DNA. They are constructed by recombinant DNA technology and then packaged in lambda phage protein coats. They are used to transfer genes to bacteria.

The *Lambda phage* is a naturally occurring virus that specializes in infecting bacteria such as *E. coli.* Phage virions literally inject their DNA genome into bacteria. Once inside a host bacterium, the cosmid replicates similar to a plasmid and integrates its DNA into the genome of the bacterium so that its genes can be expressed along with those of the bacterium [Baez 2005].

A *phasmid* is an engineered, linear (dsDNA) molecule whose ends are nucleotide sequences taken from the lambda phage. The middle of the phasmid is a sequence taken from a plasmid, to which is attached the desired DNA sequence one wants to insert into the DNA of the target bacteria. Similar to cosmids, they are assembled using recombinant DNA techniques and then packaged in lambda phage protein coats. Interestingly, both the lambda phage and plasmid replication functions are intact in a phasmid. Importantly, phasmids contain the lambda phage genes for "lysis," the process whereby this lambda phage dissolves its host's cell membrane at eclosion. Depending on the conditions, a phasmid can act either as a phage or a plasmid and, hence, its name.

2.3.5 Satellites and Virusoids

A satellite is a *subviral agent* composed of nucleic acid molecules that depends for its reproduction on the coinfection of a host cell with a *helper virus.* In other words, just as a planet can have moons orbiting it, a virus can have a satellite orbiting it. There are several kinds of satellites such as *satellite viruses* and *satellite nucleic acids*, including satellite dsDNAs and satellite ssRNAs.

All five known satellite viruses have ssRNA genomes. An example of a satellite virus is the tobacco mosaic satellite virus for which the well-known *tobacco mosaic virus* (TMV) is its "helper." The genome of *satellite nucleic acids* does not code for a protein coat. Instead, it hides in the protein coat of its virus helper. All known satellite DNAs are single stranded (ss), but satellite RNAs can be either single stranded (ss) or double stranded (ds). Satellite ssRNAs are also called *virusoids.* They are among the very smallest, nonliving, self-replicating, dependent, obligate, cellular endoparasites. They have circular loops of ssRNA containing about 350 Nts, and they can only reproduce in cells that have been infected by their helper virus because they use some of the RNA of the helper virus to reproduce. Their helper virus is typically an RNA virus that causes a plant disease and has about 4500 Nts.

Baez [2005] is of the opinion that a virusoid is a nonliving, molecular parasite of its helper virus. He points out that their association is not always so simple. Sometimes the helper virus is unable to reproduce unless the virusoid is present. Then symbiosis occurs instead of parasitism.

Baez stated: "An interesting theory about the origin of virusoids is that in plants infected with both viruses and viroids, the viroids got encapsulated in the viruses and later lost their ability to reproduce independently."

2.4 THE ORIGINS OF LIFE

2.4.1 Introduction

In order to consider the origin of life on Earth, we must first review the *scientific evidence* for the age of Earth rocks and minerals, rocks from the moon, and meteorites. Fossilized life is found in rocks, and in order to date a fossil, we must date the rock in which it is imbedded. Rocks and minerals are dated using the radioisotopic decay of elements found in the rock. For example, if a given sample of a rock cooled from magma at a relative time, $t = 0$, with an initial concentration of radioactive uranium 238 (^{238}U), this element would undergo radioactive decay to a final product, which is the isotope, lead 206 (^{206}Pb). As ^{238}U decays, there are 13 intermediate radioactive daughter products formed (e.g., radon, polonium, and other unstable isotopes of uranium), as well as 8 alpha particles (helium nuclei; charge = +2, mass = 4) and 6 beta particles (electrons; charge = −1). Neutrons are also released. Radioactive decay is a completely random process, and the events in the decay process are described by the Poisson probability function. The rate of radioactive decay depends only on the element decaying. It is basically independent of temperature, pressure, and concentration, as long as the relative concentration of the radioisotope is low.

Let us consider the simple mathematical relation between the *parent isotope* (here, ^{238}U) and the ultimate *daughter* product (^{206}Pb) formed in a given volume of rock after an elapsed time, t_o. Let $N_{Pb}(t_o)$ = number of ^{206}Pb atoms at time t_o. Let N_{U0} = the number of ^{238}U molecules at the start of the radioactive decay period (i.e., $t = 0$), and $N_U(t_o)$ is the ^{238}U remaining at $t = t_o$. It can be shown that:

$$N_U(t_o) = N_{U0} \exp(-t_o/\tau) \tag{2.1}$$

where τ is the radioactive decay *time constant*. For the $^{238}U/ ^{206}Pb$ system, $\tau = 6.45 \times 10^9$ years, or 6.45 billion years. As the ^{238}U decays, there is a 1:1 increase in ^{206}Pb atoms. This can be expressed mathematically as

$$N_{Pb}(t_o) = N_{U0}\,[1 - \exp(-t_o/\tau)] + N_{Pb0} \tag{2.2}$$

The second term is any small amount of the daughter isotope that may be present at $t = 0$. Graphs of the preceding two equations are shown in Figure 2.5. Because the rock being tested is assumed to be a closed system, it is easy to show that

$$[N_U(t_o) + N_{Pb}(t_o)] = [N_{U0} + N_{Pb0}] \tag{2.3}$$

In more general terms:

$$[D(t_o) + P(t_o)] = [D_0 + P_0] \tag{2.4}$$

where D is the daughter isotope, and P is the radioactive parent isotope. It can be shown as follows:

$$t_o = \tau \ln\left[\frac{\dfrac{P(t_o)+D(t_o)}{P(t_o)}}{\dfrac{P_0+D_0}{P_0}}\right] = \tau \ln[P_0/P(t_o)] \tag{2.5}$$

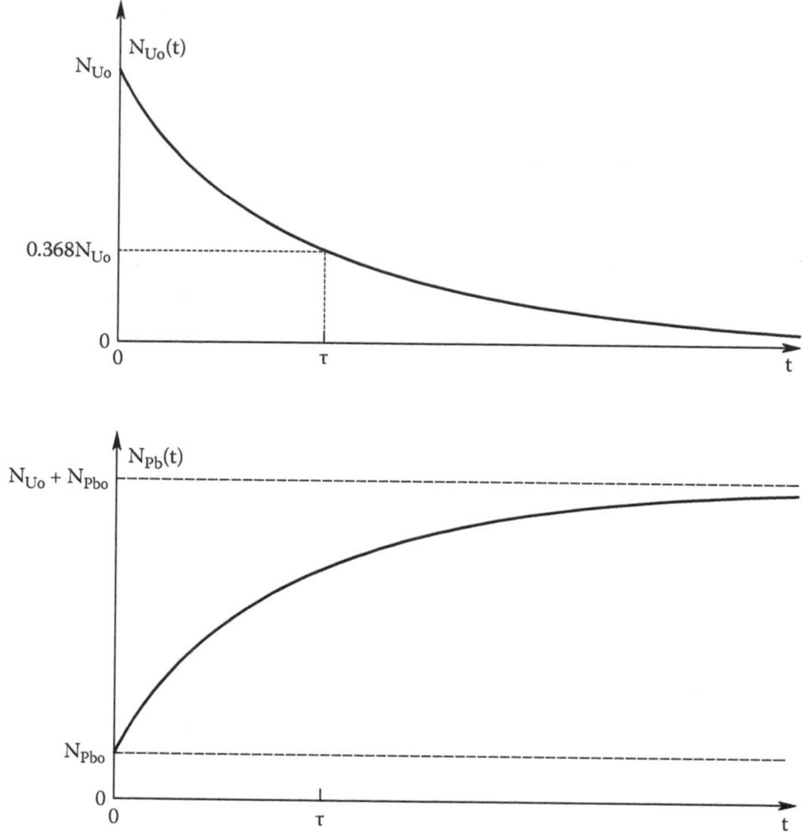

FIGURE 2.5 Plots of Equations 2.1 and 2.2 describing isotope decay and decay product growth versus time.

If D_0 can assumed to be zero, then

$$t_o = \tau \ln[1 + D(t_o)/P(t_o)], \text{ (rock age)} \tag{2.6}$$

A critical assumption in radio-dating rocks and minerals is that once the originally molten rock has cooled below a critical temperature called the *closure temperature*, no exchange of parent and daughter isotope is possible between the rock and its surrounding environment, making the rock a closed compartment. Different parent radioisotopes, their decay product daughter isotopes, and their half-lives ($T_{1/2}$) are listed in Table 2.1 in the following text. The half-life is the average time taken for a given number of radioisotope molecules to decay to half that number. It is easy to show that the decay *time constant* τ used in the preceding equations is given by $\tau = T_{1/2}/\ln(2) = 1.4427\ T_{1/2}$.

The mass of isotopes in a weighed sample of rock is determined by use of a *mass spectrometer* (MS). The MS was developed after 1918 and has evolved into several designs of extremely sensitive analytical instruments that can precisely measure the atomic mass of ions, charged radicals, and small ionized molecules, as well as the amounts present in the sample [Northrop 2002].

2.4.1.1 Carbon Dating

Carbon 14 (^{14}C), because of its short half-life, is used to date organic archaeological artifacts, generally not more than 70,000 years old. Cosmic rays from the sun and outer space strike ^{14}N atoms in the Earth's atmosphere, converting them to ^{14}C. ^{14}C is an unstable carbon isotope that emits a high-energy electron (β particle) in its decay process. It enters the geochemical carbon cycle, and some

TABLE 2.1
Half-Lives of Radioisotopes Used in Dating Rocks, Minerals, and Meteorites

Radioactive Parent	Stable Daughter	Half-Life ($T_{1/2}$) (years)
Potassium 40 (^{40}K)	Argon 40 (^{40}Ar)(gas)	1.25×10^9
Rubidium 87 (^{87}Rb)	Strontium 87 (^{87}Sr)	48.8×10^9
Thorium 232 (^{232}Th)	Lead 208 (^{208}Pb)	14.0×10^9
Uranium 235 (^{235}U)	Lead 207 (^{207}Pb)	0.704×10^9
Uranium 238 (^{238}U)	Lead 206 (^{206}Pb)	4.47×10^9
Samarium 147 (^{147}Sa)	Neodymium 143 (^{143}Nd)	106×10^9
Rhenium 187 (^{187}Re)	Osmium 187 (^{187}Os)	43.0×10^9
Carbon 14 (^{14}C)	Nitrogen 14 (^{14}N) (gas)	5730

ends up as $^{14}CO_2$, which is taken up by plants during photosynthesis. Then it gets into the food chain and eventually in all living things. All living things contain a constant ratio of ^{12}C–^{14}C, about 10^9:1 [Gore 1999]. At an organism's death, radioactive ^{14}C exchange due to metabolism ceases and any ^{14}C in the tissues continues to decay to ^{14}N. At death, the organic artifact is assumed to have "closure"; that is, its ^{14}C is not replaced from the environment. What is examined in the artifact being carbon-dated is the ratio of ^{14}C to ^{12}C. Archaeological dating with ^{14}C is useful from the recent past to the ice ages (Pleistocene epoch). It assumes that the rate of ^{14}C production in the Earth's atmosphere and, hence the flux of cosmic rays striking the Earth, has been constant over the past 70,000 years, giving a constant ratio of $^{14}C/^{12}C$ in the atmosphere and in all living creatures. It also assumes that there has been no ^{14}C exchange with the artifact after it died, either in gas form or as CO_2 dissolved in water.

As an example, if a piece of pine wood from an ancient gravesite is dated by the ^{14}C method and found to contain only half as much ^{14}C as an equal mass of wood from a piece of recently cut pine, the estimated age of the artifact is about 5,730 years.

Extrapolation from the ages of rocks on Earth, meteorites found on Earth, and Moon rocks indicates that Earth was formed about 4.5–4.6 billion years ago. Indeed, the age of Earth as extrapolated from the results of extensive isotope dating of "old" Earth rocks, meteorites, and Moon rocks is put at about 4.55 billion years, ± 2% [Stassen 2005]. The Apollo Moon rocks were dated using two methods and found to be between 4.4 billion and 4.5 billion years old. The majority of 70 well-dated meteorites have ages of 4.4–4.6 billion years. Four very old lead ore (galena) samples were found to be 4.54 billion years old. Rocks in western Greenland have been dated by four independent radiometric dating methods and found to be 3.7–3.8 billion years old [USGS 2004].

Clearly, after it condensed, Earth was very hot; it gradually cooled to a spherical shape with a solid crust over a molten core. Oparin [1953] summarized the conditions hypothesized to be on the recently solidified planet: "… the Earth must have had the following appearance: a central molten core, abounding in native metals, covered by a membrane of primary igneous rocks and, finally, all enveloped by an atmosphere consisting principally of superheated aqueous vapor, some nitrogen and other heavier gasses. The carbon, which is of special interest to us, was in the central molten core in the form of carbides of iron and of other metals, and was separated from the atmosphere by the layer of primary igneous rocks."

Oparin goes onto state that it was highly probable that Earth's primary atmosphere was devoid of free, elemental oxygen (O_2). Probably, Earth's elemental oxygen was bound up as oxides, in particular, in its atmosphere as superheated water vapor. "Carbon made its first appearance on the Earth's surface not in the oxidized form of carbon dioxide but, on the contrary, in the reduced, in the form of hydrocarbons." Oparin described many spontaneous chemical reactions that are known

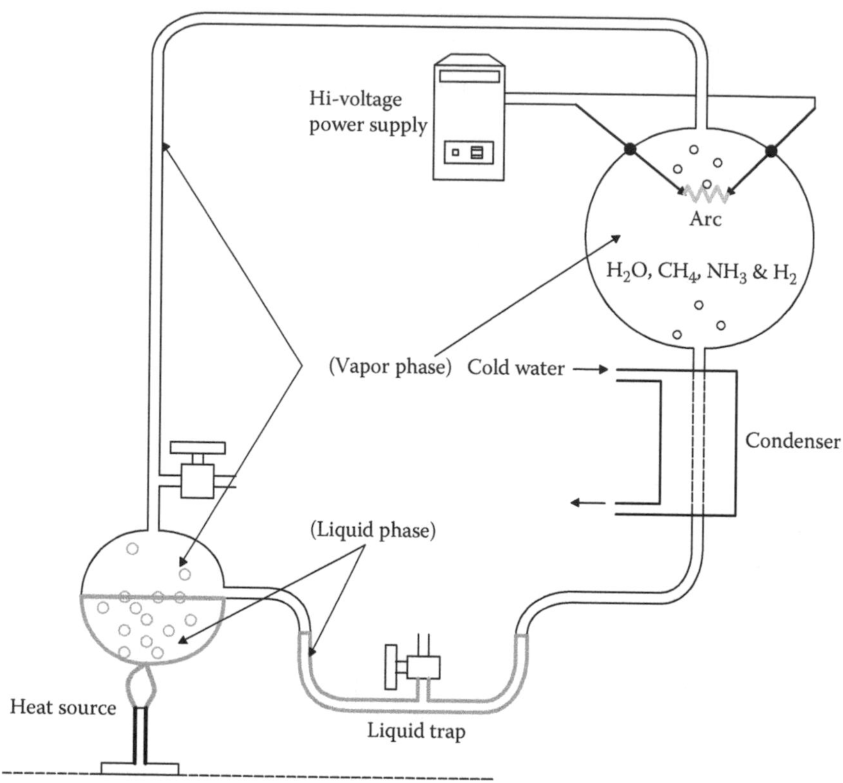

FIGURE 2.6 A schematic of the Miller–Urey apparatus. See text for description.

to take place at high temperatures between hydrocarbons and water vapor. For example, at 1000°C, two methane molecules (CH_4) spontaneously form one molecule of acetlyene (CH≡CH) plus three molecules of H_2. Acetylene reacts with water vapor at 300°C to form acetaldehyde (CH_3COH). Oparin stated: "Considerable quantities of various oxidation products of hydrocarbons, such as alcohols, aldehydes, ketones, and organic acids must have originated as a result of such [chemical] transformations on the Earth's surface." He goes on: "As the temperature of the Earth had cooled off sufficiently to allow the formation [condensation] of drops of liquid water, torrents of boiling water must have poured down upon the Earth's surface and flooded it, thus forming the primitive ebullient oceans. The oxygen and nitrogen derivatives of the hydrocarbons already present in the atmosphere were carried down by these torrential rains and the oceans and sea, at the moment of their first formation, contained, therefore, the simplest organic compounds in solution. ...alcohols, aldehydes, acids, amines, amides, etc., continued to react with each other as well as with the elements of the aqueous environment, giving rise to a prodigious number of all sorts, and even more complex, organic compounds."

Following Oparin's [1938] conjectures, Miller [1953] and Miller and Urey [1959] experimented with the abiotic (and anaerobic) synthesis of organic molecules in the laboratory. Figure 2.6 illustrates Miller's apparatus. Miller filled it initially with water, methane, ammonia, and hydrogen, but no O_2. He was trying to emulate the atmosphere of early, recently formed Earth. He boiled the liquid phase, ran the vapor phase through a continuous spark discharge, condensed it, and repeated the process in a continuous, closed cycle. After a week, he used paper chromatography to show that his reactor had made several new organic molecules as well as several amino acids. Since Miller's pioneering work, other biochemists and biophysicists have tried variations on his procedure. Many of the molecules associated with life were formed: 17 of the 20 AAs used in protein synthesis and all of the purine and pyrimidine components of nucleic acids. However, no ribose, necessary for

DNA and RNA, was found. One aspect of this chemical scenario that needs further investigation is whether biopolymers could self-assemble on catalytic surfaces [Kimball 2004a]. This research was not intended to make life, only to show that the biochemical components of life and metabolism could result from extreme anaerobic conditions of temperature, pressure, and ionizing sources (e.g., natural lightning and UV radiation).

Thus, there is general agreement [Orgel 1973; Crick 1981] that a billion years or so after planetary formation, perhaps as many as 3.9 billion years ago, the stage was set for the appearance and survival of primitive life on Earth. The primordial "stew kettle" was ready.

2.4.2 The First Life

Although there may be general agreement on the geochemical events on early Earth, some of which are based on laboratory experiments that attempted to reproduce conditions of substrate chemicals, temperature, pressure, electrical discharge, etc., there is general disagreement on what caused life to appear. Some scientists believe that life arose spontaneously about 4 billion years ago and has evolved to what we see today. Other persons of faith believe that God created life, about 4 billion years ago, and through the mechanisms of evolution, which he (or she) also created, life has evolved throughout the millennia to our present biosphere. These beliefs, of course, contradict the biblical account given in Genesis I in the Bible and contradict the position of "Young Earthers," who, on faith, believe that all life-forms (including humans) were created in one short period on Earth, some 6,000 years ago.

In the following text, we will discuss two current, plausible, nonfaith theories for the appearance of the first life on Earth. These are the theories of *Directed Panspermia* (DP) and *Abiogenesis (or Biopoesis)*.

The theory of DP presented by Crick [1981] and Orgel [1973] argues that it would have been almost impossible to start life on Earth, given the biochemical complexity of even the simplest prokaryotes and Archaea today. Sadly, DP puts off the problem of understanding and describing how life did originate and evolve on Earth. DP argues that life evolved or was created elsewhere in the universe and came to Earth either by cosmic dust, meteorites (on their interior), or by a "spaceship." This life must have been very primitive and very rugged to withstand a journey in the vacuum and cold of space, and the heat of reentry if the life-form was in a meteorite.

One argument for DP is that all life today (Prokaryotes, Archaeans, and Eukaryotes) uses a nearly universal dsDNA genetic code read out by RNA. That there is little variation in this basic design is said to argue against development of life on Earth, which supposedly would have led to more variability in the genomic code for proteins. Or, was the initial design so robust that no other genomic design could effectively compete? In any case, if one subscribes to DP, one still has to explain how life arose elsewhere in the universe.

Abiogenesis theorizes that primitive life was self-started on early Earth by *really* fortuitous chemical reactions. Conditions on the recently cooled planet first gave rise to organic monomers. Then over the millennia, biological polymers arose, including RNA and structural phospholipids. Here is the big jump; the organic substrates developed self-replication and, finally, some sort of protective cell membrane.

In the 1970s, Manfred Eigen at the Max Planck Institute examined the transient stages between the molecular chaos in the prebiotic "soup" and the transient stages of a self-replicating *hypercycle* between the molecular chaos of the prebiotic soup and simple, macromolecular, self-replicating systems. It was postulated that in a hypercycle, the information storage system (ISS) (probably RNA) produced an enzyme that catalyzed the formation of another information system, in sequence, until the product of the last ISS aided in the formation of an ISS, similar to the first system. Perhaps, ribozyme-like enzymes mediated the catalysis required in the hypercycle. Still missing is the development of the *enclosure* of the hypercycle system by a membrane and the coding of proteins by nucleic acids.

One slant on abiogenesis concerns *the RNA World*. This term was coined in 1986 by Walter Gilbert writing on the then-recent observations about the catalytic properties of RNA *ribozymes* [Kimball 2005q]. The RNA World hypothetically took place in a span of about 500 million years, from 3.5 billion to 3.0 billion years ago. During this period, proteins and protein enzymes did not yet figure in biochemical reactions, and RNA was postulated to carry both the genetic information storage and perform the catalysis necessary in primitive, self-replicating, organic chemical systems. Gilbert noted that neither DNA nor proteins were required in such a primitive organic chemical system. A major difficulty in accepting the RNA World scenario is modeling the transition from RNA-based information storage and catalysis to DNA information storage and protein catalysis. One such **RNA → DNA + Protein** molecular scenario has been proposed by Line [2005].

We note that DNA is much more suitable for genomic information storage; it is chemically more stable than RNA. So, did some sort of early *reverse transcriptase* copy original RNA information to DNA molecules? Modern rRNA is found in association with many proteins, but it is the rRNA that catalyzes the assembly of polypeptides, not the proteins.

We have several difficulties in accepting the validity of the RNA World hypothesis. Neither nucleotides nor ribose were found in the Urey–Miller reactor experiments. Somehow, they would have to be assembled from their components: deoxyribose; the bases **A**, **T**, **G**, and **C**; and phosphates. This assembly involves the enzyme-directed expenditure of chemical energy to form *covalent bonds* between the bases. Only 8% of purines will bind to the 1′ carbon of ribose spontaneously, under anhydrous conditions. Pyrimidine bases will not spontaneously bind to the 1′ carbon of ribose anhydrously. Bases do not react spontaneously with ribose in aqueous media. The base cytosine easily undergoes hydrolysis and has no plausible prebiotic synthesis scenario. Phosphate ions are rare because they precipitate with calcium and carbonate ions. Thus, it appears to us that the RNA World hypothesis is built on shaky ground.

The Oparin theory of *coacervates* describes how at the correct pH and ionic milieu, colloidal organic particles form a closed surface (the process of coacervation) around a central volume. This coacervation formed a protective, protocell membrane. By isolating the inner volume, the chemical reactions therein were protected and could have better access to the molecular substrates needed, which penetrated the *membrane*. There lies the rub; how did these primitive boundaries keep out the "bad" molecules and let in the "good" ones? Such mechanisms are seen in modern, eukaryotic cells. Complex protein receptors and "gatekeepers" are embedded in the cell membranes to regulate ionic and molecular in/out traffic. Note that there is a quantum leap in complexity between a coacervate "cell" and the actual, sophisticated cell membrane structures seen in contemporary archaean, prokaryote, and eukaryote cells.

Another puzzler in contemplating the "spontaneous" development of cells is the nucleic acid–enzyme problem. Perhaps, earliest life used available metal carbides and oxides as catalysts of their chemical reactions. In a modern cell, the genomic readout and cell reproduction are promoted by protein and RNA enzymes whose structures are very specific, and in turn, are coded in the DNA genome of the cell. Did primitive, short DNA molecules code randomly at first until a protein structure was found that increased the survival probability of a particular DNA gene? And what about the role of RNA in early protocells? Which came first, the mRNA or the DNA? Life today generally uses D-isomer sugars (relative to D-glyceraldehyde) and builds proteins out of L-amino acids. (One exception is seen in the molecular *chirality* preferences of contemporary Archaeans.)

Kimball [2004a] points out that the (present) synthesis of DNA and RNA requires protein enzymes; nucleic acids cannot be made without proteins, and proteins cannot be made without nucleic acids (the classic *chicken or egg first* paradox). Kimball writes that we now know that certain RNA molecules called *ribozymes* incorporate two major features necessary for life: storage of information (in their Nt base sequence) and the ability to act as catalysts.

Kimball states: "While no ribozyme in nature has yet been found that can replicate itself, ribozymes have been synthesized in the laboratory that can catalyze the assembly of short oligonucleotides into exact complements of themselves. The ribozyme serves as both the template on which short lengths of RNA 'oligonucleotides' are assembled following the rules of base pairing and the catalyst for covalently linking these oligonucleotides. In principal, the minimal functions of life might have begun with RNA and only later did proteins take over the catalytic machinery of metabolism and DNA takes over as the repository of the genetic code. Several other bits of evidence support this notion of an original 'RNA world': Many of the cofactors that play so many roles in [contemporary] life are based on the [sugar] ribose; for example: ATP, NAD, FAD, coenzyme A, cyclic AMP, GTP. In the cell, all deoxyribonucleotides are synthesized from ribonucleotide precursors. Many [contemporary] bacteria control the transcription and/or translation of certain [DNA] genes with RNA molecules, not protein molecules." [This, and all other quotes from J. Kimball's *Biology Pages*®, are used with his kind permission.]

An interesting speculation, but why does all contemporary life now use dsDNA for information storage and relegate RNA to executive roles?

Clearly, the study of biochemical evolution is highly speculative. There is a fossil record of early prokaryotes found in sandstone dating about 3.8 billion years ago. Unfortunately, fossils tell us nothing about a cell's molecular biology. We only have evidence from comparative cell biochemistry, the genome structures, and the proteins of various contemporary life-forms. This evidence is used to make informed conjectures as to what life-form evolved to what over the millennia to create evolutionary "trees."

Some of the earliest traces of early prokaryotic life are found fossilized in rocks isotopically dated at about 3.8 billion (3.8×10^9) years of age. These fossilized cells resemble present-day *cyanobacteria*. Modern cyanobacteria are prokaryotes that have an elaborate and highly organized system of internal membranes which function in photosynthesis. Blue-green cyanobacteria capture photons and use them to drive their metabolic engines. Chlorophyll IIA and some accessory pigment molecules (phycocyanin and phycoerythrin) are embedded in their photosynthetic lamellae. These pigments give the cyanobacteria a wide spectrum of colors: red, yellow, green, blue-green, deep blue, and violet cyanobacteria are known. Cyanobacteria can be single celled or colonial. Depending on the environment and the species, they can form filaments, sheets, or hollow balls. They may even form spores under harsh environmental conditions. Cyanobacteria also form complex symbiotic relationships with many fungi, forming lichens. Other cyanobacterial symbioses include certain protists, marine sponges, diatoms, echiuroid worms, angiosperms, cycads, pteridophytes, liverworts, and mosses. Cyanobacteria reproduce by binary fission, budding, or fragmentation.

Lindell et al. [2004, 2005] have reported that the marine phage, *Podovirus* P-SSP7, infects the marine cyanobacterium, *Prochlorococcus*. This in itself is not remarkable, but the fact that the phage virus carries photosynthesis genes in addition to the genes it needs to reproduce itself in a cyanobacterial host is. Of course, the phage virion does not practice photosynthesis or any other type of metabolism; for that matter, it is not alive. The phage's photosynthesis genes have overlapping start and stop codons and are transcribed along with essential phage capsid genes, suggesting that they are integral parts of the viral DNA genome. Both the phage and the cyanobacterium carry the ***hli*** high light-inducible genes for proteins that protect the bacterium's photosynthetic machinery from light damage. They both also carry ***psbA***, a gene encoding the *photosystem PSII* core reaction protein ***D1***. Cyanobacteria photosynthesis uses the *Photosynthesis II* (PS-II) process. Experimental evidence indicates that the viral genome boosts photosynthesis in the host cell, giving the cyanobacteria more energy (ATP) and thus a longer time for the phage to reproduce inside it, ultimately leading to more phages on viral eclosion. One wonders how many other viruses bring "gifts" to their hosts to enhance host stamina in support of viral reproduction.

Probably the most important contribution to life on Earth by this 3.5-billion-year-old class of bacteria is its ability to produce *molecular oxygen* (O_2) by photosynthesis. Thanks to this photosynthesis in the *Archaean* and *Proterozoic* eras, the O_2 concentration in Earth's atmosphere began to rise about 2.2 billion years ago. It rose slowly because the free O_2 formed oxides with exposed, reactive minerals in Earth's crust. Iron oxides were formed, which show up today as red layers in sandstone about 2 billion years old. This mineral buffering of O_2 went on for perhaps 1.4 billion years until there was little exposed elemental iron left to oxidize. At that point, about 800 million years ago, the atmospheric O_2 partial pressure rose abruptly to about 20%. It spiked again to a peak of about 37% about 400 million years ago, no doubt owing to the flourishing of photosynthetic algae in the oceans and an abundance of plants in the *Devonian, Carboniferous*, and *Permian eras.* In support of this hypothesis, the geological record shows a drop in atmospheric CO_2 at the same time as plants photosynthesized.

The O_2 partial pressure dropped during the *Triassic period* to about 15% (the CO_2 level rose again), then after some fluctuations rose to about 24% during the *Cretaceous/Cenozoic eras* and finally dropped to its modern level of 21% [RSBS 2006; Palaeos 2002]. Atmospheric O_2 supported creatures with aerobic metabolisms and permitted an ozone layer to form, protecting life from mutation-causing ultraviolet photons from the sun.

Another ecologically important property of certain cyanobacteria is their ability to convert (fix) atmospheric N_2 gas to nitrates and ammonia, bioavailable forms of which aid growth in chlorophyll-containing plants. We can also thank early photosynthetic cyanobacteria for supplying the basic genome for the plant (eukaryote) chloroplast [Sato 2002; Brinkman et al. 2002; Hasebe et al. 2003].

An important theory is that the chloroplasts of plants originated through endosymbiosis from cyanobacteria with eukaryotic cells. Chloroplasts are members of a class of subcellular organelles known as *plastids* having specialized functions. Raven and Allen [2003] have commented: "... plastids contain DNA, RNA and ribosomes, supplying a structural and biochemical basis for non-Mendelian, cytoplasmic inheritance of plastid-related characters. Subsequent molecular genetic studies have demonstrated the ubiquity of plastid genomes and confirmed that their replication, transcription and translation closely resemble those of (Eu)bacteria."

> Molecular phylogenetic studies now make it abundantly clear that the closest bacterial homologs of plastids are indeed cyanobacteria, supporting earlier conclusions from the comparative biochemistry of photosynthesis. Only cyanobacteria and chloroplasts have two photosystems and can split water, to make oxygen, as a source of reducing power. [Quoted with permission.]

As we have noted earlier, one of the earliest single-cell life-forms to appear on Earth were the *Archaebacteria.* The *Archaea* have two phyla: *Crenarchaeota* and *Euryarchaeoata.* This interesting group of prokaryotes still has the ability to live in extreme environmental conditions; their metabolic enzymes are unique in being able to function at extreme temperatures. There are over 250 named species of *Archaea.* Their biochemical adaptations are unique. Certain Archaeans can metabolize H_2, S, and CO_2, materials found in the early Earth's crust and atmosphere [Kimball 2004c]. *Methanogens* take in molecular hydrogen gas and water and metabolize them to methane gas (CH_4) and more water. A salt-loving Archaean, *Halobacterium,* contains *bacteriorhodopsin,* a purple, photon-trapping pigment that the halophile uses for energy. Trapped photon energy pumps protons (H^+ ions) to the outside of the cell membrane. When these protons flow back inside, they provide energy for the synthesis of ATP, used to power other chemical reactions in the bacterium. This important process is called *chemiosmosis* [Kimball 2007b,c].

This is not photosynthesis as practiced in the chloroplasts of higher plants, but a first step in trapping photons for energy. The earliest Archaebacteria probably appeared on Earth over 3 billion years ago [Armstrong 1997].

Still another candidate for early life on Earth is found among the putative *Nanobacteria,* or nanobes. They were first described by Folk [1992, 1993] in Italian hot spring minerals as spherical, calcified fossils, about 25–200 nm in diameter. Folk called them *Nannobacteria.* NASA scientist Chris Romanek had heard a presentation on nanobes by Folk and decided to look for their fossils in a meteorite of Martian origin. He found "fossil nanobes" embedded in the meteorite, which had been radioisotopically dated at about 4.5 billion years. The fossils were 20 to 200 nm across. Was this evidence of microbial life on Mars? Most skeptics replied that prokaryotic life as we know it today would require more than a 200-nm diameter "cell" to contain all the nucleic acid, enzymes, and a few mitochondria that a living, minimal, prokaryote cell would need. Others were not so sure. In 1997, scientists at the University of Queensland discovered fossil nanobes in ancient Australian sandstones. Then, Kajander and Ciftcioglu [1998] found (presumably living) coccoid (spherical) nanobacteria in mammalian blood. The nanobes ranged in size from 20 to 200 nm. They assumed that they had discovered new living bacteria and named them as *Nanaobacterium sanguineum.*

Nanobes were subsequently found in (living) human dental calculus and arterial plaque. They could be cultured and grown under cell culture conditions and were shown to produce carbonate hydroxyapatite on their "cell" envelopes. This mineral accretion was related to the possible accumulation of atherosclerotic plaque and renal calculi [Kajander & Ciftcioglu 1998]. In optimal cell culture conditions, their doubling time was about 3 days. Similar to viruses, gamma radiation did not seem to affect their viability; over 1 Mrad had no effect! Also, their growth could be inhibited by *tetracycline, 5-flourouracil,* and *cytosine arabinoside* but not by *chloramphenicol, rifampin, kanamycin, erthromycin,* or 0.1% *sodium azide,* which inhibits respiration [Kajander & Ciftcioglu 1998; Cisar et al. 2000].

Two factors led other workers to question the conclusions of Kajander and Ciftcioglu. They are the following: (1) The putative (calcified) bacterial envelope was considered too small to enclose the necessary molecular machinery to run even a primitive prokaryotic cell, and (2) a DNA genome, RNA, and cytoplasmic protein from nanobes had not yet been isolated and described. The size argument can be rebutted by pointing to the soil *ultramicrobacteria* described by Bae et al. [1972], which have diameters as small as about 80 nm. Other marine ultramicrobacteria have also been described [Button et al. 1993]. The DNA, RNA, and protein question is more problematic, however.

In 2000, Cisar et al. published a stunning rebuttal to the assumption that nanobes are "alive." We quote their abstract in the following text:

> The reported isolation of nanobacteria from human kidney stones raises the intriguing possibility that these micro-organisms are etiological agents of pathological extraskeletal calcification. [Refs.] Nanobacteria were previously isolated from FBS [fetal bovine serum, a culture medium] after prolonged incubation in DMEM [another culture medium]. These bacteria [sic] initialized biomineralization of the culture medium and were identified in calcified particles and biofilms by nucleic acid stains, 16S rDNA sequencing, electron microscopy, and the demonstration of a transferable biomineralization activity. We have now identified putative nanobacteria, not only from FBS, but also from human saliva and dental plaque after the induction of a 0.45-μm membrane filtered samples in DMEM. Although biomineralization in our "cultures" was transferable to fresh DMEM, molecular examination of decalcified biofilms failed to detect nucleic acid or protein that would be expected from growth of a living entity. In addition, biomineralization was not inhibited by sodium azide. Furthermore, the 16S rDNA sequences previously ascribed to *Nanobacterium sanguineum* and *Nanobacterium sp.* were found to be indistinguishable from those of an environmental organism, *Phyllobacterium mysinacearum,* which was previously detected as a contaminant in PCR. Thus, these data do not provide plausible support for the existence of a previously undiscovered bacterial genus. Instead, we provide evidence that biomineralization previously attributed to nanobacteria may be initiated by nonliving macromolecules and transferred on "subculture" by self-propagating micro-crystalline apatite. [Quoted with permission of the U.S. National Academy of Science.]

Cisar et al. were not able to describe the mechanisms by which nanobelike structures grew in various media. In fact, no one to date has explained the growthlike multiplication of the hollow, hydroxyapatite nanospheres. Some physicochemical chain reaction must be involved, perhaps depending on some unknown organic catalyst.

In conclusion, it appears that nanobacteria are not alive in the sense we have described earlier. Until definitive evidence is found that they have some sort of nucleic acid genome, proteins, and a metabolism, we are left with the position that nanobacteria are not alive in spite of the fact that their putative, hydroxyapatite "cell envelopes" self-replicate. At present, it appears that the self-replication of the calcified envelopes is a purely abiotic, physical phenomenon.

2.4.3 Discussion

The time of the (scientific) origin of life on Earth is something we will never know with certainty. Life probably appeared 3.6–3.7 billion years ago. This protolife lived in a reducing atmosphere (no O_2) and gained the energy for its chemical reactions by metabolizing H_2, H_2S, S, Fe, CH_4, etc. Photosynthesis appeared first in cyanobacteria about 2.5 billion years ago and spread to plants and other unicellular creatures, gradually causing the pO_2 on Earth to rise to its present level. Contemporary life-forms still exist that are obligate anaerobes. Many cells have both anaerobic and aerobic metabolic pathways.

We can date rocks in which early fossilized prokaryotes are found, and later fossils show more advanced single-cell life-forms. Workers led by Miller and Urey in 1953 have demonstrated in the laboratory that in the reducing atmosphere of early Earth, using liquid water and basic organic substrates, and under conditions of heat, pressure, and electrical discharge, key organic molecules could be generated that theoretically could support anaerobic life. These feasibility studies stimulated speculation that the coacervate theory of Oparin in 1938 may need reexamination. There is evidence that the first self-replicating, information-storing molecules may have been RNA. There is only speculation regarding how RNA shifted its role of acting as a self-replicating information carrier to DNA in all life known today. Yes, there are RNA viruses, viroids, and satellites, but they are not alive. They lack cell membranes and metabolisms outside of their host cells. Clearly, understanding the puzzle of genomic and biochemical evolution is a challenging problem. If life of some form or other is found on Mars or the moon *Io,* it will really challenge our science.

2.5 CELL AND ANIMAL DEATH

2.5.1 Introduction

We think we know death when we see it; for example, a road-killed squirrel or a plant that has lost all its leaves and whose stem has turned brittle. However, the appearance of death does not always signify death. The plant may send up new shoots in the spring from its roots; however, the squirrel is gone. How about bacteria? When is a bacterium dead; or is it a spore? Consider the state of *dormant life*: seeds, spores, and the *tardigrade.* Add water and wait. If a seed fails to sprout, a spore fails to germinate, or if a tardigrade fails to revive, it is evidently dead.

True cell death is easy to recognize. The cell membrane ruptures (lyses), and the internal cytosol that surrounds and feeds the organelles and nucleus mixes with the cytochemically hostile external environment. The cell is incapable of metabolism, and it no longer can make the ATP, proteins, and RNAs it needs to survive as a closed entity and reproduce. Its highly organized molecular subsystems become organic flotsam and jetsam that may be taken up by other cells as sources of energy, or for molecular substrates.

Death in metazoan (eukaryote) animals is characterized by a loss of *organismic homeostasis* and then the inability of the creature to metabolize energy. An animal can have dead cells in its body or have lost parts of its body and be perfectly "alive." It is when damage to one or more critical organ systems reaches a level where it prevents the animal from performing *coordinated, organismic homeostasis* that we see a *cascade* of physiological system failures leading to total organismic death. As you will see, because of the interdependence of animal physiological systems, as well as their ability to function autonomously in some cases, death in higher animals and humans is sometimes very difficult to define, or pronounce. Modern medical life-support technology has made it possible to postpone death and create medical and moral dilemmas about death.

2.5.2 Cell Death by Apoptosis (Programmed Cell Death)

Apoptosis is the genomically regulated form of cell death that permits the safe elimination of senescent or damaged cells when they have fulfilled their intended biological functions or have been damaged in certain ways. It is also known as *programmed cell death* (PCD). Examples of PCD can be found throughout the eukaryote animal and plant kingdoms. Apoptosis is necessary for embryonic development in animals and normal replacement of tissues in adult life. Its dysregulation in humans can result in oncogenic, inflammatory, autoimmune, and neurodegenerative diseases [Hajra & Liu 2004]. Infectious agents, including viruses, can exploit apoptosis in the host to evade the immune system. A simplified schematic diagram of the complex apoptotic biochemical pathways in a eukaryotic cell is shown in Figure 2.7. The broad arrows should be interpreted as "causes" or "leads to." Many other molecular pathways have been omitted for simplicity.

When a cell is induced to undergo apoptosis, it shrinks. It then develops bubblelike blebs on its outer membrane. The genomic material (DNA and histones) in its nucleus is broken down. Its mitochondria also break down, releasing the protein *cytochrome* **c**. The phospholipid, *phosphatidylserine,* which is normally hidden within the plasma membrane, is exposed to the surface. Phosphatidylserine is bound by receptors on *phagocytic cells* (e.g., macrophages and dendritic cells) that ingest the cell fragments. The phagocytic cells of the immune system secrete cytokines (e.g., *Interleukin-10* and *TGF*-β) that inhibit the immune system's inflammatory response.

Under what conditions do cells undergo apoptosis? The answer is, "under many conditions." PCD is necessary in normal embryological development. Some examples include the metamorphosis of *Lepidoptera* (a caterpillar changes to a butterfly), the resorption of the tadpole's tail as it metamorphoses into a frog, the removal of tissue webs between the fingers and toes of a fetus, and the resorption of excess or irrelevant neural connections (synapses) in the developing brain.

Other examples of normal, programmed apoptosis include the constant turnover of adult epithelial cells in the skin, the constant replacement of cells in the lining of the stomach and gut, the replacement of olfactory sensory cells, and the periodic replacement of endometrial cells in the uterus following menstruation.

Apoptosis is also needed to destroy cells that represent a threat to survival of the organism. When cells are infected with viruses, the viruses can leave fragments of their coat proteins on the host cell surface. These signs of infection are recognized by *cytotoxic T lymphocyes* (CTLs), which induce the newly infected host cells to undergo PCD before they can produce new viruses. Once an infection is cleaned up by the immune system, the excess immune system cells themselves are removed by CTLs, which also induce PCD in each other and in themselves [Kimball 2004b]. It should be clear that such a cascade of PCD must be very strictly regulated.

Cells with *DNA damage* also undergo PCD. DNA damage can come from many external sources, such as *ionizing radiation* (gamma, alpha, and neutrons), heat, UV radiation, and oxidizing chemicals. It can have a host of bad effects, including the development of cancer and birth defects. Cells respond to DNA damage by increasing their production of the *transcription factor protein* **p53**, a potent inducer of apoptosis [Kimball 2007d,e; Whitmarsh & Davis 2000]. Ionizing radiation and certain antimetabolite drugs are used on cancerous cells in hopes of damaging them to the point of inducing their PCD mechanism.

The mechanisms of cell apoptosis are triggered by the shift in a balance between *positive signals* (i.e., the need for continuing survival) and the input of *negative signals* that induce PCD. Some examples of positive signals are the presence of *Interleukin-2* (**IL-2**), an essential factor for immune system lymphocyte proliferation and growth, and *neural growth factor.* Negative signals include increased levels of oxidants within the cell, DNA damage (see preceding text), and accumulation of proteins with bad tertiary structures (this could be the result of gene damage). The presence of PCD activator molecules include: *tumor necrosis factor alpha* (**TNF-α**), which binds to the TNF receptor (**TNFr**); *lymphotoxin* (aka **TNF-β**), which also binds to the **TNFr**; and *Fas ligand* (**FasL**), a

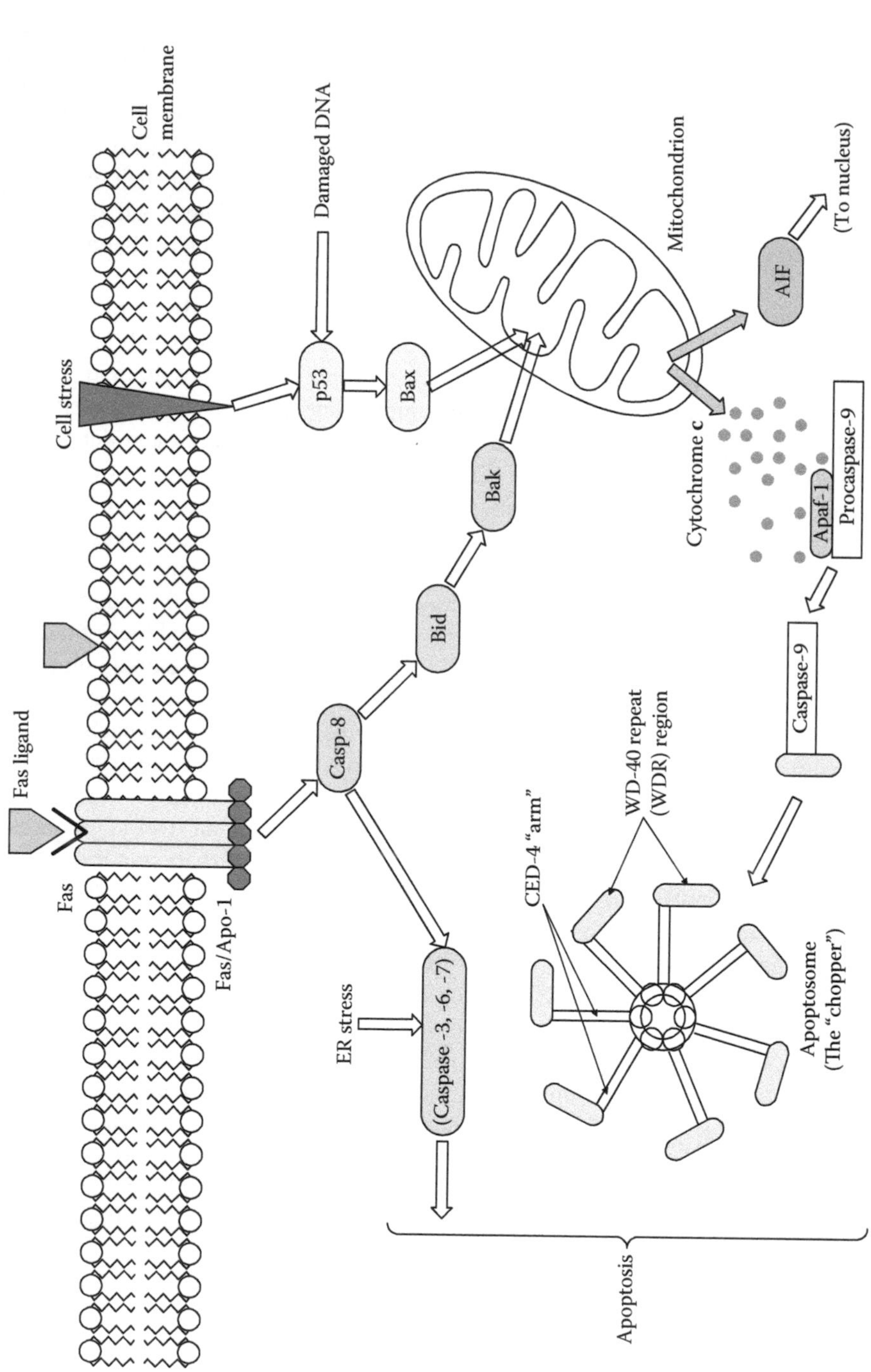

FIGURE 2.7 (See color insert.) A schematic overview of some of the molecular pathways leading to apoptosis in eukaryotic cells. The seven-spoked apoptosome molecular complex is shown.

molecule that binds to a cell surface receptor protein called **Fas** (aka **CD95**). **FasL** can be expressed on cell surfaces along with **Fas**.

There are three different mechanisms leading to PCD [Kimball 2004b]:

1. One triggered by signals internal to the cell.
2. A second initiated by *cell death activators* binding to certain cell surface receptors, which include **TNF-α**, **TNF-β**, and **Fas**.
3. A third that may be triggered by dangerous oxidizers inside the cell, which cause molecular damage.

Let us first examine the internal, *mitochondrial PCD pathway*. In a healthy cell, the outer mitochondrial membranes have the surface receptor protein **Bcl-2**. **Bcl-2** in turn is bound to the protein *apoptotic protease-activating factor-1* (**Apaf-1**). Internal damage to the cell (e.g., from oxidizers) causes **Bcl-2** to release **Apaf-1** and a related protein, **Bax**. **Apaf-1** and **Bax** penetrate the mitochondrial membranes and cause the enzyme **cytochrome *c*** to leak out into the cytosol. The released **cytochrome *c*** and **Apaf-1** bind to form a seven-spoked multimeric molecular complex (see Figure 2.7), the center of which binds to seven molecules of **procaspase-9**. This wheel-like molecular complex is known as the **apoptosome**. Somehow, a molecular signal activates the **procaspase-9** in the **apoptosome** to form **caspase 9** [Zou et al. 1999]. (Note that some 13 different caspases have been identified in cells.) **Caspase 9** is an enzyme specialized in cleaving proteins, generally at an **aspartic acid** AA. When **caspase 9** cleaves, it activates other caspases in an expanding cascade of proteolytic activity that leads to digestion of the cell's enzymes and structural proteins in the cytoplasm, thence to degradation of chromosomal DNA, apoptosis, and ultimately to phagocytosis of the cell by other cells such as immune system macrophages. (See the paper by Hill et al. [2003] for a good description of the caspases and the **apoptosome**; a glossary of the many, many abbreviations and acronyms used to describe the molecules in the regulatory pathways of apoptosis can be found at: www.biosource.com/content/apop/glossary.html.)

There are other intracellular proteins in the **Bcl-2** family. **Bcl-2** and **Bcl-X$_L$** by themselves tend to inhibit apoptosis, at least partly by blocking the release of **cytochrome *c*** from mitochondria. Other members of the **Bcl-2** family are not PCD inhibitors but instead promote procaspase activation, leading to apoptosis. Some of these PCD promoters such as **Bad** function by binding to and inactivating the PCD-inhibiting members of the family, whereas others such as **Bax** and **Bak** stimulate the release of **cytochrome *c*** from mitochondria. If the genes encoding **Bax** and **Bak** are inactivated, cells are remarkably resistant to most apoptosis-inducing signals. This indicates the importance of these two proteins in apoptosis control. **Bax** and **Bak** are themselves activated by other apoptosis-promoting members of the **Bcl-2** family, such as **Bid**. Clearly, these relations are extremely complex and are only now beginning to be understood.

Another important family of PCD regulators is the **IAP** (inhibitor of apoptosis) family of proteins. It belongs to a specialized group of proteins called **E3 ubiquitin protein ligases** that transfer **ubiquitin** protein "labels" onto ***caspases***, thereby blocking PCD. **IAP**s are thought to inhibit apoptosis in two ways: (1) They bind to some **procaspases** to prevent their activation, and (2) they bind to **caspases** and inhibit their activity. **IAP** proteins were first discovered as being produced by certain insect viruses (the host cell produces them from the viral genome). They prevent the host cell from killing itself before the virus has had time to fully replicate. When mitochondria release **cytochrome *c*** to activate **Apaf-1**, they also release a protein that blocks **IAP**s, thus greatly expediting the PCD process.

The Ubiquitin-Proteosome Pathway (**UPP**) is the principal mechanism for protein catabolism in a eukaryotic cell's cytosol and nucleus. This pathway is highly regulated and affects a wide variety of cellular processes and substrates. Defects in the **UPP** can result in the pathogenesis of several important human diseases. Hershko, Ciechanover, and Rose received the 2004 Nobel Prize in Chemistry for their work on elucidating the **UPP**. The **UPP** is important in a wide variety of cellular processes. It is involved in regulating cell mitosis, DNA repair, the biogenesis of cell organelles, immune response and inflammation, antigen processing, modulation of cell surface receptors, ion channels and the secretory pathway, embryogenesis, the regulation of transcription, endocytosis, and of course, apoptosis. We shall consider the latter application here.

Ubiquitin (**Ub**) is a small (76 AA) protein found only in eukaryotic cells. (However, **Ub**-like proteins are also found in Eubacteria and Archaea.) The **Ub** AA sequence is highly conserved along the eukaryotic evolutionary tree branches. For example, there is only a three-AA difference in the **Ub** from yeast to humans. **Ub** is found throughout the cell and can exist in free form or complexed with other proteins. **Ub** functions to regulate protein turnover in a cell by regulating the breakdown of specific proteins. This regulatory role is important because by quickly destroying a protein (enzyme) that regulates another function, it can change the rate of a chemical reaction in the cell and also the expression of specific genes. It does not destroy proteins itself; it serves as a molecular tag to mark a protein that will be destroyed through ATP-dependent enzymatic hydrolysis by a specialized protein complex, the ***proteosome***. Special **E1, E2**, and **E3** enzymes (many types of each exist), using energy supplied by ATP, attach **Ub** to a protein selected to be destroyed. The **E1** and **E2** enzymes also catalyze the attachment of **Ub** molecules to itself by covalent bonding, forming short **Ub** chains that are attached to the "victim" protein. A "library" of **E3** proteins *(**Ub**-ligases)* is responsible for victim selection for **Ub** tagging. When **Ub** tags onto a protein, it is called *ubiquitination* or *ubiquitinylation*. To underscore the complexity of the **UPP** system, special deubiquitinylating enzymes are involved in unbinding **Ub** from proteins and disassembling **Ub** chains.

Once obsolete or unnecessary proteins are tagged with at least four **Ub** molecules, they are destroyed by the large, about 1,000 kD, proteins called the **26S proteosomes.** The process is shown schematically in Figure 2.8. A proteosome is a very efficient molecular protein shredder, but its destructive machinery is guarded. So, it cannot run amok in the cell and destroy useful proteins; it needs the four **Ub** molecules to work. What is not known well are all the signals used by the ***E***-enzymes in selecting their victim protein molecule and how the signals work. There is evidence that the N-terminal AA of a protein may send one signal; it has been called the **N-degron**.

Certain AA sequences in the victim appear to be degradation signals. The "PEST" sequence is one such signal. PEST stands for proline, glutamic acid, serine, and threonine. One example of PEST regulation of protein destruction is the transcription factor, **Gcn4p**, which has 281 AAs. Its PEST sequence is found at positions 91–106. The normal half-life of this protein is about 5 minutes. However, if only the PEST sequence AAs are removed, the half-life increases to about 50 minutes, a tenfold increase. Signals to tag a protein with **Ub** may also lie buried in the hydrophobic core of the folded molecule. This is why partially folded or abnormal proteins may be more prone to degradation. An important question is, how do the **E3**-proteins "recognize" an abnormality as such? Is it because a PEST-like AA sequence that signals "tag for death" is exposed to the **E**-complex?

The *core particle* (CP) of a proteosome complex is cylindrical in shape; its active sites are hidden inside the tube. The outside of the proteosome does not induce hydrolysis of proteins; it is biochemically benign. The two molecular "caps" on the ends of the proteosome CP regulate entry of the victim protein into the center hydrolysis chopping-chamber. The victim protein is reduced to pieces from 3 to 23 AAs in length, which are further broken up by other enzymes, and the free AAs are "recycled" in the cell's metabolism to make new proteins. The core particle is made of a stack of four rings; each ring is made from seven protein subunits (not unlike the apoptosome's "wheel"). The *regulatory particle* (RP) molecules cover each end of the core particle. Each RP is made of 14 (!) different protein subunits (none of the same as used in the CP). Six of these proteins are ATPases, and some subunits recognize (have affinity for) ubiquitin. (ATPase is an enzyme that cleaves the two outer phosphate bonds on ATP, releasing the chemical energy in the bonds to drive other biochemical reactions; AMP is the result.)

Clearly, there must be a delicate balance between the normal destruction of ubiquitinylated proteins in a cell and the mass destruction of cell proteins by caspases in the apoptosome. The **UPP** process is clearly not apoptosis but can contribute to apoptosis. A major question that must

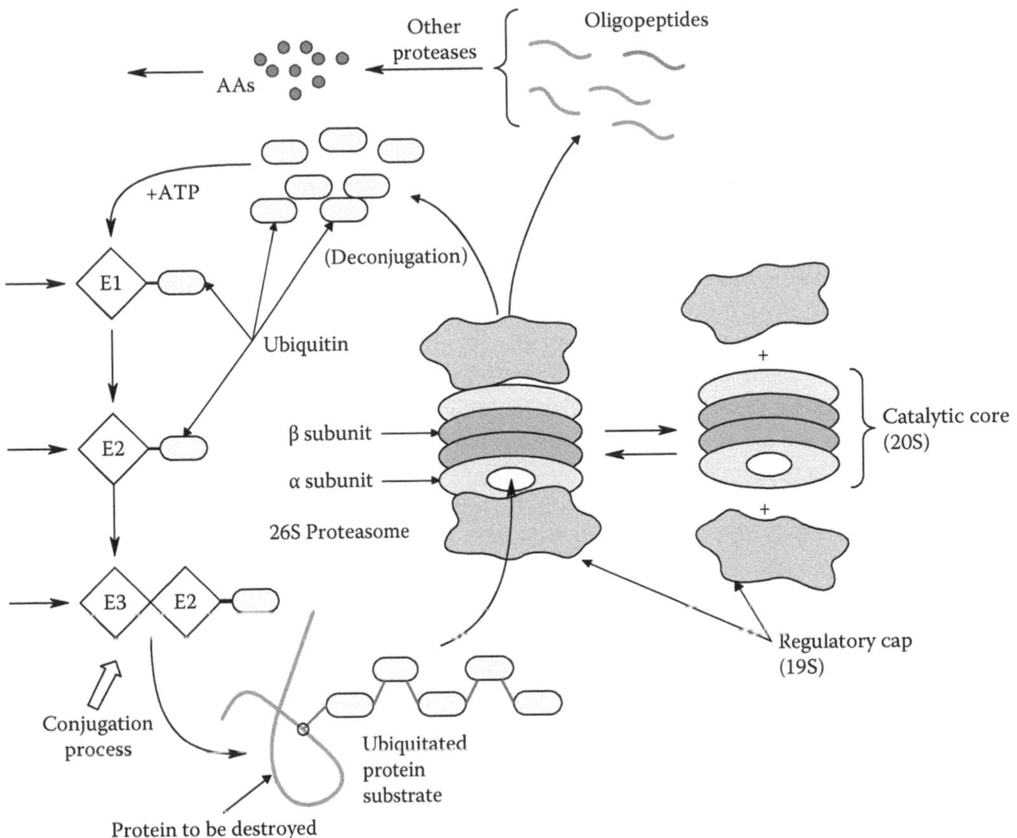

FIGURE 2.8 (See color insert.) Schematic illustrating the pathway where a protein tagged by polyubiquitin is broken down by a proteosome.

be answered is, how do **Ub**, the **E**-proteins, and the proteosome interact with caspases and the apoptosome?

According to Friedman and Xue [2004]: "The proteosome appears to play an important role in suppressing inadvertent activation of proapoptotic proteins in cells that need to live, and participates in eliminating anti-apoptotic proteins in response to apoptotic stimuli. After the initiation of apoptosis and the activation of caspases, the proteosome is then disabled to allow the build-up of proapoptotic proteins, which tilts the balance of life and death further toward the point of no return". [Quoted with permission.]

External signals can also trigger the PCD pathway. Recall that the cell's surface membrane has the integral membrane receptor proteins, **Fas** and **TNFr**, embedded in it. (An **Fas** receptor is shown in Figure 2.7.) When the external, complementary *death activator proteins* (**FasL** and **TNF-α or TNF-β**, respectively) bind to these two receptors, the event triggers a signal into the cytoplasm that activates **caspase 8**, which, similar to **caspase 9**, initiates a cascade of proteolytic activity that ends with the death and phagocytosis of the cell. Note that some 11 highly regulated caspases have been recognized in human cells. They normally are dormant and have to be activated to kill cells from within.

Neurons can undergo PCD without the aid of caspases. *Apoptosis-inducing factor* (**AIF**) is a protein normally located in the intermembrane spaces of mitochondria. When the neuron receives its "time-to-die" signal, this leads to **AIF** being released (not unlike the triggered release of **cytochrome *c*** in the first PCD pathway). An **AIF** complex migrates to the cell's nucleus where it binds to the DNA and triggers its destruction. Cell death follows. The DNA is ultimately chopped into oligos as small as 200 bps. It has been shown that in vitro, massive potassium loss from the neuron cytosol is associated with apoptosis. Blocking potassium channels in the neuron membrane with *tetraethylammonium* (TEA) reduced the rate of neuron PCD [Sage 1997].

Some pathogens contain a means of blocking apoptosis so that the infected cells do not do PCD and remain as "hosts," allowing the pathogen to replicate. According to Kimball [2004b], some viruses that cause cancers use molecular "tricks" to prevent apoptosis in the cells they have transformed. For example, the human *papilloma virus* (HPV) has been implicated in causing cervical cancer. It produces a protein (**E6**) that binds to and inactivates the critical *apoptosis promoter protein*, **p53**. The *Epstein-Barr virus* (EBV) is the cause of infectious mononucleosis and is associated with some lymphomas. Once inside their host cells, EBVs produce a protein similar to **Bcl-2** and also induce the cell to make more of its own **Bcl-2** apoptosis inhibitors. Melanoma (a dangerous type of skin cancer)-transformed cells avoid apoptosis by inhibiting the expression of the gene encoding **Apaf-1**. Lung and colon cancer cells secrete elevated levels of a "decoy" molecule that binds to **FasL**, blocking it so that it cannot bind to **Fas** receptors on the cancer cell that trigger PCD. More **FasL** molecules are produced at the surface of a *cytotoxic T-lymphocyte* (CTL) of the immune system when it binds to a cancer cell's altered cell surface. Normally, this excess **FasL** would bind to the cancer cell's **Fas** receptors and trigger its PCD. However, the decoy molecules from the cancer cell inactivate the **FasL** molecules. As a further tribute to the "cleverness" of cancer cells, we note that other types of cancer cells secrete excess **FasL**, which can bind with the CTL's own **Fas** receptors, causing it to undergo apoptosis. (Note that the CTL is somehow protected from its own **FasL**.)

The body's use of **FasL** to actively prevent PCD of certain tissues is seen in "immunologically privileged sites," which include the anterior chamber of the eye and the testes. High densities of **FasL** bound on the surfaces of the target cells kill any CTLs that draw near to it. Because the **FasL** is bound, it cannot bind to the cell's own **Fas** receptors (good thing).

In HIV/AIDS, the HIV virions infect a few of one type of their target host cells, the *CD4+ T-lymphocytes* of the immune system. Far more CD4 cells die than are virally infected; the cause is apoptosis. HIV-infected CD4 cells express an HIV gene that makes a protein, **Nef**, which causes the infected cell to make high densities of membrane-bound **FasL** at its surface which cannot bind with its own **Fas** receptors, preventing the infected cell from undergoing PCD. However, when the HIV-infected CD4+ cell, which is bristling with virally induced **FasL**, encounters another uninfected lymphocyte closely, the uninfected cell's **Fas** binds to the infected cell's **FasL**, and the uninfected CD4+ cell undergoes apoptosis. Removal of its host cells by this mechanism actually works against the HIV virus, but it surely kills its host animal (if untreated) by weakening its immune system!

In conclusion, we note that apoptosis, or PCD, relies on a highly regulated, complex, interactive biochemical system with multiple inputs. PCD is necessary for normal embryonic development and the maintenance of a healthy mature animal or plant. PCD is also a very active medical research topic. If we can learn to manipulate its complex biochemical control systems, we may find an efficient means of fighting cancer by inducing the PCD of cancer cells alone. The HIV also can produce abnormal stimulation of apoptosis in certain immune system cells. Perhaps that can be blocked, inhibiting the release of HIV virions.

Organ transplant rejection is also due to transplant cell apoptosis induced by the host's CTLs and other immune cells. A cell's normal **UPP** system, which is concerned with internal protein housekeeping, interacts with the PCD system. Both systems destroy proteins, are regulated by proteins, and are incredibly complex. Fortunately, there is a finite number of protein species involved.

2.5.3 Cell Necroptosis

A second morphologically distinct way that individual cells die is called *necroptosis*. In some cases, when the apoptotic pathway for PCD is blocked, the cell somehow reverts to a caspase-independent, genetically regulated death mode called n*ecroptosis,* which is one identified means of cell necrosis. Similar to apoptosis, necroptosis can be induced when a cell's membrane-bound *death receptors* (DRs) bind with PCD trigger molecules such as **TNF** and **FasL.** Necroptosis is generally characterized by cell membrane and organelle disruption, mitochondrial impairment, cell swelling, and cell lysis with a following severe inflammatory response by the immune system. (Recall that in apoptosis, cells *shrink* and disintegrate internally into smaller "parts".)

Necroptosis is known to figure in many pathological conditions, for example, in *ischemia* from stroke or myocardial infarction, trauma, and possibly some forms of neurodegeneration. Teng et al. [2005] have shown that the chemical 5-(1H-indole-3-methyl)-2-thio-hydantoin, called *Necrostatin-1*, effectively blocked DR-induced necroptosis. Degterev et al. [2005] have demonstrated the effectiveness of necrostatin-1 in delaying ischemic brain injury in mice by necroptosis. These authors point to the potential therapeutic value of necrostatin-1 in stroke and myocardial infarction.

2.5.4 Metazoan Death: Failure of Critical Organ Systems

Various means of genetically controlled cell death (PCD) were described earlier. Everyone will agree that a cell has died when its membrane has ruptured. It no longer is able to practice homeostasis, and its metabolism has ceased to function.

The definition of death when considering humans (at the peak of the evolutionary tree) is far more complex. A complex, metazoan, vertebrate animal, a human is composed of interacting, interdependent, regulated organ systems. Perhaps, the most critical organ systems are the *central nervous system* (CNS), *respiratory system, cardiovascular system* (CVS), *hepatic system,* and *renal system*. The CVS is responsible for the circulation of oxygenated blood, hormones, and cell nutrients throughout the body; it also removes metabolic waste products from the body's cells and delivers CO_2 to the lungs and nitrogenous waste to the kidneys for removal. The portal circulation delivers digested food from the intestines to the liver. All aspects of the cardiovascular system are under autonomic nervous regulation from the brain. Nerve signals slow or speed up the heart, constrict or dilate local blood flow in arteries and veins to ensure tissue perfusion in a prioritized manner (the brain comes first).

Major impairment of the CVS, including exsanguinations from trauma or major blood vessel rupture from an aneurism, or heart stoppage from heart attack or electrocution, leads to massive tissue anoxia and irreversible failure (death) of other critical organs such as the brain, kidneys, liver, etc. In the absence of O_2, aerobic cell metabolism throughout the body grinds to a halt. If a heart attack, electrocution, or cold-water drowning has recently stopped the heart, prompt application of cardiopulmonary resuscitation (CPR) may be able to start the heart beating again. If the heart is in ventricular fibrillation, electrical cardiac defibrillation may restart normal sinus rhythm. Time is critical; the brain can remain viable for about 4–6 minutes under conditions of zero blood pressure at body temperature before irreversible damage to the CNS begins to occur.

Because of the close coupling and strong functional interdependence of mammalian organ systems, certain causative factors can precipitate what is called *multiple organ dysfunction syndrome* (MODS) leading to organismal death [Johnson et al. 2004]. Some causes of MODS include severe allergy; bacterial endotoxins (e.g., from *Streptococcus sp.* causing toxic shock); massive viral infection (e.g., Ebola); venoms from certain snakes, mollusks, and spiders; and plant toxins such as *abrin, ricin,* etc. In MODS, one can witness a serial cascade of organ system failures or simultaneous parallel system failures because of concurrent assault on all their tissues or a scenario involving catastrophic serial/parallel failures.

A physician can pronounce a person dead when he or she is found to have zero blood pressure and has had no electrical heart activity for a long time. At death, because cell metabolism throughout the body shuts down, the body temperature approaches the surrounding (ambient) temperature, and signs such as a complete failure to respond to sensory inputs, dilated and fixed pupils, lividity, pallor, and eventually *rigor mortis* are present. Oxygen-sensitive cells begin to die and release potassium ions into the interstitial fluid and blood. Clearly, persons with these signs are dead.

A major area of medical and ethical controversy today is the establishment of a definition of "death" when a person is in a coma. The coma may be the result of trauma to the head (brain), drug overdose, electrocution, CNS ischemia from a stroke, drowning, or infection. Persons in a severe coma can be kept "alive" by artificial ventilation and feeding through an nasogastric (NG) tube. Such persons still have a functional brain stem if their heart continues to beat, receiving neural cardioregulatory signals. Such persons may be able to breathe on their own, as well. However, their cerebrum and cerebellum may be irreversibly damaged. Signs of severe higher brain damage may include the lack of response to sensory stimuli, the lack of eye movements, fixed pupils, and the inability to eat, chew, and swallow. Persons in this state of CNS damage are said to be in a *persistent vegetative state* (PVS). One of the signs of PVS is diagnosed by recording a zero electroencephalogram (EEG) over the entire surface of the skull. However, the normal EEG is recorded as low-frequency, low-amplitude voltages generally lying in the range of tens of microvolts. The EEG is the result of groups of neurons near or on the surface of the brain (cortex) firing nearly synchronously in waves of activity. The summated neural action potentials are picked up through the poorly conducting skull and scalp. If neurons in the cerebrum are still alive and have lost the ability to fire synchronously, EEG recording from the scalp surface will not show this. The electrical signals from single live neurons would be buried in the EEG recording system noise. Thus, we see that even the brain, similar to the entire body, has interconnected, functional subsystems, some of which still may function at a basic level, preserving cardiac action and respiration, whereas other parts may be temporarily "shut down" or be irreversibly dysfunctional.

PVS patients are generally considered "alive" by the law, even if they have no future as functional human beings. There is controversy as to whether PVS patients have a right to die as dictated by their "living will," which specifies that they do not want to be kept in a PVS with near-zero probability of ever recovering. In one recent, widely publicized and politicized case, a long-term PVS patient capable of spontaneous respiration was caused to die by removing her NG feeding tube and IV saline. Removal was at the request of her former husband, who said that even though she had left no living will, it had been her desire not to be kept alive in a PVS. No proactive means, such as injecting an overdose of an opiate, was used. Dehydration and starvation caused the rest of her body's systems to irreversibly shut down, and her final death. Some persons are opposed to terminating the life of a PVS patient on religious grounds, regardless of the patient's wishes.

True brain death is clearly the death or irreversible dysfunctionality of all of the brain's component parts. The true brain-dead patient provides *no* autonomic brain stem (medullar) neural signals to regulate the heart rate, circulation, body temperature, or respiration. The heart may beat by itself in the immediate absence of neural control, but it eventually will stop in 2–10 days after loss of neural control.

The clinical diagnosis of true brain death generally involves the following signs: (1) There is no recordable EEG signal anywhere on the scalp, and repetitive sensory stimulation does not produce a cortical evoked potential. (2) There are no spontaneous respirations. An apnea test in which respiratory support is withheld for few minutes does not result in any respiratory efforts. (3) All extremities are flaccid; that is, there is no muscle tone and no reflex activity in any of the limbs (arms, legs, hands, or feet). (4) The pupils are dilated and unresponsive to changes of light intensity. There are no eye movements. (5) There is no response to noxious stimuli anywhere on the body. (6) There are no signs of brain stem activity: These include a total lack of eye movements and no corneal reflex (touching the cornea does not cause eyelid motion [blinks]). There are also no gag or cough reflexes. There is no vestibulo-ocular reflex, in which the injection of cold water in the ear canal causes ocular nystagmus. In addition, MRI can be used to examine brain metabolism, and a radio-opaque dye can be injected, and x-ray can be used to see what parts of the brain lack circulation. The preceding tests are generally repeated at least a day later to add redundancy to the diagnosis.

In addition to the testing outlined earlier, the examining physicians also must be sure that the loss of brain function *has not been due to:* (1) a recent overdose of opiate or barbiturate drugs, (2) recent hypothermia (as caused by cold water drowning), or (3) recent hypoglycemia (as caused by an insulin overdose). These factors may cause a possibly reversible, deep coma.

Once a committee of physicians has declared total brain death, if the patient is an organ donor, this is the time that organs would be harvested. Life support (ventilation, NG feeding, IV fluids) is then removed, and the patient's other organ systems are allowed to fail. A "dead brain" has never been known to recover.

It should be noted earlier that the *apnea test* is generally performed by interrupting artificial ventilation of the patient for several minutes to see if the rise in blood pCO_2 triggers spontaneous breathing. The persistent rise in pCO_2 in the blood, and the drop in pO_2, may further damage borderline nervous tissues that may be trying to recover from some initial anoxia. It appears that this is an ill-advised and possibly destructive test.

From the viewpoint of a physiologist, perhaps a less destructive test of respiratory competence could be done by giving the patient a few breaths of air with increased CO_2 concentration while he or she is on the ventilator and by seeing if there is a transient rise in heart rate and blood pressure. A transient CO_2 input test is not intended to test for spontaneous breathing. Increased blood pCO_2 would act on a functional *vasomotor center* in the *medulla* of the brain to produce a powerful vasoconstrictor effect that is transmitted through the sympathetic vasoconstrictor nervous system. The resulting constriction of arterioles and veins increases peripheral resistance, and thus, a transient increase in blood pressure (BP). At the same time, neural signals from a functional medulla would increase the heart rate [Guyton 1991]. The absence of the transient increase in BP and heart rate would argue for a dysfunctional medulla. So, it really is not necessary to disconnect a patient's ventilator for minutes, causing possibly harmful anoxia and high pCO_2 in the blood.

In conclusion, we have seen that criteria for death in man and other mammals can be difficult to establish because of the ability of some organ systems, including the CNS, to continue functioning at a reduced level following trauma, infection, heart attack, or stroke. In questions of brain death, the decision is based on a battery of neurological tests given to ensure brain death has indeed occurred. A persistent vegetative state should not be confused with brain death. The difference is that in a PVS, the patient's brain stem's autonomic "housekeeping" functions work, whereas other parts of the brain required for consciousness, thought, sensory perception, and motor control are irreversibly dysfunctional. A PVS patient may have spontaneous respiration and is still able to maintain thermal and cardiovascular regulation, and ion balance with their kidneys. (The latter mechanism is implemented by feedback systems involving the secretion of the hormone vasopressin by the posterior pituitary neurosecretory cells and the hormone aldosterone by the adrenal cortical cells. These hormones act on the kidneys to regulate sodium and potassium ion concentrations in the blood.) Through artificial respiration and NG tube feeding, the organ systems (e.g., heart, lungs, liver, kidneys, etc.) of such patients can be kept alive for years. Is keeping a person's body alive in an irreversible PVS ethically correct?

Tests for brain death vary from country to country and also from hospital to medical center in a country. They should be standardized, based on medical physiology.

2.6 SUMMARY

In this chapter, we have considered some of the toughest problems in biology today that have no clear, scientific answers: What is life; how can it be defined? Also, when and how did it first arise on Earth? The quantum leap in cellular complexity from prokaryotic cells (bacteria, Archaeans) to eukaryotic cells (with membrane-enclosed true nuclei and other organelles) took place about 2.8 billion years ago. We can only speculate how this may have happened. It probably occurred slowly over a half billion years on either side of the 2.8 billion year marker. Although life-forms have evolved complexity, we still have living examples of the earliest single-celled prokaryotes with us today.

We have also considered death. Genetically programmed cell death (PCD), or apoptosis, was seen to be necessary for embryonic development in eukaryotic metazoans and required for maintenance of healthy organs and tissues. Other forms of PCD called necroptosis and cell necrosis use some of the cell machinery in the apoptotic/caspase pathway. We saw that some viruses can inhibit a cell from PCD when it is infected so that the virus can reproduce with maximum efficiency inside the cell. Certain cancers do the same thing. Consequently, there has been much recent interest by cell biologists in learning the details of the PCD control system. The goal is to signal cancer cells to undergo apoptosis and not their healthy neighbors.

Finally, we examined the criteria for human death, including brain death. The fact that medical science can keep the rest of a body's organ systems functioning by various means of life support in the presence of a totally dysfunctional brain has created ethical, moral, religious, and medical controversy. A brain-dead person is dead; he or she is not in a coma and will never recover consciousness. A person in a persistent vegetative state, by definition, has lost the use of all higher brain functions, even though his or her functional brain stem continues to perform autonomic regulation of the CV system, breathing, and the kidneys. Such a person, too, by definition, will never recover consciousness. The differential diagnosis between a PVS and a person unconscious in a coma is difficult. Because people have recovered from comas after long periods of time, it is prudent not to rush in disconnecting life support from PVS patients.

In general terms, cell death is characterized by a total loss of cell metabolism and homeostasis. Cells rupture, and their cytosol reaches molecular equilibrium with the extracellular milieu. In metazoan animals, death occurs when there is sufficient dysfunction of one or more of the interdependent organ systems so that system interaction fails, cell metabolism fails, and homeostasis fails. Ultimately, cellular decomposition occurs.

CHAPTER 2 HOME PROBLEMS

2.1 What is the size of the typical chloroplast genome? How many genes are present?

2.2 What is the size of the human mitochondrial genome? How many genes are present?

2.3 Which prokaryote (not an obligate parasite) has the fewest genes? Give the gene number.

2.4 Which virus (RNA or DNA) has the smallest genome?

2.5 An ancient gravesite is discovered in a pyramid. Carbon-14 dating of the sarcophagus shows that it contains 0.634 as much ^{14}C as a contemporary sample of the same kind of wood taken from a living tree. How old is the sarcophagus?

2.6 Give some examples of metazoans that use *dormancy* to survive environmental hardship.

2.7 List the major types of cells (and their locations) in the human body that continually renew their populations by apoptosis.

2.8 Calculate the osmotic pressure in mmHg at 37°C of "normal saline solution" (9 g/L NaCl).

2.9 A recipe for artificial aqueous humor contains the following substances dissolved in water to make 1 L of solution: 0.77 g dextrose, 0.26 g urea, 0.0504 g lactic acid, 0.185 g ascorbic acid, 131 mM/L NaCl, and 20 mM/L $KHCO_3$. Calculate the osmotic pressure of this solution in mmHg at 37°C, also in pascals.

2.10 There are a number of biochemical regulatory pathways that can activate programmed cell death (apoptosis). Describe them. What is the role of the protein p53?

2.11 Describe how certain viruses and cancers can inhibit apoptosis.

2.12 Certain viruses have "jumped species"; for example, have gone from a bird host to a human host. What makes this interspecies jump possible?

3 Review of Basic Cell Anatomy and Physiology

3.1 INTRODUCTION

This chapter is about cells, the fundamental units of life on Earth. As you will see, the molecular biology of even the simplest of cells is, paradoxically, complex. In the following sections, we will review the anatomical features, the molecular biology, and the physiological properties of cells belonging to the three domains of life: the *Prokaryotes,* the *Archaeans,* and the *Eukaryotes.* One common feature of *all* life is that the double-stranded DNA (dsDNA) molecule is used to store the information required for self-reproduction (of the cell and the organism in which it resides), as well as the information required to do metabolism, maintain homeostasis ("molecular housekeeping"), and execute the cell's function. DNA codes protein and RNA synthesis, and sequences developmental events, including programmed cell death (apoptosis). Another common feature of all cells is their cell membranes, described in the following section. Cell membranes have a common architecture, but molecular details vary between the domains and kingdoms of life.

3.2 THE CELL MEMBRANE

As noted earlier, a universal feature of all cells is their outer, limiting (plasma or cell) membrane. Recall that eukaryotic cells also have a variety of *internal* membranes (e.g., nuclear, ER, mitochondrial, etc.). A cell's outer membrane serves as a protective interface between the variable extracellular environment and the internal molecular machinery of the cell. It also serves as a platform for the attachment and insertion of many different membrane-associated protein molecules, which have many diverse functions. Membrane proteins allow a cell to preserve the internal milieu required for metabolism and reproduction by regulating the rates that substances enter and leave the cell.

The outer membrane in nearly all cells is made from a two-layer "sandwich" of phospholipid molecules. The number and kinds of proteins associated with the cell membrane depend on the type of cell (neuron, heart, liver, etc.), the type of organism (bacteria, plant, animal, etc.), and where the membrane is located in the cell (outer, nuclear, ER, Golgi, etc.). The exact chemical composition of the outer membrane itself also varies between members of the domains and kingdoms of life, but the basic molecular architecture remains constant [see Karp 2007, Chapter 4].

All membranes contain lipids from three main categories: *cholesterol, phosphoglycerides,* and *sphingolipids.*

Most eukaryote membrane lipids are constructed around a D-glycerol backbone to which a phosphate group is attached by an *ester linkage*; hence, they are called *phospholipids.* Two fatty acid (-**F**) chains are also joined to the D-glycerol by *ester linkages.* (An ester linkage is formed when a hydroxyl group combines with a carboxylic acid group with a loss of H_2O.) Another group is generally linked to another oxygen of the phosphate group. This group can be as simple as a –**H** (forming a *phosphatidic acid*) or can be a *choline* ($-\mathbf{CH_2\text{-}CH_2\text{-}N^+(CH_3)_3}$), forming p*hosphatidyl-choline* (aka *lecithin*) or *serine*

$$(-CH_2\text{-}\underset{\displaystyle COO^-}{\underset{|}{C}}H\text{-}N^+H_3),$$

forming *phosphatidylserine* or *ethanolamine* (**$-CH_2$-CH_2-N^+H_3**), forming *phosphatidylethanolamine* (aka *cephalin*), or *inositol* forming *phosphotidylinositol*. Other membrane lipids include *sphingolipids*, which are derived from the aminoalcohol, *sphingosine*. These include the *ceramides* and *sphingomyelin*. Some of these membrane molecular subunit structures are diagrammed in Figure. 3.1.

We noted that each type of cellular membrane has its own unique lipid composition. Some examples of the fatty acid molecules used in plasma membrane structures, and the number of carbons (k) in each, are the following: *myristic acid* (14), *palmitic acid* (16), *stearic acid* (18), *oleic acid* (18; double bond [db] between carbons 9 and 10), *linoleic acid* (18; dbs between 9-10 and 12-13), *Linonenic acid* (18; dbs between 9-10, 12-13, and 15-16), *Arachidonic acid* (20; dbs between 5-6, 8-9, 11-12, and 14-15), and *Eicosapentoic acid* (20; dbs between 5-6, 8-9, 11-12, 14-15, and 17-18). Eicosapentoic acid is also called an *omega-3 fatty acid* because it has a db 3 carbons from its carbon-20 methyl termination.

The basic molecular structural subunit of the bilayer membrane used in eukaryotic cells and eubacteria generally uses two hydrophobic fatty acid chains of 16 to 18 carbon atoms, joined to a molecule of D-glycerol (the hydrophyllic head) by *ester linkages*. The basic outer membrane has two layers of many closely packed fatty acid + *hydrophyllic head* subunits arranged so the fatty acid chains are directed inward toward the center of the membrane, and the hydrophyllic heads are directed away from the membrane. The hydrophillic heads either make contact with the cell's cytoplasm or the extracellular fluid. The lipid bilayer is generally impermeable to water-soluble substances such as glucose, ions, urea, etc. However, fat-soluble substances such as steroid hormones, alcohol, CO_2, and O_2 have high permeabilities [Guyton 1991].

Eubacteria have a three-layer *cell-covering* structure: On the outside is a slimy, hydrated *polysaccharide capsule*. Next is a cell wall made from *peptidoglycan* [Meroueh et al. 2006]. At the inside of the cell wall is the bacterial bilayer plasma membrane. Inside of the plasma membrane is the bacteria's cytoplasm, DNA, and the ribosomes used for protein synthesis.

The outer cell walls of *Archaeans* are of varied compositions but do not contain the peptidoglycan polymer. They may contain proteins, glycoproteins, or polysaccharides. *Methanogen* cell walls contain *pseudopeptidoglycan,* which uses different sugars and a different molecular architecture from eubacterial peptidoglycan.

The outer membranes of Archaeans are unique in composition. Instead of using unbranched fatty acids, their plasma membranes have lipid subunits that use a pair of hydrophobic, branched *isoprene chains*, joined to a molecule of ***L-glycerol*** by *ether linkages*. See Figure 3.1. (An ether linkage is formed when two hydrocarbons are joined by an oxygen atom [e.g., **R–O–R′**]).

However, in both Archaean and other cells' membranes, both the L- and D-glycerol "heads" have a phosphate group (**-$PO_4^=$**) attached. Because of its isoprene chemistry, the Archaean plasma membrane is more physically robust, enabling Archaeans to inhabit more physically and chemically challenging environmental niches.

Another polar/lipid molecule found in eukaryotic cell membranes along with the phospholipids is *cholesterol* (see Figure 3.2 and Figure 3.3B). Cholesterol molecules have several functions in cell membranes: They mechanically strengthen their phospholipid neighbors and decrease the permeability of the membrane to small, water-soluble molecules. Cholesterol also prevents crystallization of hydrocarbons and phase shifts in the membrane. *Glycolipids (glycososphingolipids)* are also found in cell membranes. They, too, have hydrophobic lipid tails that point inward in the membrane, and their hydrophilic heads are oriented away from the membrane. They can form clusters in the membrane, and one of their roles is to serve as the site of receptor protein binding, including

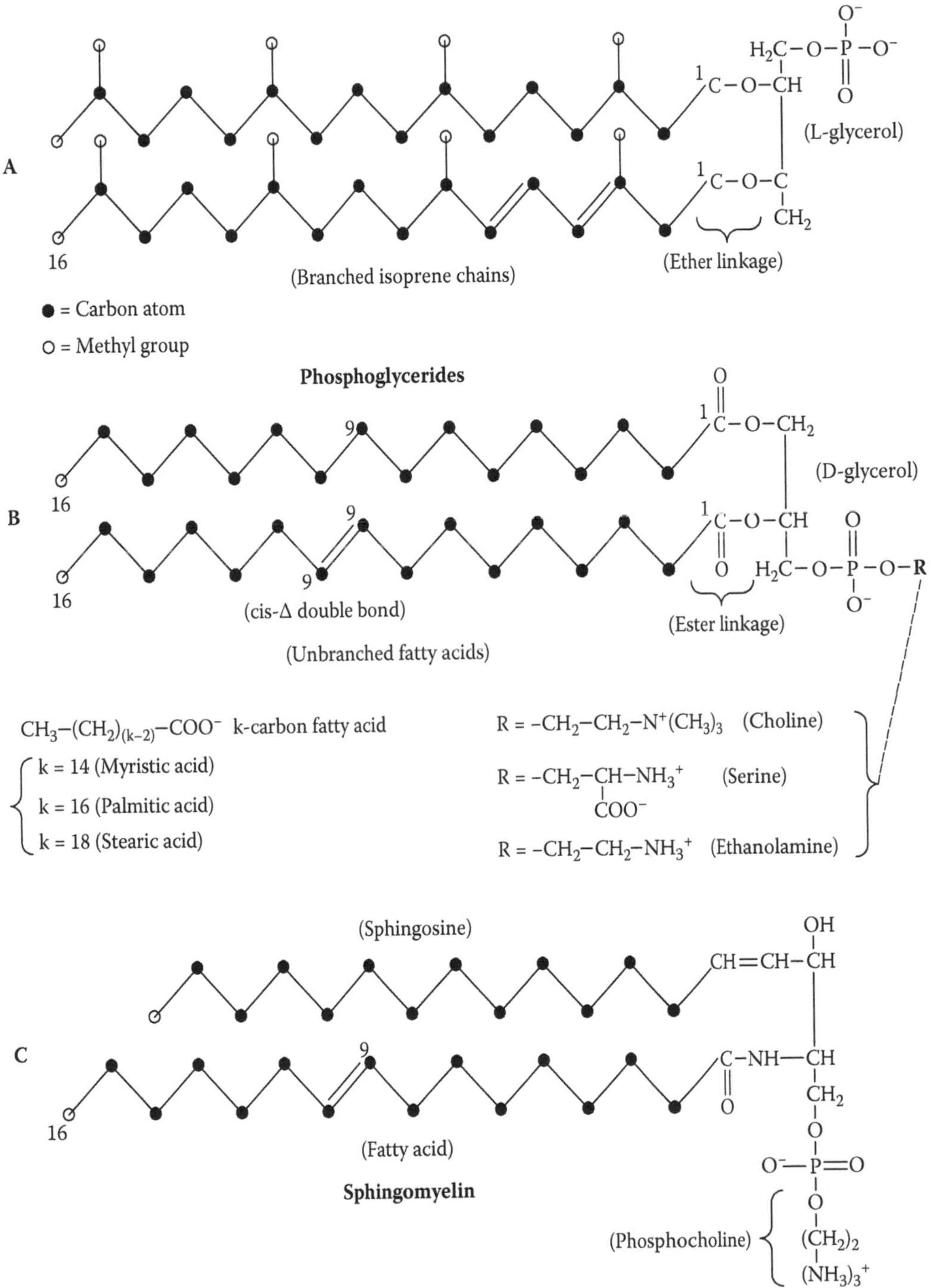

FIGURE 3.1 (**A**) Basic architecture of an Archaean plasma membrane phospholipid subunit. Note ether linkages from isoprene chains to L-glycerol. (**B**) Basic architecture of the phospholipid subunit found in Eukaryotes and Eubacteria. Note ester linkages to D-glycerol. (**C**) A sphingomyelin plasma membrane subunit. See text for discussion.

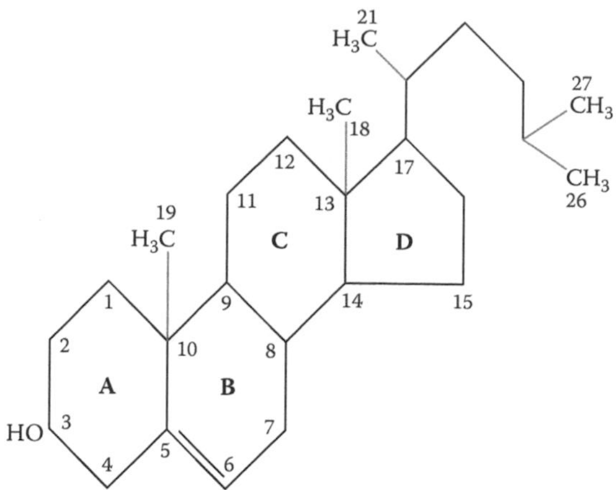

FIGURE 3.2 A 2-D schematic of a cholesterol molecule.

bacterial cell toxins such as tetanus toxin and cholera toxin [Childs 2003b]. Note that plant cells strengthen their cell walls with cellulose molecules but still retain the basic eukaryote design for the plasma membrane.

Figure 3.3B illustrates a simplified schematic cross section of a eukaryote plasma membrane, showing the lipid bilayer with cholesterol molecule inclusions. The "typical" plasma membrane is about 6 to 7.5 nm thick (the apparent thickness seen in electron micrographs depends on fixation techniques). Note that the lipid bilayer is not a rigid structure, but at its "operating temperature" it is fluid. Its subunits can flow to accommodate shear stresses and can even self-repair small punctures. It is interesting to note that the self-repairing property of the plasma membrane allows cells to be penetrated with fine, glass micropipette electrodes (with tips $\leq$ 0.5 μm diameter) in order to measure the electrical potential difference across the cell membrane or inject substances into the cytoplasm electrophoretically. Also, in genetic engineering, plasmids and other macromolecules can be forcibly injected into cells using techniques such as micropipettes, *electroporation* (see Chapter 10, Section 10.2.4.7), or a *"gene gun"* (see Section 10.2.4.6), and the membrane will generally reseal around the breach [see Section 10.2.4].

Figure 3.3A illustrates the details of a basic, *phosphatidyl choline* plasma membrane subunit molecule; the polar head is shown in detail. The interested reader can find more details about the molecules that make up plasma membranes and their various structures in Karp [2007], Christie [2007], Diwan [2006], and Atkins [1998].

A living cell necessarily must import certain molecules used for metabolic energy and export the molecular products it makes, including the "waste" by-products of its metabolism. It also must adjust its internal osmotic pressure and ion balance by importing/exporting selected ions, and sometimes water molecules, across its cell membrane. All of these transmembrane transportations are carried out by specialized *transmembrane proteins* (TMPs), which are actually imbedded in and penetrate through the plasma membrane. Other integral membrane proteins are fixed to the outer or inner surfaces of the plasma membrane. The TMPs have a broad range of activities; some are *ion pumps,* i.e., they actively exchange an internal ion for a different external ion (or vice versa) powered by the breakdown of ATP to ADP. (Ion pumps are described in detail in Chapter 7, Section 7.2.5.2.) Other TMPs are *receptors* for external signal molecules such as hormones, still others are *gates,* allowing the inward, or outward, diffusion of certain ions when the transmembrane potential reaches a certain value or a signal gating molecule is present. A special eukaryote TMP, the glucose receptor, allows the inward diffusion of D-glucose, which is required for cell metabolism. In insulin-sensitive cells, this hormone acts parametrically to increase diffusion through the glucose TMPs. TMPs are discussed in more detail in Section 3.7 and in Chapter 7, Section 7.2.5.

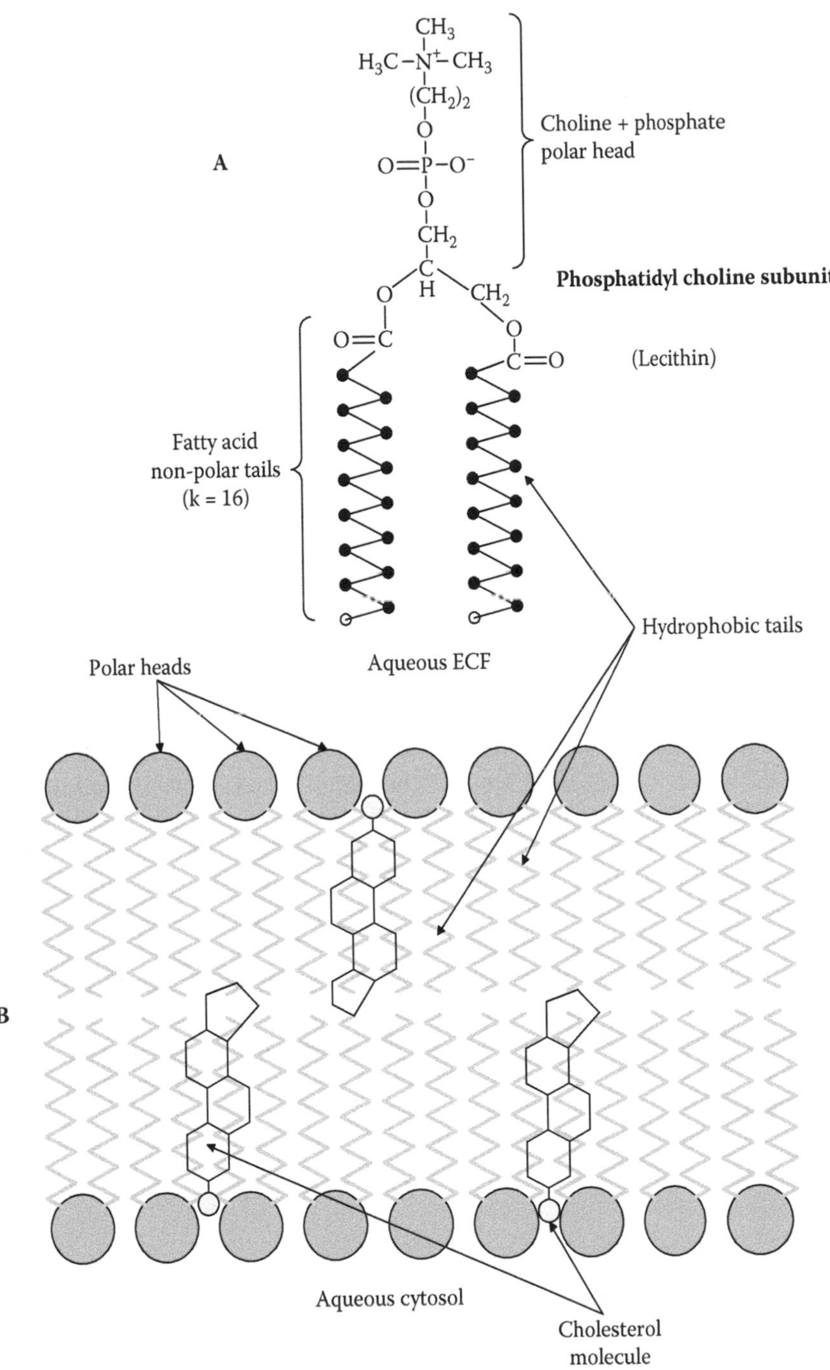

FIGURE 3.3 Schematic cross section through a plasma (unit) membrane. (**A**) Illustrates the (hydrophyllic) polar head structure of the basic unit. The zig-zag lines are the fatty acid, nonpolar (hydrophobic) "tails." (Two each joined by a *phosphatidyl choline* molecule.) (**B**) The cross-sectional schematic shows the polar heads facing the aqueous cytosol and extracellular fluid; the hydrophobic tails face the interior of the bilayer. Three cholesterol molecules are shown embedded in the tails.

3.3 INSIDE THE CELL MEMBRANE

The insides of all cells are complex biochemical "factories" in which a great number of regulated chemical reactions occur simultaneously. For example, in cells practicing oxidative metabolism, substrates such as D-glucose and molecular oxygen are converted in multistep reactions to energy-containing molecules such as ATP and NADH, which are used by the cell to power other reactions. *All* cells also contain a dsDNA genome, which contains the instructions for cell self-replication as well as its life processes. The dsDNA genome can be in the form of a circle (a plasmid) in the case of prokaryotes, and is in the form of pairs of linear sections of DNA (chromosomes) in eukaryotes. All cells also contain the molecular machinery (RNAs and ribosomes) required to translate their genomes into the proteins and RNA molecules they require for their existence. (Nucleic acids—DNA and RNA—are discussed in detail in Chapter 6.)

All cells have a fluid filling called the *cytosol*. In eukaryotic cells, the cytosol *excludes* intracellular organelles such as the *nucleus, Golgi membranes, cytoskeleton, ribosomes*, and *mitochondria*. It contains various ions and has about 20 to 30% free proteins. The *cytoplasm* of eukaryotes is the cytosol together with organelles, excluding the nucleus.

It is interesting to consider the extremes of unicellular life. One candidate for the *largest living unicellular organism* is the green algae, *Caulerpa sp. Caulerpa* is found in warm, shallow, tropical seas throughout the world. Some species grow to be about 1 m long. It has fernlike fronds that grow up from a horizontal, cylindrical stem with anchoring appendages. This immense, single-cell organism is supported by a cellulose cell wall that encloses its plasma membrane, providing mechanical support. *Caulerpa* also has internal structural stiffening formed by a lattice of thin cellulose rods. One way *Caulerpa* has overcome the problem of internal/external mass transport is by being very thin; its fronds are only about 5 μm thick, giving it a huge surface-area-to-volume ratio.

One of the smallest known bacteria is *Haemophilus influenzae*. It is capsule shaped and measures about 200–300 nm × 500–2,000 nm. Another small bacterium is *Mycoplasma genitalium,* about 400 nm in diameter. *Mycoplasma* has only 580,073 bps of DNA in its genome, which code only 483 proteins and 37 RNAs. The elementary body (analogous to a spore) of *Chlamydia trachomatis* is also quite small, about 300 nm in diameter. It has 1,042,519 bp of DNA encoding 936 genes. (*Chlamydia* causes chlamydia, a common sexually transmitted disease.)

Clearly, the prize for the largest bacterium should go to *Epulopiscium fishelsoni* that lives in the guts of the brown surgeonfish that live in the Red Sea. Individual specimens have been measured at 80 μm thick × more than 0.6 mm long! They are visible to the naked eye [Sizes 2006].

Multicellular, eukaryotic animals have also produced some extremely large cells. These include the (haploid) unfertilized ostrich egg (for sheer mass) and spinal motoneurons in the blue whale (axons several meters long).

In the following sections, we will review the characteristics peculiar to the three major domains of cells, the *Prokaryota,* which include the *Eubacteria* and *Archaea,* and the *Eukaryota* (plants and animals).

3.4 PROKARYOTIC CELLS

3.4.1 Introduction

Prokaryotic cells lack a membrane-enclosed nucleus or any other membrane-enclosed, internal organelles. The two prokaryotic domains are the *Archaea* = *Archaebacteria* and true bacteria = *Eubacteria*. There are significant anatomical, biochemical, and genetic differences between the *Eubacteria* and the *Archaebacteria* that warrant assigning them separate domains. Prokaryotes lack membrane-bound cell compartments such as vacuoles, endoplasmic reticula, mitochondria, and chloroplasts. Both prokaryotes and eukaryotic cells have ribosomes. However, prokaryotic cells' ribosomes are smaller (70S) than eukaryotic cells' (80S).

Reproduction of unicellular prokaryotes is generally asexual, using a process known as *binary fission*. The chromosome is duplicated and attaches to the inner cell membrane; then the cell divides in two. Some prokaryotes do exchange dsDNA rings between cells, allowing a mixing of genomes.

Although prokaryotes are unicellular, they can be *colonial,* attached by specialized, external cell membrane proteins to each other and a substrate. The colonies are formed by cells that remain attached following cell division but can act as individuals if separated from the colony. In unfavorable environmental conditions (e.g., loss of water), some bacteria can go dormant in the form of spores.

3.4.2 Archaeans

Archaeans deserve their own classification domain because of their unique molecular biological features. They include the following groups: *Methanogens, Extreme Halophiles, Hyperthermophiles,* and *Thermoplasma.* These creatures have evolved to live under extreme conditions of temperature, salinity, or low pH. They have some unique structural and biochemical properties not found in the modern Eubacteria.

The cell shapes of Archaeans ranges from spheroidal (coccus), to baccillus, to tubular (elongate baccillus) forms. They may or may not have flagella. The cell walls of Archaeans *do not* contain peptidoglycans. Their plasma membranes are bilayers made from L-glycerol and branched, 20-carbon, *isoprene chains* (see Section 3.2). Isoprene is the basic building block or subunit of a class of biopolymer molecules called *terpenes.* An isoprene molecule is as follows

$$\begin{array}{c} \quad CH_3 \\ \quad | \\ H_2C{=}C{-}\underset{H}{C}{=}CH_2 \end{array}$$

Unlike Eubacteria, the DNA of Archaeans is associated with histonelike proteins and contains *introns* (non-protein-coding regions) similar to eukaryotes. Archaeans also use eukaryote-like transcription machinery, including the enzyme, *RNA polymerase,* and *TATA binding proteins.* They also use eukaryote-like translation machinery such as *initiation factors, ribosomal proteins,* and *elongation factors.* From the molecular biology of modern Archaeans, it appears that they may lie evolutionarily between the Eubacteria and eukaryotes.

The Archaean *Methanogen* bacteria are anaerobic and use H_2, methanol, formate, carbon monoxide, or acetate as electron donors for the reduction of CO_2 to make *methane gas* (CH_4) [Kenyon 2006]. Methanogens may prove economically valuable as a means of generating methane as an alternative fuel under controlled bioreactor conditions.

3.4.3 Eubacteria

Probably the first form of cellular life on Earth appeared about 3.8 billion (3.8×10^9) years ago and was some form of very primitive prokaryote. It is known that prokaryotes have diversified greatly throughout their tenure on Earth. They certainly have had the time to do it. The first protocells presumably evolved into modern Eubacteria.

Before the age of genomics/proteomics, bacterial classification relied on basic features such as shape, size, and habitat: *Cocci* are spherical, *Bacilli* are rod shaped, and *Spirilla* have a corkscrew appearance. Other classification features include the ability to form spores, method of metabolism (anaerobic, aerobic, photosynthetic, etc.), source of metabolic energy, and reaction to the *Gram stain.* (The Gram stain is a simple means of determining the type of bacterial cell wall. Bacteria are first stained with a dye called *crystal violet,* and then washed with a solvent such as alcohol or acetone. If the stain is washed away, the bacteria are called *Gram-negative;* if the stain remains, they are in the *Gram-positive* group.) Recently, bacterial genome sequencing and sequencing of their 16S ribosomal RNA has provided guidance on bacterial evolution and has refined their classification.

Eubacterial classification includes *Proteobacteria,* subdivided into five clades: α, β, γ, δ, and ε. Proteobacteria are all *Gram-negative* but come in all shapes and have a variety of metabolisms. For example, the β-Proteobacteria include sulfur bacteria and iron bacteria. The γ-Proteobacteria include *E. coli, Salmonella sp, Yersina pestis* (plague), *H. influenzae*, and *purple sulfur bacteria.* The ε-Proteobacteria include *Helicobacter pylori,* which cause stomach ulcers, etc.

The *Gram-positive classification* includes economically and medically important aerobic bacilli such as *B. anthracis, B. thuringiensis*, and *Lactobacillus sp.* Anaerobic Gram+ bacilli include *Clostridium tetani and C. botulinum.* Gram-positive *cocci* include *Staphylococci sp. and Pneumococci sp.*

Other bacterial classifications are as follows:

1. *Spirochetes*, which cause syphilis and Lyme disease.
2. *Mycoplasmas*, which are obligate intracellular parasites and have the distinction of being the smallest (0.1 μm) living organisms with few genes. For example, *M. genitalium* has 580,073 base pairs of DNA encoding only 517 genes.
3. *Chlamydiae* are also intracellular parasites.
4. *Cyanobacteria* (blue-green algae) are photosynthetic; they use photon energy to convert $CO_2 + H_2O$ to O_2 gas and other metabolic products. They use Photosystem-II as well as PS-I [Kimball 2004c].

Eubacteria have a morphologically simple cell construction. The outer surface of a Eubacterium is generally coated with hydrated polysaccharides. This slime coat allows the bacterium to cling to substrate surfaces and to form colonies. Inside the slime coat, there is a tough, *peptidoglycan* cell wall. This wall is relatively porous to movement of small molecules (metabolites, water, waste, and proteins for the slime coat). The innermost layer is the bacterial *plasma membrane,* described in Section 3.2. Inside the plasma membrane are bacterial ribosomes and a simple circular dsDNA plasmid genome.

The genome of prokaryote *Eubacteria* generally lacks the histone proteins and introns found in eukaryotic DNA. There are exceptions on the genome structure: The Lyme disease pathogen, *Borrelia burgdorferi,* and the mold *Streptomyces sp.* have linear chromosomes, rather than rings. Other prokaryotes have many small DNA plasmids (rings) in their cytoplasm in addition to their main chromosome.

Bacteria have evolved many interesting forms of metabolism—the chemical processes by which they extract chemical energy from their environment. *Cyanobacteria* perform oxygenic photosynthesis using photons and CO_2, similar to eukaryote plant cells. An ancient bacterial form that has been on Earth for over 3 billion years, they are responsible for introducing oxygen gas into the atmosphere during the Archaean and Proterozoic Eras. Their importance is enhanced by the theory that they became endosymbionts in certain eukaryotic cells and then evolved into *chloroplasts,* found in all modern (eukaryotic) plants. Modern cyanobacteria are aquatic and photosynthetic. They come in a variety of colors because of variations in their light-absorbing, photosynthetic pigment molecules: violet, deep blue, blue-green, green, yellow, and red. Cyanobacteria form symbiotic relations with a wide group of life-forms, including certain protists, marine sponges, and echiruid worms, and also with a wide variety of photosynthetic hosts. They live singly and in colonies. They reproduce by binary fission, budding, or fragmentation. They have had a lot of evolutionary time to adapt to many, many niches [rvt 2005].

Iron and sulfur are also used as metabolites by certain prokaryotes. *Purple sulfur bacteria* are also photosynthetic; they use hydrogen sulfide gas (H_2S) for the electrons needed to synthesize NADPH and ATP for chemical energy. In this process they reduce sulfur according to the overall reaction shown in Equation 3.1. Note that $\mathbf{CH_2O}$ is formaldehyde! Both hydrogen sulfide and formaldehyde are poisonous to most other life-forms. Bacterial photosynthesis uses only Photosystem I for the conversion of photon energy into chemical energy [Todar 2004]

$$2H_2S + CO_2 \rightarrow CH_2O + H_2O + 2S \tag{3.1}$$

Iron bacteria are *anaerobic lithotrophs*; i.e., they use elemental iron (Fe) as a source of energy. They remove electrons from iron (oxidize it) and push them through an electron transport system that will produce ATP by *electron transport phosphorylation.* The iron bacteria *Gallionella sp.* and *Thiobacillus ferrooxidans* oxidize ferrous iron (Fe^{++}) to ferric iron (Fe^{+++}) to harvest electrons. *T. ferrooxidans* can also oxidize sulfur; in this case, it begins with H_2S gas and ends with $SO_4^{=}$ ions. Iron bacteria are not toxic to humans or animals. The red ferric iron produced by their metabolism is largely insoluble and is deposited in the slimy, gelatinous material surrounding their cells. It is often found on the pumps and pipes in deep, drilled water wells.

Another unique genus of bacteria is the *Spiroplasmas.* They are thought to have evolutionarily diverged from Gram-positive Eubacteria *(Clostridia sp.)* billions of years ago; they generally have a unique helical (corkscrew) cell shape that lacks the standard bacterial cell wall and slime coat. Instead, they have a cholesterol-containing unit membrane, which places them in the Prokaryote class, *Mollicutes* with *Mycoplasma sp* and *Acholeplasma sp. Mollicutes* are the smallest and simplest free-living and self-replicating life-forms. They are motile and chemotactic and have internal cytoskeletons [Trachtenberg 2004].

Spiroplasma sp. move by undulating their corkscrew bodies by contracting internal motor fibers. Trachtenberg, Gilad, and Geffen [2003] have described the biomechanics of motion in *S. melliferum* and have also given average dimensions for this spiroplasma cell. The mean helical repeat (pitch), P, is about 0.85 μm, the mean cell diameter is $d \cong 0.2$ μm, the coil diameter $(D + d) \cong 0.6$ μm, and the overall length is variable, seen to be about 3.5 μm in a figure in the reference. Depending on the contractile states of its internal motor fibers, the shapes of spiroplasmas can range from a tightly coiled helix to a long, thin rod.

The genome sizes of *Spiroplasma sp.* are relatively small, ranging from 580 kbps to 2.2 Mbps, depending on the species [OARDC 2003a]. An interesting property of spiroplasmas is their ability to infect a broad spectrum of eukaryotic metazoans, including plants, insects, and mammals [Maramorosch 1981; OARDC 2003b]. Injection of a tick-associated *Spiroplasma sp.* into newborn mice, rats, hamsters, and rabbits caused cataracts in the mice [Clark 1964] as well as CNS lesions. Maramorosch [1981] raised the question as to whether spiroplasmas could be an etiological agent for human cataracts and other human diseases.

From the published research of Bastian and Foster [2001] and Bastian et al. [1984], a certain amount of controversy has surrounded the role of spiroplasmas in the etiology of *Transmissable Spongiform Encephalopathy* (TSE) (aka mad cow disease), Creuetzfeldt-Jakob disease, kuru, and other prion-related diseases. (See Chapter 7, Section 7.3.4 for a description of the prion diseases.) Bastian has presented evidence that TSE can be induced in mice by infecting them with a *Spiroplasma sp.* His theory goes against the conventional wisdom that TSE is spread by host infection with misfolded prion protein, $\mathbf{PrP^{Sc}}$, in what appears to be some sort of autocatalytic process (e.g., $\mathbf{PrP^{Sc} + PrP^{C} \rightarrow 2\,PrP^{Sc}}$).

In a recent paper, Alexeeva et al. [2006] presented the results of their research, which demonstrated that there was no evidence of *Spiroplasma* or other bacterial 16S rRNA genes in the brains of hamsters with scrapie. It thus appears that although *Spiroplasma sp.* may be associated with the onset of TSE in many species, TSE has other causes, including the infection of the host by misfolded prion protein (PrP^{Sc}) molecules. Beware of "single-cause" thinking in molecular biology.

3.5 EUKARYOTIC CELLS

3.5.1 INTRODUCTION

Around 2.7 billion years ago, a quantum jump occurred in the biochemical and morphological sophistication of cells. The appearance of *eukaryotes,* whose domain includes the kingdoms *Ani-*

malia, Fungi, Plantae, and *Protista,* signaled a major surge in evolutionary speed and complexity associated with the rise in the use of oxidative metabolism by cells because of the appearance of molecular O_2 in the Earth's atmosphere, thanks to *Cyanobacteria.*

We can only speculate on the ancient natural origins of eukaryotes. Some theories are as follows: (1) Eukaryotes may have resulted from the complete fusion of two or more cells, the cytoplasm forming from an Eubacterium and the nucleus from an Archaean (or possibly, a virus). (This is known as the *serial endosymbiosis theory* [SET].) (2) Eukaryotes may have developed from Archaeans and acquired their Eubacterial characteristics from the protomitochondrion that entered as a commensal endoparasite. (3) Still another hypothesis is that eukaryotes and Archaea developed separately from a modified *Eubacterium.* These theories have roots in the important books by Margulis [1970, 1981] and have been discussed and elaborated on by Mohn [1998] and Vellai and Vida [1999] and also in Chapter 1 in the text by Cooper [2000].

Two other theories concern the origin of the endomembrane system and mitochondria in eukaryotes: The *Phagotrophic Hypothesis* claims that the use of internal membranes originated with the evolution of *endocytosis* and later was specialized. Mitochondria were acquired by ingestion, similar to plastids. In the *Syntrophic Hypothesis,* the protoeukaryote relied on free-living mitochondria for food/energy, and eventually mitochondria became *commensal* inside eukaryotic cells (not parasites). Note that, in modern eukaryotic cells, a sizeable portion of the genes required for mitochondrial function reside in the cell's nucleus. Perhaps the internal membranes arose later owing to mitochondrial genes.

Eukaryotic cells use a plasma (nuclear) membrane to enclose a nucleus containing linear, not circular, dsDNA chromosomes. Also enclosed by membranes are various cytoplasmic organelles such as mitochondria and chloroplasts, as well as cytoplasmic vesicles and vacuoles that permit ingestion of materials by *endocytosis* and the exportation of materials by *exocytosis. Eukaryotes* also have internal cytoskeletons composed of microtubules and microfilaments. The dsDNA chromosomes of a eukaryotic cell are physically separated during cell division by a microtubular spindle apparatus. Various tubular and sheetlike extensions of the nuclear membrane form the *endoplasmic reticulum* (ER), which figures in protein transport in the cell. The *ribosomes* where proteins are assembled are attached to the *rough ER.* The newly synthesized proteins enter the lumens of the rough ER, where they undergo posttranslational modification and modification by *chaperone proteins,* they then are transported to their functional locations in the cell by a variety of means.

Eukaryotic cells generally have a smaller surface-to-volume ratio than prokaryotes, and thus tend to have lower metabolic rates and longer generation times. Cell reproduction by eukaryotes is generally by *mitosis,* in which the diploid chromosomes are copied and go into the daughter cell. Sexual reproduction is another attribute of eukaryotes; they use *meiosis* to produce haploid cells, which fuse with other haploid cells, mixing genomes and creating a genomically diverse diploid cell that can reproduce by mitosis.

3.5.2 Single-Cell Eukaryotes (Protists)

All protists have a membrane-covered nucleus; most have mitochondria, and many have chloroplasts; that is, they can derive energy from photosynthesis. It has been hypothesized that mitochondria in eukaryotes were derived from aerobic *α-proteobacteria* that once lived as commensal prokaryotes within cells. The chloroplasts may have had a similar start from commensal, photosynthetic bacteria living in early eukaryotic cells [Margulis 1970, 1981]. Many protists are unicellular; however, some live together in cooperative colonies. The kingdom *Protista* can be classified into the following phyla:

- Euglenozoa
- Alveolates
  - Ciliates
  - Sporozoans
  - Dinoflagellates

- Diatoms
- Golden algae
- Brown algae
- Water molds
- Red algae
- Slime molds
    - Cellular slime molds
    - Plasmodial slime molds
- Protists without mitochondria
- Choanoflagellates

We now review some features of the foregoing groups. The *Euglenozoa* are unicellular; many use a single *flagellum* for propulsion. There are about 1,600 species, which include *Euglena,* which has chloroplasts; *Trypanosoma brucei*, which causes sleeping sickness; and *Trypanosoma cruzi*, which causes Chagas' disease.

The *Alveolates* all have a system of saclike vacuoles on the inner surfaces of their membranes (hence *alveoli*) as well as close homology in their genomes. The *Ciliates*, as the name suggests, are covered with rhythmically beating cilia, which they use for propulsion. They feed by sweeping a stream of water containing food particles through a "mouth" and "gullet" into a food vacuole. They eliminate wastes and pump out excess water using contractile vacuoles to do exocytosis. Ciliates include *Paramecium, Stentor,* and *Vorticella*.

The phylum *Sporozoa* consists entirely of parasites and is thus a major cause of diseases. They include the genus *Plasmodium*, which causes malaria, also *Toxoplasma gondii,* a parasite that causes toxoplasmosis. *Dinoflagellates* include about 1,000 species. Most contain chlorophyll a and c and are unicellular. Although classed as eukaryotes, their chromosomes lack histone proteins. They have two flagella with the eukaryotic (9 + 2) core design. Population explosions of certain marine Dinoflagellates are responsible for poisonous "red tides" that kill fish and render shellfish unfit to eat.

Diatoms are ecologically important in marine and aquatic ecosystems. Their cell walls are composed of two overlapping discoidal shells with radial symmetry that are largely made of silica. These single-celled organisms carry out photosynthesis and are major producers of O_2 and biomass in fresh and saltwater. They are at the bottom of the food chain and supply nutrients to heterotrophic protists and small filter-feeding animals. These creatures, in turn, are food for larger invertebrates and vertebrates. Their inorganic skeletons are used in certain polishing compounds (diatomaceous earth).

Photosynthetic *Golden Algae (Chrysophyta)* are found in fresh water; they are mostly unicellular. They have the unique property of being able to switch from photosynthesis under low-light conditions and become heterotrophic, feeding on bacteria or other smaller protists.

The marine rockweeds and kelps are multicellular organisms formed from *Brown Algae (Phaeophyta).*

Water Molds (Oomycetes) are economically troublesome organisms. They are not fungi but are genomically closer to the brown algae. They have cellulose cell walls. Two species (*Saprolegnia* and *Achyla*) parasitize fish. Downy mildews damage grapes and other fruit. The infamous Irish potato blight was caused by *Phytophthora infestans,* and *Phytophthora ramorum* damages oak trees.

Red Algae are photosynthetic marine protists; most are multicellular but some are unicellular. About 4,000 species are known. They use chlorophyll a; the structure of their chloroplasts resembles those found in cyanobacteria. The *Agar* laboratory culture medium for microorganisms is extracted from red algae.

The *Mycetozoa* (slime molds) are single, amoeba-like cells that congregate into a single, slimy mass that feeds on rotten vegetation. *Cyclic AMP* secreted by the aggregating cells serves as a chemical attractant. The life cycle of slime molds is surprisingly complex for protists. Let us start with a *haploid spore* produced in the *fruiting body* by *meiosis.* Once germinated, the spore can become *either* an amoeba-like *myxamoeba or* a flagellated swarm cell. In this stage, these cells can divide into large populations of independent haploid cells. Two myxamoeba or two flagellated swarm cells can fuse *(plasmogamy)* followed by nuclear fusion *(karyogamy)*. (Note that a myxamoeba and a flagellated cell do not fuse.) The combination of their cell contents represents a primitive form of sexual reproduction; the resultant cell is diploid zygote. This zygote next divides and forms an aggregate of cells called a *plasmodium* that moves together as a multicellular, slimy mass, seeking nutrients; as many as 100,000 diploid cells can be involved. When nutrients become scarce, the plasmodium grows fruiting bodies (fructifications) that disperse spores. The spores are spherical and are 5 to 15 μm in diameter. There are four types of fruiting bodies; some are sessile, whereas others have stalks. The plasmodium itself can be yellow or orange-brown. Indeed, the plasmodium of the slime mold, *Fuligo septica*, has earned it the common name of the *dog vomit slime mold.*

The protists without mitochondria have apparently lost them in the distant past. They include the obligate intracellular parasites, the *Microsporidia. Microsporidia* contaminate water supplies and cause diarrhea in humans. *Giardia intestinalis* is a waterborne protist without mitochondria that causes diarrhea in humans. Another gut parasite is the protist, *Entamoeba histolytica,* which causes amoebic dysentery. It, too, lacks mitochondria.

Finally, we examine the *Choanoflagellates,* single-celled, fresh and saltwater protists that have a single flagellum used for propulsion. The flagellum also forces bacteria-containing water into the collar of microvilli that surround its base; these are used for food. Kimball [2006c] speculates that *Choanoflagellates* are the closest protistan relatives to eukaryotic metazoan life. He notes that they express genes for several proteins used for cell-to-cell C^3 in metazoans, including *cadherins* and *protein kinases,* enzymes used in cell-to-cell signaling.

In summarizing this section, we note that the *Protista* exhibit a broad range of morphological, genomic, and biochemical features that place them somewhere between the bacteria, archaea, and higher eukaryotes. Because of the mixture of features this kingdom exhibits, there are numerous alternative classification schemes for protists [Sidwell 2006].

3.5.3 Multicelled Eukaryotes

The *Metazoa* are multicelled eukaryotes that include *invertebrates* and *vertebrates*; the *plants*, including *green algae (Chlorophyta)*; and the *fungi.*

The development of metazoan life probably was enabled in part from the protistan molecular mechanisms that allowed close associations of protistan cells. Metazoan life requires cell adhesion and cell functional specialization. Cell-to-cell binding is a complex subject and is described in Section 3.7.1. Intercellular command, communication, and control (C^3) also had to evolve to permit complex multicellular organisms to function as coordinated entities. The whole study of cell membrane receptors and signal substances is a complex, active area of investigation, not only for our understanding of the fundamental properties of life, but also to learn how to control cell development and fight diseases such as cancer.

The amazing range of diversity and specialization in the cells of adult vertebrates is awesome. From a zygote, then stem cells, there develops a bewildering array of cell types, all from the same genome, including but not limited to neurons of various types, including sensory cells, skin, muscles (smooth, striated, and cardiac), bone, connective tissue, and specialized organs and glands (liver, pancreas, kidney, adrenal glands, testes, ovaries, etc.). Some of the common properties of eukaryotic animal cells are examined in the following section.

3.6 ANATOMICAL COMPONENTS OF EUKARYOTIC CELLS

3.6.1 The Cytosol

As we observed earlier, the cytosol is the liquid outside the nucleus and the mitochondria and inside the cell membrane. It contains water and just about every molecule inside a cell that is not bound to a membrane or inside the nucleus or mitochondria; including proteins, amino acids, various RNAs, glucose, ATP, ADP, small ions including H^+, K^+, Na^+, Mg^{++}, Ca^{++}, HCO_3^-, Cl^-, $SO_4^=$, OH^-, etc. The cytosol is 20 to 30% dissolved (hydrated) proteins, and its pH is about 7.0 (neutral) in human cells.

The cytosol also surrounds the cytoskeleton, which is made from fibrous proteins, and also bathes the mitochondria, endoplasmic reticulum (ER), Golgi apparatus, etc. In prokaryotic cells, *all* chemical reactions occur in the cytosol. Eukaryotic cells maintain their internal osmotic pressure by adjusting the milliosmolarity of the cytosol by pumping out certain ions and molecules through the plasma membrane or cell wall so that water will not enter the cell, swelling it. They also can regulate the rate of water flux across their cell membranes with *aquaporin* transmembrane proteins. All this is part of the "molecular housekeeping" all cells must practice.

3.6.2 The Nucleus and Nucleolus

One of the necessary attributes of eukaryotic cells is a membrane-covered nucleus. The cell nucleus has an inner and an outer bilipid membrane. The volume between these membranes is continuous with the endoplasmic reticulum (ER). The nuclear membranes are perforated with *nuclear pore complex* (NPC) transmembrane proteins that control passage of molecules and ions in and out of the nuclear volume.

The cell's *chromosomes* are located inside the nucleus. Each *chromatid* consists of one long, helical molecule of dsDNA complexed with *histone* proteins. (Nucleic acids are treated in detail in Chapter 6.) Eukaryotic chromosomes are visible with special stains under the light microscope in *prophase* and *metaphase* of *mitosis.* They appear as homologous, *paired* (diploid), linear structures joined together at a point called the *centromere*, so they form a skinny letter "X." In humans, a centromere may contain about 10^6 Nt base pairs. Most of this DNA is "filler," repeat sequences of about 170 bps joined in tandem about arrays.

The diploid (**2n**) chromosome count of certain organisms is not necessarily consistent with the organism's position in terms of evolutionary complexity. For example: Ant (**2n** = 2), chicken (78), corn (20), crayfish (200), dog (78), *Drosophila* (8), horsetail plant (216), human (46), mouse (40), *Xenopis* frog (36), yeast (32), etc. The complete set of chromosomes in cells is called the *karyoptype.* The karyotype of the human female contains 23 pairs of homologous chromosomes, that is, 22 pairs of *autosomes* and one pair of X chromosomes. Human males have 22 pairs of autosomes plus one X and one Y chromosome (**2n** = 46).

Whereas the genome is located in the dsDNA, the *histones* and *transcription factor (TF) proteins* are involved in regulating gene expression. There are five kinds of histones. Two copies each of H2A, H2B, H3, and H4 form a core around which is wrapped about 147 DNA base pairs (bps). This unit is called a *nucleosome.* Twenty to 60 DNA bps link one nucleosome to its neighbors. Each linker region is complexed with a single molecule of histone H1. The importance of histones in eukaryotic cell biology is underscored by their high degree of conservation across eukaryotic species. For example, H4 in cattle differs from H4 in the pea plant by only two AAs out of a total of 102. The binding of histones to dsDNA depends heavily on the AA sequences of the histones rather than the nucleotide sequences of the DNA [Kimball 2005n].

One of the important functions of histones is to fold and compact the 46 dsDNA molecules (human chromosomes) into the small volume of the nucleus. (The typical nucleus is about 10 μm in its major about dimension.) The approximately 3.1×10^9 nucleotides in the human genome, if stretched out end-to-end in alpha helix form, would measure over 2 m in length! Nature has solved this compact packing requirement by having the nucleosomes associate into 30-nm-diameter fibers and then having the fibers further fold and twist in a complex manner yet to be fully understood. During gene expression, the fibers untwist around the gene of interest and the histones expose the DNA helix for pmRNA reading.

The nuclear dsDNA with its complexed histone proteins is called *chromatin.* Regions of (histochemically) dense chromatin are called *heterochromatin.* Less optically dense regions are described as being composed of *euchromatin*. Heterochromatin is associated with parts of a chromosome where there are few or no genes, such as the *centromere* and two *telomeres*. Those genes present in heterochromatin are generally *turned off* (not being transcribed). The DNA in heterochromatin regions exhibits increased methylation of cytosines in *CpG islands*. It also has increased *methylation* of lysine-9 in histone H3, which provides a binding site for a protein *(Heterochromatin Protein 1)* that blocks access to the gene by the *transcription factors* used in gene expression.

It has recently been demonstrated that the heterochromatin in the fruit fly, *Drosophila sp.*, contains from 230 to 254 protein-coding genes, as well as Nt sequences that code for RNAs, including small siRNAs that neutralize transposable elements (transposons). *Transposons* are short, virus-like DNA sequences that can move around (or be moved around) in the genome and are capable of disrupting gene function. The genome of the female fruit fly is over one-third heterochromatin (about 60 megabases, Mb). The male has over 100 Mb because its **Y**-chromosome of about 40 Mb is all heterochromatin. Also embedded in the fly's heterochromatin are 32 *pseudogenes*. These are truncated, inactive copies of genes. Pseudogenes are inactive because they coded for proteins no longer needed by the organism, or they may have been duplicated elsewhere in the genome and are redundant. Humans have about 20,000 pseudogenes [Science 2007].

The genes associated with euchromatin are generally active and are characterized by decreased methylation of their cytosines in *CpG islands* of the DNA, increased acetylation of histones, and decreased methylation of lysine-9 in H3. Regions of euchromatin are separated from adjacent heterochromatin by *insulator* sections of DNA that act to functionally isolate adjacent genes. The nucleosomes of euchromatin form "bead necklaces" of active genes and are found near the nuclear pore complex TMPs that penetrate the nuclear envelope. Thus, mRNA can leave the active genes in the nucleus and migrate to ribosomes in the cytoplasm, where protein assembly occurs.

Modification of histone molecules by the addition of certain groups and radicals to their AAs can affect gene expression. *Methyl groups* (**$-CH_3$**) can be added (then removed) from lysines and arginines. Phosphate groups can be added to serines and threonines, and *acetyl groups* (**CH_3CO-**) to lysines. Kimball [2005n] observed that

- "Methylation, which also neutralizes the [+] charge on lysines (and arginines) can either stimulate or inhibit gene transcription in that region. Methylation of lysine-4 in H3 is associated with **active** genes while methylation of lysine-9 in H3 is associated with inactive genes. These include those **imprinted genes** that have been **permanently inactivated** in somatic cells (e.g., one X-chromosome in human females).
- And adding phosphates causes the chromosomes to become more—not less—compact as they get ready for mitosis or meiosis. In any case, it is now clear that histones are a dynamic component of chromatin and not simply inert DNA-packing material." (Quoted with permission.)

In human cells (between cell divisions) in *interphase,* 10 chromosomes have loops forming a dense spherical mass known as the *nucleolus*. The nucleolus is the site where the genes for coding three of the four kinds of rRNA (28S, 18S, 5.8S) used in the assembly of the large and small subunits of *ribosomes* are located. Once formed, the rRNA molecules combine with the various ribosomal proteins used in the assembly of the large and small subunits of the ribosomes. The *nuclear pore complex* (NPC) proteins transport these molecular complexes out of the nucleus into the cytoplasm. Each NPC may have as many as 50 *nucleoporin* protein subunits. A nuclear pore can open to form a channel about 25 nm in diameter through which the ribosomal subunits can leave the nucleoplasm and enter the cytoplasm.

The nuclear pores also allow the import into the nucleus of key nuclear proteins. Each protein may have a characteristic sequence of AAs (a sort of molecular PIN number) that makes it eligible for transport into the nucleus. Transport in and out of the nucleoplasm is an active process; that is, it requires chemical energy such as in ATP. It also involves specialized carrier molecules and docking sites in the entry part of the NPC molecules.

Proteins that must be *imported* into the nucleus include all of the histones, all of the ribosomal proteins, all of the transcription factors required to activate and suppress gene expression, and all of the *splicing factor proteins* needed to process pmRNA into mRNA (i.e., edit out the *intron* code sequences). Molecules *exported* from the nucleus through the NPCs include (1) edited mRNA, (2) all the rRNA + protein ribosomal subunits, (3) transfer RNA (tRNA) accompanied by proteins, and (4) transcription factors that are being recycled. Both the RNAs and proteins being exported contain a *nuclear export sequence* to make them bind to the correct carrier molecules.

Imagine what a busy place (in a molecular biological sense) a stem cell's nucleus (and its NPCs) is during embryonic development!

3.6.3 Ribosomes

The ribosomes are the molecular machines that assemble polypeptides according to the genetic code. They work with the information coded on messenger RNA (mRNA), which was transcribed from a gene and then edited to remove the *intron* Nt sequences. A eukaryotic ribosome has a diameter of about 20 nm and appears approximately spherical. Ribosomes are ubiquitous to all life; they are found in prokaryotes, archaea, eukaryotes, and even in mitochondria and chloroplasts. All ribosomes have a (heavy) molecular subunit and a small (lighter) subunit. The subunits are made of associations between rRNA molecules and proteins. In eukaryotes, the large (60S) subunit has three rRNA molecules in it and about 49 proteins. The smaller (40S) subunit has one rRNA and about 33 protein molecules. They are complex "nanomachines."

Ribosomes that synthesize proteins, which are used locally in the cytosol, are located in the cytosol; e.g., the enzymes used in *glycolysis*. Proteins that are destined for use in and on the cell plasma membrane, destined for export through the cell membrane or for use in the *lysosomes*, are made in ribosomes attached to the cytosolic face of the ER, As the linear polypeptide is formed, it is extruded into the lumen of the ER for "final assembly." Before these proteins are transported to their final destinations, their structures are further "edited" in the *Golgi complex*.

The 13 proteins used in the inner membrane of the mitochondria are made by ribosomes within the mitochondrial membrane. These **m**-ribosomes are smaller and simpler in structure than those of pro- and eukaryotic cells. The large subunit (39S) has one rRNA and 48 proteins; the smaller subunit (28S) has one rRNA and 29 proteins [Kimball 2003a].

(See Chapter 5, Section 5.6.2 for a detailed description of how proteins are expressed and synthesized by ribosomes.)

3.6.4 Endoplasmic Reticulum

The *endoplasmic reticulum* (ER) is a structure in the cytoplasm of eukaryotic cells composed of an interconnected network of thin tubules, folded membranes, vesicles, and sacs. It is divided into two subtypes: *smooth ER* (SER) and *rough ER* (RER). In different types of cells, the SER has different biochemical and physiological roles. The large surface area of the SER membranes allows storage sites for membrane-bound enzymes and their products. In muscle cells, this ER is called the *sarcoplasmic reticulum* (SR); the inner volume of the SR stores Ca^{++} ions required for muscle contraction, its membranes contain ATP-driven calcium pumps that concentrate the Ca^{++} ions and which are responsible for muscle relaxation. In liver cells (hepatocytes), the SER is very convoluted with many small tubes and vesicles. Hepatocyte SER contains many enzymes; it is where glycogen is metabolized and detoxification of drugs occurs. Another example of metabolic specialization in the SER is found in adrenal cortical cells. There enzymes metabolize steroids and synthesize steroid endocrine hormones. The side chain of cholesterol is cleaved in the mitochondria, the product is transferred to the SER and further modified, and then it is taken up again by the mitochondria for final molecular assembly [Childs 2001].

The RER holds the ribosomes that synthesize proteins which are destined for a transmembrane location or exocytosis. The RER is so called because the presence of attached ribosomes gives it a

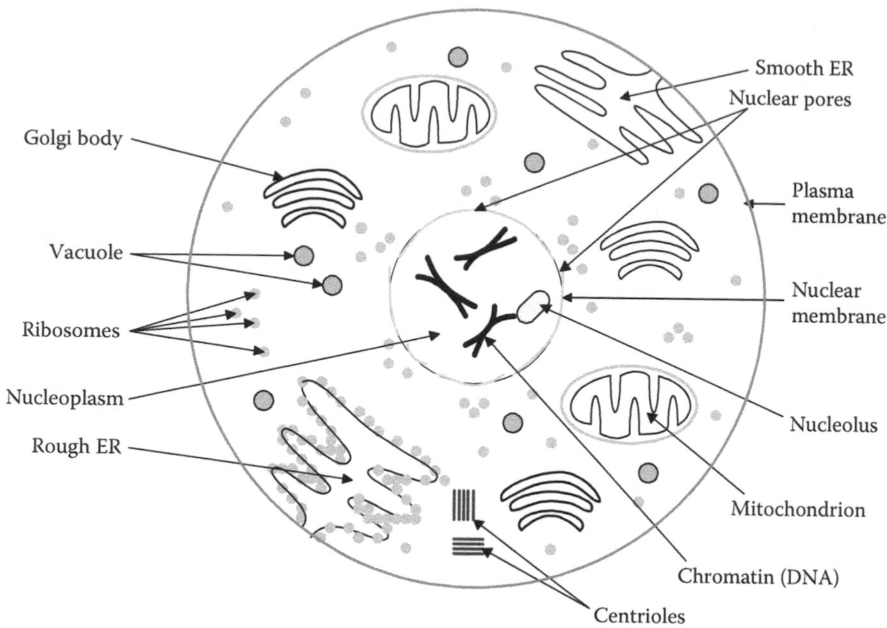

FIGURE 3.4 (See color insert following page 44.) Schematic cross section of a typical eukaryotic cell showing its nucleus and cytoplasmic organelles.

"rough" appearance in electron micrographs. The ribosomes associated with the RER are bound to the outer surfaces of the RER's sacs, or *cisternae*. The proteins they assemble are stored in the sacs and may undergo posttranslational processing. Some of the details of the complex molecular interactions between ribosomes, the polypeptides they synthesize, and the RER can be seen in the online lecture by Childs [2003e]. (How proteins destined for exocytosis or use in or on the cell's outer membrane are manipulated is described in detail by Kimball [2003h].)

Figure 3.4 illustrates a schematic of a representative eukaryotic cell showing the nucleus, SER, RER, Golgi apparatus, and lysozymes.

3.6.5 Golgi Apparatus

Some of the proteins synthesized in the RER undergo further processing in the Golgi apparatus (complex) (GA). In electron micrographs, the GA appears like a stack of pancakes—hollow, membrane-bound *cisternae*. They are typically located between the ER and the cell membrane [Kimball 2003h].

Because the GA is the site of many complex biochemical reactions, it contains many protein enzymes. It also contains molecular mechanisms that sort out the processed proteins and direct them to their destinations, which can be into *lysozymes* or embedded in the cell membrane (transmembrane proteins such as ion and water channels) or secreted by exocytosis. It also has enzymatic machinery for reclaiming the processing proteins for molecular recycling. One major molecular activity is the *glycosylation* (adding sugar molecules) of polypeptides, forming *glycoproteins*.

In epithelial mucus-secreting cells, *mucopolysaccharides* (*proteoglycans*) are made in the GA by adding an abundance of carbohydrates to a protein "skeleton." In plant cells, the GA manufactures the cell plate and the cell wall. Large protein substrate molecules are cut up to form smaller peptide hormones and neurotransmitters in the GA. For example, the precursor protein *proopiomelanocortin*, which has 256 AAs, is cut into *ACTH* (adrenocorticotrophic hormone, 39 AAs), α and β *MSH* (melanocyte-stimulating hormone), β-*endorphin*, and others [Kimball 2004h]. This subdivision of proteins is one reason the number of possible proteins is much greater than the number of protein-coding genes.

Transition vesicles pinch off from the ER surface and carry integral membrane proteins, soluble proteins requiring postprocessing, and processing enzymes to the *cis-Golgi* (the closest GA section to the ER). There, their protein coat (*coat protein II*) is removed, and they fuse with the cis-Golgi membrane and release their molecular load for processing. Amazingly, the *cis*-Golgi exports processed or partially processed molecular products in *shuttle vesicles* to intermediate GA layers. Finally, proteins and Golgi products ready for export are put into various types of transport vesicles, which pinch off from the outer face of the *trans*-Golgi and carry the products to their final destinations. This can be the *secretory vesicles* (e.g., for hormones) of lysozymes [Childs 2003f].

The remarkable GA is under nuclear control; it is broken down and disappears at the onset of mitosis, and by telophase it reappears, ready to work.

3.6.6 Lysosomes

As we have seen in the previous section, lysosomes are 0.5- to 2-μm-diameter spherical bodies that bud off the *trans-GA*. They contain over 36 kinds of hydrolytic enzyme molecules, which are used for intracellular digestion of defective or foreign objects in the cystosol. Such foreign particles include but are not limited to defective mitochondria that have been engulfed by *autophagosomes*; objects that have entered the cell by endocytosis, including bacteria and viruses taken in by immune system cells; and antigens and antigen–antibody complexes.

Lysosomes contain *proteases* that break down proteins, *lipases* that cut up lipids, *nucleases* that digest nucleic acids, and *polysaccharidases* that break down carbohydrates. They are active in embryonic development in removing structures such as tadpole tails and fetal finger webs.

The interior of a lysosome is acid, pH 4.8, which is maintained by proton pumps and CLC chloride channels [Childs 2003g].

3.6.7 Peroxisomes

Other specialized single-membrane vesicles in the cytoplasm, *peroxisomes,* are not made by the GA. They superficially resemble lysosomes, but their role is *oxidation* of certain molecules; they contain oxidative enzymes such as D-*amino acid oxidase*, *ureate oxidase*, and *catalase*. The enzymes and other proteins destined for peroxisomes are synthesized by ribosomes in the cytosol. Each contains a *peroxisomal targeting signal* (PTS) AA sequence that binds to an individualized receptor "ferry" molecule that takes the protein into the peroxisome and then returns for another target molecule (a sort of destination "barcode"). Two PTS sequences have been identified: A 9 AA sequence at the protein's N-terminal and a tripeptide at the carboxy (C) terminal [Kimball 2005o].

Peroxisomes function to remove toxic substances from cells. They are common in liver cells where toxic substances accumulate. In hepatocytes, they break down excess fatty acids and break down hydrogen peroxide with *catalase* (H_2O_2 is a byproduct of fatty acid oxidation); they also break down excess *purines* (AMP, GMP) to *uric acid* and participate in the synthesis of *cholesterol*, *bile acids*, and *lipids* used to make *myelin* [Childs 2003g].

3.6.8 Mitochondria

The basic mitochondrion structure is illustrated schematically in Figure 3.5. It has inner and outer bilayer lipid membranes studded with transmembrane proteins. The outer membrane encloses the entire organelle; it has integral proteins called *porins* that have relatively large pores (2 to 3 nm) which can pass all molecules and ions up to about 5,000 Da mass [Kimball 2007c]. Larger molecules traverse the mitochondrial membranes by an active transport process. The outer membrane has enzymes attached to it that catalyze reactions such as oxidation of epinephrine, adding carbons to fatty acid chains, and breaking down the AA *tryptophan*.

The inner mitochondrial membrane is folded into flat invaginations called *cristae*; the inner volume of a mitochondrion inside the *cristae* is called the *matrix*. The inner membrane is highly

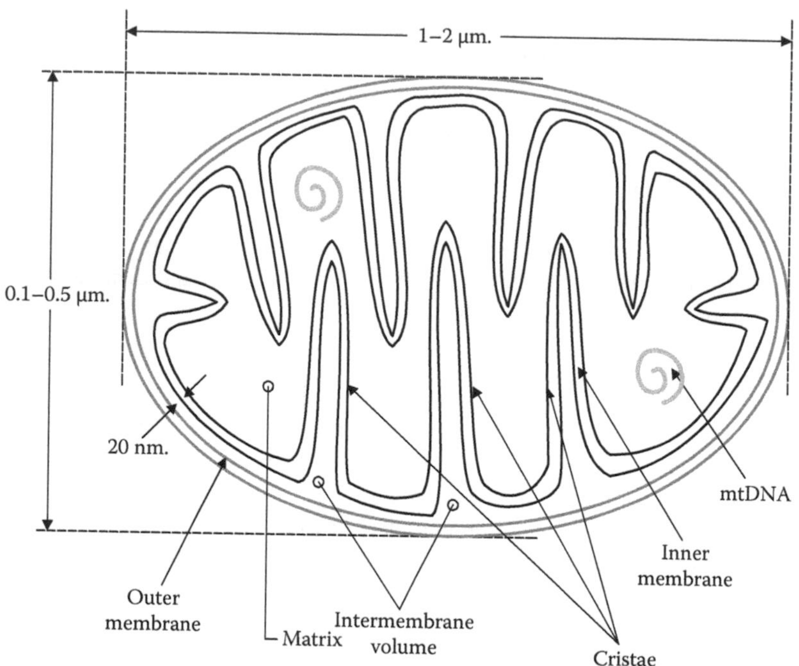

FIGURE 3.5 Simplified schematic cross section of a mitochondrion. Dimensions are approximate.

invested in proteins; in fact, it has a very high protein-to-phospholipid ratio, over 3:1 by weight, which is about 1 protein molecule per 15 phospholipid molecules; over 100 different proteins are found there. The principal functions of the inner membrane proteins include (1) the production of ATP in the matrix, (2) the redox and electron transfer reactions of oxidative metabolism (the citric acid cycle), and (3) regulation of the passage of metabolites in and out of the matrix (by transport proteins).

Although the aerobic metabolism that produces ATP is a major mitochondrial function, mitochondria also are necessary for other cellular metabolic tasks. Some of these are cell-type-specific tasks; for example, in liver cells, mitochondria can detoxify ammonia resulting from protein metabolism. These tasks include but are not limited to

- Apoptosis (programmed cell death)
- Cellular proliferation
- Glutamate-mediated excitotoxic neuronal injury
- Heat production (a byproduct of oxidative metabolism)
- Heme synthesis
- Regulation of the cellular redox state
- Steroid synthesis.

The raw materials for the production of ATP in the mitochondrial matrix are pyruvate and NADH produced by glycolysis in the cell's cytosol. Pyruvate molecules are actively transported into the matrix. There, they react with *coenzyme A* to form *acetyl CoA,* which feeds into the Krebs cycle. The Krebs cycle produces, among other products, three molecules of NADH and one molecule of $FADH_2$, which participate in the *electron transport chain.* Free energy from NADH and $FADH_2$ is transferred to molecular oxygen in the multistep electron transfer chain.

The protein enzymes in the inner membrane, including *NADH dehydrogenase, cytochrome-**c** reductase,* and *cytochrome-**c** oxidase,* produce the electron transfer. The released free energy is used to pump protons (H^+ ions) into the intermembrane compartment. This high concentration of protons in the intermembrane volume of the cristae can escape through the *ATP synthase* enzyme complex back into the matrix. This flow of protons from high to low concentration powers the synthesis of ATP from ADP and inorganic phosphate, P_i, an important biochemical process called *chemiosmosis.* (Note that chemiosmosis also occurs in the chloroplasts of plants.) If protons "leak" back into the matrix without providing energy for ATP synthesis, their energy is released as heat (much as electrons in an electric circuit produce heat when they pass through a resistor [Joule's law]).

In some respects, mitochondria act like cells within cells. They have their own ribosomes and a small dsDNA genome that codes for some of their many proteins. Their DNA is circular, rather than the linear form found in a eukaryotic cell's chromosomes. A human mitochondrial DNA ring has 16,569 base pairs coding 37 genes, and there are 5 to 10 dsDNA rings in the matrix in structures known as *nucleoids.* Thus, mitochondria have redundant information storage. Human mtDNA codes for subunits 1, 2, and 3 of the enzyme *cytochrome oxidase*, subunits 6, 8, and 9 of F_o *ATPase*, *apocytochrome-**b*** subunit of *CoQH2-cytochrome-**c** reductase*, and also seven *NADH-CoQ reductase* subunits. All the many remaining mitochondrial proteins are coded by the cell's nuclear DNA. Mitochondria also have their own tRNAs and rRNAs (16S, 12S, and 5S).

Genomic studies on maternal ancestry use mtDNA, which is inherited matrilineally. In particular, the noncoding HVR-1 segment (568 bps) joined to the noncoding HVR-2 segment (574 bps) is used for DNA inheritance studies. The incidence of mutations in the two HVR regions is about 100 times that of the rest of the mtDNA genome.

A large mitochondrion may divide by fission into two daughters. First, the nucleoids duplicate, and then the inner and outer membranes furrow and pinch off, forming two new units. Old, worn-out or excess numbers of mitochondria are destroyed and their molecules recycled by autophagy. The autophagic process starts with ER membranes wrapping around the mitochondrion. Then, vesicles arrive from the Golgi complex containing *hydrolase* enzymes; these vesicles fuse with the autophagic vesicle, forming a lysosome. The mitochondrion is then literally disassembled into its constituent small molecules [Childs 2003c].

3.7 THE ROLES OF TRANSMEMBRANE PROTEINS

3.7.1 Cell Adhesion Molecules

Most eukaryotic cells carry on the outer surface of their cell membranes specialized cell *adhesion molecules* that allow them to bind closely with other cells or to their extracellular substrate. These cell adhesion molecules are what make metazoan embryonic development, and indeed all multi-celled organisms, possible. When embryonic stem cells differentiate, they aggregate with similar, specialized, differentiated cells to form organs and specialized tissues. All cell adhesion molecules are *integral membrane proteins* that penetrate the membrane and have extracellular, transmembrane, and cytoplasmic domains. See Figure 3.6 for a schematic illustration of some representative *cell adhesion molecules* (CAMs).

Obviously, it is the extracellular portion of the *cell adhesion molecule* that forms a molecular bond to some extracellular molecule. The transmembrane and cytoplasmic domains of the CAMs form the mechanical anchor to the cell. Three kinds of extracellular binding can occur: (1) *homophilic binding,* in which a CAM binds to another CAM of the same type on another similar cell; (2) *heterophilic binding,* in which a cell's CAM binds to a different kind of CAM on another cell type; or (3) to an intermediate *linker molecule* that forms a bridge or link between the cell's CAM and another CAM. These linkers are called *catenins.*

There are many different CAMs; however, they can be organized into one of four different families. They are listed Table 3.1.

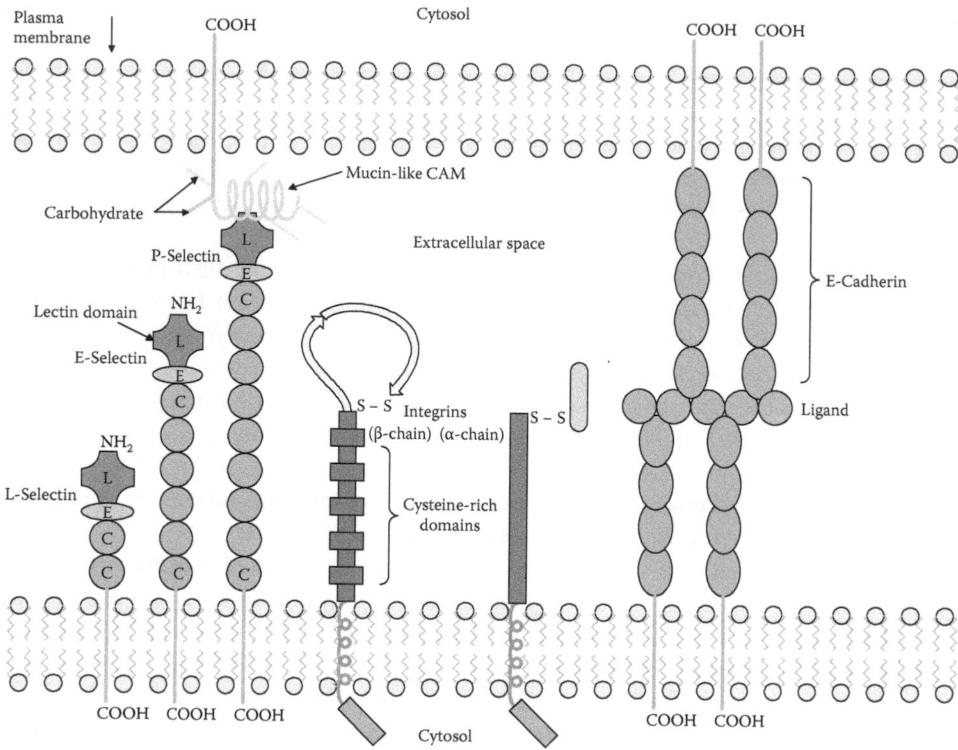

FIGURE 3.6 (See color insert.) Schematic illustration of some cell adhesion molecules in the intercellular space between two cells' plasma membranes. There are many types and subtypes of cell adhesion molecules whose roles are to attach cells to other cells, and cells to substrates.

TABLE 3.1
Cell Adhesion Molecule Families

CAM Family	Ligands	Interactions
Cadherins	Cadherins	Homophilic
Selectins	Carbohydrates	Heterophilic
Integrins	Extracellular matrix, I g superfamily proteins	Heterophilic
Immunoglobulin-like CAMs	Integrins	Heterophilic
	I g superfamily proteins	Homophilic

Cadherins' homophilic binding to other cells' cadherins requires calcium ions. Cadherins play a critical role in organizing embryonic cells into tissues and organs.

Selectins are associated with immune system function. They are expressed on the surfaces of leukocytes and endothelial cells. In the presence of Ca^{++}, selectins form relatively weak bonds with oligosaccharide ligands on the surfaces of cells. *P-selectins* allow the initial binding of leukocytes to endothelial cells. Local chemical signals at the site of inflammation cause the endothelial cells to express *P-selectin.*

Integrins are composed of a large group of *heterodimeric glycoproteins;* both the alpha and beta protein subunits participate in binding. Integrins facilitate signal transfer between a cell's cytoskeleton and nearby cells (the *extracellular matrix*), allowing each to affect the orientation and structure of the other. Integrins can become active in response to inflammatory signaling molecules, causing leukocytes to be bound to inflamed tissues, where they can fight the infection locally.

3.7.2 Transport of Substances Across the Cell Membrane

3.7.2.1 Diffusion and Osmosis

One of the requirements of being alive is that every cell has to regulate the inward and outward flow of molecules across its cell membrane. In fact, this homeostatic regulation is a cardinal property of life. Ions, molecules, and "substances" can cross the cell membrane in five basic ways: (1) by simple, *nonselective diffusion* down a concentration gradient, (2) by *osmosis* of water molecules, (3) by *selective, facilitated diffusion,* (4) by *active transport* of selected ions or molecules, and (5) by cellular *vesicles* doing *endo-* and *exocytosis*, ingesting the substance, or pushing it out of the cytoplasm into extracellular space.

Small molecules such as O_2, CO_2, and H_2O can diffuse through pores, or small defects in the lipid bilayer of the cell membrane. Lipid-soluble molecules such as steroids can also diffuse through the lipid bilayer. Mathematically, passive diffusion can be described by Fick's law in one (x) dimension:

$$\frac{\partial m}{\partial t} = -kA\frac{\partial C}{\partial x}\ \mathrm{kg \cdot sec^{-1}} \tag{3.2}$$

Where $m(x, t)$ is the mass diffusing, k is Fick's constant with the basic dimensions L^2T^{-1}, A is the area over which m diffuses, and $\partial C/\partial x$ is the concentration gradient in the $-x$ direction. When a *thin* membrane permeable to the diffusing molecule separates solutions having concentrations C_o and C_i, the 1-D diffusion equation can be written:

$$dm/dt = -k\,A\,\Delta C = -k\,A\,(C_i - C_o)\ \mathrm{kg.s^{-1}} \tag{3.3}$$

A more general 3-D form of Fick's diffusion equation is

$$\frac{\partial C}{\partial t} = k\nabla^2 C(x, yz) \tag{3.4}$$

Here $C(x,y,z)$ is the 3-D volume distribution of the concentration at time t, and

$$\nabla^2 \equiv \left[\frac{\partial^2}{\partial x^2} + \frac{\partial^2}{\partial y^2} + \frac{\partial^2}{\partial z^2}\right]$$

is the *Laplacian operator.* These diffusion equations are used in formulating mathematical models of cellular biochemical and biophysical processes, as well as in physiological systems compartmental modeling.

Osmosis is the diffusion of water across a *semipermeable membrane* (SPM) as the result of *osmotic pressure.* Osmotic pressure is the net pressure exerted on an SPM separating two different solutions. By definition, a semipermeable membrane allows the free diffusion of water but excludes the passage of ions and molecules on either side of the SPM. The net movement of water through a cell membrane depends on the *difference in osmotic pressure* between the cytosol and the extracellular fluid. Pure water, by definition, has $P_{osm} \equiv 0$. P_{osm} is measured in pressure units such as Pascals (newtons/m^2) or mmHg. (1 kPa = 7.50 mmHg.) All solutions, by definition, have $P_{osm} < 0$. Water molecules move by osmosis from solutions of low concentration to more concentrated solutions; that is, water moves toward the more negative P_{osm}. That is, nature tries to balance the concentrations on either side of the membrane by osmosis. Our interest is in the osmotic pressure experienced by a cell. The cytosol on the cell's interior is obviously the solution on the inside of the cell membrane in which ions, proteins, and other molecules to which the membrane is relatively impervious are dissolved. Outside the cell in the extracellular fluid there are many of the same inorganic ions but different proteins and lipids at different concentrations. Clearly, an osmotic pressure gradient exists across the cell membrane.

The osmotic pressure of a solution (relative to water) is measured by calculating its *total osmolarity* in *osmoles per liter of solution.* An *osmole* is the total number of gram moles of nonionizable and nondiffusable substances, such as sucrose, plus twice the total number of gram moles of all ionizable substances such as NaCl. If $CaCl_2$ is in solution, the osmolarity is thrice the total number of gram moles because one molecule of $CaCl_2$ produces three ions.

For example, let us calculate the osmoles of normal saline (0.9% by weight NaCl) solution. 9% NaCl is 9 g of NaCl in 1 l H_2O. The *gram molecular weight* (gmw) of NaCl is 58.5 g. So, 9 g in solution is 154 mM (milliMoles) of NaCl. However, the osmotic pressure is due to both the Cl^- and the Na^+ ions, so the total effective osmolarity is 2 × 154 = 308 mOsm. The osmotic pressure of a solution in mmHg is $P_{osm} = -19.3\ \Sigma$(milliosmoles) at 37°C. So, the osmotic pressure of normal saline is −5,944.4 mmHg, or −792.5 kPa.

Now consider a solution in which 1 g of D-glucose is dissolved in a liter of water. Glucose does not ionize; its gmw is 180, so the solution's concentration is 5.5556 milliosmoles = mM, and its osmotic pressure is −107.2 mmHg. Now, if the glucose is on side A of a semipermeable membrane and the normal saline is on the B side, osmotic pressure will force water to pass through the membrane from side A (glucose) to side B (saline), diluting the saline.

Note that true osmotic pressure is affected by ionic interactions in solutions. Less than 100% ionization in aqueous solutions causes the calculated osmotic pressure to be higher than that measured. For example, the calculated milliosmolarity of human plasma is about 303, but this results in an effective milliosmolarity of 283, calculated from osmotic pressure measurement. Thus, a correction factor of about 0.93 is involved. Because this correction factor also applies approximately inside cells as well as without and osmotic effects are determined by relative rather than absolute osmolarities, the correction factor is generally ignored. Guyton [1991] calculated the total (corrected) osmotic pressure at 37°C of *human blood plasma* (5450 mmHg), *interstitial fluid* (5430 mmHg), and *cytosol* (5430 mmHg). These figures were based not only on ions in solution (Na^+, K^+, Ca^{++}, Mg^{++}, Cl^-, HCO_3^-, $HPO_4^=$, $H_2PO_4^=$, and $SO_4^=$) but also a host of organic molecules including *glucose, urea, proteins, amino acids,* etc.

Most cells can adjust their internal osmotic pressure by actively "pumping" sodium ions across their membranes. The normal milliosmolar ratio of external to internal K^+ concentration is 4/140; it is 140/14 for Na^+ ions. The inside of a cell is generally electronegative, so sodium ions try to "leak in" down both potential and concentration gradients, the same for Ca^{++} ions, whose concentrations are about 2 mOsm in the extracellular fluid and about 0.001 mOsm in the cytosol for nonmuscle cells. By adjusting the rate of actively exchanging internal Na^+ for external K^+, a cell can adjust the osmotic pressure across its cell membrane as ΔP_{osm} occurs naturally in the interstitial fluid.

3.7.2.2 Aquaporins

Aquaporins are a recently discovered class of transmembrane proteins (TMPs) that have the unique property of regulating the passage of water molecules through the plasma membrane. These complex TMPs have the property of excluding all large molecules and ions such as Cl^-, Na^+, K^+, Ca^{++}, Mg^{++}, etc., from passing through the membrane; that is, they are H_2O specific. The aquaporin channels also repel and exclude protons in the form, H_3O^+. Water molecules pass though aquaporin channels serially (i.e., in single file) at rates up to about 10^9 molecules/second [Agre et al. 2002].

To date (2007) there are 13 known mammalian aquaporins (AQPs), with 6 types located in human kidneys. Aquaporins have also been found in prokaryotes (AQP-Z in *E. coli*) [DeLamarche et al. 1999] and eukaryotes, including plants, insects, and mammals. (Bacteria also have a TMP channel protein, GlpF, related to AQPs but which selects glycerol instead of water.) There are probably more than 35 versions of AQPs found in the model plant, *Arabidopsis thaliana* [von Heijne 2003]. In plants, AQPs are critical for water absorption in the roots and for maintaining the water balance throughout the plant in times of drought.

The general structure of an aquaporin TMP has been described and consists of a complex of six peptide-linked transmembrane subunits [ALSNEWS 2002; Bowen 2005]. Walz et al. [1997] described the detailed protein structure of AQP-1 using *Xenopus* oocytes and erythrocytes. They found that the AQP-1 molecular complex has six peptide-connected transmembrane domains (tilted α-helices) with the N- and C-ends inside the cell. The structures of genes coding for many AQPs are now known.

Water can cross a cell's plasma membrane by penetrating "defects" in the plasma membrane bilayer or through aquaporin water channels. Some aquaporin-like TMPs also transport glycerol and a few other molecules (e.g., CO_2, ammonia, and urea). As we have seen in Section 3.7.2.1, water flows in a direction dependent on the osmotic pressure gradient across the cell membrane. The Δ-osmotic pressure across the cell membrane depends on what molecules are dissolved in the cytosol and in the ECF. However, aquaporins operate independently of the *molecular pumps* that transport osmotic substances across cell membranes (see Section 3.7.2.3).

Clearly, aquaporin permeability must be regulated for proper physiological functioning of individual mammalian cells and complex organ systems such as kidney water resorption and also to maintain water balance in all other cells in the domains and kingdoms of life.

Resorption of water from the filtrate in the collecting ducts of mammalian kidneys has long been known to be regulated by *antidiuretic hormone* (ADH) (aka *vasopressin*) from the posterior pituitary gland [Guyton 1991]. High ADH concentration causes the kidneys to retain (reabsorb) water. Binding of ADH to receptors on the surface of collecting duct epithelial cells triggers the second-messenger, cyclic-AMP mechanism to make cAMP within the cells. The cAMP in turn stimulates transcription of the *aquaporin-2* (AQP-2) gene and also causes the cytoplasmic store of AQP-2 molecules to be inserted into the cell membrane of the collecting duct cells facing the lumen of the collecting duct. Thus, an increased density of AQP-2 allows more water from the collecting duct filtrate to enter the epithelial cells surrounding the collecting duct. This serves to concentrate the filtrate in the collecting duct as well as adding water to the cytosol of the epithelial cells. Another aquaporin, AQP-3, is located on the basolateral membranes of collecting duct epithelial cells. The AQP-3 TMPs transport intracellular water passed through the ADH-regulated AQP-2 units into the extracellular fluid, and thence to the blood, increasing blood volume.

McCaffrey [2004] described how AQP-0 found on cells in the lens of the vertebrate eye functions. In the mature lens, this AQP is located mostly at cell-to-cell contact points, where opposing aquaporins match up on opposing cell membranes and serve to bind the cells together structurally through exposed, shared proline residues, essentially serving the function of cell adhesion molecules. Researchers found that in the mature lens, the pores of an aligned pair of AQPs were too small to pass water. However, in the embryological development of the lens, before the cells pack tightly and AQPs pair up, AQP-0s can and do pass water for the developing lens cells. Table 3.2 lists some known human aquaporins.

Aquaporin gating in plant cells is mediated by phosphorylation/dephosphorylation of certain serine residues in the aquaporin molecular complex, as well as protonation of specific histidine residues. During drought stress, plant root AQPs close as the result of dephosphorylation of two conserved serine residues. Also, during flooding, the AQPs close in response to protonation of a fully conserved histidine residue [Törnroth-Horsefield et al. 2006].

There is no doubt that many more human AQPs will be discovered in the future, along with genetic diseases that affect the functioning of these ubiquitous proteins. Regulation of AQP function was seen, in the case of kidney collecting duct cells, to depend on a rapid turnover of membrane-bound AQP-2s coupled with their rapid synthesis mediated by the hormone vasopressin. In plants, AQP function depends on their controlled enzymatic dephosphorylation, phosphorylation,

TABLE 3.2
Distribution of Some Known Mammalian Aquaporins

Aquaporin	Site of Action	Function
AQP-0	Eye: lens cells	Fluid balance within lens
AQP-1	Erythrocytes	Osmotic balance
	Kidney: proximal tubules	Urine concentration
	Eye: ciliary epithelium	Aqueous humor secretion
	Brain: choroid plexus	Cerebrospinal fluid secretion
	Lung: alveolar epithelial cells	Alveolar hydration state
AQP-2	Kidney collecting duct epithelial cells	Regulation of blood volume using ADH
AQP-3	Kidney: collecting ducts	Export of water into blood: ADH
	Tracheal epithelial cells	Secretion of water into trachea
AQP-4	Kidney: collecting ducts	Reabsorption of water
	Brain: ependymal cells	CSF fluid balance
	Brain: hypothalamus	Osmosensing?
	Lung: bronchial epithelium	Bronchial fluid secretion
AQP-5	Salivary gland	Saliva secretion (water component)
	Lacrimal glands	Tear secretion
AQP-6	Kidney	Function unknown
AQP-7	Fat cells	Exports glycerol out of adipocytes
	Testis and sperm	Functions unknown
AQP-8	Testis, pancreas, liver, etc.	??
AQP-9	Leukocytes	??

and protonation. There is no doubt that many other interesting AQP-regulatory mechanisms will be discovered, leading to a better understanding of cell physiology and the development of new drugs.

3.7.2.3 Ion and Molecular Pumps

Ion pumps are complex transmembrane protein structures that use metabolic energy (generally provided by the hydrolysis of ATP) to transport ions or molecules into or out of a cell against a concentration gradient or an electrical potential gradient across the cell's membrane. Broadly stated, pumping operations are necessary for cellular ionic homeostasis and are also used for such physiological operations as generating stomach acid and causing muscles to relax. The neuromuscular systems of animals could not function without ion pumps.

Many types of ion and molecular pumps are found in cell membranes. These pumps are large, complex, multimeric TMPs that exhibit specificity for the particles pumped. They are generally powered by the free energy in the ATP phosphate bond; ATP is cleaved into ADP + P_i by the enzymatic action of the TM pump proteins, and the released free energy allows substances to be transported against a concentration gradient, and in some cases, against an electrical potential gradient as well.

There are several classes of pump (aka carrier) protein: *Uniport carriers* carry just a single solute (molecule or ion) across a cell membrane. An example is the ***GLUT1*** glucose carrier. *Symport* (cotransport) *carriers* bind two dissimilar substances together and transport them together across a membrane in the same direction. One substance, generally an ion, may drive the transport of the second substance "uphill" against its concentration gradient. One important example is the *glucose-Na^+ symport* pump found in the epithelial cell surfaces facing the lumen of the kidneys' proximal collecting ducts. This symport system serves the useful service of resorbing glucose lost in the renal filtrate (on its way to becoming urine). Amino acids in the filtrate are also absorbed by into the epithelial cells by other symport pump proteins.

Another class of pump is the *antiport carrier*, which exchanges one type of particle for another type across the cell membrane. Generally, *antiporters* use "ping-pong" kinetics in which one particle binds to the pump and is transported across the membrane; then the *antiporter protein* binds to the other particle and carries it back to where it picked up the first one. One example of an antiporter is the *adenine nucleotide translocase exchanger* located in the inner mitochondrial membrane. This antiporter effects a 1:1 exchange of ADP for ATP; an ADP^{3-} is carried into the mitochondrial matrix in exchange for an ATP^{4-} being carried out.

In 1997, Jens Skou shared a Nobel Prize in Chemistry for describing how the cell membrane Na^+, K^+-ATPase *antiporter ion pump* works in neuron membrane. The Na^+, K^+-ATPase antiport pump is found in most animal cells; it catalyzes ATP $\rightarrow$ ADP + P_i, and uses the free energy to pump three Na^+ ions out of the cell against a concentration and electrical potential gradient, and in exchange, transports two K^+ ions into the cytosol. According to Kimball [2006b], almost one-third of all the chemical energy generated by the mitochondria in animal cells is used to run the Na^+, K^+-ATPase antiport pump; such is its importance in maintaining cellular ionic homeostasis.

Parietal cells lining the stomach use an H^+, K^+-ATPase pump to carry hydrogen ions through their cell membranes into the stomach in exchange for potassium ions in a 1:1 manner; it is also powered by ATP hydrolysis. This pump takes H^+ ions from an intracellular concentration of about 4×10^{-5} mM to a concentration of about 0.15 M in the stomach acid!

Thyroid gland cells and mammary gland cells have a Na^+, I^--*symporter* that pumps iodide ions into the thyroid cells for the manufacture of the hormones *thyroxine* and *3-iodotyrosine* (3-IT). Mammary gland cells use this symporter to put I^- into milk for infant nutritional needs.

The Ca^{++}, H^+-ATPase antiport pumps, located in the membranes of the *sarcoplasmic reticulum* (SR) of muscle fibers, are necessary for muscle relaxation. These pumps are also called SERCA, for sarco (endo)plasmic reticulum **Ca^{++}-ATPase**. They can constitute as much as 90% of the protein in the SR of certain muscles. For each ATP cleaved, SERCA transports two Ca^{++} ions through the SR membrane into the lumen of the SR tubule where it is stored [Guyton 1991]. In exchange for the two Ca^{++} ions transported into the SR, the SERCA also transports two or three protons (H^+) in the opposite direction [Goodsell 2004]. A Ca^{++}, Na^+- pump can also be found in cell membranes. Driven by Na^+ ions entering down a concentration and a potential gradient, its role is to pump excess Ca^{++} out of the cytosol. The detailed chemical reactions of the SERCA reaction cycle, and the molecular structure of the SERCA pump, are described by Diwan [2005].

A feature of all transmembrane pumps is that molecular configurational changes are involved with pumping ions or molecules. The configurational changes derive their mechanical energy from chemical energy from ATP hydrolysis. Life would be impossible without transmembrane pumps and many other kinds of TMPs.

3.7.2.4 Regulated Ion Channels in Membranes

Ion channels are transmembrane gates for specific ions and small molecules that, when activated, allow specific ions to *passively* flow in or out of membranes following concentration gradients and electric field gradients. Ion channels have "filters" that give them ionic specificity; for example; the K^+ channels can exclude Na^+ ions and preferentially pass potassium. However, some channels are pores; that is, they remain open permanently, accounting for the steady-state "leakiness" of a cell membrane.

Ion channels can be opened or activated by a change in the electric field across the membrane (*voltage-gated channels*) or by specific molecules such as *neurotransmitters*, or Ca^{++} ions (ligand-gated channels). Ligands can act to open channels from outside the cell or from inside the cell membrane. Two examples of external ligands are the *neurotransmitters acetylcholine* (ACh) and *gamma amino butyric acid* (GABA). ACh binds to certain cholinergic sodium channels, such as found in the subsynaptic membrane (SSM) of muscle and cholinergic neuron synapses, and causes the channel's conductivity to suddenly increase to admit only sodium ions. This increase in the inward sodium current density ($\mathbf{J}_{Na}$) in the SSM causes *depolarization* of the cell membrane, which in turn can lead to muscle contraction or initiation of a nerve action potential (spike). When GABA binds to $GABA_A$ synapses in the CNS, the channel admits Cl^- ions and inhibits the generation of a nerve impulse by driving the local transmembrane potential negative.

Internal ligands obviously originate within a cell. In olfactory sensory neurons and also photoreceptors, *second messenger molecules* such as *cyclic AMP* and *cyclic GMP* regulate the transduction of odorants and photons. The energy molecule ATP is the internal ligand required to open the channel that allows Cl^- and HCO_3^- ions to exit the cell. Although the free energy from ATP hydrolysis is required to open this channel, it is not a pump; the ions diffuse through the open channel following their concentration gradients.

In sensory neurons, ion channels can be activated by mechanical shear stress in a membrane, heat, cold, pH, certain external chemicals and odorants, photons, etc. Opening a specific ion channel in a sensory neuron generally results in a change in the transmembrane potential of the cell in the vicinity of the opened ion channels. In some cells, this can be a *depolarization*; that is, the transmembrane potential that is, becomes less negative inside. In vertebrate retinal photoreceptors, light causes *hyperpolarization* of the cell membrane; that is, the insides of rods and cones become more negative in response to absorbing photons. Mechanical shear force in a mechanosensor's membrane can open ion channels, leading to generator potentials and nerve impulse generation. Some of the common human mechanosensors are *Golgi tendon organs, Meissner corpuscles, Pacinian corpuscles*, and *Cochlear hair cells* (the transducers used in hearing).

The critical ion channels in nerve and muscle membranes control the inward or outward passage of critical ions such as Na^+, K^+, Cl^-, and Ca^{++}. Whatever the modality of ion channel control, ion channel proteins such as pump proteins are believed to operate by physical changes in their internal dimensions; that is, they are either open or shut. Their conductive states may be binary, but the net effect can be graded because thousands of channels may be involved, some open and some shut at any given time.

Note that the internal dimensions of the internal pore in the channel, as well as the charge distribution of the AAs surrounding the lumen, affect the channel's *conductance.* (Note that *conductance* is an electrical term.) It is defined for electrical circuits as $G \equiv I/V$ siemens. The *instantaneous ionic conductance* for a single potassium channel, for example, is defined as

$$g_K \equiv \frac{I_K}{(V_m - E_K)} \text{ Siemens} \tag{3.5}$$

where I_K is the measured instantaneous potassium ion current through the channel (measured by patch-clamp technique). E_K is the potassium *Nernst potential:* $E_K \equiv (RT/zF)\ln\{[K^+]_o/[K^+]_i\}$ [Kandel et al. 1991]. R is the SI gas constant, T is the Kelvin temperature, $z = +1$, V_m is the instantaneous transmembrane voltage, and F is the Faraday constant. At 25°C, $(RT/zF) = 26$ mV. For squid giant axon, $E_K = 26\ln(20/400) = -77.9$ mV. (See Section 3.7.2.5 for a detailed description of the Nernst potential.)

Similarly, the instantaneous sodium ion conductance is given by

$$g_{Na} \equiv I_{Na}/(V_m - E_{Na}) \tag{3.6}$$

The sodium Nernst potential is about $\mathbf{E}_{Na} = 26 \ln(440/50) = +56.5$ mV.

Another important class of ion channels is the *chloride channels* (ClC) (aka CLC) that pass nature's most abundant anion. Early physiological models describing transmembrane potential and nerve action potentials focused on voltage-gated sodium, potassium, and calcium channels; chloride was assumed to "leak" to satisfy electroneutrality and Donnan equilibriums. Various types of ClC channels have been found in almost all organisms, ranging from bacteria to mammals. Nine alpha ClC Cl^- channels have been described in mammals; these can be subdivided into those in the plasma (outer cell) membrane and those in organelle membranes in the cytoplasm. ClCα channels have a *homodimeric* structure in which each transmembrane subunit forms a regulated Cl^- pore.

The two pores per receptor give a CLCα-0 receptor *three conductance states:* The lowest, G0, has both pores closed. G1 has one pore open, and G2, the highest conductance, has both pores open. When the ClC channel is active, both pores open and close randomly at a rate that depends on certain physical and chemical factors around the channel protein. The ClCα-0 channel was discovered in the electric organ of the *Torpedo* ray in 1980. ClCα-0's structure was described in 1990, and *patch-clamp recording* of a single ClCα-0 channel's current showed that the inward Cl^- current of an open pore was about 0.5 pA. The pore states appear to be either open or closed, and the actual individual ClCα channel current waveform appears to be a 3-level, random square wave. There are 2- or 3-s intervals of receptor activity interspersed with 2- or 3-s intervals of closed pores (G0). The opening of the two pores in a ClCα channel appears to be a statistically independent random process. The dwell, or open time, of each pore determines the average chloride current passed by the channel. The pore gate can be described by the simple rate process:

$$\mathbf{C} \underset{k_o}{\overset{k_c}{\leftrightarrows}} \mathbf{O} \tag{3.7}$$

Where **C** is the closed state of a pore and **O** is its open state. k_o is a *voltage-dependent opening rate constant* and k_c is the (fixed) closing rate constant. It can be shown [Pusch 2004] that the *open probability* is given by

$$p_{open} = k_o/(k_o + k_c) \tag{3.8}$$

Thus, the probability of the pore being closed is

$$p_{closed} = 1 - p_{open} = k_c/(k_o + k_c) \tag{3.9}$$

The pore state is also a function of pH inside and outside the cell membrane. A low (acid) pH on either side increases p_{open} but by different molecular mechanisms [Pusch 2004].

ClCα-0 lacks the ionic specificity of voltage-gated sodium channels, for example. ClCα channels will pass other small anions. CLCα-0 will conduct Br^- ions less well than chloride, but bromine is not a physiologically significant ion, in any case. Iodide is, however. ClCα-3 favors iodide ions over chloride over bromide; it may have some significance for thyroid cells. It has also been found that p_{open} is increased by a high external concentration of Cl^-. Low $[Cl^-]_{ext}$ markedly reduces the pore's voltage dependence and hence lowers pore conductance. Mammalian muscle ClCα-1 behaves quite similarly to *Torpedo* ClCα-0 [Maduke et al. 2000].

In whole-cell chloride current recordings of the initial (fast) peak chloride current to an applied step of transmembrane voltage V_m from zero, ClCα-0 exhibits a linear relation between I_{Cl} and V_m; that is, it has a constant conductance that is not voltage dependent. In other words, under conditions of constant temperature, pH, and ionic concentrations, the ClCα-0 channels behave as a group as a linear chloride membrane conductance, even though at the single-channel level, A single ClCα-0 channel has the fascinating stochastic gating behavior of its two chloride pores. Other ClCα channels exhibit "rectifying" current-voltage behavior; for example, human ClCα-1 has a very low conductance for $V_m > 0$ and a relatively high conductance for $V_m < 0$ (hyperpolarizing V_m). ClCα-4 has the opposite behavior; it has high conductance for $V_m > 0$ (depolarization) and very low conductance (rectification) for $V_m < 0$. Other types of ClCα channels also exhibit voltage-dependent conductances [Fahlke 2004].

The control of ClCα channel gating is just beginning to be understood. Several factors affect pore activity, including transmembrane potential, chloride concentrations on both sides of the membrane, intracellular $[Ca^{++}]$, temperature, and pH. Note that extracellular chloride concentration is about 120 mM in frog muscle; internal $[Cl^-]$ is about 3.5 mM. The more depolarized (positive) the membrane potential, the higher the open pore probability. At symmetrical in/out $[Cl^-]$, the single-pore conductance (G1) of ClCα-0, ClCα-1, and ClCα-2 is about 10 picoSiemens (pS), 1 pS, and 3 pS, respectively, per channel [Chen 2005].

The structure of ClCα proteins is still under study. One model for the ClCα-0 channels shows an amazing 18 helical regions per ClCα subunit [Babini & Pusch 2004; Estévez & Jentsch 2002]. The ClCα molecules appear far more structurally complex than the 7TMRs described later.

It would be interesting to examine the 3-state pore current waveform of ClCα receptors in the frequency domain. That is, compute power density spectrums of $i_{pore}(t)$ as a function of the variables known to affect p_{open}, namely, V_m, $[Cl^-]$, pH, T, etc. Such information on the dynamics of pore opening and closing would be useful in understanding the regulation mechanisms of ClC channels.

Voltage-gated cation channels are found principally in the membranes of electrically active cells such as nerve axons, muscle membranes, and the electric organs of certain fish. The major voltage-gated cation channels are for Na^+, K^+, and Ca^{++}. We will first examine the classic voltage-gated Na^+ channel found in nerve axon membrane and whose biophysics were first modeled in the classic Hodgkin–Huxley equations for nerve spike generation (see Chapter 4, Section 4.5.4, this text) [Hodgkin & Huxley 1952; Northrop 2001].

Voltage-gated sodium ion channels are large, complex, transmembrane proteins that are necessary for the electrical signaling activity in nerves and many muscles. There are about five subtypes of voltage-gated sodium (Na_v) channels. The Na_v channel protein is arranged as a *homotetramer;* that is, each of the four subunits has six transmembrane subunits [Yu et al. 2005]. The channel has three states: *resting* (closed), *open,* and *closed* (not resting; recovering). Na_v channels form selective, voltage-gated pores through which extracellular sodium ions can enter a cell driven by the sodium ion concentration gradient and initially by the electric field gradient across the cell membrane. The Na_v channels are generally closed (and resting) under resting conditions of transmembrane potential V_{m0}, which can range from about −60 to −90 mV (inside of cell negative), depending on the cell type. Individual Na_v channels can open randomly and transiently but with very low probability under (resting) V_{m0} conditions. When V_m is caused to depolarize (go positive), a change in configuration occurs in the sodium channel protein S4 subunits when V_m reaches a *threshold voltage,* $V_{m\phi}$. At this threshold voltage, a fraction of the Na_v channels open in the membrane, each allowing a peak sodium current of about 6×10^6 ions to rush into the cell [Anderson et al. 2005]. This influx further depolarizes the cell, and even more N_{av} channels open, causing a rapid depolarization that can cause V_m to actually go positive by tens of millivolts, reaching a peak in about a millisecond. This rapid inward I_{Na} of many Na_v channels causes the nerve action potential (AP) or spike to begin, or the muscle membrane AP to occur. Following the Na^+ ion inrush and V_m depolarization in about 2 ms, the Na_v channels close, and $I_{Na} \rightarrow 0$.

Kandel et al. [1991] define the action potential threshold transmembrane voltage, $V_{m\phi}$, as *that voltage at which the net transmembrane current is positive (inward) depositing a net positive charge inside the cell.* $V_{m\phi}$ is generally a few millivolts more positive than V_{m0}. It also depends on the rate of change of V_m. One way of examining the excitation of an electrically excitable membrane is by stimulating it with a controlled current source (positive ion inward) pulse and observing V_m [Northrop 2001].

As noted previously, an N_{av} channel can exist in three molecular states: (1) *resting* (closed) at V_{m0}; (2) *activated* (open to Na^+ ions) as the result of V_m depolarization—if the depolarization is brief, the channel reverts to the resting state; and (3) if depolarization is maintained, the channel assumes its *inactivated* (closed) state. Once inactivated, the channel cannot be activated (opened) again until $V_m \rightarrow V_{m0}$ and it reverts to its resting state. Thus, I_{na} is always a transient pulse through a particular patch of active membrane [Kandel et al. 1991]. In squid axons, an outward, voltage-dependent potassium current is critical to the recovery of V_m to V_{m0}. Chloride currents through CLC channels also figure in the reestablishment of V_{m0} following an action potential.

Voltage-gated Ca^{++} channel proteins belong to the same gene family as the Na_v and K_v channels. The Ca_v channels use the same homotetrameric architecture as the Na_v channels. The S4 subunits act as the ΔV_m sensors and the S5, S6 subunits determine the ion conductance and selectivity. Changes in only three AAs in the pore loops of domains I, III, and IV will cause a Na_v channel to have Ca^{++} selectivity! There are *currently* 10 subtypes of Ca_v channels described [UPHAR 2006]. Ca_v channels were first discovered in 1953 by Fatt and Katz, who were working with crustacean muscle and found that it still generated action potentials when Na^+ ions were omitted from the perfusion medium. It was subsequently discovered that Ca_v channels are present in all electrically excitable cells, and now they have been found in low numbers in nonexcitable cells such as immune system cells [Dolphin 2006]. Ca_v channels have been subdivided into 10 subgroups: L-types (4), T-types (3), P/Q (1), N (1), and R (1). The details of the structure, physiology, and pharmacology of the 10 Ca_v channels have been reviewed by Catterall [2000].

There is a profusion of different *voltage-gated K^+ channels* in nerve, muscle, and other cells. In excitable nerve membranes they are responsible for the repolarizing phase of the action potential. Triggered by depolarization (V_m going positive), they open and allow potassium ions to pass out of the cell, allowing $V_m \rightarrow V_{m0}$.

Similar to the Ca_v channels, the typical K_v channel has a *homotetrameric* structure with each subunit having six transmembrane regions, with both carboxy and amino terminations on the inside of the cell. Another way of describing the K_v channel architecture is *tetrameric 6TM*. The subunits are arranged around the voltage-regulated central pore. The S4 subunits serve as (molecular) voltage sensors, and the pore involves the S5 and S6 units. More than 22 different genes code K_v channel proteins, and additional varieties are made by alternative spicing or *heteromultimerization* [Yellen 2002]. Similar to other voltage-activated cation channels, K_v channels are quite K^+ specific.

It is known that intracellular calcium ion signals regulate a broad array of cellular functions, ranging from muscle contraction to longer-term control of gene transcription, the cell cycle, and cell death. The mechanisms of intracellular (cytoplasmic) calcium ion regulation in cells have been widely studied, and transmembrane proteins located in the plasma membrane (Ca^{++} release-activated Ca^{++} [CRAC] channels) and the membrane of the endoplasmic reticulum (STIM) have been implicated in this regulation. Unfortunately, the molecular details of the regulation of *store operated Ca^{++} entry* (SOCE) into the cytoplasm remain to be worked out [Venkatachalam et al. 2002; Grant 2007]. Needless to say, this regulation is thought to be of "enormous complexity" by workers in the area [Penner & Fleig 2004].

3.7.2.5 The Basis for the DC Transmembrane Potential

At the core of all neural and muscular electrical communication systems is the evolution of ion-selective, voltage-activated, and chemically activated transmembrane gate proteins. As we have seen earlier, some gates are opened by chemical signals (e.g., neurotransmitter molecules); others are opened by changes in transmembrane potential; still others open because of second-messenger action on the inside of the cell. In the steady state, at V_{m0}, these gates are generally closed, but some "leak," however. That is, they have a finite, but low, conductance to specific ions. In addition to gates, we have also introduced the various types of transmembrane *ion pumps* (cf Section 3.4.2.2). Thus, cell membranes are electrophysiologically complex structures. In the steady state they are semipermeable to certain small ions and impermeable to larger charged molecules, both inside and outside the cell. Furthermore, the charge on certain AAs and proteins depends on the pH of the medium they are in (cytosol and extracellular fluid).

The picture is further complicated by the concentration differences of key ions outside and inside the cell. Table 3.3 gives ion concentrations for the much-studied squid giant neuron axon.

From chemical thermodynamics, it can be shown that the chemical energy (or work) of moving a single (e.g., Q^+) ion from one concentration inside a cell to a different concentration outside the cell is

$$W_{cQ} = RT \ln \frac{[Q^+]_o}{[Q^+]_i} \text{ joules/ion (SI units)} \tag{3.10}$$

TABLE 3.3
Distribution of Ions across the Axon Membrane of the Squid Giant Neuron

Ion	Cytoplasm (mM)	Extracellular Fluid (mM)	Nernst Potential[a]
K^+	400	20	−76.9 mV
Na^+	50	440	+55.8 mV
Cl^-	52	560	−61 mV
Ca^{++}	30 – 50 μM	10	+149 mV
IP^-	385	—	—

Note: IP^- is the net concentration of negatively charged ions in the cytoplasm.

[a] The Nernst potential is the electric potential in volts required to exist across a simple membrane permeable to just one species of ion in order to prevent those ions from diffusing from high concentration through the membrane to lower concentration (the membrane is impermeable to water).

The electrical energy required to move the ion Q^+ through the electrical potential, V_m, across the membrane is (from basic physics)

$$W_{eQ} = \mathbf{z}\,\Im\, V_m \text{joules} \tag{3.11}$$

In the foregoing equations, R is the SI gas constant, 8.3145 J/(°K g-mole); $\Im$ is the Faraday constant, 96,485 C/g-mole; **z** is the number of charges on the ion considered (+1 in this case); T is the Kelvin temperature, $[Q^+]_o$ is the molar concentration of Q^+ ions outside the neuron's cell membrane, and $[Q^+]_i$ is the concentration in the cytoplasm.

At equilibrium, the chemical potential energy equals the electrical potential energy (there is no net force on the ion). Thus

$$W_{eQ} = \mathbf{z}\Im V_m = W_{cQ} = RT \ln \frac{[Q^+]_o}{[Q^+]_i} \tag{3.12}$$

which yields the well-known *Nernst equation* for the EMF of ion Q^+:

$$E_Q = \frac{RT}{\mathbf{z}\Im} \ln \frac{[Q^+]_o}{[Q^+]_i} \text{ volts} \tag{3.13}$$

For potassium ions in the squid giant neuron (note **z** = +1 for cations):

$$E_K = \frac{RT}{\mathbf{z}\Im} \ln \frac{[K^+]_o}{[K^+]_i} \text{ volts} \tag{3.14}$$

Using the foregoing values, we find $E_K = -76.9$ mV at 25°C. E_{Na} similarly is found to be +55.8 mV, and $E_{Cl} = -61.0$ mV. (Note that the valence number **z** = −1 for the chloride anion.) Note the direction of Nernst battery polarities in the simple electrical circuit model for a unit area of axon membrane,

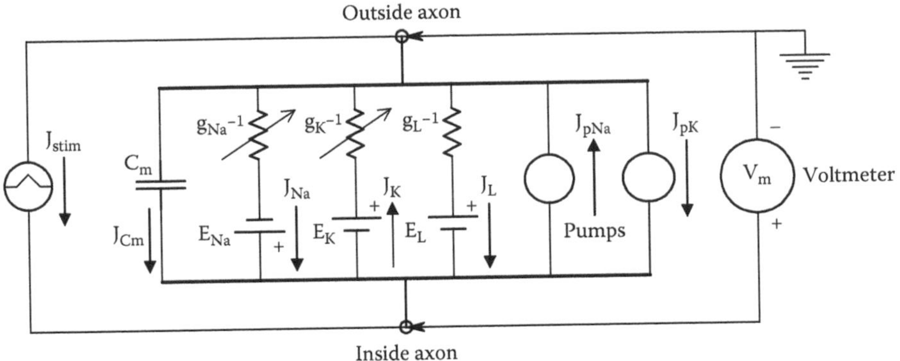

FIGURE 3.7 Simplified lumped-parameter electrical circuit used to model a unit area of nerve axon membrane.

shown in Figure 3.7. (Note that the energy of 1 electron-volt, eV = 1.603×10^{-19} joule, and that ionic concentrations are used to approximate *ionic activities* in the Nernst equation.)

The lipid bilayer of the unit membrane behaves like an electrical insulator sandwiched between two conductive "electrodes" (the polar heads) and hence forms an effective capacitor, C_m, albeit a "leaky" one. For example, the capacitance for squid giant axon membrane was estimated to be about 1 μF/cm² [Katz 1966].

So, how do we treat a membrane that has several ionic species mixed together at different concentrations on either side of it? In nerve and muscle cells, the important mobile ions are generally Na^+, K^+, Ca^{++}, Mg^{++}, H^+, Cl^-, HCO_3^-, and $SO_4^=$. Note that electrons are noticeably absent. A resting neuron or muscle cell membrane is relatively impermeable to all these ions, whereas those that leak into the cell, such as Na^+, are forced down a concentration gradient and an electrical potential gradient (see Table 3.3). Clearly, the net effect of all the ions inside and outside the cell contributes to its transmembrane potential. These ions also include a number of large, *impermeable charged molecules* inside the cell having negative charges. In the steady-state, ion leakage, regulated water transport through *aquaporins,* and ion pumping through the membrane maintain constant ion concentrations inside the cell.

It was observed many years ago that a cell's plasma membrane, and the proteins imbedded in it, are subjected to a large average electric field, ***E***. For example: Let the resting potential, V_{m0} = −65 mV, and take the lipid bilayer as about 7.5×10^{-9} m thick. Then assuming a uniform dielectric constant, $E = V_m/d = 0.065/(7.5 \times 10^{-9}) = 8.67 \times 10^6$ V/m, or 8.67×10^4 V/cm. If the lipid dielectric is assumed to be 5 nm thick, and its dielectric constant is 5.65, then E = 1.30×10^5 V/cm across the lipids. This high average electric field affects the 3-D configurations of the transmembrane ion channel proteins.

Electroneutrality also exists inside the cell in the steady state; that is, the mean total number of charges (+ and −) inside the cell in a fixed macrovolume must be equal, assuming random charge distribution. However, the charge distribution is anything but random, considering charged protein macromolecules fixed to the inner cell membrane and to mitochondrial, nuclear, and Golgi complex membranes. The small ions listed earlier can move, however, to effect electroneutrality.

A pragmatic, nonlinear electric circuit model which has been used to model both membrane resting potential and action potentials makes use of ion *channel electrical conductances,* g_Q, which can be measured using microelectrodes and the patch-clamp technique. In general, by Ohm's law, at any V_m

$$g_Q \equiv \frac{J_Q}{(V_m - E_Q)} \text{ Siemens/m}^2 \tag{3.15}$$

where E_Q is the Nernst potential for the Qth ion species. Using the simple electric circuit model for a unit area of nerve axon membrane of Figure 3.7, we can calculate the dc membrane potential. (For dc, the membrane capacitance is treated as an open circuit.) With $J_{stim} = 0$ and $V_m \equiv V_{m0}$, we can write a *node equation* for V_m. First, it is convenient to make Norton equivalents from the series Thevenin circuits. Direct currents leaving the V_m node (inside of the membrane) are taken as positive, and positive ions go in the direction of the current. Thus, we can write the node equation (Equation 3.16), using Kirchoff's current law, and solve for V_m:

$$V_m(g_{Na} + g_K + g_L) - g_{Na}|E_{Na}| + g_K|E_K| - g_L|E_L| = 0 \tag{3.16}$$

$$\downarrow$$

$$V_m = \frac{g_{Na}|E_{Na}| - g_K|E_K| + g_L|E_L|}{(g_{Na} + g_K + g_L)} \tag{3.17}$$

In this simple dc model, we regard the Nernst voltages as fixed, and g_{Na} and g_K are voltage-variable conductances. The resting membrane's Norton dc cation current densities are given by

$$J_{Na0} = -g_{Na0}(V_{m0} + E_{Na}) = g_{Na0}(-0.065 - 0.055)\ \text{A/m}^2\ (\text{inward}, < 0) \tag{3.18}$$

$$J_{K0} = g_{K0}(V_{m0} - E_K) = g_{K0}(-0.065 + 0.075)\ \text{A/m}^2\ (\text{outward}, > 0) \tag{3.19}$$

Note the sign conventions; J_{Na} is an *inward* (negative) current density through the membrane. We have used SI units here; the voltages are in volts, and conductances are in siemens/m^2. We have combined chloride current with other minor ion currents to form a *leakage current density,* J_L. g_{L0} is assumed to be a constant, independent of V_m. Thus:

$J_{L0} = g_{L0}(V_{m0} - E_L) = g_{L0}(0.065 - 0.069)$ A/m^2 (conventional current flows out, but chloride ions are negative, so J_L is effectively inward, or is negative.) (3.20)

3.7.2.6 Cell Signaling by Changes in Transmembrane Potential

We have already seen that multicellular organisms communicate intercellularly by chemical signal molecules. Metazoan animals also have evolved the ability to communicate intercellularly electrically (and electrochemically) between sensory neurons, motor neurons, neurosecretory cells, muscle cells, and glands. The basis for this electrical communication is the resting or steady-state transmembrane potential difference between the cells' cytosol and the extracellular fluid. As we have seen, all living cells have a *resting transmembrane potential* (V_{m0}) that is on the order of several tens of millivolts, with the inside of the cell being negative with respect to the exterior. What is communicated by electrically active cells are transient changes in their transmembrane potentials [$\Delta V_m = V_{m0} - V_m(t)$]. The ΔV_m signal in neurons can induce a variety of physiological changes in the receiving cells, including but not limited to muscle contraction, muscle relaxation, secretion of endocrine hormones and signal substances, production of light, modulation of heart rate, blood pressure, gut motility, etc.

Specialized sensory neurons that sense physical and chemical quantities convert these inputs to changes in their transmembrane potentials. Upon receiving the input stimulus, most sensory cells *depolarize*; that is, the transmembrane potential becomes less negative. A few others, such as vertebrate rod and cones, *hyperpolarize*; that is, their transmembrane potential goes more negative relative to the resting potential value. Sensory cells are specialized neurons; certain sensors can respond to photons (light), temperature changes, strain (caused by stretch, pressure, and tissue deformation), electric fields, magnetic fields, and a variety of chemical substances including airborne and waterborne odorant molecules and pheremones. In mammals, there are also receptors that signal pain, a more abstract, synthesized input.

Neurons form an animal's rapid communication system. *Afferent neurons* carry signals from sensors to the central nervous system (CNS), and *efferent neurons* carry signals from the CNS to activate muscles (motor neurons), give feedback to sensory arrays, and cause certain glands to secrete. *Autonomic* efferent neurons also form the communications link whereby an animal's CNS regulates the heart rate, blood vessel diameters, and blood flow through the kidneys.

Neurons carry signals in the form of *action potentials* (aka nerve spikes). These are transient changes in transmembrane potential that propagate along nerve axons at speeds that can range from about 0.1 m/s to as large as 20 m/s for squid giant axon. Nerve spike conduction velocity increases as the square root of the axon diameter and is also increased by the presence of a *myelin sheath* on the axon [Northrop 2001]. (Certain CNS neurons can also propagate depolarization signals for short distances along cell processes without spiking; we will not consider them here.)

An action potential can be initiated naturally at a specific site on a neuron cell body called the *spike generator locus.* Spike initiation requires a rapid inward flow of positive charges, generally sodium ions, in an area of *active membrane.* (One source of these positive ions can be from dendrites receiving excitatory synaptic inputs.) Once initiated, the spike propagates down the axon and reaches the end of the neuron, where it may couple to another neuron, skeletal muscle fiber, heart muscle fiber, or gland tissues, etc., by *synapses.* When the spike arrives, depending on the type of neuron, it can cause the release of a chemical neurotransmitter, or the transient voltage signal can directly induce an electrical potential change in the postsynaptic cell's membrane through an electrical synapse called a *gap junction.*

Gap junctions are intercellular channels about 1.5 to 2 nm in diameter formed from four or six copies of one of a family of transmembrane proteins called *connexins.* The connexins penetrate both the presynaptic cell membrane and the adjacent postsynaptic membrane. Because ions can flow through them, gap junctions conductively couple the two cells. There are two types of electrical synapses, *nonrectifying* and *rectifying.* As the name implies, rectifying junctions allow ions to pass in one direction, whereas two-way transmission occurs in nonrectifying junctions. The nonrectifying gap junction electrical coupling is faster than chemical synapses and is also called an *electrotonic* (ET) coupling junction [Kandel et al. 1991]. ET coupling is widely found in the CNS and in certain sensory systems such as the retina.

However, most non-CNS neurons (such as spinal interneurons and motor neurons) communicate with others by chemical synapses. When a spike arrives at a chemical synapse, it causes neurotransmitter molecules to be released from the terminating boutons. The neurotransmitter molecules then diffuse rapidly across the *synaptic cleft* separating the bouton from the *subsynaptic membrane* (SSM) of the postsynaptic cell. The SSM has protein transmembrane ion gate molecules in it that have receptors for the neurotransmitter. When the neurotransmitter docks with the receptor protein, it causes a conformational shift in the ion gate molecule that allows a specific ion, such as sodium, to rush through the SSM and depolarize the postsynaptic cell (neuron or muscle).

Nerves seldom spike at rates in excess of 10^3 /s. The upper limit on nerve spike rate is set by a number of factors, including the speed at which the transmembrane voltage can return below the spike threshold voltage, $V_{m\phi}$, and the ability of the electrically activated sodium ion gate molecules to "reset" to their prespike, "resting" state.

The propagation of nerve impulses is the fastest means of internal communication developed by animals. Because of its electrical nature, it has fascinated neurophysiologists, electrical engineers, and biophysicists. Unlike molecular conformation shifts occurring in transmembrane ion channels, the voltages and currents associated with nerve impulse origination and propagation are relatively easy to measure. Most of the early electrophysiology on neurons was done on invertebrates. The squid giant neuron axon has been a favorite subject because of its size, and the fact that the axon membrane was exposed (not covered with myelin). The research on the squid axon led to the initial discovery that the ion currents crossing the membrane acted to initiate and propagate the nerve impulse in an autoregenerative process.

Conduction of nerve spikes was found to occur in either direction on the axon. Normal (orthodromic) conduction is generally from the site of spike initiation on or near the cell body *(soma)*; however, antidromic conduction can be artificially initiated on the axon near its synapses. Antidromic conduction has been used to map neural pathways in the spinal cord.

The Hodgkin–Huxley [1952] model for nerve impulse initiation and propagation was a landmark step forward in quantitative biology and physiological modeling. Using a simple set of linear and nonlinear ordinary differential equations and algebraic relations, it describes in simple terms the generation of a nerve spike in a "unit area" of active nerve membrane (e.g., axon). Essential to the H-H model are nonlinear relations describing the behavior of voltage-dependent Na^+ and K^+ conductance channels in active nerve membrane.

We note that the H-H model has withstood the rigors of time; as knowledge has grown about the vast numbers of voltage-activated and chemically activated ion channels in membranes, they have been modeled and added to the basic H-H model. The mathematical formulation of the basic H-H model is covered in detail in Section 4.5.4 of this text on modeling complex biochemical systems and in Appendix C.

3.7.3 Electrical Synapses: Gap Junctions

Electrical synapses are important for two reasons, speed of communication and biochemical simplicity. It is not surprising that they are found in the mammalian CNS as well as in selected invertebrate neuromuscular systems where speed of reaction is critical.

The anatomy of a typical electrical synapse is shown schematically in cross section in Figure 3.8. The gap between the plasma membranes in a gap junction is about 3.5 nm. Six hexameric *connexin protein* subunits in each plasma membrane mate with each other to form a gap junction channel called a *connexon,* which connects the two cytosolic compartments. The electrical synapse between two neurons can contain up to thousands of single connexon channels.

The diameter of the lumen of an open *connexon* passageway is about 2 nm. It can pass ions and small molecules up to 1,000 Da. Such small molecules can include nutrients (glucose), metabolites, second messengers, ATP, cyclic AMP, etc. In a few neurons, connexons are *rectifying*; that is, they pass certain ions more easily in one direction than the other. This rectification is a property of the connexin molecular subunits used to form the two aligned connexons. At least 21 connexin genes may code for gap junction proteins in the human genome [Söhl et al. 2005].

Gap junctions were discovered in crayfish giant fiber neurons and in lobster cardiac ganglion neurons in the 1950s [Furshpan & Potter 1959; Wanatabe 1958]. Since the work on invertebrates, they have been found in all metazoan life-forms. For example, cardiac muscle cells synchronize their contractions by electrical signals through gap junctions between cells, allowing them to contract in tandem and in sequence. Ciliated tracheal epithelial cells are interconnected by gap junctions that act to coordinate their unidirectional waves of beating used to clear the pulmonary passages. Near mammalian birth, uterine smooth muscle cells increase their numbers of gap junctions to better coordinate uterine contractions during delivery. Evans and Martin [2002] stated that: ... "gap junctions have now been found in all tissues except in striated muscle where the cells have fused during development. …Indeed, the list of cells not utilising this mode of cellular interaction and signalling is confined to erythrocytes, platelets and sperm."

Communication by gap junctions between the many types of cells having nonexcitable membranes is beyond the scope of this section. We will review the role of gap junctions in neuron and muscle cell signaling. Furshpan and Potter [1959] discovered what proved to be a fairly rare, rectifying type of gap junction in the crayfish's giant motor synapse. It was possible for them to place glass micropipette electrodes in both the pre- and postsynaptic axons near the gap junction. They found that depolarizing (+ ion) current flowed from the presynaptic fiber, through the connexons, into the

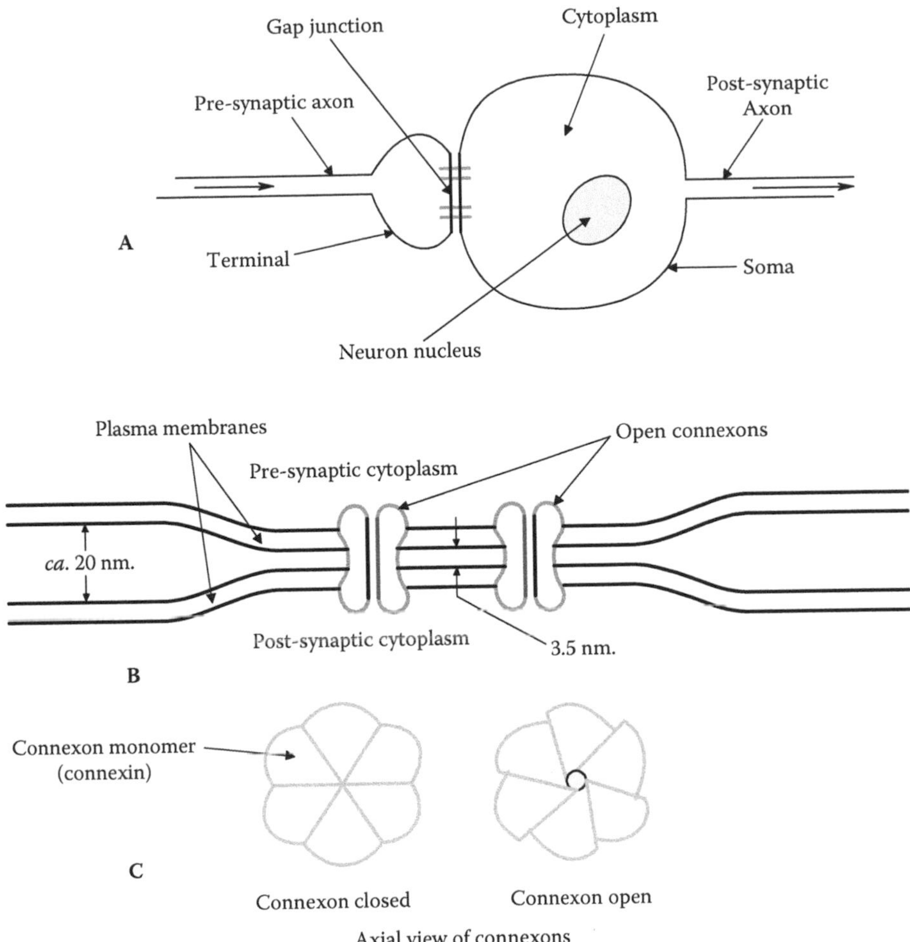

FIGURE 3.8 (**A**) Schematic illustration of an electrical gap junction between a presynaptic bouton and a postsynaptic neuron cell body (soma). (**B**) Cross section of a gap junction synapse between a pre- and postsynaptic neuron. Two open connexon proteins that penetrate both neuron's plasma membranes are shown. (**C**) Axial schematic view of a closed and an open-center connexon. Note that when the connexon is open, its six "orange section" molecular subunits are displaced to form the open center through which select ions can pass.

postsynaptic membrane, depolarizing it and initiating a postsynaptic spike with a small (about 400 μs) delay. When they depolarized the postsynaptic fiber, negligible depolarization was seen in the presynaptic fiber. This showed that cations could pass from pre- to postsynaptic cytosol but not from post- to presynaptic cytosol. That is, the connexons between the crayfish neurons were *rectifying*, i.e., allowing a one-way flow of cations.

Enter complexity again. It has been observed that connexons are not always open. Nonrectifying gap junctions close in response to lowered (acid) pH and to elevated $[Ca^{++}]_i$. Rectifying connexons are voltage operated; depolarization of the presynaptic opens them, and hyperpolarization (by depolarizing the postsynaptic membrane) causes them to close. The molecular design of a gap junction requires that the connexin proteins be "smart" molecules. Each subunit must recognize the other five subunits in order to assemble an effective hexameric connexon channel through the plasma membrane; also, each protruding, presynaptic subunit must mate up with the protruding postsynaptic connexin subunits so that the pores in both are aligned.

In their review paper, Söhl et al. [2005] stated that "neurons in the adult mammalian brain mainly form electrical synapses with other neurons." They cite anatomical, neurohistochemical, and electrophysiological evidence for gap junctions in brain areas such as the cerebellum, superior olive, hippocampus, cortex, thalamus, and in all types of retinal neurons.

The human brain has about 10^{11} neurons and perhaps 10 to 50 times as many glial cells intimately associated with them. There are three principal types of glial cells in the mammalian nervous system: (1) *Oligodendrocytes*, which participate in forming the myelin sheaths around nerve axons; (2) *Schwann cells* found in the peripheral nervous system form myelin sheaths around axons; and (3) *Astrocytes* have many dendritic processes; they make contact with neurons and capillaries and induce endothelial cells to form the blood–brain barrier. Kandel et al. [1991] commented: "Glial cells are probably not essential for processing information, but they are thought to have several other roles." They go on to list seven non-information-processing roles of glial cells. This picture has changed in 15 years.

It has now been found that certain CNS neurons form gap junctions with certain types of glial cells. The functional significance of these connections is yet to be described. In response to neural firing, associated glial cells evidently communicate with other glial cells by gap junctions and chemical signaling [Bullock et al. 2005]. Bullock et al. state:

> Astrocytes are now known to communicate among themselves by means of glial transmitters and neuromodulators as well as by gap junctions. Moreover, astrocytes can detect neurotransmitters that are released from neuronal chemical synapses. These transmitters are delivered via synaptic vesicles into the synaptic cleft and diffuse to perisynaptic astrocytes. Additionally, neuotransmitters can be released outside the synapse and detected by perisynaptic glia. In response, astrocytes can regulate communication between neurons by modifying synaptic transmission through the release of neurotransmitters and neuromodulators. Thus there may be a parallel system of information processing that interacts with neuronal communication but propagates over much slower time scales through a functional reticular network of non-neuronal cells. This functional reticulum results from gap junction coupling and the omnidirectional communication that is mediated by chemical messengers released from astrocytes over much slower time scales. Such may be the case in the human brain. (From Bullock et al. 2005. *Science* 310: 791–793. Excerpted with permission from AAAS.)

Thus, it may very well be found in the future that glial cells, in addition to their neurochemical support roles, have a major role in modulating CNS function. That is, they do figure significantly in neural information processing and possibly in memory.

3.7.4 Chemical Synapses

3.7.4.1 Introduction

As we have described earlier, the other way nerves communicate their spike signals to other neurons, muscles, and gland tissues is by chemical synapses. There are two major categories of chemical synapses: *excitatory* and *inhibitory*. Excitatory synapses act to cause a positive action; that is, the propagation of postsynaptic nerve spikes, the contraction of muscles, or the secretion of glands. Inhibitory synapses act on postsynaptic neurons to inhibit the generation of postsynaptic spikes that are stimulated by the firing of excitatory neurons. In crustacean neuromuscular systems, inhibitory synapses can prevent or reduce muscle contraction by direct action on the muscle membrane itself [Fatt & Katz 1953]. In this section, we will describe the mechanism by which excitatory and inhibitory chemical synapses work.

Figure 3.9 illustrates schematically the major features of a chemical synapse. An action potential propagates down the presynaptic nerve axon and the membrane depolarization invades the terminal branches that are terminated in *synaptic boutons* (knobs). Arrival of the depolarization wave induced by the action potential causes voltage-gated calcium channels to open in the bouton's membrane. The influx of Ca^{++} causes exocytosis of vesicles filled with *neurotransmitter molecules.* The vesicles discharge their contents in an all-or-none manner. In *cholinergic neurons,* each vesicle holds about 1,000 to 3,000 molecules of the neurotransmitter, *acetylcholine* (ACh). Each vesicle is said to hold one *quantum* of neurotransmitter.

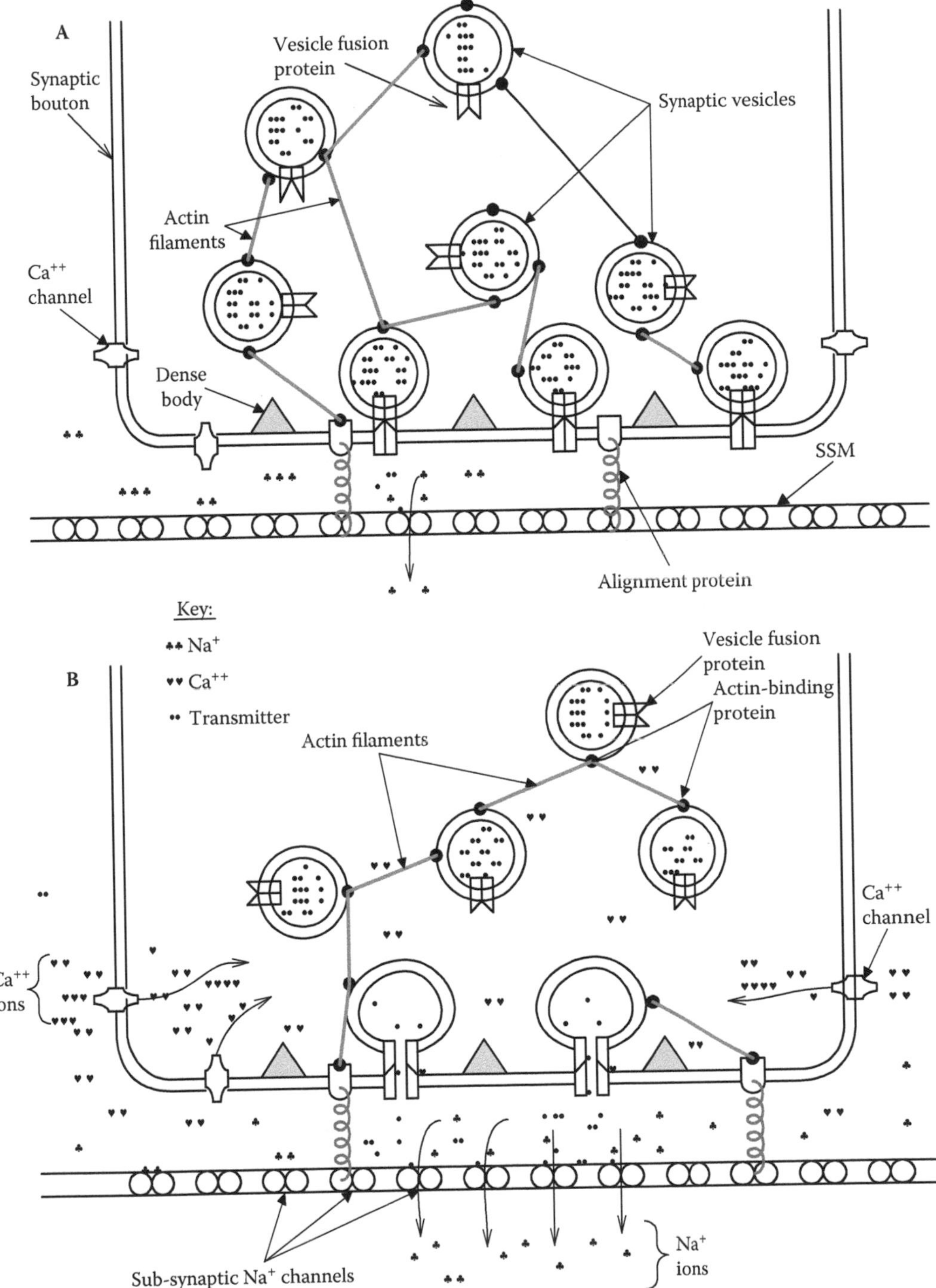

FIGURE 3.9 (**A**) Schematic cross section of a bouton forming a chemical synapse. **SSM** is the subsynaptic membrane across which the neurotransmitter (NT) molecules diffuse when released. The dots in the synaptic vesicles represent NT molecules. (**B**) Schematic of the same synapse immediately after arrival of a presynaptic nerve spike. The paired circles in the plasma membrane represent NT-activated Na^+ ion gates; two NT molecules are required to cause a transient opening of these ion gates. [Adapted from Northrop 2001. With permission from Taylor & Francis.]

The released neurotransmitter molecules diffuse across the *synaptic cleft* and bind to receptor molecules on the subsynaptic membrane. These receptors are *ligand-gated, specific ion channels.* There are several different neurotransmitters and several types of ion channels. There are enzymes associated with every chemical synapse that act to destroy the released neurotransmitter. If not destroyed, the receptors in the postsynaptic membrane would remain bound to the neurotransmitter molecules and the synapse would be effectively blocked.

Acetylcholine is an important excitatory neurotransmitter in neuromuscular junctions and nerve-nerve excitatory synapses. ACh binds to ligand-gated sodium ion channels, causing sodium ions to rush in through the SSM and depolarize it. If the net depolarization reaches the transmembrane potential firing threshold, a postsynaptic nerve spike or a muscle action potential (and twitch) will be generated. The transient positive-going V_m is called an *excitatory postsynaptic potential* (EPSP). At a neuromuscular junction, about 150 quanta of ACh are released at the arrival of a motor neuron spike [Kandel et al. 1991].

Gamma aminobutyric acid (GABA) is an example of an inhibitory neurotransmitter substance found in inhibitory CNS synapses. The binding of GABA to SSM "type A" receptors allows ligand-gated chloride channels to open, causing the subsynaptic transmembrane voltage to rapidly approach the chloride Nernst potential. This clamps $V_m \approx E_{Cl}$, effectively preventing V_m from reaching the firing threshold of the postsynaptic cell. The GABA type B receptor is a *second-messenger system.* GABA binding to the B receptor activates an *internal* **G**-protein and an internal second messenger causes potassium channels to open more slowly, driving V_m toward E_K (about −80 mV). This hyperpolarization is generally summed in the postsynaptic neuron with the depolarizing effect of excitatory inputs. The negative-going V_m caused by GABA B-binding is called an *inhibitory postsynaptic potential* (IPSP). Neurotransmitters are presented in Table 3.4 [Kimball 2005p].

Many animals and organisms have evolved poisons and toxins that interfere with neurotransmission, including jellyfish, mollusks, spiders, snakes, lizards, and bacteria. The poisons from animals are used for defense and for prey capture. Some of these poisons block synaptic neurotransmitter-esterases, others bind to subsynaptic ligand sites and block neurotransmitter action, and others prevent neurotransmitter release. For example, bacterial *Botulinum toxin A* binds to ACh releasing boutons and prevents ACh release, causing flaccid paralysis. *Tetanus toxin,* also from a *Clostridium*

TABLE 3.4
Some Mammalian Neurotransmitters and Neuromodulators

Neurotransmitter	Acts On
ACh	Skeletal muscle Preganglionic neurons of the autonomic NS Postganglionic neurons of the parasympathetic NS Certain CNS neurons
Glutamic acid	CNS excitatory synapses
GABA, glycine	CNS inhibitory synapses.
Norepinephrine	Postganglionic neurons of the sympathetic branch of the autonomic NS Certain CNS synapses
Dopamine	Certain CNS synapses
Serotonin (5HT)	Certain CNS synapses
Histamine	Certain CNS synapses
Peptide hormones:vasopressin (ADH), oxytocin, gonadotropin-releasing hormone, angiotensin II, cholecystokinin, substance P, 2 en-kephalins (met-enkephalin, leu-enkephalin)	Various target sites in the body

bacteria, selectively binds to and enters spinal neurons, where it blocks glycine release and causes spastic paralysis [Hardman et al. 1996]. (The neurotransmitter glycine causes inhibition by increasing chloride conductance.)

3.7.4.2 Neural Transmembrane Chemical Signal Receptors

By now, you, the reader, are aware that membrane proteins are extremely diverse and plentiful. One reference cites 608 protein transmembrane structures [PDBTM 2006]. Transmembrane molecular receptor (TMMR) molecules are one factor that makes metazoan life possible. For an organism to act as an entity rather than an amorphous collection of cells, the cells must have a command, control, and communication (C^3) system to coordinate *organismal homeostasis* (as opposed to *cellular homeostasis*) and behavior. The organismic-level C^3 is carried out in part by a nervous system in higher metazoans, as well as by hormonal and chemokine signals.

Neural and hormonal signals are transduced by cells' transmembrane signal receptor (TMSR) molecules. That is, the TMSR molecules are *molecular transducers,* converting the binding of the signal molecule to another molecular event within the cell. Receptors generally have very high specificity for their signal molecules (ligands), which bind to the extracellular part of the TMSR and cause the inner portion of the receptors to activate (or suppress) cell activities such as expression of specific proteins.

The structures of many TMSRs are known; they are composed of two or more protein subunits that operate collectively. The genes coding many receptors and the chromosomes where they are located have been listed [SenseLab 2006]. TMSRs have three structural domains: *extracellular* (contains a binding site for the target ligand), *transmembrane* (the molecules that penetrate the cell membrane), and the *intracellular* domain. The event of a ligand binding to the extracellular domain is communicated through the transmembrane domain to the intracellular domain.

3.7.4.3 G-Protein-Coupled Receptors

A major class of TMSR is called the ***G****-protein coupled receptors* (GPCRs), or *seven transmembrane receptors* (7TMRs), also known as *heptahelical receptors.* This class of TM receptor is ubiquitous in eukaryotic cells. 7TMRs can be activated by photons, odorants, neurotransmitters, hormones, certain glycoproteins, and chemokines. The 7TMRs possess seven linked, helical peptides that penetrate the lipid bilayer of the cell membrane. This class of receptors has a free end and three loops that protrude into the extracellular fluid; also, a free end and three loops project into the cytosol. The extracellular parts of the 7TMRs can be *glycosylated.* The extracellular loops also contain two cysteine residues that enable stable disulfide bonds to form, making the 7TMRs more chemically stable.

G-proteins are called that because they bind to the ***g****uanine nucleotide phosphates,* GDP, and GTP. The details of the actual transduction mechanism of GPCRs are not completely understood. It is thought that once the ligand binds within the transmembrane domain, the GPCR conformation shifts and somehow activates the **G**-protein, which is bound to the inner domain of the receptor. The inactive **G**-protein has a heterotrimeric (α, β, γ) structure; it is associated with the C-terminal (inner) end of each 7TMR. There are about 23 different Gα subunits in mammals, 5 β, and 14 γ subtypes, which have been found in a wide variety of combinations [McCudden et al. 2005; Preininger & Hamm 2004].

The activated **G**-protein's Gα subunit exchanges *guanine diphosphate* (GDP) for GTP. The Gα-GTP splits from the G$\beta\gamma$ dimer. Both these altered **G**-proteins detach from the inner 7TMR and can induce intracellular signaling events. Once the **G**-protein is activated, the 7TMR can uncouple the ligand and go back to its "ready" state, or it can stay active and activate another nearby **G**-protein.

The activated **G**-protein initiates the production of a *second messenger* (SM) molecule. One common SM is *cyclic AMP* (cAMP), which is produced by the enzyme *adenyl cyclase* (AC) using ATP. AC is bound to the inner cell membrane near the 7TMR. Another SM molecule is *inositol 1,4,5-triphosphate* (IP_3). The SM molecule, in turn, can trigger a sequence of intracellular events such as phosphorylation and activation of enzymes or the release of calcium ions into the cytosol from stores inside the ER. The SM, cAMP, activates the *transcription factor* (TF), *cAMP response element binding protein* (CREB), which is bound to its response element, **5′TGACGTCA3′**, in the promoters of those genes that are supposed to respond to the signaling ligand. Once CREB is activated, it turns on gene transcription, causing the cell to make the desired proteins and RNAs required by the signal ligand that has bound to the 7TMR on the cell's surface.

Another layer of complexity in the picture of cellular signal transduction is introduced by the observation that the β-*arrestin* proteins modulate second-messenger biochemical signals, making the description of second-messenger networks even more complex. The arrestins regulate aspects of cell motility, chemotaxis, apoptosis, and probably many other cellular functions [Lefkowitz & Shenoy 2005]. But what regulates the arrestins?

(The reader interested in pursuing the topic of **G**-proteins in depth should read the review paper by McCudden et al. [2005]).

3.7.4.4 Summary

In concluding this section, we stress the incredible complexity and diversity of intercellular chemical communications by listing the following categories of membrane *receptors* (the numbers of each type of receptor known are in parentheses). The entries are for receptors for *neurotransmitters, neuromodulators, hormones*, and *signal substances.*

> 5-Hydroxytryptamine (5HT) receptors (14), Acetylcholine receptors Muscarinic (5), Acetylcholine receptors, Nicotinic (8), Adenosine receptors (4), α_1-Adrenoreceptors (3), α_2-Adrenoreceptors (3), β-Adrenoreceptors (3), Angiotensin receptors (2), Atrial natriuretic peptide receptors (2), Bombesin receptors (3), Bradykinin receptors (2), Calcitonin, amylin, CGRP & adrenomedullin receptors (4), Cannabinoid receptors (2), CC-Chemokine receptors (11), CX3C-Chemokine receptors (5), Chemotactic peptide receptors (3), Cholecystokinin & Gastrin receptors (2), Corticotropin releasing factor receptors (2), Dopamine receptors (5), Endothelin receptors (2), Glycine receptors (1), $GABA_A$ receptors (2), $GABA_B$ receptors (1), Galanin receptors (3), Glutamate receptors, ionotropic (4), Glutamate receptors, metabotropic (8), Glycoprotein hormone receptors (3), Histamine receptors (3), Ins(1,4,5)P3(IP3) receptors (3), Leukotriene receptors (3), Lysophospholipid receptors (5), Melanocortin receptors (5), Melatonin receptors (3), Neuropeptide Y receptors (6), Neurotensin receptors (2), Neurotrophin receptors (4), Opioid & opioid-like receptors (4), P2X receptors (transmitter-gated channels) (7), P2Y receptors (G-protein-coupled) (6), Peroxisome proliferator activated receptors (3), Prostanoid receptors (8), Protease-activated receptors (4), Ryanodine receptors (3), Somatostatin receptors (5), Steroid hormone receptors (4), Tachykinin receptors (3), Thyrotropin-releasing hormone receptors (2), Vasoactive intentinal peptide & pituitary adenylate cyclase activating peptide receptors (3), Vasopressin & Oxytocin receptors (4). [SenseLab 2006. Quoted with permission.]

The description of cell receptor proteins, how they work intracellularly, and how they are regulated is a fascinating, complex discipline of molecular biology and physiology. Cell receptor protein functions probably hold the answer to what regulates stem cell differentiation, and hence, embryonic development.

3.8 REVIEW OF CELL REPRODUCTION

3.8.1 The Cell Cycle: Mitosis

The process of eukaryotic cell mitotic division is a complex and carefully regulated event. Cells such as *stem cells* of various types (see Chapter 10, Section 10.4), embryonic cells, mature epithelial cells, and corneal cells divide continuously; other cells such as neurons and erythrocytes do not divide once they have matured. All cells require external molecular signal cells to trigger division. For example, cells such as fibroblasts are stimulated to divide (enter the *cell cycle*) by external regulatory signals from platelets associated with tissue damage.

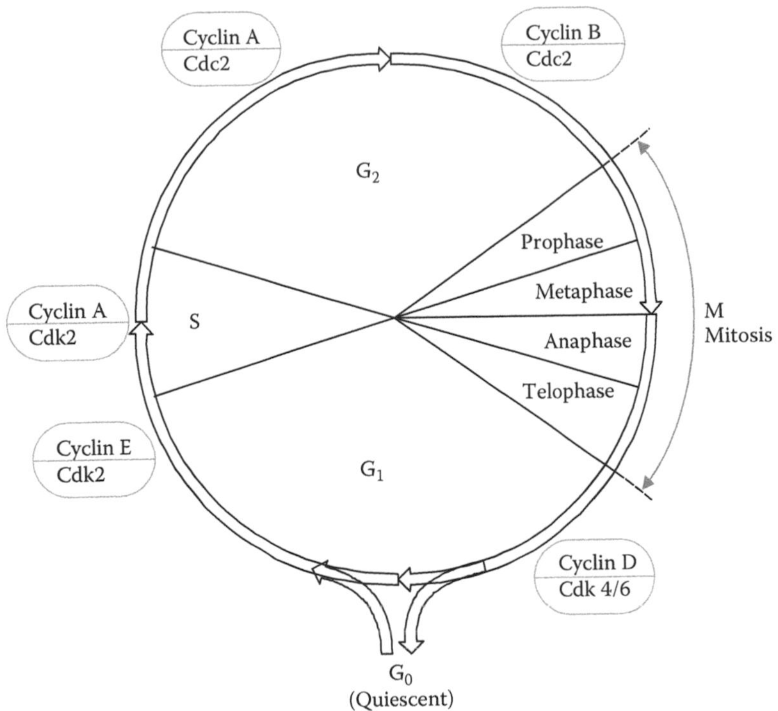

FIGURE 3.10 Generalized time line for the mitotic cell cycle showing when regulatory molecules are active.

The cell cycle leading to cell division has five distinct morphological and biochemical stages. These are

- **G_1**: Growth and preparation for chromosomal (genomic) replication. This is also called *G1 interphase*. The G1 chromosomes each have one *chromatid* (**n** = 1).
- **S**: DNA replication and synthesis of *centrosomes*. Also part of *interphase*.
- **G_2**: The cell is preparing for mitosis. By *G2 interphase,* the cell has duplicated its DNA and is **2n.** Each chromosome now has two chromatids joined by a *centromere*.
- **M**: The mitotic events occur (*prophase, metaphase, anaphase, telophase, and cytokinesis*).
- **G_0**: Quiescent phase; no further entry into the cell cycle. Some cells can be stimulated to reenter the cell cycle and divide by external signals.

Special cytosolic proteins called *cyclins* regulate the cell cycle internally: *Cyclin D* (a G_1 cyclin), *cyclins E and A* (S-phase cyclins), *cyclins A and B* (M-phase cyclins). In addition, there are *cyclin-dependent kinases* (Cdks). (A *kinase* is an enzyme that phosphorylates a target molecule.) These include *Cdk4* (used in G_1), *Cdk2* (used in S-phase), and *Cdk1* (used in M-phase). There is also the *anaphase-promoting complex* (APC) (aka the *cyclosome*), which triggers biochemical events that allow the *sister chromatids* to separate. APC also degrades the mitotic *cyclin B*. Figure 3.10 illustrates the cell cycle schematically and shows when the cycle regulating molecules is active.

Mitotic events in the cell cycle visible in the light microscope are also described with traditional nomenclature:

- *Interphase:* Chromosomes are not clearly discernable in the nucleus, although the nucleolus may be visible. A pair of centrioles may be visible. The **2n** cell is engaged in metabolic activity preparatory to mitosis; its mass may double. (Plant cells lack centrioles.)

- *Prophase: Chromatin* in the nucleus becomes visible as chromosomes. (Chromatin is a mass of uncoiled DNA and the histone proteins associated with it.) The *nucleolus* disappears. Centrioles begin moving to opposite ends of the cell, and *microtubule fibers* are visible extending from the centromeres. Some fibers cross the cell and form the *mitotic spindle*. (Microtubule fibers are made from α- & β- *tubulin* proteins. See Chapter 7, Section 7.2.4.2.)
- *Prometaphase:* The nuclear membrane dissolves at the beginning of *prometaphase*. Proteins attach to the *centromeres*, creating the *kinetochores. Microtubules* from the *centrioles* attach to the kinetochores and begin moving the chromosomes.
- *Metaphase:* Spindle fibers align the chromosomes along the *metaphase plate* that bisects the former nucleus. This geometry ensures that in the next phase when the chromosomes are separated, each new daughter nucleus will receive one complete set of chromosomes.
- *Anaphase:* The paired chromosomes are separated at the kinetochores and pulled to opposite ends of the cell by the *kinetochore microtubules*. This motion results from the coordinated shortening of the kinetochore microtubules and the polar microtubules anchored to opposite ends of the cell. This motion is called *karyokinesis*.
- *Telophase:* The *chromatids* arrive at the opposite poles of the cell and new membranes form around the daughter nuclei. The chromosomes disperse and are no longer distinct. Also, spindle fibers disperse, and the actual partitioning of the cell begins.
- *Cytokinesis:* The final event is the formation of the two daughter cells. *Actin* protein fibers circling the cell's "equator" contract, eventually pinching the cell in two. Each daughter cell initially has a haploid (**n**) nucleus. The chromatids double (to **2n**) in S-phase. G2 interphase daughter cells have doubled (**2n**) chromatids.

These steps are illustrated schematically in Figure 3.11. The blue chromatids represent DNA from the sperm; pink is from the egg. (For simplicity, no X, Y chromatids are shown.) Because of their stiff cellulose cell walls, plant cells cannot pinch in two like animal cells. Instead, they form a new cell plate between the daughter cells and then eventually form two primary cell walls.

Mammalian cells cannot divide indefinitely. There is a master control that limits the number of mitotic cell divisions to about 125. The limit is set by the presence of a repeat Nt sequence at the two ends of each chromosome called a *telomere*. In young human cells, the telomere Nt sequence 5′ **TTAGGG** 3′ is repeated thousands of times. Every time the cell divides, it looses about 16 of these repeat sequences. When the telomere repeats are gone, the cell can no longer divide. (See Kimball 2006d for the details of how telomere shortening occurs and how it inhibits mitosis.)

The enzyme *telomerase* can reconstitute the telomeres by adding the telomere repeat sequence to the 3′ end of the DNA. Telomerase is a ribonucleoprotein; in mammals, it contains a single snoRNA strand with the Nt sequence **AAUCCC** that guides the insertion of **TTACCC** on the end of the DNA. The protein component of telomerase is *telomere reverse transcriptase* (TERT), which catalyzes the insertion.

Telomerase is generally not found in adult somatic cells. However, it is found in cells that must divide many times, including *embryonic stem cells* and cells involved with *gametogenesis*. It is also present in unicellular eukaryotes and cancer cells. It has been found that 85 to 90% of cancer cells have somehow regained the ability to synthesize high levels of telomerase throughout their cell cycle, ensuring their ability to keep reproducing.

Worn-out and damaged cells are removed from the population by *programmed cell death* (PCD), or *apoptosis*, which was described in detail in Chapter 2, Section 2.5.2.

3.8.2 Meiosis

Meiosis is a cell-division protocol used in animal embryogenesis to produce haploid sperm and eggs from diploid, primordial germ cells. It also promotes *genetic diversity*. Meiosis consists of two

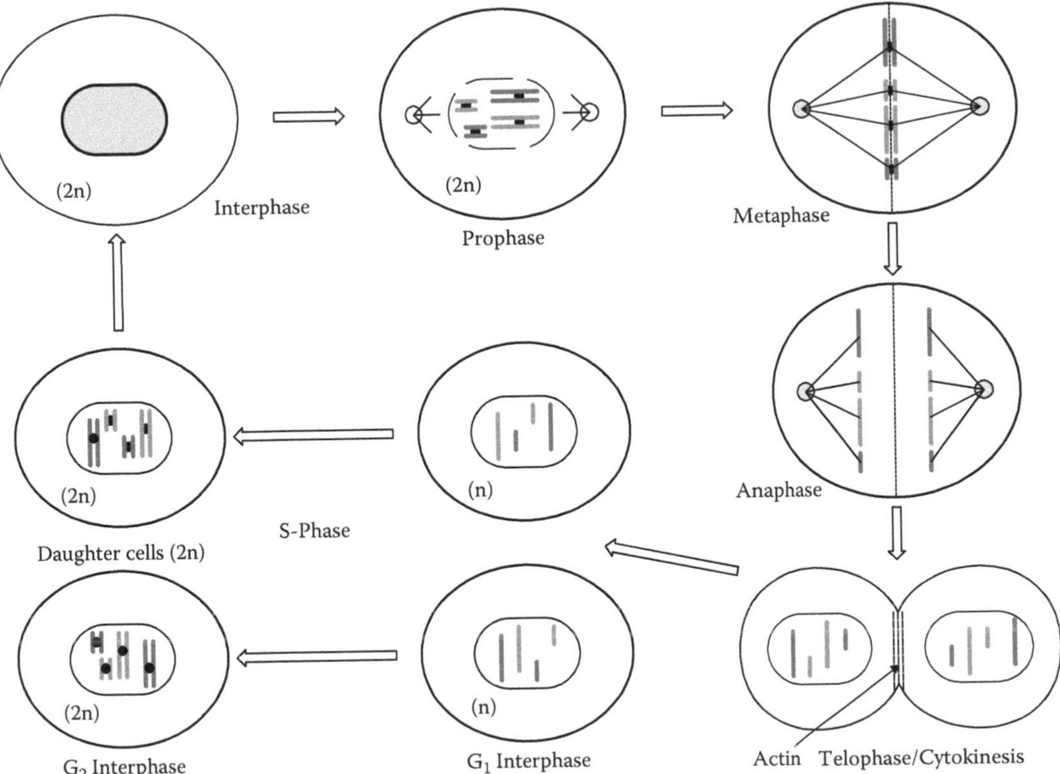

FIGURE 3.11 (See color insert.) A schematic of the cell cycle showing what happens to the chromatids in mitosis. See text for further description.

sequential nuclear divisions that reduce the number of chromosomes in the cell nucleus from **2n** to the haploid number, **n**. Chromosome shuffling also occurs. First, two diploid cells are produced, and then in a second division four haploid cells result.

Figure 3.12 illustrates the process far more eloquently than a verbal description. Note that there are important differences between meiosis and mitosis, although they both involve tubulin "traction motors" to separate the genetic material. Meiosis is subdivided into *Meiosis I* (M1) and *Meiosis II* (M2). Referring to Figure 3.12 (top left), a diploid (**2n**) progenitor cell in *meiotic interphase* duplicates its chromosome pairs and becomes **4n.** The tetrads formed are held together by proteins called *cohesins.*

In *meiotic Prophase I,* the chromatin condenses and organizes (the cell in the figure has **n** = 2). During chromosome duplication in *Prophase I, crossing over* can occur in which chromatids in a tetrad actually cross, stick together, and swap genetic material between chromatids, as illustrated in Figure 3.13.

Late meiotic prophase I is also called *Diakinesis.* The nuclear membrane breaks down, and the *centromeres* of the paired *chromatids* become attached to spindle fibers at their *kinetochores.* The dividing cell now progresses to *Metaphase I.* Pairs of chromatids line up at the center of the cell on the *metaphase plate.* In the figure, the red/blue colors show where crossover has occurred.

In *Anaphase I,* the spindle traction motors exert force and separate the joined pairs of chromatids (dyads). Each centromere *does not* divide in meiotic anaphase; they *do* in mitosis. In *Telophase I,* cytokinesis occurs, yielding two daughter cells each with four chromatids. A nuclear membrane may form here in some species.

The second stage of meiosis has features similar to mitosis. In *meiotic Prophase II,* the chromosomes condense and any nuclear membrane breaks down. In *Metaphase II,* the dyads line up

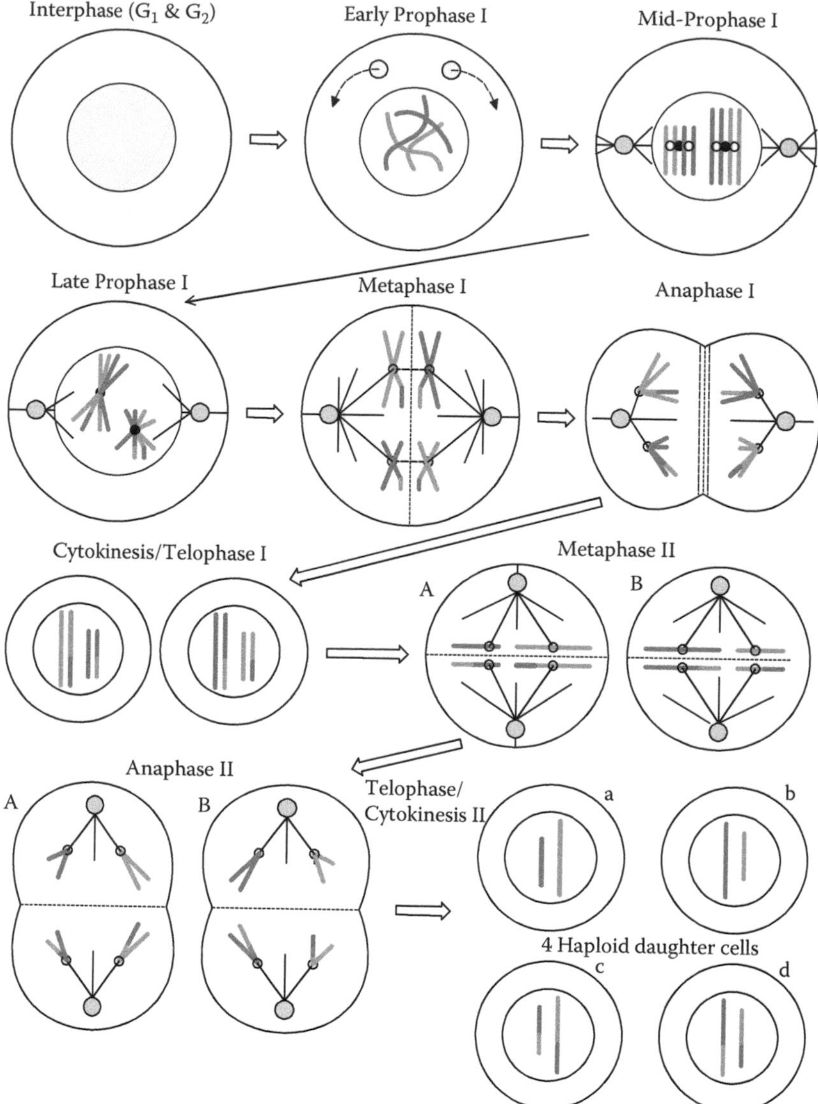

FIGURE 3.12 (See color insert.) Schematic of how meiosis produces four, haploid daughter cells. The fates of the chromatids are shown. See text for further description.

parallel along the metaphase plate connected to spindle fibers attached to the centrioles. The dyads are pulled apart at their centromeres and single chromatids are pulled toward the centrioles in *Anaphase II.* In *Telophase II,* nuclear membranes again can form, and cell division occurs, yielding a total of four haploid (germ) cells (**n = 2** in our example) from one progenitor cell.

Note that meiosis "shuffles" the genomic deck in producing sperm and eggs. The full meiotic process in humans can theoretically produce $2^n = 2^{23} = 8{,}388{,}608$ different types of haploid gamete genomes (without crossing over). With crossovers, the number becomes astronomical; in fact, there is a vanishingly small probability that any of the billions of sperm a man produces in his lifetime will have an identical gene content. The same goes for the hundreds of eggs that mature over the lifetime of a woman [Kimball 2006e].

It would be naive to expect that processes as biochemically and physically complex as *spermatogenesis* and *oogenesis* are without error. Perhaps 10 to 20% of all human *zygotes* contain chromosome abnormalities. These genomic errors are the most common cause of early spontane-

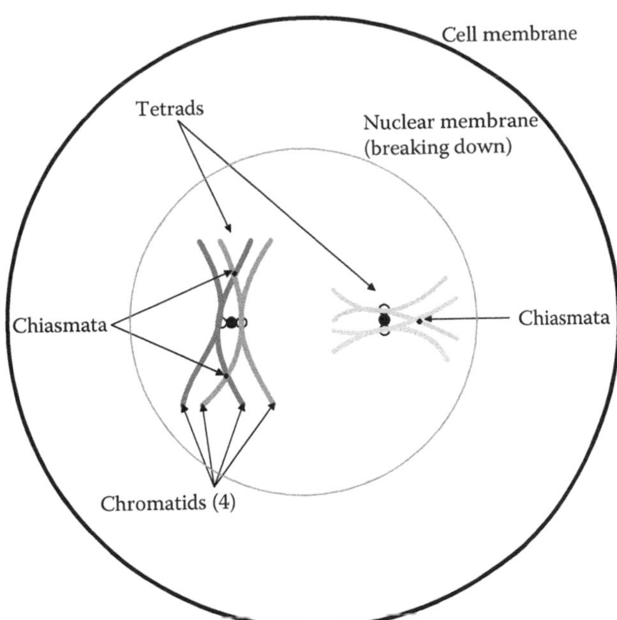

FIGURE 3.13 (See color insert.) Schematic illustration of *diplonema*. See text for further description.

ous abortion (35% of the cases). According to Kimball [2006e], chromosome errors in zygotes arise from errors in meiosis, usually Meiosis I. The errors occur with 90% greater frequency in oogenesis (versus 10% in spermatogenesis). They also increase in frequency as a woman ages. The most common abnormality is *aneuploidy,* the loss or gain of a whole chromosome. This failure of the tubulin traction motors to separate homologues during Meiosis I, or sister chromatids during Meiosis II, is called *nondisjunction.* Zygotes missing one chromosome cannot develop, except for females with only one X-chromosome (the second one is normally deactivated). On the other hand, three of the same chromosome (i.e., *trisomy*) is also lethal except for human chromosomes number 13, 18, and 21. *Down syndrome* is the result of trisomy of chromosome 21. If there is trisomy of the X-chromosome, two of the three Xs are inactivated and the embryo is viable.

Oogenesis, which occurs in the ovary, is a special case of meiosis. Each of the human oogenetic divisions is unequal. The result of the first division of a **4n** *primary oocyte* is a **2n** *secondary oocyte* and a **2n** *polar body.* The polar body is a nonfunctional cell that is eventually broken down. When Meiosis II occurs, the secondary oocyte forms the haploid oocyte and also a haploid (**n**) polar body. The **2n** polar body undergoes Meiosis II as well and forms two more haploid (**n**) polar bodies. Thus, accompanying the mature, haploid oocyte are three haploid polar bodies. Only the mature egg can be fertilized by a sperm, however. The creation of polar bodies may be nature's way of saving cytoplasm. Know that oogenesis is under a complex hormonal regulatory system, which we will not discuss here.

Women are born with about 2×10^6 primary oocytes. By puberty, there are about 400,000 left in the ovaries. Each month, about 10^3 primary oocytes will mature, but most will die and be broken down enzymatically. A woman will ovulate about 400 to 500 ova during her lifetime, about once every 4 weeks.

Sperm production in the testes is a more lively cellular process. Mature human males produce about 10^8 sperm per day in each testicle. The Sertoli cells surround, nourish, and protect the developing sperm cells. *Primordial Germ Cells* (PCGs) divide by mitosis to form *type A_1 spermatogonia* (A_1S), a type of *stem cell.* Thus, the A_1S cells can divide by mitosis to reproduce themselves and also transform into type A_2S cells, also believed to be stem cells. The A_2S cells in turn can make A_3S cells, thence A_4S cells. An A_4S cell has three possible fates: (1) it can form another A_4S cell; (2) it can undergo apoptosis (PCD); or (3) it can differentiate into the first committed stem cell type, the *intermediate spermatogonium* (IS). ISs are destined to become sperm; each divides once by mitosis into two *type B spermatogonia* (TBS). Each TBS also divides by mitosis to form two *primary spermatocytes* (PSs). Each *primary spermatocyte* undergoes Meiosis I and forms two *secondary spermatocytes* (SSs). Each SS undergoes Meiosis II and forms two haploid *spermatids*. The spermatid cells then differentiate into *spermatozoa*, leaving behind a syncytium of *residual bodies*, which are broken down and molecularly recycled. During his lifetime, a human male can produce 10^{12} to 10^{13} sperm. (The biochemical details of *spermiogenesis*, the differentiation of the spermatid cells into sperm, can be found in Gilbert [2003]. It is far more complex, compared to oogenesis.)

Following the initial mitosis of the A_1S cells, the daughter cells of all subsequent generations form a *syncytium* with each other. That is, each cell communicates with others of its generation with numerous cytoplasmic bridges, about 1 μm in diameter, between adjacent cell membranes. Successive generations by mitosis or meiosis produce clones of interconnected cells. One purpose of the syncytia is to cause each cell generation to mature synchronously. It takes about 65 days for a spermatogonial stem cell to develop into sperm. Figure 3.14 illustrates schematically the steps in human sperm development. Note that one *type A_1 spermatogonium* can produce 256 sperm cells. Also, six stages of mitotic divisions precede Meiosis I and II.

What we have not considered in this section is meiosis in plant cells. Plant life cycles differ from those of animals by adding a phase (the haploid gametophyte) after meiosis and before production of gametes. They are no less important, but space prevents a detailed description of plant life cycles here.

3.9 SUMMARY

In this chapter, we have reviewed the domains of life, including the *Archaea*, the *Protista*, and the *Eukaryotes*. We have seen that tracing what ancient life-form has evolved into what present life-form is a considerable challenge, now made easier by our ability to quantitatively examine an organism's genome and proteins. We have seen that *all* life involves cells. *All* cells have cell membranes. The information required for *all* living organisms to reproduce, metabolize, grow, etc., is stored in their DNA genomes.

The evolution of complex eukaryotic, metazoan life was seen to follow the development of molecular systems permitting cell adhesion and intercellular communication systems (molecular and electrical), as well as transmembrane proteins capable of maintaining individual cells' molecular environments (e.g., ion pumps, ion gates, molecular gates, etc.). Molecular intercellular communication required the development of membrane receptor proteins, some of which gate ions, and others activate internal second-messenger molecules. One of the bigger challenges in biology today is gaining an understanding of how an organism's genomic control system works, especially in development.

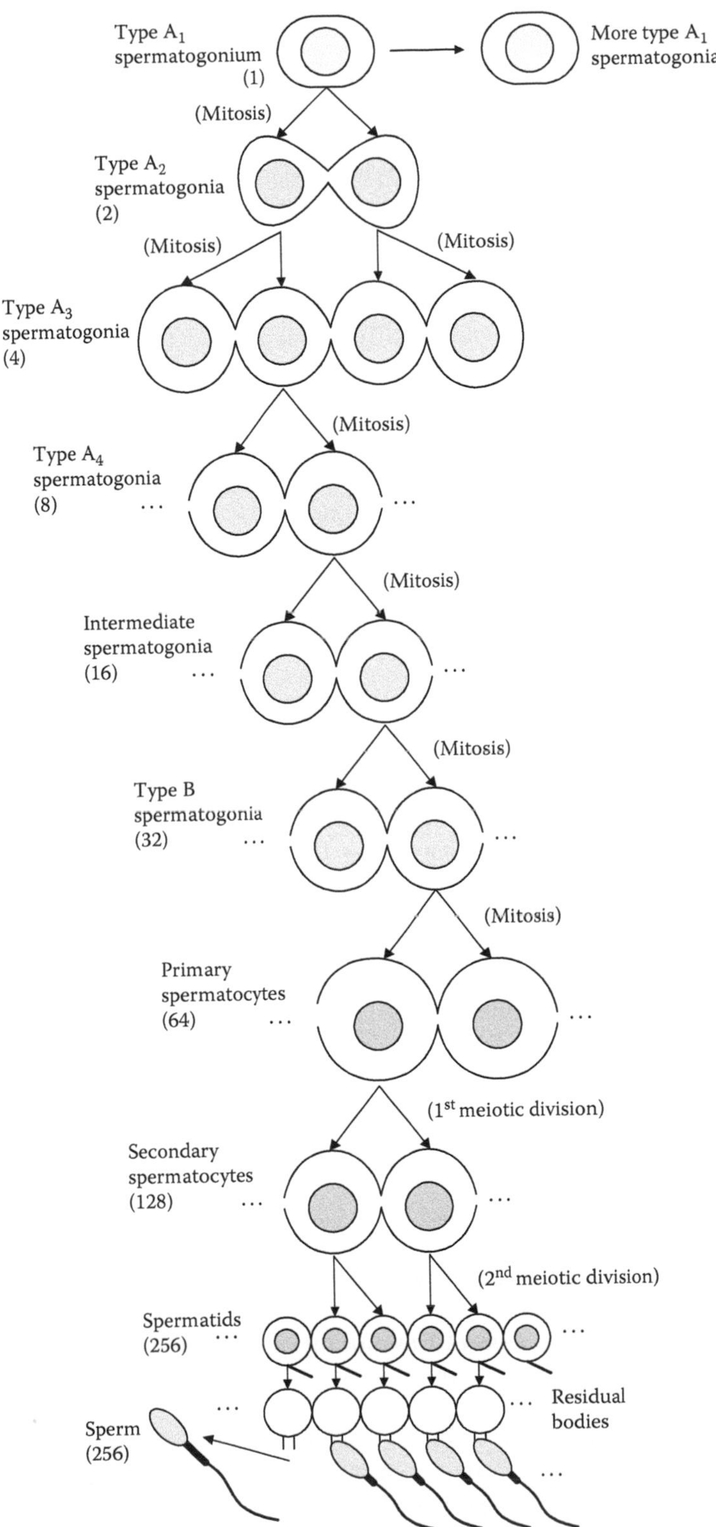

FIGURE 3.14 Schematic of the multiple cell division steps in the production of sperm. Note that meiosis only occurs in the final stage of spermatid production. One spermatogonium cell theoretically can yield 256 spermatozoans.

CHAPTER 3 HOME PROBLEMS

3.1 Most eukaryotic cells have one nucleus; however, some have two or more nuclei. Give at least two examples of normal human cell types that have multiple nuclei. What advantages does multinuclearity confer on a class of cells?

3.2 Mature red blood cells have no nuclei. What advantage does this confer on their role as components in the circulatory/respiratory systems?

3.3 The kinetics of neurotransmitter (NT) release, receptor binding to NT, and enzymatic breakdown of NT at a chemical synapse are modeled in the following text. Each presynaptic action potential causes the all-or-nothing release of NT from a number of presynaptic vesicles into the synaptic cleft. The NT diffuses out of the cleft and is destroyed enzymatically (T is the deactivated transmitter), as well as combining with the subsynaptic receptors. Two NT molecules are required to open each subsynaptic ion channel. An enzyme breaks down the bound NT molecules with rate constant k_3. The voltage change of the epsp generated is assumed to be proportional to the density of open ion channels. That is, $v_m(t) = k_v\,\mathbf{x}(t)$. The chemical reactions are

$$\alpha \longrightarrow T \xrightarrow{k_4} \bar{T}$$

$$2T + R \underset{k_2}{\overset{k_1}{\rightleftarrows}} R * T_2 \xrightarrow{k_3} 2\bar{T} + R$$

Let: y = density of NT molecules (T) in cleft,

x = fraction of Na^+ channels opened by combining with 2 NT molecules,

$1 - x$ = fraction of ion channels closed,

$\alpha(t) \equiv a_o\,\delta(t)$: Input rate of NT caused by an action potential arriving at the bouton. (Approximate 8(t) in simulation with a narrow, rectangular pulse.)

R = unbound (muscarinic) receptor channels,

T = NT,

R^*T_2 = Open ion channel (two NT molecules bound to a channel receptor.)

The mass action ODEs are nonlinear:

$$\dot{y} = -k_4 y + \alpha(t) + k_2 x^2 - k_1 y^2 (1 - x)$$

$$\dot{x} = k_1 y^2 (1 - x) - (k_2 + k_3) x$$

Simulate the above reactions with *Matlab* or *Simnon*. Let $k_1 = 2$, $k_2 = k_3 = k_4 = 1$, and $a_o = 200$. Alpha pulse width = 200. Plot x and y. Note that each t unit = 1 ms. Use an Euler integrator with a 0.001 step size. Observe how the time course of the epsp (**x**), $v_m(t)$, varies with k_3.

3.4 Individual cells practice *homeostasis*. Describe the internal parameters and molecular and ionic concentrations of a cell that are autoregulated. Where does the energy come from that enables homeostasis?

3.5 What is the *diplotene* stage (of the first meiotic prophase) of egg development (oogenesis)?

3.6 Describe gene transcription in a growing mammalian oocyte following ovulation (assuming no fertilization occurs). What gene products are made? What are their destinations and functions?

3.7 Describe how an eukaryotic cell maintains the appropriate osmotic pressure across its cell membrane.

3.8 List the differences between prokaryote and eukaryote cells.

3.9 What mechanisms insert *integral proteins* in the cell membrane? List the different types of integral proteins in a "typical" mammalian neuron and give their functions.

3.10 The cell membrane and its integral proteins experience a considerable electric field because of the steady-state dc transmembrane potential V_M. Calculate this electric field **E**, in V/m, assuming $V_M = 100$ mV, inside negative.

3.11 The human CNS has about 10^{11} neurons. Estimate the percentage of brain neurons that communicate with electrical synapses (versus chemical neurotransmitters).

3.12 We know that the bacterial neurotoxins *Botulinum A* and *Tetanus toxin* act on the mammalian nervous system with extremely low LD_{50} concentrations (about 1 ng/kg for botulinum, 1 ng/kg for tetanus). What survival value does this extreme toxicity to *Clostridium botulinum and Clostridium tetani* confer on mammals?

4 Introduction to Physical Biochemistry and Biochemical Systems Modeling

4.1 INTRODUCTION TO CHEMICAL REACTIONS

4.1.1 Introduction

This chapter is about biochemical reactions and how we can describe them quantitatively. Biochemical reactions are incredibly complex; many branching reaction pathways and "cycles" exist where a molecular species is transformed by multistep, enzyme-catalyzed reactions, and one of the end products turns out to be one of the starting molecules. In the following text, we provide the reader with some tools by reviewing some selected major biochemical pathways. These examples amply illustrate the complexity of biochemistry in action. First, however, we consider chemical reactions and chemical kinetics.

We write chemical reactions as a statement of molecular balance, or stoichiometry, when molecules react together. On the left side of the reaction equation are the reactants or substrates, and on the right are the products of the reaction. The reaction, as written, merely tells us how many molecules of each chemical participate in the reaction. It does not tell us the rate at which the reaction proceeds (kinetics), or whether it will proceed at all, or whether or not there are intermediate steps.

Noncatalyzed chemical reactions occur when molecules of one species randomly collide with molecules of another species. The medium in which the collisions occur may be a gas or liquid, or even a vacuum. If the molecules having the collision have sufficient kinetic energy and are positioned correctly with respect to one another, chemical bonds will be broken and other bonds will be formed, forming new molecular species or products. It is the breaking and reestablishing of chemical bonds that is the origin (or sink) of heat energy released by the reaction.

4.1.2 Chemical Bond Energy

A chemical bond is what physically binds atoms to other atoms, enabling molecules (made of two or more atoms) to exist without flying apart from heat energy or collisions with other molecules or photons. That chemical bonds can be repeatedly broken and reformed, however, is what makes chemistry (and life) possible.

All chemical bonds contain stored or *potential energy*. The amount of energy input required to break the bond is exactly the same as the energy released when the bond is formed. There are many forms of energy: electrical, electromagnetic (including heat and photons), mechanical, and chemical. The basic dimensions of *work* or *energy* from classical mechanics is (force × distance); note that force is (mass × acceleration), so *all* forms of energy have the fundamental dimensions of ML^2T^{-2}. The SI (or old MKS) unit of energy is the *joule*. In the cgs system energy is measured in *ergs*, and chemical energy is generally measured in *kilocalories* (kcal, or Calories with a capital *C*), which is also the amount of heat energy required to raise the temperature of 1 kg of pure water by 1°C. The gramcalorie is the heat energy required to heat 1 g of pure water by 1°C; it is 1/1,000 Calorie. The Btu, or *British thermal unit,* is the amount of heat energy required to raise the temperature of 1 lb of water by 1°F. It can be shown that 1 Cal = 4186 joules. In photonic and photochemical systems, energy is also given in *electron-volts* (eVs). This is the kinetic energy an initially stationary electron acquires when it is accelerated between two electrodes having a potential difference of 1 V. One eV is equal to 1.602×10^{-19} J.

The energies associated with chemical reactions can be measured by using precisely measured masses of reactants in a *bomb calorimeter* (which contains the reaction), surrounded by a known volume of water. The steady-state temperature change in the water is measured to determine the net energy associated with the reaction in the calorimeter.

Kimball [2003d] gave an example of chemical bond energy using the electrolytic decomposition of water. The chemical equation is

$$\mathbf{2H_2O \rightarrow \underset{(gas)}{2H_2} + \underset{(gas)}{O_2}}, \quad \Delta G = +118 \text{ Cal} \tag{4.1}$$

The net *free energy* balance in breaking apart two moles of water is found to be 118 Cal. This energy is called the *Gibbs free energy* of the reaction, G. *G is positive when the reaction requires a net energy input to occur.* In fact, more than 118 Cal are required to decompose the water, but some of the excess energy input is absorbed as the atomic bonds for H_2 and O_2 gases form. Note that the electrical energy input to the electrolysis process is the product of the voltage across the electrodes (assumed to be held constant) times the time integral of the electrode current over the time taken to totally decompose the water. The time integral of the current is the total charge in coulombs required to decompose the water. Kimball [2003d] gave the numbers:

- The total bond energy of the H-O bonds is 110 Cal.
- The total bond energy of the H-H bonds is 103 Cal.
- The total bond energy of the O=O bonds is 116 Cal.
- The decomposition of two molecules of water requires breaking four H-O bonds, and thus a total equivalent electrical energy input of 440 Cal for two moles of water.
- On the plus side, the formation of two moles of H_2 gas yields 2 × 103 = 206 Cal.
- The formation of one mole of O_2 gas yields 116 Cal.

$$\text{Thus, the energy balance: } (\underset{\text{(consumed)}}{+4 \times 110} \underset{\text{(released)}}{- 206 - 116}) = +118 \text{ Cal.}$$

Note that the + sign means it takes a net 118 Cal energy *input* to electrolytically decompose two moles of water to O_2 and H_2 gas, or 59 Cal/mole of water.

Now, let us recombine the two moles of H_2 and one mole of O_2 by *burning* the gases together. It is found that when this combustion takes place, heat equivalent to 118 Cal is released as water is formed according to the following reaction:

$$\mathbf{2H_2 + O_2 \rightarrow 2H_2O}, \quad \Delta G = -118 \text{ Cal} \tag{4.2}$$

The net free energy, ΔG, is *negative* because, by convention, *heat is released* by the reaction (combustion). Note that these examples were simple one-step reactions.

Biochemical reactions, such as cellular respiration in mitochondria, are generally multistep, and they require many enzymes (catalysts) and many bonds to be broken and made. The net free energy

TABLE 4.1
Average Chemical Bond Energies in Cal/Mole for Various Bonds Made or Broken in Biochemical Reactions

Bond	Average Bond Energy, Cal/Mole
C-H	98
O-H	110
C-C	80
C-O	78
H-H	103
C-N	65
O=O	116 = (2 × 58)
C=C	145 = (2 × 72.5)
C=O	187 = (2 × 93.5)
ADP-P	About 7.3
Peptide	About 69

Note: Not all possible biochemical bonds are given in the table; sulfur–(C, H, O) and nitrogen–(O, H) have been omitted.

Source: From Kimball, J. 2003d. *Bond Energy.* //users.rcn.com/jkimball.ma.ultranet/BiologyPages/B/Bond-Energy.html.

produced when one mole of D-*glucose* is oxidized by six moles of O_2 gas can be shown to be −686 Cal. The net reaction can be written as follows:

$$\mathbf{C_6H_{12}0_6 + 60_2 \rightarrow 6CO_2 + 6H_20}, \quad \Delta G = -686 \text{ Cal} \tag{4.3}$$

Average chemical bond energies per mole are given in Table 4.1.

4.1.3 Types of Chemical Bonds

Chemical bonds are what hold molecules together and to each other. They are important in biochemistry because they determine the active tertiary 3-D structures of proteins, protein enzymes, hormones, nucleic acids, etc. A chemical bond is an attraction between atoms caused by a sharing of one or more electrons between them, or a complete transfer of electrons. There are five kinds of chemical bonds: *noncovalent, ionic, hydrogen, covalent, and polarized covalent.*

Noncovalent bonds are individually low-energy (weak) bonds but are used in parallel in biochemical systems for strong intermolecular bonding. Noncovalent bonds enable antibodies to bind to antigens, protein enzymes to their substrates, transcription factors to bind to one another and to DNA, and protein hormones to bind to their cell-surface receptor proteins, etc.

Ionic bonds arise from atoms with low electronegativity (their outer electron shells are mostly empty) reacting with elements with high electronegativity (mostly full outer shells). In this case, there is a complete transfer of some electrons. Sodium chloride is an example of an ionically bonded molecule; Na gives up its one outer shell electron to Cl, which needs only one electron to fill its outer shell. In aqueous, ionized form, a salt solution contains Cl^- and Na^+ ions. The sodium is positively charged because it has lost an electron, and the chlorine is negatively charged because it has an extra electron it picked up from the sodium in the solid state. The attraction between sodium and chlorine is largely electrostatic. Ionic bond strengths are in the range of 4 to 7 Cal/mol.

Hydrogen bonding exists between a hydrogen atom in one molecule and a small atom of high electronegativity in another molecule. That is, it expresses an intermolecular force, not an intramolecular force. In a water molecule, the hydrogen end is positively charged and the oxygen end is negatively charged, Thus, one (liquid) water molecule's two hydrogens electrostatically attract the oxygen of other H_2O molecules. A hydrogen atom covalently bonded to a very electronegative atom (e.g., N, O, or P) shares its partial positive charge with a second electronegative atom. The energy in hydrogen bonds is about 5 Cal/mol. These bonds are frequently found in proteins and nucleic acids and serve to preserve and stabilize molecular structure. It is noted that H-bonds in proteins could also H-bond with the oxygens in the surrounding water, so the relative energy in protein–protein H-bonds is less than the 5 Cal/mol of protein–H_2O, and H_2O–H_2O H-bonds.

Covalent bonds are the strongest chemical bonds. They are formed by the *sharing of a pair of electrons* between two atoms. Single covalent bond energies range from about 50 to 110 Cal/mol, with about 80 Cal/mol being typical. There are single, double, and triple covalent bonds. Once formed, covalent bonds rarely break spontaneously. (The thermal energy of a molecule at 298 K is only about 0.6 Cal/mol.) Protein *peptide bonds* are covalent. In organic molecules, nitrogen can form four covalent bonds, oxygen can form one covalent bond, and sulfur can form one covalent bond. Peptide bonds connect all polypeptide and protein amino acid primary sequences in a linear chain. The amino nitrogen covalently bonds to the carboxyl carbon. The -OH from the carboxyl and the -H from the amino form water when the peptide bond forms. Enzyme catalysts and energy from molecules such as ATP are required to cleave covalent bonds naturally.

Polarized covalent bonds are asymmetrical covalent bonds in which the sharing of the electron pair is unequal, with the electrons statistically spending more time around the less electronegative atom (both oxygen and nitrogen are electronegative). Water is an example of a molecule with polarization. Polarized covalent bonds can form anywhere that oxygen or nitrogen atoms are covalently bonded to hydrogen atoms.

All life is carbon based because carbon is tetravalent. That is, it has four stable electron vacancies in its outer shell, enabling it to bind to another carbon atom with one, two, or three bonds or to oxygen with one or two bonds and to a variety of other atoms and radicals, for example, **-CH_3**, **-CH_2-**, -**COOH**, **-OH**, **-NH_2**, **-S-**, **-C_6H_4OH** (tyrosine), **-C_6H_5** (phenylalanine), etc.

4.1.4 Chemical Kinetics and Mass Action

Chemical kinetics describe the rates at which reactants disappear and products appear in chemical reactions under idealized conditions. Chemical kinetics are based on the *law of mass action*, which, in turn, is based on the assumptions that the reacting and product molecular species are well mixed and free to move in a closed vessel (i.e., they are not bound to a surface similar to a mitochondrial membrane or a cell surface), and there are no diffusion barriers. Furthermore, we assume that the molecules and ions in the vessel are in solution in weak concentrations and move randomly because of thermal energy and momentum exchanges due to collisions with themselves and one another. The law of mass action is based on the probabilities that molecular species will collide and react; that is, bonds will be broken and made. Basically, it states that the *rate* at which a chemical is produced in a reaction (at a constant temperature) is proportional to the product of the concentrations of the reactants. Thus, a chemical reaction described by mass action kinetics can be expressed as a set of first-order *ordinary differential equations* (ODEs). These ODEs are generally nonlinear and can have time-varying coefficients.

For example, if two reactants combine and form a product, ***C***:

$$\mathbf{A} + \mathbf{B} \rightarrow \mathbf{C} \tag{4.4}$$

Then by mass action the rate of appearance of ***C*** is given by the simple first-order nonlinear ODE:

$$\frac{d[C]}{dt} = k[A][B] \tag{4.5}$$

Also, it is easy to see that both the reactants behave as

$$\frac{d[A]}{dt} = \frac{d[B]}{dt} = -k[A][B] \tag{4.6}$$

In these simple ODEs, brackets [*] denote a concentration, such as moles/liter. k is the *reaction rate constant,* given by the *Arrhenius relation* [Maron & Prutton 1958]. Note that k has temperature dependence given by Equation 4.7:

$$k = k_o \exp(-E_a/RT) \tag{4.7}$$

where k_o is a positive constant (prefactor), E_a is the *activation energy* of the reaction in joules per mole (recall that E_a is lowered by enzymes), T is the Kelvin temperature, and R is the SI *gas constant* = 8.31441 J/(mole K).

Another way of writing Equation 4.5 is to let a = the initial concentration of **A**, b = the initial concentration of **B**, and x = the number of moles of **C** made at time t (a "running" concentration). Thus, for $t \geq 0$:

$$\dot{x} = k(a-x)(b-x) = k[ab - x(a+b) + x^2] \text{ (moles/liter)/sec, where } \dot{x} \equiv dx/dt \tag{4.8}$$

Note that a reaction may be *reversible.* For example, a product, **C**, can spontaneously dissociate back into the two reactants, **A** and **B**. Reversibility is represented by Reaction 4.9 in the following text. Note that the forward reaction rate constant, $\boldsymbol{k_1}$, is in general different from the reverse constant, $\mathbf{k_{-1}}$.

$$\mathbf{A} + \mathbf{B} \underset{k_{-1}}{\overset{k_1}{\rightleftarrows}} \mathbf{C} \tag{4.9}$$

Now we can write three ODEs based on mass action:

$$\frac{d[A]}{dt} = k_{-1}[C] - k_1[A][B] \tag{4.10}$$

$$\frac{d[B]}{dt} = k_{-1}[C] - k_1[A][B] \tag{4.11}$$

$$\frac{d[C]}{dt} = -k_{-1}[C] + k_1[A][B] \tag{4.12}$$

Note that the ODE for [A] is identical to that for [B]. Also, the rate constants k_1 and k_{-1} do not have the same units. k_{-1} has the units of T^{-1}, and k_1 has the units of T^{-1} $mole^{-1} = (MT)^{-1}L^3$.

Now let us examine a reaction in which two molecules of **B** must combine with one molecule of **A** to make one molecule of **C**:

$$\mathbf{A} + 2\mathbf{B} \xrightarrow{k} \mathbf{C} \tag{4.13}$$

Mass action tells us that the rate of increase of the molar concentration of the product, **C**, is given by the ODE:

$$\frac{d[C]}{dt} = k[A][B]^2 \tag{4.14}$$

Note that the concentration of **B** is squared in the ODE.

In another example, consider the physiologically important formation and decomposition of *carbonic acid*. This reaction figures in the regulation of the acid–base balance in body fluids and the elimination of CO_2 as a metabolic waste gas [Guyton 1991, Chapter 30].

$$\mathbf{H_2CO_3} \underset{k_{-1}}{\overset{k_1}{\rightleftarrows}} \mathbf{H_2O} + \mathbf{CO_2} \tag{4.15}$$

We assume that water is present in excess; let **x** = the concentration of carbonic acid, and **g** = the concentration of dissolved CO_2 gas. Thus, the chemical kinetic ODE is

$$\dot{x} = -k_1\,\mathbf{x} + k_{-1}\,\mathbf{g} \tag{4.16}$$

As a final example, we consider the important *Michaelis–Menton reaction* (MMR). The MMR is representative of a typical enzyme-catalyzed reaction, where a substrate molecule, **S**, combines with an active site on the enzyme to form a *complex*, **S*E**. In the complex, the enzyme cleaves off part of S to form the product, **P**. (The remainder of **S** is not shown in the reaction.) **P** is released from the intact enzyme. The product, **P**, is assumed to be removed from the compartment by first-order kinetics with rate constant, k_3. The MMR is expressed as a chemical equation:

$$\mathbf{S} + \mathbf{E} \underset{k_{-1}}{\overset{k_1}{\rightleftarrows}} \mathbf{S} * \mathbf{E} \xrightarrow{k_2} \underset{\downarrow k_3 \atop *}{\mathbf{P}} + \mathbf{E} \tag{4.17}$$

Let **e** = the running concentration of free enzyme at time $t > 0$, **s** = the running concentration of the substrate, **c** = the running concentration of the complex, and **p** = the running concentration of the product. Because the system is closed and the enzyme is conserved, we can write the auxiliary equation, $\mathbf{e}_o = \mathbf{e} + \mathbf{c}$, where $\mathbf{e}_o$ is the total (initial) concentration of enzyme at $t = 0$. Note that the reactants (**S**, **C**, and **P**) all have initial conditions when doing a simulation. The mass action ODEs for the MMR are

$$\dot{s} = -k_1(e_o - \mathbf{c})\,\mathbf{s} + k_{-1}\,\mathbf{c} \tag{4.18A}$$

$$\dot{c} = k_1(e_o - \mathbf{c})\,\mathbf{s} - (k_{-1} + k_2)\,\mathbf{c} \tag{4.18B}$$

$$\dot{p} = k_2\,\mathbf{c} - k_3\,\mathbf{p} \tag{4.18C}$$

Note that the ODEs for **s** and **c** are nonlinear; the **p** ODE is linear.

It is clear that complete dynamic solutions of the three Michaelis–Menton ODEs are best done by computer, using a program such as Simnon® (a product of SSPA, Sweden, and a registered trademark of Department of Automatic Control, Lund Institute of Technology, Sweden) or Matlab/Simulink (Matlab and Simulink are trademarks of MathWorks, Inc., Natick, Massachusetts).

From enzyme conservation, we see that $\dot{c} = -\dot{e}$. In the *steady state*, we can set $\dot{s} = \dot{c} = \dot{p} = \mathbf{0}$, because no concentration is changing. Now we can solve for the steady-state concentration of the complex, $\mathbf{c}_{ss}$:

$$c_{ss} = \frac{k_1 e_o s_{ss}}{k_{-1} + k_2 + k_1 s_{ss}} = \frac{e_o s_{ss}}{K_M + s_{ss}} \tag{4.19}$$

where $\mathbf{K_M \equiv (k_{-1} + k_2)/k_1}$ is the well-known *Michaelis constant*, and $\mathbf{s}_{ss}$ is the steady-state concentration of the substrate (for $t \to \infty$).

Under conditions of constant substrate concentration (**S** is replaced as fast as it is used up), $\mathbf{s} = \mathbf{s}_o$, and in the steady state, $\mathbf{p}_{ss} = (\mathbf{k}_2\, \mathbf{c}_{ss})/\mathbf{k}_3$, and we can finally write

$$p_{ss} = \frac{k_2 e_o s_o}{k_3 (s_o + K_M)} \tag{4.20}$$

Although the kinetics of the MMR are a good model for some in vitro, enzyme-catalyzed bioreactions, they are based on mass action kinetics, which in turn are predicated on low concentrations of molecules free to move and collide randomly in a large volume compared to molecular size. Intracellular biochemical systems very often violate these conditions. The reaction volume is often small. Enzymes (or reactants) may be immobilized on membranes. Reactants and products may diffuse into or out of the reaction volume, introducing more kinetics for diffusion. Molecules in the reaction volume may not be well mixed. Finally, the effectiveness of the enzyme in the MMR can be modulated in time by *allosteric* (feedback) *regulation* and *competitive inhibition*.

4.2 IMPORTANT COUPLED REACTIONS

4.2.1 Introduction

A characteristic of life is that its sustaining biochemical reactions are coupled. That is, these vital reactions tend to be *concatenated,* that is, joined one to another in a complex topology. There are exceptions; for example, *shunt reactions* that take "shortcuts" from one chemical species to another. The linear topology makes closed-loop parametric control of the reactions simpler through the regulation of enzyme concentrations and effectiveness. Modulation of the effectiveness of one enzyme at the beginning of a reaction chain affects all the products in the chain. What is least understood in physical biochemistry are the control pathways. Some affect the expression of genes coding enzymes; others involve direct interference with an enzyme's function by a reaction end-product molecule.

Progress is rapidly being made, however, in understanding and describing quantitatively these complex regulatory networks. Available on the Internet, the *Kyoto Encyclopedia of Genes and Genomes* (KEGG) contains all known metabolic pathways and some of the known regulatory pathways in about 100 hypermedia diagrams (click on a section of a diagram and download details of that section). The metabolic pathways were derived in part from the Roche Applied Science *Biochemical Pathways* charts, Parts 1 and 2, and the *Metabolic Maps* book by the Japanese Biochemical Society. A KEGG diagram is intended as a reference map of all *chemically feasible* pathways, not a consensus of known pathways in different genomes. "The genome-specific pathways are then automatically generated by matching the enzyme genes in the gene catalog with the enzymes on the reference pathway diagrams" [JGI 2004].

In the following text, we describe two of the more important biochemical energy-producing pathways.

4.2.2 Glycolysis

Probably one of the more-studied and better-known systems of coupled reactions is *glycolysis,* the first step in the complex reactions of cellular metabolism. Glycolysis is the initial reaction pathway for the conversion of the monosaccharide, D-glucose, into chemical energy. It describes the *anaerobic* catabolism of glucose. Glycolysis is also known as the *Embden–Meyerhoff Pathway.*

Glycolysis is found in *all* cells (with some variations). In eukaryotes, glycolysis occurs in the cytosol; a "soup" of ions, reactants, products, enzymes, coenzymes, redox, and energy-source molecules (ATP, GTP). The "input" molecule to the glycolysis pathway is *D-glucose.* D-Glucose itself can come from enzymatic transformation of other monosaccharides that, in turn, can arise from enzymatic cleavage of disaccharide sugars (e.g., sucrose) or the enzymatic digestion of starches or cellulose.

In the *first step* of glycolysis, one molecule of D-glucose is *phosphorylated* by the enzyme *hexokinase* (HK), using the energy from one phosphate bond in ATP. The products are *glucose-6-phosphate* (G6P), ADP, and free hexokinase. The *second step* in glycolysis is the conversion of G6P to *fructose-6-phosphate* (F6P) by the enzyme *phosphoglucose isomerase* (PGI). In the *third reaction*, F6P is given a second phosphate group by the enzyme *phosphofructokinase* (PFK) to form *fructose-1,6, biphosphate* (F1,6-BP). ATP is also used in this reaction for its energy. Next, in the *fourth reaction*, the hexose sugar F1,6-BP is lysed by the enzyme *aldolase* to form two *triose sugars* called *dihydroxyacetone phosphate* (DHAP). (Now we have two, identical, parallel pathways. All the reactions described in the following paragraph take place at the same time. For example, one molecule of D-glucose leads to two GADP molecules, etc.)

In the *fifth reaction*, in each of two parallel pathways, a DHAP molecule is acted on by the enzyme *triose phosphate isomerase* to form *glyceraldehyde-3-phosphate* (GADP). In the *sixth reaction*, a GADP is acted on by the oxidoreductase enzyme *glyceraldehyde-3-phosphate dehydrogenase* (GAP) to become *1,3-biphosphoglycerate* (1,3BPG). This reaction is important because the hydrogen displaced by the phosphate radical is used to *reduce* a molecule of *nicotinamide adenine dinucleotide* (NAD^+) to NADH. The *seventh reaction* in glycolysis yields metabolic energy: 1,3BPG reacts with the enzyme *phosphoglycerate kinase* (PGK) to form *3-phosphoglycerate* (3PG), plus an enzymatic coupling of a phosphate to ADP, producing one molecule of ATP. In the *eighth reaction*, the enzyme *phosphoglyceromutase* (PGAM) makes 3PG into *2-phosphoglycerate* (2PG). In the *ninth reaction*, 2PG is acted on by *enolase* to form *phosphoenolpyruvate* (PEP). In the *tenth step* in glycolysis, PEP is converted by *pyruvate kinase* (PK) into the end product, *pyruvic acid* (PA) or pyruvate. In this last step, substrate-level phosphorylation makes another ATP molecule from ADP. Now, because there are two parallel DHAP → PA pathways, the glycolysis system of coupled reactions yields a total of $2(2 - 1) = 2$ ATP molecules per glucose molecule input [Farabee 2001].

In yeasts, an extension of the glycolysis pathway leads to *fermentation.* In *reaction eleven*, PA is converted to *acetaldehyde* by *pyruvate decarboxylase.* One molecule of CO_2 gas is produced in this reaction. In *reaction twelve*, acetaldehyde is converted to *ethanol* by the *alcohol dehydrogenase* enzyme. NADH is converted to NAD^+ in this final step. Note that one molecule of glucose produces two molecules of CO_2 and two molecules of *ethanol.*

Other variations of fermentation exist; certain bacteria can produce *methanol* or *methane gas* by similar steps. *Lactobacillus* sp. converts two PAs to two *lactic acids.* Figure 4.1 illustrates schematically the 10 reactions that comprise glycolysis, as well as one form of fermentation.

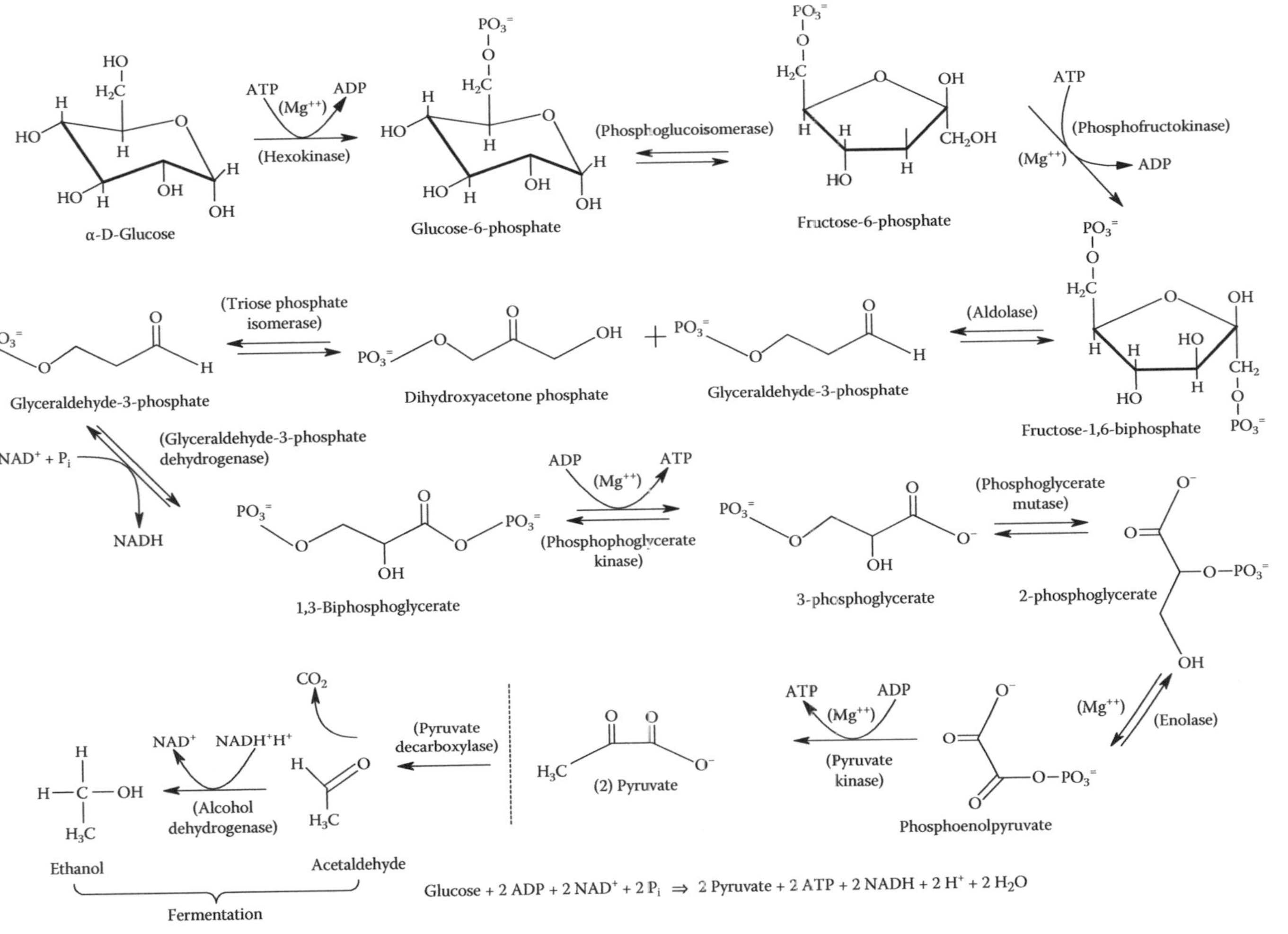

FIGURE 4.1 Schematic of the glycolysis pathway showing one fermentation pathway. One glucose is changed to two pyruvates. Two ATP and two NADH molecules are produced per glucose metabolized. In the fermentation, one pyruvate produced one molecule of ethanol at the expense of one NADH molecule.

The overall glycolysis reaction is [Saccharomyces 2005]:

$$\textbf{D-Glucose} + 2\,\mathbf{NAD^+} + 2\,\mathbf{ADP} + 2\,\mathbf{P_i} \rightarrow 2\,\textbf{Pyruvate} \\ + 2\,\mathbf{NADH} + 2\,\mathbf{ATP} + 2\,\mathbf{H_2O} + 2\,\mathbf{H^+} \tag{4.21}$$

Thus, two pyruvates are supplied to the oxidative citric acid (Krebs) cycle for every glucose molecule metabolized in glycolysis. *Pyruvate kinase* is the enzyme that attaches the P_i (phosphate) to ADP to make ATP.

4.2.3 The Citric Acid Cycle

The second set of coupled reactions we will describe occurs in mitochondria; it is also known as the *tricarboxylic acid cycle (TCA Cycle)* or *Krebs cycle.* The CA cycle is a circular closed-loop series of chemical reactions. Certain inputs to the CA cycle come from the cytosol. The CA cycle is an important component of the reactions that break down carbohydrates, fats, and proteins into CO_2, water, heat, and chemical free energy stored in ATP (refer to Figure 4.2; the enzymes involved are not shown for simplicity; they are cited in the following text, however).

Step one in the CA cycle is when the four-carbon, *oxaloacetic acid* (OAA) condenses with *acetyl-CoA* (2 carbons) to form *citric acid* itself. The catalyst enzyme is *citric acid synthetase* (CAS).

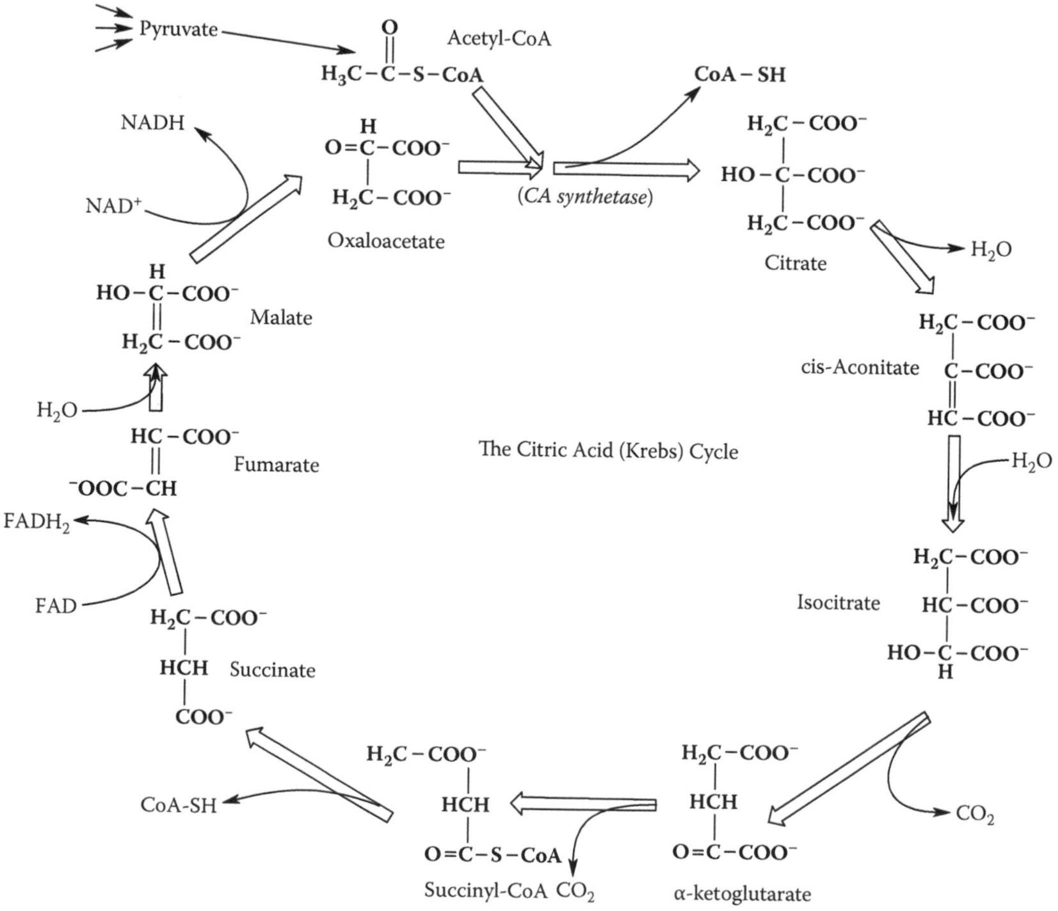

FIGURE 4.2 Illustration of the citric acid (Krebs) cycle. Note that the input is the 3-carbon pyruvate molecule. See text for further description.

The acetyl-CoA is made by reacting *pyruvic acid* with *coenzyme A* (CoA), causing the reaction NAD+ → $NADH^+ + H^+$.

In *step two*, CA is transformed to *cis-aconitic acid* (CAA) by the enzyme *aconitase*. The CAA remains bound to the *aconitase* for the next step.

In *step three*, also catalyzed by *aconitase*, the CAA is hydrated to make *isocitric acid* (ICA).

In *step four*, ICA is oxidized to make *oxalosuccinic acid* (OSA). The enzyme *isocitrate dehydrogenase* causes two hydrogens and two electrons to be transferred to NAD^+ to form NADH + H^+. *Isocitrate dehydrogenase* also catalyzes the *fifth reaction*. Its product is the five-carbon *alpha-keto-glutaric acid* (αKGA). The carbon molecule lost in the decarboxylation of OSA leaves the reaction as CO_2 gas.

In the *sixth reaction*, three chemical transformations take place: oxidation, decarboxylation, and thiol ester synthesis. αKGA is transmogrified into *succinyl coenzyme A* (S CoA) by the *α-keto-glutarate dehydrogenase enzyme complex* in the only *nonreversible step* in the entire CA cycle. (The nonreversibility of the αKGA → SCoA *reaction* prevents the entire CA cycle from operating in the reverse direction.) A molecule of CO_2 is lost in this complex reaction. Also, one molecule of NAD^+ is reduced to NADH + H^+.

In the *seventh reaction*, SCoA is hydrolyzed, releasing free energy, which is trapped by the formation of a molecule of ATP. SCoA acts as its own enzyme to form *succinic acid* (SA).

In the *eighth reaction*, the enzyme *succinate dehydrogenase* catalyzes the *oxidation* of SA to *fumaric acid* (FA). The hydrogen acceptor is the coenzyme FAD, instead of the more usual NAD^+. $FADH_2$ is formed.

The enzyme *fumarase* catalyzes the formation of *malic acid* (MA) from FA in *reaction nine*.

In *reaction ten*, the final step in the CA cycle, MA is *oxidized* by *malate dehydrogenase* to form the ketone *oxaloacetic acid* (OAA). The coenzyme NAD^+ causes the transfer of two electrons and two hydrogens to form NADH + H^+. This brings us back to the beginning of the CA cycle.

The sum of all the linked reactions in the CA cycle can be written:

$$\begin{gathered}\textbf{Acetyl-CoA} + \textbf{3 NAD}^+ + \textbf{FAD} + \textbf{GDP} + \textbf{P}_i + \textbf{H}_2\textbf{O} \rightarrow \\ \textbf{GTP} + \textbf{CoA-SH} + \textbf{3 (NADH} + \textbf{H}^+) + \textbf{FADH}_2 + \textbf{2 CO}_2\end{gathered} \tag{4.22}$$

The CA cycle is not a closed cycle. The inputs to the CA cycle can come from many other metabolic pathways. For example carbohydrates, glycerol, the AAs **Ala**, **Cys**, **Gly**, **Ile**, **Ser**, **Thr**, and **Trp** can be metabolized to *pyruvate*, which can be changed to *acetyl-CoA*. Also, even- and odd-chain fatty acids and the AAs **Leu**, **Lys**, **Phe**, **Thr**, **Trp**, and **Tyr** can be transmogrified directly to acetyl-CoA. *Succinyl-CoA* molecules can be entered into the CA cycle by the metabolism of fatty acid chains and the AAs **Ile** and **Met**. There are also other molecular substrate entry points to the CA cycle (these complex pathways are shown in exacting detail in the Roche Applied Sciences *Biochemical Pathways Wall Charts,* 3rd ed.). The CA cycle also can export intermediate product molecules into other biochemical pathways. For example, *succinyl-CoA* is used in the biosynthesis of *heme*. The complex molecular structure of *coenzyme A* (CoA) is shown in Figure 4.3. Note that *CoA* is built from the purine *adenine* and is coupled to a group such as *succinyl* by the sulfur molecule at the end of *CoA's* "tail." *Adenine* turns out to be a very popular adjunct molecule in various energy compounds such as ATP, NAD^+, FAD^+, and $NADP^+$.

The energy balance in the CA cycle is complex. Not all energy is put into ATP molecules. The reduced *coenzymes, nicotinamide adenine dinucleotide* (NADH) and *flavin adenine dinucleotide* ($FADH_2$), are energy-rich molecules. Further oxidation of one mole of NADH yields three moles of ATP, and oxidation of one mole of $FADH_2$ produces two moles of ATP.

Eukaryotic aerobic mitochondrial respiration produces (about 34) additional molecules of ATP for each molecule of glucose that enters glycolysis. The net ATP production is based on the following factors:

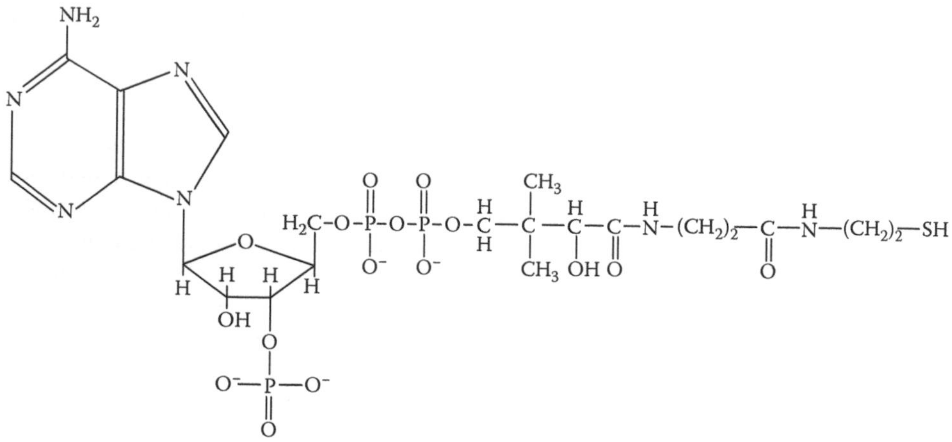

FIGURE 4.3 2-D structure of the coenzyme A molecule.

1. The electrons and hydrogen from each of the two NADH molecules produced in *glycolysis* provide sufficient free energy to produce about two ATP molecules/NADH in the mitochondria.
2. The electrons and H^+ from the $FADH_2$ produced in the CA cycle provide sufficient energy to produce about two ATP/$FADH_2$ by *chemiosmosis* [Kimball 2007c].
3. The H^+ and electrons from the two NADH produced in the CA cycle and the NADH produced by oxidation of *pyruvate* to *acetyl CoA* provide enough free energy to produce about three ATP/NADH by chemiosmosis. Note that the glycolysis cycle produces two pyruvates/glucose input. H^+ ions passing down a concentration gradient interact with the enzyme *ATP synthase* to create new ATP. See the *Glossary* for a further description of chemiosmosis. The details of the construction of the mitochondria and its electron transport systems were covered in Chapter 3, Section 3.6.8.

4.3 OTHER METABOLIC PATHWAYS

4.3.1 The Chemistry of Organic Redox Reactions

Many *redox* (reduction/oxidation) reactions are important in storing and releasing energy in biological systems. *Photosynthesis* involves the *reduction* of CO_2 into sugars and the *oxidation* of water into molecular oxygen. The reverse reaction, *respiration*, *oxidizes* sugars to produce CO_2, water, heat, and high-energy molecules. Biochemical redox reactions are regulated by a variety of parametric control mechanisms involving enzymes.

Our first example of a *reduction/oxidation* (redox) chemical reaction is the common carbon-zinc flashlight battery, or dry cell, aka voltaic cell. The dry cell was invented in 1867 by French scientist Georges Leclanché. It has an extremely simple construction: A thin, cylindrical zinc metal can is filled with an electrolyte paste, which is a mixture of ammonium chloride (NH_4Cl) as a source of H^+ ions, manganese dioxide (MnO_2) as an oxidant, water, and a neutral binder. A thin porous material soaked in an aqueous mixture of the two salts separates the electrolyte paste from the inside of the Zn can. An electrode consisting of a thin carbon rod is inserted in the center of the electrolyte-filled can (not touching the bottom) and is held in place by a nonconducting top seal to the can. The carbon rod has a metal cap to ensure good electrical contact. The carbon center electrode is the positive terminal of the cell (anode), the bottom of the can is the negative terminal (cathode). By U.S. convention, current leaves the positive terminal and flows through an external resistive load back through a wire to the zinc electrode. However, electrons leave the zinc electrode and flow in the external circuit toward the positive carbon anode.

The zinc anode reaction is an *oxidation:* $\mathbf{Zn \rightarrow Zn^{++} + 2\,e^-}$.

The electrons enter the carbon cathode and cause the *reduction:*

$$\mathbf{2\,NH_4^+ + 2MnO_2 + 2\,e^- \rightarrow Mn_2O_3 + H_2O + 2\,NH_3} \tag{4.29}$$

Adding the two reactions, we find the overall reaction of the dry cell:

$$\mathbf{Zn(s) + 2\,NH_4^+(aq) + 2\,MnO_2(s) \rightarrow Zn^{++}(aq) + Mn_2O_3(s) + 2\,NH_3(aq) + H_2O(l)} \tag{4.30}$$

The open-circuit EMF of a zinc/carbon dry cell is about 1.55 V at room temperature. Note that from electrochemistry, the EMF will increase linearly with Kelvin temperature. By harnessing the electron flow in the redox reactions in the cell, it can supply electrical power: P = (1.55) I W. I is the current in coulombs/second.

Note that a Zn/C dry cell cannot be efficiently recharged. Other types of cells can be recharged by reversing the electron flow in the external circuit. Rechargeable cells include the well-known lead/acid (automobile) battery, solid-state lithium cells, and the Edison cell; they are called *secondary electrochemical cells.*

A convenient mnemonic acronym for redox reactions is *OIL RIG.* OIL stands for oxidation is loss *of electrons* (i.e., the oxidized matter loses electrons). RIG stands for reduction is gain *of electrons* (i.e., the reduced matter gains electrons). In an electrochemical cell, the electrons can be transferred in a copper wire between half-cells. In a biological redox system, free electrons can move through specialized organic carrier molecules such as *cytochrome* molecules.

Some organic redox reactions are different from ordinary electrochemical redox reactions because they do not actually involve the transfer of electrons in the electrochemical sense of the process; instead, protons can be involved [Sweeting 1997]. Two general redox scenarios are illustrated schematically in Figure 4.4. In the top diagram, compound **E** loses two electrons as it is oxidized. The electrons are passed directly to compound **F**, which is reduced. In the lower diagram, the redox reaction is coupled to and driven by the NAD^+, which is the oxidizing agent for compound $\mathbf{GH_2}$. The reduced form, $\mathbf{NADH + H^+}$, now acts to reduce **B** to $\mathbf{BH_2}$.

In a biochemical redox reaction, there is simultaneous oxidation and reduction of two, participating molecular species. An *oxidizer molecule* removes electrons from a second, *oxidized molecule.* The *oxidized molecule* is a *reducing agent* in the redox reaction; it gives up electrons and *reduces* the oxidizing molecule.

The second example of an organic redox reaction is the spontaneous oxidation of oxalic acid by permanganate ions in aqueous solution. The oxidation product $\mathbf{CO_2}$ leaves as a gas. The permanganate ions are reduced to divalent manganese ions.

$$\mathbf{2\,MnO_4^- + 5\,H_2C_2O_4 + 6\,H^+ \rightarrow 10\,CO_2\uparrow + 2Mn^{++} + 8\,H_2O} \tag{4.31}$$

Note that the net charge on both sides of the reaction is 4+.

Atoms, molecules, and ions that have a large affinity for electrons make good oxidizing agents. Fluorine gas, F_2, is the strongest common oxidizing agent, attacking stable materials such as glass and asbestos. Ozone (O_3), hydrogen peroxide (H_2O_2), chlorine (Cl_2), and oxygen (O_2) are all strong oxidizers. Recall that *reducing agents are electron donors.* Strong reducing agents include elemental K, Ba, Ca, and Na.

A third example of a complex, coupled redox reaction is the oxidation of the cofactor *nicotine adenine dinucleotide* (NADH) by the complex *electron transport chain* in a mitochondrion [King 2005d]. The *net reaction* can be written:

$$\mathbf{2\,NADH + O_2 + 2\,H^+ \rightarrow 2\,NAD^+ + 2\,H_2O} \tag{4.32}$$

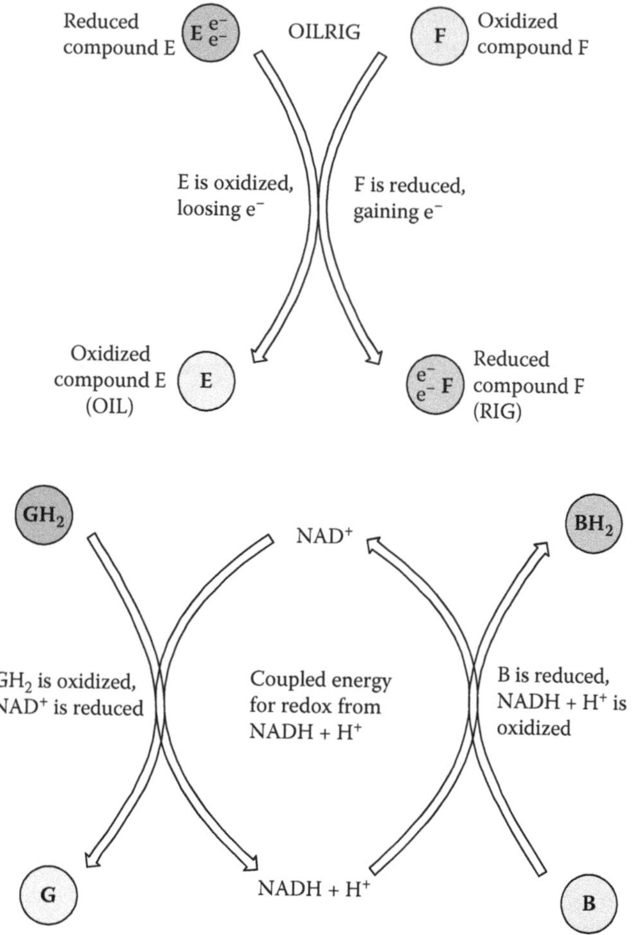

FIGURE 4.4 Two redox scenarios found in biochemical systems. See text for further description.

Because 2 NADH are oxidized, they each lose an electron; oxygen is the oxidizer (which is reduced), that is, two oxygen atoms each gain an electron so they can combine with four protons to form two water molecules.

A fourth example of an organic redox reaction is the biological conversion of *ethanol* to *acetaldehyde.* The reaction is catalyzed by the protein enzyme *alcohol dehydogenase.* (NAD^+ is the oxidized form of *nicotinamide adenine dinucleotide.* NADH is its reduced form.)

$$\mathbf{CH_3CH_2OH + NAD^+ \rightarrow CH_3CH = O + NADH + H^+} \tag{4.33}$$

This is an oxidation reaction that results in the removal of two electrons plus two protons from the ethanol and adding them to the NAD^+, converting it to the reduced form NADH + H^+. Note that atomic hydrogen can be considered to be an electron combined with a proton, sic:

$$\mathbf{H \Leftarrow \Rightarrow H^+ + e^-} \tag{4.34}$$

Note that as a result of the oxidation of any organic compound, there will be an increase in the number of bonds to oxygen or an increase in the number of double or triple C–C bonds.

A fifth example is the net redox reaction for aerobic respiration in mitochondria. This redox reaction can be written:

$$C_6H_{12}O_6 + 6\ O_2 \rightarrow 6\ CO_2 + 6\ H_2O, \quad \Delta G^0 = -686\ \text{Cal/mole} \tag{4.35}$$

The sugar is oxidized to CO_2; O_2 is the oxidant. The entire process in vivo is complex and involves many enzyme-mediated steps, some involving e^- transfer.

A sixth example of an overall redox reaction is the multistep, enzyme-mediated generation of methane gas by the anaerobic reduction of CO_2. This reaction occurs in certain bacterial species.

$$4\ H_2 + CO_2 \rightarrow CH_4\uparrow + 2H_2O, \quad \Delta G^0 = -8.3\ \text{Cal/mole} \tag{4.36}$$

H_2 is the reductant that transfers electrons to the oxidant, CO_2. On the other hand, if we burn methane in oxygen:

$$2\ O_2 + CH_4 \rightarrow CO_2 + 2\ H_2O, \quad \Delta G^0 = -193.5\ \text{Cal/mole} \tag{4.37}$$

Here, methane is the reductant that transfers electrons to oxygen, the oxidant, which is reduced.

As a seventh and final example of an organic redox reaction, consider the simple colorimetric test for ethanol. A driver suspected of drinking is asked to blow his or her breath into an acid solution of potassium dichromate. Alcohol vapor on the breath dissolves in the solution where the chromate ions oxidize it. The chromate ions are reduced to Cr^{+++}, which gives the solution a green color, providing a qualitative test for alcohol on the breath. The net reaction is

$$CH_3CH_2OH + 2\ Cr_2O_7^{2-} + 16\ H^+ \rightarrow 3\ CH_3COOH + \underset{\text{(green)}}{4\ Cr^{+++}} + 11\ H_2O \tag{4.38}$$

Analysis of this reaction is facilitated by writing it as two separate half-reactions. The sum of the half-reactions equals the total reaction (Equation 4.38).

$$3\ CH_3CH_2OH + 3\ H_2O \rightarrow 3\ CH_3COOH + 12\ H^+ + 12\ e^- \tag{4.39}$$

(Electrons are released by the oxidation of methanol.)

+

$$2\ Cr_2O_7^{2-} + 28\ H^+ + 12\ e^- \rightarrow 4\ Cr^{+++} + 14\ H_2O \tag{4.40}$$

(Electrons are gained by the reduction of bichromate.)

To summarize organic redox reactions, we note that *reducing agents* remove oxygen from another substance or give hydrogen to it. *A reducing agent is oxidized*, and it loses electrons (OIL). *Oxidizing agents* give oxygen to another substance or remove hydrogen from it; *an oxidizing agent is reduced*, and it gains electrons (RIG).

4.3.2 Review of Carbohydrate Metabolism

You have already been introduced to one aspect of *carbohydrate* (CHO) metabolism in the sections on glycolysis and the citric acid cycle earlier. In this section, we will summarize how dietary CHOs reach the cells and how liver cells store glucose in the form of the polymer *glycogen* and then release it when needed.

Cellulose is a CHO made by plant cells; it is a linear polymer of β-*(1→4)*-D-*glucopryanose* units in $^4C_1$ conformation. The cellulose polymer consists of from 2,000 to 14,000 residues. It forms natural crystals (cellulose I_α) in which intra- and intermolecular hydrogen bonds make it mechanically rigid and insoluble in water [Chaplin 2005]. Plants use this ubiquitous molecule for a structural material.

Humans cannot digest cellulose. The animals that do (e.g., ruminants, odd- and even-toed ungulates, beavers, termites, etc.) have colonies of specialized bacteria in their digestive systems that break cellulose down to water-soluble CHOs. Humans obtain dietary CHOs from sugars (mono- and disaccharides), starches and, of course, from internally stored *glycogen*. The sugars include the disaccharides sucrose (cane sugar) and lactose and the monosaccharides fructose and glucose. Glucose is the "entry energy molecule" for glycolysis, thence the CA cycle.

The first enzymatic step in the digestion of CHOs is the conversion of the higher polymers to simpler, soluble forms that can be transported across the walls of the intestinal epithelial cells and then eventually delivered to the circulatory system for distribution to body cells. The breakdown of polymeric CHOs begins in the mouth with the enzyme *ptyalin* (an *α-amylase*). This enzyme hydrolyzes starch into the disaccharide *maltose* and other small polymers of glucose containing three to nine glucose molecules, such as *maltotriose* and *α limit dextrins* that are the branch points of the starch polymer. About 3 to 5% of the starches eaten are hydrolyzed in the mouth before being swallowed. However, starch digestion continues in the stomach, although the salivary α-amylase enzyme is eventually inactivated by the stomach acid having a pH less than about 4. Perhaps 30 to 40% of the starches in the stomach are broken down before the stomach contents (*chyme*) pass into the small intestine.

There, pancreatic *α-amylase* completes starch hydrolysis into maltose and other small glucose polymers before the chyme has passed beyond the duodenum or upper jejunum. The microvilli brush border of the epithelial cells lining the small intestine contains the enzymes *lactase, sucrase, maltase*, and *α-dextrinase* whose roles are to split the disaccharides *lactose, sucrose*, and *maltose* as well as other small glucose polymers. Lactose splits into a molecule of galactose and a molecule of glucose. Sucrose breaks into molecules of fructose and glucose. Maltose splits into two glucose molecules. The resultant monosaccharides are transported across the epithelial cell walls and eventually turn up in the blood of the *hepatic portal vein* where they are transported to the liver parenchymal cells and, later, they enter the metabolic pathways of glycolysis, the citric acid cycle, or are converted into *glycogen,* fatty acids, amino acids, etc. Glucose and triglycerides are distributed to other cells in the body by the circulatory system. Figure 4.5 illustrates the steps that liver cells use to synthesize the glucose storage polymer, glycogen. Up to 10% of liver weight can be glycogen.

Two pancreatic endocrine hormones control the synthesis and breakdown of liver glycogen: *insulin* and *glucagon.* Insulin is secreted by the *pancreatic beta cells* in response to rising blood glucose levels. It activates the glucose transmembrane transport proteins in insulin sensitive cells, letting glucose diffuse into these cells at a higher rate. It also activates the important liver cell enzyme *glucokinase.*

Insulin exerts other hormonal effects beside its major one of facilitating the diffusion of extracellular glucose molecules into insulin-sensitive cells. It inhibits *gluconeogenesis;* it does this by decreasing the quantities and activity of the liver enzymes necessary for gluconeogenesis. It also acts to decrease the rate of release of amino acids from muscle and other nonliver tissues, thus reducing the available pool of precursor molecules used in gluconeogenesis, and promotes the conversion of excess intracellular glucose into fatty acids. As discussed in the following section, the fatty acids are converted to triglycerides in low-density lipoproteins, which are transported to adipose tissues where they are made into fat. A high portal insulin concentration, $[I]_p$, inhibits the enzyme *glucose phosphatase* in liver cells, an enzyme which removes the phosphate group from *glucose-6-phosphate* (G6P), converting it to glucose that can diffuse from the liver back to the blood. High $[I]_p$ also activates the liver enzyme *phosphorylase.* Phosphorylase makes G6P from glucose that has diffused into liver cells, thereby trapping the glucose inside the liver cells. G6P is enzymatically converted to *glucose-1-phosphate,* then to *uridine diphosphate glucose,* and then to the *glycogen* polymer. Thus, elevated blood insulin concentration increases the rate at which liver glycogen is formed and also inhibits its breakdown.

The hormone *glucagon* is made by the *pancreatic alpha cells* of the islets of Langerhans. Its release rate is stimulated by low blood glucose concentration ([BG]). Normal plasma concentration of glucagon is between 100 and 200 ng/L. One of the major effects of glucagon is to raise the rate at which glycogen is broken down, releasing glucose stored in the liver cells back into the blood

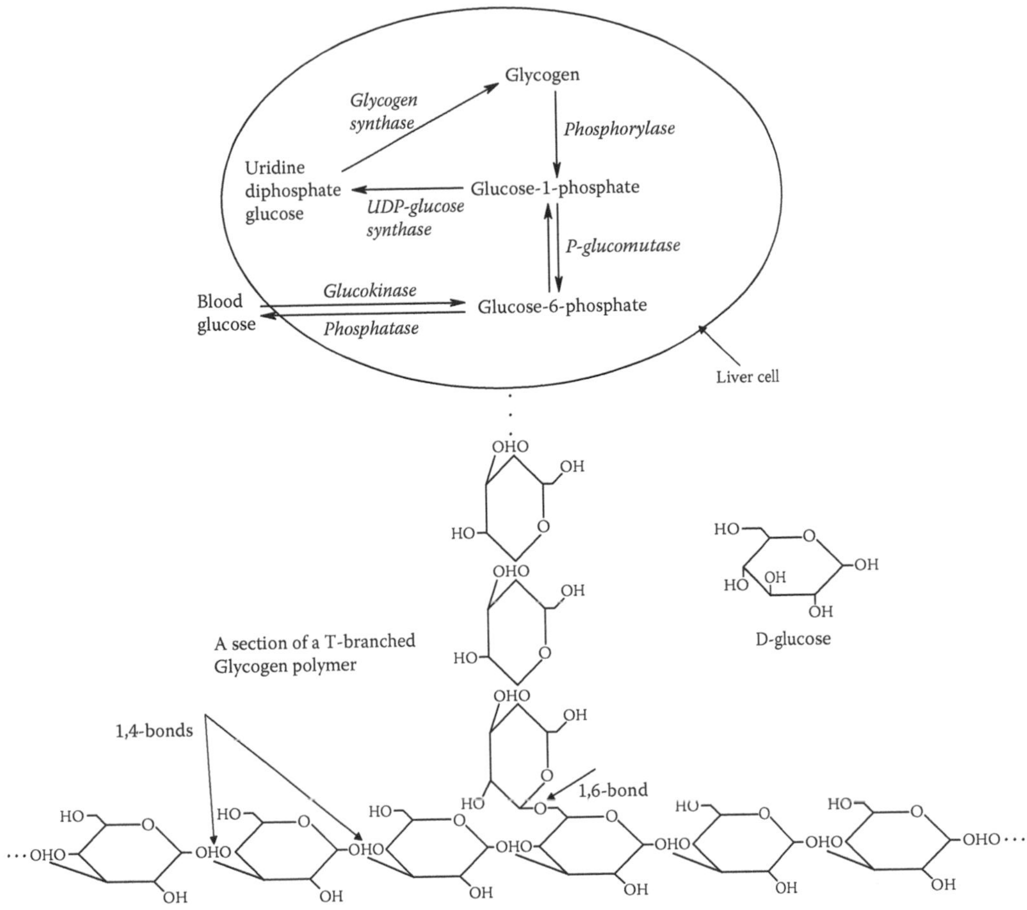

FIGURE 4.5 *Top:* metabolic pathways in liver cells in which the polymer glycogen is formed and broken down and released into the portal blood as D-glucose. *Bottom:* 2-D structure of a section of a branched glycogen molecule, showing the bonds linking the sugars.

and extracellular fluid (ECF). Glucagon also stimulates the process of gluconeogenesis, which also raises [BG] over the long term. After combining with glucagon receptor proteins on the hepatic cell membrane, the enzyme *adenyl cyclase* is activated. Adenyl cyclase catalyzes the formation of *cyclic AMP* that activates the following series of reactions: *protein kinase regulator protein* → *protein kinase* → *phosphorylase **b** kinase* → *phosphorylase **b*** → *phosphorylase **a***. Phosphorylase **a** is the active enzyme that cleaves a glycogen unit, forming *glucose-1-phosphate* (G1P). G1P is converted to G6P by another enzyme, and then *phosphatase,* activated by glucagon, removes the phosphate group and the resulting D-glucose diffuses out of the liver cell. Note that the outward flow of glucose occurs during periods of fasting as the blood glucose regulator attempts to maintain normoglycemia.

Gluconeogenesis is activated by high *glucagon concentration* ([GUN]). Several metabolic pathways are involved, and many enzymes are necessary for the multistep reactions. The details of these reactions are beyond the scope of this text. However, high [GUN] causes free amino acids in the blood to be taken up by liver cells where they are converted to glucose. High [GUN] also activates the enzyme that converts *pyruvate* to *phosphoenolpyruvate;* this conversion is a key step in gluconeogenesis. Note that the gluconeogenesis processes are catabolic. Under extreme conditions of starvation, the body, in order to maintain normoglycemia, first converts fats then proteins to glucose through the gluconeogenesis pathways.

4.3.3 Review of Lipid Metabolism

There are two sources of lipids in the body: those made in the body (endogenous) and those from diet (exogenous). Fat deposition in the body serves as a source of long-term, metabolic energy storage. In considering lipid metabolism, we will describe first how dietary fats enter the metabolism, how key lipids are synthesized in humans, and also how they are broken down for energy.

Over 40% of the daily energy requirements of humans in developed countries come from dietary fats (DFs). DFs are emulsified in the gut by *bile salts*. Bile salts are synthesized from cholesterol in the liver and stored in the gallbladder. In the small intestine, the bile salts combine with fats to form *micelles,* spherical globules about 3.5 nm in diameter. The sterol nuclei of 20 to 40 bile salt molecules combine around a fat globule; the hydrophilic polar groups of the bile salt molecules project outward and cover the outer surface of a micelle, making the micelle water soluble. The bile salt micelles act as a transport medium that carries the free fatty acids (FAs) and monoglycerides to the brush borders of the intestinal epithelial cells. There the FAs and monoglycerides are absorbed, and the bile salts are released back into the *chyme* for recycling as "ferry" molecules [Guyton 1991].

Most of the cholesterol in the diet is in the form of cholesterol esters (a combination of one molecule of free cholesterol and one molecule of a fatty acid). Both the cholesterol esters and the phospholipids are hydrolyzed by *lipases* from the pancreatic duct, for example, *cholesterol ester hydrolase* and *phospholipase A_2*. About 40% of the triglycerides combine with bile salts before absorption; 60% can be absorbed directly.

After entering the small intestine's epithelial cells, FAs and monoglycerides are taken up by the cells' *smooth endoplasmic reticulum* (SER). Here, some of these lipid molecules are resynthesized into new triglycerides that together with absorbed cholesterol, absorbed phospholipids, and small amounts of newly synthesized cholesterol and phospholipids are formed into globules with the hydrophobic fats facing the center and the polar portions on the surface. Small amounts of β-lipoprotein, synthesized in the SER, also are on the surface of each globule. The globules are excreted from the brush-border cells by *cellular exocytosis* into extracellular space, thence are taken up by the lymphatic system as chylomicrons. Eventually, the lymphatic chylomicrons enter the blood circulation, where they find their way to cells that use them in their metabolic processes.

Endogenous fatty acid (FA) synthesis occurs in cellular cytoplasm, whereas oxidative metabolism producing ATP occurs in mitochondria. Synthesis of fats involves the oxidation of NADPH and utilizes an activated two-carbon intermediate molecule, *acetyl-CoA*. Oxidation of fats involves the reduction of $FADH^+$ and NAD^+, and also *acetyl-CoA*. The synthesis of *malonyl-CoA* is the first committed step in fatty acid synthesis, shown below:

$$\underset{\text{bicarbonate}}{\mathbf{HCO_3^-}} + \underset{\text{acetyl CoA}}{\mathbf{H_3C{-}\overset{\displaystyle O}{\overset{\|}{C}}{-}SCoA}} \xrightarrow[\text{ATP} \rightarrow \text{ADP} + \text{P}_i]{ACC} \underset{\text{malonyl-CoA}}{\mathbf{{}^-OOC{-}CH_2{-}\overset{\displaystyle O}{\overset{\|}{C}}{-}SCoA}} \tag{4.41}$$

ACC is the protein enzyme *acetyl-CoA carboxylase.* The enzyme ACC in turn regulates the rate *malonyl-CoA* is synthesized, hence the rate of fatty acid synthesis. Both *acetyl-CoA* and *malonyl-CoA* are transferred to an *acyl carrier protein* (ACP) by the action of the enzymes *acetyl CoA transacylase* and *malonyl-CoA transacylase.* The attachment of these enzymes to ACP allows them to enter the *fatty acid synthesis cycle.*

All of the steps of fatty acid (FA) synthesis using the *acetyl CoA* and *malonyl-CoA* substrates are mediated by the homo-dimer enzyme *fatty acid synthase* (FAS). Four other enzymes are involved; two of the steps are reductions in which NADP is oxidized to $NADP^+$. Palmitic acid $[CH_3(CH_2)_{14}COOH]$ is the primary saturated FA synthesized by FAS. (A saturated FA has no double bonds between its carbon atoms.) After palmitate is released from the enzyme, it then undergoes many separate desaturation and/or elongation reactions to produce other FA molecules, generally in the membranes of the ER.

Desaturation occurring in the SER involves four broad specificity fatty acid *acyl-CoA desaturases* that induce unsaturation at C4, C5, C6, or C9. The electrons transferred from the oxidized FAs during desaturation are transferred from the desaturases to *cytochrome **b5*** and then to *NADH-cytochrome **b5** reductase*. These electrons are uncoupled from mitochondrial oxidative phosphorylation and thus do not figure in the synthesis of ATP. Because these enzymes cannot introduce sites of saturation beyond C9, they cannot synthesize either linoleic or linolenic acids. Thus, these unsaturated fatty acids must come from diet; hence, they are called *essential fatty acids*. Linoleic acid is a substrate for the synthesis of *arachidonic acid,* a precursor for the *eicosanoids,* a family of FAs critical to the inflammatory response and immune system function [King 2006a].

Triglycerides (aka *triacylglycerols*) are molecules of glycerol to which three FAs have been esterified. The FAs in triacylglycerols are mostly saturated. The synthetic pathway for triacylglycerols begins with *dihydroxyacetone phosphate* (DHAP) and involves five enzyme-mediated steps. The long-chain FAs that are attached to glycerol are generally *stearic acid,* oleic *acid*, or *palmitic acid.* Stearic and palmitic acid are fully saturated, whereas oleic acid has one double bond in the middle of its 18-carbon chain. Triglycerides are broken down for their metabolic energy. They also play a role in the development of heart disease and stroke; that is why physicians routinely test triglyceride levels.

In addition to triglycerides, the body also makes and uses *phospholipids* (PLs). They are synthesized by the esterification of an alcohol to the phosphate of *phosphatidic acid.* King [2006] describes six phosphatidyl PLs.

The general form of *phosphatidyl glycerol* (PG) is shown below:

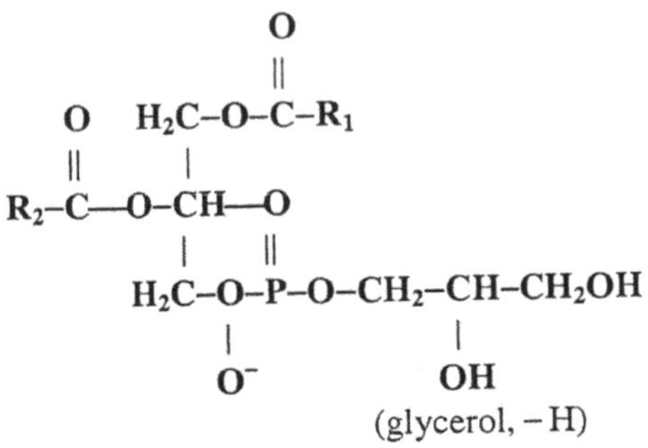

```
                  O
                  ||
      O  H2C–O–C–R1
      ||     |
R2–C—O–CH—O
             |    ||
         H2C–O–P–O–CH2–CH–CH2OH
                  |                |
                  O⁻              OH
                             (glycerol, – H)
```

PG is synthesized from *CDP-diacylglycerol* and *glycerol 3-phosphate.* PG molecules are found in high concentration in mitochondrial membranes and as components of pulmonary surfactant.

Another class of phospholipid is *phosphatidylcholine* (PC) (see Chapter 3, Figure 3.3A). PCs are also called lecithins. The lecithin *dipalmitoyl* is also a component of pulmonary surfactant. It contains *palmitate* at both carbon 1 and 2 of glycerol and is the major phospholipid found in the extracellular lipid layer lining the pulmonary alveoli. The general form of PC is given as follows:

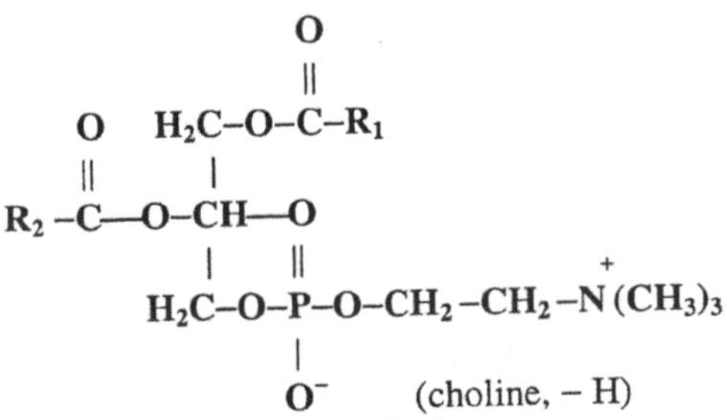

```
                  O
                  ||
      O  H2C–O–C–R1
      ||     |
R2 –C—O–CH—O
             |    ||                +
         H2C–O–P–O–CH2–CH2–N(CH3)3
                  |
                  O⁻     (choline, – H)
          (Phosphatidylcholine)
```

$H_3C\text{-}(CH_2)_{14}\text{-}COOH$

(Palmitic acid)

In addition to the lipids described earlier, there are many other physiologically important classes of lipids. These include, but are not limited to:

- *Plasmalogens:* These are *glycerol ether phospholipids*. One particular choline alkyl ether plasmalogen is *platelet activating factor* (PAF), effective at concentrations as low as 10^{-11} M. PAF functions as a mediator of acute inflammatory reactions, immune hypersensitivity, anaphylactic shock, platelet aggregation, and the release of serotonin from platelets.
- *Sphingolipids:* The core of sphingolipids is the long-chain amino alcohol *sphingosine.* They are composed of a polar head group and two nonpolar tail molecules. Sphingolipids are components of all membranes and, as *sphingomyelins,* are particularly abundant in the myelin sheaths that protect neurons (see Figures 3.1 and 3.3).
- *Cerebrosides:* They have a single sugar group (e.g., galactose) linked to *ceramide.* Galactocerebrosides are found predominantly in neuron cell membranes.
- *Sulfatides:* They are sulfuric acid esters of galactocerebrosides.
- *Globosides:* They are *cerebrosides* that contain additional carbohydrates.
- *Gangliosides:* Similar to globosides except they contain NANA (aka *sialic acid*) in varying amounts.
- *Eicosanoids:* A physiologically important class of lipid that includes the *prostaglandins* (PGs), *thromboxanes* (TXs), and *leukotrienes* (LTs), active in the immune system. Eicosanoids have 20 carbon atoms; the structures of most eicosanoids are derived from the C20 fatty acid, *arachidonic acid.* The eicosanoids have a wide spectrum of effects on inflammatory responses affecting the eyes, joints, and skin and on the duration and intensity of fever and pain. They also affect blood pressure through vasodilation or constriction of vessels and inhibit or activate platelet aggregation and thrombosis. All mammalian cells (except erythrocytes) synthesize eicosanoids, which act on tissues locally around the site of synthesis [King 2006a].

4.3.4 Review of Lipid Catabolism

We have outlined above how lipids are synthesized in the body and some of their physiological roles in the preceding text. Now we will address lipid catabolism. In times of starvation, the body can mobilize fat stores, break these molecules down, and oxidize them for biochemical energy (ATP, NADH, etc.).

There are *two reasons* that fats are more efficient forms of energy storage: First, they are more highly reduced than carbohydrates (CHOs); that is, they have more C-H bonds in which the electrons are supplied by the carbon, rather than C-O bonds in which the electrons come from the oxygen. Hence, during fat catabolism, there are more electrons transferred to O_2 per gram of fat than per gram of CHO. *Thus, more free energy is released for lipids per gram than for CHOs.* Secondly, because fats are highly insoluble in water, there is no water of hydration associated with stored fat, unlike for glycogen, for example. *A gram of hydrated fat yields more than six times the energy than does a gram of hydrated CHO.* For example, a 70-kg man with 11% body fat has about 10^5 Cal of energy stored as fat but only about 600 Cal stored in glycogen.

Triglycerides stored in *adipocytes* are mobilized for catabolism by hydrolyzation by the enzyme *hormone-sensitive lipase. Epinephrine* and *norepinephrine* trigger this process. The hydrolyzed fatty acids (FAs) diffuse out of the fat cells into the blood where they are bound to albumins and are carried to the target cells where the lipids are to be catabolized. Once inside the target cells, the FAs are converted to *fatty acyl-CoA* and activated by members of the *fatty acyl-CoA synthetase* enzyme family. The activated fatty acyl-CoA is transported to the mitochondria where fat catabolism takes place. A molecule of *fatty acyl-CoA* has the structure **R-CO-S-CoA**. **R** is the fatty acid chain. The enzymes *carnitine acyltransferase I* and *II* mediate the carnitine cycle whereby the fatty acyl-CoA is transported into the *matrix* of the mitochondria [Hackert 2001].

The catabolism of fatty acyl-CoA to *acetyl-CoA* occurs in the mitochondrial matrix in a series of *beta-oxidation* steps. The coenzymes FAD and NAD+ drive these reactions. The final step in beta-oxidation involves the action of *thiolase* that catalyzes the cleavage of the *fatty acyl-CoA* into another *fatty acyl-CoA* whose acyl chain is shortened by two methylene groups plus the desired *acetyl-CoA*. Note that this process must be repeated six times for a 16-carbon fully saturated FA (e.g., palmitate). The yield is eight *acetyl-CoA* molecules that can directly enter the citric acid (CA) (Krebs) cycle of oxidative metabolism. It can be shown that the CA cycle oxidation of the acetyl-CoA molecules from a 16-carbon fatty acyl-CoA yields 129 ATP molecules. Other FAs having other states of saturation yield different numbers of ATPs. For example, oxidation of a fully saturated 18-carbon FA yields 146 ATPs; for an 18-carbon FA with one double bond between two carbons, oxidation yields 144 ATPs because one fewer $FADH_2$ is generated. 141 ATPs are produced by oxidizing an 18-carbon FA with two double bonds. Beta oxidation of FAs with an odd number of carbons ends up producing a three-carbon end product, *proprionyl-CoA*. It enters the CA cycle by acquiring an additional carbon to yield *succinyl-CoA*.

Fat metabolism is seen to be dependent on the CA cycle to process the acetyl-CoA produced by beta oxidation. Without a balanced diet containing CHOs, FA metabolic activity can be slowed.

4.3.5 Review of Amino Acid Metabolism

In this section, we will describe the synthesis and catabolism of individual AAs. (The role of AAs in the formation of proteins is described in Chapter 7.) There are 20 AAs involved in protein synthesis in the human body. Half of them are considered *nonessential*, that is, they can be synthesized by cells in adequate quantities to meet the body's needs. The remainder is called *essential*. They must come from diet because the body cannot synthesize them in adequate quantities to meet metabolic needs. *The nonessential AAs include* alanine (**Ala**), asparagine (**Asn**) aspartic acid (aspartate **Asp**), cysteine (**Cys**), glutamic acid (glutamate **Glu**), glutamine (**Gln**), glycine (**Gly**), proline (**Pro**), serine (**Ser**), and tyrosine (**Tyr**).

The 10 essential AAs are arginine (**Arg**), histidine (**His**), isoleucine (**Ile**), leucine (**Leu**), lysine (**Lys**), methionine (**Met**), phenylalanine (**Phe**), threonine (**Thr**), tryptophan (**Trp**), and valine (**Val**). Figure 4.6 illustrates the planar structures of the 20 major AAs. The AA carboxyl group is shown in the nonionized form.

The AAs *glutamic acid* and *aspartic acid* are synthesized from their ubiquitous α-keto acid precursors (*α-ketoglutarate, oxaloacetate,* respectively) by a simple one-step, reversible transamination reaction mediated by the enzymes *glutamate dehydrogenase* and *aspartate dehydrogenase,* respectively. NADPH + H^+ is used to drive the forward reaction; NAD^+ drives the reverse reaction (e.g., **glutamate → α-ketoglutarate +** $\mathbf{NH_4^+}$). **Asp** can also be derived from the AA asparagine through the action of the enzyme *asparaginase*.

Muscle cell metabolism of D-glucose produces *pyruvate;* it is transmogrified to the AA *alanine* by the action of the enzyme *alanine transaminase* (ALT). The excess Ala enters the circulation and finds its way back to liver cells. Entering hepatocytes, Ala is transformed back to pyruvate that then can be made into D-glucose by *gluconeogenesis*. The glucose, in turn, can be carried by the blood back to myocytes where the cycle is completed.

The sulfur-containing, essential AA *methionine* is the starting point for the three-step biosynthesis of the sulfur-containing AA *cysteine*. First, ATP and the enzyme *methionine adenosyltransferase* react to form the important substrate molecule *S-adenosylmethionine* (SAM). SAM is changed to *homocysteine* that then combines with the AA *serine* to form *cystathionine*. Finally, the *cystathionine* is broken down into the AA *cysteine* plus a molecule of *α-ketobutyrate*.

Tyrosine is produced in cells by hydroxylating the essential AA *phenylalanine*. The enzyme *phenylalanine hydroxylase* and the coenzyme *H_4-biopterin* are necessary. The reaction produces *tyrosine* plus *H_2-biopterin*. *H_2-biopterin* is then recycled by being reduced by NADH + H^+ to *H_4-biopterin* plus NAD^+.

The AAs *ornithine* and *proline* are biosynthesised from a *glutamic acid* precursor. While not used in protein synthesis, *ornithine* participates in the urea cycle.

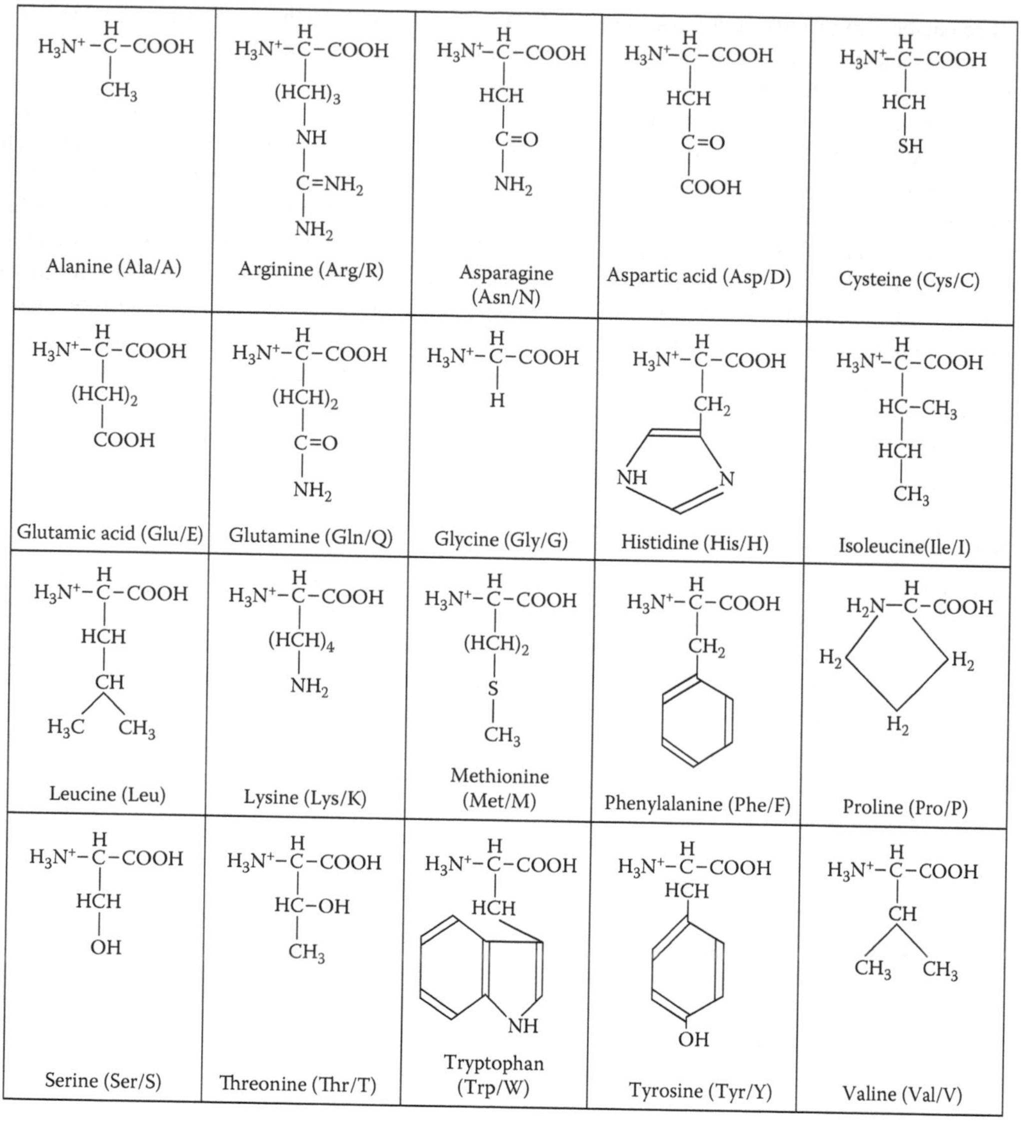

H H_3N^+–C–COOH CH_3 Alanine (Ala/A)	H H_3N^+–C–COOH $(HCH)_3$ NH C=NH_2 NH_2 Arginine (Arg/R)	H H_3N^+–C–COOH HCH C=O NH_2 Asparagine (Asn/N)	H H_3N^+–C–COOH HCH C=O COOH Aspartic acid (Asp/D)	H H_3N^+–C–COOH HCH SH Cysteine (Cys/C)
H H_3N^+–C–COOH $(HCH)_2$ COOH Glutamic acid (Glu/E)	H H_3N^+–C–COOH $(HCH)_2$ C=O NH_2 Glutamine (Gln/Q)	H H_3N^+–C–COOH H Glycine (Gly/G)	H H_3N^+–C–COOH CH_2 NH N Histidine (His/H)	H H_3N^+–C–COOH HC–CH_3 HCH CH_3 Isoleucine(Ile/I)
H H_3N^+–C–COOH HCH CH H_3C CH_3 Leucine (Leu)	H H_3N^+–C–COOH $(HCH)_4$ NH_2 Lysine (Lys/K)	H H_3N^+–C–COOH $(HCH)_2$ S CH_3 Methionine (Met/M)	H H_3N^+–C–COOH CH_2 Phenylalanine (Phe/F)	H H_2N–C–COOH H_2 H_2 H_2 Proline (Pro/P)
H H_3N^+–C–COOH HCH OH Serine (Ser/S)	H H_3N^+–C–COOH HC–OH CH_3 Threonine (Thr/T)	H H_3N^+–C–COOH HCH NH Tryptophan (Trp/W)	H H_3N^+–C–COOH HCH OH Tyrosine (Tyr/Y)	H H_3N^+–C–COOH CH CH_3 CH_3 Valine (Val/V)

FIGURE 4.6 The 20 amino acids used in assembling peptides (and proteins) in man. Both three-letter abbreviations and the corresponding one-letter AA codes are given.

The biosynthesis of *serine* begins with the glycolysis intermediate *3-phosphoglycerate*. Several enzyme mediated steps are involved.

The AA *glycine* is made from *serine* in a one-step reaction catalyzed by *serine hydroxymethyltransferase*. Glycine is important because, in addition to protein synthesis, it is a substrate for the biosynthesis of *purine nucleotides, creatine, serine, glutathione,* and *heme.*

4.3.6 Review of Amino Acid Catabolism

Amino acids undergo chemical breakdown or *catabolism* as part of the natural order of things. Their breakdown products, in general, lead to molecules that can be used by cells for metabolic energy (e.g., glycolysis or the citric acid cycle) to make ATP and other high-energy molecules. We have previously cited the fate of the AA *alanine* in liver cells. It is deaminated and changed to *pyruvate*, which is recycled back into D-*glucose.*

The kidney tubules contain the enzyme *glutaminase,* which converts the AA *glutamine* to *glutamic acid (glutamate)* and NH_3^+ that is excreted in the urine. The *glutamic acid* is then converted to *α-ketoglutarate,* a substrate in the CA cycle. Another enzyme *asparaginase* is found throughout the body's cells; it deaminates AA *asparagine* to form ammonia and the AA aspartate. *Aspartate* is transaminated into *oxaloacetate,* which participates in the CA cycle.

The nitrogen-rich AA *arginine* is hydrolyzed into *urea* and *ornithine* by the enzyme *arginase. Ornithine* is transaminated to form *glutamate semialdehyde,* which can be converted to *glutamate* or serve as a precursor for the AA *proline* biosynthesis.

The catabolism of *proline* is basically a reversal of its synthesis. The *glutamate semialdehyde* resulting from *arginine* and *proline* breakdown is oxidized to glutamate that then can be converted to *α-ketoglutarate,* and thus back to the CA cycle.

Serine is converted to *glycine,* and then *glycine* is oxidized to $CO_2 + NH_3$. *Serine* can also be enzymatically converted to pyruvate by deamination.

According to King [2006b], there are at least three pathways whereby *threonine* can be catabolized: (1) *Threonine dehydrogenase* converts Thr to *α-amino-β-ketobutyrate,* which is either converted to *acetyl-CoA + glycine* or spontaneously degrades to *aminoacetone* and is finally converted to *pyruvate.* (2) Pathway 2 involves *serine/threonine dehydrogenase* yielding *α-ketobutyrate* that is broken down to *propionyl-CoA* and, finally, into the CA cycle as *succynyl-CoA.* (3) *Threonine aldolase* is utilized in the third pathway. Both *acetyl-CoA* and *pyruvate* are produced.

The small AA *glycine* is catabolized by conversion to *serine* by the enzyme *serine hydroxymethyl transferase.* The *serine* can be converted to *3-phosphoglycerate* or to *pyruvate* by *serine/threonine dehyratase.*

The catabolism of the sulfur-containing AAs, *cysteine* and *methionine,* is also complex and has alternate pathways. A minor pathway for *cysteine* catabolism is in liver cells where a *desulfurase* produces H_2S and the useful *pyruvate.* The major mechanism for *cysteine* breakdown is by *cysteine deoxygenase* leading into many complex steps to end products such as *taurine,* a bile salt precursor, *thiocysteine, sulfite, sulfate,* and *pyruvate. Methionine* is catabolized into *cysteine* and *α-ketobutyrate,* which is in turn transmogrified into *succinyl-CoA.*

The branched-chain AAs (BCAAs), *valine, leucine,* and *isoleucine,* use the same enzymes for the first stages of their catabolism, which begins in muscle cells. The first step for all three AAs is transamination by the enzyme *BCAA aminotransferase.* α-ketoglutarate is the amino receptor producing three different *α-keto acids.* These acids are oxidized using a common *BC α-keto acid dehydrogenase;* three different CoA derivatives are produced. The product from valine is *propionyl-CoA,* a precursor for *succinyl-CoA.* Isoleucine eventually is broken down to the ubiquitous *acetyl-CoA* as well as *proprionyl-CoA,* and *leucine* ends up as *acetoacetyl-CoA* and *acetyl-CoA.*

Phenylalanine is changed into *tyrosine* by *phenylalanine hydroxylase. Tyrosine* is a substrate for the synthesis of the hormones, *thyroxine, 3,5,3'-triiodotyrosine, dopamine, norepinephrine,* and *epinephrine. Tyrosine* can also be broken down to *fumarate* and *acetoacetate.*

Lysine catabolism is complex; its ultimate end product is *acetoacetyl-CoA. Lysine* also is a substrate for eventual *carnitine* synthesis. *Carnitine* is required for the transport of fatty acids into the mitochondria for oxidation.

The AA *histidine* is a substrate for the allergy substance *histamine.* The first step in *histidine* catabolism is catalyzed by *histidase,* which removes the α-amino group and introduces a double bond in the ring forming *urocanate.* The end product of histidine breakdown is *glutamate.*

Tryptophan is ultimately catabolized to *acetoacetate.* It participates in a number of important side reactions; it is a substrate for the biosynthesis of the hormones *serotonin (5-hydroxytryptamine)* and *melatonin.* Molecules of *tryptophan* also can form the AA *alanine* (see the previous text) and, in the liver, can form *nicotinic acid,* which leads to limited amounts of NAD^+ and $NADP^+$.

As this section shows, biochemistry is incredibly complex. Just examining a single category of molecules such as amino acids illustrates the vast network of interconnected causal pathways between their anabolism and catabolism and shows that every step is controlled by an enzyme that, in turn, must be regulated so that everything "stays in balance." (For perspective on this complexity, the interested reader is encouraged to obtain the two Biochemical Pathways Charts from *Roche Applied Science* [see Bibliography and References section of this text] and study them.)

4.4 PHOTOSYNTHESIS

4.4.1 Introduction

In this section, we will summarize the complex chemical pathways found in photosynthesis and examine its energy relationships.

Certainly, photosynthesis is one of the miracles of life. It has been called the most important chemical pathway, period, and not just for plants but for the entire planetary ecosystem. Plants, green algae, and cyanobacteria utilize the energy of sunlight plus water and carbon dioxide molecules to manufacture carbohydrates, ATP, and molecular oxygen (O_2); these *producers* are also called *autotrophs*. Autotrophs permit aerobic, *consumer* or *heterotrophic*, life to exist. Heterotrophic life-forms (such as humans) metabolize sugars and carbohydrates using oxygen, and make CO_2 and nitrogenous waste that plants use, closing the cycle.

4.4.2 The Pigments of Photosynthesis

Photosynthesis uses the photon energy of light to drive its reactions [Kimball 2007b]. The photon energy is trapped by a large variety of photosynthetic pigments, depending on the species of autotroph. The pigments are generally bound in orderly molecular arrays to protein molecules in the *thylakoids*, which orient them optimally to capture photon energy. This local grouping of pigment and protein molecules in a chloroplast is called a *reaction center.* A large number of pigment molecules (100–5,000) held together is called an *antenna,* in analogy to a radio antenna that traps electromagnetic radiation of far-longer wavelengths than visible light photons. The *antenna pigment* molecules are predominantly *chlorophyll* **a**, *chlorophyll* **b**, and various *carotenoids*. The photon energy absorbed is shuttled by molecular resonance energy transfer to two specialized *chlorophyll* **a** molecules at the reaction center of each photosystem. When either of the two *chlorophyll* **a** molecules (aka. P_{700}) absorbs this resonance energy, an electron is produced and is transferred to an acceptor molecule, leaving a *hole* (in a semiconductor sense) in the donor *chlorophyll* **a**. Somehow, electrons from water participate in an oxidation reaction in which the hole is filled, and diatomic oxygen is produced. P_{700} chlorophyll is common to all eukaryotic autotrophs because of its essential role in the reaction center in photosynthesis. The accessory pigments such as *chlorophyll* **b** and carotenoids are not essential. Brown algae and certain diatoms use *chlorophyll* **c** as a substitute for *chlorophyll* **b** [Markwell & Namuth 2003].

Other photosynthetic bacteria use other variants on *chlorophyll* **b***:* For example, purple and green bacteria use *bacteriochlorophyll* that absorbs IR light between 800 and 1,000 nm. These bacteria do not produce O_2 but do *anaerobic photosynthesis. Green sulfur bacteria* have *chlorobium chlorophyll. Red algae* have additional photon-trapping pigments called *phycobilins.* Phycobilins appear red or blue because they do not absorb light in these wavelengths. Figures 4.7 and 4.8 illustrate the 2-D molecular structures of *chlorophylls* **a**, **b**, **c1**, **c2**, and **d.** Note that chlorophyll includes a structure reminiscent of a heme ring with Mg in the center instead of Fe. The small circles at the corners of the bond lines represent tetravalent carbon atoms.

Chlorophyll **a** is found in all chloroplasts; chlorophyll **b** occurs mostly in the chloroplasts of land plants; chlorophyll **c1** and **c2** are found in various algae; and **d** is unique to cyanobacteria [Eggink et al. 2001].

Figure 4.9 illustrates the primary molecular structures of the α and β carotene pigments and *xanthophyll* that differs only slightly from β-carotene. Carotenes are chemically classified as *terpenes;* β-carotene is also the dimer of *retinol* (vitamin A). There are also γ, δ, and ε carotenes in the

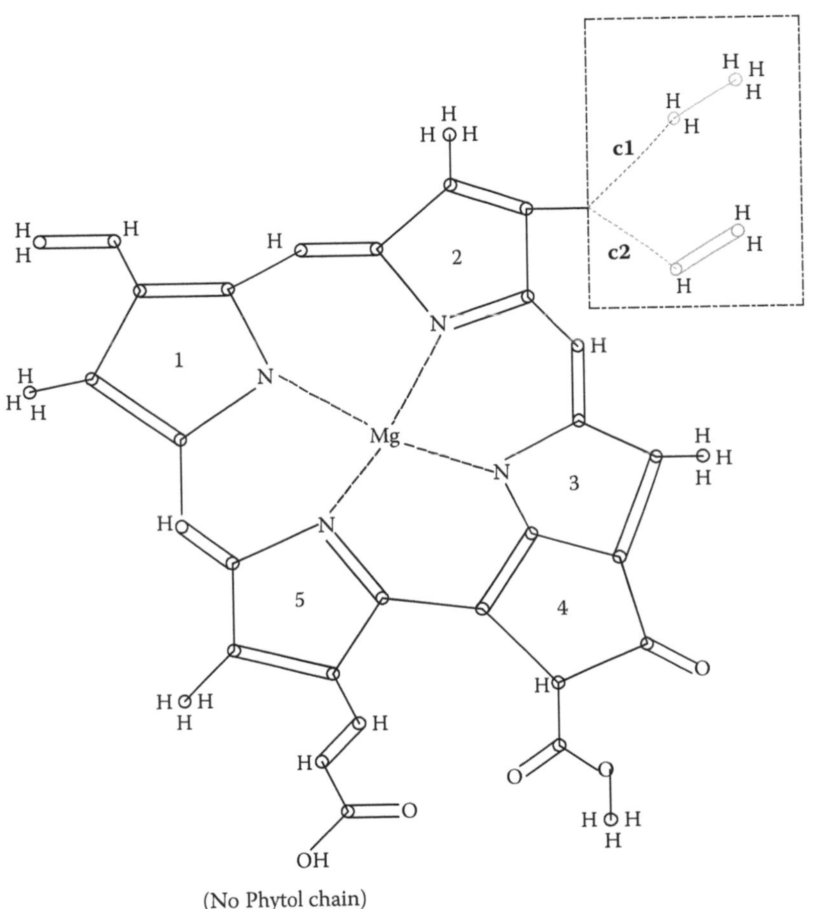

FIGURE 4.7 A chlorophyll molecule showing the **c1** and **c2** structures. Magnesium, rather than iron, is the metal in the center. There is no phytol chain.

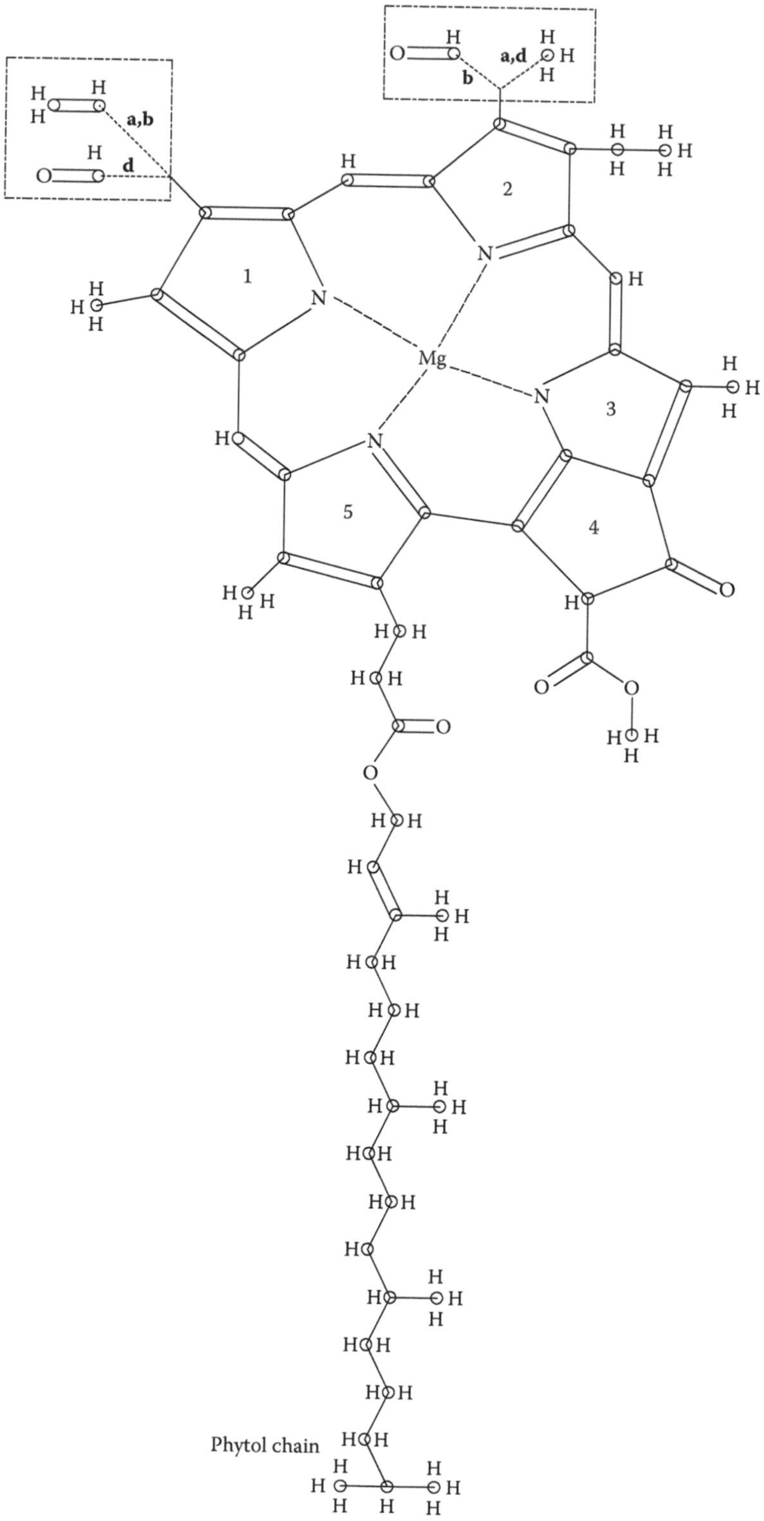

FIGURE 4.8 2-D structure of the chlorophyll **a, b,** and **d** molecules. The phytol chain is attached to the 5-ring in these molecules. (The small open circles are carbon atoms.)

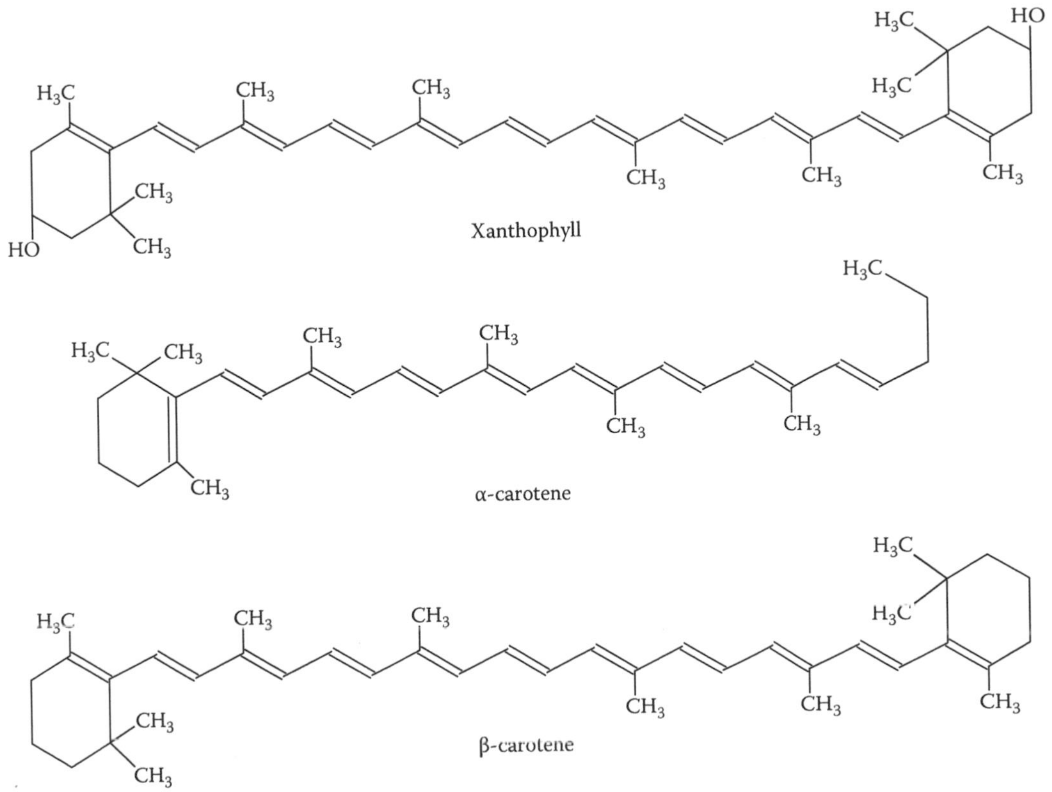

FIGURE 4.9 The photopigments *xanthophyll, α-carotene,* and *β-carotene.*

family. Carotenes are unoxidized carotenoids. Carotenoids in which some of the double bonds have been oxidized to single bonds are known generally as *xanthophylls.*

The most common form of photosynthesis is carried out by higher plants and green algae. Photon energy from sunlight is trapped by certain photopigments (*chlorophylls* and carotenoids) and, through a complex chain of reactions, *reduces* CO_2 gas to carbohydrates and sugars. Electrons from this reduction ultimately come from water, which is converted to O_2 gas and protons (H^+). (We will describe the redox reactions of photosynthesis in more detail in the following sections.) Chlorophylls absorb light maximally in the blue and red wavelengths, and carotenoids absorb blue-green light. The green and yellow light not absorbed is either reflected or transmitted, giving plant leaves a general green hue.

UV light with wavelengths below 400 nm is not very effective at producing photosynthesis, and wavelengths below 330 nm can damage all types of cells by creating free radicals that damage nucleic acids and proteins.

4.4.3 Structure of the Chloroplast

Photosynthesis takes place in chloroplasts. Figure 4.10 illustrates a highly schematic cutaway of a typical chloroplast organelle. A "typical" chloroplast is a flattened ovoid in shape, about 5.5 μm in length and about 1.2 μm thick in the direction of grana stacks. It is covered with a double unit membrane about 5 nm thick. Inside are stacks of hollow *thylakoid disks;* each hollow disk is about 15 nm thick and about 500 nm in diameter. There are 2–2.5 nm between the outsides of adjacent disks in a stack [Heslop-Harrison 1963]. The insides and outsides of the thylakoid membranes are covered with protein molecules [Kaftan et al. 2002]. A stack of thylakoid disks is called a *granum.* The photosynthetic molecules are located inside each thylakoid disk. The *stroma* is the fluid volume

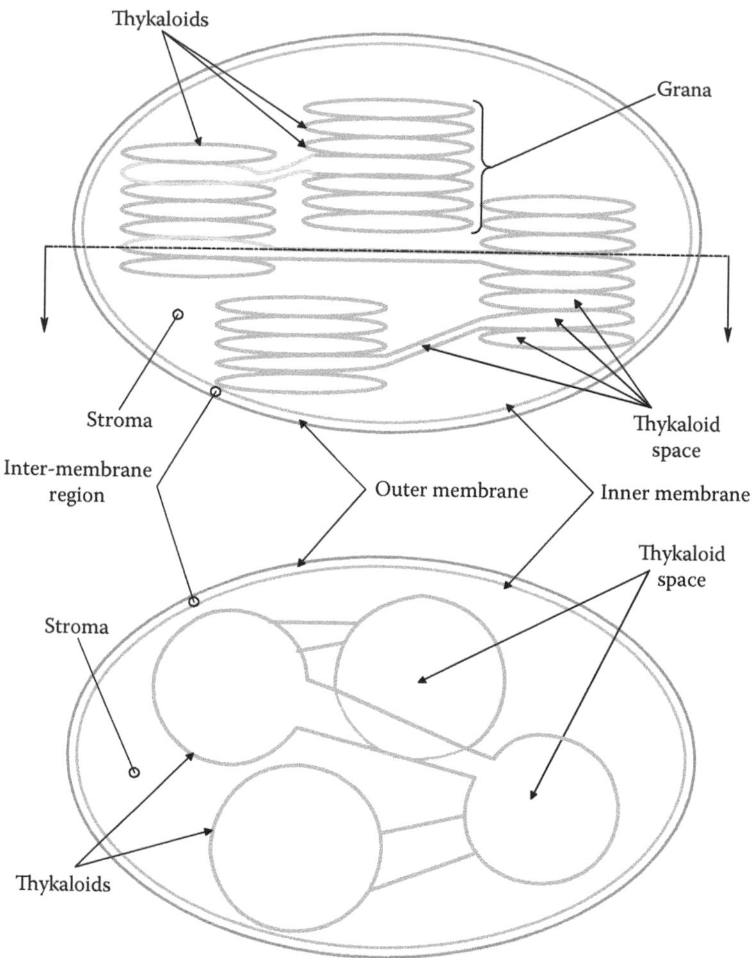

FIGURE 4.10 Simplified schematic drawing of a chloroplast. The *grana* are stacks of hollow *thylakoid disks,* in whose membranes are located the photopigments and associated molecules such as *plastoquinone* and *plastocyanin.* The interiors of the thylakoids are interconnected and form a *thylakoid space* separated by membranes from the *stroma* volume inside the chloroplast outer membrane. A chloroplast is a topologically complex structure.

inside the chloroplast double membrane but outside the thylakoid disks. All chlorophyll is located in the thylakoids and is associated with photosystem I, II, or with *antenna proteins* supplying chemical energy into these photosystems.

Various thylakoid disks in adjacent grana are connected by thin, hollow thylakoid membrane tubules that are attached to multiple lamellae in the stroma, which surround the thicker thylakoid disks. These tubules and lamellae provide a continuous thylakoid (interior) volume between disks.

Four types of molecules are embedded in the thylakoid membranes: *Photosystem I molecules* consisting of chlorophyll and carotenoid molecules. *Photosystem II molecules* contain other chlorophylls and carotenoids. Also embedded are *cytochromes* **b** and **f** and the enzyme *ATP synthase.* Figure 4.11 illustrates photon trapping by Photosystem I and II molecules embedded in the thylakoid membrane and the fate of pumped protons (**H**$^+$) (i.e., chemiosmosis) in which they drive the enzymatic production of ATP.

The *stroma* contains all the enzymes needed to do the *dark reaction* (Calvin cycle) of photosynthesis, that is, the conversion of CO_2 into carbohydrates and other molecules (see Section 4.4.5). Also in the stroma are multiple copies of the chloroplast's DNA genome. Similar to mitochondria, chloroplasts contain a *partial* genome for their reproduction and function. The chloroplast genome

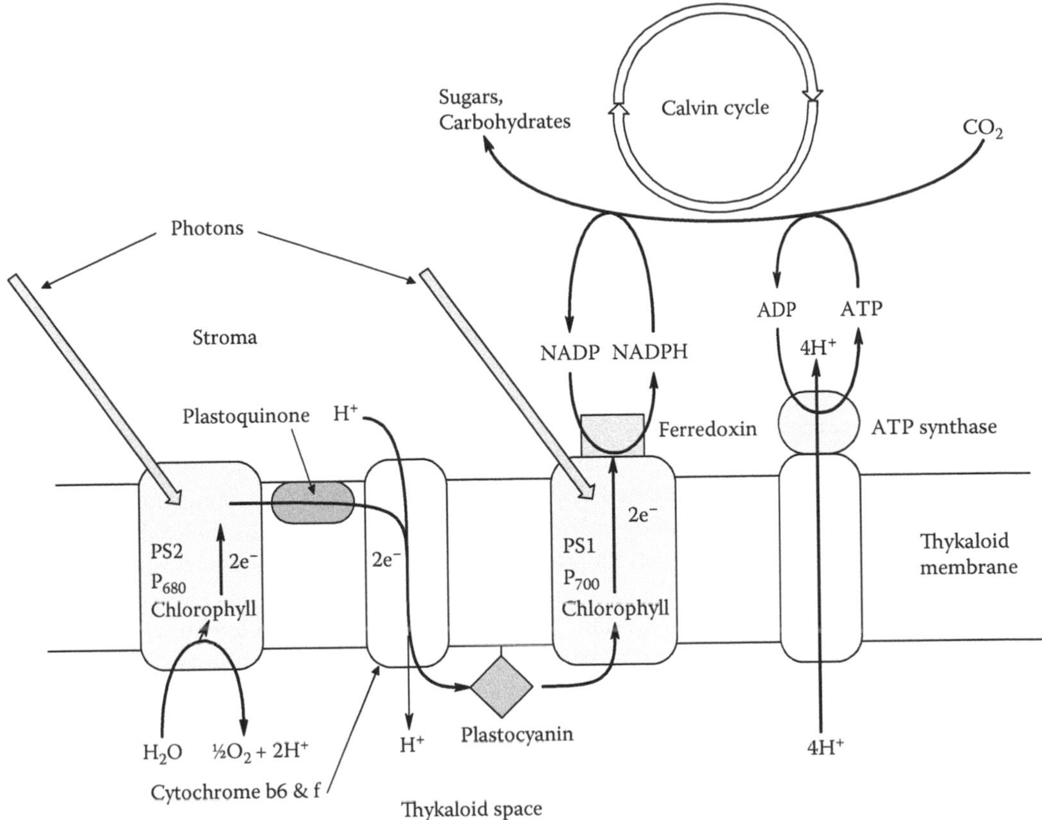

FIGURE 4.11 (See color insert following page 44.) Schematic showing the details of photon capture by Photosystem I and II molecules embedded in the thylakoid membranes of chloroplasts; also the transfer of electrons used to make NADPH. Both NADPH and ATP are used in the Calvin cycle. See text for a further description of photosynthesis. **H**$^+$ pumping and the use of high **H**$^+$ concentration to drive the generation of ATP by chemiosmosis is also shown.

is a circular plasmid of superhelical dsDNA having two inverted repeats separating one small single copy and one large single copy. The structure is highly conserved in all plants. There are between 120 and 160 kilobases. There are about 100 protein-coding genes, 30 tRNA genes, and 4 rRNA genes. Most chloroplast proteins are coded for by the *nuclear genome,* synthesized in the plant cell's cytoplasm, and transported to the chloroplasts [Schlindwein 2006].

It is generally suspected that mitochondria and chloroplasts at one ancient time were free-living organisms and somehow became internal, commensal, then symbiotic organelles in heterotrophic and autotrophic cells, respectively. The fact that mitochondria and chloroplasts have their own (partial) genomes, and rely on many more genes stored in the cell's nuclear genome, deepens the mystery of their evolution as indispensable cytoplasmic organelles.

4.4.4 Chemical Reactions and Energy in Photosynthesis

The light-induced chemical reactions that occur in chloroplasts are totally dependent on the structure and arrangement of the participating macromolecules. This is a clear example of function following form in biology.

The overall chemical equation for the photosynthesis of D-glucose is

$$\mathbf{6\,CO_2 + 6\,H_2O + Photon\ Energy \rightarrow C_6H_{12}O_6 + 6\,O_2} \tag{4.42}$$

In order for this reaction to happen, 24 electrons must be removed from water where they have been held by the oxygen redox potential (0.82 eV) and pumped "uphill" to carbon atoms that they partially reduce to carbohydrate with a redox potential of –0.42 eV. The potential difference, $\Delta E = 0.82 - (-0.42) = 1.24$ V. Allowing 24 moles of electrons to pass through this potential gives us a Gibbs free energy yield of $\Delta G = 24 \times 23.062 \times 1.24 = 686$ kcal (Cal). This is the photon energy that must be put into the photosynthetic system to make one mole of D-glucose.

4.4.5 The Calvin Cycle

Plants make D-glucose as a by-product of the *Calvin cycle*. The Calvin cycle is driven by chemical energy in the form of ATP and NADPH. Photon energy trapped during the *light reactions* makes ATP and NADPH.

The first step in the Calvin cycle is one molecule of atmospheric CO_2 gas combines with the phosphorylated, 5-carbon sugar *ribulose biphosphate* (RuBP). The indispensable enzyme *ribulose biphosphate carboxylase oxygenase* (RUBISCO) catalyzes this reaction. RUBISCO may be the most abundant protein on earth [Kimball 2003f]. Next, the resultant 6-carbon compound breaks down into two molecules of *3-phosphoglyceric acid* (3PGA). The two 3PGA molecules are further phosphorylated by ATP and then receive a proton from NADPH and release a phosphate group to form *glyceraldehyde 3-phosphate,* aka *phosphoglyceraldehyde* (PGAL).

The Calvin cycle reactions release ADP, $NADP^+$, and P_i that are recycled for use again in the light reactions. Most PGAL is converted back into RuBP to keep the Calvin cycle going. Some PGAL is exported from the chloroplasts and serves as the starting substrate for the synthesis of the 6-carbon sugars, *D-glucose* and *fructose,* as well as *starch, cellulose,* and some AAs. This synthesis takes place in the cytoplasm of the plant cells.

Similar to the citric acid (Krebs) cycle, the Calvin cycle is not a closed process; molecules can enter and leave the loop. Each "turn" of the Calvin cycle fixes one molecule of CO_2, so it takes six turns to make one molecule of glucose. The Calvin cycle is illustrated schematically in Figure 4.12. Each CO_2 molecule entering the Calvin cycle requires two NADPH molecules and three ATP molecules. Each molecule of O_2 released by the *light reactions* supplies the four electrons needed to make two NADPH molecules. The *chemiosmosis* driven by the four electrons as they pass through the *cytochrome **b6/f*** complex liberates enough energy to pump 12 *protons* (H^+) into the thylakoid compartment of a granum. Kimball [2004g] noted that from the stoichiometry of photosynthesis, it must take 14 protons passing through the *ATPase* to make three molecules of ATP. He speculates that the missing two protons may come from *cyclic phosphorylation***.** The interested reader should see Kimball's description of one possible molecular scenario for cyclic phosphorylation.

Note that CO_2 fixation has two alternate forms, C_3 and C_4. C_3 is the classical Calvin cycle described earlier. In the C_4 CO_2 fixation pathway, also called the *Hatch–Slack cycle,* CO_2 is added to one molecule of *phosphoenolpyruvate* (PEP), forming the four-carbon oxaloacetate (OAA). Instead of RUBISCO, the enzyme *phosphoenolpyruvate carboxylase/oxygenase* (PEPC) is used. The OAA is usually converted to *malate,* off which the CO_2 is again split off enzymatically. Now, this CO_2 is bound to RuBP and enters the regular Calvin cycle. The ability to do C_4 photosynthesis is an adaptation found in plants (e.g., sugarcane) in the tropics, which have more solar energy to deal with [Sage 2002; Sengbusch 2003].

4.4.6 The Light Reactions

The photosynthetic light reactions are a complex, concatenated series of physical and chemical reactions that involve large complexes of molecules embedded into the thylakoid membranes in chloroplasts. Photosystems I and II are described in the following text.

Photosystem I (PS1) has been completely worked out in the *cyanobacteria* (blue-green algae). Other plants probably use very similar PS1 architectures. PS1 is a homotrimer, with each subunit in

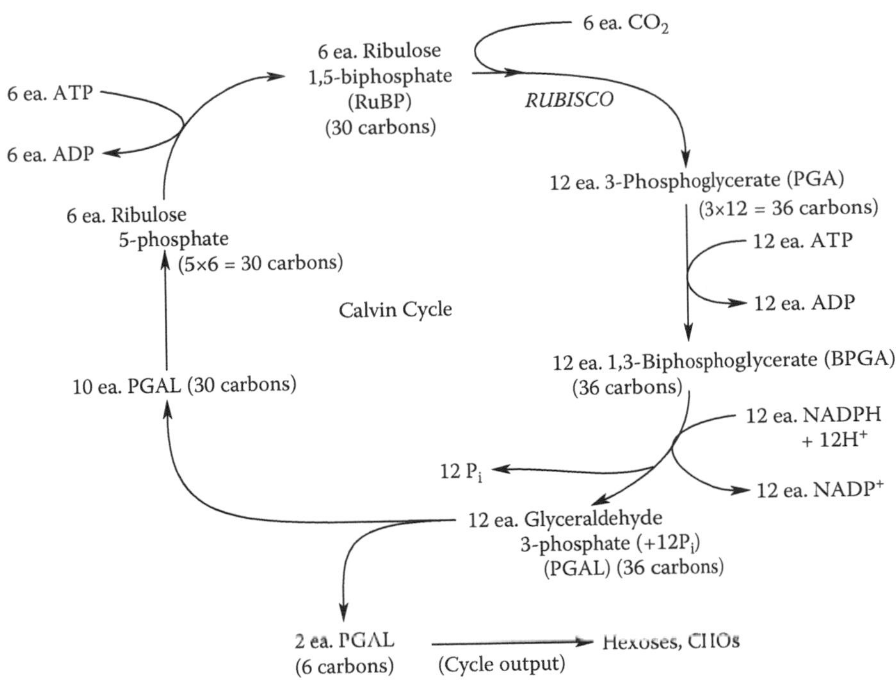

FIGURE 4.12 Schematic of the Calvin cycle. See text for a more detailed description.

the trimer consisting of 12 different protein molecules bound to 96 molecules of *chlorophyll* **a**, plus two molecules of the *reaction center chlorophyll* P_{700}, plus four accessory molecules closely associated with them, plus 90 molecules serving as *antenna pigments.* Also in the subunit are 22 *carotenoid molecules,* four *lipid molecules,* three clusters of Fe_4S_4, and two *phylloquinones.* (Hence, our frequent use of the adjective, "complex.")

PS1 catalyzes light-induced charge separation similar to that in PS2. Light energy is absorbed by an *antenna* and transferred to a reaction center chlorophyll where charge separation is initiated. In PS1, electrons are eventually transferred to *nicotinamide adenosine dinucleotide phosphate* (NADP) molecules, the reduced form of which can be used for carbon fixation. The oxidized reaction center chlorophyll picks up another e^- from the *cytochrome* ***b6*** complex.

Photosystem II (PS2) is also a complex of more than 20 protein molecules bound to 50 or more *chlorophyll* ***a*** molecules plus two molecules of the *reaction center chlorophyll,* P_{680}; close to two accessory molecules; plus two molecules of *pheophytin* (chlorophyll without the Mg^{++}); plus about six *carotenoid* molecules and two molecules of *plastoquinone.* PS2 is the molecular complex where water splitting and O_2 production occurs. Upon oxidation of the reaction center chlorophyll in PS2, an electron is pulled from an adjacent tyrosine AA in the surrounding protein. This e^- is immediately replaced by one from the water-splitting complex. Electrons flow from the PS2 reaction center to electron-carrying molecules (plastoquinones) in the thylakoid membrane and from there to another membrane–protein complex, the *cytochrome* ***b6*** system. The net electron transfer through PS2 and PS1 results in water oxidation, yielding O_2 gas and NADP reduction. Two quanta of photon energy for this complex process are required for each e^- transported through the entire PS2–PS1 chain.

Chemiosmosis is the term applied to the acidification of the interior thylakoid space by protons accumulated during PS1 and PS2. The pH during active photosynthesis is about 4. The energy stored in the proton gradient is harnessed in the synthesis of ATP as the protons pass into the stroma through the *ATP synthase* enzyme. The resultant ATP from chemiosmotic energy is used to drive reactions in the Calvin cycle, as is NAPDH [Kimball 2003g]. We have seen that a type of chemiosmosis also occurs in mitochondria.

4.5 THE CHALLENGE OF MODELING COMPLEX BIOCHEMICAL SYSTEMS

4.5.1 INTRODUCTION

The adjective "complex" describes a system that has many states or components that are causally interconnected, forming an entity that is difficult to describe, understand, and model. The causal interconnections can be excitatory or inhibitory, time variable, and nonlinear, and in some cases noise can modulate the connections. Often a complex system can be subdivided into a group of smaller, less complex subsystems that can be modeled individually. In some cases, this reductionism can lead to erroneous conclusions about overall system behavior. The dynamic properties of a large, complex system are generally the result of the total number of states and their nonlinear and time-variable interactions.

What is the rationale for mathematically modeling and simulating the complex behavior of biochemical systems? The first goal might be a desire for academic accuracy; to prove to ourselves that by matching model results with known, measured system behavior (i.e., state concentrations measured in vitro or in vivo) that we understand the system under study and have described it accurately, physically, chemically, and mathematically, in our model. This is called *model verification.* The second goal, having verified the model, is *prediction.* We can provide the model with novel inputs and see what behaviors it predicts. Some predictions can be confirmed experimentally in vivo or in vitro. For example, a robust mathematical model of cell biochemistry can thus provide an in silico test bed for new drugs and therapies before any attempt to use animal models is made.

Inspection of the *Biochemical Pathways Charts, Parts 1 and 2,* published by Roche Applied Science, leaves even the knowledgeable observer in awe at the complexity and interconnectivity of the known mammalian biochemical reaction networks. These charts summarize the many known biochemical pathways, cycles, and shunts in biochemical metabolism, synthesis, and catabolism that are amenable to modeling.

Biochemical systems modeling in the past has often yielded to the temptation of *reductionism.* That is, there was a trade-off between computational ease (capability) and model complexity. Results of such reductionist modeling have often proved useful, but by eschewing complexity, the overall, interesting, nonlinear behavior of certain major biochemical systems has been avoided. The defining goal of modern biochemical modeling is to be able to model whole-cell systems. In particular, stem cells engaged in embryological development, proliferating cancer cells, or autoimmune reactions.

Biochemical systems are challenging to model because:

1. They are of great size and complexity. Every reaction step generally has an associated enzyme. Enzymes are generally proteins made by gene expression. They may undergo regulated, posttranslational modification. Gene expression is itself regulated.
2. The topology of the biochemical C^3 networks involves many feedback and feed-forward paths. Some of the feed-forward signals can be excitatory, others inhibitory. Feedback is generally inhibitory, or negative, and there can be multiple, nested loops. Regulation of biosynthetic pathways is generally *parametric.* That is, to effect regulation, the reaction rate constants are altered by signal-induced changes in enzyme concentrations or effectiveness. The limited availability of a critical enzyme (or lack thereof) can introduce a *rate-limiting step* in a reaction pathway.
3. System dynamics are dictated by molecular and ionic diffusion, mass action kinetics, and time lags associated with protein synthesis.
4. In biochemical systems, signals are nonnegative molecular or ionic concentrations. Once a signal is received by a receptor molecule, it is generally destroyed enzymatically.

Simplification of a complex biochemical system model can come from an a priori knowledge that certain rate constants in a mass-action-based model are comparatively negligible. For example, the reverse rate constant in the second step of a Michaelis–Menton reaction (shown earlier) can often be neglected.

4.5.2 Generalized Approaches to Modeling Biochemical Systems

We have already seen that simple biochemical systems can be modeled with sets of first-order, nonlinear ODEs based on chemical *mass action* (MA) *kinetics* and simple diffusion. The concentrations of reactants (or the number of reactant molecules in a *compartment*) are usually taken as the state variables. Examples of this approach are given in the analysis of speculative models for biochemical oscillators in Section 4.5.3.

In 1988, Voit described a mathematical formalism useful in the analysis of complex biochemical system that meshes with the ODEs describing mass action and diffusion. Voit called this the *Power-Law Formalism* (PLF). The PLF requires that the states be nonnegative (easy for concentrations or densities) and that the functions describing the process be differentiable. Symbolically, we can write for the *i*th state:

$$\dot{\mathbf{X}}_i = \mathbf{F}_i^+(\mathbf{X}_1, \ldots, \mathbf{X}_n) - \mathbf{F}_i^-(\mathbf{X}_1, \ldots, \mathbf{X}_n), \text{ molecules/second. } 1 \le i \le n \qquad (4.43)$$

Mathematically, this equation states that the rate of appearance of reactant $\mathbf{X}_i$ is equal to its net rate of appearance influenced by all system states minus its net rate of diminution influenced by all system states. In Equation 4.43, Voit observed that $\mathbf{F}_i^+$ and $\mathbf{F}_i^-$ *are logarithmically differentiable functions of all system states* and describe gain and loss rates of the states, $\mathbf{X}_i$. $\mathbf{F}_i^+$, and $\mathbf{F}_i^-$ are approximated nonlinearly by "...first expressing their logarithms as functions of the logarithms of their arguments, second, developing a Taylor series, third, retaining only the constant and linear terms of these expansions, and finally, translating the truncated Taylor expansions back into the Cartesian space." The result is a nonlinear local representation for $\mathbf{F}_i^+$ and $\mathbf{F}_i^-$, each in the form of a product of power-law functions:

$$\mathbf{F}_i^+ - \alpha_i \prod_{j=1}^{n} \mathbf{X}_j^{g_{ij}} = 0 \qquad \mathbf{F}_i^- - \beta_i \prod_{j=1}^{n} \mathbf{X}_j^{h_{ij}} = 0 \qquad (4.44)$$

The *S-system* defining equations are thus

$$\dot{\mathbf{X}}_i = \alpha_i \prod_{j=1}^{n+m} \mathbf{X}_j^{g_{ij}} - \beta_i \prod_{j=1}^{n+m} \mathbf{X}_j^{h_{ij}} \qquad (4.45)$$

This structure is called an **S**-*system* to indicate its applicability to saturable and synergistic systems.

Note that $\mathbf{X}_1, \ldots, \mathbf{X}_n$ are the system's states (dynamic concentrations of metabolites) or *dependent variables*. $\mathbf{X}_{n+1}, \ldots, \mathbf{X}_{n+m}$ are *fixed* concentrations of metabolites (independent variables). g_{ij} and h_{ij} are *kinetic orders;* unlike MA systems, they may be nonpositive and have noninteger values. α_i and β_i are nonnegative rate constants.

An alternate form for Equation 4.43 allows for multiple F-functions; the *i*th state ODE is written:

$$\dot{\mathbf{X}}_i = \sum_{k=1}^{p} \mathbf{F}_{ik}^+(\mathbf{X}_1, \ldots, \mathbf{X}_n) - \sum_{k=1}^{q} \mathbf{F}_{ik}^-(\mathbf{X}_1, \ldots, \mathbf{X}_n) \qquad (4.46)$$

Each F-function in the two sums is then approximated by a product of power-law functions using the Taylor series form. The final form is called the *Generalized Mass Action* (GMA) *System* form. Again, for the *i*th state:

$$\dot{\mathbf{X}}_i = \sum_{k=1}^{p} \alpha_{ik} \prod_{j=1}^{n} \mathbf{X}_j^{g_{ijk}} - \sum_{k=1}^{q} \beta_{ik} \prod_{j=1}^{n} \mathbf{X}_j^{h_{ijk}} \tag{4.47}$$

Thus, every *S*-system is a special case of a generalized mass action system, and MA systems can be transmogrified to S-systems.

Polisetty, Voit, and Gatzke [2006] have described their version of a GMA model using the *S*-system notation. Polisetty et al. stated: "In the GMA formulation within BST [Biochemical Systems Theory, Voit & Radivoyevitch 2000], the change in each dependent pool (state variable) is described as a difference between the sums of all fluxes entering the pool and all fluxes leaving the pool." They go on to point out that each flux is individually linearized in logarithmic coordinates, which in Cartesian space corresponds to a product of power-law functions which contains those and only those variables that directly affect the flux, raised to an exponent called its *kinetic order.* The product also contains a rate constant that determines the magnitude of the flux or speed of the process. The general mathematical formulation of any GMA model uses the ***n*** flux ODEs:

$$\dot{\mathbf{X}}_i = \sum_{p=1}^{k} \left\{ \gamma_{ip} \prod_{j=1}^{n+m} \mathbf{X}_j^{\xi_{ijp}} \right\} \quad (i = 1,\ldots n)\ (\gamma_{ip} \text{ can be positive, negative, or zero.}) \tag{4.48}$$

Where γ_{i1}, ..., γ_{ik} are rate constants corresponding to k reactions of production/consumption of state $\mathbf{X}_i$, and ξ_{ijk} are kinetic orders for state *i* in reaction *k* involving state *j*. In cases where state *j* does not have any influence on a given power-law term, $\xi_{ijk} \equiv 0$ [recall, $a^0 \equiv 1$, $a \neq 0$]. The number of reactions in one ODE, *k*, may be different for each state.

Polisetty et al. go on to say: "The power-law terms in [Equation 4.48] are the result of a straightforward Taylor approximation, which is applicable to an essentially unlimited variety of underlying processes and may include different types of interactions, activation, inhibition and processes associated with dilution and growth."

Other workers have further embellished the **S**-system approach. Antoniotti et al. [2002] described "... *XS-systems,* computational models whose aim is to provide users of S-systems with the extra tool of an automaton modeling the temporal evolution of complex biochemical reactions. The automaton construction is described starting from both numerical and analytical solutions of the differential equations involved, and parameter determination and tuning are also considered. A temporal logic language for expressing and verifying properties of XS-systems is introduced and a prototype implementation is presented." [Quoted with kind permission of Springer Science & Business Media.]

Due to the unique nonlinear structure of *S*-systems, the *steady-state equations* are linear when they are viewed in logarithmic coordinates, thus allowing application of constrained linear optimization methods such as the *simplex algorithm*. This feature is the basis of the *indirect optimization method* (IOM) that has proved useful in the simulation of metabolic networks with the goal of directed production of metabolites [Vera & Torres 2003].

Although much attention is paid to **S**-systems, we submit that in most cases of small (1–10 states) and medium-sized (11–32 states) systems (the small/medium boundary is arbitrary) the direct use of MA and diffusion ODEs is simpler and gives the same results when dynamic solutions are required. The trees are not hidden by the forest.

4.5.3 Models for Biochemical Oscillators

A biochemical oscillator is a nonlinear biochemical system in which the concentrations of certain reactants can vary periodically in the steady state in the absence of any periodic inputs. In this section, we will consider dynamic models for intracellular biochemical oscillators. These models are instructive because they illustrate the art of modeling small, isolated, chemical reaction systems having feedback. They are generally simpler to analyze than trying to model glycolysis or photosynthesis, etc. Note that interacting assemblies of muscle cells and neurons can also form multicellular oscillators [Kiss et al. 2007], such as (natural) cardiac pacemakers.

Biochemical oscillators appear to be one basis for natural circadian (diurnal) behaviors in organisms ranging from bacteria to humans. Their oscillations affect the concentrations of hormones and signaling substances that affect behavior, as well noncircadian events such as menstrual periods in primates [Foster & Kreitzman 2005]. It is believed that the mammalian *central circadian oscillator* is located in the suprachiasmatic nucleus of the hypothalamus. Interacting feedback loops affecting gene transcription and translation interact to regulate this master mammalian clock. Other, peripheral tissues such as the liver and pancreas also contain clocks. These cyclical concentration rhythms are generated at a subcellular level by stable, limit-cycle oscillations that involve coupled biochemical reactions. In one scenario, interacting transcription/translation-based feedback loops involving "clock" or "period" genes periodically express a critical protein that affects behavior. For example, in mice, three period genes have been found (**mPer1**, **2**, and **3**) for which their respective mRNA concentrations are expressed cyclically in vivo. [Fujimoto et al. 2006]. There is evidence that even master circadian "clocks" require exogenous entrainment (sync) signals, such as natural photoperiod, to maintain a precise phase-locked period over many cycles. A clock without entrainment signals tends to oscillate with a longer period, but it still is periodic. Later, we examine some simple mathematical models of biochemical oscillators.

Many coupled biochemical reaction systems have *parametric feedback* inherent in their architectures. The effectiveness of critical enzyme catalysts is often modulated by molecular concentrations from within or without the system, altering reaction rates. The modulation can be from other enzymes, product concentration, or substrate concentrations. Control of transcription/translation of an enzyme is one means of affecting its concentration. Regulation of vital chemical reactions by parametric control is a fundamental property of life.

The kinetics of biochemical systems with endogenous feedback can be approximated by systems of nonlinear ODEs based on mass action. Large, nonlinear systems are capable of bizarre and counterintuitive behaviors, including sustained, oscillatory, *stable limit-cycle* behavior, in which the concentrations of reactants and products vary periodically within bounds around average values. A stable limit cycle is a bounded, steady-state oscillation in the amplitudes of states in a nonlinear system having constant (or zero) inputs. A stable limit cycle is entered regardless of initial conditions [Ogata 1970, Chapter 12]. The concentration waveforms in a nonlinear, biochemical oscillator are generally not sinusoidal. In fact, they can be pulsatile (see the Hodgkin–Huxley model for nerve pulse generation and the observed oscillations in insulin secretion by isolated pancreatic beta cells) or resemble a distorted sine wave.

Some biochemical and physiological systems that exhibit limit-cycle behavior are listed in Table 4.2.

The fact that coupled biochemical systems can exhibit stable limit-cycle behavior was demonstrated by Chance et al. [1964]. Working with anaerobic yeast cultures, in vitro, Chance observed that there were periodic fluctuations in the concentration of NADH, a reactant in the glycolysis pathway whereby sugars are converted to alcohol. The fluctuations in [NADH] were measured as nearly sinusoidal variations in the UV-induced fluorescence of NADH that had about 5-minute period. More recent studies on glycolytic oscillations have shown that the period of glycolytic oscillations in yeast can be modulated by the [NAD+]/[NADH] ratio without affecting the instability of the system [Madsen et al. 2005].

A fascinating biochemical oscillator that has recently been given attention is the cyclical secretion of insulin by "resting" pancreatic beta cells. This oscillation has been studied in vivo in monkeys and in man and in vitro with single beta cells, clusters of cells, and entire islets using dog and

TABLE 4.2
Some Typical Endogenous Biological Oscillators and Their Approximate Periods

System	Period (Approximate)
Human ovarian cycle	28 d
Circadian rhythms (behavioral)	24 h
Cell cycle	30 min to 24 h
Gonadotropic hormone secretion	1 h
Insulin/Ca++/glucose oscillations	10 min; also 50 to 110 min
cAMP oscillations	10 min
Glycolytic NADH oscillations	1 min to 1 h
Protoplasmic streaming	1 min
Calcium oscillations	seconds–minutes
Smooth muscle contractions	seconds–hours
Cardiac pacemakers (human)	0.4 to 2 s
Insect fibrillar flight muscle	0.005 to 0.1 s
Neural membrane potential oscillations. (Modeled by the Hodgkin–Huxley equations)	0.002 to 1 s

mouse pancreases. The slow, in vivo, oscillations of plasma insulin and glucose concentrations have periods of 50–110 minutes [Schaeffer et al. 2003].

Faster insulin and intracellular Ca^{++} concentration ($[Ca^{++}]_i$) oscillations have also been measured in mouse beta cells. The $[Ca^{++}]_i$ oscillations are in phase with the insulin concentration oscillations; their periods are on the order of 5–10 minutes. Gilon et al. [2002] hypothesized: "Because a rise in β-cell $[Ca^{++}]_i$ is required for glucose to stimulate insulin secretion, and because glucose-induced $[Ca^{++}]_i$ oscillations occur synchronously within all β-cells of an islet, it has been proposed that oscillations of insulin secretion are driven by $[Ca^{++}]_i$ oscillations."

Thus, the bounded oscillations in $[Ca^{++}]_i$ may in fact drive the insulin secretion mechanism by parametrically modulating the sensitivity of β-cells to glucose. But what causes the oscillations in $[Ca^{++}]_i$ to occur? It may turn out that the lower-frequency component of insulin and glucose oscillations in vivo is related to a different mechanism involving time delays, and perhaps the glucagon control system. Perhaps there are two, distinct, cross-coupled oscillator mechanisms leading to the two, superimposed limit cycles.

In 1961, Spangler and Snell (*Nature.* 191(4787): 457–458) proposed a simple, *theoretical,* chemical–kinetic model architecture as a possible mechanism to describe certain "biological clocks." Their model was based on the concept of *cross-competitive inhibition.* The four, hypothetical chemical reactions shown in the following are based on mass action and diffusion. Eight system state equations given in Equations 4.50A–D are based on the coupled reactions of Equations 4.49A–D in the following text.

$$\mathbf{(A)} \xrightarrow{J_a} \mathbf{A} + \mathbf{E} \underset{k_{-1}}{\overset{k_1}{\rightleftarrows}} \mathbf{B} \xrightarrow{k_2} \mathbf{E} + \mathbf{P} \xrightarrow{k_3} * \tag{4.49A}$$

$$\mathbf{nP} + \mathbf{E'} \underset{k'_{-4}}{\overset{k'_4}{\rightleftarrows}} \mathbf{I'} \tag{4.49B}$$

$$\mathbf{nP'} + \mathbf{E} \underset{k_{-4}}{\overset{k_4}{\rightleftarrows}} \mathbf{I} \tag{4.49C}$$

$$(\mathbf{A}') \xrightarrow{J'_a} \mathbf{A}' + \mathbf{E}' \underset{k'_{-1}}{\overset{k'_1}{\rightleftarrows}} \mathbf{B}' \xrightarrow{k'_2} \mathbf{E}' + \mathbf{P}' \xrightarrow{k'_3} * \quad (4.49D)$$

The hypothetical oscillator works in the following manner: Substrates **A** and **A'** diffuse into the reaction volume at constant rates $\mathbf{J}_a$ and $\mathbf{J}_a$**'**, respectively. **A** and **A'** combine with enzyme/catalysts **E** and **E'**, respectively, and are transformed reversibly into complexes **B** and **B'**. **B** and **B'** then dissociate to release active enzymes and products **P** and **P'**. **P** and **P'** next diffuse out of the compartment to effective zero concentrations with rates $\mathbf{k}_3$ and $\mathbf{k}_3$**'**, respectively. The two Michaelis–Menton-type main reactions are entirely independent. However, the system is *cross-coupled,* therefore dependent and possibly unstable because of the reversible combination of the *products* with the complementary free enzyme. The entire system is described by eight state equations (mass action ODEs) in which **a** = *running concentration* (RC) of **A, b** = RC of **B, p** = RC of **P, i** = RC of **I, a'** = RC of **A', b'** = RC of **B', p'** = RC of **P', i'** = RC of **I'.** There are no ODEs for **e** and **e'**. We assume the enzymes are conserved in the compartment, that is, enzymes are either free or tied up as complexes; they are not destroyed or created. Thus: $\mathbf{e}_0 = \mathbf{e} + \mathbf{b} + \mathbf{i}$, and $\mathbf{e}_0' = \mathbf{e}' + \mathbf{b}' + \mathbf{i}'$, where $\mathbf{e}_0$ is the initial (total) concentration of enzyme **E,** and $\mathbf{e}_0$**'** is the initial (total) concentration of **E'.** Note that **n** molecules of **P** in the compartment combine with one molecule of **E'** to reversibly form one molecule of complex **I'**, etc. Thus, when **I'** breaks down, **n** molecules of **P** are released with one molecule of **E'**, etc.

Using the notation detailed earlier and mass action kinetics, we write the eight ODEs and two algebraic equations (for enzyme conservation) describing the coupled, Spangler and Snell system:

1. $\dot{a} = \mathbf{J}_a - \mathbf{k}_1\,\mathbf{a}\,\mathbf{e} + \mathbf{k}_{-1}\,\mathbf{b}$ (4.50A)
2. $\dot{b} = \mathbf{k}_1\,\mathbf{a}\,\mathbf{c} - \mathbf{b}(\mathbf{k}_{-1} + \mathbf{k}_2)$ (4.50B)
3. $\dot{p} = \mathbf{k}_2\,\mathbf{b} - \mathbf{k}_3\,\mathbf{p} - \mathbf{k}_4'\,\mathbf{n}\,\mathbf{p}' + \mathbf{k}_{-4}'\,\mathbf{n}_{\mathbf{i}'}$ (4.50C)
4. $\dot{i} = \mathbf{k}_4\,\mathbf{p}'^{\,n}\,\mathbf{e} - \mathbf{k}_{-4}\,\mathbf{i}$ (4.50D)
5. $\mathbf{e}_0 = \mathbf{e} + \mathbf{b} + \mathbf{i}$ (4.50E)
6. $\dot{a}' = \mathbf{J}_a' - \mathbf{k}_1'\,\mathbf{a}'\,\mathbf{e}' + \mathbf{k}_{-1}'\,\mathbf{b}'$ (4.50F)
7. $\dot{b}' = \mathbf{k}_1'\,\mathbf{a}'\,\mathbf{e}' - \mathbf{b}'\,(\mathbf{k}_2' + \mathbf{k}_{-1}')$ (4.50G)
8. $\dot{i}' = \mathbf{k}_4'\,\mathbf{p}^{n}\,\mathbf{e}' - \mathbf{k}_{-4}'\,\mathbf{i}'$ (4.50H)
9. $\dot{p}' = \mathbf{k2'b'} - \mathbf{k}_3\,\mathbf{p}' - \mathbf{k}_4\,\mathbf{p}'^{n}\,\mathbf{e} + \mathbf{k}_{-4}\,\mathbf{n}_{\mathbf{i}}$ (4.50I)
10. $\mathbf{e}'_0 = \mathbf{e}' + \mathbf{b}' + \mathbf{i}'$ (4.50J)

To verify that this system is capable of sustained, limit-cycle oscillations, we simulated the system's behavior at "turn-on" using *Simnon* software. Note that the initial conditions used for states **b** and **b'** were **4** units. Other states have zero initial conditions. The initial amounts of free enzymes, $\mathbf{e}_0$ and $\mathbf{e}_0$**'** = **5** units. We used the parameters listed in the program `snell2.t` given in Appendix A.

Model system oscillatory behavior indeed did occur, as shown in Figure 4.13. To verify that the system has an SS limit cycle, we plotted **dp/dt** versus **p(t)** and **dp'/dt** versus **p'(t)** (phase-plane plots) in Figure 4.14. A stable, closed trajectory in the phase plane indicates a stable, finite-amplitude, steady-state limit cycle. That is, bounded, continuous oscillations occur. Double-letter variables ending in 'p' in the `snell2.t` program are the primed variables in the ODEs. Also note that in the simulation, we have "rectified" the concentration states. In nature, obviously there are no nega-

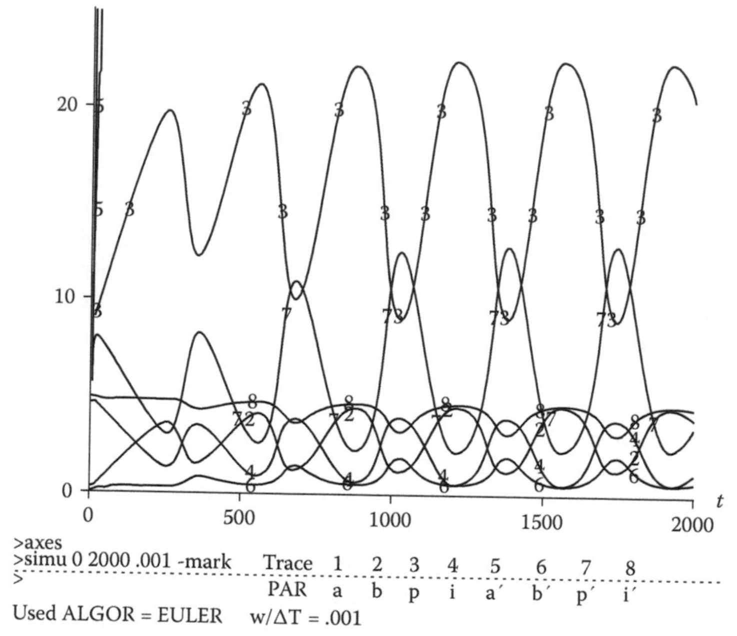

FIGURE 4.13 Plot of sustained limit-cycle oscillations generated by the *Simnon* metabolic oscillator model Snell2.t in the text. See text for further description.

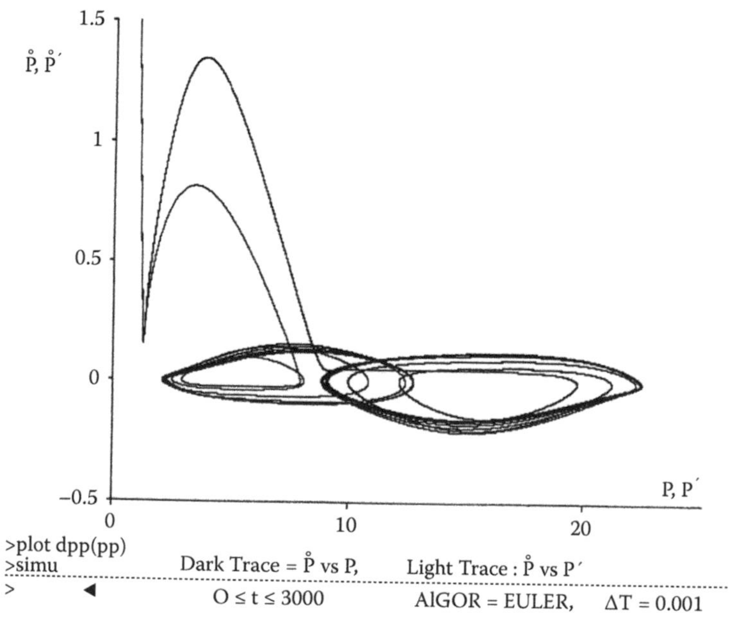

FIGURE 4.14 Phase-plane plot of *p* versus *p* generated by the sustained limit cycle of the Snell2 model. Note that the closed pattern denotes a steady-state oscillation of the *p* parameter. (Note that all of the other parameters have limit-cycle oscillations as well.)

tive concentrations; if the system ICs or constants were ill-chosen in the model, it is possible that a concentration could go negative, creating a paradox. (In the Simnon program `snell2.t`, the ODEs are written as text; no graphic interface is required.)

In an innovative project encompassing both genetic engineering and mathematical biology, Elowitz and Leibler [2000] designed a synthetic biochemical network called the *repressilator* in the *E. coli* lac⁻ strain MC4100 genome. (The **Lac** operon is described in detail in Chapter 6, Section 6.4.2.) This genetically modified (GM) bacterium exhibits steady-state limit-cycle oscillations in three protein concentrations. It uses three genes in a sequential (nonlinear) negative feedback topology. Each gene produces a protein, and each of which inhibits the expression of the next gene in the cycle. As the concentration of one of the three produced proteins increases, the expression of the next gene in the cycle is inhibited, and the concentration of the second protein falls, permitting increased expression of the third gene. As the third protein increases in concentration, the first gene now becomes inhibited, allowing expression of the second gene, etc. Note there are probably transport lags or delays in the living system due to protein synthesis mechanisms, etc., which even though not used in the Elowitz and Leibler (E & L) model could certainly contribute to system instability.

The E & L oscillating genomic network periodically induced the synthesis of a *green fluorescent protein* (GFP) marker in individual GM bacteria as a sign of the oscillation. The GFP oscillations have periods of about 150 minutes, which are slower than the cell division cycle (about 1 hour), so the oscillator (with its initial conditions) has to be transmitted from generation to generation of GM *E. coli*. How this transfer of oscillator initial conditions occurs through bacterial generations is not known. Some sort of epigenetic transfer may be involved.

A *plasmid* containing the GM repressilator oscillator and another, coupled "reporter" plasmid that makes the GFP were introduced into target bacteria. The repressilator itself is a cyclic negative feedback loop composed of three repressor genes and their corresponding promoters. Molecular details of the repressilator plasmid design and the reporter can be found in the Elowitz and Leibler paper. These authors noted that their GM bacterial "clock" displayed noisy behavior, possibly because of stochastic fluctuations of its components.

Elowitz and Leibler constructed a continuous, kinetic, mathematical model of the in vivo repressilator system that exhibited limit-cycle oscillations in protein concentrations and mRNAs under certain boundary conditions and initial conditions. Using their notation: $\mathbf{p}_i$ = repressor protein concentration, and $\mathbf{m}_i$ = the corresponding mRNA concentration. (Ribosomal action was not specifically simulated.) To simplify their model, they considered only the symmetrical case in which all three repressors are identical except for their DNA-binding specificities. Thus, the kinetics of the system were represented by *six*, coupled, first-order, nonlinear ODEs. (Details of their simulation are found in their paper.)

$$\dot{m}_i = -m_i + \frac{\alpha}{(1+p_j^{n_i})} + \alpha_0 \qquad i = \mathbf{1}\text{ (lac)}, \mathbf{2}\text{(tetR) or }\mathbf{3}\text{(cl)}: j = \mathbf{1}\text{(cl)}, \mathbf{2}\text{(lac) or }\mathbf{3}\text{(tetR)} \quad (4.51\text{A})$$

$$\dot{p}_i = -\beta(p_i - m_i) \quad (4.51\text{B})$$

where the number of protein copies per cell per minute produced from a given promoter type during continuous growth is α_0 in the presence of *saturating amounts* of repressor (owing to the leakiness of the promoter), and $(\alpha + \alpha_0)$ in its absence. β is the ratio of the protein decay rate to the mRNA decay rate. The exponent $\mathbf{n}_i$ is a Hill coefficient (the fraction is a Hill hyperbola, modeling gene expression inhibition by the preceding gene product).

See Figure 4.15A for a signal flow graph representation of the ODE set earlier, which was taken from the E & L paper. Note that the linear part of this system's loop gain has six real poles. All $\mathbf{n}_i = \mathbf{2}$ in the figure. The larger the $\mathbf{p}_k$ value, the smaller will be the output of the hyperbolic Hill nonlinearities.

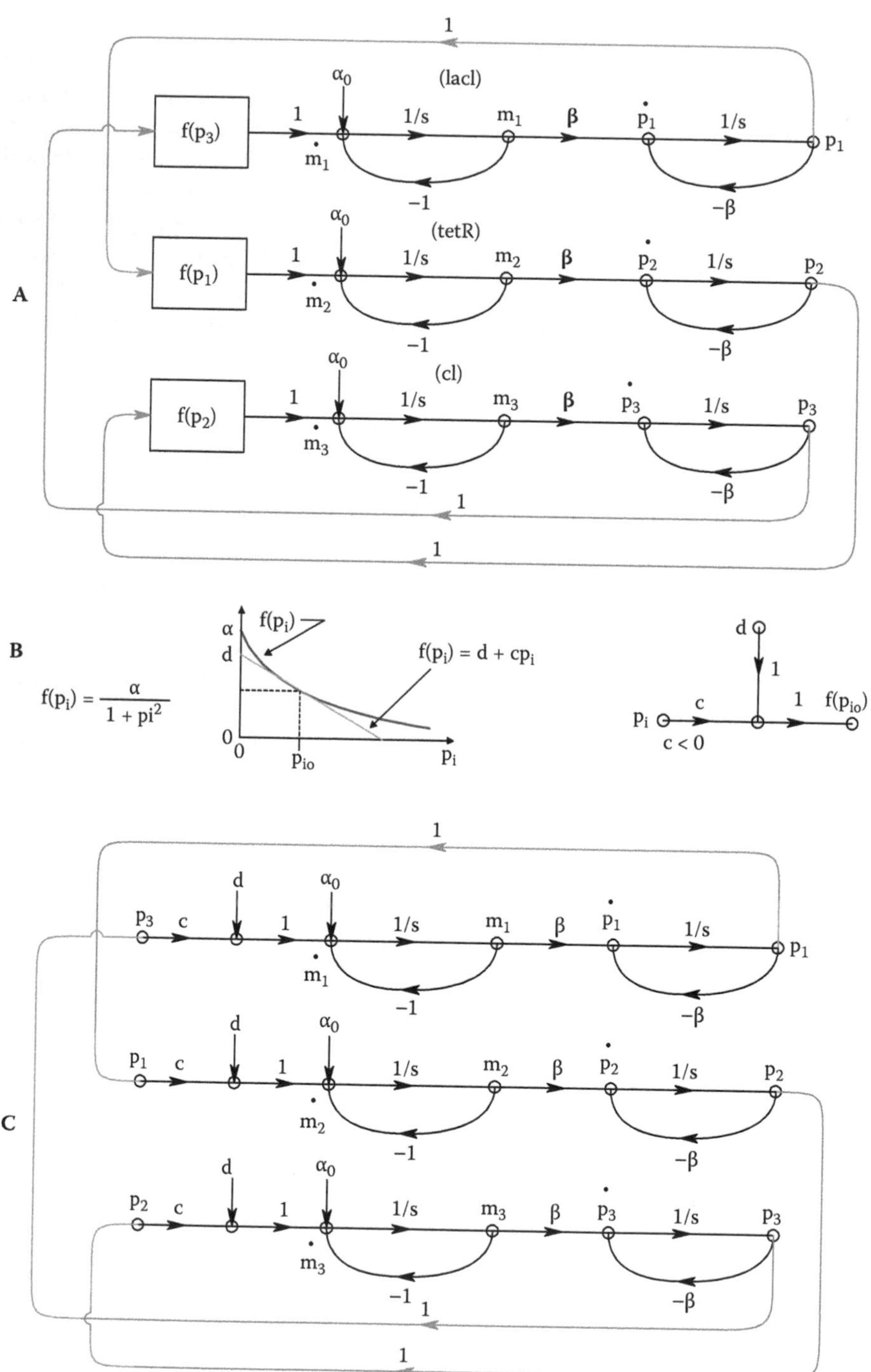

FIGURE 4.15 (**A**) A linearized model for the *repressilator* biochemical oscillator shown in signal flow graph format. The hyperbolic nonlinearity, $f(p_i)$ is shown in (**B**) where it is linearized around an operating point. (**C**) Illustrates a completely linear SFG for the repressilator system having six real poles in its loop gain. See text for further description.

Figure 4.15B illustrates a linearization of the Hill hyperbolic functions. We assume an average operating point, $\mathbf{p}_{io}$, and draw a tangent to the $\mathbf{f}(\mathbf{p}_i)$ curve at $\mathbf{p} = \mathbf{p}_{io}$. The tangent defines a linear transfer function, $\mathbf{g}(\mathbf{p}_i) = \mathbf{c}\,\mathbf{p}_i + \mathbf{d}$ based on the assumption that $\mathbf{p}_i \approx \mathbf{p}_{io}$. The linear gain, $\mathbf{c} = \mathbf{d}[\mathbf{f}(\mathbf{p}_{io}]/\mathbf{dp}_i = -2\alpha\mathbf{p}_{io}/(1 + \mathbf{p}_{io}^2)^2$. Thus, for small changes of $\mathbf{p}_i$ around the operating point, $\mathbf{p}_{io}$, the output of the linearization function is

$$\mathbf{g}(\mathbf{p}_i) = -2\alpha\mathbf{p}_i\,\mathbf{p}_{io}/(1 + \mathbf{p}_{io}^2)^2 + \mathbf{d} \tag{4.52}$$

Inspection of the linearized repressilator system shows in Figure 4.15 that its loop gain in Laplace transform format is

$$\mathbf{A}_L(\mathbf{s}) = \frac{-\left|c^3\beta^3\right|}{(\mathbf{s}+1)^3(\mathbf{s}+\beta)^3} \tag{4.53}$$

Thus, we see that the linearized system approximation has a negative loop gain and six real poles. Application of root-locus theory shows that the system will exhibit unstable oscillations for $|c^3\,\beta^3|$ above some critical value [Northrop 2000].

A number of other workers have elaborated on the repressilator model set forth by Elowitz and Leibler; for example, see Drennan and Beer [2006].

In another simulation of the reactions of a putative biochemical oscillator, we considered a model biochemical system suggested by Chance, Pye and Higgins (*IEEE Spectrum.* August, 1967. pp 79–86). This model is based on one Michaelis–Menton (M–M) equation and three other rate equations:

$$(\mathbf{B}) \xrightarrow{J_0} \mathbf{B} + \mathbf{E}_a \underset{k_4}{\overset{k_1}{\longleftrightarrow}} \mathbf{B} * \mathbf{E} \xrightarrow{k_2} \mathbf{E}_a + \mathbf{P} \xrightarrow{k_3} \mathbf{D} \tag{4.54A}$$

$$\mathbf{E}_a \underset{pk_6}{\overset{k_5}{\longleftrightarrow}} \mathbf{E}_i \tag{4.54B}$$

Reactant **B** diffuses into the reaction volume at constant rate, $\mathbf{J}_0$. There it reacts in an M–M reaction by binding to active enzyme to form the complex, **B*E. B*E** is converted with rate constant $\mathbf{k}_2$ to a product, **P**, which disappears with first-order kinetics with rate constant, $\mathbf{k}_3$. In an ancillary reaction, the active enzyme is *reversibly* converted to an inactive form, $\mathbf{E}_i$. Let $\mathbf{b}$ = RC (running concentration) of **B, c** = RC of complex **B*E, p** = RC of product, **P**, $\mathbf{e}_{a0}$ = initial concentrations of enzyme that is conserved. Thus, the system equations can be written:

1. $\dot{b} = \mathbf{J}_0 + \mathbf{k}_4\mathbf{c} - \mathbf{k}_1\,\mathbf{b}\,\mathbf{e}_a$ (4.55A)
2. $\dot{c} = \mathbf{k}_1\,\mathbf{b}\mathbf{e}_a - (\mathbf{k}_4 + \mathbf{k}_2)\,\mathbf{c}$ (4.55B)
3. $\dot{p} = \mathbf{k}_2\mathbf{c} - \mathbf{k}_3\,\mathbf{p}$ (4.55C)
4. $\dot{e}_a = -\mathbf{k}_5\mathbf{e}_a + \mathbf{k}_6\mathbf{p}\,\mathbf{e}_i + \mathbf{k}_2\mathbf{c} - \mathbf{k}_1\mathbf{b}\,\mathbf{e}_a$ (4.55D)
5. $\dot{e}_i = \mathbf{k_5}\,\mathbf{e}_a - \mathbf{k_6}\,\mathbf{p}\,\mathbf{e}_i$ (Not used as-is.) (4.55E)
6. $\mathbf{e}_{a0} = \mathbf{e}_a + \mathbf{c} + \mathbf{e}_i$ (Conservation of enzyme) Thus: (4.55F)
7. $\mathbf{e}_a = \mathbf{e}_{a0} - \mathbf{c} - \mathbf{e}_i$ (Equation 7 is now substituted into Equation 5 and we have the final ODE for $\mathbf{e}_i$.) (4.55G)
8. $\dot{e}_i = \mathbf{k}_5\,(\mathbf{e}_{a0} - \mathbf{c} - \mathbf{e}_i) - \mathbf{k}_5\mathbf{p}\,\mathbf{e}_i$ (4.55H)

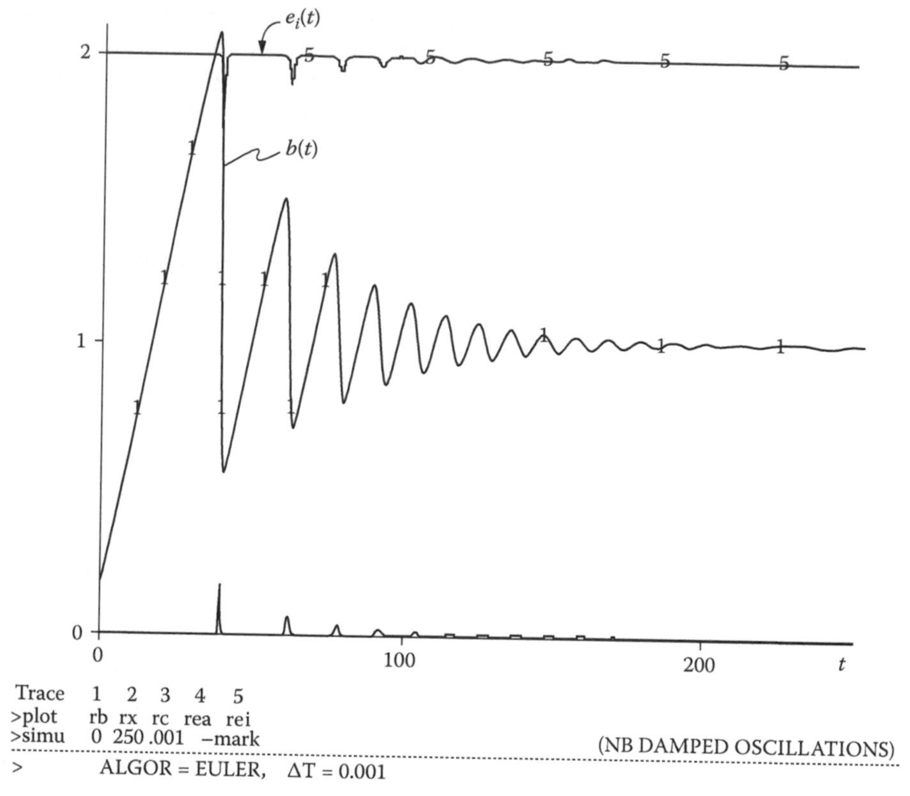

FIGURE 4.16 Results from a *Simnon* simulation of the putative biochemical oscillator proposed by Chance et al. (1967). This result, and many others run, all gave damped (nonlimit cycle) behavior.

The five ODEs above were simulated using a wide range of ICs and parameters using the Simnon® program, `CHANCE3.t`, given in Appendix B. Again, we ensured that the states were nonnegative because there are no negative concentrations. For a number of simulations with decidedly different ICs and parameters, we observed damped, periodic response—never sustained oscillations. See Figure 4.16 for a typical result. We did not try introducing time delays in the system; delays could have destabilized the system.

Biochemical oscillators are found throughout living systems. They presumably form the basis for endogenous "biological clocks" that modulate hormone secretion, diurnal behavior, behavior in response to tides, monthly behavior, etc. Most biological clocks are *entrainable*. That is, they use a synchronizing signal such as sunrise, the time of low tide, etc., to ensure their oscillations are phase locked to environmental conditions even for several cycles in the short-term absence of synchronizing signals.

The fact that coupled chemical systems have parametric feedback, parametric feed-forward (autocatalytic behavior), and transport lags (phase shifts) and are nonlinear often leads to bounded limit-cycle behavior. If a closed-loop biochemical system were to become frankly unstable, the organism with that system in its genome would quickly be eliminated from existence by natural selection. Small, limit-cycle oscillations are not lethal, but large ones can be lethal.

4.5.4 The Hodgkin–Huxley Model (1952) for Action Potential Generation

One of the great leaps forward in neurophysiology was the elaboration of the basic molecular mechanisms underlying the generation and propagation of nerve action potentials, or "spikes" on nerve axons. Most of the early experiments leading to the understanding of nerve spike generation and propagation were done on the giant, unmyelinated nerve axons of the squid, *Loligo* sp., in the mid-20th century. Squid giant axons were attractive because their large size (up to 1 mm in diameter) made it easier to replace their (internal) axoplasm with an artificial ionic media of any desired composition. Of course, the composition of their external bathing solution could be made up in any desired manner, as well.

In 1952, Hodgkin and Huxley published a unique and innovative dynamic mathematical model describing the generation of the nerve axon action potential based on the electrophysiology and then-known molecular biology of squid axon membrane. We describe this model in the following text and its significance.

The cell membrane of the squid axon was found to have a relatively high distributed capacitance across it of about 1 μF/cm^2. This capacitance is the result of the thin lipid bilayer in the axon's unit membrane. The lipids act as the dielectric of an exceedingly thin (about 75 nm) capacitance. The net ionic conductance inside and outside the axon is relatively high, compared to that across the lipid bilayer.

The membrane is studded with protein molecules that penetrate it between the inside and outside of the axon; these proteins offer selective passage for major ionic species to enter or exit the axon interior, depending on the potential energy gradient across the membrane for a given ionic species. For example, the inside of a squid axon has a resting potential from 60 to 70 mV, inside negative [Bullock & Horridge 1965, Chapter 3]. The resting or equilibrium concentration of sodium ions is higher outside than inside the axon. Thus, both an electrical field and a concentration gradient across the membrane provide thermodynamic potential energy that tries to force Na^+ inward through the membrane. The net inward potential energy in electron volts (eV) for Na^+ ions is the sum of the transmembrane electrical potential, V_m, and the *Nernst potential* for sodium ions. This potential energy is given by the well-known relation [Katz 1966]:

$$E_{Na} = V_m - (RT/\boldsymbol{F})\ln\left(\frac{[Na^+]_o}{[Na^+]_i}\right)\ \text{eV} \tag{4.56}$$

where V_m is the transmembrane resting potential in volts, – 0.070 V for squid at 20°C (note V_m also depends on temperature); R is the SI gas constant, 8.314 J/mole K; ***F*** is the *Faraday constant,* 9.65 × 10^4 C/mole; T is the Kelvin temperature; $[Na^+]_o$ and $[Na^+]_i$ are the concentrations of sodium ions outside (460 mM) and inside (46 mM) the membrane, respectively, in moles/liter for squid axon. $E_{Na} < 0$ means the net force on Na^+ is *inward.* That is, a sodium ion outside the cell membrane has a total potential energy of E_{Na} electron Volts acting to force it inward. For squid axon at 293 K (20°C), we calculate E_{Na}:

$$E_{Na} = -0.070 - (8.314 \times 293 / 9.65 \times 10^4)\ \ln(460/46) = -\ 0.128\ \text{eV} \tag{4.57}$$

Similarly, we can calculate the potential energy, E_K, acting to force a K^+ ion *outward* through the resting membrane:

$$E_K = -\ 0.070 - (8.314 \times 293/9.65 \times 10^4)\ \ln(10/410) = +\ 0.024\ \text{eV} \tag{4.58}$$

Na^+ ions generally enter the axon through special voltage-gated Na^+ channels. With the membrane unexcited and in its steady-state condition, there is a very low, random leakage of Na^+ ions inward through the membrane. Potassium ions also leak out because the net potassium potential energy is dominated by the concentration gradient (the inside having a much higher potassium concentration than does the outside, that is, $[K^+]_i \gg [K^+]_o$). Other small ions leak as well, being driven through the membrane by the potential energy difference between the membrane's resting potential and the ion's Nernst potential.

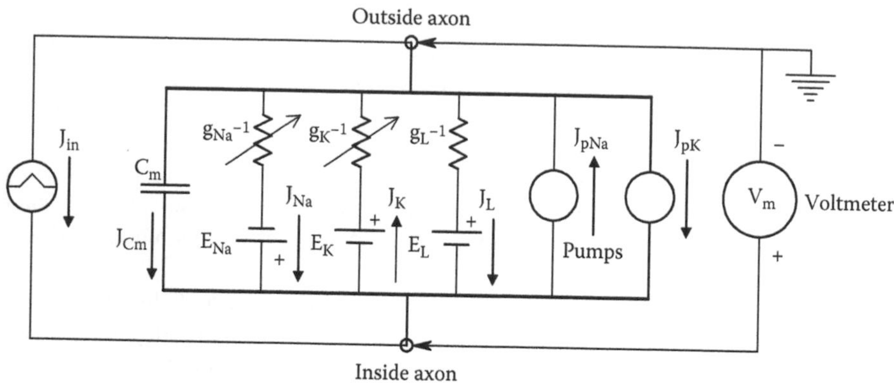

FIGURE 4.17 Lumped-parameter equivalent circuit of the unit area of the H–H model's active neuron membrane. The ion pumps do not figure in the H–H model's formulation.

At steady-state equilibrium, a condition known as *electroneutrality* exists in unit volumes inside and immediately outside the axon. That is, in each volume there are equal numbers of positive and negatively charged ions and molecules. Inside the axon, a significant percentage of negative charges are bound to large protein molecules that cannot pass through the membrane because of their sizes.

Not specifically pertinent to the Hodgkin–Huxley kinetic model for nerve spike generation is the fact that nerve membranes (on soma, axon, and dendrites), and those of nearly all other types of cells, contain molecular "pumps" driven from the energy in ATP molecules, which can eject (pump) Na^+ from the interior of the cell against the potential energy barrier, E_{Na}, often as an exchange operation with K^+ being pumped in. *Ion pumps* are ubiquitous in nature and are responsible for the maintenance of the steady-state transmembrane resting potentials of nerves, muscle cells, and all other cells. (Ion pumps are described in detail in Chapter 7, Section 7.2.5.2.)

To try to describe, summarize, and model the transient electrical events associated with squid nerve axon membrane, a simple, parallel, electrical circuit was developed [Hodgkin & Huxley 1952; Katz 1966] that includes *specific ionic conductances* for Na^+, K^+, and "other ions" (leakage) for a 1 cm^2 (unit) area of axon membrane. Figure 4.17 illustrates the simple circuit model. $\mathbf{V}_{Na}$ and $\mathbf{V}_K$ are the Nernst potentials for sodium and potassium ions, respectively, and $\mathbf{V}_L$ is the equivalent Nernst potential for *all* "leakage" ions, including chloride. A specific ionic conductance is defined by the specific ionic current density divided by the difference between the actual transmembrane potential and the Nernst potential. Thus, we can write:

$$g_K \equiv J_K/(V_m - V_K) \text{ Siemens/cm}^2 \tag{4.59A}$$

$$g_{Na} \equiv J_{Na}/(V_m - V_{Na}) \text{ Siemens/cm}^2 \tag{4.59B}$$

$$g_L \equiv J_L/(V_m - V_L) \text{ Siemens/cm}^2 \tag{4.59C}$$

For example, in resting squid axon, $g_{Na} = J_{Na}/(-0.070 - 0.058)$. The sodium ion current density, J_{Na}, is negative inward by definition, so g_{Na} is positive as it should be.

The mathematical model devised by Hodgkin and Huxley (1952) to describe the generation of the squid axon action potential began by writing a node equation for the equivalent circuit of Figure 4.17 using Kirchoff's current law:

$$\mathbf{J}_{in} = \mathbf{C}_m \, \mathbf{dv/dt} + \mathbf{J}_K + \mathbf{J}_{Na} + \mathbf{J}_L \tag{4.60A}$$

$$\downarrow$$

$$\mathbf{C}_m \, \dot{v} = \mathbf{J}_{in} - \mathbf{J}_K - \mathbf{J}_{Na} - \mathbf{J}_L \tag{4.60B}$$

$$\downarrow$$

$$\dot{v} = (\mathbf{J}_{in} - \mathbf{J}_K - \mathbf{J}_{Na} - \mathbf{J}_L)/\mathbf{C}_m \tag{4.60C}$$

By definition, the *transmembrane voltage change,* $\mathbf{v(t)} \equiv [\mathbf{V}_{m0} - \mathbf{V}_m\mathbf{(t)}]$. ($\mathbf{v}$ is in millivolts.) If $\mathbf{v} < 0$, $\mathbf{V}_m$ is depolarizing (going positive from the resting $\mathbf{V}_{m0} = -70$ mV). The leakage current density is assumed to obey Ohm's law

$$\mathbf{J}_L = \mathbf{g}_{Lo} \, (\mathbf{v} - \mathbf{V}_L) \quad \mu A/cm^2 \tag{4.61}$$

$\mathbf{g}_{Lo}$ is taken as 0.3 mS/cm², $\mathbf{V}_L \equiv -10.613$ mV.

From previous studies with the squid axon membrane under electronic voltage clamp conditions (where $\mathbf{V}_m$ is forced to assume set values), where the potassium current was measured with the sodium channels blocked, it was observed that $\mathbf{g}_K = \mathbf{J}_K/\mathbf{V}_m = \mathbf{g}_K(\mathbf{V}_m, t)$.

From chemical kinetics considerations, Hodgkin and Huxley assumed that *four* "particles" must simultaneously occupy specific sites on the potassium gate protein in order to open it. Thus, based on chemical kinetics, they wrote:

$$\mathbf{J}_K = \mathbf{g}_{Ko} \, \mathbf{n}^4 \, (\mathbf{v} - \mathbf{V}_K) \quad \mu A/cm^2 \tag{4.62}$$

where $\mathbf{n}$ is the K^+ *activation parameter;* it is taken as the probability of a K-gate opening particle being at the active site. $\mathbf{g}_{Ko}$ is taken to be 36 mS/cm², $\mathbf{V}_K = 12$ mV. $\mathbf{n}$ was given by the first-order, nonlinear ODE:

$$\dot{n} = -\mathbf{n} \, (\alpha_n + \beta_n) + \alpha_n \tag{4.63}$$

α_n and β_n are exponential functions of membrane potential:

$$\alpha_n = 0.01(\mathbf{v} + 10)/[\exp(0.1\mathbf{v} + 1) - 1] \tag{4.64}$$

$$\beta_n = 0.125 \exp(\mathbf{v}/80) \tag{4.65}$$

The *sodium current* is *activated* once $\mathbf{V}_m$ reaches a depolarization threshold voltage. (Note the threshold voltage does not appear explicitly in the H–H model.) The sodium current or conductance *deactivates* spontaneously once its ion channels have been opened. Hodgkin and Huxley observed that sodium channel activation had kinetics that suggested *three* "particles" were required (not necessarily the particles for potassium channel activation).

Deactivation of the sodium channels suggested *monomolecular kinetics,* where a single event of probability $(1 - \mathbf{h})$ causes inactivation. Thus, the probability of Na^+ channels being open was $\mathbf{m}^3 \, \mathbf{h}$, and the sodium current density was written:

$$\mathbf{J}_{Na} = \mathbf{g}_{Nao} \, \mathbf{m}^3 \, \mathbf{h} \, (\mathbf{v} - \mathbf{V}_{Na}) \quad \mu A/cm^2 \tag{4.66}$$

where $\mathbf{g}_{Nao} = 120$ mS/cm² and $\mathbf{V}_{Na} = -115$ mV. The nonlinear ODEs for sodium activation and deactivation are, respectively,

$$\dot{m} = -\mathbf{m}(\alpha_m + \beta_m) + \alpha_m \tag{4.67}$$

$$\dot{h} = -\mathbf{h}(\alpha_h + \beta_h) + \alpha_h \tag{4.68}$$

The four, nonlinear, voltage-dependent parameters used in Equations 4.67 and 4.68 are

$$\alpha_m = 0.1(\mathbf{v} + 25)/[\exp(0.1\mathbf{v} + 2.5) - 1] \tag{4.69}$$

$$\beta_m = 4\exp(\mathbf{v}/18) \tag{4.70}$$

$$\alpha_h = 0.07\exp(\mathbf{v}/20) \tag{4.71}$$

$$\beta_h = 1/[\exp(0.1\mathbf{v} + 3) + 1] \tag{4.72}$$

We wrote a simple *Simnon* program, HODHUX.t, to compute $\mathbf{V}_m \equiv -(\mathbf{v} + 70)$ mV and the various current densities and conductances versus time. Note that the current densities are in μA/cm^2, voltages are in mV, conductances are in milliSiemens/cm^2, and the capacitance is in millifarads/cm^2. The strange treatment of units and voltage and current signs follows from the original 1952 H–H paper.

We will examine how the nonlinear dynamic system model of Hodgkin and Huxley responds to depolarizing and hyperpolarizing input current densities. Of particular interest is the system's sensitivity to the rate of change of the *injected current density,* $\mathbf{J}_{in}$. Also, you will see that the system acts as a nonlinear current-to-frequency converter. Such steady-state oscillatory behavior is seen in nature in certain *pacemaker* neurons.

Using Hodgkin and Huxley's original sign convention, currents (seen as positive ion flow) *entering* the axon from outside, such as $\mathbf{J}_{Na}$, are *positive.* $\mathbf{J}_K$ *leaving* the axon lumen is *negative.* $\mathbf{V}_m$ is shown as we would measure it; the resting $\mathbf{V}_{m0} \equiv -70$ mV. $\mathbf{J}_{in} < 0$ depolarizes the membrane, that is, it drives $\mathbf{V}_m$ positive. ($\mathbf{J}_{in}$ is a negative ion [or electron] current density.)

Figure 4.18 illustrates a single simulated action potential (AP) produced when $\mathbf{J}_{in} = -2$ μA/cm^2 for 5 ms. Also plotted are the scaled membrane capacitance current density, potassium ion current density, and the sodium ion current density. Scaling details are given in the figure caption. In Figure 4.19 we plot the *auxiliary parameters* **n**, **m**, and **h** when the AP is generated. The AP is scaled by 1/70 in order that the same vertical scale can be used for **n**, **m**, and **h.** Note that the *sodium activation parameter,* **m**, falls rapidly after reaching its peak near unity. Finally, in Figure 4.20, the *specific ionic conductances* $\mathbf{g}_K$, $\mathbf{g}_{Na}$, and $\mathbf{g}_{net}$ are plotted with $\mathbf{V}_m(t) + 70$ mV. Note that $\mathbf{g}_K$ dominates the recovery phase of $\mathbf{V}_m$ following the spike.

A little-appreciated property of living squid axon membrane is its propensity to fire on *rebound* from a prolonged, induced hyperpolarization of $\mathbf{V}_m$. Hodgkin and Huxley [1952] called this phenomenon *anode break excitation.* Figure 4.21 illustrates in the H–H model what happens when $\mathbf{J}_{in}$ is positive (note that $\mathbf{J}_{in} = \mathbf{J}_{Na}$), forcing $\mathbf{V}_m$ more negative, hyperpolarizing the membrane. Rebound firing occurs when the positive $\mathbf{J}_{in} \to 0$; $\mathbf{V}_m$ overshoots $\mathbf{V}_{m0}$ enough to cause a spike to occur for $\mathbf{J}_{in} = 2$, 3, and 5 μA/cm^2. There is not enough rebound in $\mathbf{V}_m$ for $\mathbf{J}_{in} = 1$ or 1.5 μA/cm^2 to cause firing, however.

Another interesting property of the H–H model is that it models the natural behavior of active neuron membrane as a *current-to-frequency converter* (or, more physiologically, an epsp-to-frequency converter). In this case, we apply a prolonged, negative, depolarizing, input current density, $\mathbf{J}_{in}$. For the model parameters given, and $\mathbf{C}_m = 0.003$ mF, we find that for $\mathbf{J}_{in} \leq -7$ μA/cm^2, the model fires repetitively at a constant frequency; the frequency increases as $\mathbf{J}_{in}$ becomes more negative. Figure 4.22 and Figure 4.23 illustrate this behavior. Note particularly that the peak-to-

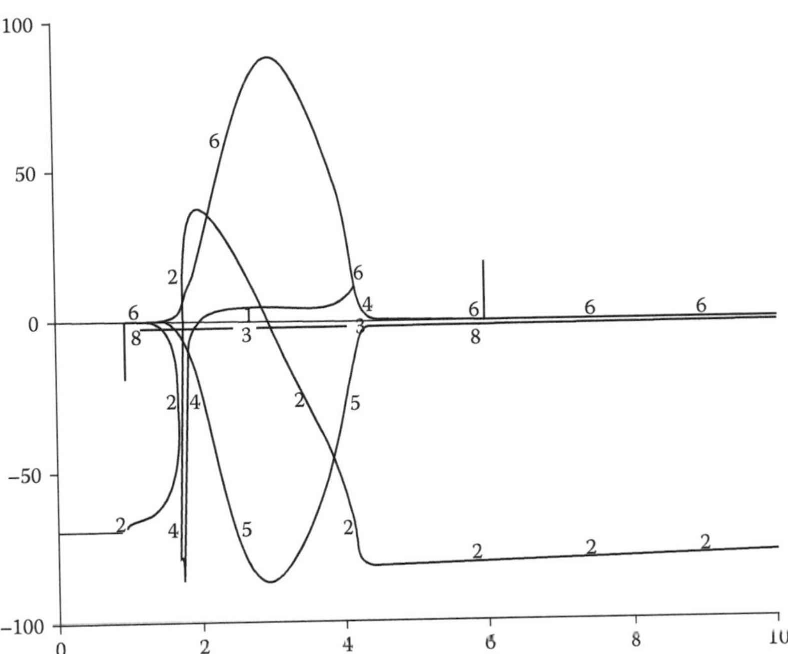

FIGURE 4.18 Results of the *Simnon* simulation of the H–H model for nerve impulse generation using Euler integration with $\delta t = 0.00001$, $C_m = 0.01$. Horizontal axis, time in milliseconds. Traces: (2) V_m in mV. (3) $J_{in} = -2\ \mu A/cm^2$ from $t = 1$ to 6 ms, else 0. (4) $J_C = 10\ J_{Cm}$ (scaled capacitance current density in $\mu A/cm^2$). (5) $J_K = J_{Km}/10$ (scaled potassium ion current density in $\mu A/cm^2$). (6) $J_{Na} = J_{Nam}/10$ (scaled sodium ion current density in $\mu A/cm^2$).

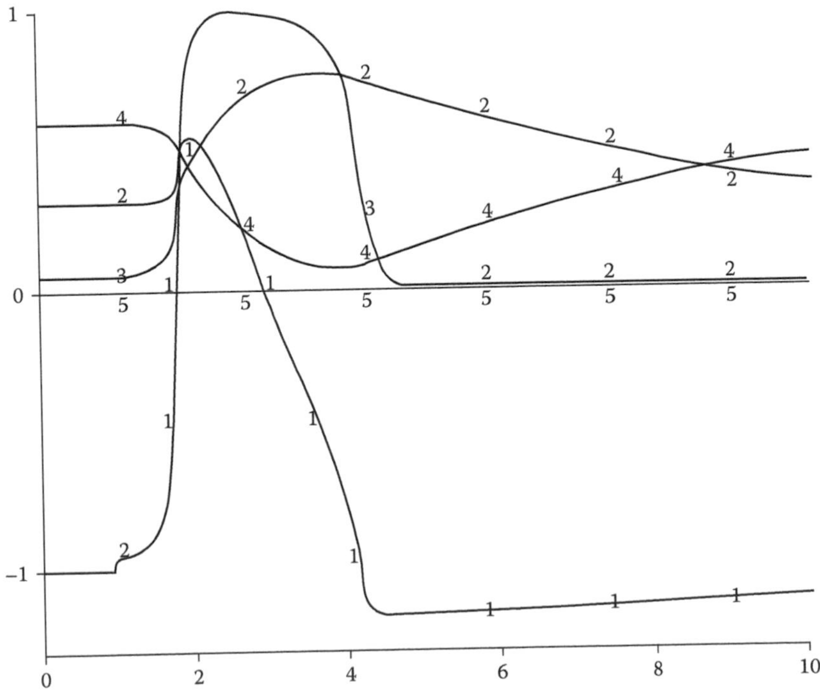

FIGURE 4.19 Further results of the Simnon simulation of the H–H model nerve action potential. Traces: (1) $V_m(t)/70$ mV, (2) $n(t)$, (3) $m(t)$, (4) $h(t)$, (5) zero volts. Horizontal axis, time in milliseconds.

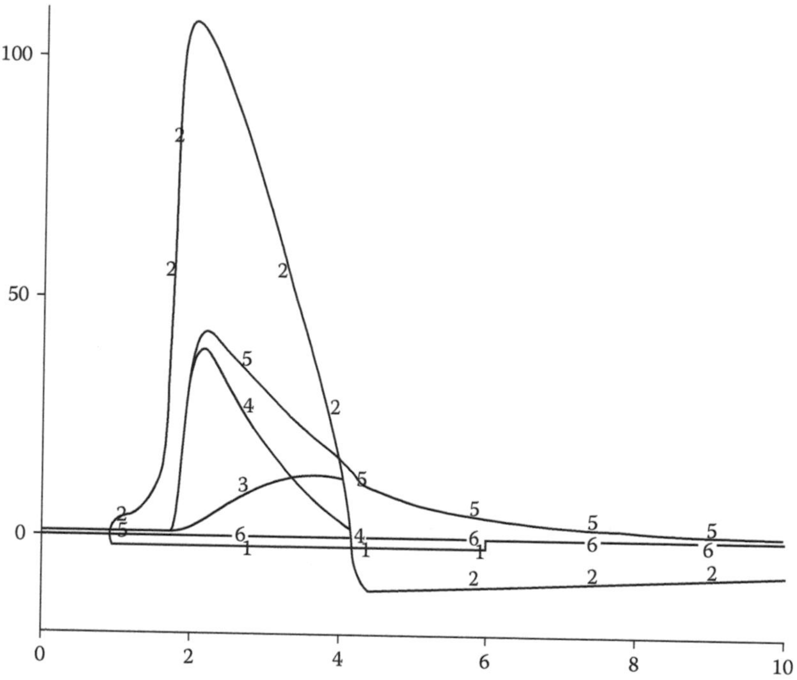

FIGURE 4.20 Further results of the Simnon simulation of the H–H model. Traces: (1) J_{in} μA/cm², (2) [$V_m(t)$ + 70] mV, (3) $g_K(t, V_m)$ mS/cm², (4) $g_{Na}(t, V_m)$ mS/cm², (5) $g_{net} = g_K + g_{Na} + g_L$ in mS/cm². Horizontal axis, time in milliseconds.

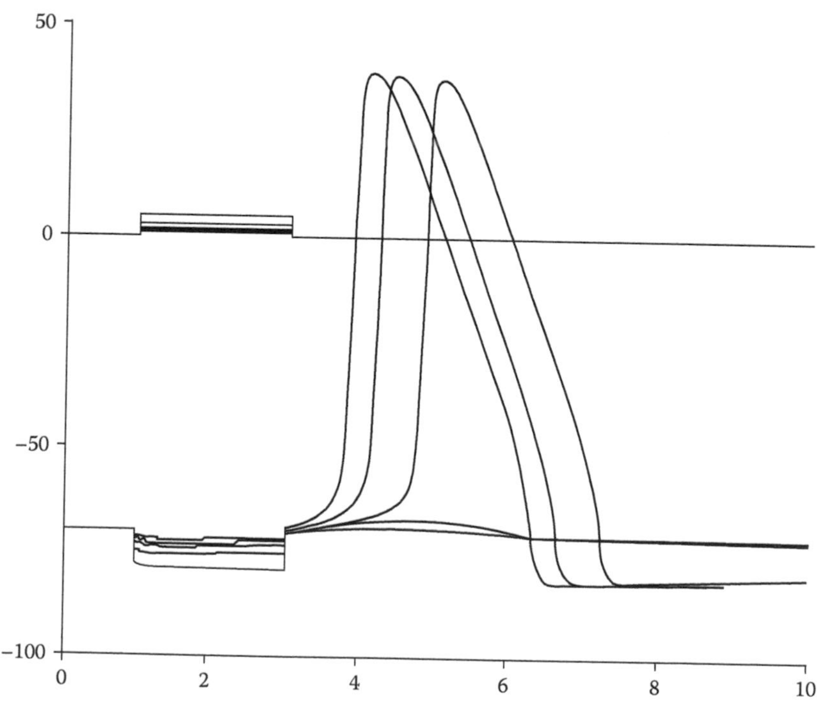

FIGURE 4.21 Simulation of anode break excitation. H–H model with $\delta t = 0.00001$, $C_m = 0.003$. Horizontal axis, time in milliseconds. Top trace: J_{in} (hyperpolarizing current densities of 1, 1.5, 2, 3, 5 μA/cm²). No spike for $J_{in} = 1$ and 1.5.

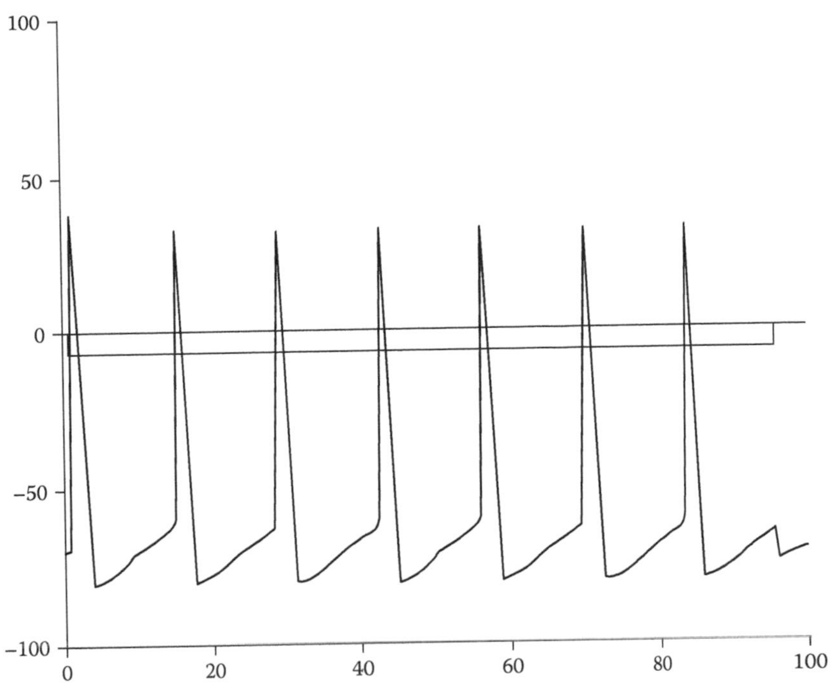

FIGURE 4.22 Simulation of the H–H model as a dc current-to-frequency converter. Euler integration with $\delta t = 0.00001$, $C_m = 0.003$. Horizontal axis, time in milliseconds. DC $J_{in} = -7\ \mu A/cm^2$. Horizontal axis, time in milliseconds.

peak height of the periodic spikes decreases as the frequency increases. At $\mathbf{J}_{in} = -\ 140\ \mu A/cm^2$, the membrane oscillation was centered around −50 mV and was nearly sinusoidal, having a peak-to-peak amplitude of about 12.5 mV (hardly nerve spikes). The steady-state frequency of the H–H model "oscillator" versus dc $\mathbf{J}_{in}$ is plotted in Figure 4.24 on log–log coordinates. An approximate model that fits the plot is $f = 57 + 6|\mathbf{J}_{in}|^{0.64}$ pps. Thus, the H–H model is not a linear current-to-voltage converter but behaves in a power-law manner over the range, $-\ 7 \geq \mathbf{J}_{in} \geq -140\ \mu A/cm^2$. This result quantifies the physiological observation that neurons fire faster in response to increased, steady-state excitatory stimulation.

4.6 SOME SIMULATION LANGUAGES FOR BIOCHEMICAL SYSTEMS

As biochemists have described more of the details of biochemical pathways, and interest in quantitative biology has grown, so have the number of software packages proliferated, specialized for the solution of biochemical kinetic problems. Several commercial software application packages can be used to solve small- and medium-sized sets of first-order, nonlinear ODEs. These include the well-known Matlab/Simulink software and Simnon. We present and discuss a list of commercial programs, as well as an extensive list of public domain (free) applications for modeling molecular biological systems.

Of all the software used for the analysis and characterization of biomedical signals and systems, the Matlab family of object programs is the most ubiquitous. (Matlab is at v. 7.4, 31 May 2007.) Most engineering schools introduce their students to Matlab early in their academic careers. It is used to solve home problems in the systems analysis area and is the designated software of certain systems engineering textbooks. Historically, it started out as a set of linkable object programs used for linear, state-variable matrix operations on LTI systems. To satisfy the need for specialized applications, MathWorks has developed over 30 different Toolboxes™ (a trademark of MathWorks, Inc.,

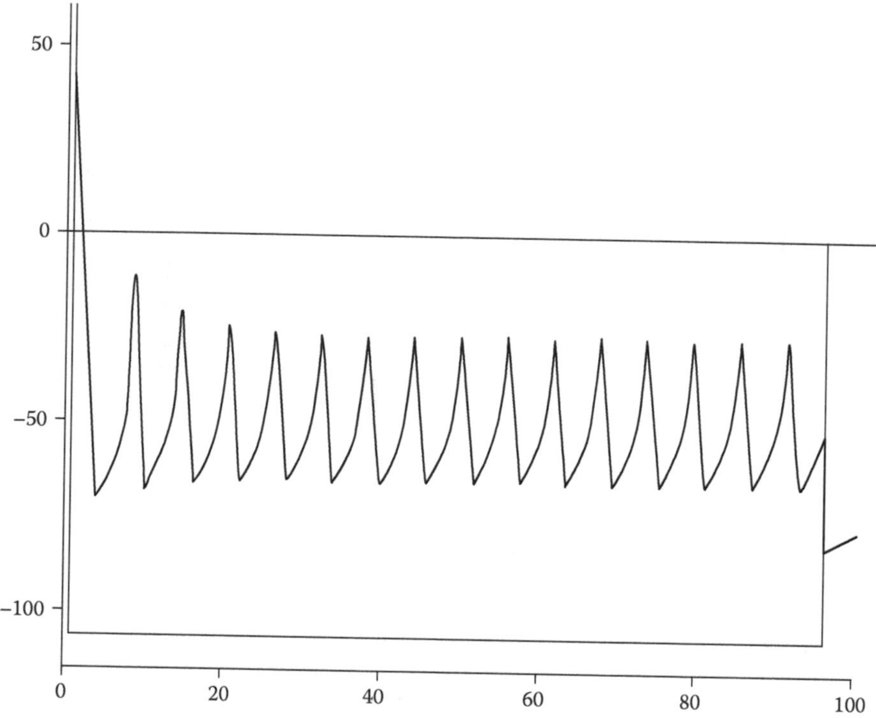

FIGURE 4.23 Simulation of the H–H model as a dc current-to-frequency converter. Euler integration with $\delta t = 0.00001$, $C_m = 0.003$. Horizontal axis, time in milliseconds. DC $J_{in} = -105\ \mu A/cm^2$. Note spikes do not reach 0 v.

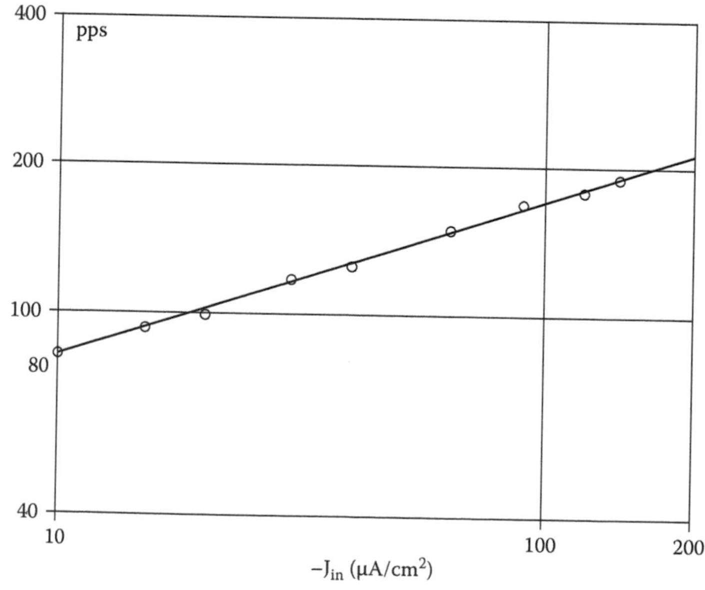

FIGURE 4.24 Log–log plot of steady-state frequency versus $-J_{in}$ for the H–H model. Same simulation parameters as in Figures 4.22 and 4.23. See text for discussion.

Natick, Massachusetts) for Matlab. Each Toolbox includes a collection of *m-files* dedicated to some specialized aspect of signal processing, system analysis, or simulation. The Toolboxes which may have the greatest application to biochemical signals and systems include *communications, control systems, curve fitting, data acquisition, neural networks, partial differential equations, signal processing, statistics,* and *system identification.*

Simulink is a GUI modeling and simulation program that runs with Matlab (Simulink is now at v. 6.6.1, 31 May 2007.) Different flavors of Matlab and Simulink run on personal computers (PCs) with Windows 95, 98 NT4.0, Me, 2000, XP, and Vista or on Macs or Unix platforms. Simulink is an icon-driven, dynamic simulation package that allows the user to represent a dynamic system by a graphical block diagram. The blocks fall into the categories of *Sources, Sinks, Discrete LTI Transfer* Functions (including delays and holds), *Continuous LTI Transfer* Functions (including time derivatives, integration, LTI state-variable systems, and delays), *Nonlinear Function Blocks* (including switching, hysteresis, etc.), *Mathematical and Matrix Operations* (including Matlab and Toolbox functions), *Signal* and *Systems, and user-defined Subsystems operations.* Simulink blocks are also organized by application; for example, there is a *Communications Blockset v2, DSP Blockset v s4, FixedPoint Blockset v3, Power Systems v2.1, Real-Time Workshop v.4,* etc.

As a system's block diagram is built, the user has to specify numerical values for the parameters in the blocks and, of course, the interconnections between them. Before the simulation is run, the user specifies the integration routine to be used, the stepsize, and start and stop times. A diverse selection of integration routines includes *R-K 2-3, R-K 4-5, Euler, Gear, Adams,* and *"Linsim"* (plain vanilla for purely linear state systems). *Gear* is recommended for stiff, nonlinear systems. Because Simulink runs in the Matlab "shell," it can make use of all of the many features of Matlab and its various toolboxes.

Although extremely versatile for the simulation of analog and/or discrete systems whose subsystems are describable as rational polynomial transfer functions or sets of LTI state equations, in the author's opinion, Simulink does not shine in the simulation of large systems of functionally nonlinear ODEs, such as found in chemical kinetics, physiology, or neural modeling. Its block diagram format becomes unwieldy, and it is clear that a program such as Simnon that accepts the algebraically written ODEs directly is easier to set up for running a simulation. Simulink is best used for systems engineering modeling and design verification.

The reader interested in learning more about Simulink and Matlab can visit the URL: http://www.mathworks.com/products/simulink.

Simnon is a venerable, time-domain, nonlinear, ordinary differential equation solver. It was developed at the Department of Automatic Control at the Lund Institute of Technology, Sweden, in the late 1980s. The program in its early versions (v1.0–v3.2) was written to run under non-Windows MS-DOS on PCs. A simple, algebraic input script was used. In 1988, the author found that Simnon was particularly well suited for simulation of *compartmental pharmacokinetic systems, biochemical kinetic systems*, and *nonlinear physiological systems.* Because its input modality is in the form of the sets of algebraic, first-order, linear, nonlinear, and time-variable ODEs that arise naturally from the analysis and modeling of these three classes of systems, there is no need to put the sets of nonlinear ODEs in transfer function (block-diagram) form. Simnon also allows the simultaneous simulation of a discrete controller along with a nonlinear analog system when it is desired to simulate a complete, exogenous, closed-loop control system used to regulate a drug concentration. One can also model systems described by sets of linear or nonlinear difference equations. Noise can be included in Simnon simulations.

Simnon 3.0, the latest Windows version, is well suited to solve sets of stiff, nonlinear ODEs. The user has the choice of one of four integration algorithms: Runge-Kutta/Fehlberg 2nd/3rd order, Runge-Kutta/Fehlberg 4th/5th (4th order has variable or fixed step size), Euler, and Dormand-Prince 4th/5th. Simnon 3.0/PCW can handle up to 10,000 states (ODEs), 100 subsystems, 50 pure time delays, 50 function tables, 32 plot variables, and 100 stored variables. Calculations are performed with double precision. Simnon uses a user-friendly, interactive GUI, and it has quality graphic outputs on monitors and to color laser and bubble jet printers. Solutions of sets of nonlinear ODEs by Simnon can be displayed in the time domain, or parametrically as phase-plane plots. Simnon data files can also be exported to Matlab. Matlab Toolboxes can then be used to do frequency- and JTF-domain operations not found in Simnon.

The author has effectively used Simnon 3.2 to model the dynamic behavior of closed-loop, nonlinear physiological systems [Northrop 2000] and to model the behavior of neurosensory systems [Northrop 2001].

The current Windows version of Simnon (v. 3.0) is available from SSPA Sweden AB, PO Box 24001, SE-400 22 Göteborg, Sweden. http:/sspa.se/software/simnon.html, email URL = simnon@sspa.se. Its price is still €99 from SSPA; add €80 for courier shipping (31 May 2007).

Teranode® Corporation offers the *Pathway Analytics*® application as a component of their *Solutions* group software (Teranode is a registered trademark of Teranode Corporation, Seattle, Washington). This is "... a pathway modeling and analysis solution for biologists and bioinformaticians to easily manage data in the context of biological pathways." *Pathway Analytics* allows researchers to create new biological models, attach raw data files generated from laboratory experiments, and automate calculations of biological kinetic parameters. It also permits the annotation of pathways with experimental or simulated observations and store simulation results for later comparison with experimental data.

The Teranode *Pathway Analytics* features permit the modeler to:

- **Graphically create and share detailed pictures of biological pathways.** Draw pathways and annotate them with text, lines, and data for viewing. Analyze and identify targets, biomarkers, diseases, or biological functions.
- **Integrate disparate sources of biological data.** Leverage publicly curated biological content from sources such as KEGG, PubMed, NCBI, and SBML or integrate with your proprietary content sources.
- **Incorporate experimental data into pathway analysis.** Import experimental data from MAGEML, Affymetrix, or *.csv file formats and integrate with experimental data to better understand interactions.
- **Build your own pathway repository.** Create and maintain a central, searchable file-set of biological content to share.
- **Automatically generate simulations of biological systems.** Convert a pathway model into a mathematical format for kinetic analysis.

Teranode also offers a *Design Suite Biological Modeler* which also does kinetic simulations from pathways; a GUI is used. Teranode Corporation was founded in 2002. It offers a wide selection of bioinformatics and biochemical simulation software that empowers scientists to automate, manage, and unify experimental data and work flow within an ever-changing R&D environment [see www.teranode.com].

There are several *free*, licensed software packages that are available online that are specialized for biochemical systems simulation. *GEPASI,* or General Pathway Simulator, runs under Windows 95® and above (it is a product of the Virginia Bioinformatics Institute, Blacksburg, Virginia). Its features are:

- *Gepasi* is free software (*user license*).
- *Gepasi* runs under Microsoft Windows 95 and above.
- Models in *Gepasi* v3 can be composed of many compartments with different volumes.
- The number of reactions and metabolites in each model is only limited by available memory.
- Simulations can be followed interactively (including adding perturbations to a time course).

- *Gepasi* characterizes steady states using *metabolic control analysis* and linear stability analysis.
- *Gepasi*'s *scan* utility provides a way for advanced exploration of a model's behavior in multidimensional parameter space.
- *Gepasi* is capable of doing data fitting (parameter estimation) with experimental data.
- *Gepasi* is capable of finding maxima or minima of any model variables with any number of adjustable model parameters.
- The results of simulations can be plotted in 2-D and 3-D directly from the program (*Gepasi* uses the excellent **Gnuplot** package).
- *Gepasi* supports SBML level 1 for model interchange with other systems biology modeling software.

See www.gepasi.org/. The *Gepasi* current version is 3.30, released 4 September 2002. The program and its applications are discussed in Mendes [1993] and Mendes and Kell [1998].

Another free software package is *Copasi*™(a product of Virginia Bioinformatics Institute and a registered trademark of Virginia Tech Intellectual Properties, Inc., Blacksburg, Virginia), or Complex Pathway Simulator. Its current features are

- Stochastic and deterministic time course simulation
- Steady-state analysis (including stability)
- Metabolic control analysis/sensitivity analysis
- Elementary mode analysis
- Mass conservation analysis
- Parameter scans
- Optimization of arbitrary objective functions
- Parameter estimation using data from time course and/or steady-state experiments
- Sliders for interactive parameter change
- Global parameters to change multiple kinetic rates at once
- Imports and exports **SBML** (export only in level 2 version 1, import all levels)
- Loads *Gepasi* files
- Versions for MS windows®, Linux, OS X, and Solaris SPARC
- Command line version for batch processing
- Runs on Windows 98 and above

The latest version is Release Candidate 1, Build 14, 15 February 2006. Visit www.copasi.org/tiki-index.php for details.

A software application called Biochemical Pathway Simulator (BPS) was developed at the University of Glasgow beginning in 2002 (BPS is a product of Bioinformatics Research Centre, University of Glasgow, Glasgow, Scotland). More about *BPS* can be found at: www.brc.dcs.gla.ac.uk/projects/bps/ links.html.

Still another simulation package is *VCell 4.2* (a product of the National Resource for Cell Analysis and Modeling, Farmington, Connecticut). *VCell* will do biochemical kinetic system modeling as well as compartmental models with membrane diffusion. Several tutorial examples are available online from www.vcell.org/ that cover such models as the Hodgkin–Huxley neuron active membrane system and facilitated calcium diffusion in the intestinal epithelial cell. *VCell* has an elaborate GUI for entering model structure and parameters.

Cellerator™ (a product of the Institute of Genomics and Bioinformatics, Jet Propulsion Laboratory, California Institute of Technology, and a trademark of the California Institute of Technology, Pasadena, California) is a Mathematica®-based (a product and registered trademark of Wolfram Research, Inc., Natick, Massachusetts) program for biosystems simulation [Shapiro & Mjolsness 2001]. Cellerator was designed to simulate the following essential biochemical processes: (1) signal transduction networks (STNs), (2) cells that are represented by interacting signal transduction networks, and (3) multicellular tissues that are represented by interacting networks of cells that may themselves contain internal STNs. Cellerator will run on any computer on which Mathematica v. 4.1 is installed (Mathematica is now at v. 5.2); it runs on Windows, Mac, Unix, and Linux systems. It is implemented as a Mathematica "notebook." It solves complete sets of ODEs predicted by the law of mass action. Cellerator uses a chemical "arrow notation" to enter the reactions that must be simulated, from which it writes the mass action ODEs.

The results of Cellerator simulations can be written using *Systems Biology Markup Language* (SBML). See www.sbml.org/ for information on this utility. Output can also be expressed as Mathematica ODEs, in C, FORTRAN, MATHML, or HTML files. It is not Matlab compatible.

Multicellular systems are represented by graphs containing a list of *nodes* (representing cells), a list of *links* (representing intercellular connections), and a *lineage tree* (a familial history of cell birth). Cell division can be made to occur in a model whenever a user-specified variable passes a threshold, and new cells are added to the graph in corresponding locations, within nodes. Genetic regulatory networks are represented by a generalization of the connectionist model [Shapiro et al. 2003]. Cellerator is not public domain software; it is the property of the California Institute of Technology. For more information on Cellerator, visit www.cellerator.org/. Cellerator can be licensed free of charge to noncommercial academic, U.S. government, and nonprofit users. Details are available at: www-aig.jpl.nasa.gov/public/mls/ cellerator.

Associated with Cellerator is an *enzyme mechanism language, kMech*. kMech is a comprehensive collection of single and multiple substrate–enzyme reactions and regulatory mechanisms that extends Cellerator function for the mathematical modeling of enzyme reactions. kMech was developed by Yang, Shapiro, Mjolsness, and Hatfield [2005]. See: www.igb.uci.edu/servers/coli/kmech.html.

xCellerator™ (a trademark of the California Institute of Technology, Pasadena, California) is a modification of the Cellerator software intended for cell signal transduction modeling. It, too, requires Mathematica. xCellerator is designed to aid biochemical modeling by automatically converting chemical reactions into nonlinear ODEs. It is compatible with the reaction syntax defined in Cellerator, but it uses a completely different algorithm to perform the translation. Some features include faster reaction translation (as much as 100 times faster than Cellerator), direct entry points limited to those needed by the user to significantly reduce the possibility of global identifier collision, user can define new reactions "on the fly," and stochastic modeling using Gillespie's SSA. For more information on xCellerator, see www.xcellerator.info/ (13 August 2005).

Calzone et al. [2006] describe a recently developed software system, *BIOCHAM*, the biochemical abstract machine (BIOCHAM is a product of Project Contraintes, INRIA, Rocquencourt, France, and is protected by the Gnu Lesser Public License). In their paper's abstract, the authors state: "BIOCHAM is a software environment for modeling biochemical systems. It is based on two aspects: (1) the analysis and simulation of boolean, kinetic and stochastic models and (2) the formalization of biological properties in temporal logic." They go on to describe BIOCHAM as providing the tools and languages for describing protein networks with a simple and straightforward syntax enabling the integration of biological properties into the model. When simulating biochemical kinetics, BIOCHAM can search for appropriate parameter values in order to reproduce a specific behavior observed in experiments (model verification). Coupled with other methods such as bifurcation diagrams, this directed search assists the modeler/biologist in the modeling process. BIOCHAM v. 2.5 is a free software available for download, with example models, at: http://contraintes.inria.fr/BIOCHAM/. Contact Sylvain.Soliman@inria.fr.

Vera and Torres [2003] have developed *MetMAP*, a structured library of Matlab functions specifically designed to model, analyze, and optimize S-system biochemical and biotechnological models.

The authors point out: "The user can develop an S-system model of a given biochemical system by including a mathematical description of all the kinetic equations of reactions. At this level the program is highly flexible and can operate with any type of mathematical representation, including the usual enzyme kinetic rate laws. There are two possible ways to control the functions of this package. One can use the Graphical User Interface [GUI], or alternately use the version that runs from an editable file called *kernel*." [Quote used with permission.]

The GUI (MetMAP-GUI) runs in the Matlab environment. Vera and Torres recommend MetMAP-GUI for Matlab beginners. Alternatively, the user can use the *editable file version* that runs from the Matlab ***m***-file editor. The advantage of using the ***m***-file editor is that it allows a more flexible use of functions and the customization of calculations. The authors illustrate MetMAP with a two-state, nonlinear kinetic system. The fact that MetMAP runs in the ubiquitous Matlab environment makes it attractive for biochemical kinetics modeling.

Quoting Vera and Torres' conclusion: "MetMAP is a flexible package of functions that performs numerical, dynamic and optimization analysis of biochemical systems, designed as an open environment that can easily be modified by users to include new features. It was designed to share all the advantages that make Matlab a powerful and flexible tool for computational analysis. Thus the graphics generated with MetMAP are of high quality and are easy to work with. Calculations are made in a fast and accurate fashion and there are no predictable limitations in the size of the systems to be analysed. MetMAP has been designed to provide complete connectivity among all the different modules and is open to be adapted by the user to fit his specific needs." [Quote used with permission.]

This software package appears to be well suited for the modeling of small and medium biochemical systems. Interested readers should contact the corresponding author at: ntorres@ull.es.

At the risk of boring our readers with a seemingly endless list of biochemical systems modeling software, we cite one last application program, the *SMBL ODE Solver Library* or SOSlib (in its current release, a product of CellDesigner, Inc., the Systems Biology Institute, Tokyo, Japan, protected by the Gnu Public License). SOSlib "...is a programming library for symbolic and numerical analysis of chemical reaction network models encoded in the Systems Biology Markup Language (SMBL). It is written in ISO-C and distributed under the open source LGPL licence." We note that SOSlib employs libSBML structures for formula representation and associated functions to construct a system of ODEs, their Jacobian matrix, and other derivatives. SUNDIALS' CVODES are incorporated for numerical integration and sensitivity analysis.

[SUNDIALS, or Suite of Nonlinear and Differential Algebraic Solvers, is a product of the Center for Applied Scientific Computing, Laurence Livermore National Laboratory, Livermore, California]. A short description of SOSlib can be found in the applications note by Machné et al. [2006]. Availability: contact: www.tbi.univie.ac.at/~raim/odeSolver/ or xtof@tbi.univie.ac.at.

A utopian objective of quantitative biology is to simulate *all* of the biochemical processes and functions of a *cell,* including metabolism, signal transduction, gene expression, DNA replication and repair, RNA functions, transmembrane transport, and physicochemical events such as cytoplasmic streaming and cytoskeletal changes ("the full catastrophe," to quote from the film, *Zorba the Greek*).

Whole-cell simulation involves many different components and variables: In *Homo sapiens,* about 10^{17-18} total molecules of 10^{4-5} species with about 10 component types and 10^{1-4} interaction modes [Takahashi et al. 2002]. The creation of an accurate set of mathematical models that describe cell behavior will lead to a more complete understanding of cell biology and also enable the investigation by means of mathematical models of ways of controlling cell proliferation (e.g., direct stem cell differentiation and growth in *regenerative medicine,* and also cause *apoptosis* of tumor cells). Takahashi and coworkers have been developing a comprehensive simulation package called *E-Cell* since 1996. The current version is *E-Cell 3.* Quoting Takahashi et al: "... E-Cell 3's simulation engine truly supports mixed-mode computation by integrating native computation modules and object model-based modules."

We note in closing this section that the majority of cellular biochemical processes are describable as nonlinear ODEs based on the format used in modeling mass action kinetics and diffusion. The ODEs will typically contain nonlinear algebraic terms to model receptor saturation, as well as time lags associated with diffusion, active molecular transport, gene expression, protein synthesis, etc., and parametric control of enzyme activity. A number of specialized computer programs have been developed in the past decade for dynamic simulation of cellular biochemical processes. Some programs are general purpose nonlinear ODE solvers (Simnon), others are stand-alone software packages specialized to solve biochemical mass action and S-system equations, and still others are specialized interfaces that adapt biochemical kinetic equations to be solved on powerful software systems such as *Matlab* and *Mathematica.*

4.7 SUMMARY

In this chapter, we have introduced the physicochemical basis for modeling the dynamics of complex biochemical systems. The concept of *free energy* of the various types of chemical bonds was presented, and the *law of mass action* was shown to be the result of idealized statistical interactions of reacting molecular species. Even though the nonlinear ODEs of mass action kinetics are predicated on well-mixed, dilute solutions, they were shown to be useful in describing biochemical reactions that generally occur under severe boundary conditions (e.g., on reacting molecular species bound to a membrane surface).

We next examined the well-known, coupled biochemical reactions of glycolysis, fermentation, and the Krebs (citric acid) cycle. We have observed that nature uses a variety of metabolic pathways to produce intracellular chemical energy stored in the cell in the form of ATP and NADH molecules. Metabolic pathways for carbohydrates, lipids, and amino acids were described in the context of organic redox reactions. Important material was reviewed on what is undoubtedly one of life's indispensable chemical pathways, photosynthesis. The Calvin cycle, photosystems I and II, chemiosmosis, and RUBISCO were introduced.

Biochemical oscillators were shown to be able to be modeled by small sets of interacting, nonlinear ODEs based on mass action. These systems were shown to be able to generate bounded limit-cycle oscillations in concentrations. We examined the venerable Hodgkin–Huxley model for nerve impulse generation and its properties as a dc current-to-frequency oscillator.

In the remainder of this chapter we reviewed some of the many software packages useful for modeling the physical biochemistry of cells. Some were seen to be based on well-known, widely used software packages such as *Matlab* and *Mathematica,* others were "home-grown" programs designed and written in certain cell biology laboratories for specific modeling purposes and distributed as licensed freeware.

CHAPTER 4 HOME PROBLEMS

4.1 Define and give some examples of a system's *phase portrait.*

4.2 As we have seen in Chapter 4, Section 4.5.3, limit-cycle oscillations can occur in certain biochemical systems modeled by nonlinear ODEs. *Define* what is meant by: (1) *a stable limit cycle,* (2) *an unstable limit cycle,* (3) *a semistable limit cycle.* Illustrate these three cases with $\dot{x}$, x phase portraits showing some initial conditions on $\dot{x}$ and x and typical trajectories. [See, for example, Ogata, K. 1970. *Modern Control Engineering (1st ed.).* Prentice-Hall, Englewood Cliffs, NJ.]

4.3 A chemical kinetic model for photoreceptor transduction has been proposed** in which the photoreceptor cell's depolarization voltage is proportional to the concentration of product C in the cell, **c**; that is, $v_m = k_6 \mathbf{c}$. The product C is made according to the following chemical reaction (see Figure P4.3):

* * See Jones, R.W., D.G. Green & R/B. Pinter. 1962. Mathematical simulation of certain receptor and effector organs. *Fed. Proc.* 21(1): 97–.

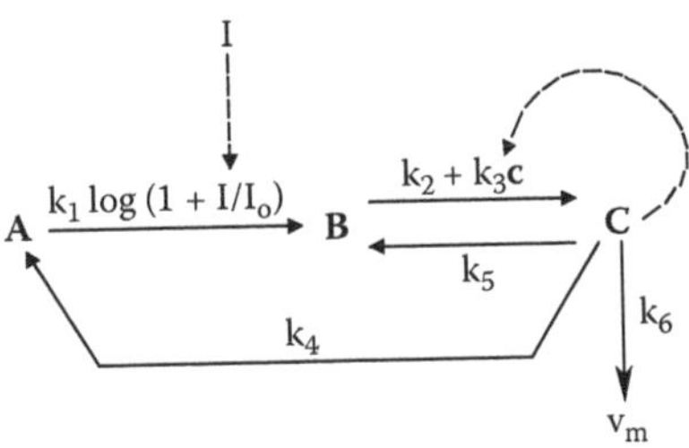

FIGURE P4.3

The conversion of A to B proceeds at a rate proportional to the log of the light intensity, $k_1 \log(1 + \mathbf{I}/\mathrm{I_o})$. The rate of conversion of B to C contains an autocatalytic term, $(k_2 + k_3\,\mathbf{c})$. In the absence of light, C is converted to A at rate k_4. From chemical mass action kinetics, we can write the three ODEs in the following (**a**, **b**, and **c**) are the running concentrations of molecules A, B, and C, respectively.

$$\dot{a} = k_4\mathbf{c} - \mathbf{a}\,k_1 \log(1 + \mathbf{I}/\mathrm{I_o})$$

$$\dot{b} = \mathbf{a}\,k_1 \log(1 + \mathbf{I}/\mathrm{I_o}) + k_5\mathbf{c} - \mathbf{b}\,(k_2 + k_3\,\mathbf{c})$$

$$\dot{c} = \mathbf{b}\,(k_2 + k_3\,\mathbf{c}) - \mathbf{c}\,(k_4 + k_5)$$

A. *Simulate* the three nonlinear chemical kinetic equations earlier: Use $\mathbf{a}(0) = 1$, other ICs $= 0$. $k_1 = 4$, $k_2 = 0.3$, $k_3 = 40$, $k_4 = 10$, $k_5 = 0.1$, $k_6 = 80$, $\mathrm{I_o} = 1$. Let $\mathbf{I} = 0.1$, 1.0, 10, 10^2, 10^3, 10^4 light units for 5 ms, and then zero. Plot the depolarization, $v_m(t)$. Does the system saturate? Plot the initial peak of v_m, and v_m at 5 ms as a function of intensity, **I**.

B. Let $\mathbf{I} = 2$. Plot $a(t)$, $b(t)$, and $c(t)$ over 15 ms.

4.4 Monomolecular binding of a hormone H in extracellular fluid to receptor sites R on a cell membrane is expressed by a two-step chemical reaction in which the complex mediates the desired end effect of the hormone. The complex can either dissociate back to hormone and receptor or break down to yield a free receptor R and an inactivated hormone molecule *H*, as shown in the reaction.

$$\mathrm{H + R \underset{k_2}{\overset{k_1}{\rightleftarrows}} H*R \xrightarrow{k_3} \underline{H} + R}$$

Let: $\mathbf{h}$ = concentration of free hormone, $\mathbf{fr}$ = concentration of unbound (free) receptors, $\mathbf{br}$ = concentration of complexed receptors (H*R). Note that in this closed system, $\mathbf{Cr} = \mathbf{fr} + \mathbf{br}$ = total receptor concentration.

A. Write the first-order ODE for $d(\mathbf{fr})/dt$ based on mass action kinetics.

B. Assume steady-state conditions ($d(\mathbf{fr})/dt \equiv 0$). Derive an expression for $\mathbf{B}(\mathbf{h}) = \mathbf{br}_{ss}/\mathbf{Cr}_{ss}$. Plot and dimension B(**h**) versus **h.** Note that **B** is proportional to the hormone's effect on the cells.

4.5 Reactants A and B combine reversibly to form product P inside a cell. P diffuses out of the cell to "zero" concentration. Assume that reactant B has a constant concentration, $\mathbf{b_o}$, inside the cell. A diffuses into the cell from the extracellular fluid where its concentration is a fixed $\mathbf{a_o}$.

$$\underset{\mathbf{a_o}}{\mathrm{A_o}} \xrightarrow{k_{da}} \underset{\mathbf{a}}{\mathrm{A}} + \underset{\mathbf{b_o}}{\mathrm{B}} \underset{k_{-1}}{\overset{k_1}{\rightleftarrows}} \underset{\mathbf{p}}{\mathrm{P}} \xrightarrow{k_{dp}} *$$

A. Write the ODEs for *a* and *p*, the running concentrations of A and P.
B. Note that the system is linear. Draw a signal flow graph for the system's ODEs.
C. Derive an expression for the steady state **p**, given $\mathbf{a}_o > 0$.

4.6 Refer to Problem 4.5. Now assume reactant **B** is made inside the cell at a fixed rate, $\dot{b}_o$. (i.e., B no longer has a fixed concentration.) A biochemical feedback system has been discovered which can be modeled by the rule:

$$\dot{b}_o = \dot{b}_{oo} - \beta \mathbf{p}$$

where: $\dot{b}_{oo}$, $\beta > 0$, and $\mathbf{p} \geq 0$.

A. Write the state equations for the system.
B. Give an expression for the net outward diffusion rate of P in the steady state.

4.7 A Michaelis-type chemical reaction in which *parametric control* is present is shown as follows. In this reaction, the end-product concentration, **p**, affects the forward rate constant according to the law, $k_1 \cong \beta/\mathbf{p}$. The enzyme E is assumed to be conserved, that is, no new enzyme enters the system. Thus, $\mathbf{e} + \mathbf{c} = \mathbf{e}_0$. R*E is a complex between a reactant molecule, R, and an enzyme molecule, E.

$$\underset{\mathbf{r}}{\mathbf{R}} + \underset{\mathbf{e}}{\mathbf{E}} \xleftrightarrow{k_2 \;\; k_1} \underset{\mathbf{c}}{\mathbf{R*E}} \xrightarrow{k_3} \underset{\mathbf{e}}{\mathbf{E}} + \underset{\mathbf{p}}{\mathbf{P}} \xrightarrow{k_4} *$$

A. Write the mass-action ODEs governing the nonlinear system's dynamics.
B. Assume **r** is constant ($\mathbf{r} = \mathbf{r}_o$). Find an expression for the steady-state concentration of P, $\mathbf{p}_{ss}$. Show that $\mathbf{p}_{ss}$ is independent of **r**. (Hint: This answer is arrived at by solving a quadratic equation in $\mathbf{p}_{ss}$. Use the approximation,

$$\sqrt{(1 + \varepsilon)} \cong 1 + \varepsilon/2, \; \varepsilon << 1.)$$

4.8 A proposed eight-state linear model for the storage dynamics for thyroid hormone (TH) and 3-iodotyrosine (3IT) is shown in the following text. Note that there are two compartment volumes in the model (extra- and intracellular fluid volume) as well as storage equilibrium reactions for TH and 3IT in both volumes. The states are the concentrations, $\mathbf{x}_1$–$\mathbf{x}_8$. In addition to the storage equilibria, TH is deiodinized to form 3IT in both volumes. To linearize the model, we have assumed that the concentration of thyroid-binding globulin (TBG) is constant everywhere at $\mathbf{x}_o = 1$ ng/mL.

A. The first state equation for the system is

$$\dot{x}_1 = -\mathbf{x}_1 (k_1 \mathbf{x}_o + k_5 + k_7) + k_2 \mathbf{x}_2 + k_7 \mathbf{x}_5 + \mathrm{u}_1.$$

Write the remaining seven state equations for the system. Assume the diffusion rates are proportional to the concentration differences between the extracellular and intracellular spaces.

B. Write the system's 8 × 8 **A** matrix in numerical form.
C. Assume $k_1, \ldots, k_{13} = 1$, $k_{14} = 10$. Let $\mathrm{u}_1(t) = 10\delta(t)$. Plot the system's states versus time to scale. Simulate the system with *Matlab* or *Simnon* (see Figure P4.8).

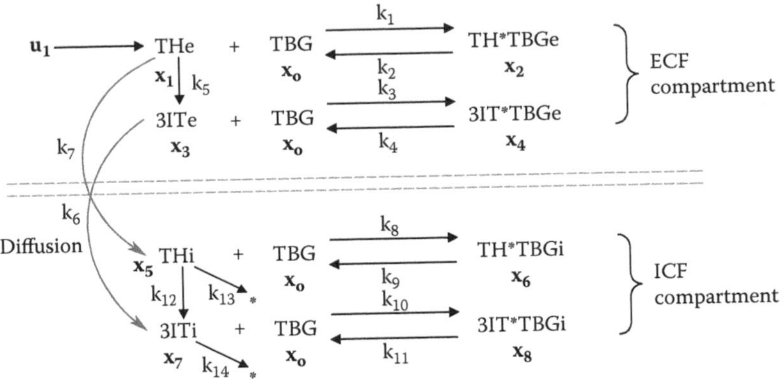

FIGURE P4.8

4.9 An inherent biosynthetic regulation of reaction end product P arises from P competing for the same enzyme binding site as does the substrate S. The reactions are shown as follows. Note that $\mathbf{e}_o = \mathbf{c} + \mathbf{q} + \mathbf{e}$. $\mathbf{e}_o$ is the total enzyme (conserved). $\mathbf{e}$ = free enzyme concentration, $\mathbf{c}$ = enzyme complexed with S, and $\mathbf{q}$ = enzyme complexed with P.

$$\underset{\mathbf{s}}{\mathbf{S}} + \underset{\mathbf{e}}{\mathbf{E}} \xleftrightarrow{k_2 \quad k_1} \underset{\mathbf{c}}{\mathbf{S*E}} \xrightarrow{k_3} \underset{\mathbf{p}}{\mathbf{P}} + \underset{\mathbf{e}}{\mathbf{E}}$$

$$\underset{\mathbf{p}}{\mathbf{P}} + \underset{\mathbf{e}}{\mathbf{E}} \xleftrightarrow{k_6 \quad k_5} \underset{\mathbf{q}}{\mathbf{P*E}}$$

From chemical mass action, we can write the ODEs:

$$\dot{c} = k_1 \mathbf{e}\,\mathbf{s} - \mathbf{c}(k_2 + k_3)$$

$$\dot{p} = k_3 \mathbf{c} - k_4 \mathbf{p} - k_5 \mathbf{p}\,\mathbf{e} + k_6 \mathbf{q}$$

$$\dot{q} = k_5 \mathbf{p}\,\mathbf{e} - k_6 \mathbf{q}$$

In the steady state, the derivatives → 0.

A. Using the steady-state assumption with $\mathbf{s} = \mathbf{s}_o$ = constant, derive expressions for $\mathbf{p}_{ss} = f(\mathbf{e}_{ss}, \mathbf{s}_o)$ and $\mathbf{e}_{ss} = g(\mathbf{p}_{ss}, \mathbf{s}_o)$.

B. Solve the quadratic equation derived from the results of (A) for $\mathbf{p}_{ss} = h(\mathbf{s}_o)$.

C. Show what happens when we assume that $k_1 \mathbf{s}_o >> (k_2 + k_3)$.

4.10 Figure P4.10 illustrates the model for a nonlinear, two-compartment, pharmacokinetic (PK) system in which compartment 2 has Langmuir saturation. $\mathbf{x}_1$ and $\mathbf{x}_2$ are the *drug masses* (ng) in compartments 1 and 2, respectively.
Compartment volume dimensions are hidden in K_{L1}, K_{21}, and σ_2. The system's ODEs are

$$\dot{x}_1 = b\,u_1 - K_{L1}\,\mathbf{x}_1 - K_{21}\,(1 - \sigma_2\,\mathbf{x}_2)\,\mathbf{x}_1 \text{ ng/min}$$

$$\dot{x}_2 = K_{21}\,(1 - \sigma_2\,\mathbf{x}_2)\,\mathbf{x}_1 \text{ ng/min}$$

A. Investigate the system's dynamics by simulation. Let $\sigma_2 = 20$, $K_{L1} = 1$, $K_{21} = 3$, $b_1 = 0.5$, $\mathbf{x}_1(0) = 0$, and $\mathbf{x}_2(0) = 0$.
Generate an impulse input u_1 at $t = 0$ and again at $t = 1$. Impulses can be approximated by narrow rectangular pulses of width 0.02 and heights u_{1o} ranging from 1 to 50. Thus, the impulse areas are $D_1 = 0.02\,u_{1o}$. At what level does $\mathbf{x}_2$ saturate?

B. Show how the first impulse at $t = 0$ can be used to find b_1 (see Figure P4.10).

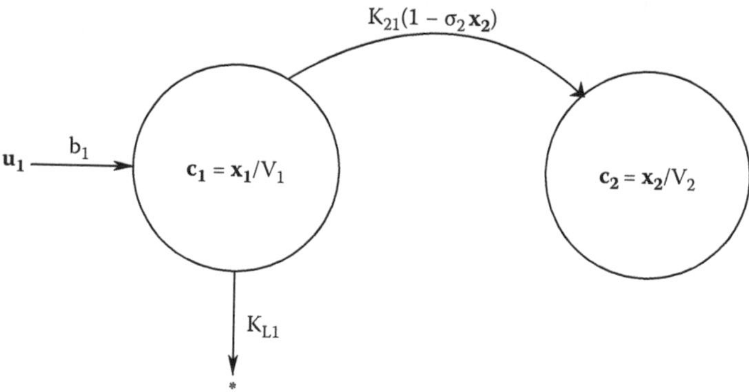

FIGURE P4.10

4.11 Many cellular biosynthetic pathways are regulated by *end product inhibition* (EPI). Describe how EPI works and give examples from the known metabolic pathways that use EPI in the biosynthesis of *heme* and the AAs, *methionine* and *threonine*.

5 The Basis of Genetic Inheritance

5.1 INTRODUCTION

How are genetic traits passed on from parent to offspring? In very simple terms, genes are the instruction book for building an organism by specifying what proteins to use and, equally important, when to use them. The genetic instruction for a trait is called the *genotype*. The physical expression of a gene is called the *phenotype*. Just as phenotypes vary, so do genotypes.

You may have heard expressions such as "They've found a new gene for breast cancer," but this is misleading. Oncogenes, or cancer genes, are simply "misspellings" of normal genes. For example, the "ATM" gene *(ataxia telangiectasia mutated)* normally controls cell division but, in an altered form, it may play a role in the formation of cancers. Everyone has this gene, but about 1% of Americans carry the defective form [LBNL 2006].

An important principle in genetic inheritance is *polymorphism*, or varying forms taken by genes at the same locus. We do not have different genes; we have different versions of the same genes. Genes can vary as alleles or mutations.

Genetic variation through alleles (normal variations of genes): Specifically, an allele is a DNA sequence variation that is common in the population. No single allele is regarded as the standard or default version. Instead, there are two or more equally "normal" alternatives. To be classed as an allele, the least common variation in a gene's sequence must have a frequency of 1% or more in the population [Twyman 2003]. These variations are also called *polymorphisms*. If the frequency is lower than 1%, the genetic variation is regarded as a *mutation*. The term *allele* is often used interchangeably with the term *gene*.

Example: An example of variation based on alleles is human eye color. Eye color is determined by a combination of alleles in several genes. Scientists once thought that a single gene pair controlled human eye color, with the allele for brown eyes dominant over the allele for blue eyes [SEP 2006]. The actual genetic basis for eye color is neither as complex as "a gene for every color," nor is it as simple as a single genetic pair. Currently, *three* allelic pairs controlling human eye color are known.

- A gene on chromosome 15 has a brown and a blue allele (and blue is recessive).
- A gene on chromosome 19 has a blue and a green allele.
- Another gene on chromosome 15 is a central brown eye color gene [SEP 2006].

Genetic variation through mutations (essentially "spelling errors"): A mutation is a change in the DNA sequence of a particular gene away from "normal." This implies that a normal allele is prevalent in the population and that the mutation is a rare and abnormal variant of the sequence [Twyman 2003].

Examples: The genetic variation that causes *Huntington's disease* is caused by a "stuttering repeat" of the base sequence **CAG** [Friedman 2001]. The repetition of this sequence inserts an extraneous glutamine into an otherwise normal protein [Blakeslee 1996]. The more often the sequence is repeated, the earlier in life the onset of Huntington's disease will occur [Jankovic 2006]. Several other known genetic diseases, including Fragile X Syndrome (a form of mental retardation), are caused by excessive repeats of a base sequence [MSGN 1994].

Dominance and recessiveness: An allele can be dominant or recessive, as can a mutation. Because our genes come in pairs (one from each parent), dominance plays an important role in determining which gene is expressed. A dominant allele "wins" over a recessive allele, so in order for a recessive allele to be expressed, it must be paired with its recessive mate.

Dominant and recessive genes interact in autosomal chromosomes (paired chromosomes, one inherited from each parent). Another form of inheritance is through sex-linked genes (genes located on the X and Y chromosomes).

5.2 MENDELIAN INHERITANCE

The concept of dominant and recessive genes, and the principles for how inheritance is determined, was established by an Austrian monk, Gregor Mendel, in the 1800s. Mendelian inheritance refers only to the combination of genes in the cell nucleus, not to cytoplasmic heredity. This is traditionally the type of inheritance that geneticists address.

Although Mendel understood that characteristics were determined by discrete units of inheritance, he did not coin the term *gene* himself. Terminology aside, his work paved the way for modern genetics. Mendel used pea plants as his model, cultivating and testing some 28,000 plants [Field Museum 2007]. He discovered that dominant and recessive genes combine in specific, predictable ratios in each generation. Mendelian inheritance can be summarized in generalizations known as *Mendel's Laws of Heredity:*

Mendel's First Law: The Law of Segregation. During gamete formation, each member of the allelic pair (of genes) separates from the other member, ensuring variation. In other words, the expression of the dominant allele does not eliminate the recessive allele in the pair but merely suppresses it. Thus, a recessive gene can be passed on through the generations, even when the phenotype is not expressed.

Mendel's Second Law: The Law of Independent Assortment. During gamete formation, each allelic pair separates independently of all other allelic pairs. This is the essence of Mendel's conceptual breakthrough—genetic inheritance happens in discrete units.

Linked Genes: During his experiments, Mendel encountered an exception to the second law. Some traits did not appear independently but always appeared together with at least one other trait (e.g., red hair and a pale complexion are often linked) [Hunter & Mitchell 1997]. Today, we know that some genetic traits are located close to one other on a chromosome and tend to be linked, or inherited together. Therefore, Mendel's Law of Independent Assortment only applies to genes on different chromosomes.

Mendelian F2 Ratios: Mendel found that mixing parents with different forms of the same trait yields a 3:1 ratio between dominant and recessive phenotypes in the offspring (see Table 5.1).

Example: Red versus white flowers in pea plants. Recessive alleles are indicated with lowercase letters.

Exceptions to the F2 rule: Despite his large sample sizes, Mendel did not find perfect 3:1 ratios in the F2 generation. Round versus wrinkled seeds occurred in a 2.96:1 ratio; yellow to green seeds occurred in a 3.01:1 ratio; red versus white flowers occurred in a 3.15:1 ratio; green versus yellow pods occurred in a 2.82:1 ratio [Moore 1972].

Much discussion has ensued regarding Mendel's ratios and whether they were "too close" or "not close enough." The well-known statistician R.A. Fischer once analyzed Mendel's experiments for statistical probability using chi-squared analysis [Novitski 2004].

A possible cause of variations off the perfect 3:1 ratio was a class of genetic elements called *transposons.* They are a type of intron that "jumps" into and alters genes. They were not discovered until the 1940s, so Mendel, and indeed Fischer, had no way of guessing of their involvement.

Transposons were discovered by Dr. Barbara McClintock, who eventually received the Nobel Prize in Medicine for her work. She was a botanist studying maize and found that some Indian corn developed streaks or variegation in its pigmentation that could not be explained through Mendelian

TABLE 5.1 (See color insert following page 44.)
Mendelian Ratios: Dominant-Recessive Alleles

Parent Generation	Phenotype:		Phenotype:	
	Genotype: ww		Genotype: RR	
F1 Generation (Offspring of Parents)	wR	wR	wR	wR
F2 Generation (Offspring of hybrid F1 generation)	RR	Rw	wR	ww

FIGURE 5.1 (See color insert.) Indian corn with streaks caused by transposons. (Photograph by Wayne P. Armstrong. With permission.)

genetics (see Figure 5.1). The transposons turned off the ability to produce pigment in some cells but not others, yielding streaks in the kernels [Armstrong 2007].

What if traits are neither dominant nor recessive? You may have observed phenotypes that seem to belie Mendel's laws. For example, a parent of African heritage and a parent of Caucasian heritage together produce children whose skin color seems to be a "blend" of the two colors.

An example from botany is a flower called *Mirabilis jalapa,* whose color alleles are neither dominant nor recessive. The red and white alleles together make a pink variant. In the F2 generation, half the offspring will show a blending of colors, whereas half will retain one color or the other from the parent generation. This 1:2:1 ratio occurs when neither allele is dominant (see Table 5.2).

5.2.1 Gene Swapping during Reproduction and Genetic Variation

Why do we look different from the other members of our families? Genetic variation is due in large part to the way genes from both parents are combined during reproduction.

Through the process of *meiosis*, one chromosome from each pair is selected and passed on to the daughter cells. Meiosis ensures that daughter cells receive one of each kind of chromosome (at

TABLE 5.2 (See color insert.)
Mendelian Ratios: Neither Allele Dominant

Parent Generation	Phenotype: Genotype: ww		Phenotype: Genotype: rr	
F1 Generation (Offspring of Parents)	wr	wr	wr	wr
F2 Generation (Offspring of hybrid F1 generation)	rr	rw	wr	ww

least where the *autosomes* are concerned). Independent assortment provides 2^n possible combinations of genes. In humans with 23 chromosomes, $2^n = 2^{23}$ (2 each of 23 chromosomes). That amounts to 8,388,608 possible combinations [Wolf 2006]. Where the sex chromosomes are concerned, a female receives two X chromosomes, but one is silenced. A male can receive only one Y chromosome because his mother has no Y chromosome to contribute.

Variation is increased by a phenomenon called *crossing over*. During this normal process, chromosome partners physically swap sections with one another during the first stage of meiosis, generating chromosomes with a "patchwork" of sections from the original pair. If only one crossover occurred within each chromosomal pair, 4^{23} combinations, or 70,368,744,000 combinations, would be possible [Wolf 2006].

The likelihood that any two genes will cross over, or *recombine*, is a function of how close together they are. The closer together two genes are, the less likely it is that they will cross over. For genes at opposite ends of a chromosome, there is about a 50-50 chance that they will cross over. It is very rare that genes right next to each other will be swapped from chromosome to chromosome [McClean 1997]. These genes are considered to be *linked* (see discussion of Mendelian principles in the preceding text).

Clearly, sexual reproduction will not produce offspring that are genetically identical to either parent. Whether through chromosome swapping, crossing over of sections of chromosomes, or through the actions of mobile genetic elements such as transposons (see the section on introns and exons, above), a vast amount of variation is possible. The amazing outcome is that we resemble our family members as much as we do.

5.3 EPIGENETIC INHERITANCE

Epigenetic inheritance is the transmission of information from a cell or organism to its descendants, aside from what is encoded in the genes. Yoo and Jones [2006] defined "epigenetic" as "mitotically and meiotically heritable states of gene expression that are not due to changes in DNA sequence."

Not only do genes themselves vary, but the genotype (and therefore the phenotype) varies on the basis of which genes are expressed. Gene silencing, or suppression, can occur through chemical modification of DNA, through structural modification of DNA, or through regulation of transcription [Zuckerkandl & Cavalli 2006].

Chemical gene suppression: DNA methylation is one way that the protein expression of a gene can be blocked. This chemical modification of DNA can be inherited *epigenetically* (without changing the DNA sequence). In humans, DNA methylation is executed by three enzymes: *DNA methyltransferase 1, 3a,* and *3b* [Epigenetic Station 2007]. A methyltransferase enzyme (with a methyl group of carbon and hydrogen) attaches to the 5-position of a cytosine ring [Yoo & Jones 2006] in a CpG dinucleotide sequence (called "CpG islands"). These "islands" are usually in a *gene promoter region* (at the beginning of a gene). This prevents the methylated gene from being expressed [Halim 1999]. Conversely, *promoter regions* that have not been methylated are correlated with active gene expression. DNA methylation has important medical implications as it may play a "crucial" role in "setting the stage for carcinogenesis" [Yoo & Jones 2006].

Methylation patterns can be analyzed through bisulfite mapping. Methylated cytosine residues are unchanged by the treatment, whereas unmethylated ones are changed to uracil [Schmitt et al. 1997].

Abnormal methylation patterns are thought to be involved in carcinogenesis because many tumor suppressor genes are silenced by DNA methylation [Smith et al. 2006]. Although many attempts to reverse the methylation of these suppressor genes (using methlytransferase inhibitors) have been unsuccessful due to toxicity, recent work has identified less toxic inhibitors that may be clinically viable [Yoo & Jones 2006; Stresemann et al. 2006].

Proteins or chemical groups that are attached to DNA and modify its activity are called *chromatin marks*. These marks are copied with the DNA. The number and pattern of such methylated cytosines influences gene amplification: low levels of methylation correspond to high potential activity, whereas high levels correspond to low activity. (See Chapter 7, Section 7.2.4.7 for a description of histone proteins.)

Structural modification of DNA: In addition to chemical modification, DNA may be structurally modified in a way that alters the density of its packing and, therefore, its transcription [Raftopoulou 2005]. Protein complexes called *histones* control the supercoiling of DNA (see Chapter 7, Section 7.2.4.7) [Garner 1987]. They can be modified by processes such as *phosphorylation* or *acetylation* [Mahadevan 2004]. Even when the structural change is temporary, because it affects transcription, it can create permanent changes in the expression of certain genes. The more densely packed the DNA, the lower the gene expression. Yoo and Jones provide this description of histone modifications: "DNA is wrapped around a core of eight histones to form nucleosomes, the smallest structural unit of chromatin. The basic amino-terminal tails of histones protrude out of the nucleosome and are subject to … modifications, including acetylation."

Chromatin is the packed mass of DNA and proteins in the nucleus. A 2006 study showed that the chromatin of stem cells has unique "bivalent" characteristics, containing regions with both active and repressed gene expression; this is considered to be a key to "stemness" [Bernstein et al. 2006].

5.4 GENETIC IMPRINTING

Geneticists used to believe that one entire chromosome out of each autosomal pair was methylated. However, the prevailing belief now is that a more subtle mechanism called *imprinting* is involved. Imprinting is the suppression of certain genes, based on their parental source [Genetics 2004].

Imprinted genes are chemically "marked" as coming from either the father or the mother. Imprinting does not affect the chemical composition of the gene. It determines whether a gene will be expressed [Woods 2000].

Abnormal imprinting has been linked to a number of hereditary and other diseases, including colon cancer, lung cancer, and prostate cancer. Environmental factors may be involved in the loss of maternal or paternal imprints [Woods 2000]. A well-known genetic disease related to imprinting is the Prader–Willi Syndrome, in which the disease gene is only expressed if it comes from the father's chromosome 15 [MSGN 1994].

5.5 NUCLEAR AND MITOCHONDRIAL DNA

Eukaryotic organisms (most living things) have cells with a nucleus. DNA in eukaryotic organisms is stored in the cell nucleus. This is known as "nuclear DNA."

Mitochondrial DNA (mtDNA) is located in organelles called mitochondria. Mitochondria are unusual in that they possess their own DNA rather than being defined by nuclear DNA. It is generally believed that mtDNA is inherited only from the mother. Human mtDNA consists of 5 to 10 rings of dsDNA, carrying 37 genes related to cellular respiration. It does not recombine as nuclear DNA does, so that changes between generations are largely due to mutation [Cann et al. 1987].

5.6 GENES AND THE GENOME: THE CENTRAL DOGMA

The genome, or sum total of an organism's genetic material, can be thought of as an instruction book for the creation, maintenance, and function of a life-form, written large in dsDNA.

Using the metaphor of the genome as a book (beautifully explained by Matt Ridley in his book, *Genome* [Ridley 1999]), the genome contains the following elements:

- *Chromosomes,* which are like the chapters in the book (the human book has 23 chapters).
- *Genes,* which are like stories in the chapters; each gene is expressed as a protein.
- *Exons,* which are paragraphs in the stories.
- *Introns,* which are like advertisements that interrupt the stories.
- *Codons,* which are like words in the paragraph; all words in the genome are three "letters" or bases long. Most of these are expressed as amino acids, which combine to make peptides, polypeptides, or proteins.
- *Bases,* which are like the letters that make up the words. There are only four letters found in DNA: **A** (adenine), **C** (cytosine), **G** (guanine), and **T** (thymine). Bases come in pairs in the double helix form of DNA. **A** pairs with **G**, and **C** with **T**. Bases are also referred to as *nucleotides* (Nts).

The *central dogma* of molecular biology is based on the principle that *the flow of genetic information is one-way, from genetic material to proteins.* Once information is encoded into protein, it cannot flow back to nucleic acids.

The central dogma is often misunderstood as having a more narrow meaning, that "DNA codes mRNA which codes proteins," but there are exceptions to this specific flow of information [Crick 1970]. Genetic information does not always flow from DNA to RNA. See Section 5.6.1 for a discussion of retrotransposons, an example of the RNA-to-DNA flow. Again, note that there are no cases in which proteins encode genetic information into RNA or DNA. DNA only carries the *information* that controls synthesis; various RNAs, ribosomes, ribozymes, and certain protein enzymes carry it out.

The genetic code really is a code. How does mRNA specify amino acid sequence? It would be impossible for each amino acid to be specified by one nucleotide because there are only four nucleotides and 20 amino acids. Similarly, two nucleotide combinations (4^2) could specify only 16 amino acids.

Each amino acid is specified by a combination of three nucleotides, called a *codon.* Four nucleotides, combined in groups of three (4^3), yield a total of 64 possible sequences or codons. All 64 have been found to exist. The order in which they appear *does* matter; **UCU** codes for serine, whereas **CUU** codes for leucine. Most codons code for an amino acid, but 3 of the 64 do not; they are "stop" codes, similar to punctuation in the instruction book. These stop codes tell the protein synthetic mechanism of the ribosome to stop synthesis and release the protein amino acid chain. Because there are only 20 amino acids normally found in proteins made by all cells, the genetic "code" is *degenerate,* with some amino acids being selected by more than one codon.

Table 5.3 "decodes" the amino acid meaning of all 64 codons in RNA, where the left column indicates the first position in the codon, the top row indicates the second position, and the right column indicates the third position.

Note that each amino acid might have up to six codons that specify it. For example, leucine will be produced by any codon beginning with **C** or **U**, no matter whether the third codon is **U**, **C**,

TABLE 5.3
Amino Acid Coding by Nucleotides (Codons)

	U	C	A	G	
U	Phenylalanine	Serine	Tyrosine	Cysteine	**U**
					C
	Leucine		Stop	Stop	**A**
			Stop	Tryptophan	**G**
C	Leucine	Proline	Histidine	Arginine	**U**
					C
			Glutamine		**A**
					G
A	Isoleucine	Threonine	Asparagine	Serine	**U**
					C
			Lysine	Arginine	**A**
	Methionine				**G**
G	Valine	Alanine	Aspartic acid	Glycine	**U**
					C
			Glutamic acid		**A**
					G

A, or **G**. Only two amino acids, methionine and tryptophan, are coded by only one codon, **ATG** and **TGG**, respectively. The DNA codon **ATG** codes for methionine *inside* a peptide sequence or signals the *start* of a sequence with methionine if **ATG** is the very first codon in the gene. [UCSD 2001]. Note that no AA is coded by five codons.

From the results of *Human Genome Project* and, more recently, the *ENCODE pilot project,* a startling fact has emerged. There are now known to be about 23,000 different protein-coding genes in the human genome divided among 44 genomic regions [doegenomes 2004; Little 2005; ENCODE 2007]. (At one time cell biologists estimated that there were close to 10^5 protein-coding genes.) However, there are well over 22,000 proteins in the human body, perhaps more than 230,000. This numerical uncertainty in the protein number is because a newly created protein can be swiftly cleaved enzymatically to form two or more daughter protein molecules. Also, enzymatic posttranscriptional modification can also add or delete functional groups on a protein's AAs such as methyl, hydroxyl, phenyl, phosphate, etc., creating new proteins [Germenis & Patrikidou 2005]. Finally, in a given class of cells, not all proteins are ever expressed, and those that are may be at different times in the cell cycle.

Interestingly, different organisms have different frequencies of codon usage. A giraffe might use **CGC** for arginine much more often than **CGA**, and the reverse might be true for a sperm whale [MIT 2006].

Table 5.4 shows the DNA genetic code from the amino acid point of view.

5.6.1 Introns and Exons

The Human Genome Project, which sequenced the entire genome, revealed that genes are only the "tip of the iceberg." The vast majority of the genome is not made up of genes at all [Ridley 1999] but by other material known as introns (non–protein-coding DNA sequences that occur in packets between and among the genes). The "average" protein contains about 333 AAs, so 1,000 DNA bases are required to code each protein, or a total of about $1,000 \times 30,000$ genes $= 3 \times 10^7$ bases are required. However, it is estimated that the entire human genome has a total of about 3 gigabases (3.1×10^9), so it appears that only about 1% of the human genome is protein-coding genes; about 99%

TABLE 5.4
Amino Acids Coded by DNA Codons (Replace T with U for mRNA Codons)

AA Name	Abbreviation	One-Letter Abbreviation	DNA Codons for Each AA
Tryptophan	Trp	W	TGG
Methionine	Met	M	ATG*
Aspartic acid	Asp	D	GAC, GAT
Glutamic acid	Glu	E	GAA, GAG
Cysteine	Cys	C	TGC, TGT
Histidine	His	H	CAC, CAT
Asparagine	Asp	N	AAC, AAT
Lysine	Lys	K	AAA, AAG
Phenylalanine	Phe	F	TTC, TTT
Glutamine	Gln	Q	CAA, CAG
Tyrosine	Tyr	Y	TAC, TAT
Isoleucine	Ile	I	ATA, ATC, ATT
Alanine	Ala	A	GCA, GCC, GCG, GCT
Glycine	Gly	G	GGA, GGC, GGG, GGT
Proline	Pro	P	CCA, CCC, CCG, CCT
Arginine	Arg	R	CGA, CGC, CGG, CGT
Threonine	Thy	T	ACA, ACC, ACG, AGT
Valine	Val	V	GTA, GTC, GTG, GTT
Leucine	Leu	L	TTA, TTG, CTA, CTC, CTG, CTT
Serine	Ser	S	TCA, TCC, TCG, TCT, AGC, AGT
Start			ATG*
Stop, release			TAA, TAG, TGA

is non–protein-coding DNA, either codes RNAs or has poorly understood or unknown regulatory functions, or even no function at all. Scientists once referred to introns as "junk DNA," but recent discoveries have indicated that this is far from the case [Gibbs 2003].

Introns may also be defined as sections of DNA that will be spliced out after transcription but before the mRNA is used. Introns are common in all eukaryotic RNAs but are rare in prokaryotes [Lynch 2002]. In contrast to introns are exons, the regions of a gene that remain in spliced (edited) mRNA [McClean 1998]. They code for proteins [Traut 1988].

Any intron bases lying within a mammalian gene are transcribed into the *premessenger RNA* (pmRNA) for the gene and then are "edited" out of the gene sequence in the pmRNA before protein synthesis occurs. The editing is done in a complex eight-step process by a set of five nuclear "caretaker" RNA molecules known collectively as a *spliceosome.* The spliceosome is composed of five subunits called *small nuclear ribonuclear proteins* (snRNP) [MCB 411 2000]. This is a wonderful and indispensable molecular machine as far as proteins are concerned, because our genome is 99% irrelevant in terms of polypeptide (protein) coding. (The details of how the spliceosome works are beyond the scope of this text.)

With what is now evident as short-sightedness and hubris, scientists once referred to intron DNA as "junk" DNA [Ohno 1972]. It now turns out that introns can contain *cis-regulatory elements* (base sequences) that determine which (and when) master control genes are activated, thus providing a clue to how embryonic development takes place. As discussed throughout this text, many introns have other epigenetic functions. Introns can also contain RNA coding sequences.

It is our opinion that about 99% of our DNA is not noise or junk, just because it does not directly code proteins. One wonder of life lies in its many biochemical regulatory mechanisms. The elaboration of the *cis-regulatory DNA base sequences* in introns promises to be a major challenge in genomics in the next decade. Evidently, the Human Genome Project will continue for some time as scientists clarify the complex molecular mechanisms of genetic regulation and developmental timing associated with introns.

There are many hypotheses of how introns arose and why they persist in the eukaryotic genome. For example, it is widely believed that introns once served as spacers between the sections of DNA that coded for specific, comparatively simple proteins. During the evolution of complex proteins, regions of the genetic code may have been shuffled to generate new sequences [Bidwai 1999].

It is also believed that many introns were originally ancient genes that have since lost their protein-coding ability and their original function [Koonin 2006], although they can now retain certain epigenetic functions [Fedorov & Fedorova 2004].

A known source of introns is *retrotransposons,* which constitute up to half of the human genome [Deininger & Batzer 2002]. About 8% of the human genome is made up of a type of retrotransposon known as *Human Endogenous Retroviruses* (HERVs) [Polavarapu et al. 2006].

Transposons and retrotransposons: An important example of an intron is the transposon, a class of genetic "parasite" that includes retroviruses [Economist 2005]. Transposons are mobile genetic elements that can move from chromosome to chromosome. When they skip the normal genetic flow of information through transcription and use reverse transcriptase to copy and splice themselves directly into DNA, they are known as retrotransposons [Salk Institute 2005].

The most common form of retrotransposon is called LINE-1. This is a "paragraph" of DNA, between 1,000 and 6,000 bases long, that includes the code for reverse transcriptase. There are as many as 100,000 unique copies of LINE-1 sequences in each copy of your DNA, and many of them are repeated [Salk Institute 2005].

Recent research shows that transposons may be a "major creative force in the evolution of mammalian gene regulation" [Mikkelsen et al. 2007]. Over 99% of the genes of *Monodelphus domestica* (a small South American opossum) are also found in humans [NIH 2007]. Opossums belong to the infraclass *Metatheria,* the marsupial branch of the mammalian class, in contrast with *Eutheria,* the placental mammals. *Eutheria* and *Metatheria* split between 125 and 144 MYA [Ogburn & Brogdon 2004]. The comparison of opossum and eutherian genes shows that "true innovation in protein-coding genes seems to be relatively rare" [Mikkelsen et al. 2007]. In other words, the differences between opossums and humans, from the point at which they diverged from the common ancestors, are due almost entirely to noncoding regions, or introns. Specifically, the differences are due to the activity of transposons [Mikkelsen et al. 2007].

Because they move around the genome so much, retrotransposons often "jump" into the middle of normal genes, damaging or mutating them. This sounds serious, but the good news is that retrotransposons are naturally suppressed in the human genome through methylation. This may, in fact, be one reason that methylation evolved [Ridley 1999]. Some researchers believe that transposons played a key role in the evolution of complexity in eukaryotes [Bowen & Jordan 2002].

Retroviruses transcribe RNA into DNA (the inverse of the "normal" process) using a special enzyme called reverse transcriptase. Interestingly, the gene for *reverse transcriptase* is the most common gene in the human genome, even though it serves no function in the human body. Reverse transcriptase was "planted" in human DNA by retroviruses as a way to enter the genome. It is as if a computer hacker planted a "back door" that would allow unauthorized entry to another person's computer.

The resulting DNA is inserted into a host cell's DNA using an enzyme called *integrase* and is reproduced along with the rest of the cell's genome. The viral DNA in the daughter cell then makes an RNA copy of itself that is covered in a protein shell and leaves the cell to infect new cells [Columbia University Press 2003].

Some retroviruses can jump between species. These are known as exogenous retroviruses. Retroviruses, especially exogenous retroviruses, sometimes destroy the cells whose DNA they alter, as with HIV, the virus that causes AIDS. Retroviruses can also cause cells to become cancerous, as with the viruses that cause some leukemias [Columbia University Press 2003].

Because reverse transcription lacks the usual "proofreading" of DNA transcription [Kimball 2007a], retroviruses mutate frequently [Drake 1993]. This allows the virus to develop resistance to antiviral pharmaceuticals quickly and is one of the main reasons why an effective vaccine for HIV has not been developed yet [Coffin et al. 1997].

Retroviruses commonly carry three genes that encode proteins found in the mature virus:

- **gag** (Group-specific antigen) codes for core and structural proteins of the virus
- **pol** (Polymerase) codes for *reverse transcriptase, protease,* and *integrase* enzymes
- **eEnv** (Envelope) codes for the proteins used to coat the retrovirus [Cann 1999]

RNA is considered more ancient than DNA [Jeffares & Poole 2006]. Some researchers speculate that the processes used by retroviruses (RNA to DNA to RNA to protein) may be the key to the evolution of DNA; at some time in the distant past, retroviruses may have evolved to create DNA from RNA templates [Biocrawler 2006].

Intron functions. As mentioned earlier, it was once believed that introns had no function, but research has increasingly revealed significant roles played by introns.

- *Some introns* can promote the expression of a neighboring gene. Introns are not required for gene expression, however, because several genes that lack introns express themselves normally [DeSouza 1999].
- *Some introns* contain genes for small nuclear RNA, which is important for the translation of messenger RNA, an intermediary between DNA and proteins [*Ibid*].
- *Nuclear introns* can also be important in a process called *alternative splicing*, which can produce multiple types of messenger RNA from a single gene [*Ibid*].
- *Some introns* play a role in the formation of cancer [Zheng 2006]; some serve as tumor suppressors [Berggen et al. 2000].

Yet another class of introns codes for *microRNAs* (miRNAs). miRNAs are short, noncoding RNAs that appear to play a role in *suppressing* the expression of certain proteins and may regulate the formation of shape and size in development. Recent studies have indicated that miRNAs can also *activate* gene expression at different times in the cell cycle [Vasudevan et al. 2007]. miRNAs exert their effects by interacting with mRNAs. There are over 150 miRNAs in humans alone. It is still uncertain what their exact roles are, but they probably have a role in regulating embryological development. It has been shown that the levels of certain miRNAs are precisely regulated in mice as their brains grow [Gibbs 2003]. This is a tantalizing hint of the research yet to come on these interesting nucleic acid molecules.

The bottom line is that genes, although an important part of the genome, are not the last word. A great deal of work remains to be done on understanding introns as well as exons.

5.6.2 Protein Synthesis

Protein synthesis is the process whereby DNA encodes for the production of certain messenger RNAs, linear chains of amino acids, and finally proteins [Access Excellence 1999]. Tables 5.3 and 5.4 showed us how the genetic code is translated for AA assembly by mRNA, tRNA, and ribosomes. The process consists of two parts, transcription and translation. The transfer of the codon code of a certain gene from DNA to *messenger RNA* (mRNA) is called *gene transcription. Translation* is the process whereby the genetic code in the mRNA is used to construct proteins.

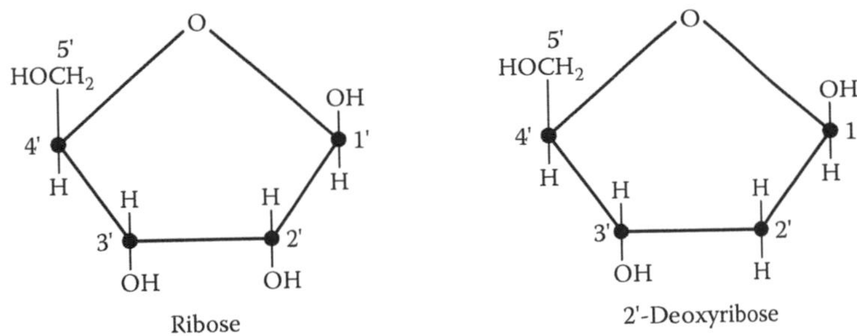

FIGURE 5.2 The ribose and 2′-deoxyribose molecules.

5.6.2.1 Transcription

Before the synthesis of a protein begins, the corresponding RNA molecule is produced. One strand of DNA is used as a template by RNA polymerase to synthesize a messenger RNA (mRNA). In RNA, uracil is used in place of thymine, but all other bases remain the same. This mRNA migrates from the nucleus to the cytoplasm. During this step, mRNA is spliced to eliminate noncoding Nt sequences. The resulting mRNA coding sequences are called *codons* [Access Excellence 1999].

The RNA *ribonucleoside* polymer is held together by covalent bonds between alternating *ribose* and phosphate groups. Ribose and deoxyribose are sugars with five-carbon rings, shown in Figure 5.2. Each subunit or monomer of RNA, similar to DNA, also has a base side chain that is composed of either *adenine, cytosine, guanine*, or *uracil* (DNA uses thymine instead of uracil). Sometimes, the base *inosine* is found attached in the third position of a tRNA anticodon. The six common nucleic acid bases are shown in Figure 5.3.

The four major types of RNA involved with protein synthesis are *premessenger RNA* (pmRNA), *messenger RNA* (mRNA) (outside the nucleus), *transfer RNA* (tRNA), and *ribosomal RNA* (rRNA). Other important classes of small RNA molecules also exist, for example, *small nuclear* RNAs (snRNAs), *small nucleolar RNAs* (snoRNAs), microRNAs (miRNAs), and the *4.5-S signal recognition particle RNA* (srpRNA). We are only concerned at this point with mRNA, tRNA, and rRNA because they are the kinds principally involved with gene expression and protein synthesis.

The *first step* in protein synthesis is the binding of some 50 different *transcription factor proteins* (TFPs) to the *promoter sites* on the dsDNA, usually on the 5′ side of the gene to be transcribed. The TFPs themselves are regulated in order to turn gene expression on and off. Phosphorylation of TFPs by *protein kinase enzymes* (also proteins) activates them, and no doubt, protein kinases themselves are regulated, ad infinitum. (Protein kinase enzymes phosphorylate other protein AA residues.) Then an *RNA polymerase* enzyme binds to the TFP complex, causing the unfolding of the DNA helix from its storage configuration and the separation of the two complementary halves of the helix in the region of the active gene. This unraveling exposes the DNA bases so that special enzymes called *DNA-dependent RNA polymerases* (RNAP) can act to form *premessenger RNA* (pmRNA) that will have a complementary base structure to that of the DNA template strand. (Actually there are three types of RNAP: *Type I* is used to synthesize rRNA, *Type II* synthesizes pmRNA, and *Type III* makes tRNA. Nothing is simple in molecular biology; enjoy the complexity.)

The RNAP II proceeds down one DNA strand moving in the 3′ to 5′ direction. In eukaryote DNA, another complex of proteins works ahead of the advancing RNAP II to remove the *nucleosomes* before the RNAP II reaches them. (Nucleosomes are an octamer of histone proteins whose role is to bind inactive dsDNA in compact form—see Glossary.) The same complex of proteins replaces the nucleosomes after the gene has been transcribed. The RNAP II synthesizes pmRNA in the 5′ to 3′ direction on the *template* (readout) *strand* of DNA. As each nucleoside triphosphate is attached to the growing 3′ end of the growing RNAP II strand, the two terminal phosphates are enzymatically removed. Transcription is complete when a STOP codon is reached, and the

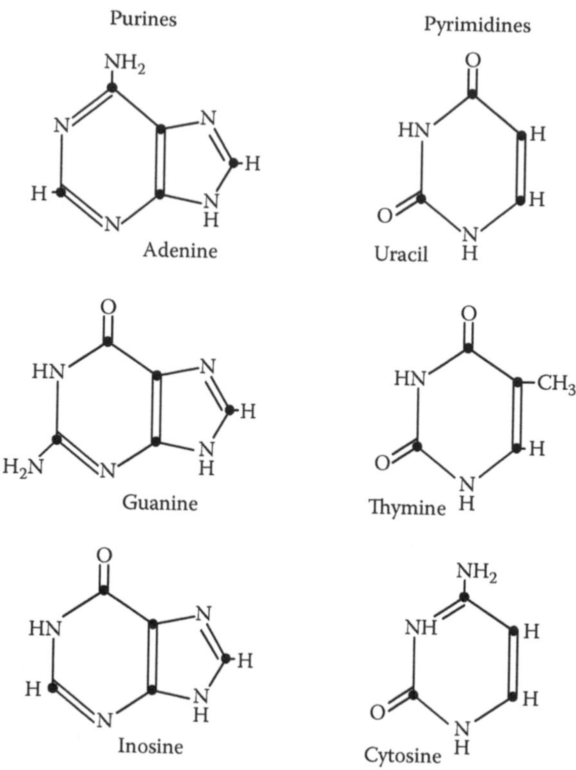

FIGURE 5.3 The three purine molecules, adenine, guanine, and inosine, and the three pyrimidine molecules, uracil (used in RNA), thymine, and cytosine.

pmRNA and RNAP II are released from the DNA strand. The process then can repeat, making more pmRNA strands [Kimball 2005c].

The uracil found in RNA pairs with adenine on the DNA strand being transcribed. Note that pmRNA comprises only about 5% of a cell's RNA (80% is rRNA, and 15% is tRNA). After the spliceosome proteins edit the intron ("noise") bases out of the pmRNA, it becomes mRNA, which carries the code for the assembly of a specific protein from the gene site on the DNA, through the nuclear membrane, to the ribosomes in the cytoplasm where the actual protein assembly occurs. Proteins are synthesized from the amino- to carboxyl (-**COOH**) direction. The process is complex, beautiful, even baroque, but it works!

Transfer RNA (tRNA) molecules can have 73 to 93 nucleotide bases. The 3′ end sequence of the tRNA always terminates with the Nt sequence, **CCA**. The 3′ hydroxyl of the ribose of the terminal **A** (adenosine) is the point of covalent attachment to the specific amino acid. About 7 to 15% of the bases of tRNA are posttranscriptionally modified after synthesis by an RNAP enzyme. For example, adenosine in the first or 5′ position of the anticodon (corresponding to the third or 3′ position of the mRNA codon) is always modified to be *inosine* (**I**), which lacks the amino group on its purine ring. Consequently, inosine can base pair with **A, U,** or **C** in the mRNA, accounting for much of the degeneracy of the genetic code. This pairing ambiguity will be considered below when we discuss the *Wobble Theory.*

The secondary structure of tRNA is a "cloverleaf" with three loops of unpaired Nts. The end loop (opposite the AA attachment) is called the *anticodon loop* where the tRNA binds to a matching mRNA codon. In addition, tRNA has a *D-loop* and a *TΨC-loop.* The loops are connected by short stretches of helical, paired RNA bases. (Figure 5.4 illustrates a 2-D molecular schematic of a tRNA molecule coded for the AA methionine.)

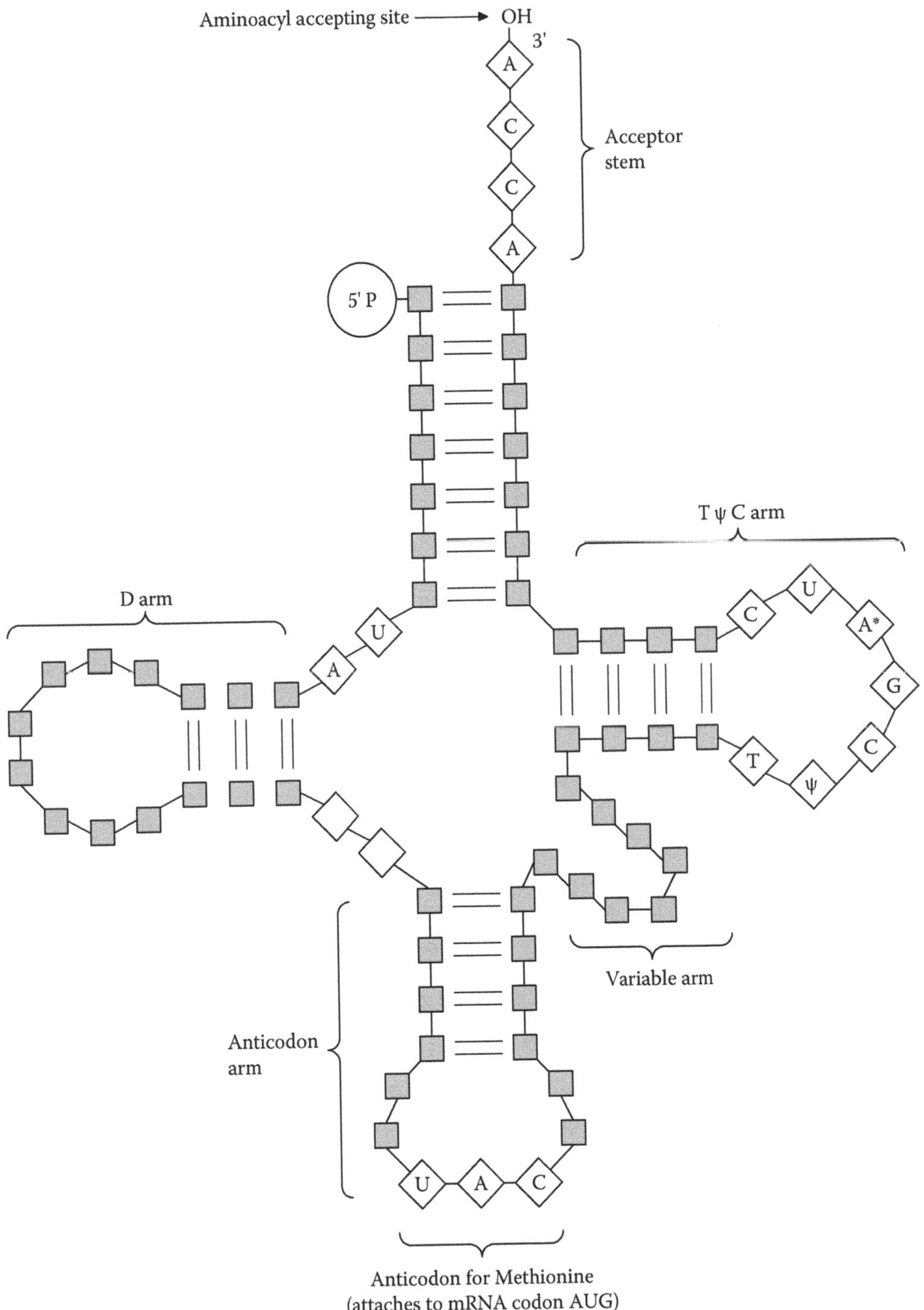

FIGURE 5.4 General 2-D structure of a transfer RNA (tRNA) molecule. Note that four nonpaired loops exist, although the *variable arm* can be a "stub" in some tRNAs. At the bottom is the all-important *anticodon loop* that matches a complementary mRNA codon read from a gene's DNA. Shown is the anticodon for the AA methionine, which is always the first AA to leave the ribosome. The TΨC arm loop incorporates some nonstandard bases. See text for further description.

TABLE 5.5
List of AAs Handled by Class I or Class II AARS Enzymes

Class I AARS	Class II AARS
Glu	Gly
Gln	Ala
Arg	Pro
Cys	Ser
Met	Thr
Val	His
Ile	Asp
Leu	Asn
Tyr	Lys
Trp	Phe

Because there are 20 different amino acids used in eukaryote proteins, in human cells there are 20 different *aminoacyl-tRNA synthetase* (AARS) enzymes that attach the appropriate AAs to the appropriate tRNA molecules. These 20 enzymes are widely different in structure, each one optimized for functioning with its own particular AA and the set of tRNAs appropriate to that AA. (For example, the AAs leucine and serine each have six tRNAs they can be attached to, whereas tryptophan and methionine each have one unique tRNA to be attached to.) The molecular recognition of the AARS enzymes must be essentially foolproof to prevent protein assembly errors. Not only must an AARS enzyme select the correct tRNA molecule to attach its AA to, but it must not attach the AA to any wrong tRNA. (*E. coli* has 21 AARSs; one for each AA except lysine, which has two. Other organisms have fewer than 20 AARSs. Special enzymes modify ambiguous AAs after they are added to the tRNA so 20 AAs are coded.)

To underscore the complexity of the AARS system, there are two classes of aminoacyl-tRNA synthetase enzymes. *Class I enzymes* are generally monomeric and attach the carboxyl of their target amino acid to the 2′ **–OH** of the terminal adenosine of the tRNA molecule. *Class II enzymes* are generally dimeric or tetrameric and attach their AA to the 3′ **–OH** of their tRNA, except for phenylalaninyl-tRNA synthetase, which uses the 2′ **–OH**. Table 5.5 shows which AAs are handled by which class of AARSs.

Figure 5.5 summarizes the reactions whereby a correct AA is attached to the terminal adenosine Nt of an appropriate tRNA molecule. In the first reaction, the enzyme AARS causes the AA's acyl group to bond to an adenosine triphosphate molecule, forming an *aminoacyl-AMP molecule.* In the second reaction, the AARS enzyme causes the aminoacyl-AMP molecule to replace the terminal adenosine molecule on the tRNA. A free AMP molecule is released, and the aminoacyl-adenosine is now bonded to the tRNA.

Each of the 20 amino acids (AAs) used in human protein synthesis has one or more corresponding tRNA molecules, which will carry the AA to a ribosome for the process of protein assembly. tRNA contains many unusual bases in addition to **A**, **G**, **C**, and **U**. Some of these unusual bases are methylated or demethylated derivatives of the four normal RNA bases; they include *inosine, dihydrouridine, and pseudouridine.* The purpose of the unusual bases may be to form critical 3-D structures that give tRNA its activity and specificity. The 5′ ends of all tRNAs are phosphorylated, and the terminal residue is usually p**G**. The base sequence at the 3′ end of all tRNAs is **CCA**. The activated AA to be put in the protein being assembled is attached to the free 3′-OH of the terminal adenosine (**A**). About half of the nucleotides in tRNA are

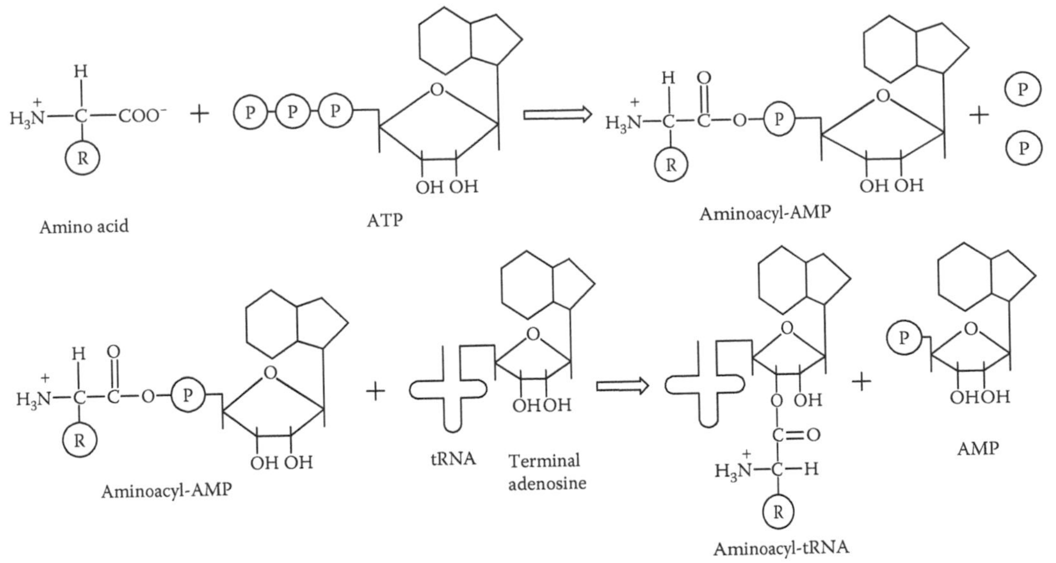

FIGURE 5.5 Schematic illustrating the reactions by which a correct AA is attached to the appropriate tRNA molecule. R is the side group of the AA. The acceptor stem of the tRNA binds to adenosine, which in turn picks up the correct AA. The specificity of the process is remarkable.

base-paired to form short pieces of double helix. Four regions of the tRNA are not base-paired. Three bases on the anticodon loop pair with a certain codon on the mRNA molecule, that is, they are complementary to that codon. The actual assembly of the protein from mRNA is done by another clever set of proteins and catalytic rRNA in the *ribosomes*. Figure 5.4 illustrates schematically one tRNA molecule that has the codon for the AA methionine (Met) bound to a specific anticodon (starting) site on an mRNA molecule. Other enzyme molecules will join the *Met* AA with a peptide bond to the next AA in the protein (in this example, valine) being assembled by the ribosomal complex. The molecular biology of this process is very complex and is discussed in the following section.

5.6.2.2 Translation

The ribosome (an organelle) binds to the mRNA at the start codon, which is recognized only by the initiator tRNA. Next, AA–tRNA complexes sequentially bind to the appropriate codon by forming complementary base pairs with the tRNA anticodon. The ribosome moves along each codon on the mRNA. Amino acids are added one by one. At the end, a release factor binds to the stop codon, terminating translation and releasing the complete polypeptide chain from the ribosome [Access Excellence 1999; Evolvingcode 2006].

Translation of the mRNA code into a protein takes place in *ribosomes*. Initially, the ribosomal *small subunit* proteins bind to an "upstream" site on the 5′ end of the gene coded in the mRNA. It proceeds "downstream" (toward the 3′ end of the gene) until it reaches the *start codon* (**AUG**) of the gene. At this point, the *small subunit* is joined by the *large subunit* proteins and a special *initiator tRNA* that always carries the AA *methionine* in eukaryotes. The *initiator tRNA* binds to the *P site* on the ribosome proteins. The region of the gene between the 5′ cap and the **AUG** codon is known as the *5′-untranslated region*.

Note that the *initiator tRNA* is the only tRNA that can bind directly to the ribosomal P-site. The P-site is called that because, with the exception of the *initiator tRNA*, it can bind only to a tRNA with the growing peptide attached (a peptidyl-tRNA molecule). The ribosomal *A-site* can only bind to an incoming aminoacyl-tRNA that matches the active codon on the mRNA [Kimball 2005b].

Next, an *aminoacyl-tRNA* (a tRNA covalently bound to its specific AA) that matches the next codon on the mRNA (following **AUG**) arrives at the ribosomal *A-site* along with an *elongation factor protein* and a source of chemical energy, *guanosine triphosphate* (GTP). The first AA (methionine) is now covalently linked to the second AA with a *peptide bond*. Figure 5.7 illustrates the three isoforms of the peptide bond. The *initiator tRNA* is released from the P-site, free to pick up another methionine AA. The ribosome complex now shifts one codon downstream (to the third codon in the mRNA gene). This shifts the #2 tRNA to the P-site and opens the A-site for the arrival of a third aminoacyl-tRNA that matches the third codon. Another molecule of GTP and another protein elongation factor (EF-G) are required to link the third AA by a peptide bond to the second AA.

The process of linking AAs continues until a *STOP codon* on the mRNA is reached (**UAA**, **UAG**, or **UGA**). The mRNA Nts from this point to its 3′ end make up the *3′-untranslated region* of the mRNA. Note that there are no AA-carrying tRNA molecules with anticodons for the three STOP codons. A ribosomal *release factor protein* does "recognize" the STOP codons and catalyzes the release of the linear protein peptide from the ribosome, which then splits back into its subunits ready for another round of protein synthesis.

For translational efficiency, there are generally several ribosomes traveling along a given mRNA molecule at a given time, all making copies of the same protein. Figure 5.6 illustrates the linking of the first three AAs at the start of protein assembly by a ribosome. Ribosomes are amazing molecular "nanomachines."

Once a linear protein is released from the ribosome, it folds and twists as hydrogen bond crosslinks form, giving it a stable, 3-D configuration that enables it to function as an enzyme, hormone, structural material, etc. This final folding to its finished, 3-D configuration is generally "supervised" by other proteins known as *chaperonins.*

The use of an intermediary between DNA and proteins allows genetic information to be "amplified" through a one-to-many relationship between DNA and RNA [MIT 2006]. Note that, when RNA is translated back into DNA, uracil is methylated into thymine. This is important because methylation protects the DNA. It makes the DNA unrecognizable to many nucleases (enzymes that break down DNA and RNA) so that it cannot be easily attacked by invaders, such as viruses or certain bacteria [Onken 1997].

Thymine also protects the DNA in another way. The components of nucleic acids, phosphates, sugars, and bases are easily water soluble, but methyl groups are water insoluble [Onken 1997].

Further, methylation marks thymine as a legitimate base, making it easier for DNA's repair mechanisms (see Chapter 6) to correct mutations. The cytosine base can spontaneously deaminate to form a uracil base, which would result in undetectable cytosine-to-uracil mutations if **U** were used routinely in DNA. As thymine is basically methyl-**U**, the cell's DNA repair mechanisms can distinguish illegitimate **U** from legitimate methyl-**U** in DNA and make the proper repair (replacing any **U** with a **C**) [Onken 1997].

5.7 THE CONTROL OF DEVELOPMENT

DNA instructs embryos in how to develop, starting from the zygote and undifferentiated stem cells in the blastocyst. But how do stem cells "know" how to organize and differentiate themselves? About 5% of embryonic stem cells serve as an "organizer" unit that provides these instructions [Kirschner & Gerhart 2005].

All animal species on Earth use one of about two dozen different basic body plans [Levine 2006]. Heads, legs, and other parts may look different, but they are all organized using a standard body plan. Body plans are divided into compartments and organized by a "compartment map" in the embryo. This compartment map (and the vestiges of body compartments) first appears in the middle stage of development, known as the *phylotypic stage* [Kirschner & Gerhart 2005].

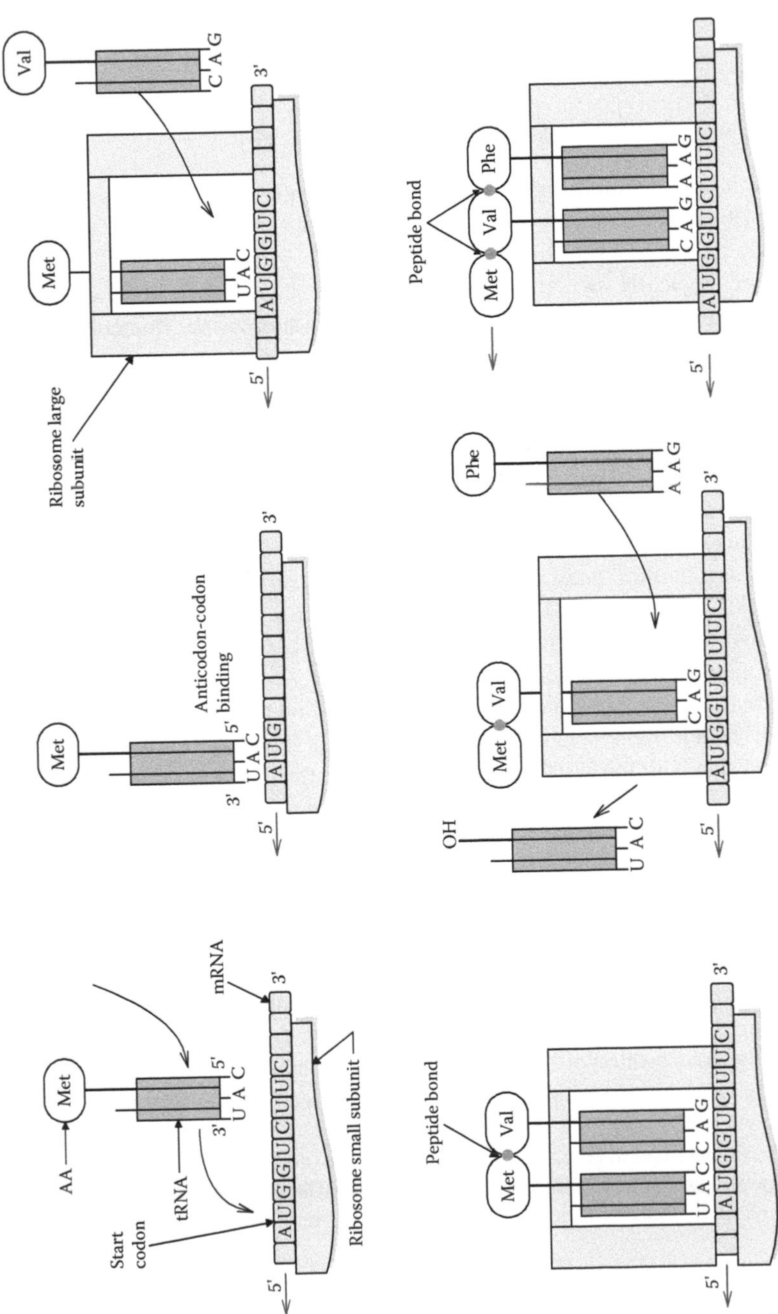

FIGURE 5.6 Schematic showing the initiation of protein synthesis with the methionine tRNA. Then, the ribosome joins two more AAs to the polypeptide. The AAs are linked enzymatically with high-energy peptide bonds.

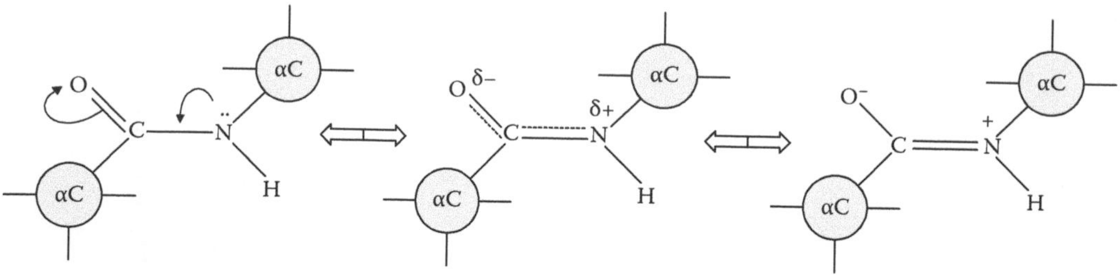

FIGURE 5.7 Three isoforms of a peptide bond formed by electron sharing.

The genes that control development of the body plan are known as *homeobox genes*. A homeobox is a special sequence found within genes that are involved in the control of development and is only about 180 base pairs long [Bürglin 2005].

Homeobox genes typically switch on cascades of other genes, such as all the ones needed to make a leg. This is similar to a computer program "calling" a subroutine. Homeobox genes encode a protein (the homeodomain) that can bind DNA. Homeodomain proteins usually act in the promoter region of their target genes, forming complexes with a high target specificity [Schier & Gehring 1993] so that no mistakes are made about which part goes where.

The *Hox genes*, a subgroup of homeobox genes, are found in a special gene cluster and provide a plan based on the body axis. They identify particular body regions such as limbs and other body segments. Hox genes are so critical to development that mutation in vertebrate Hox genes usually results in spontaneous abortion [University of Idaho 2007; Lodish et al. 2003].

The most common animal body plan on Earth is that of the tetrapod (four legged), which includes all mammals, birds, amphibians, and reptiles. Some animal groups such as snakes, cetaceans, bats, and most birds do not appear to be four legged, but they are modified tetrapods, with wings or flippers replacing legs [UCMP 2007].

The phrase "phylogeny recapitulates ontogeny" is based on the observation that all vertebrates pass through an early stage that resembles a fish embryo. Our "fish-like" early stages are directed by the same Hox genes that built early fishes—some of the first members of our branch of the animal kingdom. By the time fishes evolved, genetic control of development was already very complicated [Levine 2006]. This complexity made it unlikely that the genetic program could be rewritten to "edit out" the fish stage when other vertebrates such as mammals arose. (Once a genetic program passes a certain level of complexity, major changes early on in the program tend to have lethal "domino effects" on the remainder of development [Levine 2006].)

Therefore, what differentiates other vertebrates from fishes must occur in *later* stages of development, modifying the early body plan that we share with developing fish embryos [Levine 2006]. Bilateral symmetry, heads, and spinal cords are common to vertebrates and have been adapted into the tetrapod body plan. In addition to the conservation of Hox genes and body plans, some scientists believe that the core processes that generate the phenotype from the genotype have been conserved throughout evolution.

> For example, the basic information processing of DNA, RNA and protein synthesis is the same in all living organisms… the developmental role of the Hox genes is the same in all bilateral metazoan phyla; and the developmental program for limb formation is the same in all land vertebrates. [Kirschner & Gerhart 2005]

The conservation of core processes, therefore, has implications not only for the control of development but also for theories of evolution.

5.8 SUMMARY

We have examined how traits are passed from parents to offspring through various components of the genome. This chapter has first focused on variation in nuclear DNA, which can be accomplished genetically through alleles (normal variations, such as eye color) or mutations. The principles of Mendelian inheritance explain to a great degree how dominant and recessive alleles interact in genotypes, yielding predictable ratios in phenotypes.

Genes are swapped during meiosis, with exponential combinations of genes possible and further variation introduced by crossing over. Several exceptions to Mendelian principles include linked genes and *transposons.*

Beyond Mendelian inheritance, we considered *epigenetic inheritance* such as chemical gene suppression; structural modification, which affects transcription; and imprinting. The difference between nuclear and mitochondrial DNA was also addressed.

Protein-coding genes make up only about 1% of the human genome. As an introduction to genomics, we discussed the central dogma of molecular biology, which is essentially that genetic information flows always from nucleic acids to proteins, and not vice versa.

The remainder of the genome consists of introns. We considered the possible origins of introns and the clues that they give us to the evolution of DNA itself. A significant component of the human genome is *retroransposons,* which make up about 25% of the genome. Retrotransposons are seen to be mobile genetic parasites including retroviral DNA. The role of *reverse transcriptase* is important to the makeup of the genome. Introns have several known functions, including the promotion of gene expression.

Once a genotype is formed, it is expressed as a phenotype through the processes of transcription and translation (see also Chapter 6). DNA and RNA differ in their use of thymine versus uracil as a base, and this is significant to their existence and maintenance.

We also have reviewed the basics of the central dogma of protein coding by DNA genes and synthesis by mRNAs, tRNAs, and ribosomes.

As a new organism is formed from its new DNA, starting with stem cells, how do stem cells differentiate into body parts? The concept is introduced that an *organizer unit* provides a body plan. Homeobox genes play a key role in controlling embryonic development and aid in setting up and placing body parts.

CHAPTER 5 HOME PROBLEMS

5.1 You are a botanist studying a recently discovered variety of orchids. You notice that, out of 1,000 plants you cultivate, all are yellow with a purple center, except for one variegated plant in which the purple runs in streaks from the center out along the yellow petals. These plants have been completely isolated from outside pollens. How do you explain this one plant? Is it caused by a recessive allele or by something else?

5.2 You are breeding poultry and you cross an all-black bantam with a bantam that has a red back and is black everywhere else.

- You know that the allele for a black back is recessive (b) and the allele for a red back (R) is dominant, and the traits are not sex linked.
- Your first breeding pair is homozygous; the father is RR (with a red back) and the mother is bb (with a black back).
- The F1 generation is all black.
- In the F2 generation, three of the offspring have black backs and one has a red back.
- To produce the F3 generation, you breed one of your F2 roosters with a red back to an F2 female with a black back. However, you do not know whether your red-backed F2 rooster is homozygous or heterozygous.
- How will you know, based on an F3 generation with four offspring, whether the father is homozygous or heterozygous? If the father is heterozygous, what colors will the four offspring be?

Parent Generation	Mother BB—Black Back		Father RR—Red Back	
F1	Rb (red back)	Rb (red back)	Rb (red back)	Rb (red back)
F2	RR (red back)	Rb (red back)	bR (red back)	bb (black back)
F3	?	?	?	?

5.3 Dogs have 39 pairs of chromosomes, compared with 23 pairs in humans. During meiosis, dog genes undergo *independent assortment.* How many combinations of genes are possible in dogs as a result of independent assortment?

5.4 How many combinations of genes are possible in humans if each chromosomal pair crosses over twice during meiosis?

5.5 Two genes on chromosome 16 cross over about 40% of the time during meiosis. Are they located near the ends of the chromosome's arms or near the center point where the arms join?

5.6 The gene for a ring finger protein is located near the center of chromosome 16, in region 16p11.2. During meiosis, is it more likely to cross over (a) the gene for vacuolar protein sorting 35, near the center on the other arm in region 16q11.2, or (b) the gene for ribosomal protein L 13, near the end of the other arm in region 16q24.1 (see Figure P5.6)?

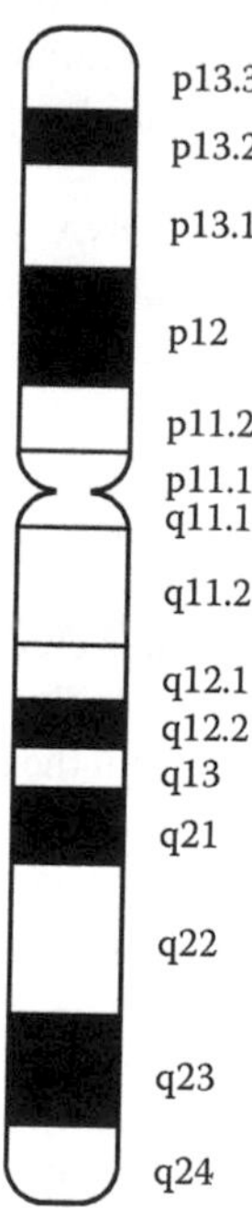

FIGURE P5.6 Chromosome 16 Ideogram. (NIH [2007]. National Cancer Institute's Cancer Genome Anatomy Project. *Distribution of SNPs by Chromosomes.* Public domain. http://gai.nci.nih.gov/cgi-bin/rh.cgi.)

5.7 According to the central dogma of molecular biology, under what circumstances does genetic information flow from proteins to nucleic acids?

5.8 Because three codons code for "stop" instructions rather than amino acids, that means there are 61 codes for amino acids. Based on the number of amino acids, is this code specific or redundant?

5.9 What AA does UGG, in that order, code for?

5.10 Is the base sequence "CTAG" from RNA or DNA? How do you know?

5.11 A certain disorder affecting muscle tone, appetite, and mental development appears only about once in every 15,000 births in a particular population. It affects both males and females equally. Statistical analysis shows that it runs in families and appears to be unrelated to environment, indicating that it is a genetic disorder. Further analysis shows that children are born with this condition only when it exists in the father's family. To what do you attribute this? (Extra credit: What genetic disease fits this description?)

6 Nucleic Acids and Their Functions

6.1 INTRODUCTION

This chapter is about the molecular biology of the nucleic acids, DNA and RNA. As we have stressed in earlier chapters, all biological macromolecules have natural, complex 3-D structures in which the interatomic binding energy is minimized; DNA, RNA, and proteins are no exception. Their 3-D molecular structures, including *chirality*, are necessary for intermolecular specificity and their biochemical and physiological activity. Deoxyribonucleic acid (DNA) is unique in that it appears in *all* cells today (Prokaryotes, Archaeans, Eukaryotes) and is used as a linear programming code for protein and RNA synthesis and control of gene expression. Protein-coding segments of DNA are called genes; they are also called *exons*.

Portions of non–protein-coding DNA Nt sequences, called *introns*, are also implicated in control of gene expression, and hence, regulation of embryological development and cellular metabolism. Some introns also contain the information for the synthesis of specific ribonucleic acid molecules. RNAs generally have the role of implementers of DNA instructions, and some serve as enzymes or catalysts.

6.1.1 DNA: THE COMMAND AND CONTROL NUCLEIC ACID

The basic organizational unit of the DNA polymer is the *2′-deoxyribose* (a sugar) molecule covalently bonded to one of two *pyrimidine bases* (*cytosine* or *thymine*) or one of two *purine bases* (*adenine* or *guanine*). The bases cytosine (**C**), thymine (**T**), adenine (**A**), and guanine (**G**) are the four *coding symbols* in a DNA molecule's linear genetic code. The bonding between the base and the deoxyribose sugar involves the C_1' carbon of the sugar and the N_9 nitrogen of a *purine base* (or the N_1 nitrogen of a *pyrimidine base* in an N-β-glycosidic linkage). The base-deoxyribose subunits are connected to one another by 3′-5′ phosphodiester linkages, as illustrated in the schematic of a dsDNA, four-base oligo, shown in Figure 6.1. Note that the 5′ methyl group on the deoxyribose binds to a phosphate's oxygen. Another phosphate oxygen binds to the 3′ carbon in the deoxyribose ring. Figure 6.1 also illustrates the cross-strand hydrogen bonds between the bases as dotted lines. These are the "glue" that holds the "rungs of the DNA ladder" together.

A *nucleoside* subunit of DNA is defined as one of the four bases attached to the C_1' position of 2′-deoxyribose, where the sugar *lacks* a phosphate group.

A *nucleotide* (Nt) is a nucleoside having one or two phosphate groups covalently bonded to the deoxyribose's 3′- or 5′-hydroxyl group's oxygen. A single strand of DNA is thus composed of a polymer formed from *phosphodiester links* between alternating nucleotide and nucleoside sugars. These links are also illustrated in Figure 6.1.

In its simplest 3-D form, the *B-DNA helix* contains two complementary polymer chains with the same number of bases organized as an *alpha helix*. The two DNA strands wrap around the helix axis with a *right-hand spiral;* the two polynucleotide chains run in opposite (complementary) directions. The radius of the alpha helix is about 1 nm, and the repeat length (pitch, or axial period) is about 3.4 nm. The bases are oriented perpendicular to the helix axis. The axial displacement

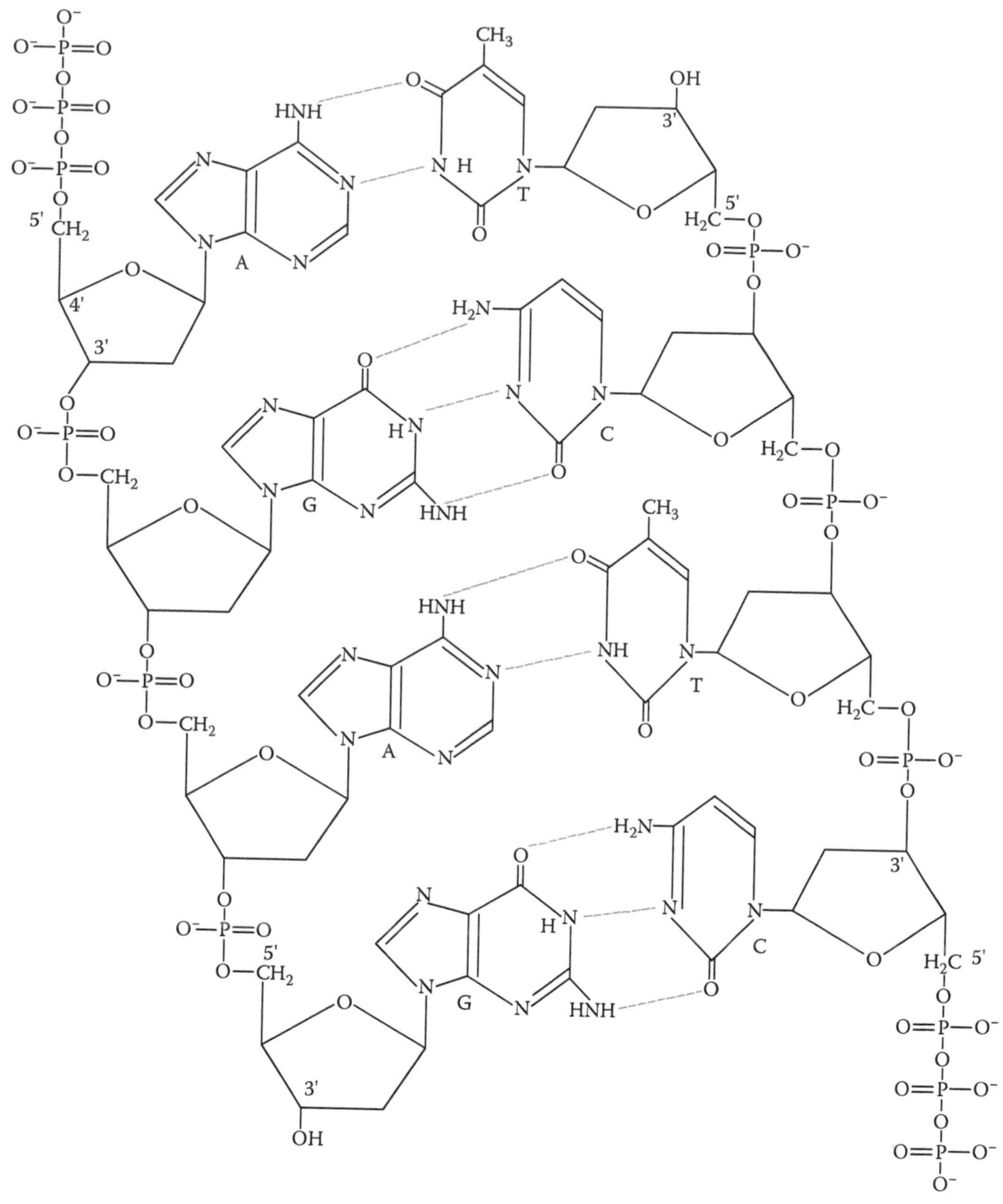

FIGURE 6.1 2-D schematic of a four-base, double-stranded DNA oligo. The dotted lines in the center represent the hydrogen bonds holding the strands together. Also, the 5′ carbon of the end ribose on each strand is linked to a triphosphate group. (Adapted from Northrop, R.B. 2003. *Signals and Systems in Biomedical Engineering*. CRC Press, Boca Raton, FL. Used with permission from Taylor & Francis.)

("steps") between bases is *about* 0.34 nm, so about 10 base pairs are found in an axial period. Figure 6.2 illustrates the α-helix of B-DNA in a 3-D side view. Note that from the side, the *B-DNA* α-helix has a large and small "groove." Most physiological DNA is of the B-form.

A-DNA is a form found when DNA is dehydrated. The small groove is much smaller relative to the large groove in the A-form (A-DNA is the "forensic form"). *Z-DNA* is unusual in that it is a left-handed helix (A- & B-DNA have right-handed helices). *Z-DNA* occurs in oligonucleotides having alternating purine and pyrimidine bases, for example, **GCGCGCGC**.

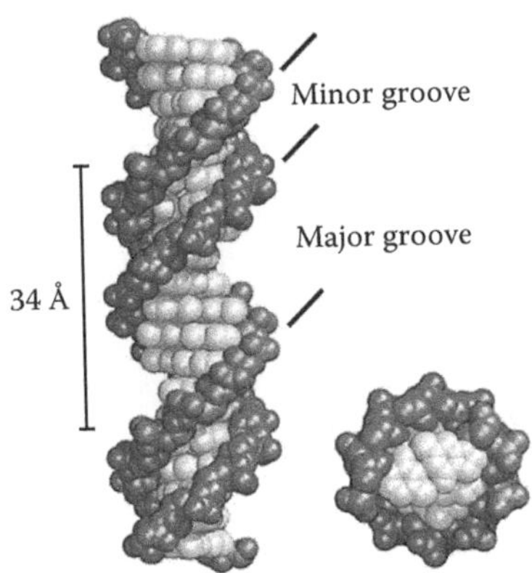

FIGURE 6.2 A 3-D rendering of a dsDNA molecule showing its helical structure and its major and minor grooves. To the right is an axial (end-on) view. (Adapted from Northrop, R.B. 2003. *Signals and Systems in Biomedical Engineering*. CRC Press, Boca Raton, FL. Used with permission from Taylor & Francis.)

The sugar-phosphate links of the two DNA strands wrap around the helical molecular axis like the railing of a spiral staircase, the "steps" of which are the hydrogen-bonded bases in the center. The adenine on one chain always pairs with thymine on the opposite strand; they are held together by two weak *hydrogen bonds*. A guanine on one side always pairs with a cytosine on the other strand; in this case, the pair is held loosely by three hydrogen bonds. Figure 6.3 illustrates 2-D schematics of pairs of purine and pyrimidine bases and the *hydrogen bonding* between them. Approximate dimensions and angles are given. Thus, because of the complementary base structure, if we know the base sequences on one strand of the DNA helix, we know it in on the opposing complementary strand. This redundancy permits enzymatic repair of damaged DNA molecules in eukaryotic cells.

The backbone of polynucleotides is highly charged: One unit (electron) negative charge for each phosphate group and two negative charges per base pair. If there are no salt ions in the surrounding aqueous medium, there is a strong electrostatic repulsion between the two DNA strands and they separate. Therefore, *counterions* are essential to maintain the double helix structure in an aqueous medium. Counterions effectively shield the charges on the deoxyribose-phosphate backbone. They may also contribute to an attractive interaction between the DNA strands.

Resting DNA in the nucleus (DNA that is not being "read" for its codes) is found in association with many specialized proteins called *histones*. This DNA is further wrapped and folded at five different length scales in order to achieve a high-density packing factor for storage in a cell's nucleus. If fully unwrapped and laid end to end, all DNA in a human cell *would be* about 2 m long [Masum et al. 2001]! Clearly, from nuclear volume restrictions, for a gene to be read, the two DNA helix strands must be locally separated. The single-strand bases must be read sequentially and then rejoined and refolded, both in DNA replication (in mitosis and meiosis) and in protein synthesis. Little is known about the in vivo molecular mechanisms that perform these complex 3-D molecular manipulations.

6.1.2 RNA: The Executive Nucleic Acid

If DNA is the command and control nucleic acid, then by analogy, RNA is the executive nucleic acid. It carries out the instructions encoded in the DNA genome.

All living cells we know today (*Prokaryotes, Archaea,* and *Eukaryotes*) use DNA for their genetic code. Cells use *ribonucleic acid* (RNA) to implement the command and control codes contained in DNA. RNA differs from DNA in that it uses the *pyrimidine uracil* instead of *thymine* as one of its four bases. RNA also has a hydroxyl (**–OH**) group instead of hydrogen on the 2′ carbon of the ribose sugar. Because of the **–OH** group on the sugar, RNA is too bulky to form a long, stable double helix. However, it does form short regions of double helix where there is some base complementation (**U** to **A**, **G** to **C**). When base complementation is not possible, the RNA forms

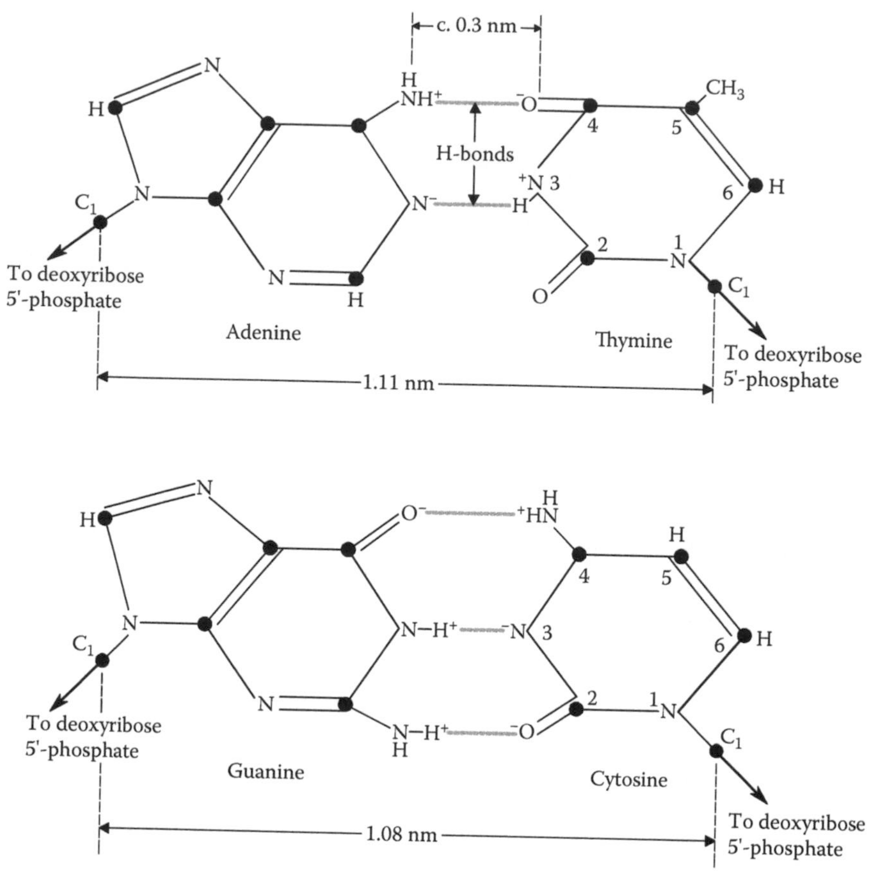

FIGURE 6.3 Details of the hydrogen bonds linking opposing purine and pyrimidine bases in dsDNA. Dimensions are given in nanometers. (Adapted from Northrop, R.B. 2003. *Signals and Systems in Biomedical Engineering.* CRC Press, Boca Raton, FL. Used with permission from Taylor & Francis.)

single-stranded *hairpin loops,* giving it a tertiary structure. Because RNA is not restricted to the more stable double helix, it can form various stable 3-D structures, as do proteins.

The first step in *gene expression* (protein manufacture) is called *transcription.* In transcription, the DNA code for a protein is converted to messenger RNA. mRNA, in association with ribosomes and a large number of enzymes, performs *translation* in which tRNA brings in amino acids, one by one, to a ribosome and assembles them in a linear chain that matches the chain of codons in the DNA and mRNA.

Cells make a number of different kinds of RNAs that figure in protein synthesis. The information to make these RNAs is encoded in the non–protein-coding parts of the cell's genome. These include *premessenger RNA* (pmRNA), *transfer RNA* (tRNA), *ribosomal RNA* (rRNA), *small nuclear RNAs* (snRNAs), and others. pmRNA is a complementary RNA copy of a DNA gene, including any intron Nt sequences embedded in the gene. *Spliceosome enzymes* (see Section 6.2.1) "edit out" the nonprotein-coding intron Nt sequences to form a pure copy of the protein-coding gene as mRNA, which carries the gene code from the nucleus to the ribosome. tRNAs read the mRNA codons; there are tRNAs for each codon in the protein code. A tRNA binds only to the amino acid (AA) it codes for. Thus, three RNA bases from the DNA gene form a codon that specifies the linear placement of an AA in the primary structure of the protein being synthesized. Figure 6.4 illustrates a "typical," trilobe tRNA molecule with the anticodon for the AA methionine (Met). Methionine is the "start" tRNA.

Note that a tRNA molecule has three major hairpin loops, including the *anticodon loop* at one end. The anticodon loop has three particular bases that match the particular codon bases in the mRNA which code for a certain AA. The distal end of the tRNA molecule has bases that cause it to have a highly specific affinity to the particular AA the codon codes for. The AAs are found in the cell's cytosol. The tRNA structure is highly conserved throughout all life-forms; only the "variable arm" exhibits variability in its length and structure; 3 to 21 Nts can be in it. The 3-D structure of tRNA has an approximate "L" shape, with each leg about 6 nm long; the anticodon site and the AA acceptor site are about 7.6 nm apart. Unusual bases are used in the construction of tRNA: The TψC arm contains *pseudouridine,* and the D arm contains *dihydrouridine.*

6.2 RNAS CODED BY DNA

6.2.1 Introduction

We have seen in Chapter 6 that various RNAs are the "executive molecules" that carry out the instructions of the DNA genome (the "planner"); they are essential for protein synthesis as well as communication between, and control of, cellular processes. The *RNA polymerase enzymes* are necessary to assemble various RNA molecules from DNA codon "templates" in a cell's genome. There is a large number of RNA classes: We have already encountered pmRNA, mRNA, and tRNA. In addition, there are *ribosomal RNA* (rRNA), *small nuclear RNA* (snRNA), *small nucleolar RNA* (snoRNA), *micro-RNA* (miRNA), and *XIST RNA*. In each class of RNA, there can be several subclasses.

All three classes of *RNA polymerase* (RNAP) are dependent upon a DNA template in order to synthesize RNA. The RNA synthesized is thus complementary to the dsDNA strand used and identical to the nontemplate dsDNA strand. The nontemplate strand is called the *coding strand* because its Nt sequences are identical to those of the RNA, except, as noted earlier, the pyrimidine uracil is used instead of thymine in RNA. RNAP-I synthesizes three of four rRNAs. RNAP- II synthesizes the mRNAs and some of the snRNAs involved in RNA splicing (editing out introns from pmRNAs). RNAP-III synthesizes the tRNAs, one rRNA, and some snRNAs [King 2005a].

In spite of the molecular complexity of eukaryotic cells, we should take solace in the fact that there is a finite number of molecular species in each cell, and generally only a few genes are expressed in a given cell at a given time.

6.2.2 Messenger RNAs

As we have seen earlier and in Chapter 6, this class of RNAs is used as the active genetic coding template by the translational machinery to determine the primary AA sequence in a protein being synthesized. Premessenger RNA (pmRNA) is first synthesized from an exposed DNA strand in the nucleus by the enzyme RNAP-II. In eukaryotes, the pmRNA can contain the Nt sequences from one or more *introns* embedded in the gene's DNA. Intron Nt sequences do not code protein and must be edited out of pmRNA before precise protein synthesis can commence. The molecular algorithms that accomplish splicing out the introns in pmRNA are complex and will not be covered here. (The interested reader should consult King 2005c and Kimball 2005c.) *Splicing* basically cuts out the string of intron Nts from the pmRNA and then rejoins the cut ends to form a purely protein-coding mRNA molecule, which then leaves the nucleus and joins a ribosome in the cytoplasm where the actual protein assembly occurs.

6.2.3 Transfer RNAs

There are more types of tRNAs in cells than there are codons. The human DNA genome uses 59 codons to code the 20 amino acids used to build proteins (counting **ATG** for *Start Coding* and the AA methionine). However, it has been found that the human genome has 497 genes coding tRNAs! About a quarter of these genes are found together on chromosome 6; the others are found on all other chromosomes except 22 and Y.

TABLE 6.1
Wobble Ambiguities in mRNA–tRNA Coding

5′ tRNA, Anticodon Base	3′ mRNA, Codon Base
A (adenosine)	U
C (cytosine)	G
G (guanosine)	C or U
U (uridine)	A or G
I (inosine)	A or C or U

The structure of each tRNA is coded by a separate gene. tRNAs typically contain only 73 to 93 Nts. As we have seen earlier, they have a curious three-loop structure: the loops are formed by unpaired Nts; the paired sections are dsRNA.

Each species of tRNA has an amino acid attached to its 3′ end one of the 20 types of AAs used in all proteins. Note that not all 59 codons in the protein code have a specific tRNA, because of *wobble* in the third Nt position of anticodons. On the distal loop of tRNA are three exposed Nts (the *anticodon*), which are complementary to the mRNA codon. With the help of ribosomal enzymes, the anticodon binds to the active codon, and the tRNA delivers its AA to be added to the peptide being assembled by the ribosomal enzyme complex.

The *wobble hypothesis,* proposed by Francis Crick in 1966, suggested that although the specificity between the first two Nts of the mRNA codon and the tRNA anticodon had to be exact, it did not need to be exact in the third Nt position. Crick proposed that nonstandard base pairing might occur between the Nt base in the 5′ position of the tRNA's anticodon and the base in the 3′ position of the codon in question. Crick postulated that the base pairing schemes shown in Table 6.1 were operative between tRNA anticodons and mRNA codons.

It appears that the so-called *wobble rules* are not followed exactly. If they were, only 31 tRNAs would be needed to code 20 AAs. However, it has been found that well over 31 tRNAs exist in cells. For example, in *E. coli* there are five distinctly different tRNA molecules with which the *amino-acyl-tRNA synthetase* (AARS) enzymes bind to the AA leucine.

The rogue tRNA anticodon base *inosine* appears to be generated from adenosine by a gene that produces a protein enzyme, *adenosine deaminase.* Even with this inosine ambiguity, the genetic code in a species is stable and almost error free. Biochemical systems do not tolerate even small errors in protein composition. The error rate in attaching the correct AA to the correct tRNA is estimated to be 10^{-4}. The AARS enzymes use two means to ensure this level of coding accuracy: (1) the AA specificity "pocket" in a specific AARS will only bond to AAs similar in size and charge, and (2) the AARS has a unique "proofreading" capability with which it detects its binding to a wrong AA and then hydrolyzes it (destroys it) before it can be attached to the end of the tRNA.

6.2.4 Ribosomal RNAs

Before being incorporated into ribosomes, rRNA is processed in the subnuclear organelle, the *nucleolus.* The eukaryote 28S, 18S and 5.8S rRNAs are processed from a long primary rRNA transcript (a "pre-rRNA") containing long noncoding spacer and flanker sequences of Nts. Psuedouridylation and 2′-O-ribose methylation of the rRNAs described earlier occur before nucleolytic processing of the pre-rRNA [Kiss 2005].

Ribosomes are ubiquitous cellular organelles, found throughout all domains of life, that catalyze the sequential polymerization of amino acids into proteins following the linear genetic code contained in messenger RNA. They are about 20 nm in diameter and can only be visualized with electron microscopes. Ribosomes are built of two complex molecular subunits that act together to effect protein synthesis. Both subunits are complexes of mRNA and many proteins; they surround and travel along a mRNA molecule as the protein is made. The larger subunit creates the peptide bonds between adjacent AAs and provides the path through which the linear polypeptide chain emerges from the ribosome. The smaller unit acts more in an executive manner; it decodes the genetic message, discriminates between correct and incorrect aminoacylated tRNA molecules, and controls the fidelity of codon and anticodon interactions and in mRNA-tRNA translocations [Yonath 2001]. A major difficulty in discovering the structures and functions of rRNAs is the fact that they are intimately surrounded by and bonded to many protein molecules.

Most of what we know about rRNA and ribosomal structure comes from the study of bacterial ribosomes. Bacteria grow rapidly and their ribosomes are easy to harvest. The bacterial large ribosomal subunit (50S) contains 33 proteins and two rRNA molecules; one of 23S (2904 Nts), the other of 5S (120 Nts). (The S-number in 50S refers to the *Svedberg constant*, which relates to the amount of necessary *g*-force created by a centrifuge to sediment the molecule from a standard solution.) The 30S smaller ribosomal subunit contains 21 proteins associated with a 16S (1542 Nts) rRNA molecule.

The eukaryotic cell mitochondrial large (60S) subunit contains about 49 proteins and three rRNA molecules; one 28S (4,700 Nts), one 5S (120 Nts), and one 5.8S (160 Nts). The eukaryotic smaller subunit has about 33 proteins and an 18S (1,900 Nts) rRNA molecule (these numbers vary slightly with species).

All mitochondrial ribosomes also have two subunits: The larger has 48 proteins and a 16S (1,650 Nts) rRNA molecule. The smaller has 29 proteins and a 12S (950 Nts) rRNA molecule [Kimball 2003a].

Much of what we know about the 3-D structures of ribosomal subunits comes from x-ray crystallography on protein and RNA crystals prepared from the purified molecular components (e.g., 70S, 50S, and 30S concentrates from bacteria). The entire process is very specialized and requires nearly monochromatic x-rays produced from a linear accelerator such as at CERN or the U.S. Argonne National Lab in Chicago. The studies reveal molecules of considerable 3-D complexity. Published pictures of a bacterial ribosomal subunits (rRNA together with proteins) appear like a dish of pasta (noodles, spaghetti, rotini, etc.) all joined together and cross-linked at various points [Harms & Schlünzen 2005].

Moore and Steitz [2002] stated the obvious: No enzyme system as complex as the modern mammalian ribosome could possibly have emerged from the "primordial soup" in a single step. They argued that the ribosomes in the first cells were very likely composed entirely of RNA. Moore and Steitz went on to point out that the functional core of the modern eukaryotic ribosome, its decoding site and its *peptidyl transferase* center, consist primarily of RNA. Also, the bulk of ribosomal proteins are found on its surface, well removed from its RNA-based functional centers. Probably the earliest ribosomes were almost certainly smaller, simpler structures than the modern molecular complex, and had more limited functions. These "protoribosomes" might have been a one-subunit complex that catalyzed peptide bond formation in an uncoded way, and thus made small oligopeptides of random sequence that were not proteins in the modern sense, and may have found some other purposes in the protocell.

Note that the protoribosome was in fact a *ribozyme,* that is, an RNA molecule capable of catalysis. The RNA in modern ribosomes also has catalytic function and may be considered to be ribozymes. The human mitochondrial genome contains 24 RNA genes; two are for 23S and 16S rRNA subunits of mitochondrial ribosomes.

Moore and Steitz made the point that the chemical composition of the ribosome's functional centers correlate crudely with their evolutionary age in this scheme — the trend being that the more recent the center, the more protein it contains. The most ancient center of them all, the *peptidyl transferase center*, has no protein in it. The decoding center, the second to be added to the ribosome, is predominantly RNA but does contain some protein. The E-site, which is a late evolutionary addition, is protein rich. It is clear that the arguments of Moore and Steitz are supportive of the RNA World theory (see Chapter 2, Section 2.4.2).

One reason certain natural and synthetic antibiotics (e.g., *chloramphenicol, clindamycin, erythromycin, clairithromycin, puromycin, viomycin, and roxithromycin)* are effective against bacteria is that they attack the bacterial 50S subunit at the peptidyl transferase cavity, inhibiting bacteria protein synthesis. Fortunately, these antibiotics do not attack human ribosomes [Yonath 2001; Rib-X Pharmaceuticals 2005].

6.2.5 Small Nuclear RNAs

About a dozen genes for snRNAs have been described, each present in multiple copies in the genome (Redundancy underscores importance?) [Kimball 2005c]. snRNAs combine with certain *U-proteins* to form snRNPs; these executive molecules have roles in editing other classes of RNA. The "U" designation was given to the snRNAs because they were found to be rich in uridylic acid.

An important example is the small nuclear ribonuclear proteins (snRNPs) that are components of the *spliceosome* that edits intron Nt sequences out of pmRNA, forming mRNA. With the exception of U6, which is transcribed in the nucleoplasm by RNA polymerase III, U1, U2, U4, and U5 are transcribed there by *RNA polymerase II*. Posttranscriptional editing of snRNAs and combination with specific U-proteins may take place in the nucleolus and the *Cajal body* in the nucleus. The Cajal body is a subnuclear organelle, first described about 100 years ago by the noted neurocytologist Ramon Cajal. Under the electron microscope the Cajal body appears to contain a tangle of coiled threads. Cajal bodies are known to contain U7 snRNA and components of the three RNA polymerase transcription complexes [Carmo-Fonseca 2002].

The details of the roles snRNPs play in spliceosome editing of pmRNA are beyond the scope of this text, but snRNPs U1, U2, U4, U5, and U6 are involved [WU 2003; Kimball 2005c].

6.2.6 Small Nucleolar RNAs

In eukaryotic cells, rRNA and snRNA are extensively modified and processed in the *nucleolus*. Much of this activity is affected by snoRNAs. It appears that much of the coding for snoRNAs lies in introns and other intergenic (non–protein-coding) regions [Mattick 2003].

Lander et al. [2001] have compiled a set of 97 known snoRNA gene sequences. Lander et al. tell us that *the snoRNAs can be* subdivided into two families: (1) *C/D box snoRNAs*, mostly involved in guiding site-specific 2′-O-ribose methylations of other RNAs, and (2) *H/ACA box snoRNAs*, mostly involved in guiding site-specific pseudouridylations. The RNA component of the ribonuclease for mitochondrial RNA processing (MRP RNA) fails to fit into either classification. snoRNAs can be considered to be a subgroup of the snRNAs but should not be confused with the snRNAs mediating mRNA splicing, i.e., the spliceosomal RNAs. It is interesting to note that snoRNAs have not been found in any prokaryotes or Archaeans [Lowe 2000].

The main functions of the H/ACA box snoRNAs are guiding rRNA pseudouridylations, and most C/D box snoRNAs mediate rRNA ribose methylations. Kiss [2005] noted: "Apparently, adoption of the snoRNA-guided modification mechanism provides many advantages for eukaryotic cells. Eubacterial rRNAs contain only a few modified nucleotides which are synthesized by specific protein enzymes." Commenting on eukaryotes, Kiss observed that a single snoRNA-associated psuedouridine synthase and *2-O*-methyltransferase (*fibrillarin*) can accomplish the site-specific synthesis of many modified nucleotides in rRNAs, snRNAs, and most probably other cellular RNAs, instead of needing to express hundreds of protein enzymes. Kiss noted that the snoRNA-guided modification system is much more flexible and has the potential to evolve very rapidly. The *direct* evolution of protein enzymes is complex and is highly dependent on the preexisting modification sites. Thus, the snoRNA-guided pseudouridylation and *2-O*-methylation systems may have continuously monitored rRNA sequences during evolution and have introduced novel advantageous functional modifications into eukaryotic ribosomes.

Many snoRNAs and snRNAs have been found to be encoded in introns. Studies have suggested that snoRNAs are possibly processed from the out-spliced introns by exonucleases [Isis 2000; Mattick 2003].

6.2.7 Short Interfering RNAs (siRNAs)

siRNAs and miRNAs were discovered in unrelated studies, but their biogenesis and assembly into RNA–protein complexes and their function in downregulating gene expression are closely related. siRNAs and miRNAs share common RNAse III processing enzymes, the *Dicer enzymes,* and closely related effector complexes (RISCs) for posttranscriptional repression of protein synthesis. On the other hand, siRNAs and miRNAs differ in their molecular origins. siRNAs are sampled more randomly from longer, primary, double-stranded (dsRNA) transcript molecules. In the cytoplasm, the Dicer enzymes split the dsRNA primary molecules. The finished siRNAs in animals are usually 21–22 Nts long, similar to miRNAs. Figure 6.4 illustrates schematically the biochemical pathways used by siRNAs and miRNAs.

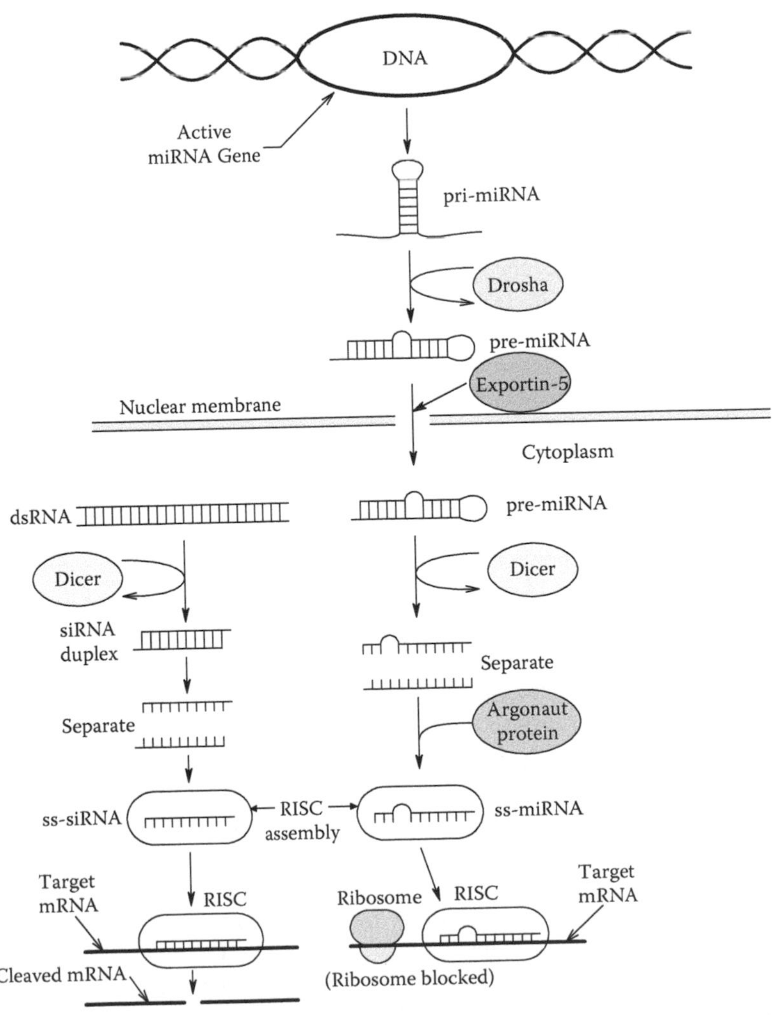

FIGURE 6.4 Biochemical synthetic pathways used by siRNAs and miRNAs.

It is also possible to introduce synthetic dsRNA into cells as plasmids that can produce siRNA–RISC complexes to inhibit a certain target gene [Cullen 2004]. Such genetic engineering has obvious medical applications.

Although they share similar preprocessing pathways, there are subtle differences between the actions of siRNAs and miRNAs. siRNAs generally inhibit expression of a specific gene, whereas miRNAs tend to be more general in their inhibition, one particular miRNA blocking expression of many different proteins. It has recently been shown that certain miRNAs can also upregulate translation as well as repress it [Vasudevan et al. 2007]. These authors proposed that miRNAs can oscillate between repression and activation during the cell cycle.

Work to clarify the differences and similarities between siRNAs and miRNAs and their roles in regulating gene expression is an ongoing challenge.

6.2.8 Micro-RNAs (miRNAs)

Micro-RNAs are a class of small, noncoding RNAs that regulate gene expression in a sequence-specific manner as required in embryonic development [Chen & Rajewski 2007; Engels & Hutvagner 2006]. miRNAs have been found throughout diverse metazoan (eukaryote) genomes, including plants [He & Hannon 2004]. They can inhibit protein expression by shutting off translation or by targeting mRNA for degradation. miRNAs were first discovered in 2001 in the widely studied worm *Caenorhabdtis elegans.* miRNA genes produce short (about 22 Nt) ss segments that fold over on themselves, forming a short section of dsRNA terminated in a loop; that is, they have a "hairpin" structure. Humans express over 460 genetically encoded miRNAs [Chuang & Jones 2007]. Also, human miRNAs make up more than 1% of the genome and may regulate over 30% of all protein-coding genes [Chen & Rajewski 2007].

As is often the case in biochemical systems, the synthesis of miRNAs is not simple. miRNA is first transcribed as a several hundred Nt long, ds, *primary transcript* called a *pri-miRNA.* In some cases, RNAPOL II mediates the transcription of pri-miRNA. The first step in posttranscriptional processing of miRNA is the excision of the about 70 Nt miRNA hairpin from the longer pri-miRNA. This cleavage occurs in the nucleus by a clever enzyme called *RNAse III,* or *Drosha.* The cleavage product is a 63 Nt, *pre-miRNA intermediate molecule* that is then exported from the nucleus into the cytoplasm. In the cytoplasm, another interesting enzyme, *Dicer,* an ATP-dependent RNAse, cleaves both strands of the pre-miRNA 21 Nt from the bound end. This cleavage leads to the formation of the 22 Nt miRNA final product. This final miRNA is incorporated into a large protein complex called the *RNA Induced Silencing Complex* (RISC). RISC directs the cleavage of mRNA targets [He & Hannon 2004]. See Figure 6.5 for a schematic illustration of these processes. The RISC composition is not known in detail, but it contains certain *Argonaute proteins* that are evidently necessary for miRNA to function as a silencer.

So, miRNA/RISC can pair exactly with an mRNA and cause its cleavage and destruction, or it can pair partially with mRNA and produce translational inhibition (block the ribosome). Perhaps as many as one-third of human genes are regulated by miRNAs. However, it is known that many essential "housekeeping" genes, such as those involved in cellular respiration, cannot be blocked by miRNAs. During embryonic development, the repression of gene expression by miRNAs appears to be one mechanism that coordinates the sequencing of cell differentiation and growth. It is possible that miRNAs play as important a role as transcription factors (TFs) in regulating development [Kimball 2007f]. Do TFs regulate the expression of miRNAs? *Quisnam ordinate ordinates?* (What regulates the regulators?)

Recent evidence cited by Chuang and Jones [2007] indicates that certain miRNAs are regulated epigenetically. That is, epigenetic mechanisms such as DNA cytosine methylation and various histone modifications can affect miRNA expression. As an example, they cite the upregulation of miR-127 in cancer cell lines after treatment with the potent DNA methylation inhibitor 5-Aza-CdR and 4-phenylbutyric acid (PBA), a histone deacetylase inhibitor. These drugs lead to more open chromatin structures (heterochromatin → euchromatin) and thus to the reexpression of genes that had been silenced epigenetically. The miR-127 downregulates mRNAs used in gene expression, in particular, the protooncogene, BCL6.

Because of the miRNA system's regulatory complexity, exactly how miRNAs function in gene regulation in embryonic development and their role in diseases such as cancer will remain a hot research topic for years to come. Their study is crucial to understanding applied regulatory genomics.

Some recent studies have shown that miRNAs are regulators of mammalian hematopoeisis [Chen & Lodish 2005]. They also are implicated in the regulation of the **Ras** cancer gene [Yale University 2005], mammalian brain differentiation [Sempere et al. 2004], and heart development. The wide extent and complexity of their regulatory actions are just beginning to be understood.

6.2.9 Ribozymes

Ribozymes are catalytic RNA enzymes that act to alter covalent structure in other classes of RNAs and certain molecules. We have already seen that they occur in ribosomes. These unique nucleotides are also found in the nucleus and the chloroplasts of eukaryotic organisms. Some viruses, including several bacteriophages, also have *ribozymes.* An optimum concentration of metal ions such as Mg^{++} and K^+ is associated with their effective functioning. *Ribozymes* generally act as "molecular scissors," cutting precursor RNA molecules at specific sites. Curiously, they also serve as "molecular staplers," which *ligate* or join two RNA molecules together.

Ribozymes are involved in the transformation of large precursor molecules of tRNA, rRNA, and mRNA into smaller final products. In their active form, ribozymes are complexed with protein molecules. For example, the enzyme *Ribonuclease-P* (RNase-P) is found in all living things. This ubiquitous ribozyme cleaves the head (5′) ends of the precursors to the tRNAs. Ribonuclease-P is a *heterodimer* containing one molecule of protein and one molecule of RNA.

Ribozymes are also contained in *Group I introns* found in rRNA genes in the mitochondrial genome of certain fungi, some chloroplast genes, and in the nuclear genome of some "lower" eukaryotes. The Group I introns "splice themselves out," making contiguous RNA. Self-splicing, *Group II introns* are found in some mRNA in the mitochondria yeast and fungi. They also edit out introns from the mRNA.

Probably the best-known *ribozymes* operate in the nuclear *spliceosome enzymes* that edit intron sequences out of premessenger RNA in the nucleus of a eukaryotic cell, forming correctly coded mRNA, ready for coding protein synthesis by ribosomes. The spliceosome complex contains five kinds of small nuclear RNA (snRNA) molecules complexed with many protein molecules. The splicing operation is evidently catalyzed by the snRNAs, not the proteins.

The ribosomal protein synthetic molecules contain four rRNA molecules and dozens of proteins. The peptide bond formation in the growing polypeptide chain has been shown to be catalyzed by the 23S rRNA molecule in the "large (50S) ribosomal subunit." The 31 proteins found in association with the rRNA in the large subunit probably help maintain the 3-D structure of the 23S ribozyme so that it functions properly [Kimball 2005g].

Synthetic ribozymes can potentially be used in genetic engineering applications to stop the expression of any gene in a sequence-specific manner and therefore may be useful in cancer therapy [Malerczyk et al. 2005] or as tools to analyze gene function in development. Two generic types of ribozymes have shown potential in genetic engineering: The "hammerhead" and the "hairpin" ribozymes [Lyngstadaas 2001]. Both types can cleave RNA **GUC** sites, but the hammerhead can also cleave at **GUA, GUU, CUC,** and **UUC** and sometimes at **AUC** sites. Ribozymes can be put into cells as synthetic oligos with a "gene gun," or the cell can synthesize them itself with engineered genes. Note that research on applications of synthetic ribozymes is an ongoing active area [Lilley 2003; Ferré-D'Amare & Rupert 2002; Soukup & Breaker 1999].

6.2.10 XIST RNA

XIST stands for *X-Inactive-Specific Transcript.* It is a large (about 17 kb) RNA coded by a gene on the X-chromosome (at locus Xq13, eight exons are involved with human XIST). XIST RNA accumulates in female somatic cells along the X-chromosome containing the active XIST gene and proceeds to inactivate nearly all the hundreds of genes on that X-chromosome. (Eighteen genes on the inactivated X are spared deactivation; these same 18 genes are also found on the Y-chromosome in males. Why these genes remain active is yet to be discovered.) The XIST RNA does not travel over to any other chromosome in the nucleus. The *Barr body* seen in a cell nucleus with light microscopy is an inactive X-chromosome covered with XIST RNA.

In early development of a female embryo, the XIST locus on her two X-chromosomes is expressed, but the resulting XIST RNA is broken down on both. Then, an unknown event occurs that makes one X-chromosome's XIST gene dominant, whereas the other is suppressed. It is known that the XIST allele on the inactive X replicates before the XIST on the active X. So, it is possible that the expressed XIST RNA on the inactive X inhibits the XIST gene on the active X by an unknown blocking factor [Spusta & Goldman 2005; Marahrens et al. 1998].

XIST-mediated shutdown is mediated by targeted methylation of the XIST regulatory sequences as well as histone hypoacetylation [Csankovszki et al. 2001]. The X-chromosome without XIST RNA is able to express all its other genes. X-inactivation by XIST in the female embryo appears to be random; either the maternal or paternal X can be inactivated in a given cell. However, in the cells of the *extra-embryonic membranes* (amnion, placenta, and umbilical cord), it is always the paternal X-chromosome that is inactivated [Kimball 2005d].

The entire XIST RNA X-blocking mechanism is very complex. Another RNA, *TSIX,* is found along with XIST and is antisense to it; it appears to reduce the level of XIST activity. The role of TSIX may be to regulate the silencing activity of XIST [Plath et al. 2004].

6.2.11 Discussion

In Section 6.2, we have seen that there are many kinds of RNAs coded by DNA. These RNAs have many roles in cell biology, ranging from specialized enzymatic actions to orchestrating the shutdown of the genes on one female X-chromosome. In each case, these RNAs carry out orders implicit in their structure coded by DNA. Many of these "executive" RNAs are active when in close association with specific proteins.

Molecular biologists are just beginning to understand the roles of these molecules in cell replication, cell metabolism, and protein expression. The fact that many RNAs complex themselves with one or more protein molecules makes understanding their regulatory architectures and functions that much more difficult. Not only does the cell's DNA code for the proteins and the RNAs but also for the enzymes that form the RNA–protein complexes.

6.3 DNA REPAIR

6.3.1 Introduction

Considering the importance of a cell's genome, eukaryotic cells are indeed fortunate to have their DNA enclosed by a protective nuclear membrane. The DNA is further protected by its alpha-helical structure and its protective, close association with histone proteins. In spite of these structural and chemical protective factors, a cell's DNA is still subject to damage by a variety of external factors. These include *ionizing electromagnetic (photon) radiations* including gamma rays and x-rays, as well as high kinetic energy, and subatomic particles such as neutrons and alpha particles. DNA damage can also occur when a nucleus is exposed to short-wavelength ultraviolet radiation (UV-B and -C). UV photons may not directly damage nucleic acids, but they can create free radicals that can break DNA bonds. Many exogenous chemicals can also damage DNA, for example, drugs used in cancer chemotherapy, insecticides, benzopyrene found in cigarette smoke, and some plant and microbial products such as *aflatoxins* produced in moldy peanuts.

Humans inherit about 3.1×10^9 base pairs of DNA from each parent. Thus, 6.2 billion (diploid) DNA base pairs can be the target of a substitution. It has been estimated that in humans and other mammals, uncorrected errors (mutations) in DNA occur at the rate of about 1 to every 40 million (5×10^6) base pairs during DNA duplication. This suggests that with 6.2×10^9 bps in a diploid cell, each new cell has about 120 new mutations. This is not as bad as it sounds because only 1% of our DNA consists of genes, so only about 1 error in 23,000 genes is relatively accurate. However, errors in the 99% remaining DNA can affect the control of gene expression, transcription, and translation [Kimball 2005f].

The types of DNA damage vary widely. The base cytosine can be converted to uracil by an amino group being replaced by an oxygen. Cancer chemotherapy drugs can cause covalent cross-links to form between bases on the same DNA strand or between DNA strands. High-energy ionizing radiation can actually break a single DNA strand or cause a double-stranded break. Finally, a failure of proofreading enzymes during DNA duplication can cause a bad base substitution (e.g., incorporation of the pyrimidine uracil instead of thymine).

DNA damage, or the failure to repair it properly, can lead to cancer, or senescence (gradual loss of cell function and failure to reproduce) leading to cell apoptosis. Uncorrected damage to (haploid) germ cells, or to cells of the embryo, produce mutations that can affect a multicelled organism's survival; that is, they can create positive or negative evolutionary (survival) pressure.

Finally, we note that prokaryotes, with their circular dsDNA genome, also possess mechanisms for DNA repair. For example, bacteria have a ubiquitous enzyme, *mycobacterium ligase* (Mt-Lig), which can perform *nonhomogenous end joining* (NHEJ) of broken DNA strands (see later text). One form of molecular complexity has been traded off for another. One bacterial Mt-Lig molecule can do the work of three necessary enzymes in eukaryotes to do NHEJ [Weller et al. 2004]. Bacteria have other enzymatic DNA repair systems, as well as Mt-Lig. Perhaps the shift to a dsDNA genome (for all life) from some sort of early RNA coding was made possible by the evolution of effective DNA repair enzymes with dsDNA.

6.3.2 DNA Repair Mechanisms

What is truly remarkable is that a eukaryotic cell's genome carries the information to make the specialized enzymes for self-repair. First, let us consider the repair of damaged or inappropriate bases. To underscore the amazing complexity of cell biology, note that each of the many types of chemical damages to bases requires its own specialized, enzymatic corrective machinery.

One frequent cause of *point mutations* is the spontaneous methylation of cytosine followed by a deamination of a neighboring thymine. There are two major means of correcting this type of DNA damage:

1. *Direct chemical reversal*, in which *glycosylase* enzymes remove the mismatched thymine and restore the correct cytosine. (There are at least eight genes coding *DNA glycosylases*, which identify and remove a specific kind of base damage. This repair is done without the need to break the DNA double helix.)
2. In *excision repair*, the damaged bases are cut out (excised) and replaced with the correct ones in a local pulse of DNA synthesis.

Excision repair can be accomplished in eukaryotes by three methods, each of which uses its own special sets of enzymes: (a) *base excision repair* (BER), (b) *nucleotide excision repair* (NER), and (c) *mismatch repair* (MMR).

Base excision repair is estimated to occur some 20,000 times a day in each nucleated cell in the human body by a DNA glycosylase. Two genes encode two enzymes that remove *deoxyribose phosphate* from the DNA spine, separating it. *DNA polymerase beta* (DNAPOL-β) is one of at least 11 DNAPOLs encoded in human genes. DNAPOL-β replaces the correct Nt in the break. The broken DNA strand is now rejoined with covalent bonds; two enzymes can do this using ATP as an energy source.

Nucleotide excision repair (NER) differs from BER is the following ways: It uses different enzymes. NER removes not just the bad Nt but cuts out a section of several Nts on each side of the bad one. Kimball [2005e] listed the steps in NER:

1. The damage is "recognized" by one or more protein factors that assemble at the location.
2. The DNA is unwound, producing a "bubble," exposing the damaged base. The enzyme system that does this is *transcription factor II H.*
3. Cuts are made both on the 3′ side and the 5′ side of the damaged area to remove the damaged section.
4. Using the intact, opposite (complementary) DNA strand as a template, the enzymes *DNA polymerases delta and epsilon* (DNAPOL-δ and DNAPOL-ε) fill the gap with the correct Nts.
5. Finally, a *DNA ligase* uses covalent bonds to join the fresh patch in the separated DNA strand [Melcher 2005].

Defects in the NER mechanisms can lead to certain hereditary DNA repair disorders often called *segmental progerias* (accelerated aging diseases) [Andressoo et al. 2006].

Mismatch repair (MMR) fixes mismatches to the normal Watson–Crick DNA base pairing (e.g., **A-T** and **C-G**). MMR can enlist the enzymes used in BER and NER, as well as its own specialized enzymes. Recognition and excision of the mismatch requires several different enzymes. An important question is: How does the MMR system recognize the incorrect Nt? In *E. coli*, we know that certain adenines in its genome become methylated soon after a new strand of DNA has been synthesized. The MMR system evidently works faster than the DNA replication, and if it detects a base mismatch, it treats the nucleotide on the already-methylated (parental) side as the correct complement and removes the Nt on the freshly synthesized, unmethylated daughter strand and replaces it. The MMR recognition system in mammals has yet to be described [Kimball 2005e].

DNA strand breaks are generally caused by photons and kinetic particles having sufficient energy to break covalent bonds. They also can be caused by certain toxic cancer chemotherapy drugs. *Single-strand breaks* (SSBs) can be repaired by the same enzymes used in base excision repairs (BER). *Double-strand breaks* (DSBs) are more serious. There are two mechanisms used by eukaryotic cells to repair DSBs:

1. Direct joining of the broken ends; This is called *nonhomogenous end-joining* (NHEJ). A specialized heterodimer protein called **Ku** aligns the broken ends of the dsDNA sections and joins them. Errors in the direct-joining process may be the source of the various DNA *translocations* that are associated with cancer development.
2. In *homologous recombination,* the broken ends are repaired using the information on the intact *sister chromatid*, the *homologous chromosome,* or the *same chromosome* if there are duplicate copies of the gene on the chromosome oriented in opposite directions. Kimball [2005e] noted that recombination between homologous chromosomes in *meiosis I* also involves the formation of DSBs, which must be repaired using the **Ku** system, etc.

6.3.3 Discussion

The repair of DNA damage is incredibly important to preserve a species' genome. Nature has developed an amazing array of enzymes to detect and repair damaged or broken DNA molecules in all cells. Wood et al. [2005] have released a list of human DNA repair genes, giving the protein enzyme activity as well as the chromosome location on which each gene is found. *Base excision repair* (BER) uses 15 genes. There are three genes coding enzymes doing *direct reversal of damage*. *One* gene is used for repair of DNA protein cross-links. *Mismatch excision repair* (MMR) uses 11 genes. *Nucleotide excision repair* (NER) has 30 known genes (some code enzyme subunits). There are 18 genes associated with *homologous recombination*. Six genes work in *nonhomologous end-joining*. DNA polymerases need 15 genes. A number of other genes associated with DNA repair or suspected DNA repair are also given. The grand total of genes associated with DNA repair is 143 [Wood et al. 2005].

One might consider this set of DNA repair genes to be "anti-evolution genes." If they worked perfectly, there would be negligible "drift" in the genomes of various species and no selective phenotypical advantage or disadvantage relative to a changing environment. However, we do see evolution, even on so short a timescale as decades. Certain insects exposed to DDT have developed enzymatic resistance to this once-powerful insecticide. Prolonged use and overuse of certain antibiotics have also led to bacteria that are "immune" to the antibiotics because the bacteria have developed enzymes or internal molecular structures that render the antibiotics ineffective. The irony here is that use of certain insecticides and antibiotics can damage DNA, and perhaps the genes that repair it, allowing for a much higher mutation rate. Antibiotic-resistant staphylococcus and tuberculosis have truly evolved, as have DDT-resistant flies. A question for the reader to ponder: Is an efficient yet imperfect DNA repair system *necessary* for evolution and evolutionary adaptation? Or is evolution caused primarily, as has been hypothesized [Mikkelsen et al. 2007], by certain introns? (See Chapter 5.) Note that introns in dsDNA are also subject to breakage and repair.

6.4 GENE REGULATION

6.4.1 Introduction

One of the more challenging problems in human genomics today is understanding the details of gene regulation, i.e., the elaboration of the biochemical pathways that control gene expression. Understanding gene regulation is pivotal in the new discipline of *Evo-Devo,* or evolutionary and developmental biology. An understanding the mechanisms and pathways of gene regulation is also necessary to program stem cells to differentiate into precise tissue types, e.g., skin, neurons, immune system cells, heart muscle, etc.

The control of gene expression involves a number of mechanisms: certain *promoter* and *repressor* molecules that interact with the DNA, RNAs, or enzyme proteins. Regulatory molecules include hormones, cytokines from the immune system, certain RNAs, end products in biosynthetic pathways, and various proteins. Gene regulators have effects on gene expression; for example, when and for how long a gene is expressed and how much protein is made. Gene controls can be graded or binary; *positive control* activates a gene in a cell, and *negative control* can repress a gene's activity or turn it off completely. Also, because gene expression is a multistep, complex process, the regulatory mechanisms can act at any step of expression. Probably the most active time for gene regulation is during embryonic development. It is then that differentiated tissues are formed beginning with totipotent stem cells. Some differentiated embryonic tissues are made to disappear through regulated apoptosis (e.g., fetal finger webs, "gill slits," and a tail). Clearly, the orchestration of gene control during development is most complex in vertebrates, in particular, humans.

Because of this great complexity, it makes sense to begin an investigation on the mechanisms of gene regulation with the simplest organisms, prokaryotic bacteria. In fact, the pioneering work of F. Jacob and J. Monod in the 1960s on gene control mechanisms in the ubiquitous bacterium, *E. coli,* has been expanded to try to describe the genetic regulatory mechanisms in eukaryotes [Struhl 1999]. Later, as an introductory example, we will describe the well-known mechanisms of **Lac** gene regulation in *E. coli.*

6.4.2 Gene Regulation in Prokaryotes

We have already seen that the bacterial genome is a double-stranded "ring" of DNA. Much of the research on gene regulation in prokaryotes has been done on the bacterium *Escherichia coli,* or *E. coli.* The genome of *E. coli, variety K12,* has been sequenced, and it has 4,377 genes contained in 4,639,221 nucleotide pairs [Kimball 2004c]. *E. coli* has the ability to sequence about 1,700 enzymes (proteins), with which it performs its vital functions; thus, it has the genes for about 1,700 different mRNAs.

Some "housekeeping genes" are active nearly all the time, at almost every stage of the bacterial life cycle. These housekeeping genes are required for continuously ongoing cell functions such as glycolysis (metabolism), replacing important transmembrane proteins such as aquaporins and ion pumps, etc. Other genes are inhibited and expressed on an ad hoc basis in response to chemical signals in the environment. This downregulation saves metabolic energy and molecular substrates until they are needed.

It has been discovered that there are three fundamental mechanisms of gene regulation in *E. coli* that use three types of specific DNA sequences. The rate of gene expression can be constant, or modulated up or down, depending on physiological and biochemical conditions. *DNA promoter sequences,* originally defined as genetic elements that determine the maximum potential rate of gene expression, are recognized by an RNA polymerase enzyme and contain all the information necessary for accurate transcriptional initiation.

Operator sequences are recognized by *repressor proteins,* which inhibit gene transcription that would otherwise occur owing to promoter action. The third mechanism of gene regulation in *E. coli* is the *positive control element sequence* (PCES). PCESs are recognized by *activator proteins* that stimulate gene transcription from the promoter. The functions of repressors and activators are modulated by specific physical and chemical conditions affecting the bacteria.

In *E. coli,* RNA polymerases (RNAPs) recognize promoters by their specific DNA sequences immediately upstream from the gene initiation site. In fact, for prokaryotic organisms, the ground state for transcription is nonrestrictive. Thus, repressors are generally required to keep gene activity at a low level, except in situations where the promoter itself is weak. Thus, for activation of specific genes, the repressors themselves must be downregulated or repressed, a sort of double-negative logic.

To make the regulatory mechanism more efficient, some *E. coli* genes are organized in *operons.* An operon is a functional group of structural genes that are regulated as a group and expressed as a group. A total of 75 different operons controlling 250 structural genes have been identified in *E. coli.* Figure 6.5 illustrates the well-studied *E. coli* **Lac** *operon.* First in the DNA is the *promoter* for the *regulatory gene* ($\mathbf{p}_i$), then the *regulatory gene* (**i**) followed by the **Lac** *operon promoter sequence* ($\mathbf{p}_{lac}$), then the *operator sequence* (**o**), then a group of *structural genes* coding certain

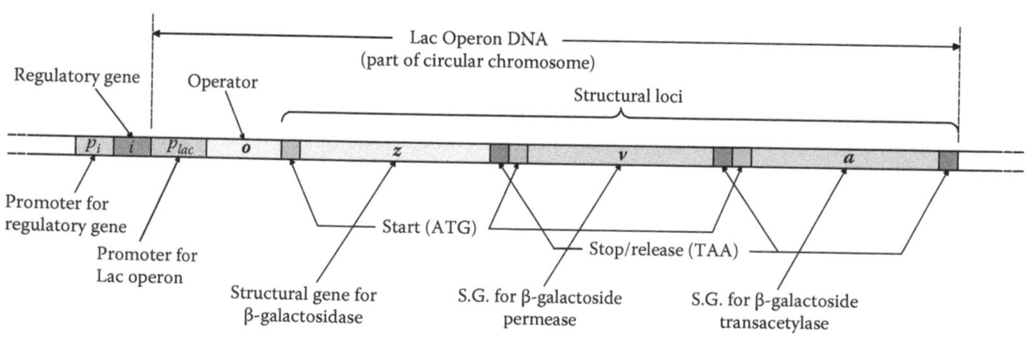

FIGURE 6.5 (See color insert following page 44.) Schematic of the *E. coli* **Lac** operon DNA. It is used to code three protein enzymes at once, once activated.

enzymes for lactose metabolism. The regulatory gene codes for a repressor protein that obstructs the promoter and hence inhibits transcription of the structural genes.

The structural genes are separated by initiation and termination signals (stop/start base sequences) and are expressed by ribosomes as a common group because they work synergistically. The **z, y,** and **a Lac** structural genes are for β-*galactosidase,* β-*galactoside permease,* and β-*galactoside transacetylase,* respectively. Note that the **Lac** gene is inducible; the presence of lactose sugar around and in the bacteria turns it on (derepresses it) [Farabee 2001]. Thus, lactose acts as a hormone or signal substance. In the absence of lactose, the **Lac** repressor protein binds to the *E. coli* DNA near the **LacZ** gene promoter. This prevents RNAP from binding, inhibiting the transcription process for **Lac**, saving energy and substrate molecules. When lactose is present inside the bacterium, it binds to the **Lac** repressor protein, inducing a conformational change that causes the repressor to dissociate itself from the DNA, allowing transcription to proceed. Then mRNA for the concatenated **z, y,** and **a** structural gene exon is made.

Regulation of transcription in eukaryotes is far more complex than that in prokaryotes; it allows for many more points of regulation. For example, in eukaryotes there are three types of RNAP that utilize different types of promoters. Each protein gene is transcribed separately, so transcription of every gene is regulated by its own promoter. Transcription is assumed to be independent of translation, and RNAP requires other protein transcription factors in order to bind to the promoter and begin transcription.

6.4.3 Some Molecular Factors in Gene Expression in Eukaryotes

Gene regulation is certainly one of the more important areas in molecular biology: It figures in developmental biology, medicine, stem cell regulation, physiology, and agriculture. Basically, gene regulation is about all of the cellular regulatory mechanisms that determine the timing of and the rate of expression of the functional products of a gene. Such products can be proteins or RNAs. Note that some gene products are expressed continuously in cells; they are necessary for cell homeostasis, respiration and metabolism, etc. The cell may also continuously secrete certain molecules as part of its physiological "job." Other gene products are regulated exogenously by hormones. Some products only appear during cell differentiation and growth. And as we have noted, in some classes of cells certain genes are never expressed.

A simplistic view of the human genome treats a gene as a continuous string of DNA codons having a *start* codon at its beginning and a *stop* codon at its termination. Most of the 30,000 human *protein-coding genes* (PCGs) have one or more noncoding *intron* base sequences inserted in them. To further complicate the picture of protein-coding genes, they have in addition to the exon base coding sequences and noncoding introns a *promoter* sequence, an *enhancer* sequence, and various *silencer* and *insulator* base sequences. The promoters, enhancers, silencers, and insulators generally lie outside the intron and exons of a PCG.

Enter complexity, again: The *transcription start site* is where a molecular complex of 12 proteins, called *RNA polymerase II* (RNAPOL II), binds to the gene. In addition, a DNA *basal promoter region* (BPR) lies about 40 bps from the *transcription start site.* The basal promoter region contains a base sequence, **TATAAA**, called the *TATA box* (aka the Goldstein–Hogness box). *TATA-binding protein* (TBP) binds to the TATA box bases, which lie about 25 bps upstream from the gene's start site; 14 other proteins bind to the TBP but not the DNA. Taken together, the complex of the 14 proteins plus the TBP are called *transcription factor IID* (TFIID). The basal promoter also contains *transcription factor IIB* (TFIIB), which binds to both the DNA and RNAPOL II. The basal promoter is found in all eukaryote protein-coding genes.

In contrast, an *upstream promoter* (UP) can be found on the DNA as many as 200 bps farther upstream of a gene's basal promoter. The UP can assume various forms for different genes.

In addition to the UP and the BPR regulatory sites, there are two other regulatory regions that can be located as many as thousands of BPs away from the gene they regulate: *Silencers* are control regions of DNA that suppress gene expression when certain transcription factor molecules bind to them; they can be viewed as a molecular "off switch" for gene expression. An *insulator* is a section of DNA, 42 bps or more, located between *enhancer* and *promoter* regions or between *silencer* and *promoter* regions of adjacent genes or clusters of adjacent genes. An *insulator* prevents the expression of a gene from being influenced by the activation or repression of its neighboring genes. All insulators contain the Nt sequence **CCCTC**. Insulators work only when bound to a *CTCF zinc finger protein* (aka **CCCTC** binding factor). Contrast this separation of gene expression to the bundling of gene expression in prokaryotes (e.g., the **Lac** operon). Clearly, there is a trade-off between efficiency and specificity in regulation.

Many, many molecular biological factors affect gene expression. There are *exogenous factors* such as steroid hormones that act at the nuclear level after being internalized into a cell, and they are signal molecules such as interleukins that trigger second-messenger signals after binding to cell surface receptors (e.g., **G**-protein-coupled receptors), turning (switching) the expression of certain genes on or off.

Methylation of DNA is a common method of gene silencing. Methylation is sometimes referred to as the *second code*, and is considered important to drug discovery, diagnostics, and cancer treatment because of its ability to silence promoter genes in particular [Bouley 2006]. DNA is typically acted on by *methyltransferase enzymes* acting on cytosine in a CpG dinucleotide sequence called a *CpG island*. Gene expression can also be downregulated by phosphorylating histone proteins found in close association with "resting" DNA. Genes can also be turned off by the methylation of histones. On the other hand, enzymes known as *histone acetyltransferases* (HATs) dissociate the DNA from the histone protein complex, exposing the alpha helix so transcription may proceed.

The reason that the regulation of gene expression is so complex is that there are so many biochemical pathways affecting it; expression of a gene can be downregulated or turned off by downregulating the concentration of any of the critical RNA and protein enzymes required for it to take place. Molecular biological regulatory mechanisms can also turn on genes by inhibiting the expression of gene products that turn genes off (a logical double-negative).

Mastery of the details of gene regulatory systems will have great impact in all walks of human life; not just can cures for chronic diseases and cancer be discovered and manipulation of stem cells in regenerative medicine be performed, but better, safer GMOs (plants, animals) will also result when we learn to improve control of gene expression exogenously. Much remains to be learned about the regulation of gene expression in embryological development. Obviously, many complex inter- and intracellular cell signaling pathways remain to be elaborated. The molecular biological events that determine how a zygote progresses to its adult form in any metazoan is a picture that is slowly being assembled.

6.4.4 Epigenetic Gene Regulation

Epigenetic processes control the way genes are expressed and their products are utilized. They are crucial in development and the ongoing maintenance of tissues and organs.

The evolution of gene regulation (as well as gene mutations) plays a pivotal role in the evolutionary process. Epigenetic effects on gene regulation are considered to be inheritable changes (at least over several generations) in gene expression brought about by modification of the DNA-associated histone proteins (chromatin). No changes occur in the nucleotide sequences of the genes in question or their regulatory Nt sequences, which include enhancers, promoters, and insulators. However, an unknown causative DNA sequence change may have occurred distal to the genes in question [Zuckerkandl & Cavalli 2006]. Zuckerkandl and Cavalli are of the opinion that epigenetic inheritance is closely associated with sectorially controlled developmental genes such as the *Hox genes*. They stated: "In sectorial regulation, the coding sequences and their primary regulatory dependencies can be either sequestered or maintained in an 'open' structure thanks to alternative chromatin conformations extending over a DNA sector that is at least tens of kilobases long." They go on to remark that in the plant *Arabidopsis sp.*, small interfering RNAs (siRNAs) are associated with establishing and maintaining inheritable states of gene activity. Also, inheritable activation of transcription is associated with histone H3 methylation (specifically, H3K4) (Kn is the *n*th lysine residue) and other specific histone K- modifications. Epigenetic gene silencing may also be associated with histone modifications such as deacetylation of H3K9, demethylation of H3K4, methylation of H3K9, and methylation of cytosine in CpG dinucleotides in DNA. How all these epigenetic modifications are regulated and the information specifying them is stored remains to be discovered. Regulatory epigenetics will certainly be an active area of research in the next few decades.

6.5 SUMMARY

Although all living organisms use DNA to carry the information that allows them to reproduce, metabolize, and practice homeostasis, they also use RNA molecules of many sorts to execute housekeeping tasks, gene expression, protein synthesis, etc. Indeed, we argue that RNAs are the executive nucleic acids, carrying out the instructions coded in the organism's DNA genome.

Some nonliving parasitic organisms (e.g., retroviruses) code their instruction set in RNA rather than DNA. As a result, their RNA must be changed to DNA inside their host cells by a *reverse transcriptase* enzyme so they can use the cells' DNA → RNA machinery to reproduce themselves, including their genomic RNA.

Points for the reader to ponder: Why does *all* life use DNA for its genetic code? Was RNA ever used genomically by cells? Perhaps DNA was naturally selected because its alpha helix is chemically more robust (stable). However, because it is robust, a host of enzymes had to arise (evolve, be developed) to repair and reproduce DNA and also read out the genes in a DNA genome. Although the prokaryotic operon is an efficient way to group and regulate structural genes with a common purpose, it was superseded by individual structural gene regulation in eukaryotes, driven by the advantage of the extra flexibility. Clearly, the biochemical C^3 networks in eukaryotes are far more complex than for bacteria. Does a very large association of stable, negative feedback biochemical regulatory networks form a more robust whole organism than a smaller network of stable regulatory networks? Bacteria have been on Earth far longer than eukaryotic metazoans, and like it or not, they have been quite successful life-forms, adapting to a broad spectrum of severe conditions (see Chapter 3, Section 3.2). They are literally everywhere. Finally, is it valid to assume that, in general, molecular biological systems tend to evolve toward biochemical complexity? And if so, why? [Bonner 1988.]

CHAPTER 6 HOME PROBLEMS

6.1 Why have estimates of the number of human protein-coding genes steadily decreased over time?

6.2 A certain free-living bacterial (prokaryote) species (not an obligate parasite) has the smallest known number of genes. Give the genus and species of this simple creature, its gene number, and genome size. Where does it live?

6.3 Give the size of the "typical" chloroplast genome.

6.4 Give the size of the human mitochondrial genome. How many proteins does it code for?

6.5 What eukaryote organism has the largest known genome? Give its size.

6.6 Is there such a thing as nonviral RNA inheritance? If so, give examples.

6.7 What is the estimated error rate in DNA duplication in eukaryotes?

6.8 (A) Describe in detail how DNA sequencing was done in the Human Genome Project. (Start by visiting: www.ornl.gov/sci/techresources/Human_Genome/faq/seqfacts.shtml.)
(B) Describe how DNA sequencing is now done more efficiently.

6.9 What is a *transcription factor* (TF) protein? Describe the functions of TFs in the nucleus. Give several examples of TFs.

6.10 Make a table of known small and micro-RNAs and their functions (and putative functions).

6.11 Discuss the evolution of robust homeostasis in metazoans as a form of genetic fitness.

6.12 Is the evolution of a CNS that has memory and that can learn the result of natural selection? What is the value of such a brain?

7 A Review of Proteins

7.1 INTRODUCTION

7.1.1 AMINO ACIDS AND THE PEPTIDE LINKAGE

Proteomics is the study of proteins, especially in biological systems. Proteins are closely associated with genetics because genes code for proteins. The field of proteomics is even more complex than that of genomics because the number of proteins is so much greater than the number of genes, and because the 3-D variations in the structures of proteins affect their specificity.

Proteomics is a "hot" field that has potential medical implications at least as great as those of genomics. In order to understand why it can be beneficial to manipulate protein function and structure for research and therapeutic purposes, it is important to understand how proteins function in the body. This chapter gives an overview of the role of proteins in hormones, tissue structures, cellular homeostasis, and the immune system. It also describes some pathologies related to proteins and opportunities for scientific advancement in this field.

Proteins are chains of amino acids joined sequentially by *peptide linkages* between adjoining amino acids (AAs). This is known as a polypeptide chain. Immediately after a chain of AAs (polypeptides) is manufactured by a ribosome, the polypeptide chain folds into a unique, complex, 3-D, *tertiary structure* that gives the protein its desired function. The current estimate is that there are about 22,726 genes encoding proteins in human cells. This represents a decrease in the estimated number of genes of about 2,000 over 6 years (2001–2007) [Southan 2004, 2005; Gingeras 2007]. (Note that the gene number is in flux, in spite of the Human Genome Project. The actual number of genes [if it *is* fixed] probably lies between 22,000 and 23,000).

In some cases, enzymes split newly made proteins up to form one or more active molecules; in other cases (as in hemoglobin) multiple protein molecules associate together to form a single active molecule. The number of distinct proteins in the human body is estimated to be well over 230,000, considerably more than the number of genes. This disparity is because of alternative splicing, RNA editing during transcription, and posttranscriptional modification by enzymes that cleave, link, phosphorylate, glycosylate, acetylate, phenylate, etc. [Germenis & Patrikidou 2005]. Thus, there is a about tenfold factor of expansion between basic polypeptides coded by genes and mature proteins with complex tertiary or quaternary structures.

In mammals, the primary linear structure of proteins is built from the 20 amino acids in the genomic protein library. All amino acids have a general structure consisting of a central α-carbon atom to which are attached (1) an α-amino group ($-\mathbf{NH_3}^+$), (2) a unique side chain or **R**-group, and (3) a carboxyl group ($-\mathbf{COO}^-$). Most AAs use the spatial isomer called L. Some use the D isomer. (The AA glycine has 3-D symmetry, so it is neither D nor L.) Figure 7.1 illustrates the general 3-D structural difference between an L- and a D-AA. D and L molecules are said to have a *chiral center*. The L and D refer to the actual 3-D geometry of the AA compared with the *chiral molecule, glyceraldehyde*. (See Figure 7.2 for a 3-D schematic of D-glyceraldehyde.)

In addition to D- or L-, AAs may be classified as positive or negative with regard to their *optical activity*. D-glyceraldehyde is a (+) *optical isomer;* that is, its molecule has *(+) optical activity*. This means that when a beam of *linearly polarized light* (LPL) passes through a solution of D(+)-glyceraldehyde, it emerges from the sample cuvette rotated to the right, clockwise, or *dexter* side. That is,

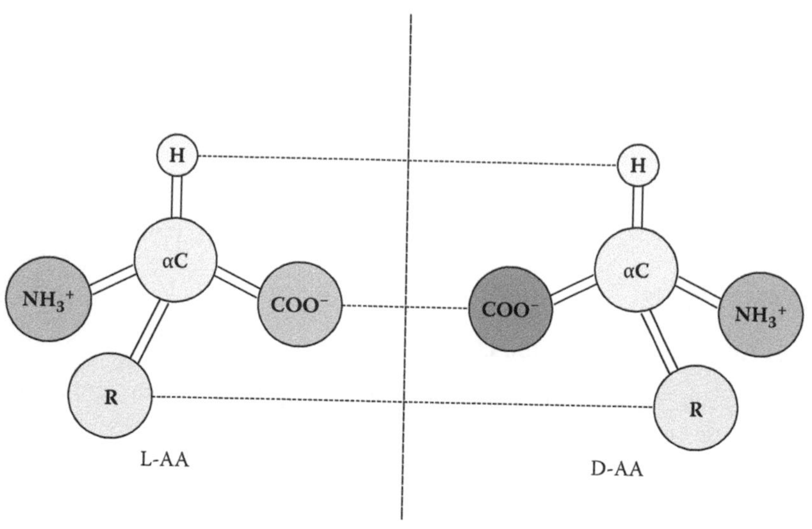

FIGURE 7.1 (See color insert following page 44.) Illustration of the D- and L-structural isomers of a generic amino acid.

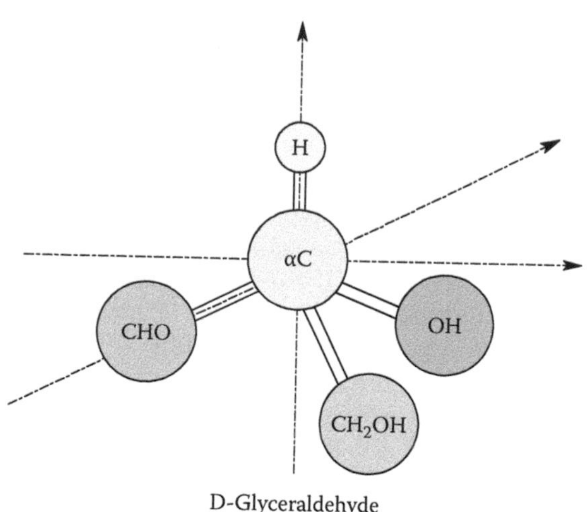

FIGURE 7.2 (See color insert.) The D(+)- glyceraldehyde molecule isomer. It is used to define D- and L-structures in other chiral biomolecules.

if one views the emerging LPL beam looking toward the source of LPL, the **E**-vector locus of the emerging LPL is rotated clockwise (CW), so that if the **E**-vector is at 12 o'clock entering the cuvette, it emerges at, say, 2 o'clock.

After passing through a vessel containing the optically active substance in solution, the number of degrees of optical rotation of the **E**-vector, θ, is given by the product $\theta = \alpha^T_\lambda$ L [C]. α^T_λ is the *specific rotation constant* of the optically active substance in solution; α is a function of temperature and wavelength. L is the optical path length through the solution, and [C] is the concentration of the optically active analyte. α^T_λ can have the units of degrees/(mole × cm) or degrees/(g/dl × cm), etc.

Note that the D- and L-designation of molecular isomers refers to their 3-D structures (shapes), not how LPL is rotated by their optical activity. *Glucose* is an example of an optically active organic molecule. D(+)-glucose, aka *dextrose,* is used universally in cell metabolisms; L-glucose cannot be metabolized by cells! However, L-AAs are generally used to build proteins. Nine of the 20 L-AAs used to build proteins are dextrorotary (at $\lambda = 589$ nm). D(−)-fructose is also called *levulose* because it is levorotary (–); that is, the emerging LPL **E**-vector is rotated CCW looking toward the source. Again, we stress that the D- and L-designation of isomers refers to their 3-D structures, not how LPL is rotated by their optical activity.

Besides being the building blocks of proteins, certain AAs are precursors for physiologically important molecules. For example, the AA *Tryptophan* is a precursor for *serotonin,* a CNS neurotransmitter. The AA *Tyrosine* is the precursor for the *thyroid hormones, melanin, dopamine,* and *epinephrine.* Glycine is one of the reactants in the biosynthesis of *porphyrins* such as heme.

The covalent linkage between adjacent AAs in a protein or polypeptide is called a *peptide bond.* It is formed enzymatically with the expenditure of chemical energy in the ribosome when the protein is assembled from the mRNA template. A peptide bond is formed between the amino group of one AA and the carboxyl group of another AA with the formation of one water molecule. (The three possible resonance states of a peptide bond were shown in Chapter 5, Figure 5.7.) The trans configuration of the peptide bond is shown; note that the bond cannot rotate, it is planar. The approximate *free energy* of opening a single peptide bond in the denatured state is –2.7 Cal/mole [Buczek et al. 2002]; however, it depends on the AAs joined by it. Note that a polypeptide chain always has a free amino end and a free carboxyl group on its other end.

The number of AAs in a polypeptide can be as few as nine (e.g., the mammalian nonapeptide hormones *oxytocin* and *vasopressin*) or several hundred (e.g., prion protein). Transmembrane proteins tend to be large (several hundreds of AAs).

The 2-D chemical structures of the 20 AAs used to build polypeptides and proteins are shown in Figure 7.3. Two of the 20 AAs (methionine and cysteine) contain sulfur molecules. The sulfur in cysteine is capable of forming strong covalent chemical bonds to heavy metal ions and can form disulfide bridge bonds across a folded, 3-D, tertiary configuration of the protein, giving it a robust chemical structure in harsh environments.

You will see later that proteins have many functions in animals: intracellularly as enzymes, in cell membranes as receptors and pumps, as hormones and cytokines, antibodies, structural materials, etc. We discuss the more important applications of proteins later. Some proteins form complexes with other proteins, lipids, RNAs, and sugars. These complexes increase protein functional diversity.

7.1.2 Primary, Secondary, Tertiary, and Quaternary Structure of Proteins

A protein or polypeptide molecule emerges from the ribosome that assembles it according to its genetic code as a linear, primary, chain structure, terminal amino group first. This initial linear chain of AAs specified by gene-coding the protein is called the protein's *primary structure.*

This chain of AAs quickly folds and twists in three dimensions and assumes a unique, stable 3-D *structure* that gives the protein its biochemical properties. In the first step of the protein reaching a stable 3-D configuration, it assumes a *secondary structure,* the most common of which are the *α-helix* and the *β-pleated sheet* configurations. The α-helix has good molecular stability because its AA side-chain groups are well separated, and each peptide linkage uses two hydrogen (H-) bonds. These H-bonds are formed down the length of the α-helix and contribute to its conformational stability. The polypeptide backbone of the β-pleated sheet is stretched so that when two or more of these extended β-strands lie together in parallel, H-bonds form between them, creating a 2-D sheet that is zigzag pleated similar to the bellows of a concertina. Alternate β-strands can run in the same direction to give a *parallel β-pleated sheet* or in opposite directions producing an *antiparallel β-pleated sheet.* Some complex proteins contain an ordered collection of secondary structures, α-helices and β-pleated sheets, and can be linked as part of the same *supersecondary structure,* as can two or more β-pleated sheets or α-helices.

A protein's *tertiary structure* refers to the complete 3-D folding and cross-H-bonding between the secondary structures. The stable tertiary structure is formed largely by unique electrostatic and molecular forces imparted by cross-strand atomic juxtapositions in its primary AA peptide sequence. Strong *disulfide bonds* can form between opposing cysteine sulfhydryl groups, making

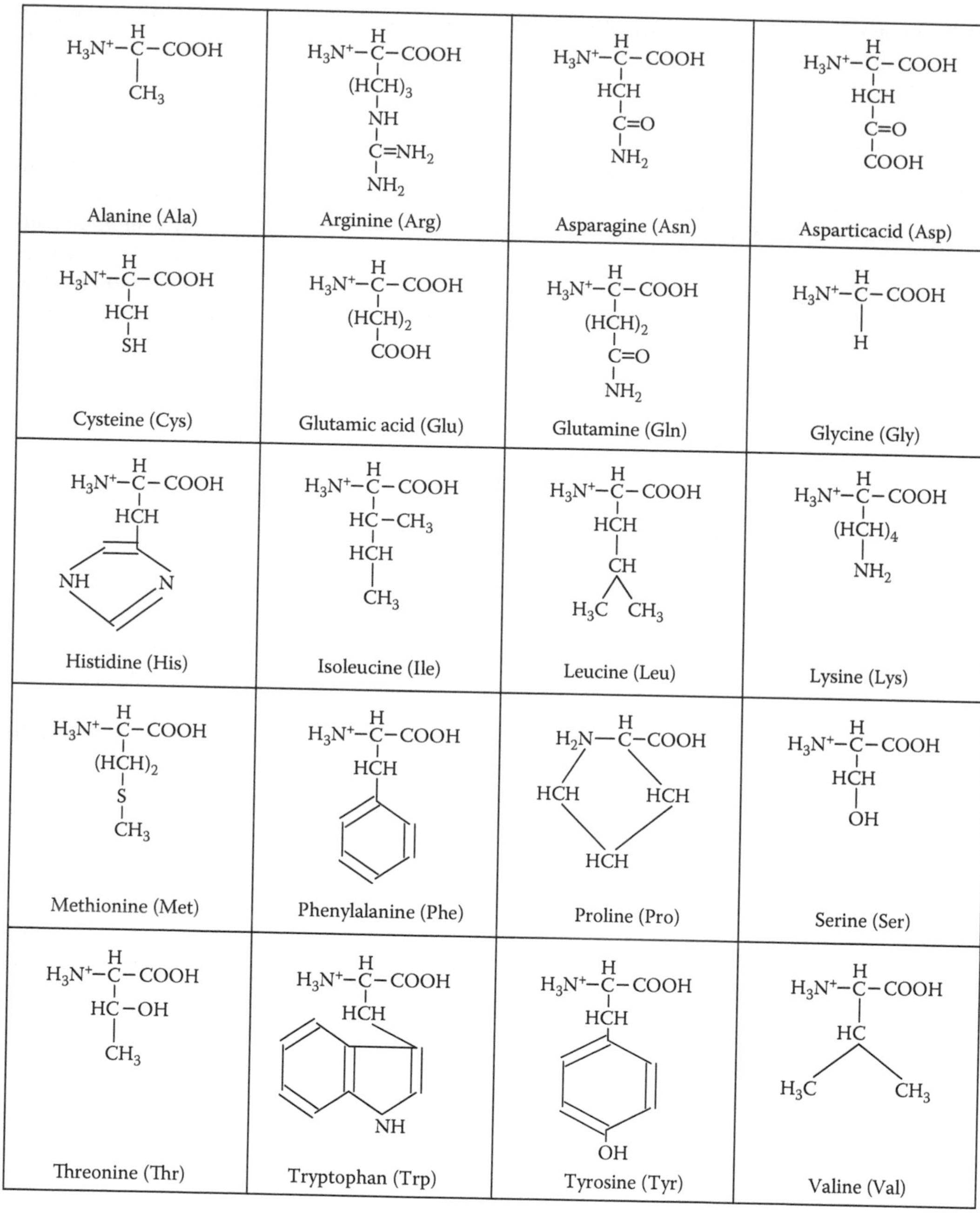

H $H_3N^+-C-COOH$ CH_3 Alanine (Ala)	H $H_3N^+-C-COOH$ $(HCH)_3$ NH $C=NH_2$ NH_2 Arginine (Arg)	H $H_3N^+-C-COOH$ HCH C=O NH_2 Asparagine (Asn)	H $H_3N^+-C-COOH$ HCH C=O COOH Asparticacid (Asp)
H $H_3N^+-C-COOH$ HCH SH Cysteine (Cys)	H $H_3N^+-C-COOH$ $(HCH)_2$ COOH Glutamic acid (Glu)	H $H_3N^+-C-COOH$ $(HCH)_2$ C=O NH_2 Glutamine (Gln)	H $H_3N^+-C-COOH$ H Glycine (Gly)
H $H_3N^+-C-COOH$ HCH NH N Histidine (His)	H $H_3N^+-C-COOH$ $HC-CH_3$ HCH CH_3 Isoleucine (Ile)	H $H_3N^+-C-COOH$ HCH CH H_3C CH_3 Leucine (Leu)	H $H_3N^+-C-COOH$ $(HCH)_4$ NH_2 Lysine (Lys)
H $H_3N^+-C-COOH$ $(HCH)_2$ S CH_3 Methionine (Met)	H $H_3N^+-C-COOH$ HCH Phenylalanine (Phe)	H $H_2N-C-COOH$ HCH HCH HCH Proline (Pro)	H $H_3N^+-C-COOH$ HCH OH Serine (Ser)
H $H_3N^+-C-COOH$ HC−OH CH_3 Threonine (Thr)	H $H_3N^+-C-COOH$ HCH NH Tryptophan (Trp)	H $H_3N^+-C-COOH$ HCH OH Tyrosine (Tyr)	H $H_3N^+-C-COOH$ HC H_3C CH_3 Valine (Val)

FIGURE 7.3 The 20 standard amino acids. Note that the AA glycine is not chiral.

the tertiary structure robust. Also, there are enzymes that add sugars to certain AAs of newly synthesized proteins that may be required for proper folding and cross-strand H-bonds.

Finally, one sometimes hears about a protein's *quaternary structure*. Many multimeric proteins are composed of two or more different polypeptide chains that are held together by the same noncovalent forces that stabilize the tertiary structures of proteins. Proteins with multiple polypeptide chains are called *oligomeric proteins*. The structure formed by the monomer–monomer interaction in an oligomeric protein is called its quaternary structure. When the component-linked chains are alike (repeat peptides), the protein is called *homomeric;* if the chains are different, the protein is *heteromeric*.

To avoid unwanted cross-bonding, the cells of eukaryote organisms contain *chaperone molecules* that stabilize newly formed polypeptides while they fold into their desired tertiary 3-D structures. A molecular chaperone is a protein enzyme that stabilizes the secondary structure a newly formed protein as it emerges from a ribosome.

About 80% of newly synthesized proteins are stabilized by *chaperone molecules* that temporarily bind to their surface until they have folded properly. Some 20% of nascent proteins are so complex that a *different group of* chaperones is required, called *chaperonins*. Chaperonins are hollow cylinders into which the newly formed protein fits while it folds. The inner wall of the chaperonin protein is lined with hydrophobic AAs that stabilize the hydrophobic regions of the polypeptide and keep it insulated from the cytosol and other proteins. Both chaperones and chaperonins use ATP as an energy source to power the folding process. (Questions for the reader to consider: Are chaperones and chaperonin molecules themselves folded by other chaperone or chaperonin molecules? How specific are chaperones and chaperonins? Is there one each for each of the genomically coded polypeptides a cell can synthesize? What is their relation to *heat shock proteins*?)

Chaperone stabilization is required because as a protein is being synthesized, it emerges amino terminal first from the ribosome into the watery cytosol, surrounded by many other proteins and free AAs. As *hydrophobic* AAs emerge, they must find other hydrophobic AAs to associate with on the synthesized polypeptide. (*Hydrophobic AAs* include **Ala**, **Cys**, **Phe**, **Ile**, **Leu**, **Met**, **Val**, **Trp**, **Gly**, and **Tyr**.) Owing to hydrophobic interactions, these associations often will be with other hydrophobic AAs on the nascent protein, not those on other proteins in the cytosol. (*Hydrophilic AAs* include **Lys**, **Arg**, **His**, **Ser**, **Thr**, **Asn**, **Gln**, **Asp**, **Pro**, and **Glu**.)

Certain physical and chemical conditions in the cytosol or surrounding a cell can lead to *protein denaturing*, in which the biochemically active tertiary structure of the protein is lost. These conditions are high temperature, extremes of pH, changes in salt concentration (osmotic pressure), and the presence of *reducing agents* that can break disulfide bonds (–S-S–) holding tertiary structure. None of these denaturing conditions has enough energy to cleave the covalent peptide bonds, so a primary polypeptide chain is left floating in the cell or extracellular volume. This denatured protein has lost all capacity to function that was endowed to it by its unique tertiary structure. In some cases, a denatured protein will spontaneously refold into its active tertiary structure, demonstrating that its unique 3-D structure is intrinsic in its linear AA sequence [Kimball 2003b]. Section 7.5.1 (Chapter 7) describes the process by which denatured proteins are "cleaned up."

The *destination of a protein* synthesized in the cells of a multicellular organism is coded in the *signal sequence* of the polypeptide emerging from a ribosome [Nobel 1999]. For certain proteins, the signal sequence is encoded in the first 15 to 30 AAs emerging from the ribosome. This AA sequence is variable from one protein to another and consists of many hydrophobic AAs. Broadly, protein destinations can be subdivided into two groups: those that remain in the cytosol and those that are attached to the *rough endoplasmic reticulum* (RER). As described by Kimball [2003c], if a signal sequence is present, translation ceases and the signal sequence is recognized and bound by a *signal recognition particle* (SRP). (The SRP is a single small molecule of RNA plus six different associated proteins. It contains binding sites for the signal sequence, the ribosome, and an SRP receptor located on the cytosol face of the RER membranes.)

The complex, consisting of the ribosome with its partially translated protein and the SRP, binds to a protein receptor on the inside surface (facing the cytosol) of the RER. The SRP now leaves and translation continues; however, the emerging polypeptide chain is extruded through a pore in the RER into its lumen. The signal sequence is now clipped off the polypeptide unless it is to be retained as an *integral membrane protein*. Once in the ER lumen, *chaperone* molecules in the lumen assist the new polypeptide to fold into its final active tertiary structure. Now various sugar residues may be added to the protein in a process known as *glycosylation*. Glycosylation occurs to some degree in the majority of eukaryotic proteins and is found on all extracellular proteins (glycoproteins). The protein's journey continues, as described in the following text.

Portions of the RER containing new proteins are pinched off, forming *transport vesicles* that carry their protein load to the *Golgi apparatus* (GA). There, the membranes of a transport vesicle fuse with the Golgi apparatus membrane, merging their contents. More glycolysation may occur within the GA; the exact pattern of glycosylation determines the final destination of the glycoprotein. Two pathways have been identified: (1) proteins glycosylated with residues of *mannose-6-phosphate* leave the GA in transport vesicles that eventually fuse with *lysosomes*, and (2) proteins that are not marked with mannose leave in transport vesicles that fuse with the inside of the plasma membrane. Some of these chosen proteins become *integral membrane proteins* that serve as signal receptors, cell recognition sites, or ion pumps, whereas other proteins are discharged from the cell by *exocytosis*. This latter class typically comprises cytokines, hormones, neurotransmitters, etc.

Proteins synthesized by free *ribosomes* (in the cytosol) have a number of intracellular destinations and functions. Some of these proteins remain in the cytosol with functions such as glycolysis, making microtubules and microfilaments. Other proteins enter the nucleus. These proteins have a sequence of 7 to 41 AAs called their *nuclear localization sequence* (NLS). The NLS has the AAs lysine and arginine as members. The nuclear membrane has protein receptors that recognize the NLS and actively transport these proteins into the interior of the nucleus, where they typically function as *histones* associated with DNA and as *transcription factors*.

Some proteins that are destined to serve in the mitochondria are encoded by nuclear DNA and translated by ribosomes in the cytosol. These proteins have a *mitochondrial stimulation factor* (MSF) signal sequence of AAs that is recognized by a special *MSF chaperone molecule*. The mitochondrial protein, bound to the MSF chaperone, is recognized by a receptor protein in the mitochondrial outer membrane, where it is internalized by the mitochondrion.

Precisely the same scenario applies to sending external proteins into chloroplasts. Chloroplast proteins also have a special set of chaperones that recognize the proteins' *AA transit sequences*. The protein–chaperone complex docks with a receptor on the chloroplasts' membranes, and the protein is internalized [Kimball 2003c].

7.2 EXAMPLES OF PROTEIN FUNCTIONS

7.2.1 INTRODUCTION

Because there can be literally hundreds of thousands of different kinds of protein molecules in every mammalian cell, we have picked and chosen certain classes of proteins we have considered important but have necessarily omitted consideration of many, many more.

Proteins serve in many diverse applications in cells, on cell surfaces, and in intercellular space as structural materials, sources of physical motion for the cell or organism, as receptors when embedded in a cell's membrane (they recognize molecular signals from other cells—hormones, cytokines, etc.), as ion pumps, as controlled ion leakage conductances in neuron membranes (control is by chemical or transmembrane electrical potential difference), as signal substances sent out by cells (hormones and cytokines), and as enzymes that catalyze biochemical reactions, support metabolism, and break up unwanted proteins and nucleic acids. We have also seen that proteins (chaperones) are required to make newly made proteins functional, and in the nucleus they participate in the repair of damaged DNA and splicing intron codons out of pmRNA from a gene containing an intron.

Modern eukaryotic life has evolved biochemical diversity based on proteins. The whole genome of a cell is about storing the information for protein synthesis (about 1% of human DNA). However, located in the remaining 99% of human DNA that does not code proteins is the regulatory information controlling the timing of protein synthesis, the amount of protein synthesis, and of course the recipe for self-replication of the genome and the cell (if it is to divide).

In the following subsections we will describe the more important roles for proteins in cells and metazoans.

7.2.2 Enzymes

7.2.2.1 The Physical Chemistry of Enzymes

An enzyme is a *catalyst.* (Most enzymes are proteins, but RNA *ribozymes* also act as catalysts.) Catalysts facilitate chemical reactions by lowering the *energy of activation* required to initiate an exergonic (spontaneous) chemical reaction. Activation energy is distinct from the energy input needed to sustain an *endergonic* reaction. An endergonic reaction needs energy added to it to keep it going, as well as to get it started. The characteristics of catalysts include the following:

- A catalyst is not consumed in the chemical reaction in which it acts; it is recycled. It is neither a reactant nor a product.
- A catalyst speeds up a chemical reaction.
- A catalyst does not change the chemical equilibrium of a reaction.
- A catalyst cannot cause a reaction to occur that is not energetically favorable.

Enzymes are almost always proteins whose tertiary (3-D) structure has a "notch" or *active site domain* that has the precise size, shape, and electrical charge distribution that exactly complements the reactants or substrates. When the substrate binds to the active site, specific bonds in the substrate are stretched and are more easily broken, lowering the activation energy for the reaction. Once the reaction occurs, the reaction products and the enzyme are released to act again.

Some enzymes can only work when in association with other proteins or molecules called *coenzymes*, or certain *cofactors* (e.g., metal ions such as Fe^{++}, Zn^{++}, Cu^{++}, Mg^{++}, and Ca^{++}). Some heavy metal ions such as Hg^{++} interfere with enzyme function. An enzyme is temperamental; it works optimally in a certain temperature range, pH, and ionic milieu.

Enzyme names either end in *ase* or *zyme.* Because there are many, many biochemical reactions, there are many, many enzymes. Enzyme use evolved because (1) they speed up reactions and (2) they allow reaction rates to be regulated up or down and even turned off through *parametric control.*

7.2.2.2 The Regulation of Protein Enzymes

Enzymes are highly specific; each catalyzed biochemical reaction generally has its own unique enzyme. Enzyme molecules are found free in a cell's cytosol as well as associated with the membranes of eukaryotic cells. The enzymes associated with cell metabolism and ATP production are bound to membranes in the cell's mitochondria and chloroplasts.

We have seen that one of the characteristics of life is *homeostasis.* The maintenance of a cell's internal steady state requires regulation by negative feedback of many biosynthetic pathways. One way the feedback is implemented is by regulating enzyme activity. This type of nonlinear feedback is called *parametric control;* it affects reaction rate constants directly and state variables indirectly. Parametric control is ubiquitous in all living systems [Northrop 2000]. In a parametric controller, functions of state variables directly affect one or more system gains (rate constants).

One way of modulating a biochemical reaction rate, and hence the amount of product produced (assuming constant substrate concentrations), is by controlling the number of active enzyme molecules available. The steady-state concentration of an enzyme is determined by the equilibrium between the rate it is made and the rate it is broken down. Protein enzymes are synthesized from DNA the way all proteins are synthesized. This rate of synthesis can be regulated genomically. Enzyme gene expression can be repressed (turned off) or derepressed (turned on) by many factors acting on the genome, including certain hormones.

Nature has developed several means of regulating the activity of functional enzymes: One means is called *allosteric regulation.* Some enzymes have specific *allosteric receptor domains* (other than the active site domain) that change the enzyme's 3-D shape to either increase or decrease its catalytic ability. Not surprisingly, the molecules that dock in the allosteric sites are called *allosteric regulators*. Another means of slowing down a catalyzed reaction is by biochemical *feedback inhibition*. Often the end-product molecules of a multistep, multienzyme reaction pathway serve as the feedback signal inhibiting the first enzyme in the pathway by allosteric means. In this case, the process is called *end-product inhibition. Precursor activation* is a feed-forward activation in which a high concentration of the initial substrate in a multistep reaction acts to increase the effectiveness (rate constant) of the last enzyme in the reaction chain by allosteric means [Kimball 2004d].

Still another means of slowing a catalyzed reaction is by *competitive inhibition*. In this system, a regulatory molecule binds to the enzyme's active site; it undergoes no reaction but instead blocks the normal substrate from binding to the enzyme's active site. The competitive inhibitor molecule may stay in place blocking the enzyme or may eventually be released, depending on the strength of the bond. Note that allosteric, end-product, and competitive inhibition systems all use parametric control.

7.2.3 Protein Hormones and Intercellular Command, Communication, and Control (C^3)

7.2.3.1 Introduction

A *hormone* is a chemical signal sent from one specialized group of transmitter cells to another group of target cells. With the exception of steroid and lipid hormones, most hormones are proteins. However, some hormones are monomolecular, such as *thyroid hormone* and *triiodotyrosine,* and are derived from a single amino acid, tyrosine. Tyrosine also initiates the biosynthetic pathways that make the catecholamine hormones and *neurotransmitters: epinephrine, norepinephrine,* and *dopamine*. The hormone *melatonin (N-acetlyl-5-methoxy-tryptamine)* and the neurotransmitter *serotonin* (5-hydroxytryptamine) are all derived from the AA, tyrptophan.

Hormones provide vital communication and control (C^2) pathways in all multicellular organisms, including plants. In animals, endocrine hormones are distributed by the blood circulation from their site of secretion to the location where they affect some cellular processes, but some hormones act locally.

Hormone secretion is generally under negative feedback regulation. The release of hormones can be stimulated (or suppressed) by the following:

- Another *stimulating* or *releasing* hormone or a certain neurotransmitter
- Plasma concentrations of certain ions or nutrients
- Neurons and mental activity
- Environmental changes, e.g., light, temperature, salinity, and pheremones (exogenous chemical signals)

Hormones *dock*, or attach with specificity, to cell membrane *surface receptor proteins*. There, their presence causes structural changes in the receptor protein leading to the initiation (or cessation) of some intracellular biochemical process. The hormone insulin causes an insulin-sensitive cell's surface permeability to glucose molecules to increase. Insulin also acts internally on hepatic cells to stimulate glucose storage as glycogen in them.

In some cells, hormone binding can initiate a *second messenger system* that actually signals the biochemical change desired. Also, some small, nonprotein hormones can actually enter the cytoplasm and penetrate the cell's nucleus, where they act to regulate transcription.

7.2.3.2 Protein Hormones

As this chapter is about proteins, we will only consider peptide and protein hormones (as opposed to the important steroid hormones, for example). The smallest peptide hormone is the tripeptide amide, *thyrotropin-releasing hormone* (TRH). It is secreted by nerve endings in the median eminence of the hypothalamus. It is *pyroglutamyl-histidyl-proline-amide*. (The amide radical is **$-CONH_2$**.)

The linear structure of the nonapeptide hormones *vasopressin* and *oxytocin* are given below; they are released by neurosecretory cells in the anterior hypothalamus region of the brain. Vasopressin is involved in the regulation of osmotic pressure in the body, and oxytocin causes uterine contractions and the expression of breast milk in nursing mothers. Note that they have seven out of nine homologous AAs.

Cys-Tyr-**Phe**-Gln-Asn-Cys-Pro-**Arg**-Gly-NH_2	**Vasopressin or *Antidiuretic hormone***
Cys-Tyr-**Ile**-Gln-Asn-Cys-Pro-**Leu**-Gly-NH_2	**Oxytocin**

Other natural mammalian peptide and protein hormones have larger molecules. They include the following: *antimullerian hormone* (AMH), *adiponectin, adrenocorticotropic hormone* (ACTH), *angiotensin, angiotensinogen, atrial-natriuretic hormone* (ANP), *calcitonin, cholecystokinin, corticotropin-releasing hormone* (CRH), *erythropoietin, follicle-stimulating hormone* (FSH), *gastrin, glucagon, gonadotropin-releasing hormone, growth hormone releasing hormone* (GHRH), *human chorionic gonadotropin* (hCG), *growth hormone* (GH), *insulin, insulin-like growth factor* (IGF), *leptin, luteinizing hormone* (LH), *melanocyte-stimulating hormone* (MSH), *neuropeptide Y, parathyroid hormone* (PTH), *prolactin, renin, secretin, somatostatin, thrombopoietin*, and *thryroid-stimulating hormone* (TSH) [Wallis 2001; Kimball 2005i]. The immune system is regulated by numerous protein *cytokines* (e.g., interleukins) secreted by immune system cells. Protein cytokines can be considered to be hormones. (See Section 7.2.6 for a detailed discussion of cytokines in the immune system.)

Every hormone docks with specific *receptor proteins* on the surfaces of the target cells. The docking is highly specific and initiates the intended physiological action. In a typical two-step hormone regulatory loop, a physiological parameter, such as the elevated concentration of an ion in the blood, acts to cause the release of a stimulating hormone, which in turn causes the release of an effector hormone that acts on the cells of an organ to reduce the concentration of the ion.

In the case of *blood glucose*, two major hormones are directly involved in its regulation. Elevated blood glucose stimulates pancreatic beta cells to release the protein hormone, *insulin*, into the portal and thence the general circulation. Insulin acts to stimulate liver cells to internalize glucose and store it as the glucopolymer, *glycogen*. Insulin also acts to facilitate the diffusion of glucose into insulin-sensitive cells, further reducing the blood glucose concentration. Another role of insulin is to stimulate adipocytes to take up glucose and convert it to lipids for storage.

Insulin release is also stimulated by high levels of the AAs arginine and lysine, as well as the gastrointestinal protein hormones, *gastrin, secretin, cholecystokinin*, and *gastric inhibitory peptide*. The hormone *somatostatin* is released by *pancreatic delta cells* in response to eating. This short-lived peptide hormone suppresses the immediate release of the glucoregulatory hormones *insulin* and *glucagon* [Northrop 2000]. (Its purpose may be to prolong the period over which food is digested.) Insulin secretion is also inhibited by high concentrations of the adipocyte-secreted protein hormone, *leptin*.

Low blood glucose has a direct effect on *alpha cells* of the pancreatic islets of Langerhans, causing them to secrete the protein hormone *glucagon*. Glucagon is every bit as important as insulin. When glucagon acts on liver cells, it reverses the process of glycogen storage, causing the liver cells to release glucose into the blood, preventing hypoglycemia. Glucagon also stimulates metabolic pathways that cause *gluconeogenesis*, i.e., the breakdown of AAs and fats to form glucose. This is a slower process than glycogen conversion and generally is seen in cases of starvation and severe, untreated diabetes.

The regulation of hormone secretion is a fascinating topic for physiologists and control engineers interested in feedback in biochemical systems. The interested reader should consult a good medical physiology text such as Guyton [1991] for details of hormone action and regulation; also see Henderson [2005]. An engineering modeling approach can be found in Northrop [2000]. Note that there are steroid and lipid hormones that function in the body as well as protein hormones. The estrogens and androgens are steroid hormones; prostaglandins and leukotrienes are eicosanoid (lipid) hormones.

7.2.4 Structural Proteins

7.2.4.1 Introduction

Proteins have important roles in forming the structural elements of cells, preserving the structure of cells, and providing motion for cells. In the cytoskeleton of eukaryotic cells, proteins and protein polymers help establish cell shape, provide mechanical strength, figure in chromosome separation in mitosis and meiosis, provide cell locomotion, and transport organelles within cells. In the following subsections we enumerate some of these structural proteins and their functions within a cell.

7.2.4.2 Microfilaments

Microfilaments are polymerized monomers of the protein *actin;* these ubiquitous fibers are about 8 nm in diameter. In striated muscle cells, they are the *thin filaments* that interact with the *myosin thick filaments,* when catalyzed by Ca^{++} ions, to provide the force for muscle contraction. Besides their use in specialized muscle cells, they also generate cytoplasmic streaming. Microfilaments generate locomotion in eukaryote calls such as the amoeba and blood leukocytes. Underneath the cell's plasma membrane, microfilaments provide mechanical strength for the cell and link transmembrane proteins to cytoplasmic proteins, preserving the juxtaposition of cell surface receptors and second-messenger systems. Microfilaments also anchor centrosomes at opposite poles of the cell during *mitosis* and pinch dividing animal cells apart during *cytokinesis.*

7.2.4.3 Intermediate Filaments

These 10-nm cytoplasmic fibers are composed of a variety of protein-derived materials: *Keratins* form hair and nails and are found in epithelial cells. *Lamins* form a reticulum of IFs that stabilizes the inner membrane of the nuclear envelope. *Vimentins* provide mechanical strength to muscle and other cells. *Neurofilaments* strengthen the long axis of neuron axons. The nucleii of epithelial cells are "shock-mounted" by a web of *keratin* IFs. Over 20 different kinds of keratins have been found, but each kind of epithelial cells may use about two kinds of keratins. Up to 85% of the dry weight of squamous epithelial cells can consist of keratins [Kimball 2004e].

7.2.4.4 Microtubules

These hollow tubules, found in both plant and animal cells, are about 25 nm in diameter. They are composed of a helical wrap of dimers of *alpha-* and *beta-tubulin* proteins (see Figure 7.4.) Tubulin tubes can grow to about 25 μm in length by polymerization of tubulin dimers at both ends, powered by the hydrolysis of *guanosine triphosphate* (GTP). They can also shrink at each end by enzymatic depolymerization. Curiously, both lengthening and shortening occur more rapidly at the *plus end* of the tubule; the less active end is called the *minus end.* Organelles such as vesicles and mitochondria are moved along microtubules by protein "motors" called *kinesins* (which move toward the plus end of the tubule) and *dyneins* (these move toward the minus end). ATP hydrolysis powers these motors. Kinesins and dyneins are used to move chromosomes along *spindle fibers* (microtubules) in the processes of meiosis and mitosis. In animal cells, microtubules originate at the *centrosome* attached to the outside of the nucleus. Just before mitosis, the centrosome duplicates, and the two centrosomes move to opposite sides of the nucleus. There they send out clusters of microtubules (the spindle fibers). (The processes of mitosis and meiosis are described in detail in Chapter 3, Section 3.8.)

7.2.4.5 Cilia and Flagella

Not unexpectedly, both cilia and flagella are built out of microtubules. *Cilia* and *flagella* are hairlike microstructures protruding from cell surfaces that move the cell (e.g., a *Paramecium*) or move some fluid cell product, such as mucus. A cell such as a *Paramecium* has many short locomotory structures called cilia. One large locomotory structure, such as found on a sperm, is called a *flagellum.*

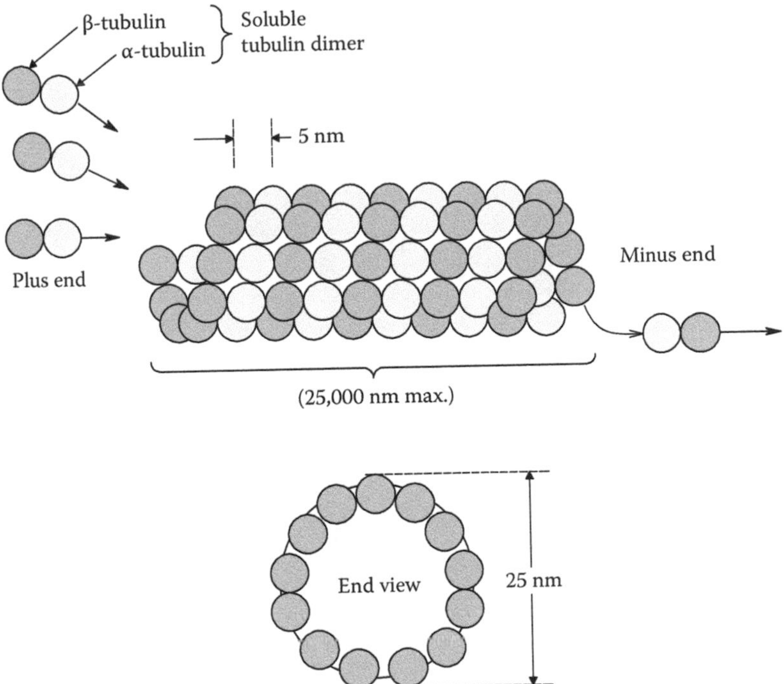

FIGURE 7.4 Schematic of a tubulin molecule. Tubulin tubes, by lengthening and shrinking, are used to move things around inside cells.

The protein structures of cilia and flagella are highly conserved throughout the kingdoms and phyla. They evidently were a successful core system that was developed billions of years ago. Their common molecular architecture is called an *axoneme.* Seen in cross section in a TEM picture, eukaryotic cilia and flagella show nine evenly spaced partial microtubules in a rough circle; in the center of the circle are two single juxtaposed microtubules, each having 13 microfilaments, forming the classical "9 + 2" axoneme pattern. The inner, core pair of microtubules is surrounded by a *central sheath.* This structure is shown in schematic cross section in Figure 7.5. The nine peripheral tubules consist of a circular, 13-protofilament *A-tubule* fused to a 10-proto filament sector of a *B-tubule.* (Note that there are 13 dimers of tubulin per turn of a microtubule molecule.) Radial spokes project centrally from each of the nine circumferential tubules to a spoke head that is attached to the central sheath. To further complicate the design, a protein called *nexin* connects the nine circumferential tubules circumferentially. The motor protein, *dynein,* is connected to the A-tubule of each of the nine circumferential tubules, forming an inner *(ida)* and outer *(oda)* dynein arm.

The waving motion of a cilium or a flagellum that generates translational force is created by the A- and B-microtubules sliding past one another in a controlled manner. Motion is generated by the *dynein* motor proteins powered by ATP. The free ends of the dynein arms make transient attachments to the B-tubule of the adjacent outer microtubule. Binding of ATP causes release of the dynein arms from the adjacent B-tubules. Hydrolysis of ATP causes reattachment of a dynein arm to a different location on the adjacent B-tubule. This results in a shear force that causes a sliding motion between adjacent outer tubule structures. The shear force causes an entire cilium to bend [Satir 1999].

Coordinated motion of dynein versus B-tubules creates a whiplike motion that moves either the cell or, if the cell is fixed, the fluid medium (such as mucus) in which the cell is immersed. Dynein molecules *(idas + odas)* are spaced about every 45 nm along the A-tubules. The beat frequency of cilia and flagella is controlled by intracellular Mg^{++}, Ca^{++}, and cAMP concentrations. Waves of dynein activity propagate longitudinally along the outer tubules, causing the remarkable but poorly understood traveling wave of cilliary bending. Cilia beat at about 10–40 times/second in coordi-

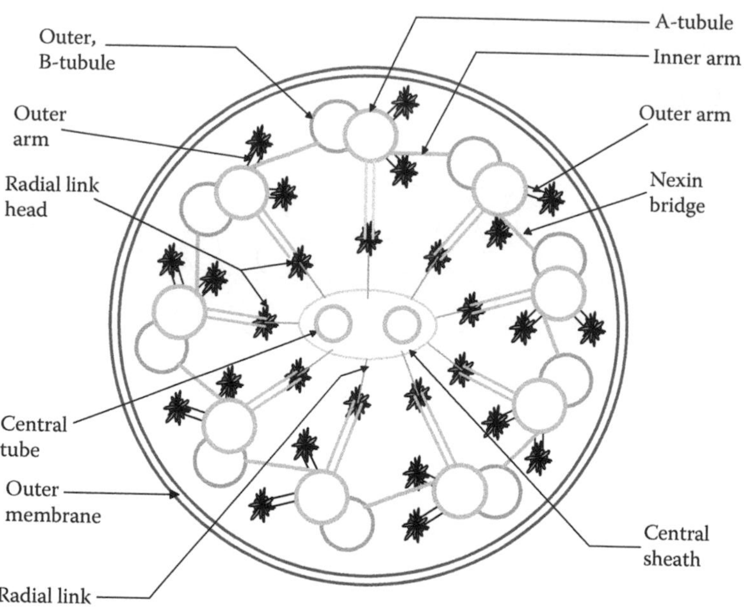

FIGURE 7.5 (See color insert.) Schematic cross section of a standard, 9 + 2 cilium stalk. Waves of complex chemical reactions create torsional forces within the cilium that cause it to wave or beat.

nated waves. In a number of species the central pair rotates as beating occurs; this "nanomotor" may act to synchronize the dynein bond cycling. The power stroke of a cilium is about ¼ of the total beat period. This means that at a 25-Hz stroke rate, the power stroke is about 10-ms duration. Thus, the bond cycling rate of the dynein *odas* and *idas* must be fast indeed.

The entire cilium or flagellum is enclosed in a unit membrane that is an extension of the cell's unit membrane.

The cellular end of each cilium or flagellum grows out of and remains attached to a complex protein structure called the *basal body*, which is embedded in the cell's cytoplasm. Curiously, basal bodies are identical to centrioles and are produced by them. Basal bodies control the direction of movement of the cilia [Childs 2003a; Satir 1999].

Not all cilia use the (9 + 2) axoneme architecture found in plants, *paramecia*, and vertebrates; some cilia on primitive eukaryotes and prokaryotes have more primitive structures. They all use microtubules, however. The cilium is a truly remarkable biomechanical "nanomachine."

7.2.4.6 Muscle Fibers

The evolution of the complex protein structures in muscle fibers gave eukaryotic animals, in particular vertebrates, the ability to move, eat, breathe, digest food, and pump blood. There are three basic types of vertebrate muscles: *striated* (skeletal), *cardiac,* and *smooth.* Smooth muscle is found in the walls of all hollow organs in the body (other than the heart). These include the esophagus, bronchial tubes, blood vessels, gut, bladder, uterus, etc.

Muscle is a unique, complex tissue having many molecular component parts. It converts chemical energy into mechanical work. When a muscle or muscle fiber contracts against an external mechanical load, it does mechanical work given by

$$W = \int_{0}^{X_o} f(x, \dot{x})dx \text{ Joules} \tag{7.1}$$

The force developed by the muscle fiber, f, is a function of the muscle's length, x, and its shortening velocity, $\dot{x}$.

The muscle's rate of doing work is simply the power P it develops in watts.

$$P = \int_0^{X_o} f(x, \dot{x}) d\dot{x} \text{ Watts} \tag{7.2}$$

The chemical energy for muscular contraction comes from ATP generated by glycolysis and the citric acid cycle [King 2006c]. Calcium ions are also necessary for muscle contraction and their loss for relaxation.

We will first examine the gross anatomy of *striated muscle* in vertebrates, then consider how an individual muscle fiber works on a molecular scale. A vertebrate skeletal muscle is made up of bundles of thousands of thin cylindrical fibers, often running the distance from origin to insertion (the full length) of the muscle. Note that not all striated muscles are attached to the skeleton: Abdominal muscles attach to fascia and have the role of supporting the lower torso and pulling the abdomen in. Extraocular muscles have origins in the eye socket and insert on the rear of the eyeball; they move the eyeballs so they can track visual objects. In long skeletal muscles, the fibers can overlap.

A skeletal muscle is signaled to contract by nerve impulses carried on the *motor neuron* bundle innervating the muscle. A motor neuron has many nerve fibers (axons) inside it; each axon is attached to a group of muscle fibers called a *motor unit,* which contracts together when stimulated by that nerve fiber. A partially stimulated muscle receives sequential nerve impulses on a subgroup of motor neuron axons. The overall tension and shortening of a muscle can be controlled by varying the number of motor neuron axons stimulated and the frequency of the nerve impulses on each axon. The overall tension developed by a stimulated muscle depends on the spatiotemporal summation of muscle fiber activity.

Each muscle fiber contains many mitochondria (to make ATP), many nuclei (a legacy from the fiber developing from the fusion of many *myoblast* cells), an extensive endoplasmic reticulum (aka the *sarcoplasmic reticulum*), and an array of *myofibrils* that are arranged in parallel and run the length of the fiber. About 75% of a skeletal muscle's total volume is made up of *myofibrils*. The nuclei and mitochondria are located just beneath the plasma membrane, whereas the *sarcoplasmic reticulum* envelops the *myofibrils*. Under the light microscope, *muscle fibrils* exhibit a banded or striated appearance, which is a clue to the protein machinery that effects shortening. Figure 7.6 illustrates the anatomy of mammalian striated muscle ranging from a *fasiculus* in a muscle, to a fiber, to a myofibril, and thence to the ultrastructure of a *sarcomere* showing the thick and thin contractile filaments. The *thick filaments* are about 15 nm in diameter and are composed of the protein *myosin*. The *thin filaments* are about 5 nm in diameter and are composed of the protein *actin* together with smaller amounts of *tropomyosin* and *troponin* proteins [Huxley 2000]. These ultrastructural components, including the myosin (thick filament) "heads," are described in the figure caption.

The functional unit of muscle contraction is the *sarcomere*. From Figure 7.6 we can see that many sarcomeres are in series in a single myofibril, and many myofibrils are in parallel in a single muscle fiber. A sarcomere is defined by dark bands perpendicular to the contractile axis of the myofibril. The perpendicular dark bands are called **Z**-*lines*. The two Z-lines of a sarcomere are actually the edge-on view of the **Z**-*disks* where all the thin actin filaments from the sarcomere join those from the two adjacent (series) sarcomeres. The thick myosin filaments define the **A**-*band* of the sarcomere. The thin actin filaments taken together form the light **I**-*band* where they do not overlap the **A**-*band*. Finally, the **H**-*zone* is the region in the middle of the sarcomere where the thick and thin filaments do not overlap.

Similar to many biological systems, the force-generating system of the sarcomere is more complex than originally thought from inspection of early electron micrographs of muscle. In addition to

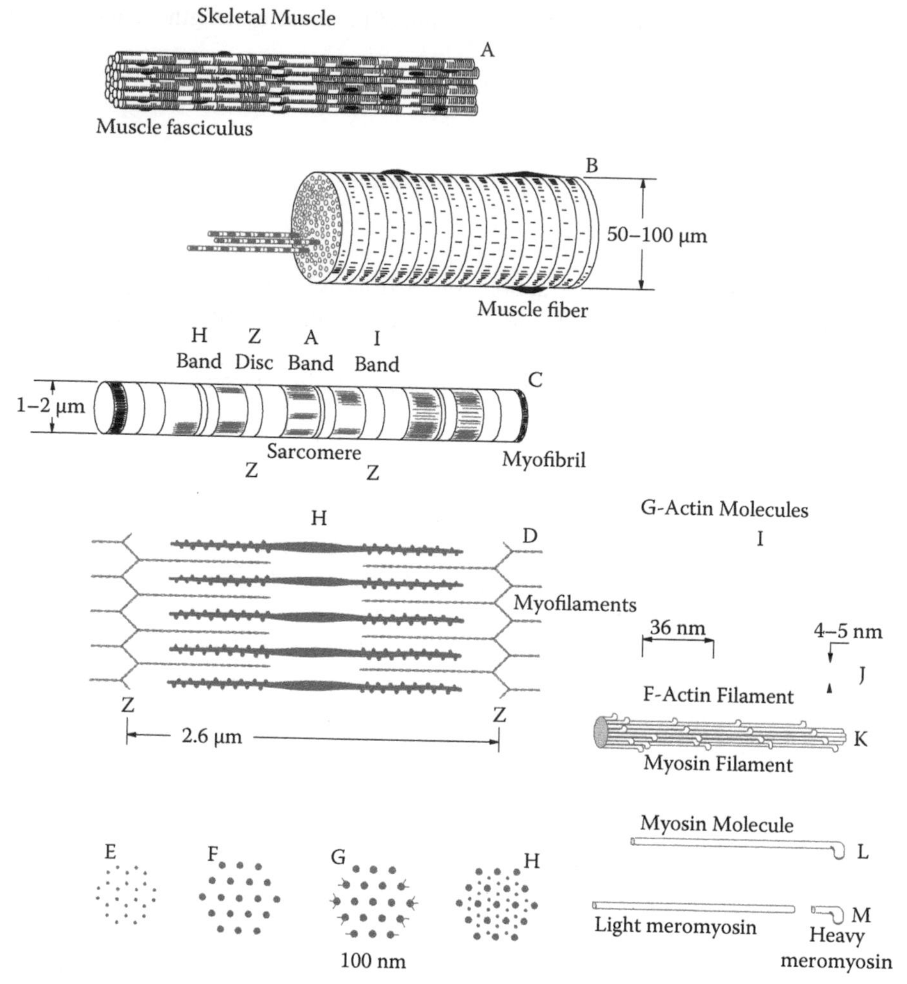

FIGURE 7.6 (See color insert.) Schematic details of the anatomy of a vertebrate striated muscle shown with expanding scales. Dimensions are approximate; they vary among species for homologous muscles and within a species depending on muscle function. The actual force generation and contraction subunit is the sarcomere with its sliding filaments of F-actin and myosin. Many other proteins are involved in muscle contraction (see text). (Adapted from Guyton, A.C. 1991. *Textbook of Medical Physiology. 8th ed.* W.B. Saunders Co., Philadelphia, PA.)

actin and *myosin,* there are eight accessory proteins: *α-actinin, β-actinin, tropomyosin, troponin,* **C**-*protein, nebulin, titin* (aka *connectin*), and **M**-*line protein.* Each molecule of *myosin* is made up from six subunits consisting of two large, heavy chains (HCs) and four smaller, light chains (LCs). The two HCs are wound together helically and have two globular headpieces. A complete myosin molecule also contains four relatively small proteins (aka light chains; LCs) that are associated with the globular headpieces. The portion of the *myosin* containing the headpiece is known as *heavy meromyosin.* A single thick filament is composed of about 400 *myosin* molecules, 200 arrayed equally on either side of an M-line.

At the end of each heavy meromyosin unit in the thick filaments are the headpiece molecules (aka crossbridges), angled away from the axis of the myosin fiber in order to be in close proximity to the adjacent *actin* LCs. As you will see, muscle contraction and force generation is the result of the regulated binding of the *myosin* headpiece molecules to the *actin* thin filaments, followed by rapid *myosin* conformational changes causing the bound actin to be moved toward the **M**-*line* in the center of the sarcomere. This action, when triggered, generates contractile force and shortens the entire sarcomere. It goes on in parallel in the hundreds of myofilaments in each sarcomere, and also in each sarcomere in each myofibril, and in parallel in each muscle fiber. The crossbridges are distributed along the myosin thick filaments. They are helically arranged around a thick filament with uniform spacing, except at the thick filament's central region.

A muscle is signaled to contract naturally when nerve impulses on a *motor nerve* innervating the muscle arrive at the nerve's motor synapses (aka *motor endplates*). The depolarization of the motor synapse releases a neurotransmitter *(acetylcholine* [ACh] from vertebrate motor nerves). The ACh diffuses across the subsynaptic gap to receptors on the subsynaptic, muscle membrane. The ACh binds briefly to the receptors and causes them to increase membrane permeability to extracellular Na^+ ions, depolarizing the muscle membrane and triggering the complex contraction process. In the synaptic cleft, the ACh is rapidly broken down by the enzyme *cholinesterase,* and its components recycled. The rapid destruction of ACh readies the system (motor synapse, receptors, and sodium channels) for the arrival of the next motor nerve impulse. The arrival of a singular motor nerve impulse causes a muscle *twitch;* it is the superposition of many twitches that builds steady muscle tension.

Figure 7.7 illustrates the normalized, steady-state, isometric tension produced by a single frog *semitendinosus* muscle fiber as a function of sarcomere length [Gordon et al. 1966; Rassier et al. 1999]. What this figure shows is that some overlap of the thick and thin filaments in a sarcomere is necessary to generate tension. At about 3.6 μm between Z-lines, there is no overlap. Again, no tension can be developed when the sarcomere has shortened to the point where the myosin heavy fibers butt against the **Z**-lines (about 1.3 μm) and shortening is mechanically stopped. Note that when a muscle develops *isometric tension,* sarcomere shortening is still possible because the force developed stretches the muscle's internal, *series elastic elements.*

To be activated, the *myosin heads* must first be activated by ATP. One molecule of ATP binds to a myosin head and is hydrolyzed to ADP and inorganic phosphate (P_i). Both ADP and P_i remain bound to a myosin head. The chemical free energy released from ATP hydrolysis is transferred to the head, activating it. Calcium ions released from the *sarcoplasmic reticulum* (SR) bind to *troponin* molecules associated with the actin thin filaments. This binding causes the shape of the troponin–tropomyosin complex to be altered. This change in configuration allows activated *myosin* head crossbridges to bind to the *actin* and release their ATP-induced mechanical energy as motion (force × distance). The contraction cycles will continue as long as there is a supply of ATP and Ca^{++} ions.

Muscle relaxation is due to the active take-up of the excess Ca^{++} by ATP-driven calcium pumps in the *sarcoplasmic reticulum* (SR). Ca^{++} is bound to the protein *calsequestrin* in the *cisternae* of the SR, which effectively decreases its concentration and favors its pumping and storage. The mitochondrial matrix also stores calcium ions under aerobic conditions. Relaxation also requires ATP to again bind with the myosin heads; this causes them to release their "hold" on the *actin/troponin/tropomyosin* filaments, as well as "recocking" the heads for the next mechanical power cycle.

Figure 7.8 illustrates the macroscale static elastic properties of a typical skeletal muscle. The isometric force developed by a muscle held at constant length and electrically stimulated (either directly or by its motor nerve) is illustrated in Figure 7.8A. Note that muscle is not Hookean; i.e., its steady-state force versus length function is not linear for either active or passive states.

Figure 7.8B illustrates the steady-state relationship between peak muscle shortening velocity when stimulated under a constant force load. This graph is called the Hill hyperbola and is modeled mathematically by the equation [$v_p = c/(f + b) - k$] (where v_p = the peak shortening velocity of a stimulated muscle fiber under isotonic conditions [constant force load]). Note that $v_{max} = (c/b - k)$ is the peak shortening velocity with f = 0. f_{max} is the critical force f at which the muscle is unable to shorten; $f_{max} = (v_{max}/k - 1)b$, given maximum stimulation. Figure 7.8C illustrates the isometric twitch response of a typical skeletal muscle fiber given a single, narrow electrical pulse to its motor neuron at t = 0, stimulating the muscle. First, the muscle exhibits a brief, *latency relaxation* phenomenon [Sandow 1947], then the tension rises rapidly as the result of the release of calcium ions from the SR in response to the electrical depolarization signal on the SR. The initial delay in con-

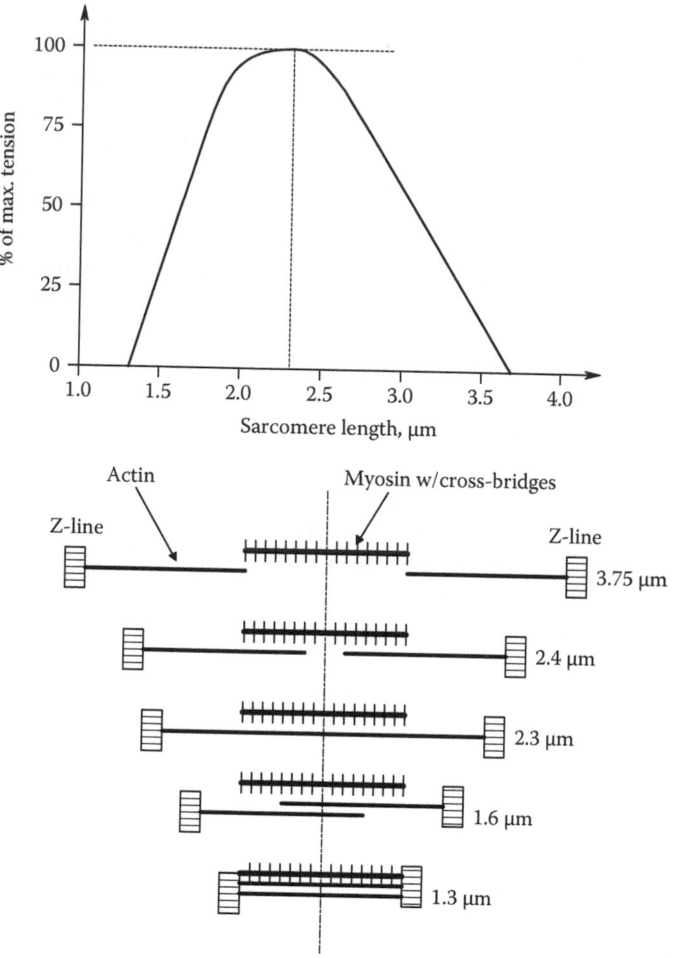

FIGURE 7.7 Top: Isometric muscle tension versus sarcomere length. Sarcomere length governs the overlap of the actin and myosin force motor fibers. Maximum isometric force occurs when the actin fibers are maximally engaged with the myosin cross-bridges. Bottom: Schematic illustration of actin/myosin fiber overlap, relative to maximum tension generation.

traction involves the time it takes the acetylcholine neurotransmitter to be released, diffuse into the motor end plate, and initiate membrane depolarization. The latency relaxation may be due to Ca^{++} binding to *titin* protein, causing a structural change in the thick filament (myosin) that transiently lowers resting muscle tension [Yagi 2003]. The tension rises rapidly as the released Ca^{++} ions diffuse into the sarcomeres and trigger contraction. The tension falls at a slower rate because of the time required to pump the Ca^{++} ions out of the sarcomeres and back into the *cisternae* of the SR. Different types of skeletal muscle have different twitch dynamics; the extraocular muscles are perhaps the fastest, whereas large postural muscles such as the *soleus* are much slower [Guyton 1991].

The ultrastructure of cardiac muscle is very similar to skeletal striated muscle, although certain of the proteins have isoforms specific to cardiac muscle. As in skeletal muscle, the thick filaments and crossbridges are made up from *myosin* molecules, and the thin filaments contain *actin, troponin*, and *tropomyosin*. Cardiac muscle cells tend to be smaller in diameter than skeletal muscle cells. They are also electrically coupled to one another through intercalated disks that contain low electrical resistance regions called *gap junctions*. This conductive coupling network, or *syncytium,* allows electrical excitation to spread from cell to cell in a coordinated, wavelike manner, triggering the spatiotemporal muscle contraction necessary for the proper pumping action of the heart.

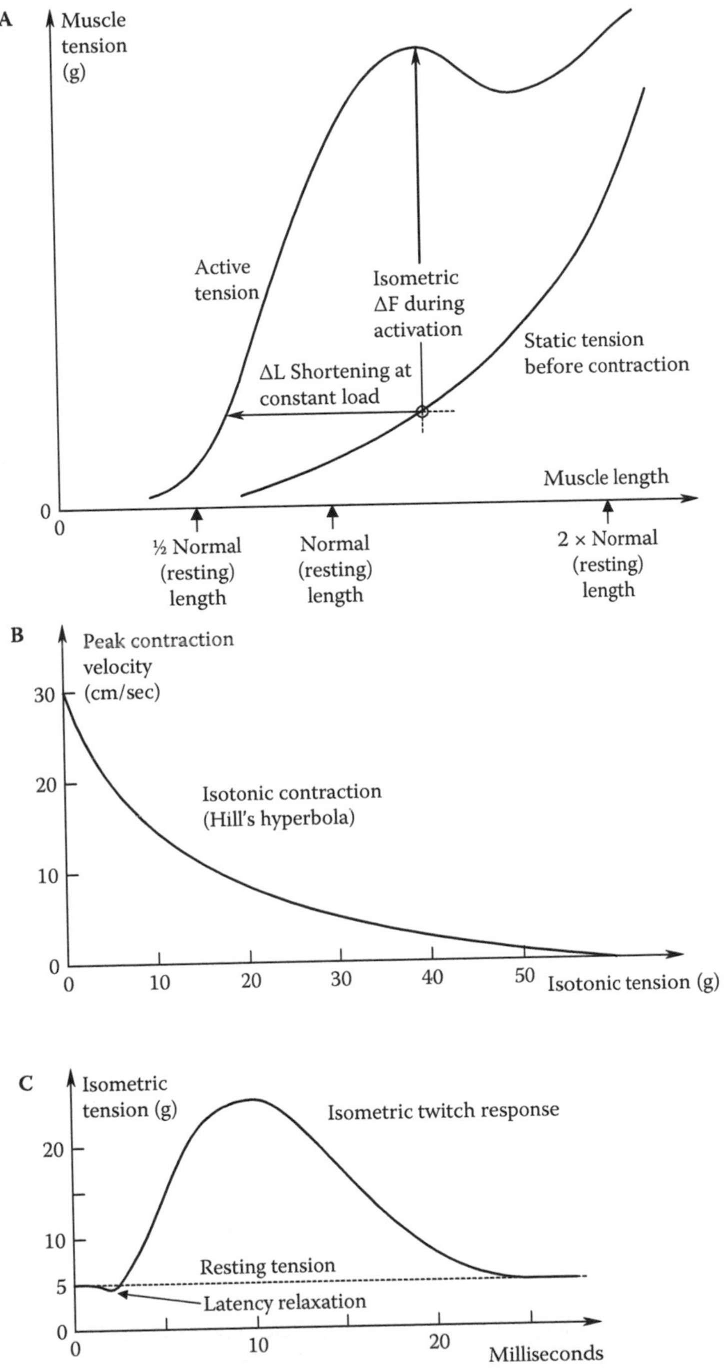

FIGURE 7.8 (**A**) Isometric force versus length plots for a typical vertebrate muscle in its unstimulated and stimulated states. (**B**) "Hills' hyperbola." Peak contraction velocity versus *isotonic* (constant) tension (force) on maximally stimulated muscle. Note that there is a maximum tension at which the stimulated muscle is unable to contract at all. Peak contraction velocity occurs, as might be expected, at zero load. (**C**) A typical isometric tension twitch in which a single muscle motor unit is stimulated by one pulse to its motoneuron axon. Note the latency relaxation (LR) before the tension rises. LR is a clue to the detailed molecular mechanisms involved with muscle activation. See text for a further description of LR.

Depolarization of skeletal muscle preceding contraction involves different ionic systems than does cardiac muscle. The relatively fast action potential associated with a skeletal muscle twitch contraction arises from the rapid opening of voltage-gated sodium ion channels in the muscle cell membrane. These fast sodium channels are triggered to open by the transient membrane depolarization induced by the motor neurotransmitter, acetylcholine, at the motor end plates. After the sodium channels spontaneously close, the skeletal muscle transmembrane potential approaches its resting level due to the action of molecular "pumps" that pump Na^+ out in exchange for K^+ in. The pumps are powered by ATP.

The action potential in cardiac muscle is quite different from skeletal muscle. After the initial depolarization of the cardiac cell membrane, it takes some 200 ms for the membrane potential to return to its resting level of about −100 mV on the inside. First, voltage-sensitive Ca^{++} channels admit extracellular Ca^{++} ions. These ions flow down a steep concentration gradient ($[Ca^{++}]_{ext}/[Ca^{++}]_{int} \cong 10$ mM/180 nM) [Shing-Sheu et al. 1984], as well as a potential gradient (about 100 mV). Because the Nernst equilibrium potential for Ca^{++} is about +100 mV, the cardiac muscle cell is strongly depolarized. This depolarization spreads by the syncytium to neighboring cells, causing their Ca^{++} channels to open, etc. The sudden rise in local intracellular $[Ca^{++}]$ causes even more Ca^{++} to be released from the sarcoplasmic reticulum, raising the $[Ca^{++}]$ at the sarcomeres to the level that causes contraction. At this level, the molecular biophysics of contraction is similar to that of skeletal muscle. The ratio of *actin:tropomyosin: troponin* is 7:1:1 in cardiac muscle, similar to that in skeletal muscle.

Cardiac muscle relaxes (i.e, stops contracting) as Ca^{++} is pumped back into the SR membranes. Cardiac muscle cells have ATP-driven ionic pumps in their sarcolemmas (outer cell membranes) that externalize intracellular Ca^{++} at the expense of $3Na^+$ ions entering the cell. This excess Na^+ is then pumped out by other pumps in the sarcolemma that exchange $3Na^+$ out for 2 K^+ in. There are also pumps in the sarcolemma that merely pump Ca^{++} out.

Cardiac muscle cells normally operate with shortened sarcomere lengths, putting them on the left-hand (ascending) side of the force versus sarcomere length curve shown in Figure 7.7. Thus, if more blood enters the heart's chambers, the walls stretch, ultimately stretching the sarcomeres to an increased length, when they can develop more force required to efficiently pump the extra blood volume. This phenomenon is called the Frank–Starling mechanism [Guyton 1991].

The final category of muscle we will review is *smooth muscle* (SM). Smooth muscle affects the contraction of hollow organs, including the bladder, the gut, the uterus, and blood vessels. Notable anatomical features of smooth muscle cells are that they lack the visible, highly organized, sarcomere structures of cardiac and skeletal muscle. Smooth muscle fibers are generally much smaller (about 2 to 10 μm diameter) than skeletal muscle fibers (about 10 to 100 μm diameter). Smooth muscle can be classified as *single-unit* or *multiunit* forms. In *single-unit SM,* the muscle fibers are organized into dense sheets or bands. The densely packed fibers run approximately in parallel; the wider portions of some fibers are packed against the narrower portions of their neighbors, etc. *Gap junctions* occur between the plasma membranes of adjacent fibers, facilitating the rapid spread of depolarization throughout the muscle's syncytium. The multiunit SM fibers have no gap junctions; individual fibers are innervated, and found mixed with connective tissue fibers.

On a molecular level, SM has *actin* (thin) filaments attached to dense bodies of the protein called *α-actinin,* a **Z**-*band* protein found in striated muscle. The ratio of thick to thin filaments in SM is much lower (about 1:15) than it is in skeletal muscle (about 1:6). SM is also rich in *intermediate filaments* that contain the proteins *desmin* and *vimentin.* Thick *myosin* filaments are also present; they must be phosphorylated for actin activation to occur.

The electrical twitch response of SM is much slower than that of skeletal or cardiac muscle. SM can be excited electrically, chemically by hormones, neurotransmitters, or drugs and in some cases by stretching (e.g., uterine SM and arterial SM). As in the case of cardiac muscle, contraction is initiated when Ca^{++} from extracellular fluid is voltage gated into the SM cytoplasm. Ca^{++} released from the sarcoplasmic reticulum triggers the contractile process. In resting SM, $[Ca^{++}]_{int} = 0.1$ μM; upon stimulation, $[Ca^{++}]_{int}$ rises to about 10 μM. As in the case of other muscles, relaxation follows the pumping of Ca^{++} out of the cell and into storage in the cell's SR. The biochemistry of smooth muscle is complex; for further reading on the subject, see Bárány [1996].

In addition to the "big three" of muscle types (skeletal, cardiac, and smooth), there are other curious types of muscle. One is *asynchronous* (fibrillar) *insect flight muscle.* Once activated by motoneuron activity, this type of muscle contracts against a mechanical load in a biochemical/mechanical limit cycle. That is, it oscillates mechanically at a rate generally slower than that of the nervous stimulation, generally driving wings for insect flight [Josephson et al. 2000].

An interesting type of smooth muscle is the "catch" muscle found in certain mollusks. The *anterior byssus retractor muscle* (ABRM) of the marine mussel, *Mytilus edulis* L., is unique in that it can contract and stay in tetanus for long periods of time without neural restimulation until it receives a neurotransmitter from a "relaxor" nerve that causes it to lose tension. Simultaneous stimulation with motor and relaxor nerve signals causes a phasic (fast) twitch in the catch muscle's isometric tension [Northrop 1964].

7.2.4.7 Histone Proteins

The five eukaryote histone proteins are an amazing evolutionary adaptation that permit a eukaryotic cell's DNA to be packed into a small (5- to 10-μm diameter) nucleus in an orderly fashion and still have readily accessible linear coding sections available for gene expression. The DNA is also organized for duplication in cell division. As we have seen, humans store about 3.1×10^9 base pairs of dsDNA in a cell's nucleus in intimate association with histone proteins. Histones (and DNA) are responsible for the structure of chromatin (chromosomes) seen in the nuclei of eukaryotic cells at prometaphase.

Histones are the highly conserved major protein components of nuclear chromatin found in eukaryotes and in modified form in *Archaea*. Bacteria lack histones. Eukaryote histone proteins are generally small, from 102 to 135 AAs; they are also very highly conserved. For example, histone H4 from cows and peas differs by only 2 AAs out of 102. Considering their job of protecting the dsDNA genome, any Nt mutation that destroys their function can lead to a cell that at best cannot express critical proteins or RNA and at worst cannot reproduce.

Histone proteins form the roughly cylindrical, *nucleosome* "beads" around which the organism's dsDNA is wound and folded in its "storage form." As discussed in Chapter 5, they also play a role in regulating gene expression. Five major histone classes are known in eukaryotes. Two each of four types of histone protein (H2A, H2B, H3, and H4) form the *octameric nucleosome* core. That is, $(H2A*H2B)_2*(H3*H4)_2$ form the nucleosome octamer. The "N tails" of the histone proteins stick out of the octamer, forming reaction sites for modifications such as methylation or acetylation of lysine residues, or phosphorylation of serines. Such modifications affect the stability of the chromatin where they occur and are involved in epigenetic control of gene expression.

dsDNA (about 146 bps) is wrapped around each 30-nm-diameter nucleosome "spool" in two tight turns [Rakyan et al. 2001], and the nucleosomes are strung together like beads along alternating linear H1 linker histone protein units. About 54 bps of dsDNA associated with H1 connect each nucleosome spool. The alternating nucleosome/DNA spools and H1/DNA units supercoil into a 30-nm-diameter *solenoid* of chromatin. One repeat unit of the solenoid involves six nucleosomes and six H1 histone subunits. The solenoids are in turn coiled onto a *scaffold,* which is further coiled into a *chromosomal matrix* about 700 nm in diameter. This is the dense prometaphase form of a single chromosome arm's chromatin (histones + DNA), called *heterochromatin* [Kimball 2007f]. Thus, a paired mitotic chromosome is about 1,400 nm across. The net result of the intimate DNA–histone association is about a 10^4-fold reduction of the length of a dsDNA alpha helix by itself.

Euchromatin is less dense under the light microscope than heterochromatin. Euchromatin is found in parts of the chromosomes that contain many active genes; it is separated from adjacent heterochromatin by *insulator* DNA Nt sequences.

Epigenetic modulation of gene expression is basically accomplished by the directed chemical modification of nucleosome histones (and also DNA) in such a way that the DNA genome is unaltered. Histone proteins can be modified by enzymatic methylation, acetylation, ubiquitinylation, phosphorylation, and ADP-ribosylation of certain AA residues to regulate such processes as gene expression, DNA repair, mitosis, and meiosis. For example, *acetylation* of lysine residues in H3 and H4 lowers the affinity of the nucleosome for its wrapping DNA, modifying the nucleosome into a more transcriptionally permissive state. Conversely, enzymatic deacetylation of H4 leads to a more condensed packaging of the DNA around the nucleosome and reduced gene expression. Thus, the modification of histone proteins in intimate association with genetic DNA can activate or silence gene transcription and, hence, protein expression. Modified or variant forms of histones have been found to play crucial roles in the organization of *centromeres, telomeres,* and X-chromosome inactivation in females.

It is known that modifications to histones are heritable both through mitosis, and in some cases meiosis, forming what is known as *epigenetic inheritance* (EI). Generally, EI obeys non-Mendelian statistics. It is thought that at some nucleosomes, inefficient erasure of epigenetic modifications at gametogenesis and early embryogenesis lead to transgenerational epigenetic inheritance [Rakyan et al. 2001; Eberharter & Becker 2004].

7.2.5 Cell Membrane Receptors and Ion Pumps

7.2.5.1 Second Messenger Systems

A major problem in biology is how all cells move molecules in and out of their cell membranes. The cell membrane is a robust isolator between the internal environment of a cell and the external milieu. In order to perform homeostasis, a cell must import molecules for energy and export metabolic waste products. Metazoan cells have also evolved signaling systems between one another and organ systems. Hormones serve as signal substances for almost all metazoan cells. We have seen that hormones can either enter the cell or dock with a receptor protein on the cell's surface and act through a *second messenger system.*

Second messenger systems are used by the following hormones in activating their target biochemical systems: *adrenocorticotropin, thyroid-stimulating hormone* (TSH), *luteinizing hormone* (LH), *follicle-stimulating hormone* (FSH), *vasopressin, parathyroid hormone* (PTH), *glucagon, secretin, the hypothalamic releasing hormones, and catecholamines.* There are several second messenger systems used by cells. The *cyclic AMP* second messenger mechanism works in the following way: First, the signal hormone binds to the receptor protein on the target cell's surface. (A target cell may have many duplicate receptor proteins for a single type of hormone.) The act of binding triggers molecular changes in the part of the receptor molecule that protrudes into the cytoplasm. This molecular activation creates the protein enzyme *adenyl cyclase. Adenyl cyclase* causes immediate conversion of cytoplasmic ATP molecules into cyclic 3′, 5′-adenosine monophosphate (cAMP). Once cAMP is formed, it activates a cascade of enzymes. That is, just a few molecules of cAMP can lead to the production of many molecules of the last enzyme in the cascade chain. The enzymes in the cascade differ from one hormone and set of target cells to another. Guyton [1991] observed: "... a thyroid cell stimulated [internally] by cAMP forms the metabolic hormones thyroxine and triiodothyronine, whereas the same cAMP in an adrenocortical cell causes secretion of the adrenocortical steroid hormones. On the other hand, cAMP affects epithelial cells of the renal tubules by increasing their permeability to water."

Another second messenger molecule is *cyclic guanosine monophosphate* (cGMP). cGMP works in much the same way as cAMP.

Calcium ions can also serve as a second messenger trigger when they enter a cell. Ca^{++} can enter the cell by electrical stimulation of transmembrane protein "calcium channels" or from a hormone or neurotransmitter acting on a calcium channel receptor protein. Once Ca^{++} ions are in the cell, they bind to a protein called *calmodulin.* Calmodulin has four Ca-binding sites; when three or four Ca^{++} ions have bound to this protein, calmodulin undergoes a conformational change that leads to a cascade of intracellular reactions not unlike what happens with cAMP. An example of calmodulin function is that it activates the enzyme *myosin kinase,* which acts on the myosin contractile protein in smooth muscle, causing contraction.

Still another hormonal second messenger system uses *phospholipase C*. When certain hormones (cytokines) are released by allergic and tissue immune reactions, they dock with membrane receptors, and these events trigger the intracellular production of the enzyme *phospholipase C,* which causes intracellular lipids to split into smaller molecules that have pervasive second messenger effects. The most important lipid broken down this way is *phosphatidyl inositol biphosphate;* it forms the important second messengers *inositol triphosphate* (ITP) and *diacyl glycerol* (DAG).

Let us examine the function of these two second messengers. ITP mobilizes free Ca^{++} ions from the mitochondria and endoplasmic reticulum, which promote all their own second messenger effects, including smooth muscle contraction and changes in ciliary action and secretion by ciliated and secretory cells, respectively. To make matters more complicated, the DAG activates the enzyme *protein kinase C,* which along with the increased intracellular [Ca^{++}] promotes cell division and proliferation. The lipid portion of DAG is the 20-carbon, *arachidonic acid,* an important precursor of the *prostaglandin and leukotriene eicosanoids.* The prostaglandins and leukotrienes are important in immune system function and in the regulation of inflammation [Northrop 2000].

In organisms that have evolved nervous systems, signals travel faster on axons to specific target cells than do chemical hormone signals, but the neurons release *neurotransmitter molecules* (e.g., *acetylcholine, epinephrine, norepinephrine, dopamine, and serotonin*), which have hormone-like action in that they bind to transmembrane receptor proteins on the target cells (e.g., muscle cells, neurons, and heart muscle). Some neurons exert their action by direct electrical coupling to the target cells. In this case, the induced transmembrane potential change causes transmembrane proteins with specific ionic conductances to transiently increase their conductances. These proteins are called *voltage-gated ion channels.*

There are two basic mechanisms whereby neurotransmitters released at a synapse cause changes in ionic currents through the cell membrane. In the first category, the neurotransmitter binds to the subsynaptic receptor molecule and causes a direct conformational change that increases a specific ionic conductance. In the second category, second messengers are involved. The α- and β-adrenergic, serotonin, dopamine, and muscarinic acetylcholine (ACh) receptors and receptors for neuropeptides and rhodopsin all are second messenger systems. In each member of this receptor class, the transmembrane receptor molecule is coupled to its effector molecules by a *guanosine nucleotide binding protein* (G-protein). Typically, the effector is an enzyme that produces diffusible cAMP, DAG, or ITP. These second messengers in turn trigger a biochemical cascade activating specific protein kinases, which lead to complex reactions that change the cell's biochemical state [Kandel et al. 1991].

Thus, we see that the second messenger systems are homologous in nerve-activated systems and endocrine hormonal systems. It is just that nervous signals are transmitted more rapidly to more specific target cells.

7.2.5.2 Ion Pumps

Note that ion pumps were described in some detail in Chapter 3, Section 3.7.2.3. We point out again that cells such as neurons have specialized transmembrane proteins that harness ATP chemical energy to continuously "pump" ions through their membranes against both an electrical potential barrier and a concentration gradient. The neuron Na–K pump is an example of an ion pump. It continuously pumps three Na^+ ions out of the cell and brings in two K^+ ions to maintain fixed ionic concentration gradients across a resting neuron's membrane. The pump is a multimeric protein complex consisting of two different polypeptides, a transmembrane catalytic subunit [α] and a glycoprotein regulatory subunit [β]. The catalytic subunit has binding sites for ATP and Na^+ on its intracellular surface and sites for K^+ and *ouabain* on its outside surface. (Ouabain is a poison that specifically and irreversibly disables the pump.) The steady-state ion flux of a neuron's Na–K pumps just balances the ion leakage through other resting ion channels (Na^+ in and K^+ out) on the neuron's membrane. Thus, these pumps maintain a neuron's internal steady-state ion concentrations and its resting transmembrane potential [Kandel et al. 1991].

Some neurons also have protein transmembrane chloride pumps that direct Cl^- ions out of the cell so that the steady-state $[Cl^-]_o/[Cl^-]_i$ is greater than the ratio that would occur if chloride ions were distributed passively by diffusion (transmembrane leakage).

Also found in muscle cells is an ATP-driven Ca^{++} pump. When muscles contract, calcium ions are released from the sacroplasmic reticulum (SR) tubules and diffuse in among the actin-myosin contractile proteins, causing contraction. Muscle relaxation is an energetically expensive process in which Ca^{++} ions in the contractile volume are pumped back into the sarcoplasmic reticulum tubules against a large concentration gradient. The normal, "relaxed" Ca^{++} concentration in the cytosol that bathes the myofibrils is about 10^{-7} molar. When the muscle is stimulated to contract, the $[Ca^{++}]$ can rise to about 2×10^{-4} molar. In about 100 ms, the Ca pumps restore the myofibril $[Ca^{++}]$ to the resting level, and the muscle fiber relaxes. Some muscles relax much faster (e.g., extraocular muscles), and others are much slower (smooth muscle and cardiac muscle). When a mammal dies, oxidative metabolism ceases. Thus, there is no energy from ATP to pump Ca^{++}. Ca^{++} that leaks into the myofibrillar space causes the temporary muscle rigidity known as *rigor mortis.*

7.2.5.3 Photon-Driven Proton Pumps

Found in *Eubacteria* and *Archaea,* transmembrane proteins known as *proteorhodopsins* (PRs) have the interesting role of trapping the energy of visible-light energy photons and using that energy to pump protons (H^+ or $H_3O_2^+$) from the bacterial cytoplasm to the outside of the bacterial plasma membrane. (PRs work in conjunction with the *retinene* molecule to trap photons of appropriate energy.)

The externalized (pumped) protons then can diffuse inward at specific sites where their *proton-motive force* (PMF) (potential energy consisting of their Nernst potential plus the cell's transmembrane potential) can (1) drive various *porter* TMPs (e.g., to pump out Na^+); (2) activate bacterial flagella, increasing bacterial mobility; and (3) diffuse inward where their potential energy drives the *ATP-synthase* TMP, which catalyzes the net reaction, **Pi + ADP → ATP**. The bacteria use this ATP intracellularly for metabolic purposes [Walter 2007].

Researchers have recently inserted the six-gene proteorhodopsin plasmid from wild-type marine bacteria into *E. coli* in order to study its regulation and biochemistry [Martinez et al. 2007]. The molecular structure of blue-absorbing proteorhodopsin (BPR) consists of about 251 AAs. Seven linked helical domains penetrate the plasma membrane. The C-terminus is on the cytoplasmic side, and the N-terminus is on the extracellular side. The AA sequence and 3-D structure of BPR have been described by Hillebrecht et al. [2006]. Giovannoni et al. [2005] have estimated that there are on the order of 10^4 PR sites in the membrane of a typical PR-containing *Peligibacter* (SAR11).

The natural role of the PR system in those bacteria possessing it is to provide extra energy by trapping photon energy and converting it to proton potential energy across the cell membrane. However, PR appears to be a supplemental energy-harvesting system; the bacteria also have other metabolic pathways. Some workers have suggested that the PR system plays the role of a *photoreceptor* in that it activates the bacterial *flagellae,* causing the organisms to move randomly (in 3-D), but with an effort proportional to the light intensity striking their PR proteins. Thus, bright light at the water surface can cause PR bacteria to swim deeper until they reach a depth where there is a lower light intensity and a reduced probability of their flagellae propelling them upward. In deeper water, there is also less probability of PR bacteria being harmed by UVB photons; UVB is absorbed by the water.

The isolation of the proteorhodopsin operon and its implantation into GM bacteria means that light-powered PR^+ bacteria may find use in more efficiently producing genetically engineered substances such as insulin.

7.2.6 Proteins in the Immune System

7.2.6.1 Introduction

The mammalian immune system is an enormously complex system, composed of many types of cells and hundreds of kinds of protein signaling molecules. Its function is to seek out and destroy invading bacteria, fungi, molds, viruses, protozoa, spirochetes, parasites, etc. That is, it protects its host animal from foreign invaders. Immune surveillance also extends under certain conditions to the detection of cancer cells and their destruction. Immune system *macrophages* (monocytes) also act to clean up necrotic body tissues damaged by injury, disease, or parasites.

As you will see, the immune system uses a number of mechanisms in its functioning. These include *humoral* (blood- and lymph-borne) molecular *factors,* including protein *antibodies* (Ab), *lectins* and *complement,* and *direct cellular attack,* including phagocytosis and lysis. *Macrophages, natural killer* (NK) cells, and *cytotoxic T-cells* (CTL) participate in the direct cellular attack of hostile invaders (the "L" in CTL is because they are a form of *lymphocyte*).

The immune system also can respond to substances that it should not react to. These responses include allergy and autoimmunity. Certain foods, drugs, etc., taken in through the oral route can trigger a variety of adverse immune responses including nausea, headaches, hives, asthma, and anaphylaxis. Allergens such as dust, pollen, mold spores, etc., can also be inhaled, giving rise to rhinitis and asthma. Other substances, such as the oil from poison ivy, on the skin cause well-known misery. For poorly understood reasons, the immune system can develop a sensitivity to certain normal cellular proteins in the body, creating *autoimmune diseases.* Two examples of autoimmune diseases are *Type 1 diabetes,* in which the immune system attacks the pancreas, and *Crohn's disease,* in which the immune system attacks cells lining the bowels, creating inflammation. The study of proteins is essential to understanding allergies and other immune disorders; much work remains to be done.

In the opinion of the authors, the human immune system is, compared to the central nervous system, the second-most complex physiological system in the adult body. Its complexity arises mostly from its biochemical pathways, rather than synaptic connections. In order to understand the vast network of complex causal interactions between its various components, we must understand its biochemical C^3 structure. These chemical interactions are, in general, of a feedback as well as a feed-forward nature. The immune system's components are normally self-regulating; i.e., the system is *homeostatic.* The components include specialized immune cells, the immunoregulatory protein *autacoids* they secrete, and the protein antibodies manufactured. The immune system responds to biochemical signals (autacoids) from immune system cells at the sites of infection, to hormones from the central nervous system (CNS), and from other tissues. Certain cellular components of the immune system are mobile and can move throughout the body's circulatory and lymphatic systems. Other immune system cells end up fixed in certain tissues and organs.

In the following section, we will summarize the cellular components of the immune system, the major chemical signals it uses (*immunocytokines,* or immunoregulatory autacoids), and how these components normally work and interact. It will be seen that immunocytokines including lymphokines, interleukins, interferons, prostaglandins, etc., form a complex hormonal regulatory network. Many immunocytokines have multiple effects on multiple target cells (a property called *pleiotropy*), adding to the difficulty of formulating meaningful mathematical models of the immune system. The problems in describing the role of immunocytokines are especially enigmatic because, in general, we have an incomplete understanding of the biochemical control mechanisms governing their synthesis and the molecular mechanisms whereby they exert their effects after binding to appropriate (or in some cases, inappropriate) membrane receptor molecules. Although dauntingly complex, certain aspects of immune system function can be mathematically modeled [Northrop 2000].

7.2.6.2 A Summary of the Cells of the Immune System

The human immune system has four major classes of *leucocytes* (white blood cells), which carry out its mission of protecting the body from infection and exogenous foreign substances (inhaled or eaten): *macrophages, T-lymphocytes, B-lymphocytes,* and *natural killer cells.* Each of these four classes of cell has several subclasses that are based on the level of cell maturity, proteins found on their cell membranes, the cells' functions, and their location in the body. Other cells such as *mast cells* and *granulocytes* also contribute to the inflammatory immune response, and *megakaryocytes* in the bone marrow make *platelets.* Platelets are 2 to 4 μm in diameter. They are not true cells in that, similar to red blood cells, they lack nuclei. However, they have membranes with receptor proteins and contain mitochondria and RNA. Platelets also contain enzyme systems that allow them to synthesize *prostaglandins*, a class of nonprotein *eicosanoid* immunocytokine. Platelets are necessary for blood clotting.

Certain immune system cells can be described as *amplifiers,* responding to foreign proteins, etc., by activating cells of like kind and other immune system components. Other cells can be described as *effectors;* they carry out offensive actions. Still other cells are *modulators;* they regulate the reactivity of the *sensor* and *effector* cells. The amplifiers, which include *macrophages* and *B-cells*, perform *antigen presentation*, which has the function of causing the proliferation of *helper T-cells, cytotoxic T-lymphocytes* (CTL), and *natural killer* (NK) *cells.* Certain messenger *immunocytokine* proteins are secreted by both the antigen-presenting cells and the receiver cells which cause the receiver cells to reproduce, or clone themselves. Each daughter B-cell and CTL retains the specificity to bind with the presented antigen. The presented antigen is generally a piece of a coat protein of an invading bacterium or virus, etc. It is by this amplification process that a specific immune response to an invading organism is strengthened.

The normal total leucocyte density in human adult blood is 5,000–10,000 cells per μl. About 5.3% of these are *monocytes,* 30% are *lymphocytes,* 0.4% are *basophils,* 2.3% are *eosinophils,* and 62% are *neutrophils.* In general, when there is an acute *viral infection* (e.g., influenza), the total leucocyte count is *decreased.* On the other hand, when a person suffers from an acute *bacterial infection,* the *total* white cell count is *increased.* Other specific diseases produce changes in the ratios of specifically stained leucocytes. For example, infection by the *trichinosis parasite* causes a signatory increase in eosinophils, their ratio rising from about 2.3% of the total leucocyte count to about 40 to 60%. Other WBC ratios stay the same, and the total WBC count increases to about 30% [Collins 1968].

In the following paragraphs, we will examine the immune system's cells in greater detail. We will describe the specialized cells of the immune system starting with *macrophages* (Mφ). The various types of *T-lymphocytes* (T-cells; so called because of their embryological origin in the thymus gland) are considered next. *Natural killer cells* (NKs) follow T-cells. NK cells come from a separate stem cell line; they have basically the same role as *cytotoxic T-cells* (CTLs) but lack the antigenic specificity possessed by CTL and B-lymphocytes. *B-cells* (so called because of their embryological origin discovered in the *Bursa of Fabricus* of birds) and their role in manufacturing antibodies are then described.

Many of the immune system cells have aliases or alternative nomenclature. Early descriptions of immune system cells were based on how they were identified by various stains when prepared for viewing using light microscopy. More recent nomenclature relies on cell function and the specific identifying proteins found on their surfaces. We will try to identify immune system acronyms as they arise, to minimize confusion.

Macrophages (Mφ) are found throughout the body. Immature Mφs are called *monocytes.* They generally circulate freely in the blood and lymphatic system. As monocytes mature, they become more sessile; i.e., they assume fixed locations in tissues and organs. Macrophages can migrate from the circulatory system and become resident in various tissues: In the liver they are called *Kupfer cells,* in the brain they are *microglia,* in the kidneys they are *mesangial cells,* in connective tissue and skin they are *histiocytes,* in the lungs they are called *alveolar macrophages*, and in the lymph nodes and spleen there are *fixed* and *free macrophages,* etc.

The immunological functions of Mφs are broad. They are involved in all stages of the immune response. First, they provide a rapid, front-line, nonspecific, cellular immune defense against invading bacteria, fungi, and parasites. Activated Mφs can engulf and internalize such invaders. Internal enzymes break down the membranes and proteins of such phagocytosed bacteria, etc., into smaller molecular subunits that can then be used to activate helper T-cells in a complex process known as *antigen presentation.*

Macrophages are the central effector and regulatory cells of the inflammatory response. They can secrete more than 100 different effector *autacoid* substances! Some of these substances, such as *hydrogen peroxide, lysozyme, neutral proteases, nitric oxide,* and so on, kill or damage the hostile target cells as well as normal cells (collateral damage) and cause inflammation. Other substances secreted also act to induce inflammation and then to repair tissues destroyed by infection and inflammation. In summary, similar to all immune system cells, Mφ are incredibly complex biochemical machines. All of their products and their actions are the subject of internal and external regulatory control by the molecular messengers of the immune system network.

Macrophages are continually produced from bone marrow progenitor (stem) cells. Their rate of production is modulated by certain immunocytokines, notably *(monocyte colony-stimulating factor)* MCSF and *(granulocyte-monocyte colony stimulating factor)* GMCSF. Those Mφ which are induced by GCSF are larger and have a higher phagocytic capacity than those induced by GMCSF, which are more cytotoxic against certain tumors, express more *major histocompatibility complex* (MHC) class II antigen molecules on their surfaces, more efficiently kill the bacteria *Listeria monocytogenes,* and secrete more (prostaglandin E_2) PGE_2 [Northrop 2000]. The very different structures and signal transduction mechanisms of the receptors for MCSF and GMCSF appear to be evidence for different differentiation pathways, and thus two different macrophage populations. Immunological studies of the coat proteins of macrophages found in different immunological scenarios (dermatitis, gingivitis, osteoarthritis, tissue graft rejection, tumors, etc.) provide further evidence for functional subtypes of Mφs. Even so, current practice is to classify all macrophages by their level of *activation.*

For modeling purposes, we have arbitrarily divided Mφ into three groups: *inactivated, primed,* and *fully activated.* In the *inactivated* condition, Mφ can either be circulating or fixed in tissues. They have low oxygen consumption, little or no monokine (immunocytokine) secretion, and a low level of MHC class II gene expression. (MHC class II is a Mφ membrane protein necessary for antigen presentation to helper T-cells.) Mature, *inactivated* Mφ do have phagocytic activity, however, can respond to chemotaxic signals, and can proliferate in respond to signals. Inactivated Mφ are primed by gamma interferon (IFN-γ) secreted from stimulated helper T-lymphocytes.

Primed Mφ have increased oxygen consumption (increased metabolism), and they exhibit enhanced MHC class II expression. Other immunocytokines such as IFN-α, IFN-β, IL3 (interleukin-3), MCSF, GMCSF, and tumor necrosis factor α (TNF-α) can also prime Mφ for selected functions. Primed Mφ can respond to secondary signals to become *fully activated.* In this stage, they cannot proliferate, and they exhibit high O_2 consumption. They achieve maximal secretion of substances for cell killing and inflammation. On the other hand, they show decreased MHC class II protein production, as well as reduced antigen presentation.

Note that there is no sharp metamorphic distinction between primed and active Mφ. Among the members of a relatively isolated population of Mφ, such as in the spleen, at any moment one can find a spectrum of Mφ development levels and capabilities. Such diversity adds interest (and confusion) to the challenge of modeling Mφ actions.

In summary, Mφ are seen to play a key role in activating the immune response. They phagocytose and destroy invading bacteria, viruses, parasites, fungi, protozoa, etc. After internal proteolysis, the Mφ present fragments of the phagocytosed invader's protein coats on their surfaces to activate natural killer and helper T-cells. NK and helper T-cells cells respond to a specific protein fragment or epitope bound to the Mφs' MHC coat protein, and they release certain immunocytokines that cause the specific responding cells to clonally reproduce themselves. Other immune cells perform antigen presentation, but the Mφ are seen to play a key role in activating helper T-cells. Mφ also secrete protein substances toxic to invading organisms.

T-lymphocytes are leucocytes originating in the bone marrow from what are called *pluripotent hemopoietic stem cells.* Their rate of production is under control by cytokine hormones from the immune system. They are called T-cells (TL) because they are "preprocessed" in the *thymus gland* (hence T-). Also, they are mostly found in lymphoid tissue, including lymph nodes, the spleen, and submucosal areas of the gastrointestinal tract. The processing of immature T-cells, or *thymocytes,* in the thymus includes selective deletion of any thymocytes that have a T-cell receptor (TCR) coat protein with an affinity for normal self-antigens on body cells. Such selective deletion is thought to prevent *autoimmunity* if portions of normal cell proteins are accidentally presented to TL by macrophages, etc.

T-lymphocytes are responsible for *cell-mediated immunity,* which is explained later. Three subsets of T-lymphocytes are characterized by their functions in the immune system and further described by the proteins found on their cell surfaces. With the exception of cytotoxic T-cells (CTLs), T-lymphocytes do not directly attack invading organisms. Instead, they serve as activators (amplifiers) or suppressors of the immune response through the protein immunocytokines that they secrete in response to *antigen* (Ag) *presentation. Helper T-lymphocytes* are generally given the abbreviation Th, *suppressor T-lymphocytes* are Ts, and *cytotoxic T-lymphocytes* are CTL.

All T-lymphocytes have unique, external *T-cell receptor* (TCR) proteins that may bind to a presented antigenic *epitope* if affinity is high. By shuffling protein subunits, the human body can make about 25 million different TCR *paratopes,* ensuring a high probability of binding to an antigenic epitope [Kimball 2006f].

In addition to the many varieties of TCR proteins, there is a five-molecule (pentameric) cell-surface protein complex, CD3, associated with each TCR. The CD3 complex can sense when the TCR has bound to an *epitope,* and then it initiates a complex sequence of intracellular events collectively called *T-cell activation.* In addition, Th lymphocytes have the *CD4 protein complex* next to their TCR proteins. CD4 is a cell surface coreceptor for *MHC Class II molecules.* (MHC II molecules are the type II *major histocompatibility protein molecules* that exist on the surfaces of all macrophages [Mφ] and antigen-presenting B-cells for cell identification.) CTL lymphocytes carry the CD8 protein complex next to their TCRs. Ts lymphocytes can also carry CD8 protein. CD8 is a coreceptor for *MHC Class I* molecules found on all body cells that do not bear MHC II molecules.

Helper T-cell activation is a complex biochemical process in which biochemical synthetic machinery in the Th cell reacts to structural changes in the intracellular portion of the CD3 molecular complex brought about by successful Ag presentation by a Mφ. The Mφ secretes *interleukin-1* (IL1), which binds to receptors on the activated Th cell's (ATh) surface. Under the combined influence of IL1 and activated CD3, two subsets of ATh cells secrete certain immunocytokines. The ATh1 cells secrete the cytokines IL 2 and 3 and *gamma interferon* (γIFN), and ATh2 cells secrete *IL3, 4, 5, and 6*; *tumor necrosis factor* α (TNFα); and *GMCSF.* These immunocytokines have profound effects on other cellular components of the immune system that will be described later.

Antigen presentation by infected somatic cells having MHC Class I + Ag on their surfaces to a cytotoxic T cell (CTL) is illustrated in Figure 7.9. When the CTL's TCR has affinity to the presented Ag, and the CD8 and CD3 molecules react, immunocytokines are released by the CTL and it is stimulated to undergo clonal proliferation. That is, it reproduces itself with the same complex TCR that has affinity for the antigenic epitope that activated it through presentation. These CTL clones can now "recognize" and bind to other somatic cells infected by the same virus or parasite and carrying the presented Ag on their surfaces. (The invader generally leaves some of its coat proteins on the cell's surface, like a criminal leaving fingerprints at the scene of a crime.) Some CTLs that bind to infected somatic cells release a protein called *perforin,* which literally bores nonrepairable holes in the target cell's membrane. Ions and water pass through the holes, causing the target cell to rupture from osmotic lysis. Another means of cell killing is thought to be by the CTL inducing *apoptosis* in the target cell. As we have already seen, apoptosis is an internally or externally triggered self-destruction mechanism that causes the cell and its contents to literally disintegrate (see Chapter 2, Section 2.5.2). By killing the infected target cell, the CTL or NK prevents any internal

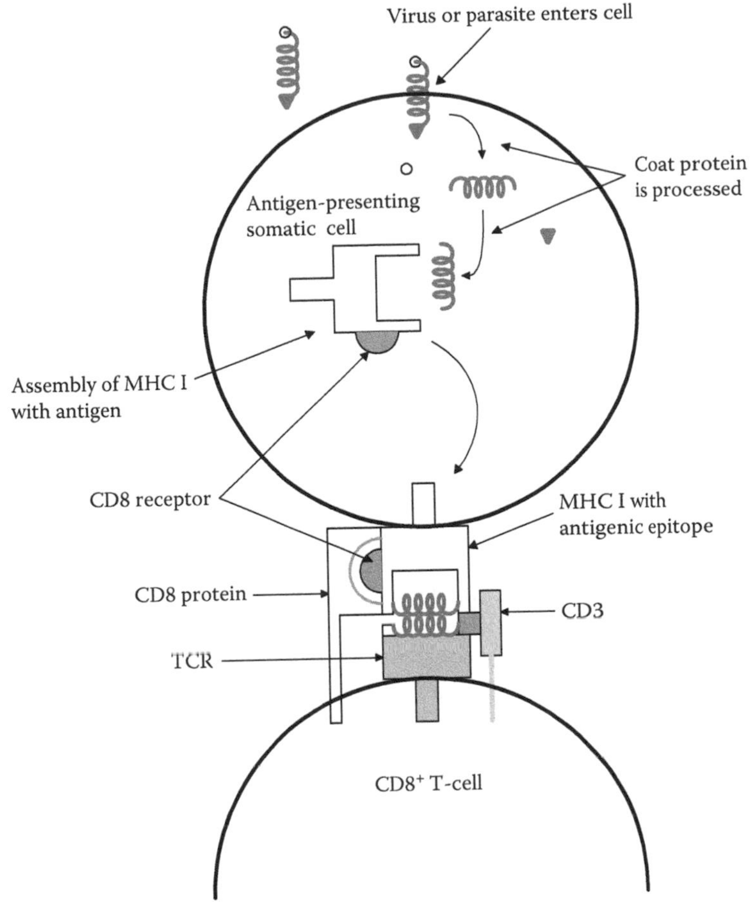

FIGURE 7.9 (See color insert.) Schematic of antigen presentation by an MHC I+ somatic cell. The antigen is presented to a CD8+ T-cell. See text for further description. (Adapted from Northrop, R.B. 2000. *Endogenous and Exogenous Regulation and Control of Physiological Systems*. Chapman and Hall/CRC Press, Boca Raton, FL. Used with permission from Taylor & Francis.)

viruses or parasites from proliferating. Once the infected cell has ruptured, other components of the immune system (macrophages, antibodies, and B-cells) attack the now externalized virions or parasites. One might wonder why there appears to be two separate methods for CTLs to kill infected cells. We speculate that redundancy ensures success, and that apoptosis, operating from within the cell, causes viral nucleic acids to disintegrate, inactivating them.

Suppressor T-cells (Ts), aka ***regulatory T-cells*** (T_{reg}), are a third class of T-lymphocyte. At one point in the 1980s, the existence of Ts cells was debated by immunologists; it is now known that their populations and functions indeed do exist, and the cytokines involved in their actions are also known. Their exact mechanism of regulation is just beginning to be understood on a molecular level. It is not unreasonable to expect that the immune system, a complex network of cells and their signaling substances (immunocytokines, including interleukins, prostaglandins, interferons, and tumor necrosis factors), has developed mechanisms to halt the excess proliferation of CTLs, NKs, plasma B cells, and antibodies, once the invading pathogen has been vanquished. This suppression of immune system actions is necessary to conserve immune system resources and prevent body damage by runaway inflammatory reactions.

Ts cells are characterized by certain cell surface proteins: CD4 and CD25 (a receptor for the IL2 α-chain) and also by the *forkhead transcription factor* FOXP3. Expression of cell surface FOXP3 protein is required for regulatory T-cell development and appears to control a genetic program directing the Ts cell's destiny. Most FOXP3-expressing Ts cells are found in the larger population of major histocompatibility complex (MHC) II, $CD4^+$ Th cells, and high levels of surface CD25 receptor protein. There also appears to be a minor population of MHC class I restricted $CD8^+$ FOXP3-expressing Ts cells. The characterization of Ts cell types is a work in progress. Additional Ts populations including Tr1, $CD8^+CD28^-$, and Qa-1-restricted T-cells have been reported [Sullivan et al. 2002].

IL2 has been demonstrated to be required for Ts function in vitro. Also, secondary stimulation of Ts cells with IL2 causes them to express IL10, a generally inhibitory, downregulatory cytokine (see Table 7.1) [de la Rosa et al. 2004]. In addition, a complex six-step sequential pathway has been described by Chess and Jiang [2004] for the activation of $CD8^+$ T-cells and their conversion into Ts cells. The MHC class Ib molecule (also known as Qa-1 in mice and HLA-E in humans) induces the $CD8^+$ T-cells to become suppressors. Chess and Jiang speculate that exogenous HLA-E may one day prove effective for treating autoimmune diseases.

As mentioned earlier, *B-lymphocytes* are so called because of their embryological origin in the organ called the *Bursa of Fabricus,* found in birds (not humans). In humans, B-cells grow from stem cells in the bone marrow and mature in lymphoid tissue of the fetal liver. They are then released into the circulatory system, where they are distributed more or less randomly throughout the body. In humans, there are normally about a total of 10^{12} B-cells. This translates into a plasma density of about 10^7 B-cells per milliliter. B-cells are morphologically identical to T-cells, so they must be identified by their coat proteins. Normally, about 10 to 20% of circulating blood lymphocytes are B-cells.

B-cells are the key effectors in the humoral arm of the immune system. They have the capability to fight infection by producing huge quantities of specific, freely circulating antibodies (Abs) with affinity to a particular antigen (Ag). The Abs bind to the Ags, inactivating them and marking them for destruction by natural killer (NK) cells, macrophages, and the complement system.

Each cell in the population of unstimulated, mature B-cells has a unique surface-bound IgM protein antibody having great affinity to bind with a specific Ag. Nature produces millions of different B-cells with unique IgM Ab molecules, each having a very different affinity to some Ag. When a specific B-cell's IgM Ab (*paratope*) binds with a soluble antigen, such as diptheria toxoid, the bound Ag is engulfed by the B-cell by the process of *receptor-mediated endocytosis.* The Ag is now digested into fragments that are picked up by MHC II molecules and then displayed on the B-cell's surface. A helper T-cell (Th) with a specific complementary TCR protein binds to the antigen-presenting B-cell, as shown schematically in Figure 7.10. As in the case of macrophages, the proteins CD4 and CD3 are involved in activating the Th, which secretes the interleukin proteins IL4, -5, and -6. The local release of these interleukins causes the B-cell presenting the antigen to reproduce clonally, copying the specific IgM antibodies (paratopes) that are bound to the antigen's epitope. The activated, mature B-cell clones (also known as *plasma cells*) then release soluble antibodies, which can bind with free or cell-surface-bound Ag. B-cell growth is also activated by IL1 from macrophages and by IL2. To make this process more complex, there are five different isotypes of a specific antibody.

The process of antigen presentation to Th by B-cells also leads to the production of special, clonally produced, circulating, inactive B- and T-cells having long lives and specificity to the particular Ag. These cells are called *memory B-cells* (M-cells), and they evidently provide a rapid and strong humoral response if the pathogen with the Ag is reintroduced.

The antibodies (Abs) produced by activated plasma B-cells have a unique protein molecular structure. They are **Y**-shaped, with the Ag binding domain (*paratope*) lying between the arms of the **Y**. Molecular diversity in encoding the binding domain structure gives the possibility of well over 10^6 different Ab paratopes in the human at a given time [Kimball 2006f]. The stem of the **Y** is made from the paired ends of two "heavy protein chains" of over 400 amino acids (AAs) in length. The arms of the **Y** are also paired; the ends of the heavy chains each pair with a light chain of over 200 AAs. (See Figure 7.11 for an antibody schematic.) As mentioned earlier, there are five classes of circulating antibodies: IgA, IgD, IgE, IgG, and IgM. (Ig stands for *immunoglobulin.*) The heavy

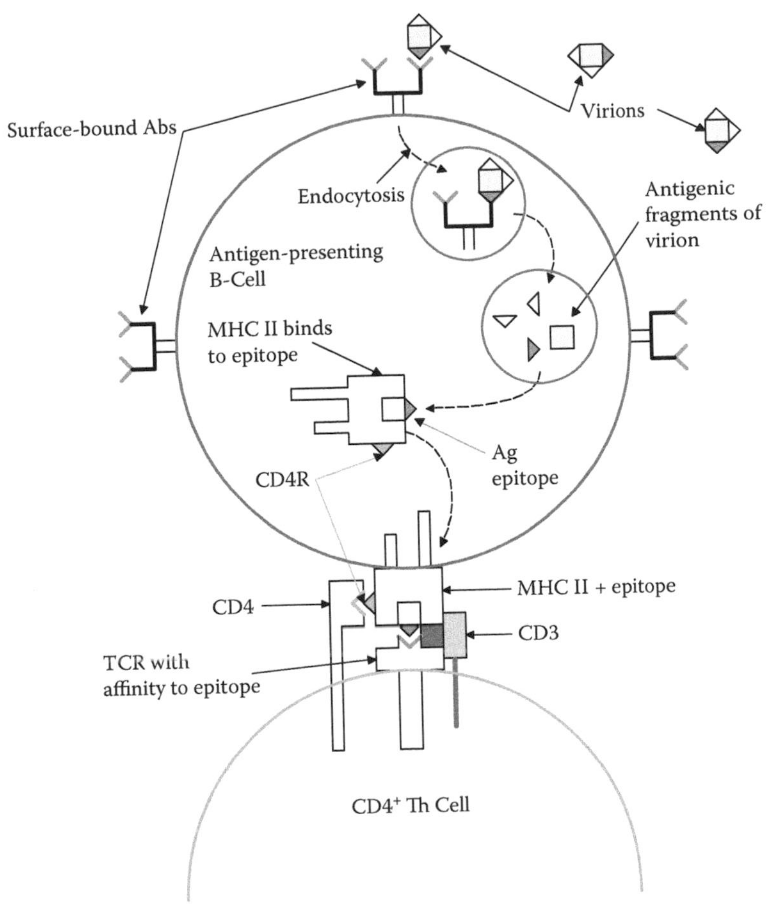

FIGURE 7.10 (See color insert.) Another mode of antigen presentation. An MHC 2+ B-leukocyte presents fragments of viral coat protein to a CD4+ helper T-cell. (Adapted from Northrop, R.B. 2000. *Endogenous and Exogenous Regulation and Control of Physiological Systems.* Chapman and Hall/CRC Press, Boca Raton, FL. Used with permission from Taylor & Francis.)

chains making up these Abs are called *alpha, delta, epsilon, gamma,* and *mu,* respectively. The light chains are called *kappa* or *lambda.* Genetically randomized variable regions of the light and heavy chains form the paratope sites, one in each arm of the **Y**. Some Abs, however, are formed from a fusion of as many as 10 light and 10 heavy chains and hence have 10 binding sites. The protein structure of the stem region of the **Y** is relatively constant and contains a *constant* ***Fc*** *binding site* through which NK cells, Mφ, mast cells, etc., can recognize and bind to Abs. If the Ab has bound to a cell-surface Ag, NK cells can bind to the Fc site and then secrete the protein *perforin,* which kills the cell. Macrophages clean the mess up.

Abs can act in three different ways to fight invading pathogens:

1. They can directly bind to the invading pathogen, inactivating it by changing its gross structure.
2. They can activate the *Complement System*, which destroys the invader. Direct binding of Ab to Ag can lead to large clusters of Abs bound to many Ags; such a cluster is said to be an *agglutination.* Agglutinations can *precipitate,* becoming immobile and prey for CTLs, NKs neutrophils, and macrophages.
3. The Fc regions of Abs bound to Ags on cell surfaces form attachment sites for NK cells, which kill the cells involved, hopefully before the pathogen can multiply inside the cell.

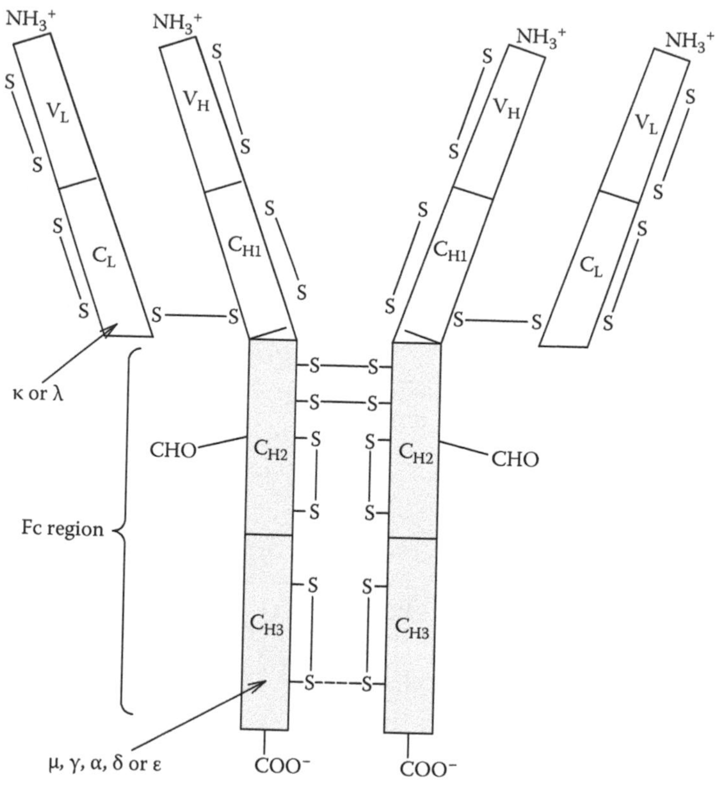

FIGURE 7.11 Schematic of a human antibody molecule. See text for further description. (Adapted from Northrop, R.B. 2000. *Endogenous and Exogenous Regulation and Control of Physiological Systems*. Chapman and Hall/CRC Press, Boca Raton, FL. Used with permission from Taylor & Francis.)

Natural *killer cells* represent a small fraction of the total leukocytes in the body. NK cells are unique because they can attack and kill cells in the body that lack normal MHC I cell surface proteins, such as cancer cells or cells infected by a virus. Thus, they can form a fast direct line of defense against cancer cells that bear altered MHC I proteins, in theory bypassing the need for antigen presentation. In a typical scenario, an NK cell uses an *activating receptor* on its surface to bind to a *critical, cell-surface glycoprotein* on a cell. If another receptor protein on the NK cell's surface binds with a normal MHC I protein on the cell surface, the NK cell is inhibited from secreting perforin or other cell-killing autacoids. NK cells may have up to 11 variations of the MHC I binding protein, ensuring that all normal variations in MHC I will be bound.

As we have already described, the traditional mode of NK cell killing involves *antibody-dependent cellular cytotoxicity* (ADCC) in which an NK binds to a site in the Fc region of an IgG antibody bound to an Ag on an infected cell's surface. When the NK binds to the Fc region, it makes contact with the cell surface and then secretes perforin, etc. It is not known whether the MHC I-induced inhibition of perforin secretion is operative in ADCC by NK cells. Presumably, *both* Fc binding and cell surface contact are required for the release of perforin, etc. Otherwise, NKs would bind to free Abs of any kind and secrete perforin at random, playing havoc with normal neighboring cells.

NK cells are stimulated by *growth hormone* (GH); *luteinizing hormone* (LH); the interleukins IL2, IL12, and IL15; and *interferons* α, β, and γ. Stimulated NK cells secrete γIFN and TNFα. IL10 inhibits the production of IL2-induced γIFN by NK cells.

Mast cells, platelets, and complement are important accessory immune effectors outside of the classical immune system structure (which includes B-cells, antibodies, T-cells, macrophages, and natural killer cells). All three effectors contribute to the inflammatory response to infection and when viewed along with the rest of the immune system underscore the functional redundancy of the immune system.

Mast cells are very important cells in allergic inflammatory reactions and, in particular, asthma. Mast cells are formed in tissues from undifferentiated precursor cells manufactured in the bone marrow and released into the blood. There are two types of mast cells: one is found in connective tissues and the other in mucosal sites.

Connective tissue mast cells are found mostly in the skin, whereas *mucosal mast cells* reside in the gut and lungs. Both kinds of mast cells are sessile. They collect around blood vessels, nerves, and lymphatic vessels. They are concentrated around potential points of entry of pathogens and foreign microparticles (e.g., pollen and mold spores) into the body.

In humans, mucosal mast cells comprise a total of 1% of the cells located in the lungs and accessory tissues. About half of the mucosal mast cells are found in the intraalveolar septa, with the other half located in the mucosa of the trachea, bronchi, and bronchioles. Mast cells contain *cytoplasmic secretory granules* of diverse morphology. Mast cells can release such substances as *histamine, proteoglycans* (*heparin* and *chondroitin sulfates*), *leukotrene-*C_4, proteolytic enzymes (*tryptase* and *chymase*), *serotonin,* IL4, IL5, IL6, TNFα, and *prostaglandin-* D_2. All of these substances contribute to the discomfort of an acute allergic reaction. Pain, edema (tissue swelling), inflammation, etc., are the result of mast cell stimulation. Activation of mast cells in the lungs can result in acute asthma, with bronchospasm (airway size reduction due to smooth muscle contraction), alveolar tissue swelling (vital capacity reduction), and thickened mucus that clogs small airways.

Mast cells are stimulated to release their granules by the binding of their high-affinity Fc receptors (FcRs) to the Fc region of IgE antibodies that, in turn, have high Fab affinities to allergen molecules such as pollens or danders. There can be as many as 5×10^5 FcRs per mast cell [Guyton 1991]. The actual trigger that initiates the release of mast cell products is the cross-linking of the bound IgE–Ag compexes attached to the mast cell's Fc receptors. Such cross-linking can occur if there are high enough concentrations of antigen and free IgE Abs and the Ag has multiple binding sites on it for the Abs. When mast cells secrete IL4, IL5, IL6, and TNFα, other immune system cells such as eosinophils and T-cells are recruited into the inflammatory scenario, which is biochemically complex.

Platelet cells have two major functions in the body: one is to form clots to stop bleeding from wounds. Clot formation is a very complex biochemical process that we will not detail here. The other function is to participate in inflammatory reactions of the immune system. There are normally from 1.5×10^5 to 4×10^5 platelets per cubic millimeter of blood. Platelets are made in the bone marrow by *megakaryocytes* and then enter the blood. Similar to mature red blood cells, they are cells without nuclei, ranging from 2 to 4 μm in diameter. Their resting shape is also discoidal; however, when activated by appropriate stimuli, they become spheroidal and develop many protruding "tentacle-like" protrusions up to 5 μm in length. They have typical phospholipid bilayer plasma membranes with many embedded receptor molecules. Their cytoplasm contains a "skeleton" of *microtubules,* largely formed from *actin* and *myosin,* proteins normally associated with muscles. Also within platelets are mitochondria, several types of secretory storage granules, and an *endoplasmic reticulum* (ER). The half-life of platelets is from 8 to 12 days. Most are destroyed by macrophages residing in the spleen [Guyton 1991].

Platelet activating factor (PAF) is synthesized by platelets, mast cells, macrophages, eosinophils, certain renal cells, and vascular endothelial cells. The 2-D PAF molecule is shown in Figure 7.12. PAF is a pharmacologically active autacoid with many diverse functions [Hardman & Limbird 1996]. It causes vasodilation, and it is over 1,000 times more effective than histamine or bradykinin in promoting edema. PAF promotes platelet aggregation in vitro and in vivo and is a chemotactic factor for eosinophils, neutrophils, and monocytes, causing these leucocytes to aggregate at the source of its release. PAF also causes the contraction of smooth muscles in the GI tract, the small airways, and the uterus. PAF certainly contributes to the edema that occurs when the immune system fights a bacterial infection introduced through the skin.

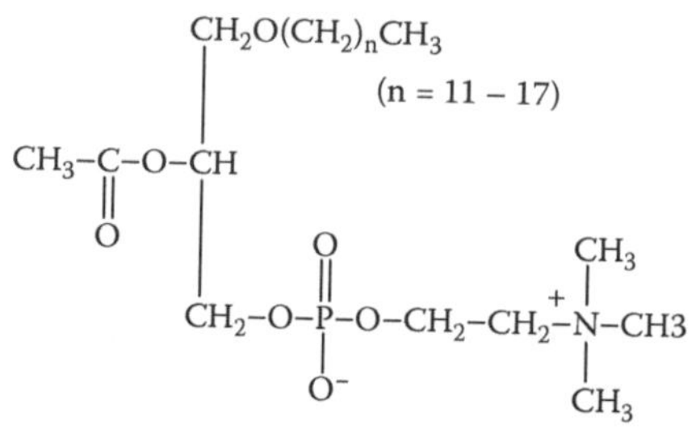

```
              CH2O(CH2)nCH3
              |            (n = 11 – 17)
              |
    CH3–C–O–CH
        ||    |
        O     |      O                  CH3
              |      ||               + |
              CH2–O–P–O–CH2–CH2–N–CH3
                     |                  |
                     O-                 CH3
```

FIGURE 7.12 2-D structure of the platelet activating factor (PAF) molecule. (Adapted from Northrop, R.B. 2000. *Endogenous and Exogenous Regulation and Control of Physiological Systems.* Chapman and Hall/CRC Press, Boca Raton, FL. Used with permission from Taylor & Francis.)

7.2.6.3 Complement

This is another remarkable, noncellular, concatenated, protein-based biochemical system employed by the immune system to fight infection. The *complement system* (C-system) consists of more than 30 large glycoproteins, most of which are soluble in blood plasma; some are bound to the surfaces of certain types of cells. When the C-system is activated, many of its component molecules are broken down and recombine with other complement molecules to form proteolytic enzymes, which further break down other complement molecules in a chain reaction. Most of the complement glycoproteins are synthesized by liver cells; however, macrophages, fibroblasts, and epithelial cells may also produce complement components [Hardman & Limbird 1996]. The organization of the complement system appears extravagantly complex; it works, however. Its complexity bespeaks the "trial and-error" processes that must operate in molecular biological evolution.

The components of the C-system have four major roles in the immune system:

1. Certain C molecules (C3b and C4b) are *opsonins;* i.e., they opsonize or coat foreign objects bound to Abs. Once opsonized, the complex of foreign object, Abs, and opsonins can be phagocytosed by cells having CR3 and CR4 opsonin receptors and then be lysed and destroyed. Portions of the foreign object can then be presented on an MHC molecule in "formal" antigen presentation.
2. Certain C molecules (C5a, C4a, and C3a) promote inflammation, increase vascular permeability, cause edema, and recruit phageocytic cells (neutrophils and macrophages) and activate platelets that secrete PAF, histamine, etc.
3. The C-system can lyse cells, similar to NK or CTL cells. In this convoluted scenario, C5b binds to the target cell surface, and thence to C6, C7, and C8. C7 and C8 undergo conformational changes that expose hydrophobic domains which penetrate the lipid bilayer of the cell's membrane. The C5b678 complex catalyzes the polymerization of the final C9 component, which, similar to perforin, makes a permanent 10-nm-diameter hole in the membrane, causing cell lysis from osmotic shock. The entire C5b6789 molecular assembly is known as the *membrane attack complex* (MAC).
4. The C-system participates in immune complex clearance. That is, the removal of groups of cross-linked Abs bound to Ags. This removal is accomplished by binding C3b and C4b covalently to the immune complex. CR1 receptors on erythrocytes (red blood cells) bind to the C3bC4bAbAg complex. Red cells carry these complexes to the spleen and liver, where they are destroyed by phagocytic cells [Mellors 1998].

There are two pathways by which C-system activation is initiated or triggered: the *classical pathway*, which begins with AgAb complexes, and the *alternative pathway*, in which activation begins at the pathogen surface. Glycoproteins in the classical pathway are prefixed with "C-."

One of the reasons the C-system is difficult to understand is that many of its proteins are cleaved (divided) to become active component molecules. In the case of C1, it is formed from one unit of C1q, two units of C1r, and two units of C1s. C1q has six binding sites for IgFc. The binding affinity is low, so that at least two bound sites are required on C1q for C1r activation. This means that two adjacent IgGs or one distorted IgM is required. Thus, random, unbound soluble IgG or IgM will not activate complement; they have to be clumped around an antigen. When activated, C1r cleaves C1s to form the *C1s protease,* which in turn cleaves C4 $\rightarrow$ C4a + C4b, and C2 $\rightarrow$ C2a + C2b. C4b combines with C2a to form the *C3- convertase,* C4bC2a, which splits C3 $\rightarrow$ C3a + C3b. Now, at the risk of really straining the reader's patience, we note that the complex C4aC2bC3b is formed and acts as *C5 convertase,* producing C5a and C5b. The complex C5b6789 is the MAC attack complex that causes cell lysis. Details of the steps in the classical and alternate pathways and their regulation are too complex to describe here; the interested reader should consult an online text such as Kimball [2005h].

Complement is clearly important in the humoral response to infection, in so far that it works with antibodies to lyse cells and promotes inflammation. The fact that dormant complement molecules are always present in the blood means that the complement response needs little time to generate a local defense against an introduced pathogen. Protein fragments from lysed pathogens can be phagocytosed and presented as antigens to Th cells by B-cells and macrophages, thus amplifying the classical responses of the immune system.

Because the activated complement proteins can cause extensive collateral damage to "good" cells, the inflammatory action of the C-system is tightly regulated in order to keep it localized. Regulatory actions are known to act at three points [Mellors 1998]:

1. A C1 inhibitor protein (C1INH) binds to free C1 and inhibits its spontaneous activation. C1INH is released upon activation by immune complexes (AgAb). It also limits the activation of C4 by inhibiting the C1r and C1s proteases.
2. *Decay accelerating factor* (DAF) found on cell surfaces, C4 binding protein (C4bp), and *Factor H* all act to speed the breakdown of C3-convertases, breaking the chain of reactions leading to MAC.
3. Two proteins found on cell surfaces, CD59 and *Homologous Restriction Factor* (HRF), inhibit the binding of C9 to C5b678 to form MAC. The factors controlling these inhibitors are not known. Because of its complexity, the complement system has never been included in mathematical models of immune system function.

7.2.6.4 Antigen Presentation

This is a complex biochemical process in which the immune system activates itself to fight specific invading microorganisms, including viruses, bacteria, parasites, molds, etc.

Antigens (Ags) are, by definition, molecules that elicit an *immune response* in the body. Ags are generally proteins and polysaccharides. Nonprotein molecules called *haptens* can also cause an immune response. This response is generally thought to be due to the hapten reacting chemically with certain tissue molecules; the product of the reaction is the actual antigen. Antigens can be of *exogenous* or *endogenous* origin. Exogenous antigens can be *inhaled* proteins from animal dander, pollen, or mold spores. Exogenous antigens can also be eaten; nuts and shellfish are common foods to which people are allergic. Another route for internalization of antigens is through the skin and mucous membranes. These can include injected antigens such as used in allergy desensitization, pollen injected by a thorn stick, bacteria entering through a cut or abrasion, parasites that are injected by insects or that burrow into the skin, and poison ivy oil. Proteins from the cell walls of internalized bacteria, viruses, spirochetes, parasitic worms, etc., can activate immune responses. *Lectin* proteins are involved in these responses.

Endogenous antigens include those that originate within the cells of the body. In the case of *autoimmunity,* a normal somatic cell protein is misidentified by some component of the immune system as a foreign invading protein. In other cases, cancer cells may have mutant proteins on their surfaces that the immune system can recognize and attack.

Antigen presentation is a crucial step in the amplification of specific immune responses. All antigen presentation involves the binding of an antigen molecule to a large protein molecule present on the antigen-presenting cell's surface. This molecule is called the *major histocompatibility complex* (MHC) molecule, or equivalently, the *human leucocyte antigen* (HLA) molecule. Part of the MHC molecule projects through the antigen-presenting cell's plasma membrane into its interior. There are two major types of MHC molecule, *Type I* and *Type II*. Nearly all nucleated cells in the body carry MHC I proteins and can present antigens derived from foreign internal proteins. The Type I MHC molecule has three major components:

1. A transmembrane protein that is exposed at the cell surface. The outermost portion is made from two alpha helices that form a groove between them.
2. A short peptide molecule attached in the groove between the alpha helices.
3. A β_2 microglobulin molecule also is attached to the α helices.

To make matters more complex (hey, it's life), humans have three subtypes of MHC I molecules, designated as *HLA-A, HLA-B*, and *HLA-C*. The genes coding the structures of these molecules are inherited, so if a person is heterozygous for HLA-A, HLA-B, and HLA-C, then we will observe six different MHC I proteins in such a person. MHC I proteins are found on all endogenous nucleated cell surfaces in the body other than macrophages, neutrophils, and certain B-lymphocytes.

The MHC I processing pathway begins inside the MHC I cell, where peptides from a virus' coating or a parasite that has entered the cell are broken up by proteolytic enzymes from proteasomes in the cytosol. The peptide fragments, between 8 and 18 amino acids, are then transported into the lumen of the rough endoplasmic reticulum (RER). Special transporter molecules move the peptide fragments inside the RER lumen, where ribosomes form the MHC I molecules around them. The assembled MHC I–peptide complex is then transferred to the cell's plasma membrane by *Golgi apparatus* [Complement 1998]. The MHC I–peptide complex is externalized through the plasma membrane; however, one transmembrane protein "root" projects back through the plasma membrane into the cell's interior. The three subunits in the externalized, active, MHC I molecule are (1) the "presented" antigenic peptide, (2) the β_2 microglobulin, and (3) the transmembrane polypeptide.

Ag presentation from cells carrying active MHC I molecules is made to leucocytes that carry the CD8 surface glycoprotein, as well as receptor proteins for the MHC I complex. CD8+ leucocytes are generally cytotoxic T-lymphocytes (CTL), although some regulatory T-cells also carry CD8. The T-cell receptor (TCR) protein for MHC I molecules and adjacent CD8 protein molecule protrude through the surface of the CTL. Antigen presentation is only possible if the TCR protein has an affinity for the presented antigenic peptide nestled in the MHC I cleft *and* the CD8 protein on the CTL binds to the MHC I side site. (See the schematic in Figure 7.9 for an illustration of this process.) Once the CD8 and TCR proteins have bound to the antigen-presenting cell's MHC I–peptide epitope complex, an internal, biochemical "message" is sent to the CTL, activating it to destroy the antigen-presenting cell. Destruction is necessary to prevent eclosion of reproduced virions or parasites from within the cell. The activated CTL either perforates the cell membrane of the infected cell with a special protein it secretes, or it sends chemical messengers to the infected cell that cause it to self-destruct by *apoptosis* (see Chapter 2, Section 2.5.2).

The affinity match between the presented antigen and a TCR protein is coincidental! That is, many billions of possible TCR configurations are randomly generated on CD8+ CTLs. The CTL receiving the message is also stimulated to divide, exactly reproducing a *clone* of CTLs with the TCR protein specific for the presented antigen. In this way the immune system greatly amplifies its ability to fight specific internal viral or parasitic invaders.

A second kind of antigen presentation is done by cells that carry the *Type II MHC protein*. The Type II MHC protein is present on the surface of mature macrophages and other phagocytic cells

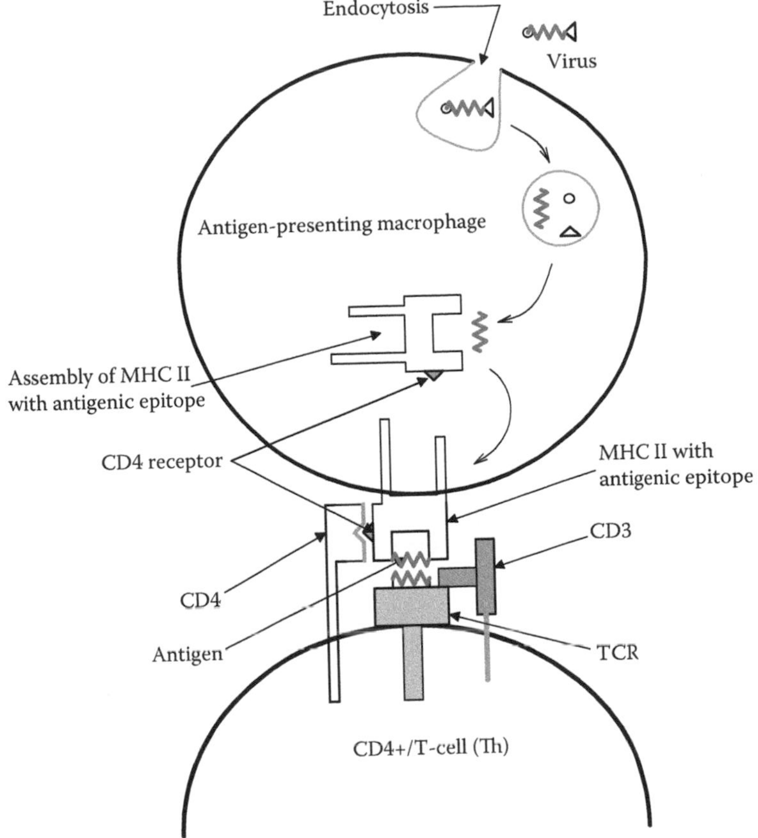

FIGURE 7.13 (See color insert.) Antigen presentation to a CD4+ helper T-cell by an MHC II+ macrophage. See text for further description of the process. (Adapted from Northrop, R.B. 2000. *Endogenous and Exogenous Regulation and Control of Physiological Systems.* Chapman and Hall/CRC Press, Boca Raton, FL. Used with permission from Taylor & Francis.)

such as neutrophils and certain B-lymphocytes (B-cells). These cells can engulf, phagocytose, or endocytose an entire bacterium or virus, as well as fragments of cell debris. Once internalized, the foreign material is broken down enzymatically to smaller peptide fragments that are taken up by MHC II molecules, which in turn migrate to the cell surface with the presented peptide fragment antigen (Ag), or epitope. When the antigen-presenting macrophage or neutrophil encounters a helper T-cell with a TCR with affinity to the presented Ag and the Th's CD4 protein binds to a specific site on the MHC II molecule, signals are passed to the Th to activate it to secrete various immunocytokines that activate the cellular arm of the immune system. In particular, *activated* helper T-cells (AThs) secrete IL2, 3, 4, 5, 6, 7, 10, 13, TNFβ, γIFN, and GMCSF. Antigen presentation by a macrophage to a Th cell is illustrated schematically in Figure 7.13.

The antigen-presenting B-cells have *B-cell receptor* (BCR) proteins bound on their outer surfaces. The BCRs are surface-bound IgG antibodies; they, too, have enormous variability in their affinities for epitopes. When a certain BCR has a strong affinity to an epitope, it binds to that epitope and is then internalized where the Ag protein is broken down, and pieces of it are bound to MHC II proteins, which are then externalized on the B-cell. If a helper T-cell (Th) has a T-cell receptor (TCR) protein with affinity to the peptide epitope nestled in the B-cell's MHC II protein, it binds to the antigen-presenting B-cell. A CD4 molecule on the Th must also bind to the side of the MHC II molecule to activate the Th, which secretes cytokines that cause the antigen-presenting B-cell to reproduce clonally with its BCRs that are specific for the Ag in question. This process is molecular "AND" logic; it is shown schematically in Figure 7.10. The clone of plasma cells grows identical BCRs with the affinity to the presented epitope. These BCRs are ultimately released as free antibodies (Abs). Because the original BCR bound to this epitope, the Abs released are also specific for it and contribute to an amplified humoral immune defense.

7.2.6.5 Autacoids: Immunocytokines, Proteins, and Glycoproteins Secreted by Immune System Cells

As you will see, under certain conditions other cells, as well as cells from the immune system, secrete cytokines that can affect immune system cells. Collectively, we call these intercellular messengers *autacoids,* a term coined by Hardman and Limbird [1996]. The word *autacoid* is from the Greek *autos* = "self" and akos = "medicinal agent or remedy." There are currently about 40 known immunocytokines (aka cytokines). The cytokines include the *interleukins* (about 28 are defined at the time of this writing, and more are being discovered every year). The immune system also uses *interferons* (3), certain *prostaglandins, tumor necrosis factors* (2), and various *cell growth stimulating factors.* In addition, there are certain *cluster of differentiation* (CD-x) *proteins* fixed on cell surfaces and many *cell surface receptor proteins.* Any one of these many and diverse humoral substances can collide with an immune system cell, where it can bind its specific surface receptor protein. Once bound, a *molecular transduction mechanism* causes the cell to internally synthesize the same or other immunocytokines, more of the same or other membrane receptors for immunocytokines, proteins used in antigen presentation (TCRs, MHC, CD3, CD4, CD8, etc.), antibodies (if the cell is a B-lymphocyte), complement components, perforin, etc., or the binding can lead to the inhibition or switching off of an ongoing biochemical synthesis by the target cell. Individual immunocytokines can posses the property of *pleiotropy,* in which a cytokine can have different effects on different target cells having receptors for it. Also, several different cytokines can have the same biological function on the same or diverse target cells.

Presumably, *suppressor T-cells* (Ts) synthesize immunocytokines (e.g., IL-10) that turn off or downregulate the cellular synthetic machinery in the immune system that was activated in the initial stages of the immune response.

It will be necessary to understand the kinetics of their production and their half-lives in vivo in order to incorporate immunocytokines into any quantitative, dynamic model of the immune network. Sadly, much of this detailed information is nonexistent or is based on in vitro studies.

We will now summarize and describe the immunocytokines in detail. The interleukins (ILs) are high-molecular-weight glycoprotein autacoids secreted by leukocytes for purposes of signaling other leukocytes and somatic cells. The ILs have diverse functions such as cell attraction (chemotaxis), inducing target cells to secrete other ILs and other immune system autacoids and to manufacture more receptors for ILs. Most ILs are stimulatory in function; there are a few, however, that inhibit or downregulate certain immune system functions.

There are about 28 interleukins at this writing (December 12, 2005). However, the ongoing research on the immune system reveals the identity of new cell surface proteins, receptors, and immunocytokines at a rate such that our knowledge base for the immune system is *never* in the steady state; new discoveries are made at an amazing rate. As of March 3, 2008, there are now 42 ILs. Thus, our knowledge about the interleukins, their origins, and functions is continually changing. The following descriptions summarize our current knowledge. Table 7.1 summarizes the known (and partially known) ILs. It gives their sources, their target cells, and their effects. To effectively signal or carry *information* to immune system cells, all interleukins (and in general, all cell signal substances, including hormones and neurotransmitters) must be eventually broken down to inactive forms and their amino acids recycled metabolically. In vivo IL half-lives are generally unknown but necessarily must be on the order of hours to days in order for the stable dynamic functioning of the immune system. A knowledge of IL half-lives is necessary to effectively formulate any kind of dynamic mathematical model of the immune system.

TABLE 7.1
Table of Human Interleukin Cytokines

Interleukin	Sources	Targets	Effects
IL1α, IL1β	Activated Mϕs:IL1α is bound to Mϕ surface, IL1β Both bind to same receptor Also, endothelial cells, B-cells, and fibroblasts	TLs, B, Mϕs, neutrophils, bone marrow cells, fat cells, bone osteoclasts, brain cells, adrenal cells, vascular endothelial cells, and smooth muscle cells	B-cell proliferation TL production of other cytokines, fever, and cachexia
IL2, aka T-cell growth factor (TGF)	Activated CD4$^+$ T-cells, CTL, and large granular lymphocytes (LGL)	IL2 receptors on Th, CTL, Mϕ, and B-cells	Proliferation of CD4$^+$Th cells; also CD8$^+$ CTL Stimulates production of IFNγ by TCs Stimulates production of NK and LAK cells, also Mϕ to secrete IL1 and TNFα Edema
IL3	Mast cells and activated CD4$^+$Th	IL3 receptors on mast, B-cell, and some mature granulocytes	Hematopoietic growth factor, aka multicolony-stimulating factor, and mast cell GF Stimulates B-cell differentiation Inhibits LAK cell activity
IL4	15- to 19-kDa glycosylated protein dimmer CD4$^+$ Th, CD8$^+$ memory T-cells, mast cells, and basophils	T-cells, B-cells, Mϕ, fibroblasts, and endothelial cells	Induces CD4$^+$ Th to differentiate into Th2 cells Suppresses development of Th1 cells Acts as growth factor for B-, T-, and mast cells Stimulates MHC2 expression on B-cells Promotes plasma B-cells to make IgE1 and IgE Abs
IL5	45-kDa protein dimmer CD4$^+$Th2, NK, and mast cells	Eosinophils, B-cells, and mast cells	Activation and differentiation of eosinophils Proliferation of immature B-cells Stimulates Ig class switching to IgA Stimulates mast cells
IL6, aka IFNβ2 and other aliases	212-AA glycoprotein, 26 kDa, made by: Mϕ, mast cells, T-cells, fibroblasts, and neutrophils	Stimulated B-cells, Mϕ, hepatocytes, CD4$^+$ and CD8$^+$ Th cells, and fibroblasts	Stimulates acute-phase protein synthesis by liver B-cell GF: induces B maturation into Ab-secreting plasma cells T-cell activating and differentiation and stimulates their production of IL2 and IL2Rs Inhibits production of TNF
IL7, aka T-cell GF	Glycoprotein, 25 kDa Bone marrow cells, thymus stromal cells	B- and T-cells	Stimulates development of pre-B- and T-cells and early thymocytes

Continued

TABLE 7.1 (Continued)
Table of Human Interleukin Cytokines

Interleukin	Sources	Targets	Effects
IL8	Activated T-cells, Mϕ, and lymphocytes	Mϕ, neutrophils, and basophils	Chemotactic substance Stimulates production of inflammatory leukotrienes
IL9	CD4+Th2 cells and some B-cells	CD4+Th cells, CD8+ T-cells, B-cells, and bone marrow precursor cells	Inhibits lymphokine production by CD4+Th cells producing IFNγ Stimulates growth of CB8+ T-cells Promotes production of immunoglobulins by B-cells and proliferation of mast cells
IL10, aka cytokine synthesis inhibitory factor	18-kDa glycoprotein made by CD4+ T-cells, activated CD8+ T-cells, and activated B-cells, Ts cells	T-cells, NK cells, Mϕs, and CTLs	Inhibits IFNγ production by activated T-cells Inhibits IL2-induced IFNγ production by NK cells Inhibits IL4- and IFNγ-induced MHCII expression on Mϕs Reduces CTL proliferation
IL11	26-kDa protein produced by bone marrow stromal cells, Mϕs, and some fibroblasts	Stimulated B-cells, Mϕ, hepatocytes, CD4+ and CD8+ Th cells, and fibroblasts	Mimics activity of IL6 Stimulates Th-dependent, B-cell immunoglobulin secretion Increases platelet production Induces IL6 expression by CD4+ T-cells
IL12: NK stimulatory factor (NKSF)	Disulfide-linked dimer; 40 and 35 kDa ea From activated B-cells and antigen-presenting cells (Mϕ and B)	Th1, CTL, NK, T-, and B-cells	Activates Th1 cells Increases rate of production of CTL, NK, and lymphokine-activated killer (LAK) cells Increases NK cytotoxicity, induction of IFNγ production by NK and T-cells Inhibition of IgE synthesis by B-cells
IL13P600	12- to 17-kDa protein made by activated Th cells	Mϕ and B-cells	Inhibits Mϕ activation and release of IL1β, TNFα, IL6, and IL8 Enhances Mϕ and B-cell differentiation and proliferation Increases CD23 expression Induces IgG4 and IgE class switching
IL14	53-kDa protein from T-cells	B-cells	Induces B-cell proliferation but inhibits their differentiation into plasma B-cells

TABLE 7.1 (Continued)
Table of Human Interleukin Cytokines

Interleukin	Sources	Targets	Effects
IL15	114-AA protein of 15 kDa From activated Mϕ, epithelial cells, and fibroblasts	NK, B-, and T-cells	Stimulates NK, B-, and T-cells Stimulates CTK and LAK cell activity and Ig production An attractant for T-cells (actions similar to IL2)
IL16: Lymphotactin LCF	Four linked, homotetrameric chains 56 kDa Made in $CD8^{+}T$ and eosinophils Released in response to histamine and serotonin	All cells carrying CD4	Attracts $CD4^{+}$ T-cells, eosinophils, and some Mϕ CD4 is receptor for IL16 Increases MHCII expression on monocytes Inhibits HIV replication in monkey and human $CD4^{+}$ T-cells
IL17:	17.5-kDa protein made by activated $CD4^{+}$ T-cells	Certain immune system cells	Induces the release of IL6, IL8, G-CSF, and PGE_2 (inflammatory cytokines)
IL18: IGIF	Mϕ	Th1-cells	Induces IFNγ production
IL19, aka MDA1	IL10 homolog Distinct population of keratinocytes Monocytes	Binds with IL20 receptors on $CD4^{+}$ T-cells	Upregulates IL4 expression in $CD4^{+}$ T-cells Induces IL10, which downregulates IL19
IL20, aka IL10D	20.1 kDa, 176 AAs IL10 homolog Distinct population of keratinocytes Monocytes	Distinct poulation of keratinocytes	Induces hyperproliferation of keratinocytes, which express IFNα
IL21	18.6-kDa T-cells	CTL, NK, and B-cells	Stimulates CTL and NK cells Suppresses growth of metastatic melanoma Augments theraputic antibody-mediated tumor lysis
IL22: IL-TIF	20 kDa IL10 homolog Activated Th1 cells, leukocytes, hematopoietic stem cells	Pancreatic acinar cells Intestinal epithelial cells	Upregulates MHC1 Ag expression No IL22R on activated B-cells
IL23A, aka IL23, P19	20.7-kDa heterodimer, IL10 homolog	Expressed mainly by dermal cells	Promotes proliferation of naïve and memory T-cells and stimulates their IFNγ production
IL24, aka MDA7	23.8-kDa IL10 homolog Lymphocytes, monocytes, Mϕ, leucocytes, melanocytes, thymus, spleen, skin, and mammary gland	Tumor cells	Selective tumor apoptosis Induces dose-dependent death in melanoma cells
IL25	Not assigned		

Continued

TABLE 7.1 (Continued)
Table of Human Interleukin Cytokines

Interleukin	Sources	Targets	Effects
IL26, aka AK155	19.8 kDa Heterodimer IL10 homolog T-lymphoblasts Activated NK Memory CD4+ T-cells		
IL27	Not assigned		
IL28A: IFNλ2	22.3 kDa Monocytes	Intestinal epithelial cells	Increased IL8 production, increased expression of antiviral proteins: myxovirus resistance A and 2′,5′-oligoadenylate synthetase No influence on Fas-induced apoptosis Decreases cell proliferation
IL28B: IFNλ3	21.7 kDa Monocytes	Intestinal epithelial cells	Increased IL8 production Increased expression of antiviral proteins: myxovirus resistance A and 2′,5′
IL29: IFNλ1	21.9 kDa Monocytes	Intestinal epithelial cells	Increased IL8 production Increased expression of antiviral proteins: myxovirus resistance A and 2′,5′
IL30, aka IL27, p28	28-kDa 4-helix bundle Antigen-presenting cells	CD4+ T-cells	Expansion of antigen-specific naïve CD4+ T-cells Promotes polarization toward TH1 phenotype with expression of IFNγ

Interferons are an important class of immune system protein. Three types of *interferon* molecules are currently known: interferon alpha (IFNα), IFNβ, and IFNγ. IFNα is thought to have 15 subtypes, which presumably act alike, so we can deal with generic IFNα. IFNα is made by virus-infected monocytes and lymphocytes, IFNβ is from virus-infected fibroblasts, and activated T- and NK cells release IFNγ, which has three different isoforms [Held et al. 1999].

All three IFNs engage in antiviral activity; they increase NK and CTL cell activity and increase MHC I expression on cells. Specifically, IFNα and IFNβ activate NK and CTLs, stimulate B-cell differentiation into plasma cells (which release soluble antibodies), induce MHC I *protein kinase*, and cause other internal biochemical events that inhibit viral messenger RNA translation, preventing viral multiplication in the target cell. IFNα is the first cytokine to be used effectively in clinical trials as an exogenous immunotherapy agent; it has been approved for the treatment of several types of human cancer. It has been demonstrated that IFNα has a direct antitumor action: It downregulates oncoproteins, induces tumor suppressor genes, has an antagonistic effect against the action of growth factors, induces cell suicide (apoptosis), inhibits angiogenesis (growth of new blood vessels that nourish the tumor), and also as increase MHC I production and tumor-associated antigens [Northrop 2000].

Because they cause immune suppression, exogenous IFNβ1a and IFNβ1b have been successfully used to treat the autoimmune disease *multiple sclerosis* (MS). Both interferons slow the progress of disability, reduce the rate of relapses, and reduce the severity and number of MRI-detected lesions. They do not cure MS, however.

IFNγ increases expression of MHC II molecules and receptors for the Fc region of antibodies (FcγR) on macrophages, increases the activity of neutrophils and NK cells, promotes T- and B-cell differentiation, and increases IL1 and IL2 synthesis. In addition, IFNγ increases type IgG2a antibodies while suppressing IgE, IgG_1, IgG_{2b}, and IgG_3 synthesis. IFNγ also inhibits proliferation of Th2 cells but not Th1 cells.

One limitation to IFN therapies is that exogenously administered IFNs have the risk of losing their effectiveness owing to the production of host antibodies [Northrop 2000]. It is ironic that a potentially effective immunotherapy may be thwarted by the host's immune system.

Two ***tumor necrosis factor proteins*** (TNFs) are known: TNFα and TNFβ. The principal effect of the TNFs is to cause inflammation. TNFα, or *cachectin,* is a 17-kDa soluble protein trimer. It is a 185 AA glycoprotein hormone cleaved from a 212 AA peptide on the outer surface of macrophages. It is the product of activated macrophages, fibroblasts, mast cells, and some T- and NK cells. Peritoneal mast cells have preformed reserve TNFα available for immediate release upon appropriate stimulation. TNFα can induce fever, either by stimulating the release of pyrogenic prostaglandins (PGs) or by causing the release of IL1, which is also pyrogenic. TNFα plays a major positive role in fighting local infections. It induces the production of acute-phase proteins, mobilizes neutrophils, activates T- and B-cells, increases the release of antibodies and complement, increases the adhesion of platelets to blood vessel walls, and increases the extravasation of lymphocytes and macrophages *(diapedesis)* to fight the infection in intracellular space. These actions result in the phagocytosis of the pathogens, local vessel occlusion, and the drainage of cells, debris, and fluid into the lymphatic system. These actions, carried out with other relevant cytokines, lead to the removal of the infecting pathogen and eventual tissue repair. TNFα also can bind to TNF receptors on tumor cells and kill them. Some tumor cells shed their TNF receptors, which become soluble and bind to TNFα, inactivating it. Excess TNFα can also cause the body to create antibodies to it, yielding a similar deactivation. The exact mechanism of infected cell killing by TNFα is not known.

High levels of TNFα cause pain, systemic edema, hyperproteinemia, and neutropenia. TNFα binds to TNF receptors on virus-infected cells, causing apoptosis and cell death. Acute infections can lead to the overproduction of TNFα, leading to fever, immune suppression, septic shock due to the loss of blood volume to extracellular space, fatigue, anorexia, and cachexia (wasting of the body tissues).

TNFβ, or *lymphotoxin,* is made by activated $CD4^+$ and $CD8^+$ T-cells. It binds to the same receptor sites as TNFα. It has similar properties as TNFα and induces apoptosis in many types of virally infected cells, tumor cells, and damaged cells. Certain endotoxins, such as those from staphococcus bacteria, cause high production of TNFs that contribute to the onset of *toxic shock syndrome.* Chronic high production of TNFs may be responsible for the cachexia observed in many chronic parasitic infections and some cancers.

The ***prostaglandins*** and other *eicosanoids* are part of a family of autacoids that includes the *leukotrienes, thromboxanes,* and *prostacyclin.* Collectively, they are called *eicosanoids* because they are synthesized from 20-carbon fatty acids that are in turn derived from the 20-carbon *arachidonic acid* (see Figure 7.14). The metabolites of arachidonic acid are varied and have diverse pharmacological effects. Subtle changes in eicosanoid structure can produce dramatic alterations in their bioactivity. Receptors for eicosanoids are highly specific for affinity and effect.

Two major biosynthetic pathways exist for the products of arachidonic acid: The prostaglandins (PGs), prostacyclin, and thromboxanes are made through the *cyclooxygenase* synthetic route. The second major pathway of arachidonic acid metabolism begins with *lipoxygenase.* This pathway leads to the biosynthesis of various leukotrienes. These pathways are shown schematically in Figure 7.15.

In discussing the immune system's use of eicosanoids as effectors and signaling substances, we will restrict ourselves to consideration of those autacoids considered by the authors as having a *direct effect* on the functioning of the immune system. First, it is relevant to remark that the

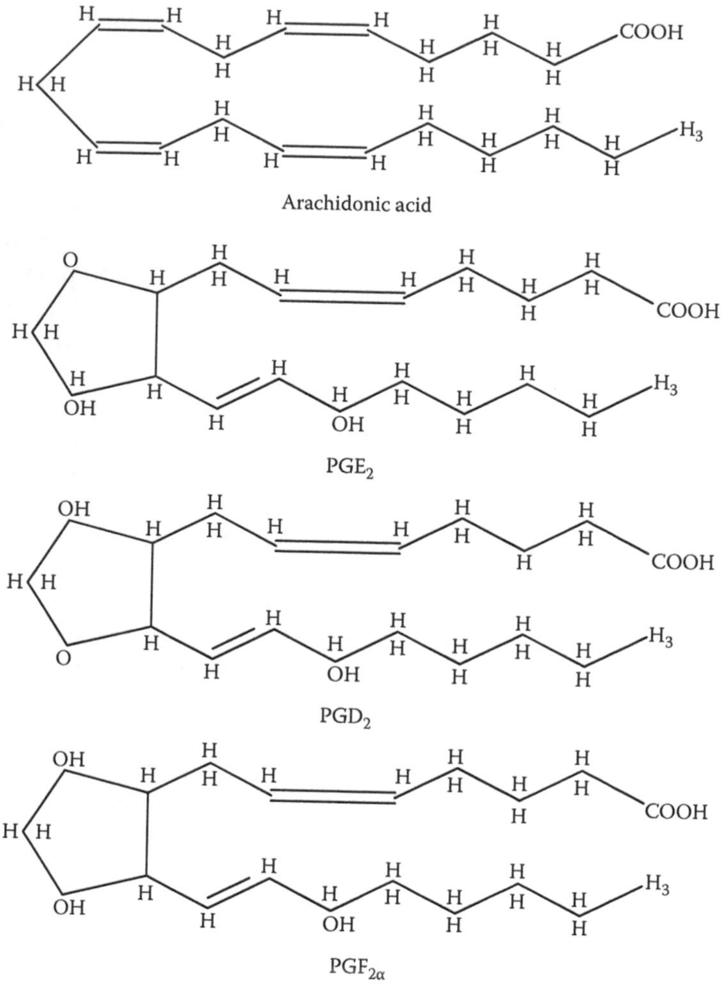

Arachidonic acid

PGE_2

PGD_2

$PGF_{2\alpha}$

FIGURE 7.14 The arachidonic acid molecule and three prostaglandin molecules derived from it. (Adapted from Northrop, R.B. 2000. *Endogenous and Exogenous Regulation and Control of Physiological Systems.* Chapman and Hall/CRC Press, Boca Raton, FL. Used with permission from Taylor & Francis.)

eicosanoid immunocytokines are short-lived. They are chemically unstable and are rapidly broken down in the body (mostly in the pulmonary circulation) to inactive forms. Their half-lives are thus on the order of minutes, and they must be continually synthesized to have effective concentrations.

PGE_2 is formed by macrophages and granulocytes; PGD_2 comes from mast cells. Both of these PGs are formed through the cyclooxygenase pathway, followed by an arcane series of biochemical reactions. PGE_2 causes the relaxation of smooth muscle, both in blood vessels and in the lungs; thus, it acts as a bronchodilator when given as an exogenous aerosol. PGE_2 and PGI_2 enhance edema formation and the infiltration of leukocytes into the site of infection/inflammation, and they potentiate the production of pain by bradykinin [Hardman & Limbird 1996].

Most importantly, PGE_2 also acts as an immune response suppressor substance. Both PGE_2 and PGD_2 decrease the rate of release of *proteases* from stimulated granulocytes, thus limiting the inflammatory reaction. In addition, PGE_2 inhibits histamine release from basophils and generally inhibits the release of activating immunocytokines from sensitized T-cells and B-cells. It also inhibits CTLs. PGE_2 simulates the release of the hormone ACTH from the anterior pituitary gland. ACTH in turn stimulates the formation of certain adrenocortical hormones, notably *cortisol.*

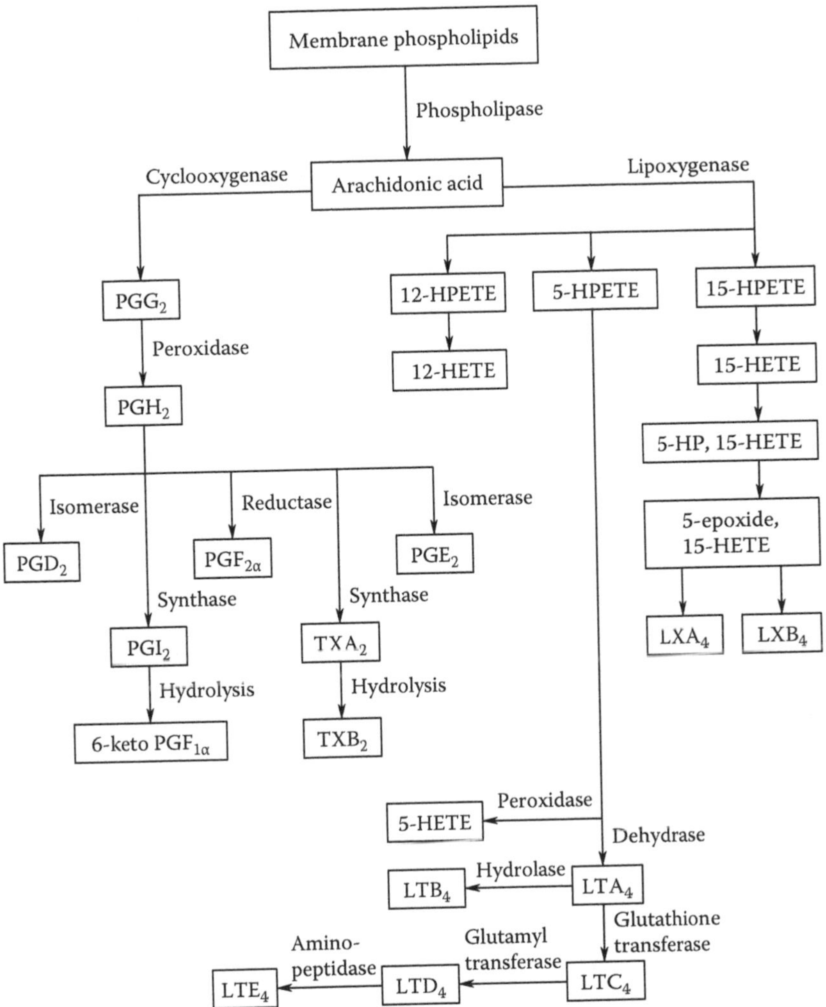

FIGURE 7.15 Flowchart describing the synthesis of prostaglandins (PGs), thromboxanes (TXs), and various leucotrienes. (Adapted from Northrop, R.B. 2000. *Endogenous and Exogenous Regulation and Control of Physiological Systems*. Chapman and Hall/CRC Press, Boca Raton, FL. Used with permission from Taylor & Francis.)

Cortisol is a potent anti-inflammatory hormone that has the following effects on the inflammatory immune response [Guyton 1991]:

1. It lowers fever, mostly by suppressing the release if IL1 by macrophages.
2. It suppresses the reproduction of T- and B-cells, reducing the number of circulating Abs.
3. It reduces the chemotaxis of macrophages, NKs, and CTLs into the inflamed area, thus reducing cell death and the release of inflammation-producing autacoids.
4. It decreases capillary permeability and, hence, edema.
5. It stabilizes and strengthens lysosomal membranes so that when cells are ruptured, there is reduced spread of proteolytic enzymes, causing reduced collateral damage to adjoining cells.

It is well known that aspirin (acetlysalicylate) acts as an antipyretic (fever reducer) and anti-inflammatory substance. Aspirin inhibits the function of the *cyclooxygenase enzymes* responsible for the conversion of arachidonic acid to the various prostaglandins. In particular, certain cytokines (IL1β, IL6, IFNα, IFNβ, and TNFα) induce PGE_2 production by the *circumventricular organs* near the hypothalamic area in the brain. PGE_2 triggers the hypothalamus to cause the body temperature to rise by readjusting the body's heat balance. Aspirin and other nonsteroidal anti-inflammatory drugs (NSAIDs) block the production of PGE_2 and hence lower fever.

We know that chronically elevated cortisol levels (as in the case of prolonged periods of stress) can lead to blood sugar imbalances [Rizza et al. 1982], decreased bone density [Merck 2007], high blood pressure [Whitworth et al. 2000], and the accumulation of abdominal fat [Epel et al. 2000]. Because NSAIDs block the production of PGE_2 and PGE_2 is a precursor of cortisol, could NSAIDs be a potential therapy for persons suffering from chronic stress?

7.2.6.6 Discussion

In this section we have summarized the immune system's cells and the more important cytokine and autacoid proteins having strong signaling or effector actions in the immune network. Clearly, most of these immunocytokines are stimulators or activators that increase the degree of immune response at a number of levels. Very few immunocytokines have definite known inhibitory actions in the immune network; these include IL10 and PGE_2. This apparent disparity between the number of activators and suppressors challenges our understanding of how the overall immune response works. All immunocytokines have finite lifetimes in vivo; they are broken down by enzymes so that their effects as immune system activators naturally decay in time. Thus, if the stimulus (e.g., a bacterial endotoxin) for immunocytokine production disappears because of successful combat by the immune system, immunocytokine production decreases and their titer is reduced enzymatically, leaving the immune network in a resting state.

Another possible natural scenario to explain why there are so few suppressor immunocytokines may lie in the concept of *anti-idiotypic antibodies.* Vigorous production of Abs for a particular antigen, **X**, leads to the demise of the pathogen associated with it. An antigen-presenting phagocytic B-cell, *by chance,* has a surface antibody with an affinity to the Abx's **X** binding site. That is, the particular, unique, IgG on the B-cell's surface has a site that looks like **X** to the Abx. The bound Abx is phagocytosed by the B-cell, broken down enzymatically, and then the **X** binding site is combined with MHC II and externalized to the B-cell's surface, where it is presented to a helper T-cell (Th). The Th is stimulated by the process of antigen presentation to release immunocytokines that stimulate clonal expansion of the antigen-presenting B-cell with its particular IgG with affinity to the Abx's. Eventually, the B-cell daughters change to plasma cells and release *soluble* Abx with affinity to Abx. These Abxs combine with the binding sites on the Abxs, inactivating them, thus damping the humoral immune response and inflammation, etc.

One can speculate that if there are enough anti-idiotypic Abxs in circulation, anti-anti-idiotypic antibodies might be made to suppress the Abxs.

In summary, the complexity of the human immune network is due to the great number of participating cells, the many immunocytokines, their often *pleiotropic* actions, and the variable causal interconnections between the IS's components. Cytokines provide the signals that lead to immune system activation or suppression. The fact that many cytokines exhibit pleiotropy makes our understanding of the mechanisms of immunoregulation more difficult. In addition to the interleukins, certain immune system cells and other cells secrete other regulatory molecules such as interferons and tumor necrosis factors—all proteins—and the eicosanoid prostaglandins (not proteins). As you have just seen from the foregoing sections, the complexity of the immune system is enormous. Its functional defects (e.g., allergies, anaphylaxis, and autoimmune diseases) are generally treated by "sledge-hammer pharmacology"—drugs that massively suppress the overall immune responses and inflammation. Only recently have drugs such as Singulaire® appeared, which selectively block leukotriene receptors to lower the inflammatory response to allergens.

TABLE 7.2
Summary of Lectin Categories and their Biological Functions

Lectin Group	Target Ligands	Biological Lectin Functions
Calnexin	Glc1, Man9	Protein sorting in the endoplamic reticulum (ER)
M-type lectins	Man8	ER-associated degradation of glycoproteins
L-type lectins	Various	Protein sorting in the ER
P-type lectins	Mannose-6-phosphate	Protein sorting post-Golgi
C-type lectins	Various	Cell adhesion (Selectins) Glycoprotein clearance Innate immunity (Collectins)
Gallectins	β-Galactosides	Glycan crosslinking in the extracellular matrix
I-type lectins	Sialic acid	Cell adhesion
R-type lectins	Various	Enzyme targeting Glycoprotein hormone turnover

7.2.7 Lectins

Lectins are a group of nonenzymatic proteins or glycoproteins that bind specifically to carbohydrates. All lectins have a specific AA carbohydrate recognition domain (CRD). They have been found in prokaryotic and eukaryotic cells (plants and animals) and in viruses in both soluble and cell-associated forms and exhibit a wide range of functions. Lectin structure is generally multimeric, consisting of noncovalently associated polypeptide subunits. A lectin may have two or more of the same subunits, as well as different subunits.

The term *lectin* was originally coined to define *agglutinins* that could discriminate among types of red blood cells. Currently, the term has expanded to include proteins that bind to sugar subunits in carbohydrates (CHOs). Most lectins can *agglutinate* some kinds of cells, forming precipitates with glycoconjugates in a manner similar to antibody–antigen reactions. As nature is full of exceptions, there are lectins that bind to cell surfaces and do not cause agglutination. It is known that some lectins will bind only to CHO molecules with glucose or mannose residues, whereas others are specific for CHOs with galactose residues. Some lectins will bind only when the particular sugar is in a terminal, nonreducing position in an oligosaccharide, whereas others can bind to sugars within the oligosaccharide molecule. The affinity between a lectin molecule and its receptor CHO may vary a lot owing to small changes in the CHO structure. Because nearly all cell membranes and walls contain some kind of glycoconjugate, the specificity of lectin binding has led to the use of synthetic lectins to isolate one glycoconjugate from a mixture of cells, bacteria, or virions [Calbiochem 2005].

In Table 7.2, we summarize lectin categories, their ligands, and functions [Drickamer 2005].

Two extremely potent plant toxins contain lectins: *ricin* from castor beans and *abrin* from the rosary pea. Davis [2000] has observed:

> These related proteins [abrin and ricin] can act as both galactose-binding lectins and base-specific RNA-*N*-glycosidase enzymes. As enzymes, they hydrolyse a specific purine base from the sugar [ribose] backbone of the large RNA subunit of the ribosome (the site of protein synthesis in a cell). This single modification inactivates the ribosome, stopping the synthesis of cellular proteins and ultimately causing cell death [by apoptosis]. This has earned this family of proteins the apt acronym, RIPs—ribosome inactivating proteins.
>
> The toxin enters the [target] cell by binding to sugars on its surface. This is followed by a process of internalization and release into the cell's cytoplasm where ribosomes are then inactivated. Just one internalized RIP molecule is needed to kill a cell [!]. The lectin subunits of RIP's account for their

extreme cytotoxicity (three orders of magnitude more potent than cyanide) – proteins which only have RNA-*N*-glycosidase activity are, in general, much less toxic than those carrying a second lectin subunit. [Quotations used with permission.]

The LD50 dose of a poison or toxin is that weight of poison/kg of body weight that is lethal to one-half the animals to whom it is administered. The LD50 depends on the animal species (e.g., mice) and the route the poison is administered (e.g., oral, inhaled, or IV). The mouse LD50 cited for abrin is 20 μg/kg [NIOSH 1995]. The Environmental Health and Safety Office at the University of Florida lists the LD50 doses of abrin and ricin at 0.7 μg/kg and 2.7 μg/kg, respectively. Assuming a 68-kg human, this would mean that a dose of about 48 μg of abrin would either kill or make the victim horribly ill. (Note that there are more toxic poisons. For example, the LD50 for tetanus toxin is 1 ng/kg, and the botulinin toxins range from 1.2 to 2.5 ng/kg [also UFL data]!)

The mammalian immune system has several other means of taking on invading bacteria, including many different *Toll-like receptor proteins* expressed on the surfaces of macrophages that bind to specific lipoproteins on the surfaces of pathogens.

Selectins are a group of membrane-attached lectin proteins that include *L-*, *E-*, and *P-selectin.* A carbohydrate recognition domain (CRD) is attached to the end of a flexible polypeptide "stalk." The stalk is anchored to the cell surface by a single transmembrane helix, which has a small C-terminal cytosolic domain. The endothelial cells of venules are stimulated by certain immune system cytokines to express P- and E-selectins. These selectins are involved in recruiting immune system macrophages and leukocytes into sites of inflammation; they determine the endothelial cells to which lymphocytes will adhere and fight infection. L-selectin is the smallest of the three vascular lectins and can be found on most leukocytes [Ehrhardt et al. 2004].

Collectins belong to the superfamily of mammalian *C-lectins* and are believed to participate in innate immunity. There have been eight collectins identified [van de Wetering et al. 2004]. *Mannose-binding lectin* (MBL), *surfactant proteins A and D, collectin liver-1* (CL-L1), *collectin placenta-1* (CL-P1), *conglutinin,* and *collectin of 46 kDa* (CL-46).

As components of the *innate immune system,* collectins play a key role in the first line of defense against invading microorganisms. The *lectin-complement pathway* has the ability to recognize the cell surfaces of bacteria and bind to them, ignoring the host's cells. Mannose-binding lectin (MBL) recognizes bacterial cell wall mannose residues and binds to them; it does not recognize D-galactose and sialic acid, which are abundant on mammalian cells. Binding of MBL to microbial carbohydrates activates members of the MBL-associated serine protease (MASP) enzyme family on the bacterial cell surface. MASPs, which form complexes with MBL, are enzymes that cleave immune system *complement* protein components and activate the complement pathway. Certain complement molecules (C_{5b6789}) then destroy the lectin-marked bacteria [Guyton 1991].

We have just scratched the surface in this discussion of the lectin family of proteins. They have numerous roles in cellular housekeeping, and in animals they figure as important components of innate immunity and are also involved in postribosomal protein sorting.

7.2.8 Glycoproteins

A glycoprotein is a polypeptide to which sugars or oligosaccharides are attached. When they attach to the amino group ($-NH_2$) on the γ carbon of asparagine, it is called *N-glycosylation.* Carbohydrates can also bond to the –OH group on the β carbon of serine and the β carbon of threonine; this is called *O-glycosylation.* The oligosaccharides can be composed from the sugars glucose, glucosamine, galactose, galactosamine, mannose, fucose, and sialic acid. Important glycoproteins include mammalian *antibodies* and *immunoglobulins*; molecules of the *major histocompatibility complex* (MHC); components of the *zona pellucida,* which surrounds an oocyte (important for sperm-egg interaction); and *glycophorin-A,* a transmembrane protein found on erythrocytes (RBCs). Each RBC has about 500,000 copies of glycophorin-A in its plasma membrane. The positions of the carbohydrate chains on the glycophorin-A molecule are genetically determined and determine whether a person has blood group M, N, or MN [Kimball 2005j]. The most common blood groups, A, B, AB, and O, are specified by the activity of specific gene products whose activities are to incorporate distinct sugar groups onto RBC membrane glycosphyngolipids, as well as secreted glycoproteins [King 2005c].

7.2.9 Blood Proteins

7.2.9.1 Introduction

Vertebrates have the advantage of a closed circulatory system in which blood efficiently transports oxygen to metabolizing tissues, hormones to target cells, immune cells to sites of infection and inflammation, CO_2 to the lungs, and molecular waste products to the liver and kidneys for further processing or elimination. In the following subsections, we discuss a few of the more important proteins associated with the blood.

7.2.9.2 Hemoglobin

One of the more important proteins in vertebrates is *hemoglobin.* Hemoglobin (Hb) is a tetrameric protein–heme complex that has been specialized to carry oxygen in the red blood cells (RBCs, aka erythrocytes) of the circulatory system to tissues in the body. The O_2 from the lungs is lightly and reversibly bound to the Fe^{++} centers of the four heme molecules in each hemoglobin molecular complex. In the capillaries, the O_2 is released by the hemoglobin and diffuses into the tissues surrounding the capillaries, where it is used in oxidative metabolism by cells. Figure 7.16 illustrates the heme ring schematically with iron; the large attached globin protein is not illustrated.

As the O_2 carried by HbO_2 is released to the tissues, the CO_2 given off by cell metabolism is taken up by the blood plasma, the cytosol of RBCs, and hemoglobin in the RBCs. *Carbaminohemoglobin* (CO_2HHb) is formed when gaseous CO_2 reacts directly with Hb proteins. The CO_2 binds loosely to certain AAs in the globin tetramer (not to the heme centers) so that it can easily be released in the lungs. About 23% of the CO_2 transported to the lungs arrives complexed with the heme proteins.

The protein enzyme *carbonic anhydrase,* located inside RBCs, facilitates the storage of CO_2 inside RBCs as bicarbonate ions (HCO_3^-) in the RBC's cytosol (in water, not the Hb). The net catalyzed reaction, shown below, is reversible, allowing CO_2 gas to be released into the alveoli.

$$\mathbf{CO_2 + H_2O \Leftarrow\Rightarrow HCO_3^- + H^+}$$

About 70% of metabolic CO_2 is transported to the lungs inside RBCs as bicarbonate, where it is converted to gas and diffuses into the inhaled air. This is probably the most important mechanism of CO_2 transport to the lungs. It is also estimated that about 7% of the CO_2 produced is transported to the lungs in the blood plasma as dissolved gas.

In summary, about 70% of CO_2 produced is transported in RBC cytosol as HCO_3^-, 23% as CO_2HHb in RBCs and 7% as dissolved gas in the blood plasma [Guyton 1991].

The heme portion of the hemoglobin molecular complex is synthesized from *succinyl-CoA* and the AA *glycine;* most of this synthesis occurs in the mitochondria in a complex sequence of eight enzymatically catalyzed reactions [Huskey 1998]: (1) succinyl-CoA combines with glycine to form δ-aminolevulinic acid, (2) next, two δ-aminolevulinic acids combine to form porphobilinogen, (3) porphobilinogen is changed to linear tetrapyrrole, (4) which becomes Uroporphyrinogen III, (5) which is changed to Coproporphyrinogen III, (6) which becomes protoporphyrin IX, (7, 8) which finally is transformed into heme with Fe^{++} in two steps. Eight enzymes are necessary for the total heme synthesis. One can only speculate regarding the molecular evolutionary pressures leading to this baroque biosynthetic pathway for heme.

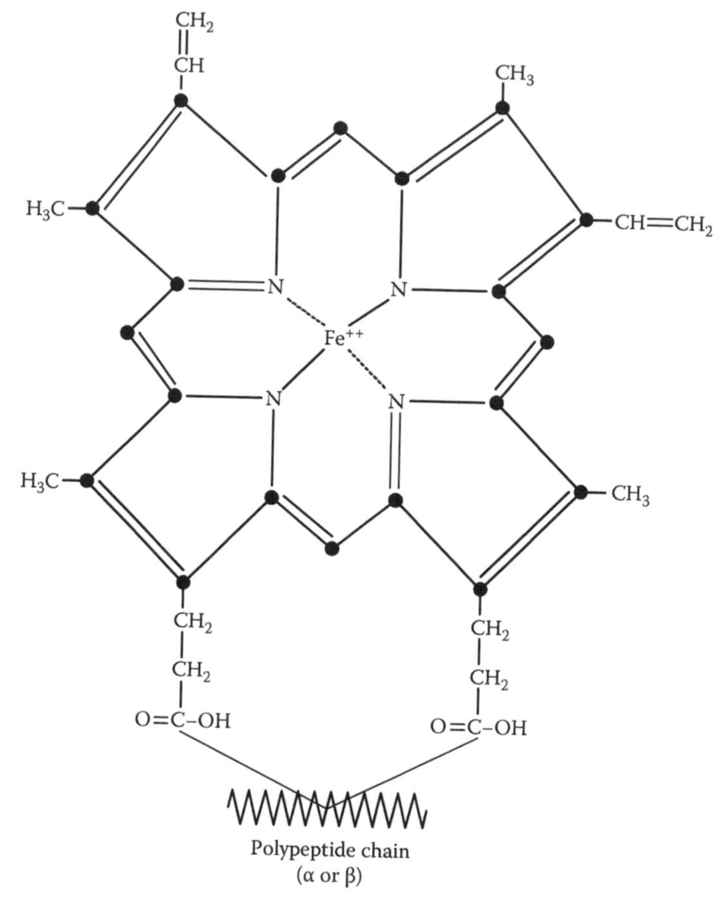

FIGURE 7.16 2-D schematic molecular structure of the heme ring. The globin point of attachment is shown.

The protein structure of mammalian hemoglobin consists of four subunits: An alpha-1*beta-1 dimer subunit and an alpha-2*beta-2 dimer subunit join to form a tetrameric hemoglobin molecule combined with four iron-carrying heme centers to form an adult Hb molecule with a molecular weight of 64.458 kDa. The alpha globin chain has 141 AAs; the beta chain has 146 AAs. Each of the four iron atoms in hemoglobin can bind loosely and reversibly to one O_2 molecule and carry them to tissues as O_2 molecules. Mature RBCs contain about 34% hemoglobin (about 2.8×10^8 Hb molecules) [Heiner 2005].

Inside an RBC, the hemoglobin can be in one of four forms:

1. *Reduced Hb* (HbR), nonoxygenated; HbR can carry CO_2 as CO_2HHb.
2. *Oxyhemoglobin,* HbO_2; Four O_2s per Hb molecule.
3. *Carboxyhemoglobin,* where carbon monoxide (CO) has replaced O_2 (HbCO).
4. *Methemoglobin,* in which the Fe^{++}'s in the heme rings have been oxidized to Fe^{+++} (metHb).

Hb is an *allosteric protein complex* that has more than one (stable) shape and can undergo reversible conformational changes in its structure in response to its chemical environmental conditions. When HbO_2 loses its O_2 to tissues surrounding the capillaries, it transforms into the *T-state* (T for "tense" configuration). Noncovalent interactions between the two globin heterodimers are strong, and the "cavity" between the β subunits is large. Fe^{++} is pulled out of the plane of the heme complex and is less available to bind O_2. Now, when alveolar O_2 binds to the T-Hb, it converts it to the high-affinity-to-O_2 R-state (R for "relaxed" Hb structure). *Oxyhemoglobin* (HbO_2) is in the *R-state*. The globins are made up of 75% α-helices; the rest is hydrophobic random coil in order to bind dissolved, nonpolar O_2 in the heme rings. R-HbO_2 has a higher affinity for O_2 than T-Hb. On the other hand, CO_2 can bind more easily to T-Hb to be carried to the lungs. When T-Hb arrives at the lungs, it once again changes to the O_2-carrying R-Hb and releases its CO_2.

To illustrate how important even small alterations in protein structure can be, consider the heritable disease *sickle-cell anemia*. Sickle-cell anemia has a variety of unpleasant symptoms, including abdominal and bone pain, jaundice, breathlessness, fatigue, rapid heart rate, and delayed growth [MedlinePlus 2007].

In sickle-cell disease, AA 6 in the beta chain is changed from glutamine to valine. This small change allows neighboring Hb molecules to stick together in long, stiff, polymer-like chains. These Hb polymers warp the RBC from its usual flattened disk to a "C," or sickle shape. Sickle erythrocytes can carry less O_2 and rupture more easily, leading to a more rapid metabolic turnover of sickle Hb. The one advantage conferred by this mutation is that the sickled RBCs are resistant to malaria parasites.

Another congenital Hb disease is *thalassemia*. Here, production of one of the two proteins (α-, β-globin) used to form Hb is reduced or repressed. In alpha thalassemia, the severity of the disease depends on the number of alpha protein genes that are defective. The loss of one alpha gene (out of four) has little effect on the blood's ability to carry O_2 and CO_2. The loss of two genes causes small RBCs and mild anemia. The loss of three alpha genes creates severe anemia, often requiring blood transfusions to survive. Three-gene deletion alpha thalassemia causes abnormal β-globin association, creating *hemoglobin H*. The loss of all four alpha globin genes is invariably fatal; most victims die in utero or shortly postpartum. Beta thalassemia is caused by damage or poor expression of the two genes for beta-globin. One-gene beta thalassemia may lead to mild anemia; two-gene beta thalassemia produces severe anemia. Note that the thalassemias are inherited (genetic) diseases [Sickle 1999].

Carbon monoxide is a very poisonous gas because it binds to HbR about 250 times as effectively as O_2; thus, a CO partial pressure of about 0.7 mmHg or a concentration of about 0.1% in air can prove lethal, tying up all the victim's Hb as HbCO. A CO poisoning victim must be treated immediately with 98% O_2 to displace the CO, plus 2% CO_2 to stimulate breathing.

Oxygenated blood leaving the heart is about 97% saturated with O_2. Venous blood leaving the capillaries is typically about 75% O_2 saturated under basal conditions. Under conditions of severe exercise, the % O_2 saturation of venous blood can drop to as low as about 28% [Guyton 1991].

The *optical pulse oximeter* (POX) is a simple, ubiquitous clinical instrument used in hospital ICUs and recovery rooms to monitor the % O_2 saturation of a patient's blood [Northrop 2002]. The operation of the optical pulse oximeter assumes that Hb is distributed between HbR (CO_2HHb) and HbO_2 and makes use of the fact that the IR optical absorbencies of HbR and HbO_2 are the same at one wavelength and different at a second wavelength. (A pulse oximeter will give erroneous readings in the presence of HbCO or metHb.) One form of POX uses light transmitted through tissues having dense capillary beds near the skin surface, such as fingertips or earlobes. Because of the pulsatile flow of blood through the capillary bed, at systole, the % O_2 saturation sensed is higher than that recorded just before the next systole. Thus, the TPOX output consists of a periodic waveform at the heart rate added to a dc level. The TPOX output is usually presented as a running average of the % O_2 saturation of Hb over each cardiac cycle. If a patient's TPOX reading drops precipitously, it generally indicates some form of respiratory distress or heart problem.

7.2.9.3 Serum Proteins and Antibodies

Although hemoglobin resides inside RBC "carriers," there are many other proteins "swimming free" in blood plasma. These are *serum albumin* and *serum globulins;* together, they make up about 6–8% of the blood. *Serum* is blood plasma minus cells, fibrinogen, and other clotting factors. Serum albumin protein is made in the liver; it has a molecular weight of about 69 kDa and accounts for about 75% of the normal *colloid osmotic pressure* (COP). *Globulin proteins* contribute about 25% to the COP. They have molecular weights of about 140 kDa [Guyton 1991].

The *alpha globulins* are the proteins that transport *thyroxine* and *retinol* (vitamin A) as complexes. *Beta globulins* include the iron-transporting protein *transferrin.* Plasma transferrin is an 80-kDa glycoprotein with homologous N-terminal and C-terminal iron-binding domains. The transferrin protein tertiary structure has an array of AAs that readilly form four bonds with each of two Fe^{+++} ions, but only with a carbonate anion cofactor. Certain cells that use iron in their biochemical reactions have transferrin receptors on their surfaces. When an iron-carrying transferrin molecule binds to its receptor, it is drawn into the cell, iron and all, by endocytosis. The cell then acidifies the endocytic vesicle containing the transferrin. In addition, a plasma membrane *oxidoreductase enzyme* reduces the Fe^{+++} to Fe^{++}, facilitating the release of the tightly bound iron from the transferrin. Then the receptor and the empty transferrin molecule are externalized. In the near-neutral pH of the blood, the receptor releases the empty transferrin molecule, which continues its job of picking up iron ions. The half-life of a transferrin molecule is about eight days, and one molecule may deliver iron up to 100 times to cells. An (unknown) transport protein moves the Fe^{++} ions into the interior of the cell. Once inside the cell cytoplasm, the Fe^{++} is bound to a low-molecular-weight carrier molecule and delivered to mitochondria, where it participates in electron transport and heme biosynthesis, or it may be stored inside the cell in another specialized protein called *ferritin.*

Ferritin proteins are found inside both prokaryotes and eukaryote cells. Ferritins are complex 24-subunit heteropolymers of H (for heavy or heart) and L (for light or liver) protein subunits. L subunits weigh 19.7 kDa, and H subunits are 21 kDa. Curiously, the tertiary structure of ferritin is a hollow sphere. Eight narrow channels penetrate into the lumen of the sphere; they are lined with hydrophilic AA residues, and six more channels are lined with hydrophobic AA residues. Evolutionary conservation of the hydrophilic channels suggests that they provide the route for iron to enter and exit. The Ferritin "bottle" can hold up to 4,500 atoms of iron as a poly-iron-phosphate-oxide complex and by some unknown mechanism and release the iron slowly as needed by the cell [Sickle 2001].

Gamma globulins include the *antibodies* made by immune system plasma B-cells and mucosal lymphocytes in immunoglobulins (Igs). These large, specialized glycoprotein protein molecules have been grouped into five classes: IgA, IgD, IgE, IgD, and IgM. IgM and IgD are antigen receptors fixed on the surfaces of B-cells. Once activated, plasma B-cells secrete IgM, which has 10 antigen binding sites; IgM also activates the complement cascade system. Plasma B-cells also secrete IgG. IgA is made predominantly in mucosal lymphoid tissues. It is present in mucous secretions of the respiratory, digestive, and urogenital tracts and in breast milk. IgE is made in response to parasite (worm) infection and to allergies; it causes the release of histamine by mast cells. A complete Ig antibody is formed from two light-chain (L-chain) polypeptides and two heavy-chain (H-chain) polypeptides. Figure 7.11 illustrates the general schematic of an Ig Ab.

The mammalian body can make about 25×10^6 variations on antibody (Ab) structure to ensure they can match the widest spectrum of antigenic *epitopes.* In times of infection, as a result of antigen presentation, clonally expanded, plasma B-cells manufacture those Abs specific for that one pathogen. An amazing feature of the human immune system is that it can manufacture about 25×10^6 different kinds of Ab molecules using fewer than 2.3×10^4 genes. This amazing diversity in Ab receptors is obtained because each Ab polypeptide chain is encoded by several different *gene segments.* The genome contains a pool of gene segments for each type of chain. Random assortment of these segments leads to a huge Ab receptor diversity. Effectively, a *Monte Carlo transcriptional system* is used.

The gene segments used for Ig diversity are 51 V_H segments (each encodes most of the N-terminal of the Ab, including the first two hypervariable regions), 27 D_H diversity gene segments (these encode part of the third hypervariable region), 6 J_H gene segments that encode the remainder of the V region, and 9 C_H segments that encode the C region of the B-cell Ab. The C-gene is subdivided into one mu segment that encodes the C region of IgM, one delta segment that encodes the C region of IgD, four gamma segments that encode the four types of IgG, one epsilon segment for IgE, and two alpha segments for the two types of IgA. All these gene segments are clustered on a complex locus on human chromosome 14. Note that for the Ab H-chains, the random assembly of 51 V_H, 27 D_H, and 6 J_H gene segments leads to about 8.3×10^3 different possible structural combinations. However, the recombination (joining) process for the gene segments is not exact; the exact points of splicing (V_H to D_H and D_H to J_H) can vary over several Nts. Also, extra Nts, called N-regions, can also be inserted at these joints. This "slop" in the V-chain system increases the number of structural combinations and, hence, Ab diversity.

A similar, combinational joining of gene segments creates structural diversity for the L-chains. The L-chains pair randomly with H-chains, creating further diversity [Kimball 2005l].

We have seen that the first 100 or so AAs at the N-terminal end of both H- and L-chains vary greatly from Ab to Ab, forming the *variable region* of the Ab. There are three *hypervariable regions* of an Ab where the AA sequence variability in the variable regions is especially pronounced. In the tertiary structure of the Ab, the three hypervariable regions of both the H- and L-chains come together to form the *antigen-binding site* of the Ab. The antigen epitope binds to the particular antigen-binding site with great affinity. In contrast, only a few different AA sequences are found on the **–COOH** terminals of L- and H-chains. These are the so-called *constant* (C) *regions* of Abs. (Refer to Figure 7.11). The complexity of Ab architecture is underscored by the following: Humans make two kinds of L-chain C regions, i.e., kappa (κ) or lambda (λ) L-chains. There are five different types of H-chain C regions: mu (μ) (the H-chain of IgM Abs), gamma (γ) chain (IgG), alpha (α) chain (IgA), delta (δ) chain (IgD), and epsilon (ε) chain (IgE). Each of these five types of H-chains appears to have no preference for pairing with the kappa or lambda L-chains [Kimball 2005k].

At times of infection, specific activated plasma B-cells produce vast numbers of plasma Igs with specificity to the invading pathogen. The V regions recognize the pathogenic epitope; the C regions trigger a coordinated immune system response (inflammation, complement, and all). When an infection has been overcome, the specific circulating Abs are broken down enzymatically and their AAs recycled. All proteins (Abs, hormones, cytokines, etc.) have finite life spans. *Memory B-cells* (MBCs) are derived from clones of the plasma B-cells that made the specific Ab. Memory B-cells store the genes for the Ab to ensure rapid deployment of the Ab in case the pathogen returns. Presumably, the mechanism that originally randomized the production of the specific Ab is in the genome of the MBC, so when the MBC begins making that Ab, its structure is deterministic; i.e., it is the desired Ab to fight the specific pathogen.

7.2.9.4 Hemocyanins in Invertebrate Circulatory Systems

Hemocyanins (Hcs) are O_2-transporting proteins found in the blood of certain arthropods (horseshoe crabs and lobsters) and mollusks. There is no heme in Hc, in spite of its name. Hemocyanin is made from individual subunit proteins, each of which is bound to two copper atoms that can reversibly bind one O_2 molecule. In deoxy-Hc, the coppers are Cu^+. In oxy-Hc, the copper is oxidized to cupric (Cu^{++}) ions, giving the molecule a blue-green color (lobster blood). The coppers are bound as prosthetic groups comprising histidine peptides. Each arthropod Hc subunit weighs about 75 kDa. Subunits may be arranged in dimers or hexamers, depending on the species. The dimer or hexamer complexes can further be arranged in chains or clusters of enormous molecular weight, over 1.5 MDa! Hemocyanin complexes float freely in the animal's blood; they are probably too big to put in cells.

Not surprisingly, molluskan Hcs have a different structure than those from crustaceans. The polypeptide chains are very large, about 350 to 450 kDa each, and each consists of seven or eight globular "functional units" connected by linker peptide strands. In cephalopod mollusks (squids and octopi), the circulating Hc exists as decamers of these large subunits, forming hollow cylindrical arrays. Gastropods have been found to have even more immense Hc complexes, about 9 MDa, which contain 160 oxygen-binding sites. Hcs probably originated in arthropods about 550–600

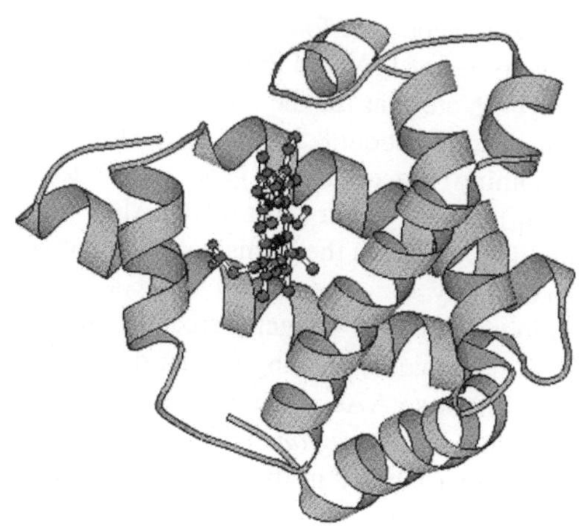

FIGURE 7.17 3-D "ribbon" schematic of a myoglobin molecule.

MYA, and earlier in mollusks, about 800 MYA, when planetary O_2 began to rise to present levels [van Holde et al. 2001].

7.2.10 Myoglobin

Myoglobin is an O_2-buffering protein found in muscle cells. Myoglobin was one of the first protein molecular structures to be elucidated. John C. Kendrew received the Nobel Prize for Chemistry in 1962 for this work. Myoglobin (Mb) is a bright red protein that gives red meat much of its red color. Intracellular O_2 is bound to Mb in a protein–heme complex. Mb has a single monomer protein with eight helices complexed with a single heme structure and surrounding it. It has a molecular weight of 17.6 kDa, about one-fourth that of hemoglobin. A 3-D "ribbon" model of the Mb molecule's tertiary (ribbon) structure is shown in Figure 7.17 [King 2005b; Scripps 1997].

The function of cytoplasmic Mb is to store O_2 close at hand to the muscle cells' mitochondria, which need it to make ATP. The O_2 released from the MbO_2 diffuses into the mitochondria through its thin membrane, where it participates in oxidative metabolism.

Mb, similar to Hb, binds strongly with carbon monoxide. *Carboxymyoglobin* is largely responsible for the asphyxiation that results from CO poisoning. Treatment with hyperbaric O_2 can help remove the carboxymyoglobin in muscle cells.

7.3 ERRORS IN PROTEIN STRUCTURE

7.3.1 Introduction

As we have seen, the natural tertiary (even quaternary) structure of a protein is absolutely necessary for its function as an enzyme, receptor, signal, pump, structural component, hormone, etc. Clearly, errors in gene sequence or in pmRNA or mRNA sequence will lead to a protein with an aberrant structure that will not perform its intended function or, at best, will perform it weakly. The primary cause of the damage to the protein may be damage to the gene's DNA that causes a *single-nucleotide polymorphism* (SNP) (see Chapter 6, Section 6.3 of this text). SNPs can cause the substitution of an incorrect AA in the peptide as it leaves the ribosome or an entire *frame shift*, resulting in a complete chain of incorrect AAs. Two consecutive Nt substitutions in the DNA or mRNA can also result in the production of a bad protein. (See Section 7.3.4 to see how serious a one-AA substitution can be in a protein.)

A host of other bad things can happen to proteins once they are in final active form. These are largely due to the effects of heat, chemicals in the environment, and ionizing radiation.

7.3.2 Nongenetic Causes of Protein Errors

The protein errors we will consider in this section are errors in competent protein structure caused by thermally caused hydrogen bond breaks and bond breaks caused by reactive by-products of oxygen metabolism (reactive oxygen species, ROS) such as the *superoxide anion radical* ($\mathbf{O_2^-}$), hydrogen peroxide (H_2O_2), and reactive hydroxyl radicals (**-OH**). The breaks disrupt the bonds tying AA residues on one peptide chain to another, altering the stable 3-D structure (and the function) of a protein. External factors such as UV photons, certain chemical agents, and high temperature can break hydrogen bonds stabilizing protein structure and also generate intracellular superoxides that cause further bond breakage and site-specific AA chemical modifications. Different AAs differ in their susceptibility to chemical change. Sulfur-containing AAs and thiol groups are specifically very sensitive sites. For example, activated oxygen can remove a hydrogen atom from cysteine residues to form a *thiyl radical* that will cross-link to a second adjacent thiyl radical to form an unwanted disulfide bridge. Or, active oxygen can add to a *methionine* residue to form methionine sulfoxide derivatives. The AA *tryptophan* is readily cross-linked to form bityrosine products. Also, *histidine, lysine, proline, argenine,* and *serine* form *carbonyl* groups on oxidation. The oxidative degradation of a protein molecule is enhanced by the presence of metal atom cofactors, such as Fe. The metal binds to a divalent cation-binding site on the protein, then reacts with H_2O_2 to form a highly reactive hydroxyl radical that rapidly oxidizes an AA residue at or near the cation-binding site. This oxidation generally inactivates the protein's function (e.g., an enzyme) and leads to the protein's destruction.

Proteins can also be damaged by *glycation*. Glycation, also called the *Maillard reaction,* is a nonenzymatic (spontaneous) reaction in which a reducing sugar (e.g., glucose) becomes attached to a protein's free *amine* groups ($-NH_2$). This bonding occurs at the free amine group of the AAs *arginine* and *lysine,* which are not involved in the covalent peptide bond. First, the double-bonded oxygen in the D-glucose's ring is lost and a double bond forms between the amine nitrogen of the AA (e.g., lysine) in the protein and carbon 1. The two hydrogens from the amine plus the oxygen form a molecule of water. The sugar so bound forms an *imine molecule* (aka *Schiff base*) in the protein. The next spontaneous step in the Maillard reaction is that the imine rearranges its molecules to form an *Amadori product.* What happens here is that the second carbon in the glucose loses two hydrogen atoms and forms a double bond with an oxygen, forming a *carbonyl group.* The number one carbon in the sugar now adds a hydrogen atom and forms a single bond with the amine nitrogen, which also takes one hydrogen (see Figure 7.18). Both the glycation and the formation of the Amadori product are reversible. However, the Amadori product can be irreversibly oxidized to an advanced glycation end product (AGE), destroying an arginine or a lysine in the protein and irreversibly damaging the molecular gestalt of the protein.

It is worth noting that sugars other than glucose are more reactive in glycation reactions. For example, *galactose* is 5 times more reactive, *fructose* 8 times, *deoxyglucose* 25 times, *ribose* (a pentose) 100 times, and *deoxyribose* 200 times more reactive than glucose.

Even treated diabetics have higher than normal blood glucose concentrations. By mass action, this leads to a higher rate of glycation and, hence, more AGEs and damaged proteins. There are a host of medical problems that are associated with high blood glucose and diabetes. These include but are not limited to atherosclerosis, cataracts, and kidney disease [Best 2006].

There are exogenous antiglycating agents that may have therapeutic value in protecting proteins from oxidative modifications. *Aminoguanidine* is one of the potent antiglycating agents that can interfere with the formation of Amadori products, preventing the formation of irreversible AGEs and protein damage. *Carnosine* has also been found to have protective properties reducing the neurotoxicity of reactive carbonyl compounds. Carnosine protects proteins from both oxidation and modification by a number of aldehydes under in vitro conditions. Both the important antiagathic enzymes *superoxide dismutase* and *telomerase* were shown to be protected in vivo and in vitro from ROS by *carnosine* [Boldyrev 2005]. You will see later that the infamous *prion protein* in its normal form can act as a *superoxide dimutase.*

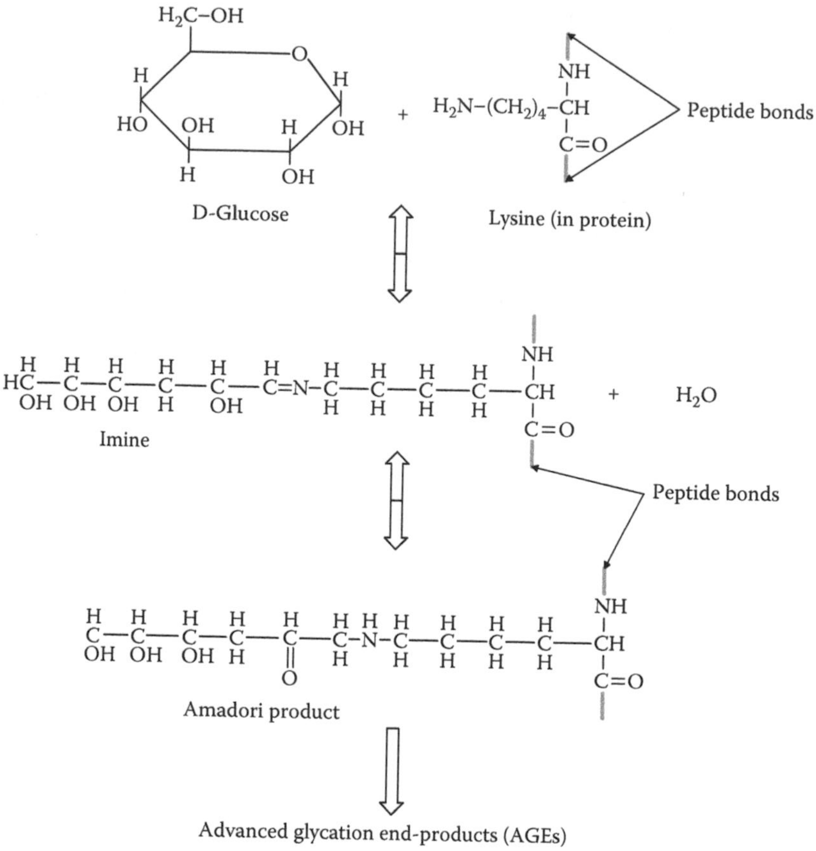

FIGURE 7.18 The spontaneous *Maillard* (glycation) *reaction* in which glucose becomes attached to an AA's free amine ($-NH_2$) groups. The resulting molecule is called an *Amadori product.*

Collagen composes about one-third of the total body protein in mammals. The inelastic, sinewy tissues characteristic of aging come from the mechanical stiffening due to molecular cross-links between collagen strands. The major protein cross-link is the arginine/lysine glucose product, *glucosepane.* Figure 7.19 (adapted from Best [2006]) illustrates how glucose forms an Amadori product with lysine in one collagen strand, then *1,4-dideoxy 5,6-glucosone* forms, followed by cyclization, then reaction with an arginine residue in an adjacent collagen strand forming a glucosepane cross-linkage. Not only do cross-links lead to inelastic tissues, but they can contribute to arteriosclerosis (hardening of the arteries), reduced kidney function, and reduced lung vital capacity.

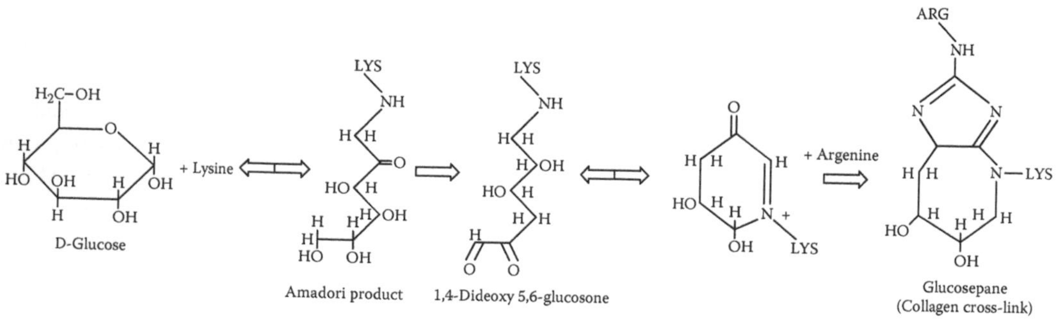

FIGURE 7.19 The pathways in forming *glucosepane*, which forms cross-links between arginine and lysine AAs in adjacent collagen strands.

Not all unwanted protein cross-links are due to glycation. *Aldehydes* produced by lipid peroxidation, for example, malon dialdehyde (MDA, or propanedial), can cross-link proteins by forming covalent bonds with AAs with terminal amine groups such as lysine.

Phosphorylation and dephosphorylation of enzymes by *kinases* and *phosphatases* at tyrosine, serine, and threonine residues plays a critical role in enzyme activation and deactivation and also cell signaling (a form of biochemical *parametric control*). Phosphorylation of tyrosine AAs can be blocked by covalent bonds formed with *peroxynitrite*. Peroxynitrite is formed by the reaction of nitric oxide (NO) with a *superoxide*.

Nature is very specific about which structural isomer of certain biomolecules to use in synthesis and metabolism. For example, D-glucose (so called because of the way its structure compares with the standard enantiomeric molecule, D-glyceraldehyde) is used exclusively in metabolism. The L-isomer won't work! Yet all proteins are initially made with L-amino acids. Their functions depend upon it. Thus, another way proteins can be damaged is by thermal energy causing a few of the normal L-isomer AAs to switch to the nonbiological D-forms. This form of protein heat damage is called *racemization*.

Another possible form of protein damage is *deamination*. Deamination can occur spontaneously and is 400 times more likely for the AA *asparagine* than for *glutamine*. This defect can generally be repaired enzymatically in the cell.

The formation of a *carbonyl radical* is another bad event for an AA residue in a protein (see the carbonyl on carbon 2 in the Amadori product in Figure 7.18). *Aldehydes* and *ketones* have carbonyl radicals. The **>C=O** bond is very stable once formed, so bad proteins with anomalous carbonyls must be removed enzymatically by *proteases, hydrolases,* etc. Best [2006] tells us that the carbonyl content of an animal's proteins increases exponentially with age. At least 30 to 50% of proteins are oxidized in old animals, including enzymes as well as structural proteins.

There is only one known *endogenous* enzymatic repair mechanism for oxides in proteins. *Methionine sulfoxide* can be restored to methionine by the enzyme *methionine sulfoxide reductase,* and a *disulfide isomerase* can restore the aberrant disulfide bridges to normal form.

7.3.3 Heat Shock (Stress) Proteins

Heat shock proteins (HSPs), aka *stress proteins,* are ubiquitous in all living cells. They occur in two forms, intracellular and extracellular. Intracellular HSPs are induced as a result of stress and play a protective role in the cell. Extracellular HSPs play a role in the inflammatory response [Wu & Tanguay 2006]. They also play a role in heart disease [Zhu et al. 2001] and cancer [Schmitt et al. 2007] in humans. They were discovered in the 1960s by exposing *Drosophila* cells to high (but not lethal) temperatures. The elevated temperature caused an increase in the expression of HSPs. Their rise in expression is transcriptionally regulated by a complex biochemical mechanism that may be related to damage of nucleic acids or other proteins. One scenario has an activated *heat shock factor* (HSF) binding to a genomic sequence motif or *heat shock element* (HSE), which initiates transcription and translation of other HSPs. Protein kinases A and C may also play a role in the regulation of promoter activity.

Increased HSP concentration is also found in cells subject to a variety of other environmental stressors, including the following: cold, oxygen deprivation [Antigenics 2007], infection [Stewart & Young 2004], inflammation [Pockley 2002], exposure to exogenous chemical toxins including chemotherapy drugs [Science Daily 2005], UV radiation, heavy metals, alcohol, oxidants [Trautinger 1996], hypoxia [Otsuki et al. 2004], or desiccation [Hayward et al. 2004]. HSPs act to protect cellular proteins from stress-induced damage such as described previously.

Because of their critical roles in intracellular protein management, there is considerable homology between HSPs across species and phyla. Mouse HSPs have about 95% homology with human HSPs, and even bacterial HSPs have 50% homology. HSPs that have chaperone activity belong to five conserved classes, given by their approximate molecular weights in kDa: HSP100, HSP90, HSP70, HSP60, and many smaller (40–10 kDa) HSPs (SHSPs).

HSPs are also present in cells at lower concentrations under nonstress conditions, where they constitute about 2% of the cell's contents. They have a number of diverse roles in a cell's protein housekeeping functions. They act as *chaperones,* regulating protein–protein interactions such as folding, and guiding new proteins into their proper 3-D shape; they also prevent unwanted protein aggregation. HSPs also act to shepherd newly made proteins around among the cytosolic compartments. Damaged proteins are transported by HSPs to centers where proteolytic enzymes cut them up into individual AAs and short peptides. Another possible role of certain HSPs is attaching peptide detritus (e.g., from viral, bacterial, or abnormal cancerous proteins) onto major histocompatibility complex (MHC) proteins. A "loaded" MHC molecule migrates to the cell membrane's outer surface, where it can present the detritus peptides to certain immune system cells that sense them as abnormal and then kill the infected cell [Goldman 2005].

7.3.4 When Good Proteins Go Bad: Prions

The physiological role of the normal *prion protein* (PrP^C) is poorly understood, perhaps because its actions are pleiotropic (the ^{C} in PrP^C stands for "cellular"). The human prion gene is located on chromosome 20. Normal prion protein is a copper-binding glycoprotein expressed by neurons, glial cells, and all other eukaryotic cells [Brown et al. 1999]. Brown et al. presented evidence that normal PrP^C exhibits copper-ion-dependent enzymatic activity as a *superoxide dismutase.* That is, PrP^C converts metabolically derived, protein-damaging superoxides in the CNS to hydrogen peroxide, which is further broken down, thus protecting neurons and glia from oxidative cell damage. Most of the PrP^C made by a cell ends up tethered to outer cell membranes by a *glycosyl phosphatidyl inositol* (GPI) anchor. The half-life of cell surface PrP^C is 3 to 6 hours. Other PrP^C molecules are found in the cytoplasm [Gains & LeBlanc 2007].

Zhang et al. [2006], in a recent study, found that PrP^C was expressed on the surfaces of 40% of adult mouse bone marrow cells. Also, Zhang and coworkers found that more than 80% of these PrP^C-marked cells were erythrocytes or their developmental precursors. Importantly, they found that PrP^C expression was necessary for hematopoietic stem cell self-renewal. The mechanism by which expressed PrP^C does this is unknown and the subject of ongoing research.

Another role for normal prion protein was elaborated by workers at the Whitehead Institute for Biomedical Research and Harvard Medical School/Massachusetts General Hospital. These workers found that PrP^C helps preserve stem cells in the blood. They reasoned (correctly) that because there is far more PrP^C in the CNS, it may affect neurogenesis from *neural stem cells* (NSCs) in the embryonic brain and, to a lesser degree in the adult brain. Neural stem cells from the brains of GM mice that expressed an excess of PrP^C differentiated into neurons in vitro more rapidly than NSCs from the brains of normal mice and far faster than NSCs from the brains of PrP^C knockout mice. All three groups of mice ended up with approximately the same number of CNS neurons [BrightSurf 2006].

In summary, the interested reader can find a comprehensive description of some of the known (and putative) physiological functions of normal PrP^C in mammals in the review article by Gains and LeBlanc [2007]. These authors discuss the role of PrP^C in cellular copper metabolism and in an antioxidant role, signal transduction and growth, Ca metabolism and the regulation of neuronal activity, nucleic acid binding, the immune response, neuron survival following an ischemic or traumatic event, and of course the role of PrP^C in *transmissible spongiform encephalopathy* (TSE) and related prion diseases, described later.

Nearly everyone has heard of *mad cow disease* (aka *bovine spongiform encephalopathy* [BSE], *transmissible spongiform encephalopathy* [TSE]) and perhaps related conditions such as *scrapie* in sheep and *chronic wasting disease* in deer and elk. In humans, related prion diseases include *Creutzfeld–Jakob disease, Gerstmann–Straüssler–Scheinker syndrome, kuru, fatal familial insomnia* (FFI), and *Alpers syndrome.* It is now known that the causative agent in these diseases is not a bacterium or a virus but rather the PrP^C *protein* folded wrong. That is, the normal, endog-

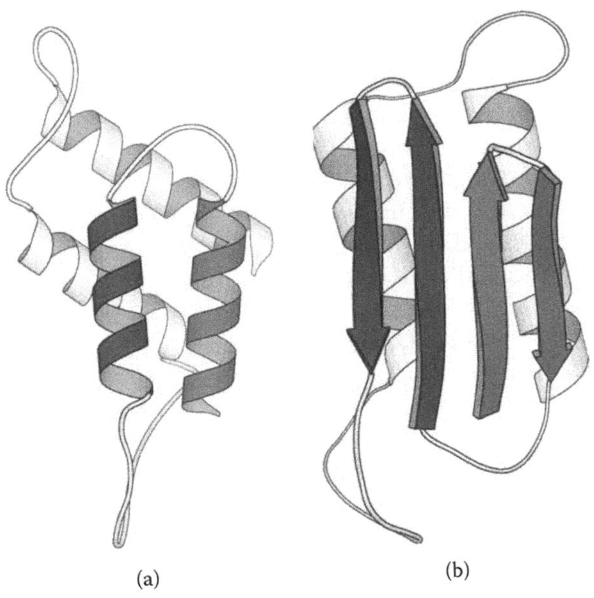

FIGURE 7.20 3-D "ribbon" schematic molecules illustrating (a) normal prion protein (PrP^C) structure and (b) misfolded, pathological, prion protein (PrP^{SC}).

enous prion protein has lost its biochemical function by assuming an alternative, "bad," ultrastable, tertiary or 3-D structure.

Figure 7.20 illustrates schematically the 3-D "ribbon" structures of normal PrP^C and PrP^{Sc}. (The ^{Sc} in PrP^{Sc} stands for *scrapie,* the name given to the prion disease in sheep.) The linear primary structure of PrP^C is about 257 amino acids long, depending on the species [OMIM 2006d]. No nucleic acid has been found to be associated with normal or bad prion proteins. By extension, bad prions in other animals are also given this abbreviation, even though their structures can be slightly different. The PrP^{Sc} 3-D molecular model is speculative; its exact structure is not yet known. The opposing beta sheets in PrP^{Sc} allow extensive hydrogen, disulfide, and other bonds to form between molecules in the opposing sheets, conferring great chemical stability. It has been found that prion diseases in mammals are both genetic *and* transmissible.

When the misfolded prion protein (PrP^{Sc}) was found to be the causative agent of BSE, or mad cow disease, little was known about the role of normal prion protein in the brain. PrP^{Sc} was found to be incredibly resistant to proteases and other natural chemical agents known to denature other proteins. It soon became apparent that the PrP^{Sc} protein's resistance to the normal biochemical proteinases that recycle AAs in the body caused PrP^{Sc} to accumulate in abnormal concentrations in extracellular space in the CNS, presenting a "spongiform" appearance in histological sections of an infected animal's brain.

PrP^{Sc} is an incredibly robust protein molecule: It is unaffected by unlimited UV radiation exposure and is resistant to autoclaving, ashing at 350°C, and protease enzymes. However, it can be denatured by strong solutions of bleach, *phenol,* KSCN, *urea, proteinase K, trypsin,* and alkali [Leiscester Microbiology 2004].

The model for intra- and interspecies transmission of PrP^{Sc} was assumed to follow the ingestion of PrP^{Sc} molecules by an uninfected animal or person. It was discovered that one way in which cattle caught BSE was to eat processed cattle food that contained waste body tissues from slaughtered, infected animals that were added to increase the protein content of the feed. The brains and spinal cords of sheep with the PrP^{Sc} disease, *scrapie,* were often used in the feed. Cooking did not denature the robust PrP^{Sc} proteins that the cattle ate. Since 1998, the use of animal protein in cattle food has been banned in most of the world. One insidious property of PrP^{Sc} diseases is the long latent period (about 1–5 years) before symptoms are expressed. Amazingly, cross-infection of CJD prions in humans can be spread by autoclaved (but not supercleaned) surgical instruments in hospitals and by contact tonometry (the eye is evidently a rich source of PrP^{Sc} proteins). In the United Kingdom, tonometry is now almost exclusively done by the air-puff method [Northrop 2002]. Unlike viral diseases, there is no vaccine or cure for prion disease; it is 100% fatal.

Chronic wasting disease (CWD) in deer and elk is endemic in the western central United States and now even affects white-tailed deer in eastern New York State. Of course, CWD in deer and elk is not spread by human-made food. It is now clear that the PrP^{Sc} proteins can appear in an infected animal's urine, feces, or placental material (afterbirth tissues) and can be ingested by a new host when that animal licks the infected material on the ground or eats contaminated plants.

Owing to its robust biochemical stability, PrP^{Sc} somehow travels from the gut to the circulatory system to the CNS, where it autocatalytically induces the $PrP^C \rightarrow PrP^{Sc}$ transition. Stanley Prusiner's research on the spread and cause of BSE and CJD indicated that the infective agent was evidently a protein (PrP^{Sc}) and not a nucleic acid (DNA or RNA). (Prusiner was awarded the Nobel Prize in Medicine in 1997 for his work on prion disease.)

Recently, in a review article, Grossman et al. [2003] presented evidence that nucleic acids (NAs) are also involved in PrP^{Sc} pathogenesis. Quoting from their conclusion: "… recent studies demonstrate a functional interaction of prion protein with RNA and DNA and thus indicate that NA participation in prion disease may be a pathophysiological result of normal prion function. As the molecular basis for NA-PrP structural transitions are better understood and insight is gained into how these events contribute to the disease, a rational basis for developing therapeutic and diagnostic strategies may be provided."… [Quoted with kind permission of Springer Science and Business Media.]

Could this PrP^{Sc} cofactor, or "chaperone," be a small, ribozyme-like NA that is resident in the infected host? Most prion researchers think not. If there is such a chaperone, perhaps this RNA affects the PrP^C structural changes that cause it fold into the bad, beta sheet form. Another theory holds that the PrP^C to PrP^{Sc} transition is aided by an unknown *chaperone enzyme, protein-X,* resident in the infected host. This X-protein could "donate" one or more stable disulfide bonds to the normal PrP^C, causing it to fold and form beta sheets, etc. [Landrau 2004].

There has been a flurry of recent research on the prion gene and its mutations, which allow it to either convert more easily to the beta sheet (infectious) form or be resistant to TSE disease. Clearly, there is *interspecies transfer of TSE.* The PrP^C protein structure is well conserved in eukaryotes. It is even found in yeast. Prion diseases in mammals are both genetic and transmissible. The prion gene is found on chromosome 20 in humans, and the 3-D (tertiary) structure of mouse PrP^C has been largely unraveled. "Knock-out" mice lacking the PrP^C gene do not come down with TSE even when inoculated with PrP^{Sc} in their brains.

Susceptibility of a host to TSE infection and the latent period are evidently determined by both the species from which the prion innoculum was taken and the host's PrP^C gene. Table 7.3 lists TSE diseases in mammals.

Note that interspecies transmission has been demonstrated for BSE → nvCJD, and other interspecies infections are possible.

Gerstmann–Straüssler–Scheinker syndrome (GSS) was shown by Prusiner and coworkers to be an inherited human TSE. A one-point mutation (one Nt base pair out of more than 750) caused codon 102 to change, substituting the AA leucine for the normal AA, proline, or valine instead of alanine at codon 117 in the patient's PrP^C gene. Prusiner et al. found these same point mutations in PrP genes from a large number of patients with GSS, providing strong evidence that this particular TSE was in fact a genetic form of prion disease. The fact that a TSE can arise spontaneously owing to a mutation in the PrP^C gene, as well as by infection with PrP^{Sc} protein, is scary.

Prusiner and colleagues also found, working with transgenic mice, that the more similar the infecting PrP^{Sc} is to the host's normal PrP^C, the more likely it is that interspecies transmission of TSE will occur. There is a low efficiency of bovine PrP^{Sc} infecting human PrP^C in vivo and in in vitro studies. There is a much higher experimental in vivo infectivity of apes with human PrP^{Sc}.

TABLE 7.3
Mammalian Prion Diseases

Name of TSE Disease	Host Animal	Prion Name	PrP Isoform
Scrapie	Sheep and goats	Scrapie prion	$OvPrP^{Sc}$
Transmissable mink encephalopathy (TME)	Mink	TME prion	$MkPrP^{Sc}$
Chronic wasting disease	Deer and elk	CWD prion	$MDePrP^{Sc}$
Bovine spongiform encephalopathy (BSE)	Cattle	BSE Prion	$BovPrP^{Sc}$
Feline spongiform encephalopathy (FSE)	Cats	FSE prion	$FePrP^{Sc}$
Exotic ungulate encephalopathy(EUE)	Nyala and greater kudu	EUE Prion	$NyaPrP^{Sc}$
Kuru	Humans	Kuru prion	$HuPrP^{Sc}$
Creutzfeld–Jakob disease (CJD)	Humans	CJD Prion	$HuPrP^{Sc}$
New variant CJD (vCJD, nvCJD)	Humans	BSE prion*	$BovPrP^{Sc}$ *
Gerstmann–Straüssler–Scheinker syndrome (GSS)	Humans	GSS prion	$HuPrP^{Sc}$
Fatal familial insomnia (FFI)	Humans	FFI prion	HuPrPSc

That a small probability of cross-infection exists at all between BSE and human PrP^C has led to numerous (temporary) import embargos on beef exports from the United Kingdom, Canada, and the United States, where BSE cases were reported.

Spontaneously occurring CJD disease in humans is inherited as an *autosomal dominant* mutation. Ten to 15% of CJD patients have inherited at least one copy of a mutated PrP gene. Some of the more common mutations are substitution of the AA lysine for the normal glutamic acid at codon 200, substitution of asparagine for aspartic acid at codon 178 when it is accompanied by a polymorphism in the gene normally encoding valine at codon 129, and a change from valine at codon 210 to isoleucine. (A detailed comparison of the prion protein orthologs in humans, mice, dogs, sheep, pigs, and cattle is available online from Winter [2005].)

Variant Creutzfeldt–Jakob disease (vCJD) is induced in humans from ingesting bovine PrP^{Sc}. Because human and cow PrP genes (aka PRNP genes) differ by 30 codons, the cow-to-human infectivity rate is very low. Still, the causality is there. Another possible etiology for acquiring vCJD is from eating the meat of deer or elk infected with CWD. Most TSE epidemiologists agree that infection is possible, but the probability is very low [Belay et al. 2004].

The State of Connecticut (which has not yet reported CWD in its deer herd, even though cases have been present in rural eastern New York State in fall, 2005) has posted advice to hunters on how to dress a deer carcass to avoid infection. They state: "Because prions of CWD accumulate in tissues, such as the brain, spinal cord, spleen, lymph nodes, tonsils, and eyes, as well as saliva, urine and feces, contact with these items should be avoided. Deer meat is not known to harbor the disease agent. Hunters should exercise the following precautions:" We give their salient points:

- Avoid shooting, handling, or consuming any animal that is behaving abnormally or appears to be sick.
- Wear [disposable] latex gloves when field-dressing deer.
- Minimize the handling of and do not eat the brain, spinal cord, eyes, spleen, tonsils, and lymph nodes.
- Debone the meat from the animal. Do not saw through bone or cut through the brain (skull) or spinal cord (backbone).
- Wash hands and instruments thoroughly after field dressing is completed.
- Instruments should be placed in 1:1 bleach/water solution for an hour and left to air-dry before reusing. [CTDEP 2004]

The last instruction may seem extreme, but recall that PrP^{Sc} prion protein molecules are impervious to most antiseptics and autoclaving. These instructions underscore the persistence of PrP^{Sc} protein in the environment and the hazard it presents from cross-species infection.

Figure 7.21 illustrates a phylogenetic tree for PrP^{C} based on homology of the PrP^{C} genes of 46 animal species. It is based on the complete nucleotide sequences of prion ORFs in the *GenBank*. The tree shows unexpected homology groupings that suggest TSE transfer may be easier between certain species. Note that the *Bos* tree is quite removed from the primate tree, yet there certainly is evidence that $BovPrP^{Sc}$ can cause vCJD in humans [Zhang et al. 2002].

One wonders how many proteins, other than PrP^{C}, "go bad" by assuming a stable, nonfunctional, tertiary structure. PrP^{C} misfolding leads to obvious CNS dysfunction and is totally lethal because of the critical role of PrP^{C} in the CNS and the fact that PrP^{Sc} is virtually immune to the normal biological mechanisms for protein breakdown. Other "bad proteins" may not be lethal or may proceed at a much slower rate than CJD and other fatal prion diseases. Also, other bad proteins may be more easily broken down by natural cellular proteolytic mechanisms.

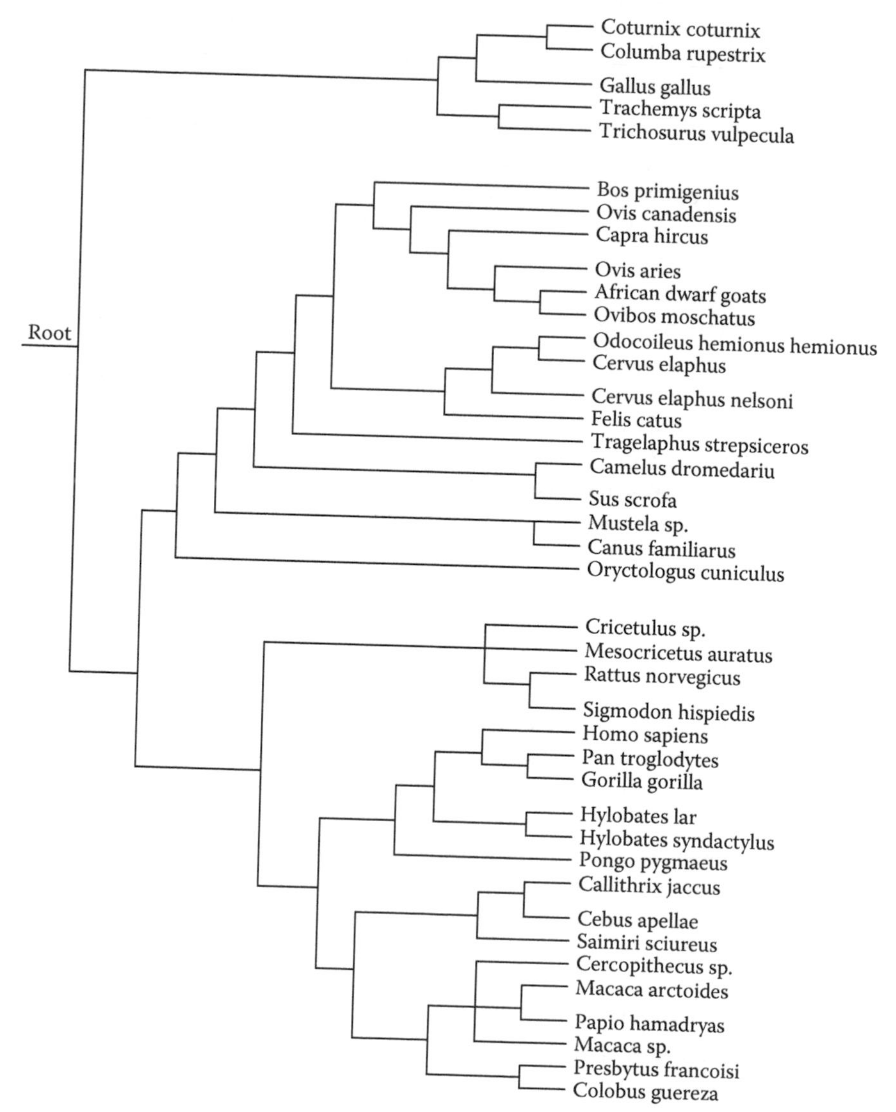

FIGURE 7.21 A phylogenetic tree for PrP^{C} protein based on protein structure. See text for further description.

7.4 POSTTRANSCRIPTION REGULATION OF GENE EXPRESSION

7.4.1 Introduction

Because protein expression is a complex multistep process, and there are literally hundreds of thousands of different proteins, there are many diverse parametric control points in the expression process, some more effective than others. These regulatory points employ posttranscriptional, posttranslational, and epigenetic mechanisms. The fluidity of gene structure is also made use of [Choudhuri 2004]. In eukaryotes, posttranscriptional regulation of gene expression can be the result of a number of possible mechanisms, including but not limited to the following:

1. Modulation of the rate of transcription.
2. Control of mRNA lifetime.
3. Alternative protein splicing.
4. Modulation of the rate of transport of mRNA from the nucleus.
5. Decreases in the length of the 3′ poly-A tail of mRNA, which can shorten its lifetime. (The direct control of mRNA concentration in the cytoplasm appears to be one rapid means of controlling the level of a gene product. Both up- and downregulation are possible in eukaryotic cells by this mechanism [Clark 2000].)
6. RNA editing in which sequences in the transcribed gene are excised and/or substituted. MicroRNAs (miRNAs) and small interfering RNAs (siRNAs) figure in regulatory networks. (See the review article by Choudhuri [2004] for an in-depth review of the plethora of mechanisms regulating eukaryotic gene expression.)
7. The mature protein, once released by its manufacturing ribosome, can be altered molecularly in a variety of ways, altering its activity.

Imagine being able to fight diseases at a specific molecular level by controlling the activity of already-expressed protein enzymes that figure in a disease. An important step toward this end has been described by Skretas and Wood [2005]. They were able to control the activity of certain proteins in *E. coli* by inserting a small synthetic protein molecule called an *intein* and then restoring activity by splicing out the intein. The molecular biology of natural inteins is complex, as you will see later.

7.4.2 Inteins and Protein Engineering

An *intein* (internal protein) is a segment of a *host protein* that is able to excise itself from its host protein, then rejoin the cut N- and C-terminal ends of the host with a peptide bond. This is similar to the behavior of transposons within genes. Inteins are generally found in conserved regions of conserved proteins. Most of the host proteins are active in nucleic acid metabolism and DNA replication and repair. Some inteins have been found in phage and virus genes, as well [Liu 2000]. Inteins are named after the organisms and the host protein in which each is located. As of 2006, more than 200 inteins have been found in more than 100 different species and in all three domains of life: *Archea, Eubacteria,* and *Eukaryotes.* To date, inteins have not been found in metazoan eukaryotes (plants and animals) [Pietrokovski 2004; Gogarten et al. 2002]. However, that does not mean they won't be found in the future. The eukaryote in which inteins have been found to date is the common yeast, *Saccharomyces cerevisiae.* Most reported inteins have two domains: a *self-splicing domain* and an *endonuclease domain.* These are known as *large inteins. Small inteins* lack the endonuclease domain, but they still can splice themselves out of their host protein. A large intein can have from 400 to 500 AAs. The endonuclease and splicing components of a large intein are genetically, structurally, and functionally separable [Belfort 2005]. Inteins are thought to share an ancestry with metazoan so-called *hedgehog proteins*, which undergo a self-cleavage reaction. According to Belfort [2005]: "It is apparent that composite elements like mobile introns, inteins and hedgehog proteins have interchanged functional domains in the course of [molecular] evolution."

An intein is transcribed and translated together with its host protein. (They are not the result of transcribing and translating DNA introns in the gene for the host protein.) Inteins posttranslationally excise themselves from a protein molecule that contains their AA sequence and rejoin the N- and C-terminal ends to make a new, shorter, intact protein called an *extein*. Following the excision of the intein, the remaining *extein* assumes a new 3-D shape and may lose its biochemical activity. If the intein is reinserted at its former location, the protein recovers its shape and activity. Thus, manipulation of an intein can *switch* a protein's activity OFF and ON. Indeed, Skretas and Wood [2005] demonstrated in *E. coli:* ... "that a number of different proteins can be inactivated by insertion of a bioengineered intein, then reactivated by the addition of thyroid hormone via ligand-induced splicing. This conditional activation was also found to occur in a dose-dependent manner. Rational protein engineering was then combined with genetic selection to evolve an additional intein whose activity is controlled by the presence of synthetic estrogen ligands. The ability to regulate protein function posttranslationally through the use of ligand-controlled intein splicing will most likely find applications in metabolic engineering, drug discovery and delivery, biosensing, molecular computation, as well as many additional areas of biotechnology." [Quoted with permission.] Clearly, the therapeutic possibilities of protein manipulation are significant, and further research must be done.

The molecular mechanism of host protein splicing is well understood, and the interested reader should see Gogarten et al. [2002] for examples.

The endonuclease activity of an intein occurs at the gene level. The *homing endonuclease* part of the intein protein is located in the center of the intein, which in turn is located within the host protein. *Homing* is defined as the transfer of a parasitic genetic element to a gene that lacks that element. The result of homing is the duplication of the parasitic genetic element when the cell divides. Homing is one path to *super-Mendelian inheritance*, which promotes a rapid spread of the genetic element in the population. The parasitic gene portion is not an intron, per se, and therefore is not spliced out of pmRNA in the transcription process.

The endonuclease ORF codes the production of endonuclease proteins, which recognize sites of 14 to 40 residues in its target gene. The target gene has no introns or intein codons in it. The intein endonuclease makes a site-specific, double-stranded break in the target DNA. Gene conversion repairs the break and generates an intein-containing product. Thus, when the target gene is expressed, the AAs of the intein appear in the target protein. The *large intein* is truly a genomic parasite; it is reproduced along with the host cell's DNA at cell division.

Here are some questions for the reader to ponder: How did inteins originate? Were they a cooperative fusion of an invasive endonuclease with a self-splicing protein? Did the intein endonuclease originally come from intron-encoded, homing endonucleases? What controls the initiation of intein splicing in proteins and intein endonuclease activity in a cell's genome? How can bioengineered, artificial, small inteins be used to control protein activity to combat cancer and other diseases?

A glimpse of the future applications of bioengineered inteins has been discussed by Belfort [2005]:

> By using the basic chemistry of the intein, mutant [engineered] intein-protein fusions can be chemically ligated to a protein or peptide with an N-terminal cysteine residue. This process has been variously termed expressed protein ligation (EPL) or intein-mediated protein ligation (IPL) [Perler 2005]. This EPL/IPL technology has already had enormous application as for example in protein stabilization and potentiation by cyclization, segmental labeling for NMR, or tagging with different reporters. Equally impressive are *in vivo* intein-based technologies, many of them based on split inteins, and many using fluorescence or light as the reporter [sensing modality]. These include monitoring of intracellular protein-protein interactions; identification of proteins specifically imported into mitochondria, endoplasmic reticulum, or nucleus; and even generation of safer transgenic plants.
>
> This staggering array of technology that has developed recently is underscored by the increasing number of papers dedicated to the subject. Looking into the crystal ball, we can envision the proliferation of

intein-based biosensors, proteomics utilizing peptide arrays, and even functional proteomics in whole animals. Nature's gift, indeed! [Quoted with permission.]

To be able to target and manipulate the activity of specific enzymes in cancer cells would indeed be wonderfully beneficial. It appears that engineered intein biochemistry will figure in future research on cancer, autoimmune diseases, allergy, and certain genetic diseases.

7.5 PROTEIN DESTRUCTION AND AA RECYCLING

7.5.1 Proteolytic Enzymes

Proteolytic enzymes, aka *proteases, peptidases, and peptide hydrolases,* are protein enzymes that catalyze the hydrolysis of the covalent peptide bonds in proteins. Recall that a peptide bond between two amino acids is formed between the amino group of one AA and the carboxyl group of the second. The carboxyl loses an **-OH** radical and the amino a hydrogen ion. Peptide bond energy is about 69 Cal/mole. In this discussion, we will call proteolytic enzymes proteases for brevity. Most proteases are expressed as inactive precursors *(zymogens)* that themselves require limited proteolysis for activation. Thus, the activator proteases must be expressed before the main proteases can operate. There are many, many proteases, some of which include the following: *carboxypeptidases, caspases, cathepsins, elastases, ficin, renin, rennin, thrombin, urokinase,* etc.

There are many proteases found outside cells. Some are used to destroy protein hormones once they have acted, e.g., in blood clotting and the complement system, and others are involved in such mundane tasks as the digestion of food. *Serine proteases* are so called because the active sites of this class of enzyme all contain the AA serine. The three pancreatic digestive enzymes, *chymotrypsin, trypsin,* and *elastase,* are serine proteases with serine as the 195th AA. Several activated clotting factors are also serine proteases: *factor 10, factor 11, thrombin, and plasmin.* The immune system's complement proteins, *Clr, Cls,* the *C3 convertases, C4b2a,* and *C3bBb,* are serine proteases. Other serine proteases also exist: *Subtilisin* is secreted by *Bacillus subtilus.* Although the enzyme *cholinesterase* breaks down the nonprotein neurotransmitter acetylcholine, it is homologous to other serine proteases [Kimball 2002].

Internal proteases are required to break down damaged proteins inside cells. Intracellular proteolysis is required because:

1. It creates a local supply of AAs for new protein synthesis.
2. It removes excess enzymes as part of downregulation processes.
3. It removes transcription factors that are no longer needed.
4. It is necessary for apoptosis (programmed cell death).
5. It is required for cell lysis.

Similar to many biochemical processes, proteolysis makes use of inhibitory control. *Serpins* are serine protease inhibitor molecules. They work by mimicking the normal 3-D structure of the protease target protein. Thus, the serpin binds to the serine protease, blocking its action. Now the protease makes a cut in the serpin chain leading to the formation of a covalent bond linking the two molecules. This produces a large allosteric change in the tertiary structure of the serpin, which leads to the destruction of the serine protease. About 20% of the proteins found in blood plasma are serpins, underscoring the importance of putting a stop to proteolytic activity when the physiological need for it has passed. The systems most affected are immune complement and blood clotting.

7.5.2 The Proteosome

The proteosome is a massively complex, large (700 kDa), intracellular protein complex that is specialized for disassembling tagged intracellular proteins. The components of the proteosome are the

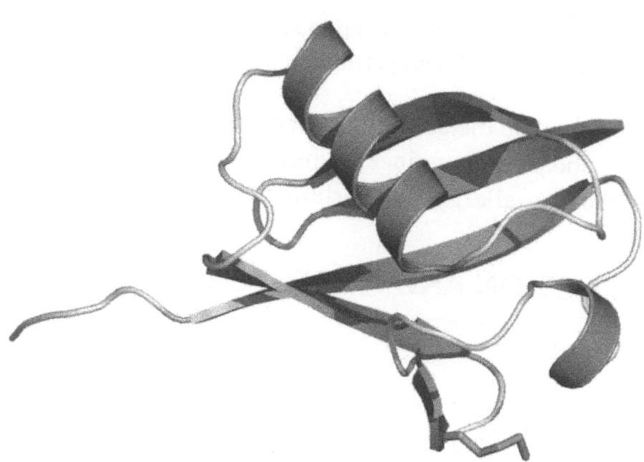

FIGURE 7.22 (See color insert.) 3-D "ribbon" schematic of the ubiquitin protein molecule.

core particle (CP) and the *regulatory particle* (RP). The CP is made from two copies of each of 14 different proteins that are assembled in groups of seven, forming a ring. Four rings are stacked on each other like tires. There are two identical RPs, one at each end of the CP. Each RP is made of 14 different proteins (none of them found in the CP). Six of these are ATPases, whereas others have binding sites for the 76 AA protein, *ubiquitin*. *Ubiquitin* (Ub) is highly conserved throughout all the kingdoms of life; its role must have evolved billions of years ago. It targets proteins for destruction by the proteosome. It is found in all human cells. Figure 7.22 illustrates a 3-D molecular ribbon diagram of ubiquitin. The intracellular *ubiquitin-proteasome pathway* (UPP) is responsible for the degradation of many cellular proteins, including abnormal, damaged, and misfolded proteins, cell-surface receptors, ion channel proteins, endoplasmic reticulum proteins, cell cycle regulators, transcriptional activators, growth and differentiation-controlling factors, and antigenic proteins targeted for presentation to immune cells on class I MHC molecules. The UPP is responsible for the cachexia (wasting) seen in cancer and other diseases, and its suppression and control is now a target for new drug development [Hall 2005].

The ubiquitination of proteins is regulated by the three E_k *ubiquitin ligase* enzymes (k = 1, 2, 3). These enzyme complexes each recognize a subset of substrates and tag them by linking the terminal carboxyl of ubiquitin with an amino group on the target protein with an *amide bond*. Amino groups are found on the R-groups of the AAs, *arginine, asparagine, glutamine, and lysine*. Ubiquitination by the ligase is itself regulated by specific hydrolytic enzymes called *deubiquitinases,* which can remove ubiquitin from its amide bond and thus rescue the tagged protein from destruction by a proteosome [Glickman & Adir 2004].

The protein destruction process begins when an ubiquitin molecule binds its carboxyl to the amino group of a lysine residue in the target protein. An additional ubiquitin molecule binds to the first ubiquitin at its lysine_{48}, forming a dimer. Additional ubiquitins (k) can bind to the lysine_{48}s of the previous ubiquitins (k – 1), forming an ubiquitin chain. The target protein–ubiquitin complex now binds to the RP, where the *target protein* (TP) is unfolded by ATPases using the energy of ATP to break bonds. The unfolded target protein moves into the central cavity of the CP, where several active sites on the two central CP rings cleave various specific peptide bonds, producing linear peptide fragments about eight AAs in length. If the polyubiquitin chains attached to the target protein are short, the proteasome may deubiquinate the target protein, releasing both the ubiquitin (for reuse) and the TP. Long polyubiquitin chains ensure that the CP will chop the TP. The octapeptide fragments leave the proteasome, where they are further broken up in the cytosol by *peptidase enzymes*. Miraculously, the ubiquitin molecules are released from the CP intact for reuse. In mammals, some octapeptide fragments may be incorporated into *Class I Histocompatibility Molecules* [Kimball 2004f].

Proteosomes also act as components of the immune system. In mammals, activation of the immune system leads to the release of *interferon gamma* (IFNγ). Acting on a cell, IFNγ causes three of the toroidal subunits of the CP to be replaced with alternate subunits. The peptide fragments generated by this modified CP are picked up by *TAP proteins* (transporter associated with antigen processing) and carried to the *endoplasmic reticulum,* where each enters the groove at the surface of a *Class I Histocompatibility Molecule.* This complex then moves through the *Golgi apparatus* and is inserted into the cell's plasma membrane, where it can be recognized by proteins on the surface of $CD8^+$ T-leucocytes. Kimball [2004f] speculated that it is no coincidence that the genes encoding the three substitute core particle rings, the TAPs, and all the MHC molecules are located together on chromosome 6 in humans.

7.5.3 Proteolysis in Apoptosis and Cell Necrosis

We have already described the major events leading to programmed cell death (*apoptosis*) in Chapter 2, Section 2.5.2. In this section, we will focus on the biochemical mechanisms used to destroy intracellular proteins in the processes of apoptosis and cell necrosis. Proteolysis in apoptosis is mostly carried out by a family of 12 to 13 *caspases* [Kroemer & Martin 2005]. Because caspase activity leads to apoptosis, the caspase molecules inside a cell are held in inactive form, and caspase activity is tightly regulated until a plurality of external or internal molecular signals for PCD are received. Caspase activation is complex and is triggered by *adaptor proteins.*

Caspases are a group of cysteine proteases that exist within all cells as inactive protein *zymogens.* These zymogens are cleaved in a cascading fashion (reminiscent of the molecular strategy used to activate immune complement) to activate them as proteolytic caspases. An initial step in activating the caspases is the release of *cytochrome* **c** molecules from damaged mitochondria. A cytochrome **c** molecule binds to a molecule of the cytosolic protein *Apaf-1.* ATP is required for this binding. A wheel-like molecular complex of seven molecules each of (*Apaf-1 + cytochrome* **c** *+ ATP)* then forms. This complex is the *apoptosome;* it recruits seven molecules of *procaspase-9* to the complex. In an arcane molecular process, the procaspase-9 is converted to active caspase-9. Caspase-9 then cleaves other procaspases, activating them, forming a chain reaction that leads to active caspase-3 and -6. These caspases are responsible for chopping key cellular proteins, leading to apoptosis. The caspase cascade of activation is complicated by the fact that certain caspases can autoactivate (e.g., caspase-3 can cleave procaspase-3, making more caspase-3) or activate other caspases (e.g., caspase-8 and -9 catalyze the proteolytic activation of caspase-3).

The complexity of the caspase system is underscored when we examine the *Bcl-2* and *IAP* families of proteins. The Bcl-2 family of intracellular proteins helps regulate the activation of procaspases. *Bcl-2* and $Bcl\text{-}X_L$ inhibit apoptosis by blocking the release of cytochrome **c** from mitochondria. The proteins *Bad* and *Bax,* on the other hand, stimulate the release of cytochrome **c**. If the genes for *Bad* and *Bax* are inactivated, cells are quite resistant to most apoptosis-inducing stimuli. *Bax* and *Bad* are themselves activated by other members of the Bcl-2 family, such as *Bid.*

IAP proteins inhibit apoptosis in two ways: (1) they bind to some procaspases to prevent their activation and (2) they bind to caspases directly to inhibit their activity. IAP proteins were originally discovered to be produced by certain insect viruses, their purpose being to put off cell apoptosis until the virus had time to replicate in the cell. Curiously, the cells developed countermeasures: When damaged mitochondria release cytochrome ***c***, they also release an *IAP blocker protein*, preserving apoptosis efficiency.

7.5.4 Autophagy

Autophagy is an evolutionarily conserved eukaryotic cellular process that is regulated by environmental conditions and by endogenous and exogenous biochemical signals. In the catabolic process of autophagy, portions of a cell's cytoplasm are sealed off in a double-membrane "bubble"

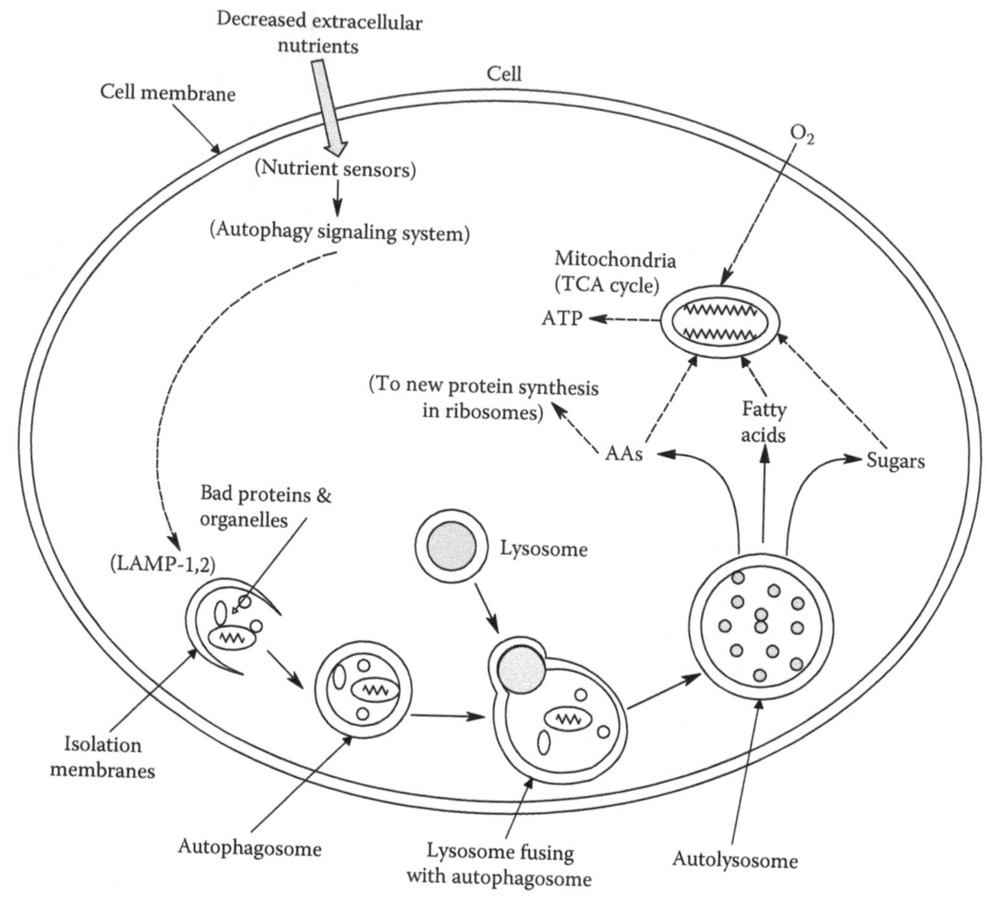

FIGURE 7.23 Schematic diagram of the *autophagy* process in a mammalian cell. This can be triggered by a lack of nutrients or the growth factor protein (GF).

(*autophagosome*) and delivered to a degradative organelle (the *vacuole* in yeast and the *lysosome* in mammalian cells). Inside the autophagosome can be proteins, cellular organelles, membrane fragments, bacteria, etc. There, hydrolytic enzymes break down the contents of autophagosome into their components (short oligopeptides, AAs, lipids, sugars, etc.) which are recycled in the cell's metabolic processes [Kimball 2005m]. The process by which the autophagosome forms is poorly understood. The source of the bilayer lipid membrane may be in the rough endoplasmic reticulum.

Autophagy occurs in yeast and other eukaryotic cells under starvation conditions, presumably to enrich the pool of substrate molecules needed for vital biosynthesis pathways at the expense of damaged or expendable macromolecules. Autophagy under starvation conditions is necessarily a stop-gap, self-limiting process for redistributing vital resources (AAs, sugars, lipids, etc.). Eventually, a starving cell will run out of substrate molecules required for metabolism, DNA repair, etc., and will ultimately die of necrosis. Broadly, necrosis occurs when the rate of autophagic substrate production plus new inputs is less than the starving cell's metabolic requirements [Edinger & Thompson 2004]. Figure 7.23 illustrates schematically the steps in mammalian cell autophagy triggered by signals resulting from lack of nutrients. This nutrient stress can be produced by birth and *growth factor* (GF) deprivation. Nutrition is temporarily interrupted at birth because the umbilical cord is lost. Lack of GF also causes limited cell nutrition, stimulating autophagy.

Mammalian cells that have been invaded by pathogens, such as the bacteria *Listeria monocytogenes, Mycobacterium tuberculosis,* etc., can surround the pathogen in the cytoplasm with an autophagosome and literally disassemble it by autophagy. This breakdown of cytoplasmic bacteria has also been called *xenophagy* because they are foreign invaders [Levine 2005]. Certain bacteria such as *Legionella pneumophila, Salmonella, Mycobacterium tuberculosis, Brucella abortus,* and *Porphyromonas gingivallis* have evolved means of blocking the autophagosomes surrounding them and can reside inside these intracellular vacuoles. *M. tuberculosis* can reside inside immature autophagosomes, but if the infected cells are treated with IFNγ or subjected to starvation conditions, the autophagic process is triggered and these bacteria are destroyed. Figures 7.24 and 7.25 illustrate schematically the fate of bacteria such as *M. tuberculosis* that are phagocytosed by a cell (e.g., a macrophage) and the fate of Group A *Streptococcus* that "hide" in a cell. Some bacteria such as *Shigella flexneri* and *Rickettsia conorii* can escape from phagosomes into the cytoplasm and multiply. Under some conditions, they are destroyed by autophagy [Ogawa et al. 2005; Rich, Burkett & Webster 2003; Gorvel and de Chastellier 2005].

The molecular mechanism of autophagy is quite complex (but then, few biochemical processes are not). Wang and Klionsky [2003] list some 43 known different protein enzymes involved with autophagy and the related *cytoplasm-to-vacuole targeting* (Cvt) pathway in yeast cells. In yeast (*S. cerevisiae*), double-membrane autophagosomes 300 to 900 nm in diameter form under conditions of nutrient deprivation. Cvt vesicles 140 to 160 nm in diameter are generated through the Cvt pathway

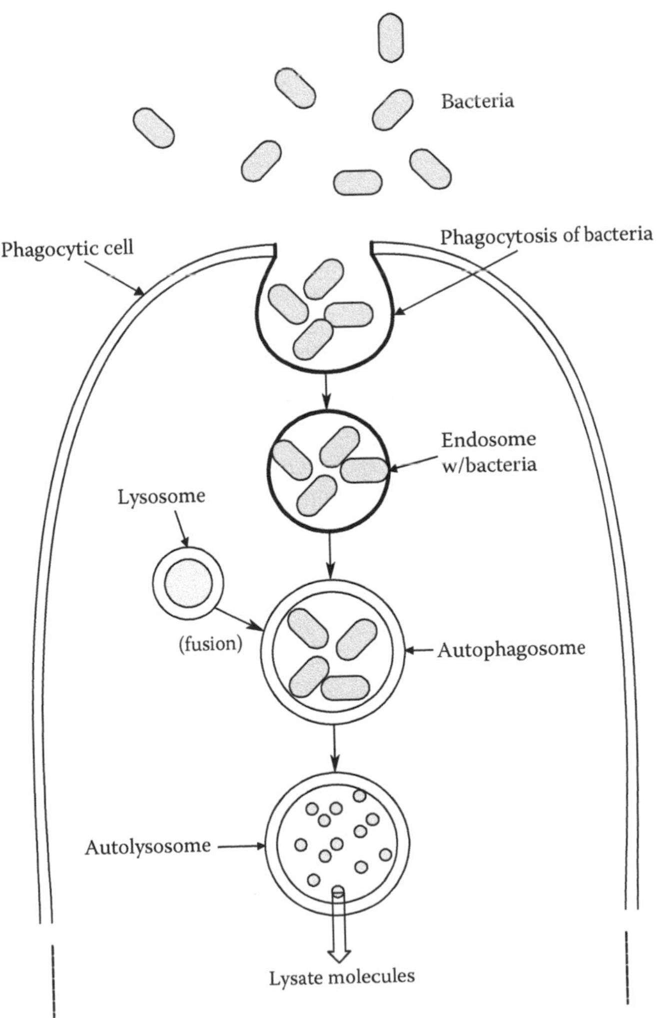

FIGURE 7.24 Schematic of the *xenophagy* process, in which a phagocytic cell ingests bacteria or virions and then "digests" them by phagosomes.

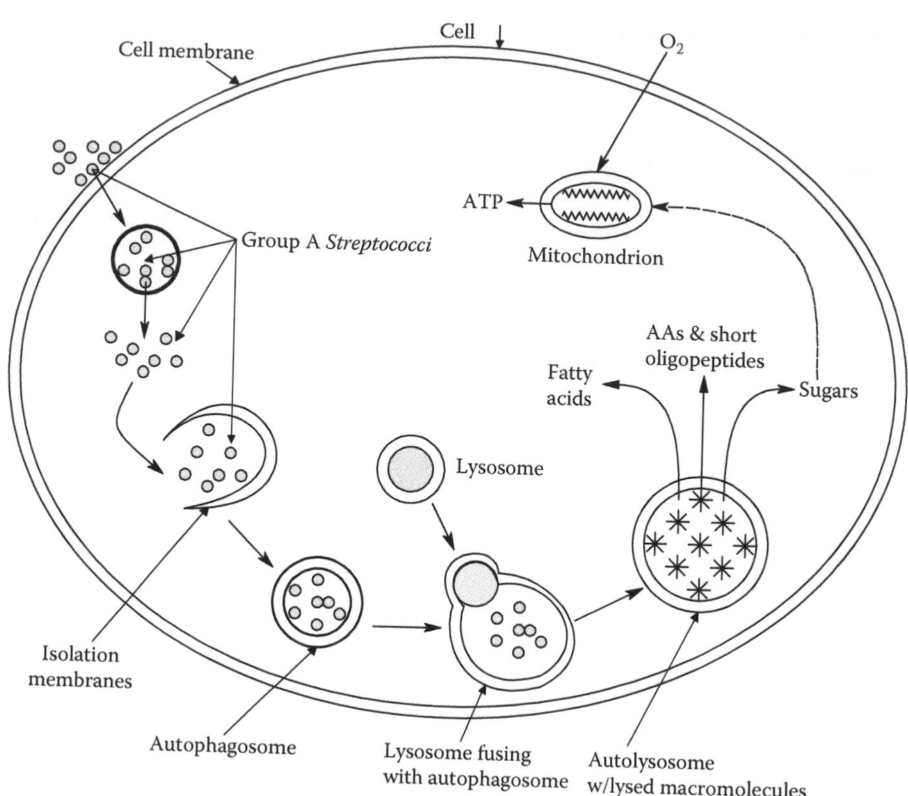

FIGURE 7.25 Another xenophagy scenario. *Streptococci* bacteria enter a cell, and isolation membranes form around them, forming autophagosomes. A lysosome complex of proteolytic enzymes fuses with the autophagosome, leading to the digestion of the bacteria. Their component molecules are recycled within the cell.

under nutrient-rich conditions. Both Cvt vesicles and autophagosomes (and their contents) fuse with lysosome vacuoles, where hydrolytic enzymes disassemble the vesicle's macromolecular contents for recycling.

Because autophagy causes a cell to literally eat itself to supply vital nutrients, the process must be highly regulated under normal conditions. Loss of regulation of autophagy has been associated with several genetic diseases; e.g., *Danon's myopathies, cancers, Alzheimer's, and Huntington's disease.*

Danon's syndrome is a particularly interesting genetic disease. The symptoms of Danon's syndrome are heart problems, mental retardation, and muscle weakness. It is a rare, X-linked disorder (at Xq24) in which a mutated, ineffective form of a protein enzyme, *Lysosomal-Associated Membrane Protein 2* (LAMP-2), is made. Normal LAMP-2 is a heavily glycosylated protein of 410 AAs that operates to regulate the autophagic process. A LAMP-2 molecule is a transmembrane protein fixed in the lysosomal membrane; its bulk lies inside the lysosome, and its short C-terminus protrudes into the cytoplasm. LAMP-2 is thought to protect the lysosomal membrane from proteolytic action within the lysosomes and to act as a receptor for proteins to be imported into lysosomes.

Danon's syndrome is a form of autophagic vacuolar myopathy in which the autolysosomes are surrounded by "secondarily-generated membranes containing sarcolemmal proteins, basal lamina, and acetylcholinesterase activity" [OMIM 2006a, 2006b]. Clearly, normal LAMP-2 is critical for normal autophagy. In LAMP-2 deficiency, a blockage of normal autophagy occurs with the intracellular accumulation of autophagic material in autophagosomes.

In some scenarios, apoptosis and autophagy appear to be causally interconnected. This interconnection can be positive or negative, supporting the concept of bistable "molecular switches" between them. Apoptosis can activate autophagy, and in some cases autophagy is required for efficient apoptosis [Wang & Klionsky 2003]. It also appears that mitochondria may figure in integrating the two types of cell death. When cells become malignant, autophagy is frequently suppressed, allowing cancer cells to proliferate [Gozuacik & Kimchi 2004].

7.6 SUMMARY

If DNA stores the plans and operational sequencing for an organism's life processes and RNA is the executive carrying out the instructions, then protein molecules are the tools by which life is implemented. Unfortunately, for human understanding of the roles proteins play, complexity rules in proteomics. Some 23,000 genes translate into over 230,000 different proteins. Understanding how a particular protein came to be involves understanding not only the protein's 4-D, 3-D, 2-D, and linear AA structures but the posttranslational factors acting on that protein. In the writers' opinion, the discipline of proteomics is orders of magnitude more complex and challenging than genomics.

The reader must appreciate that proteomics, and indeed all of molecular biology, even life itself, depends on the interactions of specific molecular 3-D structures and molecular charge distributions that enable molecules to have highly specific chemical reactions. Genetic mutations or posttranslational damage can cause specific protein enzymes to malfunction, causing *genetic diseases* (cf, Section 7.3 and Chapter 8) such as the prion diseases. In considering protein enzymes and membrane receptors, it is clear that the molecular specificity between a given signal protein and its receptor generally relies on only limited regions of both molecules. Also, specificity can be decreased (or increased) by even single AA substitutions in a critical structural region. In fact, the potential for molecular "cross-talk" exists between signal proteins and receptors, leading to *pleiotropy*.

In this chapter we have reviewed the various classes of proteins in animals. Plants are no less important, and as eukaryotes, also make use of proteins, including the world's most plentiful protein enzyme, RUBISCO (cf Chapter 4, Section 4.4.5), and many of the proteins found in animal cells.

CHAPTER 7 HOME PROBLEMS

7.1 Describe the molecular mechanisms a cell uses to maintain the osmotic pressure of its cytoplasm.

7.2 Describe, using block diagrams, the sequence of physiological operations the human body uses to restore the normal blood and extracellular fluid potassium concentration when a person is given a sublethal IV bolus dose of concentrated KCl solution. The person is allowed to drink water and urinate, as needed.

7.3 Describe, giving examples from prokaryote cells and eukaryotes, how the design of the "advanced," "9 + 2" cilium or flagellum may have evolved from more primitive cilia and flagella designs.

7.4 Describe *chaperonins.* Where and in what species are they found? What is their relation to *heat shock proteins?* List some that have been identified in humans. [To begin, see Wegrzyn, R.D. and E. Duerling. 2005. Molecular guardians for newborn proteins: ribosome-associated chaperones and their role in protein folding. *Cell. Mol. Life Sci.* 62: 2727–2738.]

7.5 Describe Group I and Group II chaperonins.

7.6 Describe the structure of the well-studied GroEL–GroES chaperonin complex in *E. coli.* How do you think this complex protein structure reaches its final functional 3-D form?

7.7 The human chaperonin CH60 (aka Hsp60) AA chain has been sequenced. What is its length and molecular weight? On which chromosome is its gene located? Where is this protein found?

7.8 What genetic diseases are associated with defects in human chaperonin-60 (Hsp60) protein? [See Ranford, J.C. and B. Henderson. 2002. Chaperonins in disease: mechanisms, models, and treatments. *J. Clin. Pathol: Mol. Pathol.* 55: 209–213.]

7.9 Describe how ionizing radiations (UV, γ-, X-, α-, and β) can damage proteins. How does this damage affect protein function?

7.10 The plant lectin protein toxins *abrin* and *ricin* induce cell apoptosis. Describe how they do this. Compare their LD50s to that of botulinus toxin.

7.11 What is meant when a certain class of cells is said to be $CD20^+$ (in general, $CDXY^+$)?

8 The Genetic Basis for Certain Inheritable Diseases: Genomic Medicine

8.1 INTRODUCTION

As man has learned more about the structure of the human genome, its command and control pathways, and the genomes of other animals, plants, bacteria and yeasts, he has also learned how to manipulate DNA and mRNA, including the insertion of recombinant genes into existing genomes. As our understanding of the human genome has expanded, so has our realization that many human diseases, conditions, and predispositions to disease are correlated with mutations of specific structural and regulatory genes.

The term *medical genomics* (aka *genomic medicine*) describes a rapidly growing, broadly interdisciplinary branch of medicine that fuses genomics, proteomics, and molecular biology with diagnostic and therapeutic medicine. In *medical genomics*, diseases having genetic etiologies are identified, studied and, in some cases, genetic therapies are developed. Genetic diseases range from the incurable (e.g., cystic fibrosis, Creutzfeldt–Jakob disease, Huntington's disease), in which death is premature, through a spectrum of symptomatic severity to those conditions in which there is a predisposition for developing a late-in-life condition such as colon or breast cancer.

Medical genomics requires *genetic diagnostics* to allow the identification of presymptomatic, preclinical, at-risk individuals having one or more known genes for a heritable disease. Such testing is possible on adult and embryonic cells because we know, for many diseases, the exact chromosome and loci for these genes [BUMC 2005] and their mutations. Genetic disease diagnosis is even done on single cells taken from a four-cell, in vitro fertilized embryo before implantation. Genetic diagnostics can find genetic determinants of the molecular basis of disease (functional polymorphisms or modifier genes) and also identify genetic determinants influencing responses to drug therapy (*pharmacogenomics*). Complicating genetic diagnosis is the fact that diseases such as breast cancer, colorectal cancer, human immunodeficiency virus (HIV) infection, tuberculosis, Parkinson's disease, Creutzfeld–Jakob disease, and Alzheimer's disease are all believed to be due to the interactions of multiple genes and environmental factors. They are thus known as *multifactorial genetic disorders*. Genetic variations in these disorders may have a pathological or protective role in the expression of the diseases [Guttmacher & Collins 2002].

Some human genetic diseases and medical conditions are the result of *single-gene mutations*. These include *hereditary hemochromatosis* (approximate incidence: 1 in 300 persons), cystic *fibrosis* (1 in 3,000), *alpha-1-antitrypsin deficiency* (1 in 1,700), and *neurofibromatosis* (1 in 3,000) [Guttmacher & Collins 2002]. Other diseases are caused by defects in two or more genes. There are a number of ways these mutations can occur and be inherited.

Some gene mutations are due to a *single-nucleotide polymorphism* (SNP). That is, a single nucleotide is substituted in a gene's codons, resulting in a single AA substitution in the protein product. According to Guttmacher and Collins, there are about 10^7 SNPs in the human genome.

Obviously, some SNPs must be innocuous; others can be very bad. Guttmacher and Collins give a diagram illustrating the effects of *point mutations* in the protein coding of a hypothetical 10-codon DNA oligo. In the "normal" scenario, the **TCA** codon codes for the AA *serine* in the protein product. In a *silent mutation,* this codon is changed to **TCC**, which still codes serine; the polypeptide product is unchanged. A *conservative missense mutation* occurs when the codon in question is mutated to **ACA** coding the AA *threonine,* which is molecularly similar to serine. A *nonconservative missense mutation* occurs when the codon becomes **CCA** that codes the AA *proline,* which is molecularly dissimilar to serine in its chemical properties. The *nonsense mutation* of a codon to **TAA**, which signals STOP to the mRNA coding the oligo, results in a prematurely truncated peptide and severely altered function. Likewise, a *frame-shift mutation* in which an extra **G** is inserted into the **TCA** codon between the **C** and the **A** in the exemplar codon creates **TCG**, which still codes serine, but all the downstream codons are shifted by one nucleotide, resulting in a string of altered codons, hence, a peptide with a string of anomalous (wrong) AAs following the serine. The latter three mutations described earlier are most serious and can lead to a nonfunctional protein product that is broken down by proteases and its AAs recycled.

Mutations in the nongene portions (about 99%) of the human genome can affect *gene regulation.* A gene can be silenced (not turned on), be turned on at the wrong time, or be permanently turned on, as in the case of the growth hormone gene in transgenic salmon (cf Chapter 10, Section 10.2.3.2). Some examples of regulatory mutations associated with disease are listed by Guttmacher and Collins [2002]: (1) the flanking region of the **FMR1** gene (causing fragile X syndrome), (2) the insulin gene flanking region (increasing the risk of type 1 diabetes), (3) the regulatory site of the type I collagen gene (increasing the risk of osteoporosis), and (4) the intronic regulatory site of the calpain-10 gene (increasing the risk of type 2 diabetes).

Guttmacher and Collins also give an example of a beneficial frame-shift mutation in the chemokine receptor gene, **CCR5.** Persons who are homozygous for this mutation are almost completely resistant to HIV1 infection, and those heterozygous for the mutation have slower onset of AIDS following initial HIV1 infection. **CCR5** is an important cell-surface molecule that enables HIV1 to enter the leukocytes it infects.

In discussing genetic etiologies for disease, the term *high penetrance* is applied to a gene that has a high likelihood that a person carrying it in mutated form will exhibit an altered phenotype (leading to disease) [Guttmacher & Collins 2002].

Genetic testing has been defined by Burke [2002]: "The analysis of human DNA, RNA, chromosomes, proteins, and certain metabolites in order to detect heritable disease-related genotypes, mutations, phenotypes, or karyotypes for clinical purposes." It should be noted that not all tests on gene structures yield positive results. This is because not all mutations in a large gene with several introns are currently detectable; also, altered posttranslational processing of polypeptides can lead to a dysfunctional protein 3-D structure. That is, the problem may lie with the gene for a *chaperone* protein that acts on the target protein.

8.2 EXAMPLES OF GENETIC DISEASES

Many years ago, physicians observed that certain diseases appeared heritable, that is, they ran in families. It waited for the genomic age for scientists to actually locate changes in specific human genes that were associated with certain diseases. The list of genomic disease genes is large and growing daily [OMIM 2006c]. *A partial list of examples includes: Hereditary nonpolyposis colorectal cancer* (mutations in the genes for **HNPCC**, **APC**, **p53**, and **K-ras**), *type 1 diabetes risk* (**MODY1**, **-2**, and **-3**), *Alzheimer's disease* (genes for **presenilin-1** and **-2, amyloid precursor protein** genes, ***apolipoprotein E ε4 allele***), *thyroid medullary cancer* (mutation in the **RET** oncogene), *sickle-cell anemia* (single specific mutation in the β-globin gene), *Huntington's disease* (autosomal dominant mutation on HD gene), *hemochromatosis risk* (two mutations of the **HFE** gene, C282Y and H63D, promote excess accumulation of iron in the body), *venous thrombosis risk* (clotting factor V

Leiden gene mutation), the *Down syndrome* (chromosome staining test to see if subject has extra copy of chromosome 21 [aneuploidy]), etc. The following text provides more detail.

Cystic fibrosis: The cystic fibrosis transmembrane conductance regulator (CFTR) gene on locus 7q31.2 on the long (q) arm of chromosome 7 has mutations. About 70% of mutations in CF patients result from the *deletion* of 3 bps in the CFTR nucleotide sequence. This deletion causes the loss of the AA phenylalanine at position 508 in the protein. Normal CFTR is large protein with a complex 3-D shape. It has 1,480 AAs and a molecular weight (MW) of 168,173 Da. Normal CFTR protein forms transmembrane chloride channels located in the cell wall membranes of cells that line the passageways of the lungs, liver, pancreas, gut, reproductive tract, and skin. That is, it is found in epithelial cells. CFTR loss also affects the regulation of other transport pathways involving Na^+ and K^+. One symptom of CF is the formation of thick, sticky mucus by epithelial cells in the lungs that cannot be cleared by cilia. This mucus traps bacteria and mold spores, resulting in chronic infections that are the leading cause of mortality among CF patients [doegenomes 2003]. Mucus from CF also blocks the normal secretion of pancreatic digestive enzymes, causing the reduced ability to digest food.

A simple, painless, noninvasive, functional diagnostic test for CF is the *sweat test*. The parasympathomimetic drug *pilocarpine* is carried into the skin of the forearm electrophoretically. It acts on the muscarinic cholinergic receptors of the parasympathetic nervous system. Applied to the skin, it causes profuse sweating and vasodilation. The sweat is collected and analyzed for its Na^+ and Cl^- concentrations. An NaCl concentration above a threshold value is a sign of CF; two separate tests are given.

A man and a woman each carrying one CF gene have a 1:4 chance of having children who express the disease (are homozygous for it). There are more than 1,200 different mutations of the CFTR gene that can lead to CF (some show milder symptoms than others). About 1 in 3,600 Caucasian children are born with CF, versus 1 in 17,000 African Americans and 1 in 90,000 Asian Americans [Padman 2004].

Breast and ovarian cancer risk: The high penetrance genes for breast and ovarian cancer are BRCA1 and BRCA2. These autosomal dominant DNA repair genes are located on chromosomes 17q and 13q, respectively. Women with faulty copies of **BRCA1** or **BRCA2** have a 50–80% chance of getting breast cancer in their lifetimes [Malone et al. 1998]. However, the mutations are rare and account for only 5–10% of all breast cancer cases. It has been suggested that a third gene **BRCA3** may be involved in the genesis of breast cancer, but the evidence for this third gene has been weak [Wooster & Weber 2003]. As it has recently turned out, there are four new genes associated with breast cancer.

Recently, Easton et al. [2007], Stacey et al. [2007], and Hunter et al. [2007] have reported on large-scale studies that have identified four new genes that significantly affect a woman's risk of developing breast cancer. In one study (Easton et al.), the genomes of 4,398 women with breast cancer and those of 4,316 controls were examined, in detail. The mutated gene with the strongest correlation with breast cancer is *fibroblast growth factor receptor 2* (FGFR2) found on chromosome 10. Evidently in its normal form, FGFR2 suppresses tumors. It was found that the presence of one copy of the mutated FGFR2 gene raises the risk of breast cancer by 20%. If a woman carries two copies of mutated FGFR2, the risk increases to 60%. FGFR2 codes a protein receptor for a *tyrosine kinase*. Three other mutated genes found statistically correlated with breast cancer are called **TNCR9**, **MAP3K1**, and **LSP1**. The function of these genes in inducing breast cancer is not known.

Creutzfeld–Jakob disease (CJD): This is a human *prion protein* disease treated in detail in Chapter 7, Section 7.3.4 of this text. CJD can be inherited (15% of cases) or acquired. *Variant CJD* (vCJD) is acquired from cattle expressing *bovine spongiform encephaly* (BSE). The majority of human cases of CJD are classified as *iatrogenic* or *sporadic* (sCJD). The gene coding the prion protein (PrP) is located at 20pter-p12 [OMIM 2006d]. Normal PrP^C has 253 AAs; its putative normal functions are discussed in Chapter 7, Section 7.3.4 and will not be repeated here. The PrP protein is an excellent example of how a small AA mutation can lead to a major structural change from a normal (PrP^C), largely alpha-helical structure to a pathological form (PrP^{Sc}) rich in beta-pleated sheets. Also remarkable is the rare documented interspecies prion disease spread between sheep and cattle and between cattle and humans.

Spontaneously occurring CJD disease in humans is inherited as an *autosomal dominant.* Ten to 15% of CJD patients have inherited at least one copy of a mutated **PrP** gene. Some of the more common mutations are (1) substitution of the AA lysine for the normal glutamic acid at codon 200, (2) substitution of asparagine for aspartic acid at codon 178 when it is accompanied by a polymorphism in the gene normally encoding valine at codon 129, and (3) a change from valine at codon 210 to isoleucine. A detailed comparison of the prion protein orthologs in humans, mice, dogs, sheep, pigs, and cattle is available online from Winter [2005].

The prion protein gene (PRNP) is also responsible for *Gerstmann–Straüssler–Scheinker disease* (137440) and *fatal familial insomnia* (600072) in humans (numbers are OMIM indices), as well as a host of fatal prion diseases in animals: *scrapie* in sheep, *transmissible spongiform encephalopathies* (TSEs) in cattle, as well as *chronic wasting disease* (CWD) in deer and elk, etc. Details about the human PRNP gene and the structure of the normal PrP^C protein can be found at OMIM [2006d].

Parkinson's disease: PD is associated with mutation on the **PARK8** gene located on chromosome 12 that encodes a protein called *dardarin.* (*Dardara* is the Basque word for tremor, a major symptom of PD.) Investigators studied the genomes of five families with a history of PD living in the Basque region of Spain and in England. They found two mutations in the **PARK8** gene, one associated with the Basque families and the other linked to the disease in the English family. There are no doubt other genes associated with PD (e.g., for α-synuclein), which is complex in its etiology, chronic, and progressive. About 50,000 Americans are diagnosed with PD every year [BrightSurf 2004].

Table 8.1 summarizes direct DNA tests for certain genetic diseases available at the Boston University Medical Center. In some cases, specific mutations are analyzed (not shown in table).

8.3 GENOMIC MEDICINE

8.3.1 Introduction

The term *genomics* was announced and defined in 1987 [Koska 1998]. In the past 20 years or so, the ability of researchers to examine the details of genes in individuals having certain apparently heritable diseases such as breast or colon cancer, CJD, Alzheimer's disease, etc., has led to the correlation of certain gene mutations with specific diseases. In some cases, there is little modern medicine can do to mitigate the effects of these adverse mutations; in other cases, biochemical therapies can mitigate the effects of the mutations but not cure the disease.

Guttmacher and Collins [2002] gave two examples of genetic diseases that have had effective therapies: *In one,* a woman has had deep venous thrombosis, earlier associated with taking oral contraceptives. (Her mother also had this condition.) The woman becomes pregnant. A genetic test shows she has a heterozygous mutation in the gene for *Factor V Leiden* that affects clotting. To avoid problems with possible estrogen-related thromboemboli, she is treated with prophylactic injections of the anticlotting agent *heparin.* No emboli developed and a healthy term infant was delivered.

In the second example, a 4-year-old boy has *acute lymphoblastic leukemia.* He is treated with chemotherapy, which includes daily oral *6-mercaptopurine.* (6-mercaptopurine is a cancer chemotherapy drug that inhibits purine synthesis and incorporation into cancer cell DNA. It is carcinogenic and teratogenic in humans.) A genetic test has shown that the boy is homozygous for a mutated gene that encodes the protein enzyme *thiopurine S-methyl-transferase*, which inactivates mercaptopurine. A decreased expression for this enzyme means that the chemotherapy must use greatly reduced amounts of mercaptopurine to prevent overdose.

Recent research has shown progress in treating genetically based autoimmune diseases. Peter Donnelly of Oxford University and a consortium of scientists analyzed the genes of some 17,000 people and found seven genes associated with type 1 diabetes, an autoimmune disease (in which

TABLE 8.1
Summary of Some Common (and Not-So-Common) Genetic Diseases That Can Be Tested For at a Modern University Medical Center

Disease/Syndrome	Gene @ Location	Mode of Inheritance
Angleman syndrome	UBE3A (ubiquitin protein ligase E3A) @ 15q11–q13	Deletion, uniparental disomy Imprinting defects Some autosomal dominant rearrangements
Bloom syndrome	DNA helicase Rec Q protein-like 3 @ 15q26.1	Autosomal recessive
Canavan disease	Aspartocyclase @17pter-p13	Autosomal recessive
Familial dysautomia	IKBKAP (inhibitor of kappa light polypeptide gene enhancer in B cells, kinase complex-associated protein) + IKAP (IKK complex-associated protein @ 9q31)	Autosomal recessive
Fanconi anemia, type C	FANCC @ 9q22.3	Autosomal recessive
Gaucher disease	GBA (acid-beta glucosidase(glucocere brosidase) @ 1q21–31	Autosomal recessive
Niemann pick disease, type A	SMPD1 (acid sphingomyelinase–sphingomyelin phosphodiesterase-1) @ 11p15.4–p15.1	Autosomal recessive
Ataxia panel	SCA1@6p23, SCA2@12q24, SCA3@14q32.1, SCA6@19p13, SCA7@3p21.1–p12, SCA8@13q21, SCA10@22q13, SCA12@5q31–33, SCA17@6q27, Friedreich ataxia@9q13, DRPLA@12p13.31	All autosomal dominant, except Friedreich ataxia is recessive
BCR/ABL (Philadelphia chromosome)	Abelson oncogene and breakpoint cluster region @t(9;22)(q34;211)	Somatic cell genetic disorder
Breast cancer	BRCA1@17q21, BRCA2@13q12.3	Autosomal dominant
CADASIL (cerebral autosomal dominant arteriopathy)	NOTCH3 @19p13.2–p13.1	Autosomal dominant
Charcot–Marie tooth disease	Myelin protein zero (MPZ) @ 1q22	Autosomal dominant
Congenital bilateral absence of the vas deferens	CFTR (cystic fibrosis transmembrane conductance regulator) @ 7q31	Autosomal recessive
Cystic fibrosis	CFTR @ 7q31	Autosomal recessive
Deafness panel	GJB3 (Connexin 26) and GJB6 (Connexin 30) @ 13q12	Autosomal recessive
Dentatorubral-Pallidoluysian atrophy (DRPLA)	DRPLA @ 12p	Autosomal dominant
Duchenne or Becker muscular dystrophy	Dystrophin @ Xp21.2	X-linked recessive
Factor V Leiden/blood clotting disorder	FV @ 1q21–25	Autosomal dominant
Familial Mediterranean fever	MEFV (Pyrin) @ 16p13.3	Autosomal recessive
Fragile X Syndrome/mental retardation	FMR-1 @ Xq27.3	X-linked recessive
Hereditary hemochomatosis/ataxia	HFE @ 6p21.3	Autosomal recessive

Continued

TABLE 8.1 (Continued)

Summary of Some Common (and Not-So-Common) Genetic Diseases That Can Be Tested For at a Modern University Medical Center

Disease/Syndrome	Gene @ Location	Mode of Inheritance
Huntington's disease/Progressive neural dysfunction	huntingtin @ 4p16.3	Autosomal dominant
Kennedy disease/spinal and bulbar muscular atrophy (SBMA)	Androgen receptor @ Xq11–q12	X-linked
Long QT syndrome/cardiac arrhythmias	LQT1 @ 11p15.5, LQT2 @7q35–q36, LQT3 @ 3p21, LQT5 @ 21q22.1–q22.2, LQT6 @ 21q22.1	Autosomal dominant Autosomal recessive
Medium-chain acyl-CoA dehydrogenase (MCAD)/ hypoglycemic coma	ACADM @ 1p31	Autosomal recessive
Methylene tetrahydrofolate (MTHFR)	Methylene tetrahydrofolate reductase @ 1p36.3	Autosomal recessive
Neuroligin (X-linked mental retardation/autism/Asperger syndrome)	NGLN3 @ Xq13NGLN4 @ Xp22	X-linked recessive
Paraganglioma (Glomus tumors)	Succinate dehydrogenase complex, subunit D (SDHD) @ 11q23	Autosomal dominant, with imprinting
Pendred syndrome/deafness	SLC26A4 @ 7q31	Autosomal recessive
Phenylketonuria/mental retardation	PAH (phenylalanine hydroxylase) @ 12q	Autosomal recessive
Prader–Willi syndrome/obesity, dwarfism, cognitive impairment	SNRPN @ 15q11	Deletion, uniparental disomy Imprinting defects Some autosomal dominant rearrangements
Prothrombin excess/risk of thrombosis	Prothrombin gene: F2 @ 11p11–q12	Autosomal dominant
Rett syndrome/neurological disorder	MECP2 (methyl-CpG-binding protein) @ Xp11	X-linked dominant
Sickle-cell anemia	Hemoglobin beta (HBB) @ 11p	Autosomal recessive
Smith–Lemli–Opitz syndrome/ microencephaly, developmental delay, syndactyly, etc.	DHCR7 (7-dehydrocholesterol reductase) @ 11q12–q13	Autosomal recessive
Spinal muscular atrophy/progressive loss of spinal cord motorneurons	SMN (survival of motor neuron) @ 5q11.2–13.3	Autosomal recessive
Tay–Sachs disease/neurodegenerative disease	HEXA @ 15q23–q24	Autosomal recessive
Thalassemia/chronic anemia	HBB (Hemoglobin beta) @ 11p15.1	Autosomal recessive
Tuberous sclerosis/skin disorders	TSC1 @ 9q34TSC2 @ 16p13.3	Autosomal dominant
Von Hippel–Lindau syndrome/ hemangioblastomas, renal cycts, etc.	VHL @ 3p25–26	Autosomal dominant
Waardenburg syndrome/hearing loss, pigmentary changes	PAX3 (paired box gene 3) WS1 & WS3 @ 2q35MITF (microphthalmia-associated transcription factor) WS2A @ 3p14	Autosomal dominant
X-linked lymphoproliferative disease/ genetic autoimmune disease	SH2D1A (SH domain protein 1A) @ Xq26	

TABLE 8.1 (Continued)
Summary of Some Common (and Not-So-Common) Genetic Diseases That Can Be Tested For at a Modern University Medical Center

Disease/Syndrome	Gene @ Location	Mode of Inheritance
Cystic fibrosis (linkage analysis[a])	CFTR @ 7q31	Autosomal recessive
Duchenne–Becker muscular dystrophy (linkage analysis[a])	Dystrophin @ Xp21.2	X-linked recessive
Familial adenomatous polyposis (linkage analysis[a])/colon cancer predisposition	APC @ 5q21–22	Autosomal dominant
Hemophilia, type A (linkage analysis[a])	Factor VIII @ Xq28	X-linked recessive
Multiple endocrine neoplasia, type 1 (linkage analysis[a])/cancer predisposition in endocrine glands	RET @ 10q	Autosomal dominant
Neurofibromatosis, type 1 (linkage analysis[a])	NF1 @ 17q11.2	Autosomal dominant
Polycystic kidney disease (linkage analysis)	PKD1 @ 16p13.3	Autosomal dominant
Wilson disease (linkage analysis[a])/ disorder of Cu metabolism	WND/ATP7B @ 13q14.3	Autosomal recessive

[a] *Linkage analysis (LA)* is not a direct diagnostic test. LA can be used to make predictions about disease status (including prenatal diagnosis) as long as there is a confirmed clinical diagnosis of the disease in the family, and that family member is available for testing.

Source: BUMC (Boston University Medical Center). 2005. Direct DNA Tests. www.bumc.bu.edu/Dept/ContentPF.aspx?PageID=2194&DepartmentID=118.

the immune system attacks the body's own pancreas). Interestingly, one of the genes associated with type 1 diabetes was also associated with Crohn's disease, an autoimmune disease in which the immune system attacks the bowels. This was reported in the journals *Nature* and *Nature Genetics* [Biello 2007].

Dr. Michael Haller of the University of Florida (and others) conducted a clinical trial on the use of umbilical cord blood to treat type 1 diabetes in children. The trial was successful in reducing the amount of insulin needed by the children, apparently by reducing the level of autoimmune attack on the pancreas. This suggests that the "adult" stem cells in cord blood can be used to treat type 1 diabetes (and possibly other autoimmune diseases). The results of the study were presented 25 June in Chicago at the 67th Annual Scientific Sessions of the American Diabetes Association [Cogdon 2007].

The mechanisms by which cord blood cells reduced the autoimmune attacks are not known, but it is possible that it is related to T- or B-lymphocytes, immune system cells produced by pluripotent stem cells in bone marrow. (See Chapter 7 of this textbook for more on the immune system.)

8.3.2 Gene Defects and Disease

The growth of metastatic tumors in humans is a biochemically and genomically complex process. It involves the sequential expression of many genes and their products that alter the normal pathways of cellular signaling, command, and control. The recent localization and identification of a cluster of potential *tumor suppressor genes* (TSGs) and *oncogenes* on the short (p) arm of human chromosome 3, particularly in the 3p21.3 region, has opened a window into the genomic causes and cures for certain cancers [Ji et al. 2005].

Ji et al. stated: "... allele loss from several distinct regions on chromosome 3p ... are the most frequent and earliest genomic abnormalities involved in a wide spectrum of human cancers including lung, breast, head and neck, ovarian, cervical/uterus, colorectal, pancreatic, esophageal, bladder and many more."

Ji et al. go on to detail the clinical and biochemical evidence for the existence of the 3p21.3 TSGs. In conclusion, they wrote: "These putative 3p21.3 TSGs are shown to be inactivated in primary tumors more frequently by alternative mechanisms such as epigenetic promoter methylation, aberrant gene splicing and truncation, deficient translation and post-translational modifications of gene products, rather than by somatic mutations observed in classic TSGs such as p53 and Rb." They noted further that the 3p21.3 genes are engaged in diverse biological processes governing many key cellular signaling and regulatory activities associated with the cell cycle (differentiation, proliferation, and death).

So, given this knowledge about the 3p21.3 TSG inactivation, how can we use this information to develop new cancer therapies? At present, we can use the knowledge that abnormalities in the 3p21.3 TSGs appear in the earliest stages of cancer development. Thus, biopsies that reveal these abnormalities can signal the likelihood of early cancer presence and indicate the need for chemoprevention. It is very likely in the next decade or so that vectored recombinant gene delivery systems will be developed to express the normal products of the damaged 3p21.3 TSGs.

8.3.3 Gene Therapy

As an example of gene therapy, we will describe a 2003 Phase I study *cystic fibrosis* (CF) patients done by a consortium of research hospitals (University Hospitals in Cleveland, Case Western Reserve School of Medicine, Children's Hospital of Denver, and Cystic Fibrosis Foundation Therapeutics, Inc.). Recall that CF occurs when the CFTR gene is dysfunctional for the chloride channel protein through which chloride ions normally enter or leave cells. The epithelial cells affected in CF line the nasal passages, the lungs, and passageways inside the liver, sweat glands, pancreas, and digestive and reproductive systems.

In 1999, Eric Alton and colleagues at the National Heart and Lung Institute of the Royal Brompton Hospital in London, in cooperation with the Genzyme Corp. of Framingham, Massachusetts, administered a normal rCFTR gene as an inhaled plasmid aerosol to eight CF patients in a Phase I study of cystic fibrosis therapy. Eight more CF patients were given a placebo aerosol. The patients receiving the rCFTR gene exhibited about 25% restored normal chloride transport; the control patients did not show any improvement. There was no improvement on sodium absorption in either group. However, this study demonstrated the feasibility of *direct gene transfection* in CF therapy.

Copernicus Therapeutics, Inc., of Cleveland, Ohio, developed a novel, nonviral, gene transfection system to insert a normal CFTR gene through the plasma membrane and into the nucleus of the genetically damaged epithelial cells. Their Phase I tests put the rCFTR gene in plasmids, which were dripped into CF patients' noses in a saline solution. Biopsies of nasal tissues showed that about 67% of the patients treated showed a meaningful increase in the transport of chloride ions in their nasal epithelium (29 April 2003 press release from University Hospitals of Cleveland). One can infer from this result that two-thirds of the nasal epithelial cells were successfully transfected with the rCFTR gene.

A problem with any topical transfection system of CF therapy is that a characteristic of lung and nasal epithelial tissues in CF patients is that they are covered with a thick, sticky mucus that the plasmid or naked rDNA must first penetrate to enter the defective cells. Removal of the thickest mucus before the application of the rCFTR gene should improve transfection efficiency. There should be few side effects to transfection therapy because it is applied topically.

When a mutated gene in cells is producing a bad protein involved in a disease, we have seen that a normal form of the gene can be introduced into the cells in order to force the cells to make normal protein. However, the cells may still be making the bad protein. One way to turn the bad product off is to introduce genetically engineered *antisense* mRNA (asmRNA) for that protein into the cell. The asmRNA binds to the cell's bad mRNA, inhibiting the ribosomal production of the bad protein. So, it should be possible to not only insert a good gene into the mutated cell but to also prevent it from expressing the bad protein.

8.3.4 GM Viruses Used to Fight Cancers

Researchers at the University of Texas' M.D. Anderson Cancer Center published amazing results of a successful treatment for *malignant glioma* (a deadly form of brain cancer) in mice [Fueyo et al. 2003]. A genetically modified (GM) adenovirus (Delta-24) was developed that targeted only the cancer cells. Mice were implanted with human glioblastoma xenografts, which expressed wild-type ***p53***, tumor suppressor protein. (The ***p53*** gene, which acts as a regulator for cell division, is missing or mutated in about half of human cancers and is dysfunctional in most of the rest.) The untreated mice lived less than 3 weeks, whereas mice treated with the GM adenovirus survived more than 4 months. Encouraging!

The *Adenoviridae* family of viruses is nonenveloped and has a dsDNA genome. They infect humans and fowl. In humans, they can cause *acute hemorrhagic cystitis* (*Adenov. 11* and *21*), *keratoconjunctivitis* (*Adenov. sp.*), and *pharyngoconjunctival fever* (*Adenov. 1–3* and *5*). Adenoviruses first dock on cell surface molecules of the target tumor cell they are infecting. A fiber knob on the end of a virion surface rod binds to the glioma cell-surface CAR protein. Then, an RGD molecule at the base of the knob shaft binds to a cell-surface, *alpha-V- integrin* molecule. The expression of alpha-V-integrin molecules is common on glioma cell membranes and uncommon on normal glial cells. The two surface bindings trigger the glioma cell to internalize the virion into an endosome. When the viral capsule reaches the nuclear membrane, it injects its dsDNA genome into the cell's nucleus. The glioma cell then sets to work making new virions according to the injected viral genome. The glioma cell is killed when the daughter virions eclose and spread to other nearby glioma cells.

The M.D. Anderson workers genetically modified an adenovirus species that they called Delta-24-RGD. In normal cells, the *retinoblastoma* (Rb) protein that prevents unlimited cell division also acts to prevent viruses that have infected the cells from replicating. It is a built-in antiviral protein. The common cold virus, for example, expresses a protein E1A that binds to Rb and blocks its functioning. In order that the adenoviruses could infect the glioma cells, the investigators deleted 24 bps from the adenoviruses E1A gene, so it cannot stop Rb from functioning, allowing the modified virus to infect and kill *only* cancer cells. A healthy cell with a normal Rb protein titer evidently can successfully defend itself against the GM adenoviruses. To improve the efficiency of the Delta-24-RGD virus in infecting glioma cells, the M.D. Anderson workers again altered the viral genome to add an RGD-4C docking region to the virus to allow the virions to dock securely to *integrin* molecules on the outside of the glioma cells.

Starting at the point where they were injected into the mouse brains, the GM adenoviruses attached to and entered the glioma cells, reproduced, and killed the cell when they eclosed. They spread out throughout the mouse's brain, only infecting and killing glioma cells. When the susceptible glioma cells were dead, the virus could no longer reproduce and died out. At necropsy, the brains of the treated and cured mice showed only empty cavities and scar tissue; no gliomas. The Delta-24-RGD viral therapy for glioma in mice was certainly effective. Phase 1 human trials are now underway. Among other experiments, the Delta-24-RGD will be tested in treatment of recurrent ovarian cancer in the abdominal cavity.

Some questions must be answered: How will the human immune system react to adenoviruses in a human brain? Also, how will the HIS react to any GM adenoviruses that escape into the rest of the body where immune system cells and antibodies have easy access to them? Will all types of normal human cells be resistant to them? We do know that wild adenoviruses can infect human tissues and cause certain diseases (see the preceding text).

Other GM adenoviruses have also been used to fight gliomas. Jiang et al. [2005] compared the effectiveness of the Delta-24-RGD adenovirus with RA55, an E1B-55 kDa GM adenovirus in killing human glioma tumors transplanted into mouse brains. The genome of the GM adenovirus contains a 903-bp deletion in its E1B gene region as well as a point mutation that generates a stop codon. These engineered changes in the E1B genome prevent the expression of p53 protein from binding to the E1B-55 kDa protein.

Human glioma cell lines were used that grew in the brains of nude mice and expressed wild-type p53 or mutant p53 protein. Jiang's study showed that Delta-24-RGD induced a more potent antiglioma effect than RA55 in vitro, also, Delta-24-RGD replicated more efficiently than RA55, both in wild-type and mutant p53 glioma cells, in vitro and in vivo. Jiang et al. [2005] stated: "… our results suggest that E1A mutant adenoviruses that target the Rb pathway are more potent anticancer agents than E1B-55 kDa mutants. ... Our data suggest that tumors that are resistant to the oncolytic effect of E1B-55 kDa mutant adenoviruses might instead be sensitive to the effect of E1A mutant adenoviruses."

Jiang et al. summarized an important goal of their research: "One of the main principles used in developing an [GM] oncolytic virus is to carefully manipulate the adenoviral genome to eliminate the ability of the virus to activate normal host cells to enter into the S-phase or to block altruistic apoptosis in normal cells, while promulgating a relatively unscathed capacity for viral replication in cancer cells, thereby improving selectivity without sacrificing efficacy." [Quoted with permission.]

To realize these goals will require time, dedicated effort, and funding; the importance of this therapeutic approach cannot be emphasized enough.

8.3.5 Recombinant Vaccines

Vaccination leads to *acquired immunity.* The principle of vaccination against viral infections goes back to 1788, when British physician Edward Jenner discovered that individuals, such as milkmaids, who had contracted cowpox from cows did not, in fact, catch the far deadlier human smallpox. He showed that a mild cowpox infection induced in humans made them resistant to the smallpox virus. He went on to take the exudate from cowpox pustules on a milkmaid in 1796 and introduced it into cuts he had made in the arms of an 8-year-old boy. The boy came down with a mild case of cowpox and recovered. Six weeks later, in a dangerous experiment, he inoculated the boy with exudate from smallpox pustules. Fortunately, the boy did not catch smallpox. Not until about 1800 did doctors of the time realize that vaccination with cowpox did work and adopted the procedure.

Jenner is credited with coining the term *virus,* and he is one of the first scientists to ever receive a government grant to support his research (a total of £30,000 from Parliament) [Harris 2007]. Smallpox is caused by the *Variola* virus, cowpox by *Vaccinia* virus; both are members of the *Poxviridae* family and are enveloped dsDNA viruses. They evidently have some coat proteins in common.

In vaccination, a protein (or proteins) derived from a virus' coat and/or envelope proteins is introduced to the human immune system, usually by intradermal injection. What is injected can be killed virions or attenuated or harmless virions carrying the coat proteins that the subject is to develop immunity against. Natural immune surveillance identifies the proteins as foreign, and the immune system makes antibodies to the foreign protein as well as *memory B-cells* [cf Chapter 7, Section 7.2.9.3] for that antigen. The traditional way to make a vaccine has been to infect chicken eggs with the virus, let it grow then harvest the virus, render it harmless, then isolate the desired proteins to make the vaccine. There is something ironic about needing millions of eggs to make vaccine for the avian flu (H5N1), which is fatal to chickens and chicken embryos. Fortunately, there is a modern way to make vaccines without chicken eggs.

The Connecticut-based *Protein Sciences Corporation* has developed a 21st century recombinant DNA means of making vaccines for viral diseases. They extracted the RNA viral genes from the H5N1 flu virus responsible for producing *hemagglutinin* on its coat and, using reverse transcriptase, recombined them with a *baculovirus'* dsDNA genome. (The *Baculoviridae* are enveloped and have a circular dsDNA genome; they generally infect insects.) Then, they grew the GM baculoviruses in cultured caterpillar ovarian cells, which produced H5N1 hemagglutinin molecules for the vaccine. For example, their FluBLØk® vaccine uses *recombinant hemagglutinin* (rHA) protein from the coats of three strains of flu virus selected by the World Health Organization and the U.S. Centers for Disease Control and Prevention for each year's vaccine formulation.

Hemagglutinin is an antigenic glycoprotein attached to the surface of the influenza viruses. About 500 molecules protrude radially outward from the viral surface and serve to bind the flu virus to the cell it is infecting. There are at least 16 different HA antigens; the first three, H1, H2, and H3, are found on human flu viruses. The bird flu virus H5N1 can infect humans at a low rate. The H1 HA protein molecule in the 1918 influenza pandemic has been shown to be only slightly mutated from the original avian virus.

After a series of successful Phase I, II, and III human trials in 2003–2005, Protein Sciences received Food and Drug Administration (FDA) permission to qualify for the "accelerated approval" mechanism. They expect to obtain market approval for FluBLØk in 2006, just in time for H5N1 to appear in Western Europe and North America. (Note that the infamous H5N1 bird flu virus is an enveloped *coronavirus* with a positive sense, ssRNA genome.)

Pandemic versions of FluBLØk have been developed from rHA cloned from highly pathogenic avian H5N1 viral strains. The H5N1 vaccine was tested in 1998 on chickens and gave a zero infection rate. Protein Sciences is also developing a vaccine against the SARS *coronavirus,* a spherical, enveloped virus whose genome is positive-stranded ssRNA.

Another non-egg approach to vaccine production has been developed by the Illinois-based *Baxter Corporation.* They are using cultured green monkey kidney cells to grow vaccine proteins for smallpox and, more recently, for H5N1 bird flu.

An interesting application of a vaccine was reported by BrightSurf [2003]. Scientists at *Caprion Pharmaceuticals* and at U.T.'s Center for Research in Neurodegenerative Diseases observed that when prion $PrP^{C} \rightarrow PrP^{Sc}$, there are newly exposed peptide side chains on the defective PrP^{Sc} molecules. One such side chain is the tripeptide, **Tyr–Tyr–Arg**. This side chain appears to be a common feature on the misfolded prions in humans, cattle, mice, hamsters, sheep, and elk. These researchers reasoned that the **Tyr–Tyr–Arg** side chain was an "Achilles' heel" for the PrP_{Sc} molecules. The terminal **Arg** AA's R-group is very hydrophilic because of its two terminal amino groups. The workers' plan was to first vaccinate and immunize mice against the **Tyr–Tyr–Arg** tripeptide "sprig," then infect them with PrP_{Sc} prions and examine the results. (Hopefully, this tripeptide sequence is not ubiquitous in other normal proteins, or else the "Law of Unintended Consequences" will be activated.)

The conventional approach to immunization requires that proteins or protein fragments from the coat or envelope of a viral pathogen be introduced to the subject's immune system. This often requires the need to inject purified viral particles or inactivated virions. An alternative, more specific approach to viral vaccination is not to use viral proteins at all. Instead, recombinant mRNA is produced from the viral pathogen that codes the viral proteins. This rmRNA is injected into the skin and taken up by immune cells that, by a complex process, produce antibodies (Abs) to the viral proteins coded by the rmRNA. It is more efficient to introduce the rmRNA directly into the immune cells than into proteins. This technique has been used successfully to vaccinate calves and sheep against the infectious *bovine virus diarrhea virus* [Cannon & Weissman 2002]. These authors concluded: "This strategy again relied on the ability of host cells to take up the [rm]RNA instead of using viral particles to introduce the RNA. This avoided the introduction of the viral coat proteins and increased the amount of antigenic protein produced through the replicative function of the viral genome. The entry of one RNA molecule led to the production of many copies of antigen encoding mRNA inside the cell thus increasing the templates available for translation." [Quoted with permission.]

8.4 EPIGENETIC THERAPY FOR CANCER

8.4.1 Introduction

Heritable changes in phenotype that do not involve DNA Nt changes in genes are called *epigenetic changes*. Such epigenetic changes are passed on through both mitotic and meiotic cell divisions and are thought to be involved in such events as X-inactivation in females and oncogenesis. At the root of epigenetic regulation of gene expression are the enzymatically caused additions of methyl groups (methylation) to cytosines in DNA at CpG islands, as well as many additions (and removals) of radicals (methyl, acetyl, phosphoryl, and ubiquitinyl) to AA residues in the histone proteins found in intimate association with DNA, especially H3.

It is known that about 37% of somatic ***p53*** gene mutations (and about 58% of germ-line mutations) occur at methyl CpGs. These mutations are strongly correlated with the development of cancers [Bannister 2006]. Other dysregulations in the expression of the oncogenes ***MYCN*** and ***Mdm2*** that affect the level of the anti-oncoprotein p53 are seen in tumors [Slack et al. 2005; Turbin et al. 2006].

The interested reader should examine the model of Ting et al. [2006] for the potential contribution of stem cell (SC) chromatin to the initiation and maintenance of aberrant epigenetic gene silencing in cancers. These authors also give a broad, detailed overview of epigenetic causes of cancers, including *cancer stem cells* (see Chapter 10, Section 10.4.5).

8.4.2 Epigenetics and Cancer

Yoo and Jones [2006], Ting et al. [2006], and Feinberg et al. [2006] are all of the opinion that the initiation and progression of cancer are controlled by *both* genetic and epigenetic events. Although *oncogenic genetic mutations* are virtually irreversible, *oncogenic epigenetic aberrations* are potentially reversible, allowing the reexpression of genes that can suppress tumor growth. Feinberg et al. state: " ... tumour cell heterogeneity is due in part to epigenetic variation in progenitor cells, and epigenetic plasticity together with genetic lesions drives tumour progression." They go on to hypothesize: " ... non-neoplastic but epigenetically disrupted stem/progenitor cells might be a crucial target for cancer risk assessment and chemoprevention."

The epigenetic alternatives to genetic causes of cancer include unbalances in global DNA methylation (hypo- or hyper-methylation) of specific genes and posttranslational radicalization of AA residues lying in histone protein "tails." It is known that overproduction of certain histone *methyltransferase enzymes* leads to the methylation of H3K4 and H3K27 residues (K is the single-letter AA code for lysine); this histone hypermethylation is frequently found in cancers and, by inference, is suspected of being causative of cancer when taken with other epigenetic changes to histones. Epigenetic changes then can mimic the effects of genetic mutations. They can lead to oncogene activation and tumor-suppressor gene silencing.

Feinberg et al. [2006] list evidence in support of an oncogenesis model in which epigenetic factors disrupt progenitor cells in a given organ or system. Such disruption leads to a polyclonal precursor population of neoplasia-ready cells. Their evidence includes:

1. In vitro studies of tumor cells demonstrate reversibility of phenotype in both leukemia and solid tumor development.
2. Comprehensive epigenetic changes that precede the initial mutations in cancer. (One change is widespread DNA hypomethylation and frequent promoter hypermethylation.) These epigenetic alterations are always found even in benign tumors and thus must precede the earliest genetic alterations.
3. Curiously, cloned mouse melanoma nuclei can be made to differentiate into normal mouse embryos. This indicates that most of the properties of these tumor cells can be reprogrammed into normal development that is epigenetically controlled.

4. Loss of *imprinting* of *insulin-like growth factor 2* is found in the colonic epithelium of patients at risk for colorectal cancer. In mice, it has been shown that this epigenetic change shifts the balance of the intestinal epithelium toward an expanded progenitor cell population.

8.4.3 Epigenetic Therapies for Cancer

Yoo and Jones [2006] stated: "With the advent of numerous drugs that target specific enzymes involved in the epigenetic regulation of gene expression, the utilization of epigenetic targets is emerging as an effective and valuable approach to chemotherapy as well as chemoprevention of cancer." They list the enzyme families involved in epigenetic regulation:

1. *DNA methyl transferases* (DNMTs) are responsible for de novo and maintenance methylation of certain DNA residues. There are five known human DNMTs.
2. HAT enzymes do histone "tail" acetylation. In this class, there are seven human *sirtuin* (SIRT) enzymes as well as GNAT and MYST classes of acetylation enzymes. Nature also supplies various *histone deacetylase* (HDAC) enzymes to remove acetyl groups.
3. Histones are methylated at serine residues by HMT enzymes.

To further complicate the epigenetic theory of oncogenesis, it has been observed that certain RNAs can also be regarded as epigenetic components involved in chromatin regulation. For example, in female mammals, the noncoding Xist RNA helps silence the inactive X-chromosome by coating the chromosome, promoting the formation of dense heterochromatin. The RNA interference (RNAi) pathway is also linked to chromatin structure.

It is known that anomalies in the RNAi system affect the formation of heterochromatin. Whether these anomalies can be directly linked to cancer remains to be determined. In the case of *micro-RNAs* (miRNAs), however, the link is stronger. Micro-RNAs affect the expression of genes linked to the cell cycle, and the expression of miRNAs is altered in cancer cells. It turns out that miRNA profiling has proved to be a very useful aid to classifying different types of cancer [Bannister 2006]. Chuang and Jones [2007] list a number of specific miRNAs that are overexpressed in various types of cancer, specifically, the miR-17-92 cluster miR-372, miR-373, miR-21, miR-155, and miR-146. Are these associations cause or effect? Clearly, suppression of these specific oncogenic miRNAs may have a role in fighting cancer. On the other hand, stimulation of the expression of known tumor-suppressor miRNAs may also be a tool in cancer therapy. Tumor-suppressor miRNAs include miR-127, miR-15a, miR-16-1, let7, and miR-145. One means of controlling the suppression of oncogenic miRNAs and also increasing the level of antioncogenic miRNAs is by epigenetic mechanisms. This regulatory strategy has already been explored by Saito et al. [2006] on miR-127 using human cancer cells.

Yoo and Jones [2006] list a number of chemicals that have putative value as epigenetic cancer chemotherapy drugs. These are summarized in Table 8.2 in the following text. Some of the chemicals are in preclinical or Phase I trials against various cancers. Others are in more advanced states of evaluation.

It is clear that in the near future, there will be a considerable work identifying specific targets for epigenetic chemotherapy drugs and evaluating their effectiveness, especially when combined with conventional chemotherapy agents and/or immunotherapy. Perhaps the epigenetic approach will be successful in restoring balance between the antitumor protein p53 and two of its immediate downregulatory TFs, the oncoproteins Mdm2 and MYCF [Oren 1999; Slack et al. 2005; Turbin et al. 2006].

No doubt new HDAC inhibitors will be developed to augment the deacetylase inhibitors listed earlier as well as new histone demethylases.

8.4.4 Some Other Diseases with Epigenetic Etiologies

In addition to various cancers, certain other medical conditions have been found to be closely associated with epigenetic dysregulation. These include but are not limited to: *Rett syndrome, immu-*

TABLE 8.2
A Partial List of Epigenetic Cancer Chemotherapy Drugs

	Dosage Range	Clinical Trials: Used On
DNA Methylation Inhibitors		
5-Azacytidine	μM	Phase I, II, III: Hematological malignancies
5-Aza-2′-deoxycytidine (Decitabine)	μM	Phase I, II, III: Hematological malignancies, cervical, nonsmall-cell lung cancer
5-Fluoro-2′-deoxycytidine	μM	Phase I
5,6-Dihydro-5-azacytidine	μM	Phase I, II: Ovarian cancer and lymphomas
Zebularine	μM–mM	Preclinical
Hydralazine	μM	Phase I: Cervical cancer
Procainamide	μM	Preclinical
EGCG	μM	Preclinical
Psammaplin A	nM–μM	Preclinical
MG98	N/A	Phase I: Advanced/metastatic solid tumors
RG108	μM	Preclinical
Vidaza	μM	Approved
Histone Deacetylase Inhibitors		
Butyrate	mM	Phase I, II: Colorectal cancer
Valproic acid	mM	Phase I: AML*, leukemias
m-Carboxycinnamic acid (CBHA)	μM	Preclinical
Oxamflatin	μM	Preclinical
PDX 101	μM	Phase I
Pyroxamide	nM	Preclinical
Scriptaid	μM	Preclinical
Suberoylanilide hydroxamic acid (SAHA)	μM	Phase I, II: Hematological and solid tumors
Trichostatin A (TSA)	μM	Preclinical
LBH589	μM	Phase I
NVP-LAQ824	nM	Phase I
Apicidin	nM	Preclinical
Depsipeptide (FK-228, FR901228)	μM	Phase I, II: CLL**, AML*, T-cell lymphoma
TPX-HA analog (CHAP)	nM	Preclinical
Trapoxin	nM	Preclinical
CI-994 (N-acetyldinaline)	μM	Phase I, II: Solid tumors
MS-275	μM	Phase I, II: Solid tumors and lymphoma
MGCD0103	μM	Phase I, II: Solid tumors

* Acute myelogenous leukemia.
** Chronic lymphocytic leukemia.
Source: Yoo, C.B. and P.A. Jones. 2006. Epigenetic therapy of cancer: Past, present and future. *Nature Reviews—Drug Discovery.* 5: 37–50.

nodeficiency/centromeric instability/facial anomalies (ICF) syndrome, alpha-thalassemia/mental retardation- X-linked (ATRX) syndrome, Rubenstein–Taybi syndrome, and *Beckwith–Wiedemann syndrome.* A common property of these disorders is that mutations in epigenomic regulators cause errors in embryological development, which then can cause phenotypic anomalies in multiple organ systems. Details of these syndromes and the epigenetic changes associated with them can be found in the comprehensive review article by Feinberg [2007].

In the case of cancers, Feinberg hypothesizes that the first steps in neoplastic progression could involve epigenetic changes that disrupt normal stem cell and/or progenitor cell programming. These oncogenic events could also affect stem cell niche cells and their role in regulating stem cells.

8.5 SUMMARY

In this chapter, we have attempted to outline how genomics and proteomics relate to disease and to disease therapy. Much of the early progress in analyzing and manipulating genomes was made on GM plants and animals. Now that most of the genes in the human genome have been sequenced, and molecular biological techniques of analysis and synthesis have been honed, it is relatively easy to examine and find genetic mutations or anomalies in patients having a well-defined syndrome such as Parkinson's disease, Alzheimer's disease, or breast cancer. In some diseases, there is "deep penetrance" of the gene defect, that is, a high correlation of the defect with the disease (e.g., in cystic fibrosis). In other diseases, such as idiopathic CJD, there may be a whole complex of interacting defective genes. Still other diseases may be correlated with defects in the control of gene expression.

Investigators now have the ability to introduce a recombinant gene into cells with mutations in order to have them express a correct form of a protein. It is also possible to turn a gene off permanently by introducing engineered, antisense mRNA for a bad protein into cells. Still other genes can be manipulated to make chemotherapy and other drugs more effective in fighting cancer and infectious diseases.

Genetically modified viruses that will only infect cancer cells make use of the fact that certain cancer cells express unique types or quantities of cell coat proteins, literally marking themselves for destruction.

Just beginning to be understood are the epigenetic molecular regulatory systems that are responsible for fighting cancers, as well as inducing them. By manipulating the methylation of the DNA of certain genes and also the radicals attached to those gene's associated histone proteins, it should be possible to design new, gentler forms of cancer chemotherapy. Clinical trials are now underway on such epigenetic chemotherapy agents.

We believe the future holds bright prospects for genomic medicine in curing chronic and, heretofore, "incurable" diseases.

CHAPTER 8 HOME PROBLEMS

8.1 Explain *copy number variation* (CNV) in the human genome. What types of genes appear to have the most CNVs and the fewest CNVs?

8.2 Discuss CNV in the human genome and its relation to diseases of the central nervous system (CNS), in particular, autism.

8.3 Describe how you would go about finding out if a certain disease has a genetic etiology.

8.4 Explain why an SNP in a gene can have a broad spectrum of effects on the animal, ranging from no effect to a fatal condition?

8.5 Describe how it is theoretically possible to prevent cells from expressing a bad (mutant) protein that causes a disease.

8.6 **cMYC** is a *transcription factor* associated with the incidence of many human cancers. Give its gene locus. Describe its molecular functions, and describe how it may be oncogenic.

8.7 *Creutzfeld–Jakob disease* (CJD) in humans is a rare, neurodegenerative disease associated with misfolding of the ubiquitous prion protein. It is estimated that about 15% of cases are heritable, hence are called *familial CJD* (fCJD). The inheritance of fCJD follows an *autosomal dominant* pattern. On which chromosome do mutations in the prion gene (PrP) occur? Discuss the hypothesis that a mutation in the chaperonin that folds the normal prion protein may be involved in fCJD.

8.8 Terrorists have developed a genetically engineered virus based on *Ebola* and H5N1 bird flu. Describe how you would develop a vaccine for the terror virus without using chicken eggs.

9 Some Instrumental Methods Used in Genomics, Proteomics, and Forensic Science

9.1 INTRODUCTION

In the past 15 years there have been tremendous improvements in the instrumental methods used to identify and analyze biomolecules including nucleic acids, carbohydrates, lipids, and proteins. These methods include, but are not limited to, *microarrays* in which the wells have a specific molecular affinity that can be used to probe for protein fragments or nucleic acid oligos in solution. Readout of a microarray is generally photonic, making use of fluorescent molecular tags.

Topics introduced in this chapter include molecular immobilization in arrays (by avidin-biotin systems), the *polymerase chain reaction* (PCR) and how it has revolutionized forensic DNA analysis (using short tandem repeats), tests for genetic diseases, fluorescent molecular tags, rolling-circle DNA amplification and padlock probes, and a review of protein analysis technology.

9.2 BIOTIN, AVIDIN, AND STREPTAVIDIN: APPLICATIONS IN GENOMICS AND PROTEOMICS

9.2.1 Introduction

The amazing thing about the chemicals *biotin* (Bi) and *avidin* (Av) (and *streptavidin*) is that they spontaneously form the strongest known noncovalent bimolecular bond. The Av*Bi reaction has a $K_a \cong 10^{15}$. The Av*Bi complex is very robust; it withstands pH between 2 and 13, brief exposures of 132°C, and it takes 15 minutes of autoclaving at 120°C to break the avidin-biotin bonds. (The Av*Bi complex is not as robust to degradation as the misfolded prion protein molecule, PrP^{Sc}, however [see Chapter 7, Section 7.3.4].) Because of the sturdiness and specificity of the Av*Bi bond, it is ideal for use in analytical biochemistry, where Av and Bi are used as molecular tags in conjunction with various signaling molecules.

9.2.2 Biotin

Biotin is a ubiquitous highly conserved *vitamin* found in all eukaryotic cells. It serves as a coenzyme involved in CO_2 transport. It has a gram molecular weight (MW) of 244.31, and its molecular structure is shown schematically in Figure 9.1. Vertebrate tissues rich in biotin include liver, kidney, and spleen. Its empirical formula is $C_{10}H_{16}N_2O_3S$, and its ACS chemical name is *cis*-hexahydro-2-oxo-1H-thienol[3,4]imidazol-4-pentanoic acid.

9.2.3 Avidin

Avidin is a *tetrameric glycoprotein* found in egg white. Its basic (monomer) gene has been sequenced and consists of 128 AAs. Its MW is 14,332, and thus Avidin's MW is 57,328. The avidin basic unit

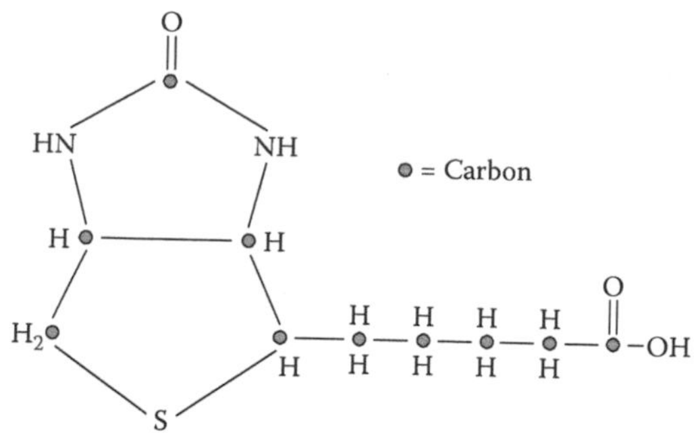

FIGURE 9.1 The biotin molecule.

protein includes nine lysine and four tryptophan residues, all of which contribute to its robust bonding to biotin. The oligosaccharide components of native avidin have not been found to be necessary for biotin-specific binding. In its 3-D form, the avidin subunit has one disulfide bond between Cys-4 and Cys-83. It is the tetrameric form of avidin that has the high affinity to biotin [Pierce Biotechnology, Inc. 2002]; one avidin tetramer molecule can bind four biotin molecules with a $K_d = 10^{-15}$. Monomeric avidin was found to have a much weaker interaction with biotin.

9.2.4 Streptavidin

Streptavidin was isolated from the growth medium of the bacterium *Streptomyces avidinii*. Streptavidin (SAv) is also a tetrameric glycoprotein; its affinity to biotin is similar to that of avidin. Sequencing of the basic unit SAv gene shows that it has 159 AAs and a calculated subunit MW of 16,500. Both Av and SAv have conserved tryptophan-lysine sequences. Avidin has Trp-Lys at positions 70–71 and 110–111. Streptavidin's two Trp-Lys pairs are at positions 79–80 and 120–121. This conserved sequence may be important in the shared affinity of Av and SAv for biotin.

9.2.5 Immobilization in Microarrays: Applications of Avidin–Biotin Systems

Microarrays are analytical devices used in proteomic and genomic research. In one simple form, a microarray is a glass microscope slide to which a rectangular array of different biomolecules (DNA oligos, mRNA, proteins, or antibodies, etc.) are attached in small, micromachined wells. A microarray is used for rapid parallel screening for molecular reactions or molecular structures of nucleic acid oligos or proteins. The array readout is generally by emitted fluorescence, but arrays have been devised that have electrical readouts, photoradiographic readouts, and surface plasmon resonance outputs. The function of a microarray is predicated on the ability to attach certain species of molecules to the array substrate (e.g., a glass slide well) in such a manner that they retain their molecular function. Reacting molecular species are passed over the fixed molecules. Fixed molecules can chemically bind to input molecules with specificity for them. The binding is generally sensed photonically and is used to identify either the fixed or reacting molecules. Microarrays were first developed and perfected for genomic research; DNA/RNA microarrays are now a mature technology. A 96-element (8 × 12) array format on a microscope glass slide is now in common use.

The analysis of protein structure and function using protein microarrays has turned out to be a considerably greater technical challenge than application of genomic arrays. The field of protein microarrays is a rapidly evolving technology. As you recall, proteins, including labeled antibodies, cytokines, cell membrane receptors, etc., have complex 3-D structures that give them their activity. Yeo et al. [2004] have reviewed various immobilization strategies that have been developed for purified proteins and Abs in microarrays. We briefly describe certain approaches to immobilization in the following subsections. Some use covalent, and others, noncovalent bonding.

9.2.5.1 Nonspecific Immobilization

In nonspecific immobilization, the biomolecules are fixed in a *random fashion* to a substrate, using either covalent or noncovalent chemical bonds. In their pioneering work, MacBeath and Schreiber [2000] tethered protein analytes *covalently* to aldehyde-derivatized glass surfaces through their lysine residues and α-amines at the proteins' N-terminals. The authors used a molecular layer of bovine serum albumen (BSA) between the glass surface and the bound peptides in order to minimize nonspecific protein absorptions to the surface and to maintain optimal spatial orientation. Upon spotting, nucleophilic groups on the analyte peptides (**–OH**, **–SH**, **–NH$_2$**, etc.) reacted covalently with the N-hydroxysuccinimide (NHS) groups on the BSA surface, producing surface-immobilized analyte proteins.

Haab et al. [2001] immobilized monoclonal antibodies (MAbs) *noncovalently* on *poly-L-lysine*-coated glass array slides. The MAbs were obtained by standard hybridoma technologies. Other workers have bound MAbs noncovalently to slides coated with nitrocellulose membranes and polyacrylamide [Yeo et al. 2004].

Carbohydrates have also been used in microarray applications. (Recall that *lectins* are an important class of glycoproteins that bind to carbohydrates and are active in immunity, cell recognition, etc. [see Section 7.2.7].) Wang et al. [2002] directly spotted various glycans from biological samples (e.g., polysaccharides, glycosaminoglycans, glycoproteins, and semisynthetic glycoconjugates) *noncovalently* onto nitrocellulose-coated glass slides. Strong noncovalent bonding occurred.

Lahiri et al. (cited by Yeo et al. [2004]) engineered a *membrane protein array* by using a number of supported lipids anchored onto a glass surface modified with γ-aminopropylsilane onto which were spotted transmembrane proteins. Simple, *noncovalent,* hydrophobic interactions between the lipids and the membrane proteins were used for immobilization.

One of the significant disadvantages of nonspecific immobilization is that there is insufficient exposure of functional domains due to the random orientations of the immobilized proteins. Another disadvantage is that noncovalent binding by hydrophobic interactions may lead to protein denaturation and loss of functional activity. An improved approach to protein binding uses *specific immobilization,* in which the analyte proteins are aligned uniformly and optimally, preserving their active sites.

9.2.5.2 Specific Immobilization

A variety of immobilization techniques have recently been developed that permit site-specific immobilization of analytes including proteins, peptides, and carbohydrates. As an example of one of these techniques, we will examine the details of the biotin-avidin arrays. These specific immobilization techniques include *Oxime/thiazolidine formation, the Diels–Alder reaction, Staudinger ligation, α-oxo semicarbazone ligation, nitrilotriacetic acid (NTA) arrays, native chemical ligation, and biotin–avidin arrays* [Yeo et al. 2004].

We will examine the last named technique: First, N-terminal *biotinylated* peptide analytes are synthesized by standard solid-phase peptide synthesis. These analytes are then spotted onto an avidin-coated glass slide. Their binding is highly specific because of the N-terminal biotin molecule's superstrong affinity to avidin. Another approach described by Yeo et al. [2004] used an intein-mediated expression system to biotinylate the C-terminal ends of recombinant proteins, which were then affixed to the avidin-coated substrate. Quoting Yeo et al.:

> The use of biotin-avidin interaction for immobilization also allows the proteins [analytes] to withstand even the most harsh conditions used for downstream screenings, thus making the protein array compatible with most biochemical assays.
>
> In order to fully realize the tremendous potential of protein and peptide microarrays, there is an urgent need to develop strategies that allow site-specific and stable immobilization of proteins, peptides and

other biomolecules onto a glass surface such that their native biological activities are retained. … Only when combining such immobilization strategies with improved arraying devices, labeling reagents, detection devices and data analysis tool[s] [software], can the potential of microarrays to perform high-throughput protein studies be demonstrated to their fullest." [Used with permission.]

9.3 THE POLYMERASE CHAIN REACTION (PCR) AND ITS APPLICATIONS IN GENOMICS AND FORENSIC SCIENCE

9.3.1 Introduction

The *polymerase chain reaction* (PCR) is a very important tool in genomics research because it allows a small pure sample of DNA to be reproduced exponentially, in vitro, to concentrations that are easily detected, analyzed, and chemically manipulated, etc. Small quantities of mRNA can also be analyzed by first converting it to DNA using the enzyme *reverse transcriptase,* then applying PCR. By attaching DNA oligos to nanospheres along with Abs, low levels of specific proteins can be quantified by applying PCR to the oligos.

The nucleic acid polymerase chain reaction was developed in 1983 by D. Kary Mullis at *Cetus Corp.* [Roche Molecular Diagnostics 2006]. In 1993, Dr. Mullis received the Nobel Prize in Chemistry for his invention of PCR technology. Over the years PCR has evolved to become indispensable in DNA forensic analysis, and as you will see later, PCR has permitted the development of ultrasensitive means to detect DNA and mRNA from bacterial and viral pathogens.

9.3.2 The PCR Process

PCR allows a particular DNA sequence (an oligo of 100 to 600 bases [33 to 200 codons] lying within a longer dsDNA molecule) to be replicated by factors of 2^{32} or higher ($2^{32} = 4.295 \times 10^9$). PCR requires the use of a pair of *primers,* each about 20 Nts long, that are complementary to a defined Nt sequence on each of the two strands of the target dsDNA. The primers are extended by a *DNA polymerase* so that a copy is made of the designated sequence. After making this copy, the same primers can be used again to make not only another copy of the target DNA strand but also of the short copy made in the first round of synthesis. Thus successive PCR cycles lead to exponential amplification of the target DNA.

Because it is necessary to raise the temperature of the sample to separate the two complementary strands of the target dsDNA, a thermally stable *DNA polymerase* (DNAP) enzyme called *Taq polymerase* is used, which eliminates the need to add new DNAP for every cycle of amplification. (Taq polymerase was isolated from a thermophile bacterium, *Thermus aquaticus,* and is stable up to about 94°C.)

The PCR process works on ssDNA. Thus, the first step in analyzing a dsDNA oligo is to heat it in solution, breaking the interstrand hydrogen bonds, yielding two complementary ssDNA molecules. In the second step, an excess of primer oligos is added and the mixture is cooled to the *annealing temperature*, allowing the primers to hydrogen-bond to the complementary Nt sequences in the target ssDNA molecules. In the third step, the temperature is increased to that optimum for the Taq enzyme to operate. Taq enzyme is added in a buffer containing $MgCl_2$ and free nucleotides (**A**, **G**, **C**, and **T**). The Taq polymerase manufactures a new strand of DNA using the primer/single-strand template and the free Nts in solution.

This three-part PCR thermal cycling is generally repeated 25 to 35 times: The target dsDNA is heated to 94°C for 30 seconds to separate the strands. The primers are added and the mixture cooled to 65°C for 30 seconds so they will anneal with the complementary Nt sequences on the ssDNA target strands. In the third step, the temperature is raised to 75°C *for 1* minute so the Taq enzyme can manufacture the new DNA strands, called *amplicons.* If each 2-minute cycle were totally efficient, the number of amplicons would be doubled each cycle, yielding $N = M2^C$ amplicons after C cycles, starting with M dsDNA target molecules. However, the efficiency of the reaction is not 100% efficient, and there can be cross-amplification "noise" due to primers binding to the wrong target sequence or wrong bases being incorporated into growing amplicons. Thus, there appears to be a practical upper limit to the "C"-value for PCR that gives the best molecular signal-to-noise ratio (SNR) in the least processing time. Figure 9.2 illustrates graphically the results of the first two PCR cycles.

As the sample is thermally cycled, at first the amplicon products increase exponentially (a log-linear plot versus cycle number). Then the rate of amplicon concentration continues to increase linearly and finally reaches a plateau (or saturates) at high cycle number (e.g., C > 35). For *end-point PCR measurements,* the cycling is stopped in the plateau region and the density of amplicons quantified by a gel reaction using ethidium bromide stain reagent.

Applied Biosystems lists some of the problems with end-point detection: *poor precision, low sensitivity, short dynamic range (< 2 logs), low resolution, nonautomated size-based discrimination only, results are not expressed as numbers, and nonquantitative nature of ethidium bromide*

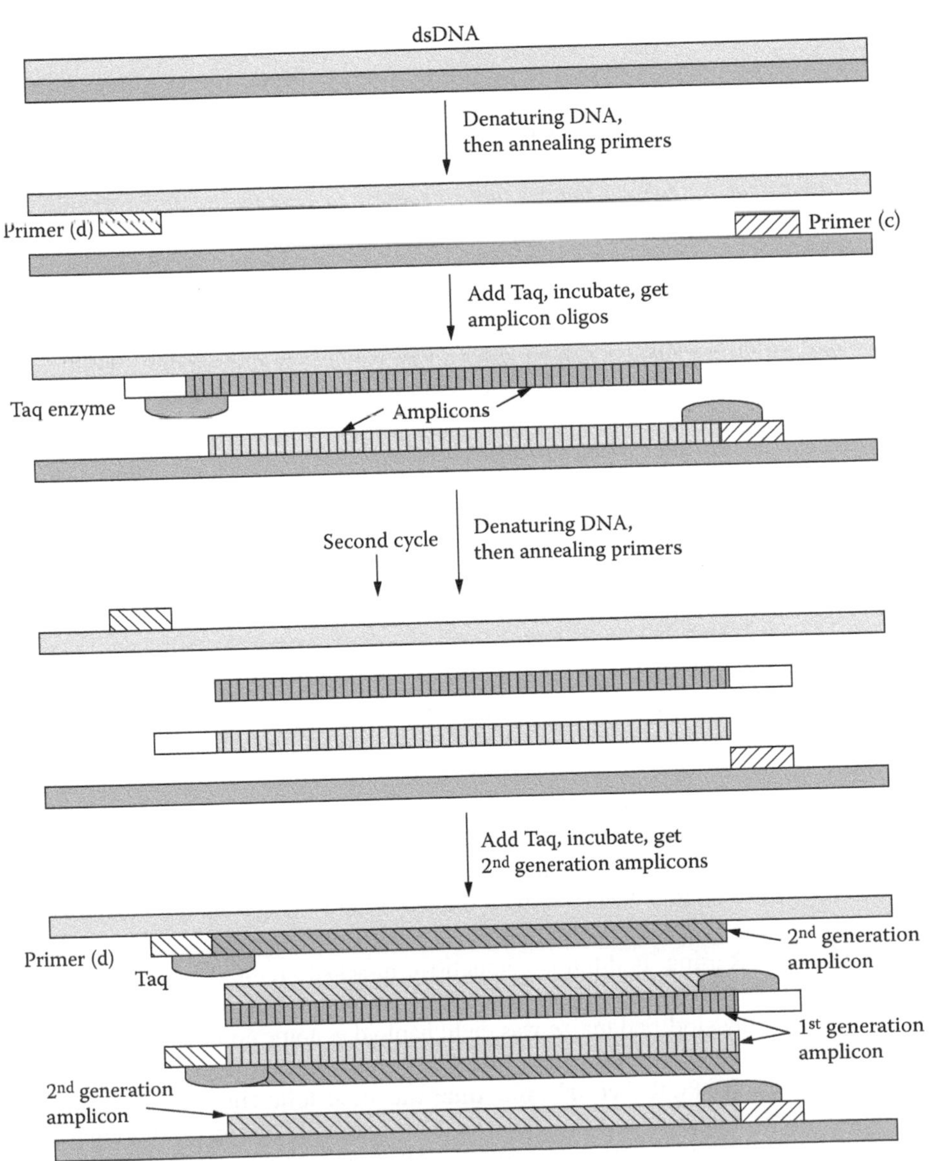

FIGURE 9.2 (See color insert following page 44.) Schematic representation of the first two cycles of the PCR process. See text for further description.

for staining [amplicons]. Note that with the 2-minute cycle period given earlier, 35 cycles would take 70 minutes! At the end of PCR cycling, one of several techniques (generally photonic) is used to quantify the amplicon product.

In order to overcome the relatively long time required to run a PCR test, *real-time PCR* (RT-PCR) was developed [Kubista et al. 2006]. Belgrader et al. [1999] devised a field-portable real-time PCR system to detect *Erwinia herbicola,* a vegetative bacterium that served as a surrogate for *Yersinia pestis,* the bubonic plague bacterium. Their device was called the *Advanced Nucleic Acid Analyzer* (ANAA), intended for field use. Its operation is based on the use of the fluorescent 5′-nuclease assay called *TaqMan®* developed by Roche Molecular Systems, Inc. A low-volume, spectrofluorometric fast thermal cycler is also used. The TaqMan probe is a quantitative homogeneous means of measuring the progress of a PCR reaction. The TaqMan oligonucleotide uses *fluorescence energy resonance transfer* (FRET), typically consisting of a green, fluorescent *reporter* dye at the 5′ end of the TaqMan probe and an orange *quencher* dye at the 3′ end. When the probe anneals (hydrogen bonds) to a complementary strand of an amplicon during PCR, Taq polymerase cleaves the probe during extension of one of the primers, and the dye molecules are separated. The separation no longer allows the quencher to suppress the reporter's green fluorescence, which is monitored. The larger the number of PCR cycles, the greater the concentration of amplicons, and hence the greater the green fluorescence of the solution.

The Belgrader et al. ANAA system began its test cycle by heating the living bacteria to 96°C to lyse them and expose their DNA. Rapid cycling was used to affect PCR detection of the pathogen's DNA. The temperature versus time cycle record appeared triangular with maxima at 96°C and minima at 56°C. Various cycle periods were tested to determine how rapidly a detection of the presence of *Erwinia* could be run with certainty and also produce quantitative information on the initial bacterial concentration. Not surprisingly, smaller initial colonies required more amplification cycles to reach the detection threshold. Using a 17-second cycle period, 5 bacteria took 9 minutes (≈ 32 cycles) to achieve threshold, 50 bacteria took 8 minutes (≈ 28 cycles), and 500 took 7 minutes (≈ 25 cycles). Evidently, fast thermal cycling gives fast results.

In 2004, the U.S. EPA Environmental Technology Verification (ETV) Program issued a report on the *Invitrogen Corp.'s* PathAlert® Detection Kit [ETV 2004], a reagent system capable of detecting smallpox. The kit is an endpoint assay rather than a real-time PCR test, providing a good deal of flexibility in the choice of platform. As of the report date, Invitrogen Corporation had conducted all its field testing of the PathAlert Detection Kit using the Agilent Bioanalyzer® 2100 and ALP.

The performance of the PathAlert® kit was evaluated by ETV. Starting with pure bacterial samples of 2×10^4 to 5×10^5 colony-forming units/mL (cfu/mL), the system was found to be 100% accurate for *F. tularensis, Y. pestis,* and *B. anthracis.* That is, there were zero false-positives and zero false-negatives. Overall specificity was tested and found to be 96% for *F. tularensis* and *Y. pestis* and 100% for *B. anthracis.* The actual speed (time/test) of the PathAlert kit was not given, but the report stated: "Sample throughput for this verification test (from DNA purification to amplified product detection) was 36 to 48 samples per day."

If we take "day" as 8 hours (480 minutes), this translates into about 10 to 13.3 minutes per PCR process. This time/test compares favorably with the real-time PCR system (ANAA) described previously by Belgrader et al. [1999].

Apparently, RT-PCR using TaqMan is competitive in sensitivity and accuracy with end-point PCR, and RT-PCR is generally faster. Yang et al. [2005] reported that the *limit of detection* (LOD) for DNA from genetically modified maize was eight haploid genome copies, and the *limit of quantification* (LOQ) was 80 haploid genome copies. These authors used both conventional PCR and RT-PCR. A 1.5-minute RT-PCR thermal cycle time was used following cell lysis (30 seconds at 94°C, 30 seconds at 60°C, and 30 seconds at 72°C). Conventional PCR thermal cycling also used a 1.5-minute period (30 seconds at 95°C, 30 seconds at 58°C, and 30 seconds at 72°C).

In order to eliminate the need for the high temperature cycling required by Taq polymerase, the use of *helicase-dependent, isothermal, DNA amplification* (HDA) has been gaining favor. Helicase is a ubiquitous enzyme that acts to temporarily unwind a portion of the DNA double helix into single strands, facilitating replication. Helicase functions at temperatures around 37°C. Its isothermal quality and simplicity offers several advantages, as noted by Vincent et al. [2004]: "These properties offer a great potential for the development of simple portable DNA diagnostic devices to be used in the field and at the point-of-care."

The HDA system devised by Vincent et al. [2004] works in the following manner:

1. Sections of dsDNA are separated by a DNA helicase enzyme and coated by a *ssDNA-binding protein* (SSB).
2. Two complementary sequence-specific DNA primers hybridize to each border of the target DNA.
3. DNA polymerases extend the primers annealed to the templates to produce two new dsDNA amplicons.
4. The two new dsDNA amplicons are then used as substrates by the DNA helicases, entering the next round of the reaction cycle. Note that unlike thermal Taq cycling, the HDA process is continuous and asynchronous.

In their experiment, Vincent et al. used *E. coli UvrD helicase* because it can unwind blunt-ended dsDNA fragments. T4 bacteriophage gene-32 protein or phage RB 49 gene-32 protein was used as the SSB. Suitable primers were used. Amplicon detection was done with agarose gel electrophoresis using ethidium bromide stain.

They amplified a 111-bp fragment from the glycogen phosphorylase gene of an oral pathogen, *Treponema denticola,* using 10^4 to 10^7 cells as starter. Also amplified was a 99-bp fragment of the *Hhal* repeat of the filarial parasite, *Brugia malayi*, isolated from human blood samples. A specific product was detected for input samples as low as 5 pg of *B. malayi* DNA. (This corresponds to about 500 copies of the genome.) The HDA reaction was run for 1 to 2 hours to endpoint in these tests; it is not fast.

Vincent et al. also investigated real-time HDA in which a 97-bp fragment of *T. denticola* DNA was amplified using a LUX® primer (from Invitrogen). After about 35 minutes, amplicon accumulation generated a typical sigmoid growth curve. In the early phase of amplification, the doubling time was about 3 minutes. Following the log-linear growth phase, the reaction slowed, entering a transition phase between 45 and 80 minutes, eventually reaching the plateau phase (end point).

The upshot of the HDA studies of Vincent et al. is that the continuous HDA process is physically simpler to implement (no T-cycling is required); however, it is also slow, constrained by the asynchronous amplification kinetics. Sixty minutes is a long time to wait for results in a field test.

9.3.3 Direct DNA Tests for Genetic Diseases

As researchers have become more adept at examining the fine structure of individual human genes, it has been possible to predict and confirm diagnoses of diseases with genetic etiologies in specific genes. These tests generally can be performed on DNA from cells removed from adults and children but also on a small number of fetal cells taken from the placenta (chorionic villus sampling). Later in pregnancy, deciduous fetal cells can be harvested from amniotic fluid taken by amniocentesis, and their DNA can be studied.

DNA from cells from a preimplantation embryo can also be used for diagnosis when in vitro fertilization (IVF) is used. Because of totipotency in the stem cells of an eight-cell embryo, all eight cells are not needed for normal development of the embryo and fetus. In IVF, one or two embryonic cells can be safely harvested to determine the sex of the embryo with a probe for Y-specific DNA. When a mother carries a dangerous X-linked trait such as *hemophilia A,* analysis of these cells permits the parents to select a female embryo for implantation to minimize the risk of this condition. Fluorescent probes for the DNA of certain chromosomes can allow the detection (using FISH, which is explained in Section 9.4.2) of an abnormal number of chromosomes such as the three No. 21s (trisomy) present in Down syndrome.

PCR is used to determine the genotype of an egg *before* it is fertilized (without destroying the egg). When *meiosis I* is completed, the egg buds off the *first polar body* (PB1) containing maternal chromosomes. The DNA in the polar body is amplified by PCR and tested. If the mother is heterozygous for a genetic disease trait, there will be a positive test for the bad allele *and* a negative result for the normal allele. This means that both copies of the mutant allele are in the polar body, no crossover has occurred, and the egg may safely be fertilized in IVF. If crossover had occurred, the PB1 would contain one mutant and one healthy allele, giving a 50:50 chance that the mutant allele will end up in the egg and be expressed, instead of in the second polar body.

An exhaustive list of the known human genetic disorders, genes, and their chromosomal locations and OMIM® links can be found online at: http://gdbwww.gdb.org/gdbreports/Chr.1.omim.html. This list covers X, Y, and all 23 chromosomes, as well as genes on unknown chromosomes. OMIM stands for Online Mendelian Inheritance in Man®, trademarks of Johns Hopkins University. The OMIM Web site entrez URL is in the bibliography [OMIM 2006c].

9.3.4 The "454" System

An example of the power of the photonic approach in genomic analysis is seen in the latest generation (2007) of genomic sequencing machines invented by *454 Life Sciences* of Branford, Connecticut (hereafter known as the 454 system). The latest version of the 454 system can read 3×10^8 Nts a day (one-tenth of the human genome). The high throughput rate of the 454 system is due to its novel, photonic-based design (chromatographic and fluorescent methods are not used). It is an array-based analyzer that uses a glass slide with an amazing 1.6 million μ-wells; each well is about 50 μm in diameter and 55 μm deep, a well volume of 1.08×10^5 cubic microns.

First, the analyte dsDNA molecule is broken up by restriction enzymes into ssDNA oligos of 200 to 300 Nts. One type of each ssDNA oligo is attached to a bead. The DNA on the isolated beads is then amplified (cloned) by PCR so that millions of identical copies of the original oligo on a bead are now attached to that bead. The ssDNA-covered beads are now mixed with DNA polymerase and loaded into the 1.6×10^6 wells on the slide. Next, smaller beads containing the enzymes *sulfurylase* and *luciferase* are packed into the wells on top of the larger, ssDNA-covered beads. Next, four solutions of pure nucleotides (**A**, **C**, **T**, and **G**) are made to flow over all of the wells, one at a time, sequentially. Thus, at any one time, there will be pure **A, C, T**, or **G** available to react with the ssDNA oligos in the wells. When a complementary Nt flows into the well, the polymerase adds it sequentially to the ssDNA oligo to make it a dsDNA oligo. Every time a free Nt is joined to the ssDNA oligo, a pryrophosphate (PP) group is released. The PP is acted on by the sulfurylase and then the luciferase, causing the release of photons. (Luciferase is used by fireflies and glowworms to make their light.) This reaction takes place in parallel because of the millions of identical ssDNA oligos attached to a given bead in a certain well. This light emitted by the addition of a single type of Nt is detected a CCD array, and the positions of the wells and oligos involved are determined.

Thus if, for example, pure *thymine* in solution is applied to the well array, only those wells containing oligos whose next uncombined Nt is *cytosine* will combine with the thymine, causing flashes. These flashes indicate that the free Nt on the reacting oligos was cytosine. It also indicates that the dark wells' oligos must have had either a thymine, guanine, or adenine that did not react with the thymine bath. The four pure Nt solutions are repeatedly applied sequentially as many times as there are Nts in the oligos. A computer "keeps score" of which cells have flashes for a certain test Nt.

The end result of the 454 system is that the exact sequence of Nts in each oligo can be determined relatively rapidly. The reassembled dsDNA oligos in the wells at the end of a run can be tested further by standard techniques if required. It is anticipated that the cost of sequencing a human genome should be less than $100,000 using improved 454 technology in 2008. It would be close to $50 million using the instrumental techniques first used in the Human Genome Project [Cohen 2007] .

9.4 FLUORESCENT MOLECULAR TAGS

9.4.1 Introduction to Fluorescence

Certain inorganic and organic molecules exhibit the phenomenon of *fluorescence.* When fluorescent molecules are irradiated with short-wavelength photon radiation (ranging approximately from UV to green), they emit longer-wavelength fluorescent photons (ranging from green to red). The details of the atomic mechanisms whereby the energy input (pump) photons are trapped and cause the radiation of fluorescence photons are beyond the scope of this text. However, we can summarize the process by saying that an input photon with energy hν is trapped by an atom and excites an electron to a higher energy state. This electron quickly loses its energy by transferring it via vibrational modes to other electrons, which, when they lose their energy, fall back to the ground state, causing the emission of a fluorescence photons with lower energy, hν'. The delay between excitation and emission of the fluorescent photon is on the order of nanoseconds [So & Dong 2002]. When the excitation stops, so does the fluorescence. Note that fluorescence is not phosphorescence.

Because the energy relaxation steps are dissipative, the wavelength of the emitted photon is longer than that of the pump photon; that is, there is a "red-shift." The difference between the emitted (λ') and the pump wavelength (λ) is called the *Stokes shift,* given by:

$$\Delta\lambda = (\lambda' - \lambda) = c(1/\nu' - 1/\nu) > 0 \tag{9.1}$$

where c is the speed of light, ν' is the Hz frequency of the fluorescent photon, and ν is the frequency of the pump photon. Also, not every excitation photon produces a fluorescence photon; every fluorescent material has an efficiency factor, η, describing the ratio of the intensity of the emitted photons at wavelength λ_F to the intensity of the excitation photons at wavelength λ_E.

All fluorescent molecules have an *excitation spectrum.* This is a plot of η versus λ_E over which fluorescent emission occurs. There is generally an excitation wavelength where fluorescence production is maximum, assuming constant excitation power. Similarly, for a fixed excitation wavelength and power, there is an emission wavelength spectrum (output power versus λ_F). Emission and excitation spectra of artificial inorganic fluorescent tags, for example, *quantum dots* (QDs), tend to be much narrower than the spectra associated with organic natural fluorophores (e.g., GFP) and synthetic organic dye fluorophores. The Stokes shift is also larger for quantum dots. QDs are essentially devoid of photobleaching (where the excitation sensitivity of a fluorophore decreases with time as it fluoresces). Bleach resistance means that the signal-to-background photon emission increases with time because the QD fluorescent output remains constant while artifactual fluorophores producing background photons gradually loose their sensitivity and produce fewer "noise" photons [Chan et al. 2005].

Pumping of fluorophores is generally done with a short-wavelength (frequency doubling) diode laser or by using incandescent radiation passed through a bandpass filter (e.g., a grating monochromator or an interference filter). Depending on the situation, fluorescence can be sensed with a vacuum photomultiplier tube, a solid-state photomultiplier, or a sensitive CCD camera. Optical filters are used to exclude pump-wavelength, Rayleigh scattered photons, as well as photons from other fluorescing molecules that might be in the preparation. Often the fluorescent light is focused by a microscope.

There are many fluorescent molecules that are used in genomics, proteomics, and molecular biology. Indeed, fluorophores are used in *ELISA, FISH, FRET, PCR,* and other methods of tagging molecules, replacing radioisotopes and enzymatically activated dyes such as those used in *ELISA.* These methods are discussed in detail later in this chapter.

One of the first natural, organic fluorophores to be used in analytic molecular biology is the *green fluorescent protein* (GFP) found in the jellyfish, *Aequorea victoria,* found along the U.S. northwest Pacific coast. The story of the isolation of *A. victoria's* GFP and its sequencing and the determination of its structure is told eloquently in the book by Marc Zimmer [2005]. Wild-type GFP is composed of 238 AAs. Its tertiary molecular structure is a cylinder with a diameter of about 3.0 nm and a length of about 4.0 nm. Eleven AA strands form the beta-sheet walls of the cylinder. Small sections of alpha helix form caps on the ends of the cylinder, and an irregular alpha helical segment also provides a scaffold for the three-AA *fluorophore* located in the center of the cylinder. GFP has a robust structure and is resistant to denaturing.

In 1992, Prasher et al. reported that they had determined the structure of the GFP protein, and had isolated its gene from the jellyfish and had cloned GFP using *E. coli* [Prasher et al. 1992]. Unfortunately, Prasher's first attempt at cloning did not produce fluorescence [Zimmer 2005]. In 1994, Chalfie et al. obtained fluorescing GFP from GM *E. coli* [Chalfie et al. 1994]. They used a PCR to amplify only the GFP gene. Their genetically engineered GFP was then used as a marker for gene expression in the worm *C. elegans.* In the time following Chalfie's seminal report, many workers have used GFP to mark a wide variety of molecules in living creatures, including proteins, enzymes, and nucleic acid oligos. GFP has been used in living eukaryotes, including the worm *C. elegans,* zebra fish, tadpoles, mosquitoes, fruit flies, rabbits, mice, etc. [Zimmer 2005].

The GFP fluorophore is a triplet of AAs, Ser65-Tyr66-Gly67, located in the center of the protein "barrel." Although this AA triplet is found in other proteins throughout nature, it generally does not fluoresce [Nikon 2006]. The wild-type fluorophore has two fluorescence excitation spectral peaks: a major peak at 395 nm and a minor one at 475 nm. GFP's emission spectrum is broad, with a peak at 509 nm (green). In the intact *A. victoria,* it has been found that the fluorescence mechanism is not as simple as shining blue light on GFP. Another photosensitive protein, *Aequorin,* is involved. Isolated aequorin fluoresces at 479 nm in vitro. In vivo, however, it does not radiate photons but rather transfers its output energy directly to an adjacent GFP molecule by a process called Förster resonance energy transfer (FRET). Ca^{++} ions are required for this process. FRET causes the GFP to emit maximally at 509 nm.

In a desire to fluorescently label more than one molecular system in an organism, wild GFP was molecularly engineered to alter its excitation and emission spectra. For example, replacement of Ser65 with Thr65 causes an increase in fluorescence intensity with little shift of the I/O spectra. Changing Try66 to His66 shifts the excitation peak to the UV (383 nm) and the fluoresence emission peak to 448 nm (blue), giving a BFP. Other AA substitutions in the fluorophore triplet give Cyan**FP** and **Y**ellow**FP** (as well as **GFP** and **B**lue**FP**).

Other fluorescent proteins have been found in nature. For example, the marine anemone *Discosoma striata* has provided workers with FPs emitting in the orange and red spectral regions. Nikon [2006] lists 7 GFPs, 3 BFPs, 5 CFPs, 5 YFPs, and 17 orange and red FPs available for research. One disadvantage of using FPs is that the molecules are relatively large and preparation time is long. FPs also tend to photobleach; that is, they gradually lose sensitivity when excited over time. The FPs have to be either inserted inside cells by genetic engineering methods or be permeable enough to penetrate the cell membrane [Selvin 2000].

Many other molecules fluoresce and can be used as tags in molecular FRET and FISH studies. The fluorescent cyanine dyes Cy2 bis (506 nm fluorescence peak, green), Cy3 (570 nm peak, orange), Cy3.5 (596 nm peak), Cy5 (670 nm peak, red), and Cy7 (767 nm peak, deep red) are used for labeling in DNA microarrays, 2-D protein electrophoresis, DNA sequencing, PCR, high-throughput screening, and flow cytometry. Cyanine dyes generally have narrow emission bands, which make multicolor screening less prone to interference by spectral overlap, and they work at biological pH ranges. They are used in FRET applications. *Fluorescein, rhodal,* and *lanthanide* derivative dyes are also used for molecular marking [Selvin 2000].

Fluorescent quantum dots (QDs) are semiconductor nanocrystals composed of atoms from groups II-VI or III-V in the periodic table (e.g., $Cd_{1-x}Se_x$, $Cd_{1-x}Te_x$, $In_{1-x}P_x$, and $In_{1-x}As_x$). They have also been synthesized using three-element alloys (e.g., $Zn_xCd_{1-x}Se$, $Zn_xCd_{1-x}S$, and $CdSe_{1-x}Te_x$). They are expressed as particles with diameters ranging from about 1 to 8.5 nm. Originally, QDs were used in optoelectronic semiconductor applications in devices such as QD lasers and superbright LEDs. Their potential in genomic and proteomic applications was uncovered in 1998 by Bruchez et al. and Chan and Nie.

QD fluorescent emission spectra are tunable by adjusting dot size and composition. Emission peaks can be made to range from 400 (blue) to 2,000 nm (NIR), given a fixed pump wavelength of 350 nm (UVA). QD emission spectra are generally sharp and narrow, far better defined than organic fluorophores. The narrow spectra mean that a number of different QDs and their ligands can be used at once and their emissions measured with minimum interference from adjacent emitters. QDs are 10 to 20 times brighter and several thousand times more resistant to photobleaching than organic dyes and FPs. For example, the dye Texas Red has a bleaching time constant of about 20 seconds, whereas QDs show no noticeable decrease in their fluorescent intensity after 2 minutes. The QD fluorescence decay time constant (to a pulse of excitation light) is about 10 times longer (20 nanoseconds) than that of organic fluorophores (2 nanoseconds) [Gao et al. 2005].

QDs generally contain from hundreds to thousands of alloy molecules and can be made as spheres, rods, or tetrapods. As prepared, QDs emerge coated with organic ligands such as *tri-n-octylphosphine oxide* (TOPO) and are hydrophobic. These new hydrophobic QDs are made hydrophilic for biological applications by coating them with amphiphilic polymers that contain both a hydrophobic segment or side chain and a hydrophilic segment or group such as *polyethylene glycol* (PEG) or multiple carboxylate groups. The carboxyls face outward to make the QDs water soluble; they also provide attachment sites for attachment ligands (e.g., biotin and avidin) and affinity molecules such as monoclonal Abs, peptides, and oligonucleotide probes [Gao et al. 2005].

The newest addition to QD fluorescent technology is *gold nanodots*. They are made up of 5, 8, 13, 23, or 31 gold atoms, each size fluorescing at different wavelengths to produce UV (385 nm), blue (455 nm), green (510 nm), red (760 nm), and IR (866 nm) emissions, respectively. The smaller gold QDs have the greater efficiencies for photon emission. The small size of gold QDs makes them better candidates than FPs or semicon QDs for use in FRET applications.

The size-tunable gold QDs (GQDs) were made by a slow reduction of $AuBr_3$ or $AuHCl_4$ gold salts in aqueous *polyamidoamine* (PAMAM) solutions. The size of the Au nanoclusters (and thus, their emission spectral peaks) was controlled by regulating the relative concentration of Au salts to PAMAM. After centrifugation and sorting, the PAMAM-encapsulated gold QDs were found to have high quantum yields (42% for the eight-atom blue QD) and narrow emission spectra [Zheng et al. 2004].

GQDs offer the advantage of small size, size tunabilty, narrow spectral emission, good brightness, imperviousness to bleaching, and nontoxicity when used in vivo in humans. GQDs are well suited for applications in proteomics and genomics.

9.4.2 Fluorescent in situ Hybridization (FISH)

FISH has been primarily used for the detection and localization of active DNA sequences on chromosomes.

FISH target molecules can also include mRNAs and proteins. FISH uses a system of binding molecules or probes to which unique fluorophores are attached. Fluorescence microscopy is used to view the fluorescing bound probe in situ.

In the FISH process, first a probe molecule is built. For DNA, it is a complementary oligo that is long enough to bind specifically to its target but not to similar short Nt sequences in the genome. If it is too large, it will impede the hybridization process. The probes are then labeled with *reporter fluorophores*.

The target DNA is generally a preparation of metaphase chromosomes from a cell (or cells) bound to the reaction vessel. The chromosomes have been denatured to separate the DNA helices into ssDNA. When the labeled probes are added to a preparation, they hybridize, forming short, labeled dsDNA. Then the unhybridized (excess) and partially attached probes are washed away. Finally, the tagged sections of chromosomal DNA are viewed under a fluorescence microscope.

FISH has found wide application in screening for certain *genetic diseases* where there are known, specific gene microdeletions or mutations. For example, FISH has been used to diagnose the following genetic diseases by detecting microdeletions: cri-du-chat, Kallman syndrome, Miller–Dieker syndrome, Williams syndrome, Smith–Mageneis syndrome, Prader–Willi/Angelman syndrome, Wolf–Hirschorn disease, steroid sulfatase deficiency and chronic myeloid leukemia [Palanisamy & Chaganti 2001], etc.

Chan et al. [2005] reported on a significant improvement in FISH technology that uses quantum dot (QD) fluorophores instead of the more widely used organic (and protein) signal molecules. Streptavidin-conjugated QDs fluorescing at 525, 565, 585, and 605 nm were incubated with biocytin, a water-soluble biotin derivative. The treated QDs were incubated with the appropriate probe oligos to attach them. QD-FISH was successfully performed on fresh mouse brain slices to detect mRNAs associated with gene expression.

Chan et al. [2005] concluded: "The direct QD-FISH method has several advantages in comparison with other approaches for FISH. First, fluorescence coming from directly QD oligonucleotide-labeled mRNA is likely to provide more accurate quantitation of relative levels of mRNA expression. An indirect detection of mRNA signal via a biotin-streptavidin interaction between the hybridized probes and streptavidin-conjugated QD can result in aggregation of the QD, altering signal intensity. Second, QD probe labeling and purification are relatively simple. After purification, probes can be stored for months or longer without affecting the hybridization or histological results. Third, the FISH protocol is simple due to the direct labeling of QD to the oligonucleotide probes and the absence of amplification steps. Finally, the high photostability of QD probes facilitates transcript analysis and quantitation due to the [fluorescent] signal stability not obtainable with conventional organic fluorophores." [Quoted with permission.]

9.4.3 Fluorescence Resonance Energy Transfer (FRET)

FRET is also called Förster resonance energy transfer after the physicist who first described and quantified the phenomenon [Förster 1948]. FRET is a sensitive biochemical/fluorescent tool that permits the analysis of the spatial relationships between two molecules interacting in vitro and in vivo. One molecule is tagged with a *donor fluorophore,* the other with an *acceptor fluorophore.* FRET can also be used to study the 3-D properties of a single protein molecule as the molecule changes shape in response to external factors such as pH and ion concentrations. The protein is tagged with a donor and an acceptor fluorophore. The transfer of captured photon energy between donor and acceptor fluorophore molecules causes a reduction in the donor's fluorescence intensity and excited state lifetime and an increase in the acceptor's emission intensity. The degree of donor/acceptor coupling can be measured ratiometrically using the two fluorescence intensities; this gives some independence from fluctuations in excitation intensity.

Instead of a second fluorophore acceptor, a nonradiating acceptor molecule called a *quencher* can also be used in FRET. Now the signal that the donor is adjacent to the quencher is a simple loss of donor fluorescence intensity.

What makes FRET a sensitive test of probe juxtaposition is the fact that the fluorescence resonance energy transfer between fluorophores is highly distance dependent. Förster [1948] showed that the *efficiency,* η, of the energy transfer is given by the sixth-power hyperbolic relation:

$$\eta = d_o^6/(d_o^6 + d^6) = I_A/(I_A + \gamma I_D) \tag{9.2}$$

where d_o is the distance, $\eta = ½$; d is the separation between the fluorophore molecules; I_A is the net fluorescence intensity from the acceptor fluorophores; I_D is the net fluorescent intensity from the excited donor fluorophores; and γ is a correction factor that compensates for the different quantum yields and filter/detector efficiencies of the donor and acceptor. d_o depends on several factors, including the refractive index of the medium, the fluorescence quantum yield of the donor in the absence of the acceptor, the dipole angular orientation of each molecule, and the integral of the overlap region between the donor and acceptor's emission spectra [Held 2005]. d_o is typically about 5 nm. Thus, if the fluorophores are greater than 8 nm apart, negligible FRET will occur between the fluorophore molecules ($\eta = 0.023$).

An example of a pair of FRET donor/acceptor molecules is CFP (donor) and YFP (acceptor). CFP is stimulated with 440-nm light and radiates fluorescent photons with a spectral peak at 480 nm. YFP can be externally excited at 510 nm and radiates with a peak at 545 nm. In FRET using CFP → YFP coupling, excitation is at 440 nm and fluorescent emission is measured at 545 and 480 nm. The optical bandpass filters are not perfect, so there are always unintended photons from the excitation source and the CFP output sensed in the 545-nm channel and unintended excitation photons and YFP short-wavelength photons sensed in the nominal 480-nm channel. Fortunately, these problems can be largely compensated for. If a 440-nm laser is used for excitation, the excitation leakage is negligible.

An example of a FRET application described by Selvin [2000] is the measurement of intracellular calcium ions. Cyan fluorescent protein-labeled *calmodulin* and yellow protein-labeled calmodulin-binding peptide were coexpressed. High Ca^{++} concentration led to binding of the calmodulin and the peptide. Juxtaposition of the fluorophores produces FRET. Another application of FRET cited by Selvin was the labeling of the muscle motor protein *myosin* at its N-terminal end with BFP and its C-terminal end with GFP. It was shown in vitro that the distance between the probes changed as myosin underwent conformational changes associated with ATP binding and hydrolysis, thereby confirming the *lever-arm model for muscle contraction* (see Section 7.2.4.6).

9.4.4 Phase Fluorometry and FLIM-FRET

Conventional FRET examines functions of the intensities of the fluorescent emissions of the donor and acceptor fluorophore molecules in order to deduce fluorophore (and hence, molecular) physical separation. *FLIM* is the acronym for fluorescence lifetime imaging microscopy. *Phase fluorometry* uses instrumental methods to measure the *fluorescence lifetime* decay time constant.

An important fluorescence parameter that can be used is the *fluorescence lifetime* of the donor and acceptor molecules. Fluorescence lifetimes can be measured by several techniques:

1. By measuring the steady-state sinusoidal frequency response of the fluorescence/excitation system; fluorescence phase and amplitude measurements allow calculation of the fluorescence lifetime.
2. By the time-domain transient response of the fluorescence system excited using very short (picosecond) excitation pulses. The time constant can be estimated directly from a plot of the averaged fluorescence waveform.
3. By *cross-correlation*, which allows direct estimation of τ given an excitation that is amplitude modulated by high-frequency random noise or sums of nonharmonically related sinusoids.

If an isolated fluorescent molecule is excited with a short rectangular pulse of light, there is a *time constant* associated with the buildup of fluorescence and its subsequent decay. If the excitation light intensity of wavelength λ_E is sinusoidally modulated in the form

$$I_E(t) = (I_{Eo}/2)\,[1 + \sin(2\pi f_m t)] \tag{9.3}$$

then the fluorescence in the steady state, to a first approximation, follows the form

$$I_F(t) = (I_{Fo}/2)\,[1 + \sin(2\pi f_m t + \phi)] \tag{9.4}$$

where $I_F(t)$ is the measured instantaneous fluorescence intensity at wavelength λ_F from the acceptor molecules, f_m is the Hz modulation frequency, and ϕ is the phase lag between the fluorescence and excitation sinusoids. Assuming linear first-order fluorescence kinetics, it is easy to show that the phase lag $\phi = -\tan^{-1}(2\pi f_m \tau)$ radians, where τ is the fluorescence time constant (time for I_F to fall to $0.368 = 1/e$ of its steady-state maximum, given a turnoff of constant excitation). First-order kinetics implies that the ODE describing the intensity of fluorescent emission is of the form

$$\dot{I}_F\tau + I_F = \eta I_E \tag{9.5}$$

where η is the FRET efficiency. Because of bleaching in organic fluorophores, η decreases with time, given constant excitation. Here we will assume that η remains constant over the time that ϕ is measured. Note that the excitation signal consists of a dc level of $(I_{Eo}/2)$ plus a sinusoidal component of peak amplitude $(I_{Eo}/2)$ and frequency f_m. The dc component is mandated by the fact that the input light intensity, by definition, is nonnegative. Thus, in the simple linear model considered earlier, the fluorescence intensity must also be nonnegative. The dc part of the fluorescence is simply $I_{Fo} = \eta I_{Eo}$. By measuring the steady-state phase lag ϕ between excitation and fluorescent emission of a single fluorophore species, it is possible to quantify the fluorescence time constant τ and study the conditions that affect it. The time constant then is given by simply

$$\tau(f) = (2\pi f_m)^{-1}\tan[\phi(f)] \tag{9.6}$$

where ϕ is generally measured at several frequencies, and an average is taken.

Because the most accurate phase measurement is at a frequency where $\phi \cong 45°$, the excitation modulation frequency is given by $f_m \cong 1/(2\pi\tau)$ Hz. For example, for a $\tau = 20$ ns, $f_m = 8$ MHz, and a 2-nanosecond fluorescence time constant requires $f_m = 80$ MHz for a 45° phase shift. (There are a variety of means of amplitude-modulating light at radio frequencies, which we will not cover here; the interested reader should see, for example, the photonics text by Sirohi and Kothiyal [1991].) The present trend is to directly modulate the output intensities of laser diodes or LEDs having appropriate emission wavelengths.

To actually measure the phase difference between two radio-frequency (RF), sinusoidal intensity signals (e.g., at frequencies between 1 and 100 MHz), it is practical to *heterodyne* the signals down to a low-kHz range where the phase can be easily measured with an IC phase detector chip. At low-kHz frequencies, the two intensity signals also can be digitized and input to a computer, which can estimate their phase difference in real time, using filtering and averaging, or cross-correlation. Calibration and zeroing of a phase fluorometer can be done using a nonfluorescent sample or utilizing the nanosecond delay inherent in light propagation over a fixed optical path length (light in air takes 3.33 nanoseconds to travel 1 m).

The heterodyning process (aka mixing) is accomplished by effectively multiplying a sine wave at frequency f_m by a reference oscillator's sine wave at frequency $f_r = (f_m - f_i)$. This multiplication (or mixing) gives two cosine waves, one with frequency $(2f_m - f_i)$ and the other at f_i. f_i is made the intermediate frequency (IF) in kHz used in the phase detector (e.g., 1 kHz). The high-frequency term $(2f_m - f_i)$ is removed by low-pass filtering and is not used. Thus, the excitation input to the phase detector is a cosine wave of low frequency, f_i. The fluorescence output signal is also converted to a cosine at f_i Hz but also has the original phase shift, ϕ. An algorithmic comparison of these low-frequency cosinusoids yields ϕ, and hence τ, by the foregoing relation. Figure 9.3 illustrates the block diagram of a generic sinusoidal phase fluorometer. In this system, only one master crystal oscillator is used; the modulation frequency, f_m, and the mixing frequency, $(f_m - f_i)$, are generated digitally using phase-locked loop frequency synthesis techniques. A half-silvered mirror is used to derive an optical signal proportional to $I_E(t)$. This signal is sensed and its dc level and ac component separated by low-pass and high-pass filters, respectively.

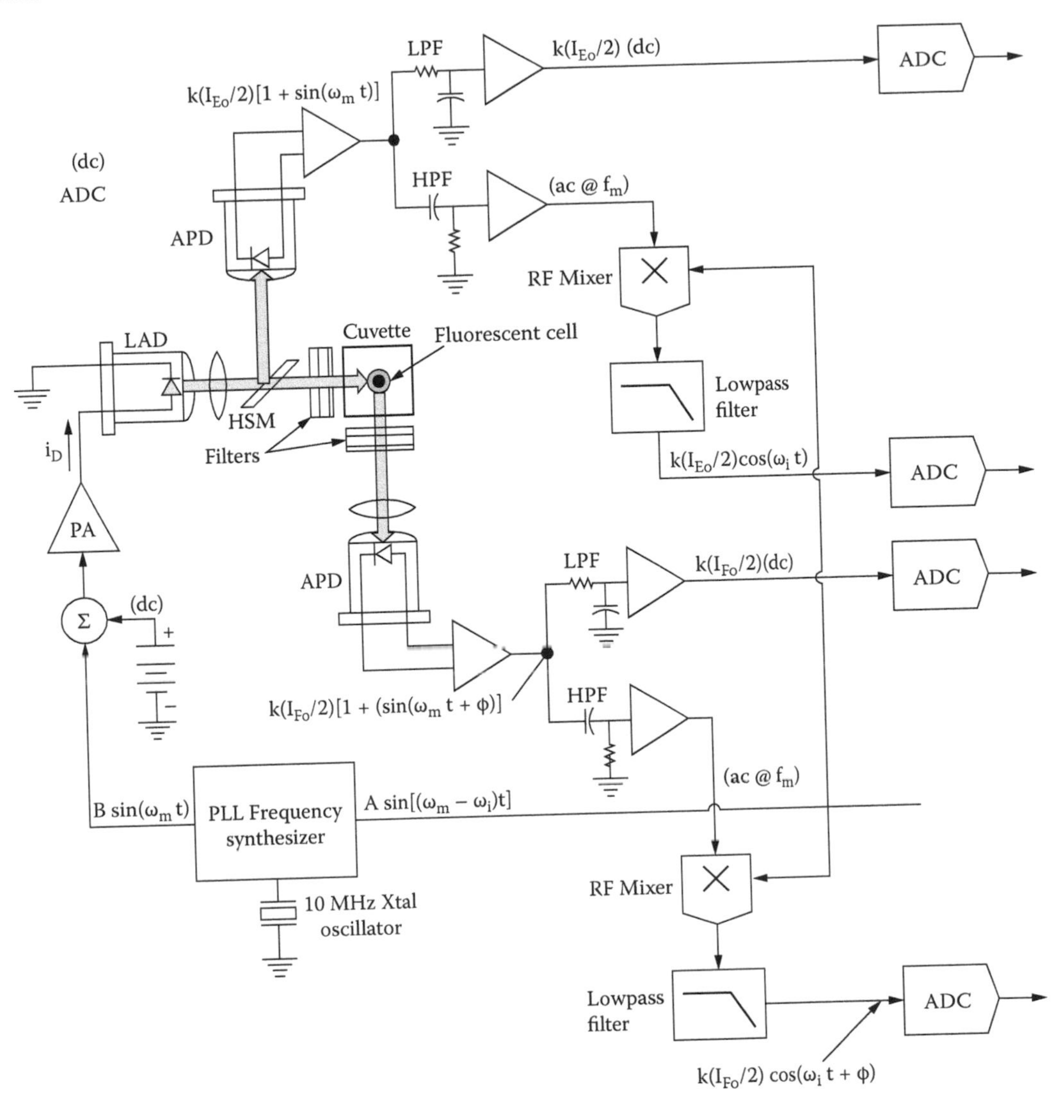

FIGURE 9.3 Block diagram of a basic sinusoidal phase fluorometer. See text for further description.

The ac component is heterodyned down to the 1-KHz working frequency, f_i, which then serves as the reference phase in sensing ϕ. Two electronically identical detection and filtering channels are used to compensate for the frequency responses of the detectors, amplifiers, filters, and mixers. If FRET is to be measured, a third detector-filter-mixer channel is added that is responsive to the acceptor fluorophore's emission wavelength. Now frequency response comparisons can be made between excitation/donor, excitation/acceptor, and donor/acceptor fluorophores. Note that the mixed-down f_i signals are digitized directly by analog-to-digital converters (ADCs), and the host computer inputs this digital information directly to do the phase detection, the τ and η estimations, as well as increment f_m.

A good example of a phase fluorometer designed to measure photosynthesis in phytoplankton was described by NASA [1996]. Unique to this system was the use of a master oscillator working at a 1-kHz baseband frequency, f_i. The laser diode was modulated at a frequency $(f_c + f_i)$ obtained by a quadrature mixing/ addition process using the carrier at f_c and baseband frequency. Final detection was accomplished by mixing the photodiode output $(f_c + f_i)$ with the carrier signal at f_c. Digitized, sinusoidal phase and amplitude information was sent to a computer for processing.

Fluorescent lifetime can also be measured directly in the time domain by exciting the fluorophore with a narrow, intense excitation light pulse with a width less than τ, and then observing the exponential decay of the fluorescence intensity. (A narrow pulse driver for LEDs was described by O'Hagan et al. [2002].) The fluorescence decay waveform can be approximated by the exponential relation, $I_F(t) = I_{Fo} \exp[-(t/\tau)]$, over a large time until $I_F(t)$ is lost in noise. The pulse excitation is repeated N times and the results averaged to reduce random noise. A fluorescence system with a single time constant will yield a linear log[$I_F(t)$] versus t plot. A break in the slope of the log[$I_F(t)$] versus t plot indicates that there are two time constants in the fluorescent system (the system's kinetics are second order). Accurate determination of τ_1 and τ_2 by the sinusoidal method now requires a detailed Bode phase and magnitude plot of the fluorescent system's steady-state frequency response. This may be impractical because long exposure to excitation energy required for a frequency response measurement can cause bleaching, changing both η and the τs.

FLIM-FRET examines the dynamics of the fluorescence coupling in dual-fluorophore FRET systems. Both donor and acceptor fluorescence dynamics are examined. The relationship between the donor fluorescence time constant, τ_D, in the absence of the acceptor and the donor time constant, τ_{DA}, in the presence of the acceptor is of interest. Clayton et al. [2004] have treated a complex FLIM-FRET system in which the fluorescence recorded emanates simultaneously from two different fluorescent systems with different time constants. Clayton et al. used linear fitting algorithms on data from steady-state single-frequency phase fluorometry to calculate the two time constants (fluorescent lifetimes).

9.5 INTRODUCTION TO FORENSIC APPLICATIONS OF DNA RESTRICTION ENZYMES AND STRS

9.5.1 Introduction

DNA (and RNA) forensic analysis is being used for an increasing number of applications, including but not limited to solving crimes (rapes, murders, etc.), demonstrating an incarcerated person's innocence, identifying victims of mass disasters or genocide, personal ID, anthropology (tracking migratory patterns of primitive humans, determining the relatedness of human populations, etc.), animal genetics, bacterial and viral forensic identification, etc. DNA from mitochondria is used to trace matrilineal ancestry because we inherit our mitochondrial DNA only from our mothers. DNA is available from nucleated cells in blood, semen, saliva (as deciduous epithelial cells from the mouth), urine (again deciduous epithelial cells), hair (with cellular "tags"), teeth, bone, and somatic tissues. Some of the difficulties encountered in forensic DNA analysis are that there may be mixed samples, samples may be contaminated, samples may be degraded by environmental conditions, and chemical contamination may inhibit the PCR process.

Many readers will be familiar with the long-running TV drama series *CSI* and its recent clones. The series is fairly technically accurate, except one gets the impression that the DNA tests are completed in minutes, rather than the 2 to 3 hours it takes to prepare the DNA, run the PCR amplification, and display the results. In this section, we will review modern forensic DNA analysis. It is perhaps a common misconception that forensic DNA analysis is carried out for the entire human genome. This would be technically nearly impossible because of the total number of Nts involved (about 3.1×10^9). Even examining each of the *about* 23,000 human genes would be overly time consuming. Instead, forensic DNA analysis in the United States is now carried out on 13 *short tandem repeat* (STR) *markers* in the human genome. These markers are found in the introns of the genome.

9.5.2 Restriction Fragment Length Polymorphism (RLFP) Analysis

Before adoption of STR forensic analysis, described later, RLFP was first adapted for DNA "fingerprinting," as described in the following text.

First, dsDNA is extracted from the sample cells using standard procedures. The extracted dsDNA is broken into fragments using *restriction enzymes* (REs). In most North American forensic laboratories, the RE of choice is generally *HaeIII.* After the extracted DNA is digested by the RE, the fragments are sorted according to size using *agarose gel electrophoresis.* Agarose gel contains nanopores through which the electrically charged dsDNA oligos can migrate, responding to the dc electric field across the flat slab. The smaller oligos have a higher average velocity of propagation through the gel and travel the farthest in a given time. Because the charged DNA fragment molecular weights are quantized, the oligos are generally separated by discrete distances at the end of their electrophoretic journeys.

In the next step in RFLP, the dsDNA fragments are denatured into ssDNA by soaking the gel in a strong alkali solution. (The alkali causes the hydrogen bonds binding the dsDNA complementary strands to separate.) Next, the ssDNA fragments are physically transferred and fixed to the surface of a thin nylon membrane in a technique called *Southern Blotting* or *Southern Transfer.*

Now, a technique called *nucleic acid hybridization* is performed. *Probe strands* of ssDNA to which a radioisotope has been attached (usually ${}_{15}P^{32}$, a beta emitter) are now incubated with the ssDNA on the nylon. After incubation, the unbound probes are washed away, and photographic or x-ray film is laid over the membrane. Wherever a group of like-labeled probes has bound with a group of matching, like ssDNA oligos, there will be a "bar code" exposure density on the film, indicating the match. The "fingerprint" autoradiogram of the DNA sample in question is then compared with those of known (or standard) DNA samples. Obviously, this is a time-consuming procedure that requires great technical skill.

Disadvantages of RFLP include the long time required to perform a test and its control, and the fact that relatively large DNA samples are required. Also, the exact sizes of the bands are unknown, and they are only relative indicators of the amount of probed DNA present. Using RFLP, it is also difficult to determine results from mixed samples, such as rape samples containing semen and vaginal cells. Because of these drawbacks, RFLP was superseded about 1991 by the superior STR-PCR technology of DNA fingerprinting, described in the following subsection.

9.5.3 Short Tandem Repeat Markers

STR markers are noncoding sections of the genome found throughout the genome. It is estimated that the human genome contains one STR sequence every 2,000 base pairs [Cullen et al. 2003]. This rate implies that there are about 1.6×10^6 STRs in the human genome! Cullen et al. identified 249 polymorphic STRs in a 3.8-Mb genomic segment encompassing the genes for the human *major histocompatibility complex* (MHC). The human MHC contains 224 genes and psuedogenes on chromosome 6p21.3. The MHC genes are associated with more diseases (autoimmune, infectious, and cancer) than any other region of the human genome.

An STR in DNA occurs when a pattern of two or more Nts is repeated and the repetitions are directly adjacent to each other. The pattern is generally from 2 to 10 pairs of bases (on a single strand of DNA), and the total size must be lower than 100 pairs of bases. An example of an STR is the tetrameric sequence

ATTA|ATTA|ATTA|ATTA, or (ATTA)$_4$.

STRs can be composed of Nt sequences of from two to five bases repeated two or more times. At every STR locus, an individual has two alleles. Thus, the 13, core STR loci used in the United States have 26 alleles. It is the variability in the STR repeat number in an allele that is of interest in forensic DNA analysis. The STR alleles comprise a very small fraction of the DNA in the human genome; a profiling STR allele is about 50 to 450 bps long, and the genome has over 3.1×10^9 bps. The number of different repeat lengths observed for all 13 CODIS STR loci was about 344 in 2001, but more are being discovered as the STR database broadens [Koupparis 2002]. [CODIS stands for combined DNA index system.]

For practical reasons, all of the 13 CODIS STR markers that are now used in the United States for DNA forensics are tetrameric repeat sequences. The designators for the CODIS STR markers are identified as follows: (The number in parenthesis is the human chromosome number on which they occur.) TPOX (2), D3S1358 (3), FGA (4), CSF1PO (5), D5S818 (5), D7S820 (7), D8S1179 (8), TH01 (11), VWA (12), D13S317 (13), D16S539 (16), D18S51 (18), and D21S11 (21).

The *Amelogenin* sex typing STR marker is also included in addition to the 13 CODIS STR markers; it uses loci on X and Y. Other known STR markers that are used in forensics are CD4, F13A1, F13B, FES/FPS, DYS19, DYS391, HPRTB, LPL, GATA132B04, D22S445, D2S1338, FIBRA, D19S433, and D16S2622.

As an example, consider the CODIS D7S820 locus on chromosome 7. The STR for this locus is $(\mathbf{GATA})_k$, where the tetramer repeat number, **k**, is known to vary between 6 and 15 in different persons' alleles. Because of the wide physical separation of STR loci on the genome, their repeat numbers, **k**, are considered to be statistically independent. For example, a 15-fold repeat of **GATA** from the D7S820 locus should not be correlated with a long STR tetramer repeat sequence on the D8S1179 locus.

Statistics on variability of the 13 STR markers used in the United States by forensic scientists is maintained by the FBI's CODIS DNA STR database, begun in 1998. It is estimated that only about 0.1% of a person's total DNA differs from one person to the next for individuals in a certain racial group (i.e., about 3×10^6 Nts). Because of the product rule in independent event statistics, an individual's STR marker uniqueness (e.g., 1-in-M) can be found by taking M = $1/p$, where

$$p = \prod_{j=1}^{j=13} p_j \tag{9.7}$$

where p_j is the probability that the jth STR pattern will repeat k times in a given demographic sample group.

Clearly, when there is DNA damage, and there are fewer than 13 loci available, the certainty of DNA uniqueness (or a match) decreases. Often, when tissue DNA samples are degraded by time, weather, fire, chemicals, etc., *mitochondrial DNA* (mtDNA) is used. A cell generally has one nucleus with one **2n** genome but may have several hundred mitochondria with many mtDNA molecules. Forensic scientists amplify the HV1 and HV2 regions of the mtDNA, then sequence each region and compare single Nt differences to a reference. Because mtDNA is maternally inherited, a directly linked maternal relative's mtDNA can serve as a match reference. STR analysis is not practical on mitochondrial rings; their genomes are generally analyzed directly. The FBI amplifies about 610 DNA bps in the two hypervariable regions of the 16,569-bp mitochondrial genome for forensic analysis [Isenberg & Moore 1999].

Once the DNA STR fragments have been tagged by sequence-specific primers and amplified by PCR, the resulting ssDNA oligos are separated and detected using electrophoresis. Either gel or capillary electrophoresis can be used. In capillary electrophoresis, the STR oligos are injected into the bottom of a polymer-filled capillary tube and are separated by their differential migration rate in an applied dc electric field. The STR oligos are detected by fluorescence from fluorophore tags that were attached to the primers used in the PCR phase of the process. Different fluorophore emission wavelengths allow multiplexed testing in which up to 10 STR markers can be copied at once. Sensitivities for the process extend down to less than 1 ng of DNA. Note that all 13 CODIS loci primers are commercially available with fluorophore tags, as well as primers for other STRs and Amelogenin (sex determination).

In gel electrophoresis, a large polyacrylamide gel plate is used to separate the amplified DNA oligos. Again, fluorophore tags on the PCR primers are read electronically, which allows quantification of tagged DNA densities.

Recently, the use of microarrays has been employed to readout STR-PCR DNA analysis. A *variable-length probe array* (VLPA) has been used to identify and quantify STR length and the number of STRs of the target molecule. The method involves hybridization of the unknown STR target sequence to a DNA microarray displaying complementary probes that vary in length to cover the range of possible STRs. A posthybridization enzymatic digestion of the DNA hybrids is then used to selectively remove labeled, single-stranded regions of DNA from the array's surface. The number of STR repeats in the unknown DNA sample is then deduced from the pattern of target DNA that remains hybridized to the microarray. A Cy5 fluorophore label permitted quantitative fluorescent detection. Note that the VLPA method eliminates the gel or capillary electrophoresis following STR-PCR [Kemp et al. 2005]. We expect that the VLPA technology will be easily adapted to field-portable, rapid STR-PCR systems for use by criminalists and the military.

9.5.4 Validity of STR DNA Evidence

Evidence from a PCR-STR test is often presented in court as a 1-in-M uniqueness match (where M is a very large number; see Equation 9.7). The problems judges and juries have with this evidence is often not the technology of STR testing but its statistical interpretation. Clearly, the 13 STR probabilities used in calculating the "match odds" by the product rule ideally should be based on very, very large samples of a genetically homogeneous population. In practice, the 13 STR match odds are different for different ethnic groups, especially ones that have been relatively isolated and have had a certain degree of inbreeding. Examples include African bushmen, Australian aborigines, Inuit, Ashkenazi Jews, etc. Statistics also will be skewed in a target population if persons are normally related and there has been inbreeding in the past (e.g., first- and second-cousin marriages). See, for example, the population genetic study of persons in Alia, Sicily, using 15 STR markers (12 were CODIS markers, 3 were not) [Calò et al. 2003]. These workers found that while comparing the Alia population with other populations from Sicily, a genetic heterogeneity within Sicily was uncovered. They noted: "Comparisons with other Italian, Mediterranean and European populations highlight the differentiation of the Sicilian population, reflecting the presence of a genetic boundary that separates Sicily from northern and central Italy and from the western Mediterranean basin."

When evaluating the validity of a DNA match or uniqueness, one should keep the following questions in mind:

1. Is the match an improbable coincidence? (A true match will result from identical twins.)
2. Is it the result of a planted sample (i.e., a frame-up)?
3. Did the accused leave the sample at the exact time of the crime?
4. What are the statistical confidence limits for the figure produced by the product rule (Equation 9.7)?
5. If the answers to questions 1 and 2 are no, and the answer is yes to question 3, is the accused guilty, or is there reasonable doubt? If two markers in 13 do not match and the other 11 do, common sense tells us that there is no way the samples came from the same individual, but the two donors were very probably related.

In summary, the simple product rule applied to CODIS data gives very optimistic results for matches; that is, M is larger than a more realistic approach (using conditional probabilities) to the statistical evaluation of the sample would produce. For example, consider the statement: "The source of blood on the shirt of the accused cannot be excluded as having come from the victim. The 13 locus STR DNA evidence is M^{-1} times more likely to have come from the victim than a random male member of the Caucasian population." For a perfect match (all 13 markers), M by the product rule can be over 10^{15}, clearly an enormous number, well more than the present population of Earth. CODIS STR probabilities for Caucasians are based on very low sample numbers compared to the total number of living Caucasian humans and do not take into account regional inbreeding. Thus, each STR marker probability estimate itself has a statistical uncertainty (variance) associated with it. These 13 uncertainties add to form a worst-case *limiting error* to the probability product estimate, M. Still, a 13 marker perfect match is a perfect match. In the absence of an identical twin as the blood source, it is a virtually certainty that *it is* the victim's blood on the shirt.

9.6 ROLLING-CIRCLE REPLICATION OF SSDNA CIRCLES

9.6.1 Introduction

Rolling-circle replication (RCR), aka *rolling-circle amplification* (RCA), of DNA oligos is a viable alternative to the PCR for analytical genomics. RCA is carried out at a constant temperature on small ssDNA plasmids (circles of genomic ssDNA) in vitro. There is no need to cycle the temperature as done in the PCR. More than a billionfold amplification of DNA oligos can be achieved with certain RCA protocols in about 1 to 2 hours, making RCA competitive with PCR for some applications. RCA was first described in the mid-1990s by Fire and Xu [1995] and also by Liu et al. [1996].

As originally described by Liu et al. [1996], single-stranded circular DNA oligonucleotides (oligos) of 26 to 74 Nts in length can serve as catalytic templates for DNA synthesis by one of several DNA polymerase enzymes (*E. coli Pol I, Klenow enzyme*, and *T4 DNA polymerase*). The DNA product is a repeating, end-to-end, multimeric, complementary copy of the ring oligo and can range from 10^3 to 12×10^3 Nts in length. Liu et al. noted that the reaction was unusual in that the synthesis proceeds efficiently despite the curvature and small size of the circles, some of which have diameters significantly smaller than that of the polymerase molecules. Also, the synthesis can proceed hundreds of times around the circle, whereas rolling replication of larger plasmid DNA circles (e.g., bacterial genomes) requires other proteins to facilitate synthesis. Figure 9.4 illustrates the RCR process as described by Liu et al. An ssDNA ring is made from N Nts. Note that a small ssDNA *primer* of about 10–15 Nts is used to initiate the RCR reaction, which requires a DNA polymerase plus dATP, dTTP, dCTP, and dGTP, from which the linear product is built. The ssDNA product emerges from the circle 5′ end first and consists of many linked copies of the ssDNA in the circle (N Nts long). These copies can be separated by the use of a restriction endonuclease that cleaves the product into many linear oligos, N Nts long, with bases complementary to those N in the ring. Liu et al. found that only one DNA primer molecule is required to facilitate the RCR process that can make over 200 complementary copies of a ring's ssDNA. The original RCR/RCA proceeds linearly, whereas PCR amplification behaves exponentially. Newer versions of RCA have been designed to amplify exponentially as well, increasing speed and accuracy.

9.6.2 RCA Protocols

In its original embodiment, the linear RCA reaction involved numerous rounds of isothermal enzymatic synthesis. DNA polymerase is used to grow a circle-hybridized primer by continuously traveling around and around the DNA circle. The plasmid generally contains ≤ 100 Nts. As the complementary ssDNA product is made, it can contain several thousand complementary repeats (cycles) of the original, small DNA plasmid. This repeat product must now be broken up for further DNA sequence analysis by *restriction endonucleases*.

The RCA method has evolved into more efficient forms. *Double-primed RCA* (aka hyperbranched, ramification, or cascade RCA) operates with a pair of different primers. One primer is complementary to a section of the ring, as used in basic linear RCR, whereas the other is targeted to the repeated ssDNA sequences of the primary RCA product.

Thus, a branching cascade of multiple-hybridization, primer extension, and strand-displacement events occurs using both primers. This reaction produces a fractal or "snowflake" product from each original plasmid. The two-primer protocol can yield up to 10^9 copies of a plasmid's DNA

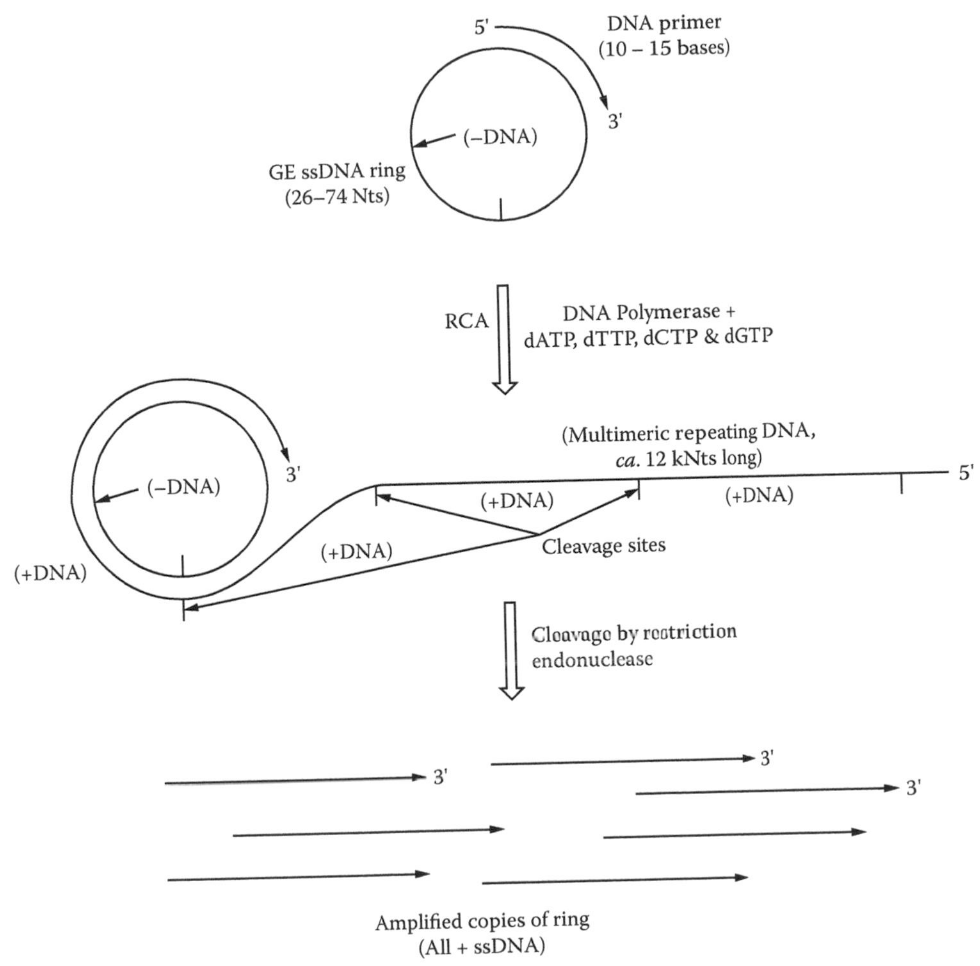

FIGURE 9.4 (See color insert.) Schematic of the rolling-circle amplification (RCA) process (for ssDNA), as described by Liu et al. [1996]. See text for further description.

in about an hour [Demidov 2005]. *GE Healthcare* (Amersham Biosciences) offers commercial *TempliPhi®* RCA kits: "From picogram [10^{-12} g] starting amounts, Templi-Phi® DNA Amplification Kits generate microgram [10^{-6} g] quantities of [circle] template DNA utilizing bacteriophage Phi29 DNA polymerase and rolling circle amplification. ... Amplified DNA can be used directly for sequencing using any chemistry and sequencing platform. Amplified DNA can also be processed by restriction enzymes and stored for future needs."

9.6.3 The JGI RCA Protocol

The DOE Joint Genome Institute (JGI) describes the use of RCA in whole-genome shotgun sequencing. They use a protocol that creates multiple gene fragments, classified as 2–4 kb, 8–10 kb, and 40 kb. Rolling-circle amplification is used in their system.

The DNA under study is extracted from cells and fragmented by the *HydroShear®* process [JGI 2004] into 2- to 4-kb oligos (in this example). In the HydroShear process, a solution of DNA is forced through a small orifice under high pressure; hydraulic shear forces fragment the DNA strands. All the small, random DNA oligos are then ligated into plasmids (using a loop of nonessential bacterial DNA called **pUC18**). The pUC18 vector allows for propagation and replication of the plasmid in *E. coli* bacteria. Each plasmid is *genetically engineered* (GEd) to have a gene for antibiotic resistance and also a LacZ gene that, when expressed, will make a blue pigment coloring the bacteria. To amplify the plasmids, they are mixed in a reactor with *E. coli* bacteria.

The plasmids are caused to enter the *E. coli* by *electroporation*. A strong electric field causes "pores" to open in the bacterial membrane, through which the GEd pUC18 plasmids enter some of the cells. The transformed bacteria are plated onto a large agar plate containing the antibiotic ampicillin. Only those bacteria that have incorporated the plasmids with the ampicillin-resistant gene grow when the plate is incubated. (When the bacteria reproduce, they also reproduce the genetically engineered plasmids.) Three possibilities exist for the plasmids incorporated into the bacteria:

1. They do not have the ampicillin resistance gene, and they are killed on the plate (don't grow).
2. They grow on the plate but are blue because the DNA oligo was not inserted into the center of the lacZ gene.
3. They are white, and they grow because the inserted fragment broke the lacZ gene. The white colonies are selected; there are about 10^6 bacteria/colony. Each bacterium contains a random fragment of the original DNA being analyzed. This select bacterial "library" is now grown further.

Next, the multiplied bacterial "library" is lysed, liberating the plasmids within. Now the RCA process is used to further amplify the bioengineered plasmids. In the protocol used by JGI, Amersham *TempliPhi reagent* is used in the RCA process. It contains dNTPs (the four nucleotides, **A**, **T**, **G**, and **C**), Φ29 polymerase (an enzyme that adds complementary nucleotides to a strand of DNA), and random primers (which give the polymerase a place to start building). The mixture of TempliPhi reagent and plasmids is incubated for 18 hours at a constant 30°C.

After RCA incubation, many "snowflake"-like DNA structures are produced, each containing multiple copies of a plasmid's DNA. This fractal-like structure is the result of the polymerase "rolling" around the circular ssDNA plasmid and also along the DNA "arms" built by the polymerase. The RCA process is terminated by the *ET terminator* sequencing mixture, which contains the following:

1. The four nucleotides.
2. ddNTPs (nucleotides with *fluorescent markers* attached). RCA sequencing terminates when the polymerase runs into a ddNTP fluorescent nucleotide.
3. One primer that targets only the insert. Only one side of the DNA strand is going to be sequenced.
4. A polymerase that adds dNTPs to a strand of DNA.

A plate with the RCAd "fractals" is thermally cycled with the ET terminator mix. In each cycle, the DNA is "unzipped" (denatured for 1 minute at 94°C), the primers attach to one strand of the DNA (annealing for 15 seconds at 50°C), and the primer sequence is then elongated by adding complementary dNTPs (4 minutes at 60°C). "Because all four normal nucleotides are present, chain elongation proceeds normally until, by chance, DNA polymerase inserts a fluorescently-tagged ddNTP instead of the normal dNTP. This halts the elongation process, leaving a DNA sequence that ends with one of the four colored ddNTPs" [JGI 2004].

In the next sequence of steps, the amplified oligos are cleaned up to remove the RCA and ET reaction products. Capillary sequencing is then used to "read" the DNA fragments.

9.6.4 Padlock Probes in RCA

Padlock probes are short, linear DNA oligos that can discriminate single-point mutations in fragments of genes.

They have one target: complementary Nt sequences at each end that can bind to an ssDNA target molecule. They are circularized by enzymatic ligation if their 5′ and 3′ ends hybridize correctly to the target DNA molecule and are adjacent. The intramolecular hybridization reaction is strongly favored over interhybridizations, which allow large numbers of different probes to be ligated simultaneously. The ligated probes become topologically locked to the target oligo. The circularized probes are processed by RCA and can give detailed information about thousands of different target Nt sequences in parallel.

Thomas et al. [1999] prepared padlock oligos that had 89 and 84 bases. The 5′ arm that bound to the target oligo had 20 bases, the 3′ arms that bound sequentially to the target were made shorter (10–15 bases), and the 54 base central region was identical on all their probes, allowing generic primers to be used for cascade RCA. The orientation of the primers is such that circularization of the probe is required for amplification to occur. By using two primers and a DNA polymerase, padlock probes can be geometrically amplified by a mechanism Thomas called *cascade rolling circle amplification* (CRCA). Thomas et al. demonstrated that as few as 10 ligated padlock probes can be detected by this method.

Figure 9.5 illustrates schematically how an ssDNA padlock probe oligo-hybridizes with an ssDNA target oligo. Its 3′ and 5′ ends are joined by a ligation enzyme, forming a ring attached by

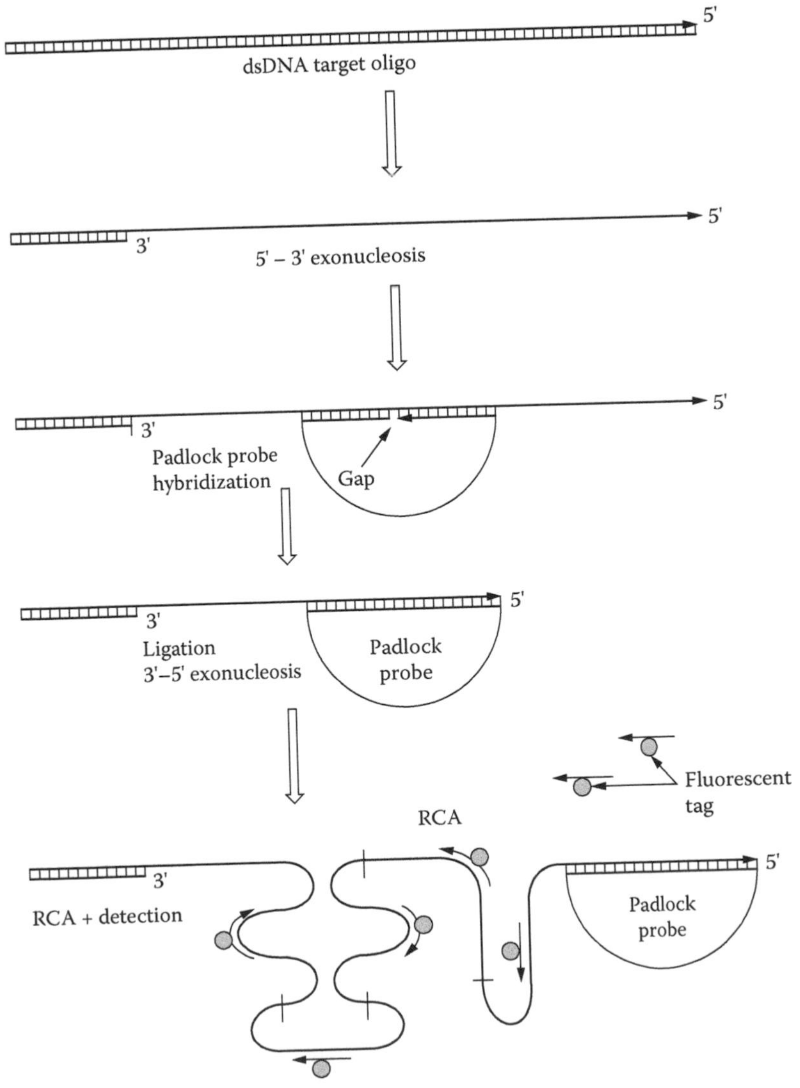

FIGURE 9.5 (See color insert.) Schematic illustration of how a padlock probe can replicate a selected oligo from ssDNA. See text for a detailed description of the process.

a few bases to the target DNA. When the RCA is initiated using Φ29 DNA polymerase, the target molecule serves as the primer because it has a 3′ end protruding past the probe circle. The padlock probe DNA then serves as the template for the RCA DNA synthesis. Many linear concatenated copies of the probe are made. Larsson et al. [2004] used fluorescent-labeled oligos complementary to an Nt sequence designed into specific probes to label those sequences in the RCA product (shown red in the figure). This ensures thousands of similar fluorescent markers on a specific RCA probe product, well marking the gene that had the sequence complementary to that on a certain padlock probe's ends. Larsson et al. used their method to search for SNP mutations in mitochondrial genomes.

9.7 METHODS FOR THE ANALYSIS OF PROTEIN STRUCTURE

9.7.1 Introduction

When we know the exact codon sequence of a DNA gene, or its mRNA, we pretty much know the linear AA structure of the protein made by that gene. (Recall that the molecules of many proteins get enzymatically altered *after* they are assembled by a ribosome.) Once made, the linear polypeptide rapidly assumes a complex 3-D (tertiary) structure, either naturally or with the aid of complex proteins known as *chaperones.* It is in this complex tertiary form that we find natural proteins inside of, on the surface of, and outside of cells. The roles of proteins as enzymes, hormones, receptors, chaperones, ion pumps, muscle components, etc., depend on the details of their tertiary form: their geometry, charge distribution, the AAs used, etc. Recall that a protein's gene only determines its linear AA sequence as it exits a ribosome; posttranscription editing can occur, and folding follows.

The *inverse problem* in proteomics is locating the gene that coded the linear structure of a particular protein, given the protein in tertiary form. To do this, one must literally take the protein apart. The first step in this process is to denature it, i.e., cause it to lose its complex 3-D structure by breaking hydrogen bonds and overcoming electrostatic forces, van der Waals forces, and hydrophobic forces that hold the polypeptide chains in their unique 3-D configuration. Cross-chain disulfide bonds must also be broken. Following denaturing, the polypeptide can be enzymatically cleaved, one AA at a time from either the C- or N-end. The AAs are identified using chromatography and/or mass spectrometry. Other kinds of proteolytic enzymes can be used cleave the subject polypeptide at certain AA residues. The biochemist now has protein fragments, or oligopeptides, to further disassemble. Also, large, very complex proteins can be crystallized and their 3-D structures deduced by x-ray crystallography.

9.7.2 Electrophoretic Separation of Proteins

The protein analyte initially is generally in tertiary form, mixed with other proteins and biomolecules. The first task is to separate the proteins from one another. This can be done with the SDS-PAGE method. (SDS is the acronym for *sodium dodecyl sulfate,* a strong detergent with a negative charge attached to it.) After incubation with SDS, cell membranes are dissolved and cell proteins are soluabilized, denatured, and given many negative charges. PAGE stands for *polyacrylamide gel electrophoresis.* When the mixture of proteins is placed on a gel in a dc electric field, they migrate at different rates toward the anode (+), depending on their negative charge distributions and their molecular weights. PAGE is first used with standard (known) protein samples having different molecular weights. SDS-PAGE is effective at separating natural proteins with different molecular weights; however, two different proteins with the same molecular weight can migrate through the gel at approximately the same rate, thwarting separation.

A tool to enhance PAGE separation is to adjust the pH of the medium containing the protein samples. All proteins have an *isoelectric point* (pI), which is the pH at which the protein has zero net charge on it. By adjusting the charges, PAGE separation may be enhanced. At a pH below the protein's pI, the molecule has a net positive charge. As the pH is made more basic than the protein's pI, it has an increasing negative charge. This process is known as *isoelectric focusing.*

9.7.3 Sequencing Protein AAs

For (denatured) polypeptides of 50 AAs or fewer, the *Edman degradation method* can be used to disassemble the peptide one AA at a time, starting from the N-terminal AA. The *alkaline phenyliso-thiocyanate* (PIT) reaction is used to react with the N-terminal AA. Next, the resultant phenylthio-carbamyl-derivatized AA is hydrolyzed in anhydrous acid. The N-terminal residue is chemically tagged to identify it. The remainder peptide is left intact, and the reaction can be repeated on its (N −1) terminal AA, etc. The result is the AA sequence of the polypeptide from the N-terminal to the C-terminal. Knowing the linear protein code, one can then search for mRNA or DNA genes that code that small protein. An example of the Edman process is shown in the reactions schematized as follows:

$$\mathbf{NH_2\text{-}AA1\text{-}AA2\text{-}AA3\text{-}AA4\text{-}\ldots + PIT \rightarrow {}^*X\text{-}AA1 + NH_2\text{-}AA2\text{-}AA3\text{-}AA4\text{-}\ldots}$$

Repeating:

$$\mathbf{NH_2\text{-}AA2\text{-}AA3\text{-}AA4\text{-}\ldots + PIT \rightarrow {}^*X\text{-}AA2 + NH2\text{-}AA3\text{-}AA4\text{-}AA5\text{-}\ldots}$$

Here, **X* is a chemical tag that the reagent attaches to the cleaved residue, which allows the cleaved AA to be identified.

Because of the size limitation of the Edman method, large denatured proteins are chopped into smaller oligopeptides by protein enzymes known as *proteases*. (See Table 9.1.) These smaller fragments can be separated by PAGE, then analyzed by Edman degradation or time-of-flight mass spectrometry. Table 9.1 summarizes the action of some commonly used proteases.

Chemical reagents can also be used to break down large denatured proteins. One example is cyanogen bromide (CNBr). CNBr cleaves the peptide bonds on the C-terminal side of methionine residues. The number of different oligos made from one polypeptide is one more than the number of methionine residues in the polypeptide. The rocket propellant *hydrazine* (**NH2NH2**) is also used for C-terminal protein digestion [King 2005c].

Determining protein structure is very important in drug design. The trend is to design "engi-neered" proteins that can competitively block certain receptors, tie up hormones with false recep-tors, or block enzymatic cascades such as caspases. Many interesting hot links to university and government research laboratories that offer specialized software for protein analysis can be found at the SWBIC 2005 Web sites.

9.7.4 ELISA Tests for Specific Proteins and Antibodies

ELISA stands for *Enzyme-Linked Immunoabsorbent Assay*. Much of the technology developed for the ELISA quantification of a specific antibody (Ab) or a specific antigen (Ag) has been carried over into the development of microarrays. ELISA allows researchers and clinicians to determine if a specific protein is present in a sample and, if so, its concentration. ELISA relies on the ability to produce *monoclonal antibodies* (MAbs) complementary to a specific Ag, including human Abs. ELISA tests can determine the presence of natural Abs to such infections as *HIV, hepatitis C, Lyme disease,* or *TB* and can also determine the levels of proteins such as *interleukins, insulin, TSH* (thy-roid function), and *human chorionic gonadotropin* (pregnancy).

In an *ELISA "sandwich" test* for an Ag protein in solution, primary MAbs to the Ag to be detected (analyte) are fixed to the test chamber walls. Next, a known amount of solution contain-ing an unknown amount of analyte molecules is added to the chamber. The Ag molecules bind to the fixed primary MAbs with high affinity. (We assume that there are fewer Ag molecules in the reaction chamber than MAbs fixed to the chamber walls.) Now a solution containing secondary MAbs to which enzyme molecules have been chemically coupled is added. The enzymes are chosen

TABLE 9.1
Some *Proteases* Used to Disassemble Polypeptides for Analysis

Enzyme	Source (Type)	Specificity	Comments
Carboxypeptidase A	Bovine pancreas (carboxyptidase)	Will not cleave at a C-terminal if AA residue = Arg, Lys, or Pro or if Pro is neighboring residue	
Carboxypeptidase B	Bovine pancreas (carboxyptidase)	Cleaves at a C-term when AA residue = Arg or Lys, not when Pro is residue	
Carboxypeptidase C	Citrus leaves (carboxyptidase)	Cleaves at all free C-terminal residues of polypeptidepH = 3.5 best	
Carboxypeptidase Y	Yeast (carboxyptidase)	Cleaves at all free C-terminal residues; more slowly at Gly residue	
Trypsin	Bovine pancreas (carboxyptidase)	Cleaves C-terminal peptide bonds when AA residue = Lys and Arg but not if Pro	Highly specific for positively charged residues
Chymotrypsin	Bovine pancreas (carboxyptidase)	Cleaves at C-terminal peptide bonds when AA residue = Phe, Tyr, and Trp but not Pro	Prefers bulky hydrophobic residuesActs slowly at Asn, His, Met, and Leu
Elastase	Bovine pancreas (carboxyptidase)	Cleaves C-terminal peptide bonds when AA residue = Ala, Gly, Ser, and Val but not if Pro	
Thermolysin	*Bacillus thermoproteolyticus* (endoprotease)	Cleaves N-terminal peptide bonds of AAs with residue = Ile, Met, Phe, Trp, Val, and Tyr but not if Pro	Prefers small neutral residues
Pepsin	Bovine gastric mucosa (endoprotease)	Cleaves at N-terminal peptide bonds w/AA residue = Phe, Leu, Trp, and Tyr but not if next to Pro	Exhibits small specificityRequires low pH
Endopeptidase V8	*Staphylococcus aureus* (carboxyptidase)	Cleaves peptide bond C-terminal to Glu	

Source: Adapted from King, M.W. 2005c. *Peptides and Proteins.* //web.indstate.edu/thcme/mwking/protein_structure.html.

because they will react with a substrate (chromogen) solution to make color, which can be measured spectrophotometrically. Typical enzymes include *alkaline phosphatase* or *β-galactosidase.* The MAb-enzyme conjugate molecules bind to exposed epitopes on those Ags already bound to the fixed MAbs. Next, the cell is washed to remove excess MAb–enzyme conjugate molecules, leaving a certain fraction of fixed Ab1–Ag–Ab2–enzyme complex "sandwiches" plus the remainder of fixed primary MAbs not bound to Ag. Now the substrate solution is added. It contains molecules that will react with the bound enzyme molecules and make color. (For example, *p*-nitrophenyl phosphate is converted to yellow *p*-nitrophenol by alkaline phosphatase. There are several enzyme-substrate combinations used in ELISA indicator reactions that make pleasant colors.) After an appropriate incubation period, the color reaction is stopped chemically. The color is read by a spectrophotometer (either absorption or transmission of two key wavelengths of light). The intensity of the color is an increasing function of the concentration of the bound Ag.

The sensitivity of the "sandwich" ELISA depends on four factors:

1. The number of fixed primary MAbs in the reaction chamber.
2. The avidity of the fixed MAbs for the Ag.
3. The avidity of the second MAb for a second epitope on the Ag.
4. The specific activity of the second MAb.

There are several variations on the ELISA test: One is the *FLISA fluorescent sandwich assay.* Instead of enzymes, fluorophores are attached to the second class of free MAbs. Now the final steps include washing out any unbound MAb-fluorophore molecules, then illuminating the cell with an excitation laser and recording the fluorescence intensity through an appropriate optical low-pass filter. This test is called FLISA *(fluorophore-linked immunoabsorbent assay).*

In *another variation on* sandwich ELISA, a standard test antigen is fixed to the cell walls. Now human serum containing a wide variety of natural Abs is added to the cell. If the patient's immune system has made Abs to the test antigen, those human Abs (hAbs) will bind to the fixed Ags' epitope. Now the serum is washed off and secondary MAbs conjugated with the signaling enzyme are added. These MAbs have affinity to the human Abs bound to the test Ag (they are called antihuman Abs [ahAbs]). Now we have the sandwich, fixed Ag-hAb-ahAb-enzyme. Color is developed when the substrate (chromogen) is added and incubated. The amount of color seen is an increasing function (generally a sigmoid curve) of the concentration of hAbs positive to the standard (test) antigen.

In *competitive ELISA assays,* only one type of antibody is used, the primary Ab (pAb). The pAb can be a purified MAb, or nonpurified natural Abs. The pAbs are fixed to the reaction vessel. Then these fixed pAbs are incubated with a known amount of solution containing the analyte antigen. The analyte Ag binds to a certain fraction of the fixed pAbs. Now a known amount of enzyme-labeled Ag is added to the chamber and incubated, the excess is washed off, and the substrate solution is added and incubated for a standard time. The higher the concentration of analyte, the more fixed pAbs will be bound, and fewer free pAbs will be available to bind to the substrate (enzyme-labeled Ag). Thus, the color intensity observed is inversely proportional to the concentration of the analyte Ag (a decreasing sigmoid curve).

Mai et al. [2002] compared the sensitivity and range of an ELISA detection method with the *AphaScreen® method* of detecting proteins marketed by *Perkin Elmer Life Sciences.* AlphaScreen makes use of tagged fluorescent nanobeads. Quoting Mai et al.:

> AlphaScreen is a bead-based non-radioactive Amplified Luminescent Proximity Homogeneous Assay. When a biological interaction brings the beads together, a cascade of chemical interactions acts to produce a greatly amplified [photonic] signal. On laser excitation, a photosensitizer in the Donor bead converts ambient oxygen to a more excited singlet state. The singlet state oxygen molecules diffuse across to react with a thioxene derivative in the Acceptor bead, generating chemiluminescence at 370 nm that further activates fluorophores contained in the same bead. The fluorophores subsequently emit light at 520–620 nm.
>
> In the absence of a specific biological interaction, the singlet state oxygen molecules produced by the Donor bead go undetected without the close proximity of the Acceptor bead. As a result only a very low background signal is produced.
>
> AlphaScreen provides a highly versatile, sensitive, time-resolved, homogeneous and miniaturizable means to efficiently perform assay development and high throughput at lower costs." [Copyright PerkinElmer, Inc. Used with permission.]

Mai et al. found that PE's AlphaScreen system and the ELISA tested both had useful detection ranges of 156 ng/mL to 5 mg/mL. However, the AlphaScreen method was found to have better linearity and accuracy.

One improvement on ELISA technology is called *Covalent Technology* (CT). In CT, covalent plates are used in the analysis chambers. The covalent plates have surface chemical groups that form strong covalent bonds to the fixed analyte Ag. Covalent plates also have spacers that put the fixed Ags off the bottom of the chamber into the solution phase, where they have a better chance of binding to free Abs. Also, the covalent fixation sites of Ags tend to have a uniform specific orientation, whereas on standard plates, the Ag molecules are fixed in a random fashion. This means that in CT, more of the fixed Ag is properly aligned (with respect to its epitope) and held in the liquid phase, where it has a greater probability of interacting with the Abs.

Another protein Ag or Ab test related to ELISA is the *Western Blot test*. The test Ag is isolated on a polyacrylamide gel by electrophoresis (i.e., PAGE). It is then transferred to and covalently fixed on a nitrocellulose surface. Specimens or control standards are introduced to discrete areas of the nitrocellulose that contain the fixed Ag. Specific Abs to the fixed Ag in the specimen will complex with the fixed Ag molecules. An additional enzyme-linked Ab is introduced and allowed to bind to the Ag–Ab complex. Reaction of the bound enzyme-labeled Ab produces a visible colored band in the discrete location where the sample was spotted, indicating the presence of complexing Abs in the specimen. Clearly, the Western Blot test is qualitative. It is easy to see, however, how it evolved into the quantitative sandwich ELISA test.

Originally, ELISA tests were carried out in large vessels such as test tubes and cuvettes; now 8 × 12 arrays are commonly used with wells holding about 50 μL. Also, the trend now appears to be the use of fluorescence detection, rather than colorimetric (or spectrophotometric) means. There are many biotech companies that offer supplies, antigens, and MAbs for ELISA and microarray technology [Uttamchandani et al. 2006].

9.7.5 Protein Microarrays

9.7.5.1 Introduction

Proteomics is the study of protein structure and function. *Genetics* tells us about genes, most of which when expressed determine the linear, 1-D AA sequences of protein or polypeptide molecules through messenger RNA. As we have seen, a quantum leap in understanding is required to predict the functional, tertiary, or 3-D structure of a mature protein, given its mRNA sequence. The inverse problem in proteomics is equally challenging; that is, given a large 3-D protein molecule, locate its gene on a chromosome, introns and all. Another key goal in proteomics is to assign functions to proteins isolated from a cell by a "shotgun" technique such as *two-dimensional polyacrylamide gel electrophoresis* (2D-PAGE). To achieve this goal, it is necessary to isolate one protein from the 2D-PAGE matrix. The resolution of 2D-PAGE separations has been improved by the use of immobilized pH gradients (IPGs) on the gel. Many thousands of different proteins can be separated on a single gel matrix, including proteins differing by a single posttranslational deamidation event [Cutler 2003]. Exacerbating the challenge of any form of proteomic analysis is the posttranslational modification of proteins by chaperone proteins, glycosylation, phosphorylation, and differential splicing.

In genomics, a very small sample of a pure nucleic acid can be grown larger by the well-known polymerase chain reaction (PCR). Unfortunately, there is no analogous amplification mechanism for protein samples. A small protein sample stays small.

Protein microarrays are an actively evolving technology [Uttamchandani et al. 2006]. They are being developed as high-throughput, labor-saving analytical systems to circumvent the labor-intensive PAGE approach. A protein microarray consists of "spots" of molecules of antibody proteins, proteins, oligopeptides, *aptamers*, or carbohydrates, immobilized in a grid pattern on a smooth, inert (e.g., glass or ceramic) surface. The advantage of microarrays is that they are small, they analyze (identify) in parallel, and they can be read out optically by fluorescence wavelength and intensity. Some versions of protein arrays have used electronic readout, or autoradiography, but such readouts either lack resolution, are too noisy, are too slow, or are technologically cumbersome and have now been superceded by photonic readouts.

The current trend in microarray fabrication is the use of glass slides (not unlike microscope slides). These can be smooth glass, a 3-D gel pad on glass, or a microwell slide. One gel pad array design used 100 × 100 pads, spaced 200 μm on centers; each pad was 20 μm square [Zhu & Snyder 2001]. Zhu and Snyder argued for the efficiency of reusable microwell slides. The microwells isolate the analytes and bound probes and reduce sample drying. These authors describe a 96 (12 × 8) microwell slide format. The microwells typically have volumes in the tens of microliters.

Different *capture agents* with specificity to various proteins are bound to the slide arrays. The ability to immobilize proteins (e.g., MAbs and analyte proteins) on the slides without altering their active form is crucial to any type of successful protein microarray design. There are many innovative, complex methods for protein binding to the microarray substrate; unfortunately, we cannot consider them in detail here. Some are: *(1) Covalent attachment using native chemical ligation, (2) leucine zipper heterodimerization, and (3) nucleic acid (plasmid) programmable protein array* [Uttamchandani et al. 2006]. Xu and Lam [2003] also illustrate some of the common chemistries used in fabricating protein microarrays by covalent attachment; six reactions are shown for nonselective covalent ligation and three reactions for selective covalent immobilization.

Detection and readout of protein microarrays has also seen a wide diversity of approaches. The paper by Rosi and Merkin [2005] presents an excellent review of microarray detection methods for both nucleic acid and proteins and summarizes published sensitivities for each approach. Their review article focuses on the use of fluorescent quantum dots, nanospheres, and other nanoparticles.

9.7.5.2 The First Large-Scale Protein Microarrays

Some of the innovators in protein microarray design were Zhu et al. [2001], who analyzed the substrate specificities of 119 yeast protein kinases using 17 different probe substrates that were attached in a microwell array. MacBeath and Schreiber [2000] also covalently tethered purified proteins to chemically activated glass slides. The proteins were studied for protein–protein interactions using fluorescently labeled protein affinity probes. These workers showed that a single protein could be detected in an array of 10,799 identical spots of another (control) protein. Sensitivity was on the order of micromoles (relatively poor by modern standards). Clearly, the first attempts at protein microarray technology were successful qualitatively, rather than quantitatively.

The first really large-scale protein microarray was developed by Zhu et al. [2001]. Quoting Phizicky et al. [2003]:

"In this experiment, 5,800 (94%) of the predicted yeast ORFs were cloned, and greater than 80% of these produced detectable amounts of protein, after purification in a high-throughput protocol.... These experiments opened a new field in which entire proteomes can be screened for binding and other biochemical assays."

The work of Zhu et al. used intact, purified yeast proteins immobilized in the array. Clearly, the creation of this protein array represented a high technical challenge.

9.7.5.3 Antibody Microarrays

These are analytical microarrays. Instead of binding intact purified protein molecules to the array surface, monoclonal antibodies, nucleic acid *aptamers* or chemical probes are spotted on the microarray, and the proteins or other analytes are added as a mixture or as purified molecules. Readouts are generally by fluorescence activity (intensity and wavelength), or even by surface plasmon resonance (SPR) [Northrop 2005], a photonic technique that can be used to track the time course of a protein–protein chemical reaction on one spot in the array.

There are two main types of monoclonal antibody (MAb) microarray: (1) those using a planar array of various antibody species (one type per spot) and (2) those using matching antibody pairs for sandwich-type assay systems.

In the simple 2-D planar array, the test MAbs are often bound to the slide in wells. Assays are performed by making a test run for calibration purposes with a standard antigen that is labeled with a fluorophore molecule. In the next run, all of the various analyte proteins have also been labeled with different-colored fluorophores. The MAbs are known, so when dye fluorescence is detected at a certain MAb cell, it shows that it has bound to the epitopes of a certain protein. Unfortunately, the same protein may have a weak affinity to one or more other MAbs and also bind to them. The weak binding to the other MAbs can block their binding sites, weakening their response and introducing noise and ambiguity into the results.

In the *sandwich microarray* (this is a physical rather than a molecular sandwich), the first set of MAbs is spotted on the lower slide, the analyte proteins with epitopes to the MAbs are introduced and bind to the appropriate MAbs in the lower array. The upper array contains bound, tagged MAbs located directly over the lower array's MAbs. Each of these labeled MAbs binds to a different epitope on the protein that has bound to the lower MAb, dramatically increasing the specificity of identification. Because of Ab–protein cross-reactions and difficult preparation, sandwich arrays do not appear practical for $N > 96$.

An alternative form of the 2-D MAb array system binds known, untagged MAbs to the array. Then, a mixture of untagged, unknown analyte proteins is added. These bind to various MAbs. Then, specific, known detector MAbs (DMAbs) carrying unique fluorophores are added. Unlike the sandwich array, these DMAbs are not fixed to an upper array grid. The DMAbs bind to other epitopes on the proteins that have attached to the known, fixed MAbs. The DMAbs can also use a detectable label molecule attached (e.g., an enzyme, radioisotope, etc.), or they are biotinylated for detection after subsequent probing with labeled *streptavidin*. For fluorescent tags, a scanned laser beam probes each cell and the resultant fluorescence is read out with a digital camera or a photomultiplier tube. Optical filters are used to block the laser's radiation and pass the longer-wavelength fluorescence. The more intense a spot of a given wavelength is, the more tagged analyte molecules are present. Optical readout is fast, precise, and easily automated.

As we mentioned earlier, antibodies have the ability to bind to a specific molecular region (or epitope) of a certain analyte protein. Unfortunately, there can be some "cross-talk"; that is, another protein type may have a similar epitope that also can bind to the DMAb in question but not as strongly as the epitope of the target protein. Note that if the human genome expresses over 220,000 different proteins, to identify and separate them using microarrays will require an awesome number of capture agents. Some protein molecules may require more than two capture agents for unique identification because of *epitopic ambiguity* (cross-reaction) [Sydor & Nock 2003; Phizicky et al. 2003]. In fact, epitopic ambiguity is a major limitation to the quantitative (and qualitative) resolution of Ab microarrays. The effect of cross-reaction on detection threshold was illustrated in tests performed by Pawlak et al. [2002]. Fluorescent antibody labeling of pure interleukin-2 had a detection threshold of about 100 pg/mL. When IL-2 was mixed with 200 pg/mL each of IL-1β, IL-4, IL-6, IL-12, IL-13, TNF-α, and IFN-γ, the detection threshold for IL-2 rose to about 250 pg/mL, a 2.5-fold *decrease* in sensitivity.

9.7.5.4 Improvements in Fluorescent Microarrays

A significant improvement in the resolution of protein microarrays was introduced by researchers at *Zeptosens®* AG, Witterswil, Switzerland. The Zeptosens innovation uses *planar waveguide technology* to illuminate the immediate fluorescent antibody layer next to the array substrate with an evanescent light field radiating from the top of the 2-D optical waveguide. Pawlak et al. [2002] described this system:

"Laser light expanded in one dimension [to make a line] to match the geometry of the incoupling grating, is coupled into a 150 nm thin film of high refractive index material (Ta_2O_5) deposited on a transparent glass support. Along the path of light propagation in the waveguide, a strong evanescent field is created at the surface, which has a limited penetration depth of about 200 nm into the adjacent medium. All molecules of one microarray area, located within this evanescent field, are excited simultaneously (typical area = 0.35 cm^2 in our case) and imaged by a CCD camera. Only surface-confined fluorescent labels are selectively excited for emission, whereas excitation does not occur in the bulk medium. This spatial discrimination and the high [local] excitation fields at the sensing surface result in a significant increase in signal-to-background ratios compared with conventional optical detection methods." [Quoted with permission.]

Conventional optical excitation methods include confocal (direct) excitation of fluorophores and use of an inclined, direct excitation beam. Optical low-pass filters are used to exclude excitation light and pass fluorescent wavelengths in conventional methods. There is always leakage of excitation light through the optical filter, and this energy presents one limitation to the absolute detection threshold of the desired flourescence photons. In the Zeptosens, negligible excitation energy interferes with detection of the fluorescence because of the rapid vertical attenuation of the evanescent excitation energy that emanates from the Ta_2O_5 planar waveguide into the reaction cell. Pawlak et al. [2002] cited the measured limit of detection (LOD) of the Zeptosens protein microarray to be about 0.8 zeptomole (0.8×10^{-21} mole) of Ab, that is, about 500 protein molecules, corresponding to a fraction of 2.8×10^{-6} of a monolayer coverage (at a dye/protein ratio of about 5) being presented and detected on a single microspot (cell). They stated: "This means that, enabled by the sensitivity of the instrument, a dynamic range of up to six decades of surface densities of immobilized fluorescent molecules can be provided and measured."

Figure 9.6 illustrates schematically the side view of one configuration of the Zeptosens *Waveguide Excitation Fluorescence Microscope* (WExFM) system. Protein analytes (e.g., proteins in a cell's membrane) are bound to the surface of the 150-nm-thick high-refractive-index ($n \cong 2.2$) Ta_2O_5 planar optical waveguide layer. Incoming excitation light is first expanded into a linear (sheet) beam and directed onto a diffraction grating etched in the waveguide surface. The grating has a periodicity of 350 nm. At a correct input angle, α, maximum light intensity is trapped in the waveguide and propagates in a monomode. Some light energy leaks out of the upper surface of the waveguide in an evanescent field. The evanescent field intensity is rapidly attenuated exponentially, dropping to an ineffective level at about 100 to 200 nm for visible light [Grandin et al. 2005]. It is this evanescent

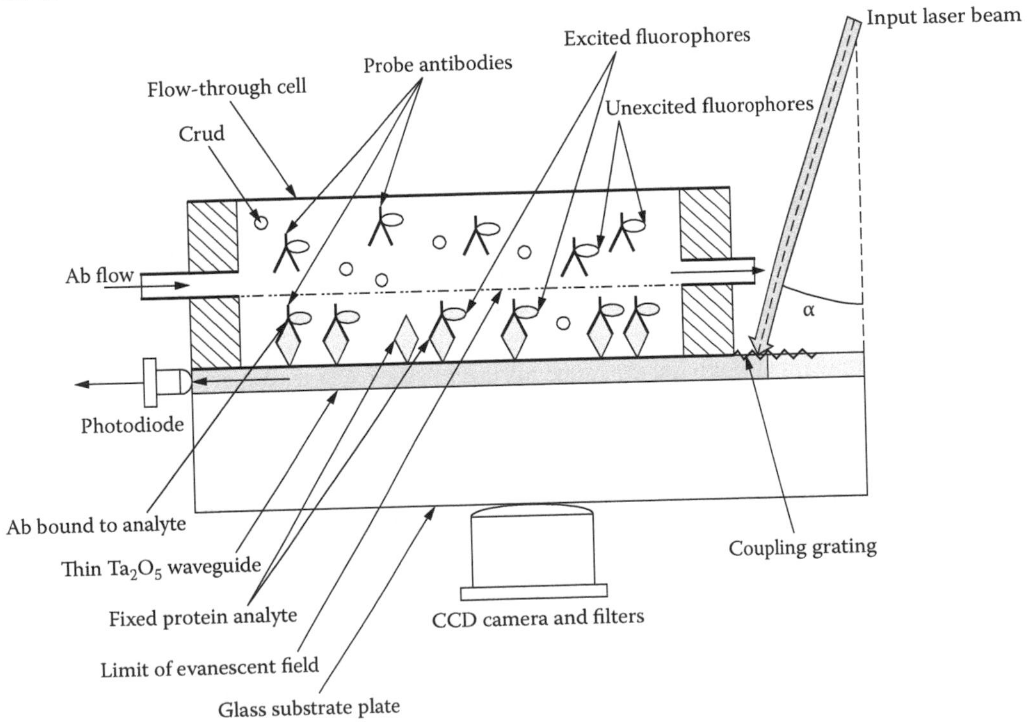

FIGURE 9.6 (See color insert.) Schematic of the Zeptosens® waveguide excitation fluorescence microscope. Evanescent photons from the waveguide excite the fluorophores. See text for further description.

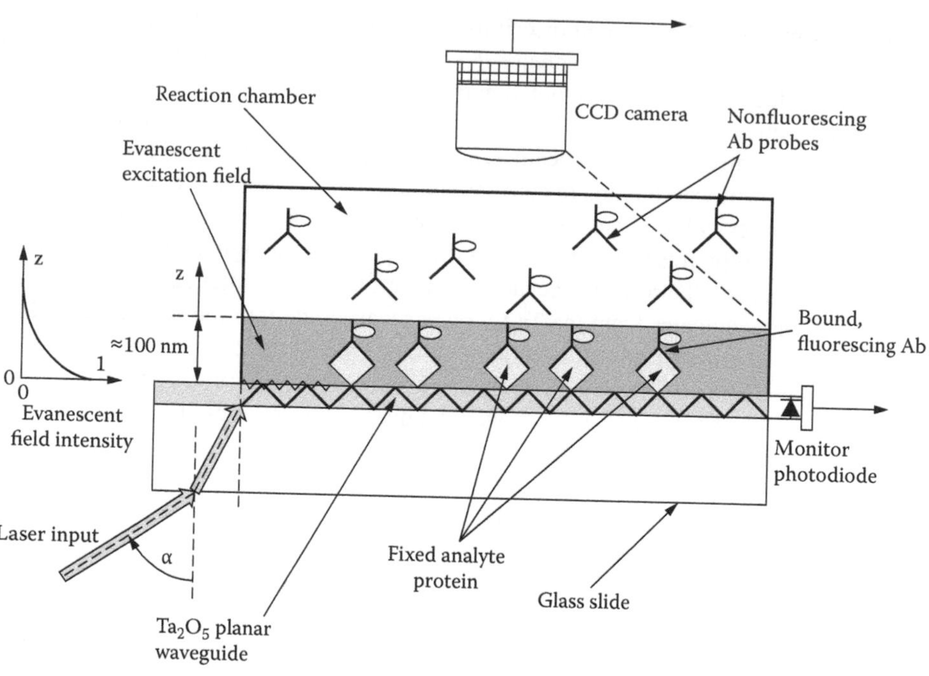

FIGURE 9.7 (See color insert.) Another version of the Zeptosens® WExFM fluorescent microscope. See text for further description.

light energy in the reaction volume that excites fluorescence in fluorophores attached to antibodies that have bound to epitopes on analyte protein molecules fixed to the surface of the waveguide. The fluorescence is detected by a microscope and a CCD camera beneath the reaction cell's glass slide. Note that considerable fluorescence light energy is lost in this scheme because of strong reflection at the water–waveguide interface (due to the difference in refractive indices). The advantage of this system is that only those fluorescent MAbs bound to analyte lie in the evanescent field and are excited; no molecules will fluoresce that lie outside the field.

Figure 9.7 illustrates the side view of another waveguide excitation system geometry. Here, the excitation light sheet is introduced into the planar optical waveguide from below. The CCD camera is directly above the reaction cell to gather the fluorescent light. Again, the input angle, α, is adjusted to maximize the light intensity in the planar waveguide, and hence, the evanescent field. Now, the high refractive index of the waveguide acts as a partially reflective mirror and directs fluorescent light upward through the thin aqueous layer to the CCD sensor.

Whichever system geometry is used, the advantage of the Zeptosens technology lies in its ability to excite only a thin layer of fluorophores, decreasing unwanted fluorescence and scatter from background artifacts in the reaction chamber or cell. Absolute JND sensitivity for streptavidin was found to be about 20 picomoles/liter in the WExFM system [Grandin et al. 2005].

9.7.6 The Immuno-PCR Method

Sano et al. [1992] devised the *immuno-PCR* (IPCR) method of quantifying proteins. This analytical system has been successfully perfected, commercialized, and marketed by *Chimera Biotec GmbH*, Dortmund, Germany, under the trade name *Imperacer*®. In general, the IPCR method of detecting proteins and other biomolecules offers a 10^2- to 10^4-fold increase in the LOD of the analyte over an analogous, conventional ELISA system! This amazing increase is accomplished by the molecular multiplication inherent in the DNA polymerase chain reaction (PCR) method. PCR increases the

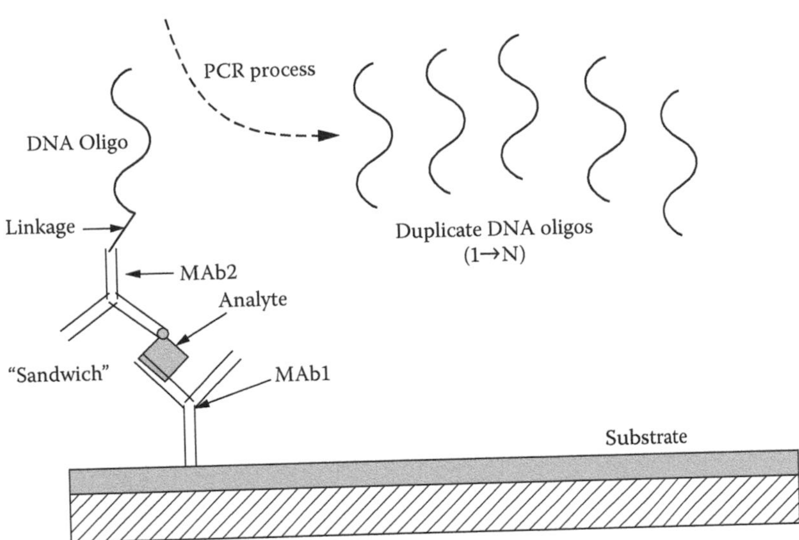

FIGURE 9.8 Schematic of the steps in the Imperacer® technology used to identify an analyte protein. See Section 9.7.6 in the text for a description of this analytical technique.

number of a given ssDNA oligo species in the reaction vessel. (As we have discussed, PCR is widely used in forensic DNA analysis, given an extremely small starter sample.)

Figure 9.8 illustrates the basic molecular steps in the *Imperacer* technology using the sandwich method. First, primary MAbs are bound to the surface of the reaction chamber. These pMAbs have a strong affinity to an epitope on the analyte protein and bind to it. Next, secondary MAbs are prepared that have an affinity to a second epitope on the analyte. These sMAbs, instead of being tagged with an enzyme or a fluorophore as in conventional ELISA, are tagged with short ssDNA oligos. The preparation is washed to remove any excess MAb-DNA molecules. Then a PCR reagent mixture is added containing special *TaqMan® probes* that enable real-time PCR detection of the amplified DNA oligo marker (cf *Glossary*). Thus, one bound sMAb molecule's DNA oligo can lead to millions of DNA copies, which are easy to sense.

In an example described by Adler et al. [2005], the *Chimera Imperacer®* serotonin-specific system was used to detect the concentration of the neurotransmitter serotonin (aka 5HT, not a protein). (Serotonin is a neurotransmitter and a vasoactive amine released by blood platelets.) The LOD was found to be 0.4 pg/mL. The LOD of a standard ELISA analysis of serotonin was 15 ng/mL. The authors claim that further refinement of the Imperacer serotonin assay gave an LOD of 3 fg! This was estimated to be the 5HT from only 10 platelets. The gram-MW of serotonin is 175.214. Thus, 3 fg of serotonin translates into a LOD of 17.1 zeptomole, or 10.3×10^6 serotonin molecules.

Adler et al. [2005] reported that the Imperacer method had comparable LOD for the detection of noradrenaline and dopamine. When Imperacer was used with ELISA, the LOD was 1.5 pg/mL for dopamine and 4 ng/mL for noradrenalin. Adler et al. commented: "... there is an increasing demand for noninvasive methods of biological sample acquisition that can be carried out by nonacademic medical personnel. Therefore, the analysis of saliva or urine is considered a promising alternative to the analysis of blood serum samples." They conclude that the Imperacer technology opens up new noninvasive diagnostic approaches with reliability and, in many cases, increased sensitivity. Imperacer technology is of particular value in the quantification of neurotransmitters and hormones, which are of particular interest in such areas as neuroscience, the field of sports medicine, and the development of ultrasensitive doping tests for athletes.

Clearly, the Imperacer technology is adaptable to microarrays and should figure in the early detection of certain tumor surface markers, thus leading to more effective therapies.

9.7.7 Bio-Bar Codes

The bio-bar code (BBC) system of detecting the presence of, and quantifying, proteins and other biomolecules also makes use of the PCR amplification of ssDNA oligos. The bio-bar code system was devised and developed by Mirkin's group at Northwestern University [Nam, Park & Mirkin 2002; Nam, Thaxton & Mirkin 2003]. These authors used the bio-barcode method to quantify the protein *prostate-specific antigen* (PSA) with an LOD of 30 attomolar (30×10^{-18} M) concentration. Elevated, circulating PSA is used as a signature test for prostate cancer, although its firm correlation with the disease has been questioned (there are false-positives) [Eastham & Scardino 2003; Pesce 2000]. PSA has also been found in the sera of women with breast cancer, albeit at much lower concentrations than in men with prostate cancer. Thus, an ultrasensitive test for PSA could serve as a breast cancer diagnosis and screening tool. The bio-barcode system appears to be such a sensitive test.

Two types of probe nanoparticles are used in the bio-barcode assay: *gold nanoparticles* (GNPs; either 13 nm or 30 nm diameter) and *magnetic nanoparticles* (MNPs; 1 μm diameter) with a polyamine surface coat over a magnetic iron oxide core. The GNPs have polyclonal Abs to PSA fixed to their surfaces along with specific "bar-code" DNA oligos used for PCR amplification and protein identification. In a separate reaction, monoclonal Abs (MAbs) to PSA were fixed on the surfaces of the MNPs. Next, sera with the PSA analyte was added to the functionalized MNPs and incubated. A certain fraction of the MAbs fixed to the MNPs bind to epitopes on the PSA molecules. In the second step of the bio-barcode assay, the GNPs are added to the solution containing the MNPs with their bound PSA molecules. The anti-PSA Abs on the GNPs bind to epitopes of the PSA already bound to the Abs on the MNPs, forming clusters of GNPs surrounding each MNP. The MNP–GNP complexes are now tethered magnetically to the vessel walls and washed. If the larger GNPs (30 nm diameter) are used, the oligos can be detected directly. When the smaller GNPs are used (13 nm diameter), the oligos are released from the GNPs by heating and isolated. The oligos are then subject to PCR to increase their number, and their concentration is quantified. The final DNA oligo concentration is an increasing function of PSA concentration. Naturally, standards are run to calibrate the procedure. Figure 9.9 illustrates schematically the salient steps in bio-bar code assay for PSA.

Nam, Park, and Mirkin [2002] list several advantages of their BBC protein detection system:

1. The analyte-binding portion of the assay is homogeneous. Thus, large quantities of functionalized MMPs can be added to the reaction vessel to facilitate the binding kinetics (by mass action) between the detection Ab and the analyte.
2. The use of GNP bio-bar codes provides a high ratio of PCR-amplifiable, ssDNA oligos to labeling Ab, which can substantially increase assay sensitivity.
3. The BBC protocol eliminates the need for complicated conjugation chemistry for attaching DNA to the labeling Abs. Bar-code DNA oligos are bound to the GNP probes by hybridization at the start of the labeling reaction and released for subsequent quantification by a simple warm wash step. Because the labeling Abs and the oligos are present on the same GNPs, there is no need to add further Abs or DNA–protein conjugates before the quantification of the bar-code DNA. In addition, the bar-code DNA is removed from the detection assay, and direct detection (or detection following PCR) is performed on DNA that is free from analyte (PSA), most of the biological sample, and the GNPs and MMPs. The isolation of the DNA substantially reduces background signal.
4. The authors argue that their BBC method has the potential for massive multiplexing and the simultaneous detection of many analytes in one solution, especially in the non-PCR form of the assay. One could prepare Abs and unique ssDNA oligos for nearly every target of interest.

Nam, Stoeva, and Mirkin [2004] also adapted the BBC method to detect an anthrax lethal factor gene DNA oligo. Instead of a protein analyte, bound complementary DNA oligos had affinity to the anthrax 12-mer oligo analyte.

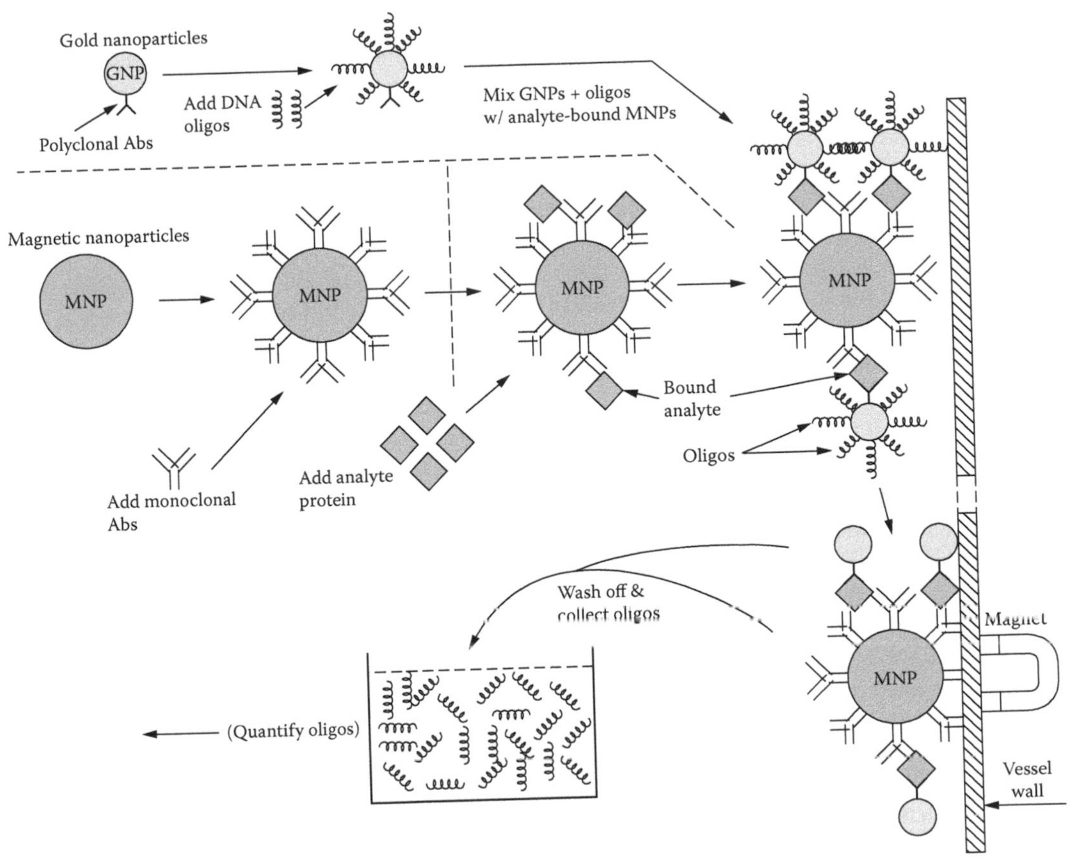

FIGURE 9.9 (See color insert.) Schematic of the bio-barcode (BBC) system for identifying and measuring the amount present of protein analytes. Magnetic nanoparticles are used. See Section 9.7.7 in the text for a description of this analytical technique.

The GNPs were coated with bar-code DNA oligos and DNA oligos complementary to another region on the anthrax oligo. Thus, in the presence of the anthrax DNA analyte, a complex was formed between MMPs plus their complementary oligos, the analyte oligo, and the GNPs with their complementary DNA oligos plus the bar-code DNA. Again, magnetic separation was used and the bar-code oligos harvested and quantified. The entire assay required 3–4 hours. The PCR amplification of the bar-code oligos and their quantification appears to be quite time dependent.

9.7.8 Structural Analysis of Protein "Crystals" Using X-Ray Crystallography

9.7.8.1 Introduction

X-ray crystallography is an analytical chemical technique that uses the diffraction and reflection of x-ray photons to describe the internal molecular structure of solid-state inorganic and organic crystals. The use of x-ray diffraction to discover the atomic structures of inorganic crystals reached technical maturity in the years between 1913 and World War II. In 1915, Sir Lawrence Bragg and his father, Sir William Bragg, shared the Nobel Prize in Physics for their use of x-ray diffraction to deduce the crystal structures of NaCl and KCl. Following the Braggs' pioneering work, x-ray crystallographers improved the collimation and spectral purity of their x-ray sources and developed mathematical analysis techniques based on 3-D Fourier transforms that are applied to the spatial intensity patterns of diffracted x-ray photons that have interacted with the atoms in the crystal lattice.

X-rays are broadband, short-wavelength, high-photon-energy electromagnetic radiation [Northrop 2002, Section 16.2]. X-rays are emitted when a high-energy collimated electron beam collides with a heavy metal anode (e.g., Cu, W, or Mo) in vacuo. Actually, x-rays, Auger electrons, and heat (IR) are produced as the electrons strike the anode of the x-ray tube. The x-rays are emitted in a fan-shaped beam from the anode and have a broad power spectrum even when a constant-energy electron beam is used. An important source of collimated high-energy x-ray photons is from synchrotron radiation sources (described in greater detail later). Synchrotron x-rays are emitted in short pulses, rather than the CW x-ray beam from an x-ray tube.

To be useful in x-ray crystallography, the x-rays must be collimated into a low-divergence beam and spectrally filtered by crystal monochomators. It is generally desirable to use an x-ray source with a wavelength less than 0.1 nm in crystallography applications because these waves/photons interact with crystal atoms having lattice spacings of the same order of magnitude.

Once the science of inorganic x-ray crystallography was established, organic chemists, biochemists, and cell biologists looked to the technique to unravel the structural mysteries of protein and nucleic acid molecules. We know that the tertiary 3-D structures of in vivo proteins are required to give them their biological activities. The knowledge of a protein's structure and its active region allows molecular biologists to understand molecular pleiotropy and pharmacologists to do intelligent drug design based on receptor blocking or stimulation. The RCSB *Protein Data Bank* (PDB) has compiled the structures of those proteins and macromolecules that have been studied using x-ray crystallography, cryoelectron micrography (cryo-EM), and in some cases nuclear magnetic resonance (NMR). (The PDB was established in 1971 at the Brookhaven National Laboratory. It initially contained data on the structures of seven molecules [Berman et al. 2000]. In 1977, the PDB was expanded to a computer-based archival file for macromolecular structures. A history of the earliest solutions for macromolecular crystal structures in available online from Martz [2005].)

As of February 7, 2006, there were some 35,026 molecular structure entries in the RCSB protein database [PDB 2006]. As of June 26, 2007 there were 44,320 proteins listed [PDB 2007]. Thus, in a 504-day interval, 9,294 new proteins were added. This is an average increase of over 18 new proteins per day! There still is a long way to go in describing proteins, considering there may well be over 220,000 proteins in the human proteome.

Protein molecule 3-D structures may be downloaded free from the PDB Web site in several forms, most of which allow 3-D rotation to see hidden 3-D features of molecules using *Chime* software. Most of the structural data have been deduced by x-ray crystallography (including synchrotron x-ray) and knowledge of the primary AA sequence.

Table 9.2 lists some of the milestone events in determining the 3-D structures of biologically important ("high profile") proteins and other macromolecules.

Knowledge of a protein's gene tells us the transcribed AA sequence, but chaperone proteins and other posttranslational events can alter the mature protein's AA sequence as well as its active tertiary form. *One problem* in doing crystallography on proteins is collecting them in purified form. *A second, very challenging problem* is growing protein crystals. *A third problem* is that bombarding a protein crystal with high-energy x-ray photons can create *free radicals* that can alter the shape of the protein being examined by breaking intermolecular bonds. We will now review these problems and some solutions.

9.7.8.2 Protein Crystals

A *crystal* is a 3-D, solid-state, ionic, atomic or molecular structure characterized by periodically repeated, conjoined, identically constituted, congruent *unit cells.* We are all familiar with crystals of sugar, salt, diamonds, quartz, other minerals, etc. Not so apparent is the fact that organic macromolecules such as proteins and nucleic acids can also form crystalline structures under appropriate conditions. (Protein crystals are our interest here.) When a protein crystallizes, it generally contains "water of crystallization," that is, water molecules tucked into the interstices of the crystal array. In

TABLE 9.2
Some Early Molecular Structure Determinations

Date	Workers	Molecule
1935	J.M. Robertson et al.	Pthalocyanins
1949	D.C. Hodgkin et al.	Penicillin
1958	J.C. Kendrew et al.	Myoglobin
1965	Phillip et al.	Lysozyme (in hen egg white)
1967	F.M. Richards and H.W. Wykoff	Ribonuclease
1968	M.F. Perutz et al.	Hemoglobin
1971	Blundell et al.	Insulin
1974	Rich et al., Klug et al.	Transfer RNA

Source: From Martz, E. 2005. *Earliest Solutions for Macromolecular Crystal Structures.* (Available online.) www.umass.edu/microbio/rasmol/1st_xtls.htm.

a protein crystal, the entire protein molecules align themselves in a periodic series of *unit cells* held together by noncovalent atomic interactions. It is the unit cell periodicity that permits the use of x-ray crystallography in deducing the 3-D shape of the molecular components of the crystal.

Growing protein crystals is an art, as well as science. A protocol that works for one protein will not work for another. Proteins are generally crystallized slowly out of aqueous solutions. The factors that are controlled are temperature, protein purity, pH, protein initial concentration, gravity, and the concentration of precipitants, such as ammonium sulfate or polyethylene glycol. In some cases, a metal ion may be required to trigger crystal growth. (For example, insulin requires trace amounts of Zn^{++} for crystallization [McRee 1993].)

Two of the most commonly used methods use water vapor diffusion (from the protein solution to a water-absorbing solution). They are known as the *hanging-drop* and the *sitting-drop methods.* (See Figure 9.10 for a schematic of the vessels used for hanging- and sitting-drop crystal growth.) Both methods use a droplet of concentrated aqueous protein solution also containing a buffer (to regulate pH) and a precipitant. The drop is placed on a slide in a sealed chamber with a larger volume (reservoir) of liquid also containing a buffer and a precipitant but in higher concentration than in the protein drop. The lid seal must be vapor tight; it can be sealed with silicone vacuum grease. Water molecules with enough kinetic energy leave the surface of the protein droplet and enter the vapor space between the solutions and preferentially (statistically) reenter the surface of the liquid in the reservoir, thus transferring water from the droplet to the reservoir, molecule by molecule. At some point when the protein concentration reaches a critical point in the droplet, its molecules begin assembling themselves into a protein crystal. Crystal growth can be sensed optically with a polarizing light microscope. After a period of *slow* growth that can last several days to weeks, the crystal reaches sufficient mass for x-ray crystallography and is carefully removed and taken to the x-ray crystallographic system.

Many NASA space shuttle missions in the 1990s carried protein crystal growth experiments done in microgravity. Horack [1997] summed up the results of these experiments, run between 1984 and 1997:

- Larger crystals in 45.4% of the cases.
- New crystal structures in 18% of the cases.
- At least a 10% increase in the x-ray crystallography brightness in 58% of the cases.
- Less thermal motion in 27.2% of the cases.

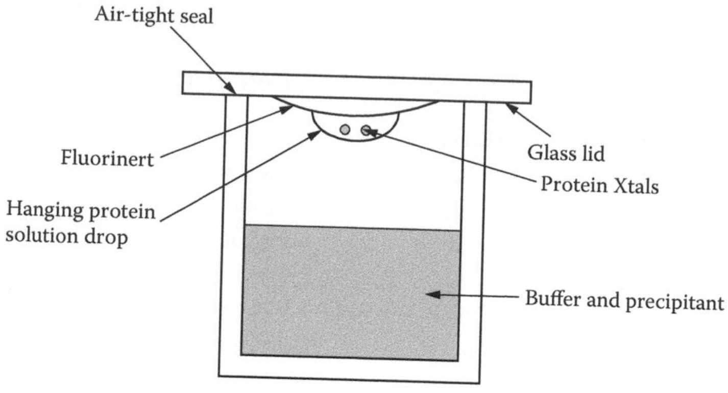

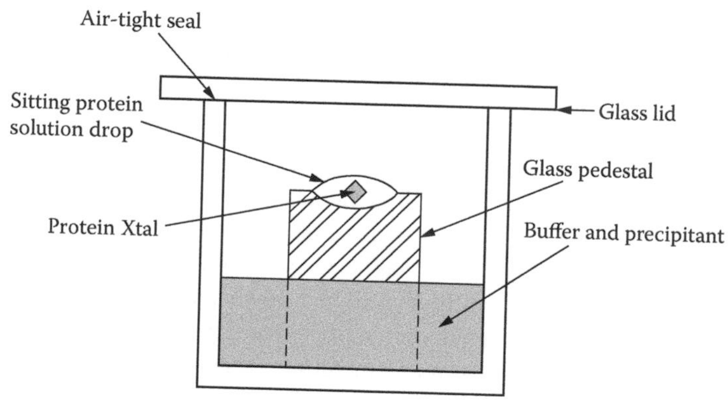

FIGURE 9.10 Schematic side view of vessels used in the hanging- and sitting-drop techniques of protein crystallization. The Flourinert® FC-70 film prevents the protein crystals from sticking to the glass in the double-hanging-drop diffusion technique [Adachi et al. 2004]. N × M arrays of small circular chambers can be used to expand the effort of crystal growth (e.g., N = 4, M = 6).

- An x-ray crystallography resolution improvement of ~0.3 Å in 42.4% of the cases.
- An x-ray crystallography resolution improvement of 0.5 to 1.0 Å in 9.9% of the cases.

To get protein samples into space without the hypergravity of takeoff affecting them, they are flash-frozen before takeoff and kept frozen until they are in orbit in the microgravity environment of the orbiting vehicle or the international space station. The LN_2 boil-off and the rate of thawing are controlled, adding yet another variable to the art of protein crystal growth in space.

Another means of slowly concentrating a protein solution until crystals form and grow is to use a liquid–liquid interface through a dialysis membrane that eliminates the vapor-phase transfer of water molecules. The dialysis membrane is permeable only to water, not protein. The osmotic pressure difference between the protein compartment and the surrounding solution is used to force the water out of the protein compartment, allowing the protein concentration to slowly rise and crystals to form [Chayen & Helliwell 2002].

However they are grown, protein crystals are delicate and must be handled carefully when they are transferred to the apparatus used to determine their 3-D structures.

9.7.8.3 X-Ray Crystallography

X-ray diffraction experiments to determine 3-D structures of proteins are carried out on single crystals. The single crystal is mounted on a *goniometer* in a variety of ways: It can be glued to a

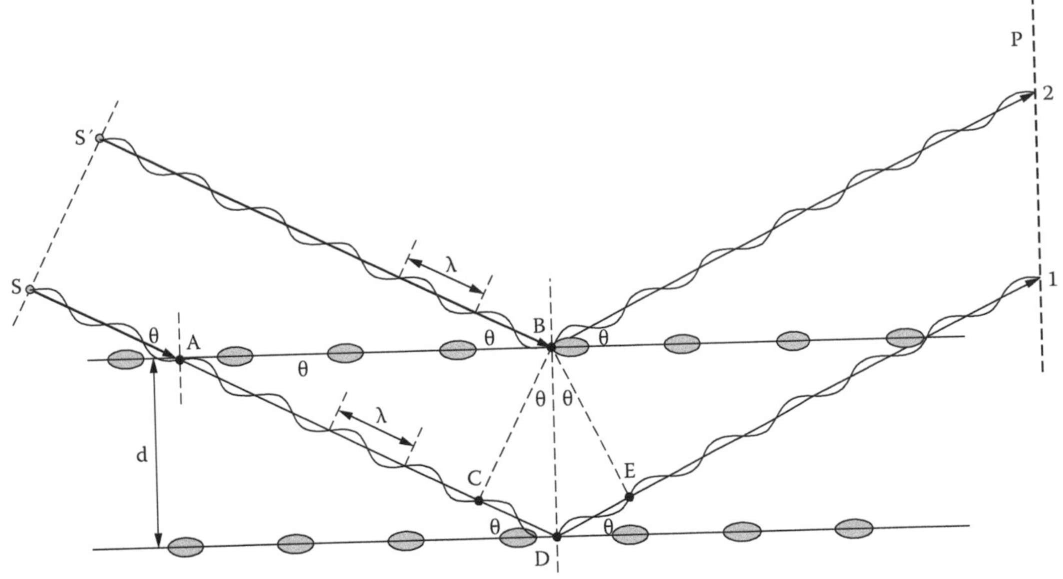

FIGURE 9.11 Side view of the ray geometry illustrating Bragg's law in x-ray crystal diffraction for a two-layered crystal.

glass fiber, put in a capillary tube, or frozen to a glass fiber using a protective oil. The goniometer is a precision mechanical device that allows the crystal to be rotated in the x-ray beam a fraction of a degree at a time.

The x-ray EM radiation has a wavelength on the order of 0.1 nm. When it is directed onto a protein crystal, the component protein molecules (and water) cause x-ray absorption, reflection, and refraction. Because all of the crystal's molecules are oriented in a 3-D periodic array, there is a superposition of the effects of reflection, refraction, and wave interference. These effects lead to a complex 3-D pattern of scattered x-ray photons detected with film or movable CCD sensors. The photon scattering is due to the electron density surrounding each atom in the crystal's component protein molecules. The spatial pattern of scattered x-ray photons contains information on the protein's 3-D molecular structure.

A simple way of illustrating the source of the intensity patterns utilizes a simple two-layer periodic structure and *Bragg's law,* illustrated in Figure 9.11. Bragg interference assumes a wave model for the x-ray radiation. Two layers of ordered reflecting electron densities surrounding atoms, ions, or molecules are shown, separated vertically by distance of d nm. Two collimated rays of x-ray radiation of wavelength λ are shown striking the first two crystal layers at an angle θ to the crystal plane surfaces and are weakly reflected at angle θ (much of the x-ray energy is lost internally in the crystal). By a simple geometric construction, it can be shown [Sears 1949] that a bright reflection occurs when the two parallel reflected waves add constructively under the condition (Bragg's law):

$$\sin(\theta) = \frac{m\lambda}{2d} \tag{9.8}$$

where m is an integer (1, 2, 3...). This requires that the distance CD + DE be exactly m wavelengths in length. Of course, a real crystal has many scattering planes, which adds to the complexity of the measurements. In addition, the molecules, atoms, or ions fixed in the crystal lattice absorb those x-ray photons striking them, further altering the pattern of x-ray photons recorded on plane P.

We see that the scattering is highly dependent on the angle with which the x-ray beam strikes the crystal and the way the crystal's periodic structure is organized. Thus, the crystal must be rotated around the beam axis in roll, pitch, and yaw with the *goniometer* to explore all its possible scattering patterns.

If we examine the 3-D pattern of the intensity of scattered x-ray photons surrounding the crystal after it has been rotated in the beam, it is possible to apply a mathematical analysis similar to 3-D discrete Fourier transforms to arrive at estimates of the electron density (ED) maxima in 3-D space for a protein molecule. The diffraction scatter intensity at a point, $I(h,k,l)$ (seen by an observer or CCD camera), is proportional to the magnitude squared of the vector sum of the wavelets striking the point, i.e., $I(h,k,l) = |\mathbf{F}(h,k,l)|^2$. (The coordinates h,k,l can be rectangular x,y,z or polar ρ,θ,φ.)

The *Patterson function,* $\mathbf{P}(\mathbf{u}, \mathbf{v}, \mathbf{w})$, was devised in 1934 to characterize the 3-D set of crystal reflection intensity amplitudes. It can be written as a 3-D Fourier transform:

$$\mathbf{P}(\mathbf{u},\mathbf{v},\mathbf{w}) = (1/V)\sum_h \sum_k \sum_l I(h,k,l)\exp[(-j2\pi(h\mathbf{u}+k\mathbf{v}+l\mathbf{w})] \tag{9.9}$$

where the tricoordinates **u**, **v**, and **w** are spatial frequencies (e.g., units of radians/micron).

This function does not produce an electron density map, per se, but rather a density map of the vectors between scattering objects in the protein crystal. Because the magnitudes squared of the reflections are used, Patterson maps of crystals with heavy atoms are dominated by the vectors between the heavy atoms. Patterson maps were made more useful in 1954 by Perutz et al. when they introduced the *Patterson difference algorithm*, which enabled the vectors between the heavy atoms in a heavy-atom-labeled protein crystal to be resolved. That is, it allowed subtraction of the light-atom vectors. X-ray crystallographers have developed many software packages to aid in the reconstruction of protein ED maps [Neuberger 2005].

Following the 3-D ED construction, and having a knowledge of the AA sequence of the protein, it is often possible to construct an exact model of the protein's tertiary structure and then compare its theoretical diffraction pattern with the measured pattern. This comparison permits subtle refinements of the model until a close match is achieved.

One problem with classical CW x-ray analysis of crystals is that the absorbed x-ray energy can generate free radicals, which can travel through the water interstices in the protein crystal and disrupt the hydrogen bonding between adjacent protein molecules in the crystal, eventually leading to breakup of the crystal structure. One way of eliminating this problem is to freeze the crystal in LN_2 or propane slush (liquid natural gas, LNG). Freezing does change the protein structure slightly, but this complication is greatly outweighed by the reduction of radiation damage to the crystal.

Still another improvement in the x-ray crystallography of proteins has been the use of pulsed highly collimated x-ray sources derived from *synchroton radiation*. Synchrotrons are charged particle storage devices. Electron synchrotrons have been around for about 50 years. In a modern synchrotron, an electron linear accelerator (LINAC) injects pulses of electrons traveling at 99.99986% the speed of light into a *booster ring*. In the booster ring, the electrons are further accelerated to energies between 1.5 and 2.9 GeV (eV × 10^9). Now the electron bunches from the booster ring are injected into the evacuated *main storage ring*, where they circle endlessly. First-generation synchrotrons used circular storage rings surrounded by magnets that kept the bunches of electrons orbiting in the circular path. In modern third-generation synchrotrons, the storage ring follows an octagonal or 12-sided path with straight segments. The electrons are kept on course inside the storage ring by being deflected at the corners between segments by magnets. Located around the periphery of the storage ring on its corners are quadrupole and sextupole bending magnets, which, when energized, deflect the circulating electrons tangentially into a beamline outside the storage ring. These bending magnets also focus the emerging electron beam, which is used in physics experiments involving electron collisions with matter. However, high-energy electrons are not used in x-ray crystallography. The synchrotron EM radiation emitted at the corners of the storage ring is of interest because it contains highly collimated, plane-polarized, high-energy, pulsed x-ray photons.

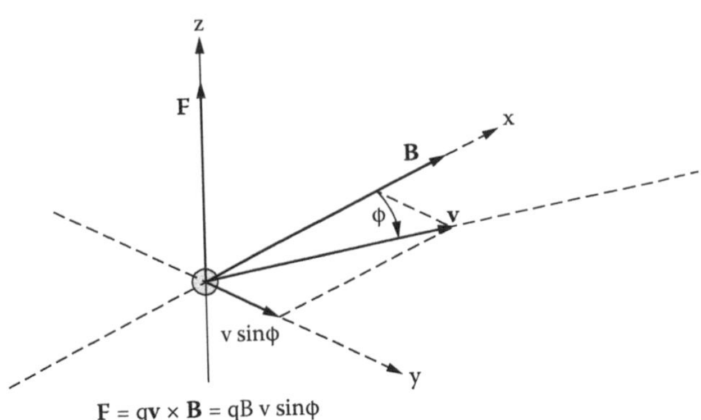

FIGURE 9.12 Vectors illustrating the force exerted on a moving charge in a magnetic field. The vector cross-product is used to find the force that is normal to the plane defined by the velocity and **B** vectors.

When a very fast, high-energy electron beam is acted on by an orthogonal magnetic field, a cross-product force is generated acting to move the electron perpendicular to the plane defined by **B** and **v** according to the vector equation:

$$\mathbf{F} = q(\mathbf{v} \times \mathbf{B}) \tag{9.10}$$

where F is the vector force in newtons, B is the vector flux density in teslas, q is the electron charge (-1.603×10^{-19} C), v is the vector electron velocity in m/s, and X is the vector cross-product operation. (Note that the electron charge is negative.) Figure 9.12 illustrates this vector relationship for orthogonal v and B. The orthogonal force on a moving electron deflects its orbit to keep it circulating inside the plane storage ring. This is a *centripetal force*, which must equal F $= m_e v_T^2/R$, where m_e is the electron's mass (9.1066×10^{-31} kg), v_T is the *tangential velocity* in m/s (same as **v** above), and R is the radius of the storage ring. It is easy to see that the radial acceleration is just $a_R = v_T^2/R$ m/s^2.

When a relativistic (ultra high-velocity) electron is accelerated, it emits synchrotron radiation. In the case of a synchrotron, a narrow beam of very intense broad-band EM radiation is emitted tangentially from the storage ring (in the same direction as $\mathbf{v}_T$ at any point on the storage ring). In third-generation synchrotrons, this radiation is enhanced by the use of *undulator magnets* around the straight sections of the storage ring. As the electron bunches pass through the alternating polarity of the undulator magnet's field, they emit a great deal more synchrotron radiation than if they had just been deflected by a single corner magnet. Figure 9.13 illustrates the geometry of an undulator magnet. It causes the passing electrons to travel in a serpentine path (in the plane of the paper), generating a very large amount of synchrotron radiation. In Figure 9.14, we illustrate the enormous spectral energy of synchrotron radiation versus the x-ray output of a classical x-ray tube. Note that the use of undulator magnets increases the synchrotron beam intensity by a factor of about 10^5. Figure 9.15 shows an octagonal storage ring emitting synchrotron EM radiation in a narrow beam. The lower panel illustrates how the synchrotron radiation is deflected, filtered, and collimated before striking a protein crystal.

The fact that synchrotron-derived x-rays are emitted in short pulses has led to a new protein crystallography discipline; that of *time-resolved x-ray crystallography* (TRXC). TRXC has been used to map functional protein motions with picosecond resolution. Hummer et al. [2004] used picosecond x-ray pulses derived from a beamline at the European Synchrotron and Radiation Facility. The time resolution of the dynamic structural determination was limited by the about 150-ps x-ray pulses (full width at half maximum FWHM). They examined hemoglobin structure after ligand dissociation. A similar system developed in Japan has been described by Adachi et al. [2005].

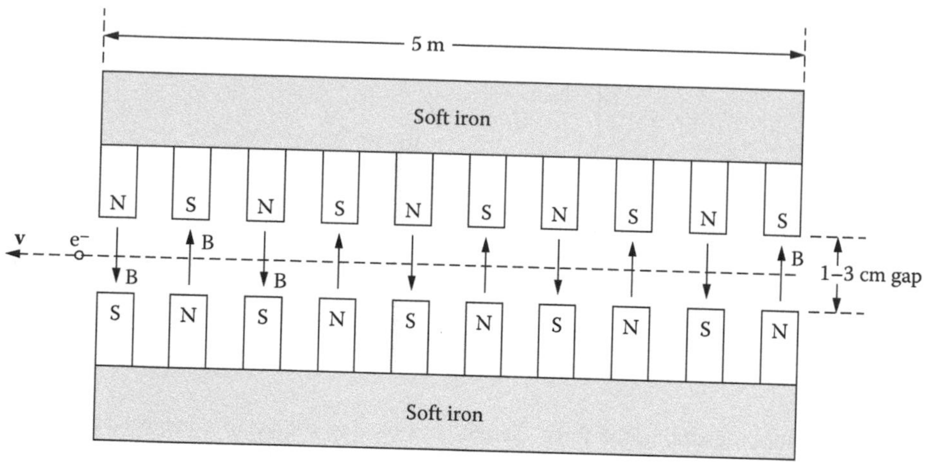

FIGURE 9.13 Geometry of an undulator magnet used to enhance the *synchrotron radiation* output of a synchrotron.

9.7.9 Cryoelectron Microscopy

Cryo-EM is a transmission electron microscopy (TEM) technique that has application in determining the gross 3-D structures of very large noncrystallized proteins. One of the innovators of cryo-EM has been Dr. Joachim Frank of the Wadsworth Institute. Frank and coworkers have focused their attention on the study of the structure and molecular physiology of ribosomes.

Cryo-EM involves freezing the protein sample in ethane slush to make the associated water into vitreous, or noncrystalline ice. (Crystalline ice [cubic ice] absorbs electrons easily, making poor contrast in the image.) LN_2 is then used to keep the sample frozen while in the EM. The advantage of cryo-EM is that it preserves the protein sample in the near-native hydrated state without the chemical distortions caused by the fixing and staining required by traditional TEM methodologies. Cryo-EM imaging can resolve molecular features below 1 nm. However, considerable computer image processing must be applied to improve the image SNR and "connect the dots." Very low electron beam currents are used in cryo-EM in order not to damage the protein object; about 6 to 19 electrons per μm^2 versus 10^5 electrons per μm^2 in EM studies of semiconductors.

Cryo-EM has been used successfully to study the functioning of the actin-myosin muscle "motor" at a resolution of 2.5 to 3.0 nm [Wells et al. 1999]. Cryo-EM was also used in a comparative study of human, *E. coli* 70S and in yeast ribosomes to investigate conformational changes associated with biological function. Pixel resolution was 0.366 nm on the object scale. A software package, SPIDER, was used to resolve the TEM images into 3-D models of individual ribosomal subunits. The final molecular density maps were displayed with IRIS Explorer [Spahn et al. 2004]. In 2000, Ushykarov et al. reported in *Nature Structural Biology 7* that they had used cryo-EM to determine the 3-D structure of *α-latrotoxin,* the neurotoxic protein in black widow spider venom at a resolution of 1.4 nm. This protein is a whopping 520-kilodalton tetramer. α-latrotoxin binds to and forces its way into neuron synapses, stimulating massive discharge of neurotransmitters in the spider's victim. It is not known how it does this.

Kumar et al. [2005] reported a 3-D reconstruction of the N-protein from *Hantavirus* using single-particle cryo-EM. In the single-particle approach, many images of the same frozen macromolecule (in many different) orientations are acquired by digital imaging, and then a sophisticated software routine puts the images together and reconstructs a 3-D model of the molecule (particle). Thus, the single-particle approach is really a particle image fusion method; resolutions obtainable from this 3-D image ensemble averaging are on the order of about 0.8 nm. Two sites working on this algorithm are The Department of Cellular and Molecular Physiology at the Yale School of Medicine and The Scripps Research Institute.

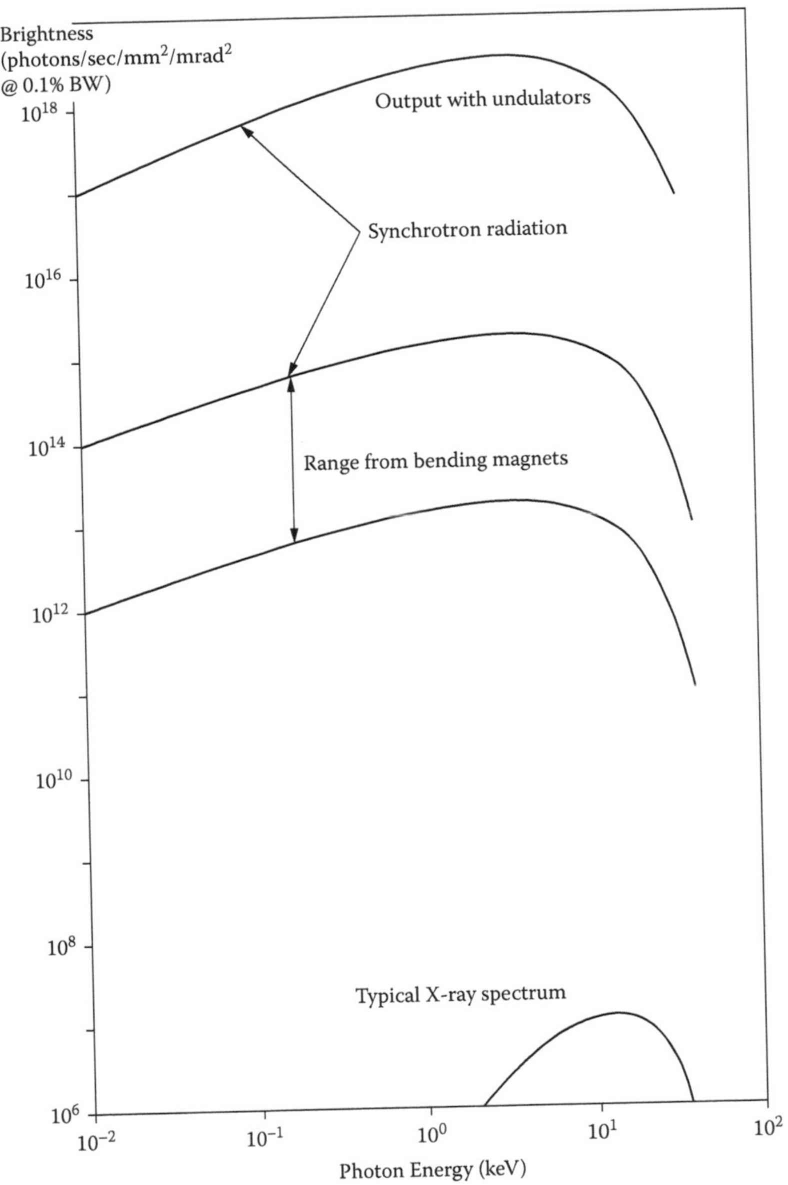

FIGURE 9.14 Comparison of the electromagnetic energy output of a synchrotron with undulators versus a typical x-ray tube. Approximately 10^{11} times as much energy is available.

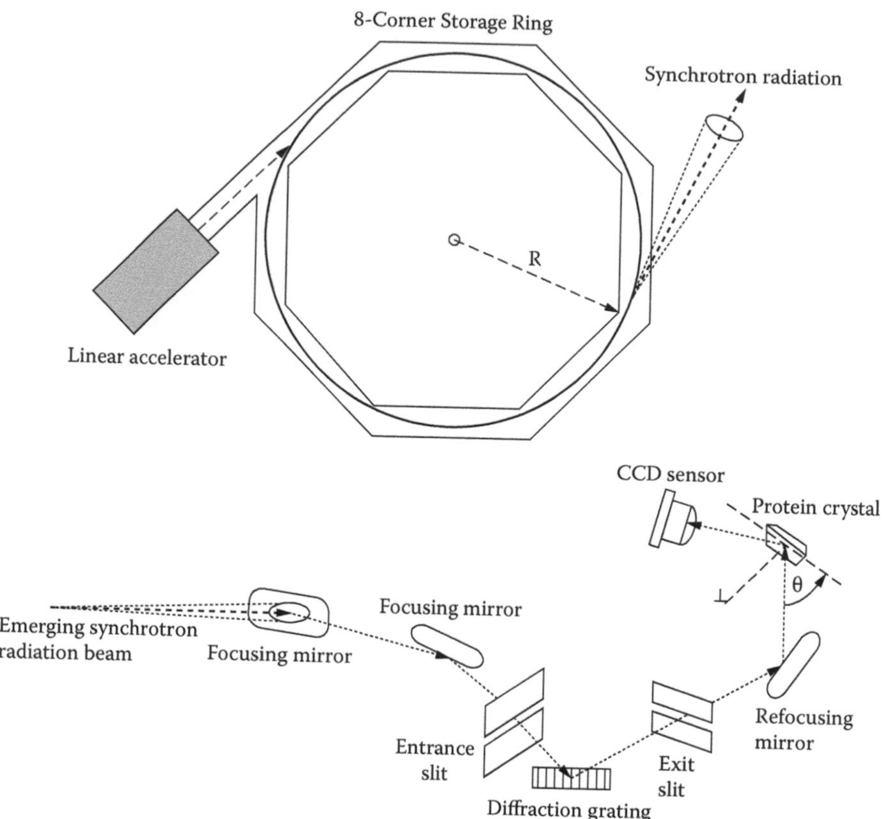

FIGURE 9.15 (See color insert.) Top: Plan view of an octagonal (eight-corner) synchrotron storage ring. The 450 GeV Large Electron-Positron Collider (LEP) synchrotron at CERN has a mean radius R of about 4.3 km. Smaller synchrotrons have lower-output beam energies. For example, a proton synchrotron at Kyoto University outputs 70–250 MeV and has a mean radius of only 3.8 m. Bottom: How the emergent high-energy beam is bent, focused, made monochromatic, and directed to a protein crystal.

9.8 SUMMARY

In this chapter we have described some of the recent technological advances that have enabled modern genomic, proteomic, and forensic DNA analysis to proceed quickly and accurately. New chemical and physical analytical systems for genomics, proteomics, and forensic science are continuously under development. Detection sensitivities for some systems extend into the attomolar range, made possible by the use of photonic systems.

The power of photonics in genomic instrumentation was underscored in our description of the new "454" system of DNA sequencing.

As we have seen in Chapter 8, applications of these analytical technologies have made possible the elaboration of the chromosomal location of the genes responsible for certain genetic diseases, as well as the mutations responsible for them.

We have also argued that problems in proteomics are an order of magnitude more challenging than those in genomics. There are roughly 9 to 10 times as many proteins as there are genes, thanks to posttranscriptional modification schemes. The function of proteins depends on their 3-D structures, which are difficult to describe instrumentally.

CHAPTER 9 HOME PROBLEMS

9.1 How fast can the latest automated analyzers determine the 13 human DNA STR markers from epithelial cells in a criminalistics investigation using some form of PCR? Give the manufacturers of the machines and the techniques used.

9.2 Describe the atomic events required for fluorescence to occur in a molecule. Illustrate your answer with a *Jablonski diagram.* What is the lifetime range of fluorescent molecular states?

9.3 Define *fluorescence quantum yield.*

9.4 Consider the block diagram of the basic phase fluorometer show in Figure 9.3 in the text.

A. Give expressions for the outputs of the two RF mixers. (Assume the mixers are ideal analog multipliers and $\omega_m >> \omega_i$.)

B. The two mixer outputs are low-pass filtered. Assume the low-pass filter outputs, $v_1(t)$ and $v_2(t)$, are in turn multiplied together, and this multiplier output, v_3, is low-pass filtered (not shown in Figure 9.3). Show that the output of this low-pass filter is of the form $v_3 = K\cos(\phi)$. Thus, $\phi = \cos^{-1}(v_3/K)$.

C. Another way to find ϕ is to calculate the cross-correlation function, $\varphi_{12}(\delta)$, between the two output signals, $v_1(t)$ and $v_2(t)$. $\varphi_{12}(\delta)$ is given by

$$\phi_{12}(\tau) = \overline{A\cos[\omega_i(t+\delta)]A\cos[\omega_i t+\phi]}$$

where the overbar denotes the process of time averaging (equivalent to low-pass filtering), and the nonphase-lagged signal is delayed by a variable time, δ. Show that $\varphi_{12}(\delta)$ is a maximum when $\delta = \phi/\omega_i$ seconds. Give that maximum.

9.5 Illustrate how an analog sinusoidal signal of frequency ω_i r/s can be delayed δ seconds by passing it through an analog, active *all-pass filter.*

9.6 Show how the fluorescence lifetime, τ, can be determined by evaluating the phase shift, ϕ, between sinusoidal excitation and fluorescent response at a number of frequencies, ω_m. [Refer to text Equation 9.5.]

9.7 The fluorescence emission of a FRET system can be modeled by the second-order linear ODE:

$$\ddot{I}_F\tau_d\tau_a + \dot{I}_F(\tau_d+\tau_a) + I_F = \eta I_E$$

Describe mathematically how multifrequency phase fluorometry can be used to estimate the donor and acceptor time constants, τ_d and τ_a, respectively. [For example, see Northrop 2003, Appendix C.]

9.8 Describe how *magnetic nanoparticles* and fluorescent quantum dots can be used to isolate and quantify a particular protein in vitro. That is, what must the particles and dots be coated with, and how should they be manipulated?

9.9 The avidin-biotin reaction can be expressed as **Av** + **Bi** ↔ **Av*Bi**. Assume this reaction takes place in vitro starting with equal moles of **Av** and **Bi** and reaches equilibrium. Give an algebraic expression for K_a, the *thermodynamic equilibrium constant* for the reaction.

10 Applications of Genomics

Coagitation

This line is straight
From here to there,
Yet not the shortest route.
The surface is unlike the air.
Solutions don't compute.
Reality is still unseen,
And also unexplainable.
The sacred cow
Is grazing now
In pastures unattainable,
For when she shone
With violet tone
Perceptive men would flee her,
Because in their reality
They'd rather be than see her.

—Adelaide P. Northrop

10.1 INTRODUCTION

In this chapter, we will consider the benefits and costs of making genetically modified organisms (GMOs), both plants and animals. We will also review the technology whereby GMOs are made.

We will also examine the costs and benefits of animal reproductive cloning (e.g., Dolly the sheep) and the creation of animal chimeras. Therapeutic cloning and somatic cell nuclear transfer is seen to be a controversial pathway to the harvesting of stem cells.

We present the controversial topics of the sources, kinds, and potential uses of stem cells in regenerative medicine. The biology of parthenogenesis is described and considered as a potential source for stem cells.

10.2 GENETICALLY MODIFIED ORGANISMS

10.2.1 Introduction

The need for GMOs has been driven by three factors: *economics* (crops and food), *the desire to produce human hormones, proteins, vaccines, and cytokines for therapeutic use,* and *research in genetic systems.* In the not-too-distant future, we also expect to see efforts in GMO production extend to the production of plants, yeast, and bacteria that produce more oil, carbohydrates, and fats for fuel use (e.g., biodiesel and ethanol) and to develop genetically modified (GM) bacteria that feed on waste and efficiently produce hydrogen or methane (again for fuel).

A *genetically modified organism* (plant, animal, or bacterium) is an organism whose genetic material has been altered using *recombinant DNA technology* in order to confer on the organism special metabolic properties not inherent in its intrinsic genome. The special properties might include the ability to produce certain human protein endocrine hormones for therapeutic medicinal purposes (e.g., *insulin, growth hormone* [GH], *parathyroid hormone* [PTH], *leptin,* etc.), also certain cytokines (*interferons, interleukins, granulocyte colony-stimulating factor* [G-CSF], *granulocyte-macrophage colony-stimulating factor* [GM-CSF], etc.), or certain enzymes (e.g., *adenosine deaminase*).

Examples of GMOs are diverse and include many plants (e.g., transgenic rice, maize, soybeans, cotton, canola, rape, etc.) as well as bacteria (e.g., GM *E.coli* that make human insulin, and those used as biological pesticides) or any of several veterinary vaccines [National Agricultural Laboratory 2000]. Although genetically modified animals are not yet used in the food supply, many exist for research purposes (e.g., mice, sheep, and goats).

GMOs are highly controversial (see Chapter 11 on ethics). Some persons object to genetic modification on moral grounds; others fear ecological "side effects" when GM plants proliferate and hybridize with similar unmodified plants through the spread of their pollen. Still other critics are concerned about more direct effects of introducing foreign proteins into the food chain (e.g., allergy and toxicity) from plants modified to be insect and herbicide resistant. Many governments have banned or strictly regulated the planting of GM plants or their importation and have required advisory labeling on GM foods. To counter some of these concerns, Monsanto Company (St. Louis, Missouri) submitted a patent application for "terminator" GM plants that produced sterile seeds in the second generation. This technology was developed by *Delta and Pine Land Company* (Scott, Missouri) and the USDA, which co-owned the patent [Engdahl 2006]. Even the terminator gene technology was viewed as controversial by those who felt it would prevent the use of seeds by Third World farmers growing these plant varieties, placing the control of those seeds into the hands of a few large corporations.

GM animals are not as common as GM plants. One example we will discuss is a GM salmon farm fish variety that was genetically engineered to grow and mature faster.

10.2.2 Examples of Transgenic Plants

10.2.2.1 The FlavrSavr™ Tomato

A relatively benign example of a GM food plant was the *FlavrSavr™ Tomato* developed by *Calgene, Inc.* (Davis, California) in 1992 (FlavrSavr is no longer a registered trademark). It had a gene spliced into its genome whose product inhibits the breakdown of its cell walls by inhibiting the natural expression of the enzyme *polygalacturonase* (PG). (This enzyme is responsible for the softening of cell walls when fruit ripens.) Thus, the FlavrSavr tomato could be picked in a riper stage and shipped to market, retaining its firm, ripe flesh, as well as vine-ripened flavor. Sounds good, but there were problems; Calgene took it off the market in 1996. These GM tomatoes were only sold in a small number of stores in California and in the Midwest.

The FlavrSavr (FS) tomato was made by introducing a recombinant gene that made antisense mRNA to the normal PG mRNA. The antisense mRNA bound to the normal mRNA for PG and effectively shut down expression of PG, preventing the tomato from softening. The antisense PG gene was introduced into the FS tomato by the following complex process: First, Calgene scientists isolated normal mRNA for PG. Once isolated, the mRNA was treated with reverse transcriptase to make *coding DNA* (cDNA) for PG. The PG cDNA was amplified, then placed between the Cauliflower Mosaic Virus 35S promoter and a termination sequence, then introduced into a plasmid in the antisense direction. The recombinant plasmid was then transferred into *E. coli* in order to multiply the number of antisense plasmids. Finally, the antisense plasmid was transferred into *Agrobacterium tumefaciens*, which was the vector used to introduce the antisense PG gene into the tomato. Once the antisense PG mRNA had bound to the normal PG mRNA, the normal mRNA could not get to the ribosome to make PG. The bound sense and antisense mRNA was quickly broken down by ribonuclease enzymes in the cell, and the Nts recycled.

In addition to the antisense PG gene, Calgene introduced a "marker" gene to identify the FS genome. This marker was a gene for resistance to the aminoglycoside antibiotic, *kanamycin*. Thus, only plants that have taken up the kanamycin resistance gene survive when grown in the presence of kanamycin, permitting the greenhouse selection of plants that also have the antisense PG gene. It was the kanamycin gene that focused the controversy.

Health concerns from the public and scientists accompanied the introduction of the FlavrSavr tomato. People were concerned that if the tomato's gene for kanamycin resistance got into other bacteria, they would develop resistance to kanamycin, diminishing its medical effectiveness. In 1993, Calgene asked the U.S. FDA to provide a ruling that GM foods are "safe," particularly with regard to the tomato's kanamycin resistance gene. The FDA gave this approval in mid-1994. However, it appeared that the results of controlled rat-feeding trials showed interesting negative effects. The FDA evidently neglected or discounted this evidence in reaching their decision. In a 28-day feeding study, four groups of 40 rats were fed one of two lines of a GM tomato, a non-GM tomato, or deionized water. Out of 20 female rats fed one of the lines of GM tomato, gastric lesions were identified in four and seven of the GM rats respectively by two expert panels. Because of these unexpected findings, the FDA requested that another rat-feeding study be done. In this study, gross and microscopic lesions were found in 2 of 15 GM rats, and no lesions were seen in the control rats. These findings were significant because such gastric "erosions" are associated with hemorrhage in humans and can be the cause of death in elderly people. In addition, the rats fed GM tomatoes grew the least, and 7 out of 40 rats eating GM tomatoes died within 2 weeks. These worrisome facts were played down by Calgene and not publicly communicated by the FDA [Soil Association 2005]. The FDA revealed these internal views in 1998 only after an FOI (Freedom of Information) Act lawsuit was filed by consumer groups, scientists, and others.

The final disposition of Calgene was based on the fact that it was a small genetic engineering company that accumulated large debts from its R&D efforts, production costs, sales in a depressed tomato market, and defense costs associated with lawsuits filed by Monsanto, which accused Calgene of "patent infringement." In the end, Monsanto bought out Calgene.

The USDA has not given up on GM tomatoes, however. The USDA Agricultural Research Service (ARS) laboratories at Beltsville, Maryland, are currently working on new strains of GM tomatoes [Weaver-Missick 2000]. At ARS, plant scientist Autar K. Mattoo has developed a transgenic tomato that ripens more slowly and also has more of the antioxidant carotenoid *lycopene*. Another GM tomato developed by Mattoo has a longer shelf life than regular tomatoes; these also bloom and bear fruit three or four times during the season, compared to two harvests for regular tomatoes. Mattoo made use of the genetic "switches" that control the timing of gene expression and the ability to direct certain recombinant proteins to the desired location in plant cells. Hopefully, the new Beltsville GM tomatoes will pass their rat tests and be more successful than the FlavrSavr in the market.

In a recent article by Coughlan [2006], a clever, unusual application of GM tomatoes was described. Workers at the Siberian Institute of Plant Physiology at Irkutsk, Russia, are using the soil bacterium *Agrobacterium tumefaciens* to shuttle a synthetic combination of HIV and hepatitis B virus (HBV) DNA fragments into tomato plants. These include fragments of genes for various HIV proteins and the gene for the HBV surface antigen protein. As they grow, the GM tomato plants then manufacture the protein antigens inside the tomatoes. When eaten, the anomalous protein Ags in the tomatoes prompt the eater to make Abs against the viruses. This system has been tested on mice, which developed Abs to HIV and HB virions. Although a clever means of making an "oral vaccine," the dose must be regulated. That is, the antigens need to be extracted from the GM tomatoes and packaged in some standard elixir or pill.

10.2.2.2 Golden Rice

An example of a beneficial GMO with apparent small environmental impact is golden rice (*Oryza sativa*) that has been genetically modified to produce beta carotene, a source of vitamin A, and iron. Rice is a staple diet in many Third World countries that have chronic problems with malnutrition leading to childhood blindness and maternal anemia. Thus, the consumption of GM golden rice should mitigate these health problems. Golden rice was developed by Ingo Potrykus of the Institute of Plant Sciences at the Swiss Federal Institute of Technology, working with Peter Beyer at the University of Freiburg. It was released in 2000. In 2005, a new golden rice was genetically engineered by the biotech company Syngenta (Basel, Switzerland). Golden Rice 2 (GR2) has 23 times more carotenoids than GR1 (≤37 μg/g) and preferentially accumulates beta-carotene (about 84% of the total carotenoids). Note that in spite of its potential to improve health in Third World rice-consuming nations, Greenpeace protested the cropping of GR1 and GR2.

GR2 was bred with local rice cultivars in the Philippines, Taiwan, and with the American rice variety, *Cocodrie*. In the latter case, the first field trials were conducted by the Louisiana State University AgCenter in 2004. Preliminary results from the LSU field tests showed that field-grown GR2 produced three to four times the beta-carotene than the GR grown under greenhouse conditions.

Work is currently under way by an international consortium to develop new improvements in golden rice (GR3?). Goals include increasing the bioavailability of provitamin A, vitamin E, iron, and zinc and improving protein quality through genetic modification [ProVitaMinRice Consortium 2006]. As long as the developers of GR3 resist the temptation to incorporate *Roundup*® (Monsanto Technology LLC, St. Louis, Missouri) resistance genes, or a recombinant insecticide gene (e.g., for Bt), it promises to be a boon to nutrition in the Third World.

10.2.2.3 Transgenic Maize

A major pest of maize (corn) is the European corn borer, *Ostrinia nubilatus,* which burrows into the corn stem, causing it to buckle and fall over. GM corn has already been genetically modified to be herbicide resistant similar to GM oilseed-rape described later. However, in this section, we will consider corn that has been genetically modified to be toxic to the corn borer. This was accomplished by inserting a gene that makes a toxin protein from a soil-dwelling microorganism called *Bacillus thurigensis,* or Bt. Borers that attempt to bore are overcome by the Bt protein, and the cornstalks are protected. Eventually, as with any chronic application of pesticide, the pests develop immunity. Thus, Bt corn is a short-term fix for high crop yield. (The Bt is inside the corn cells, including the corn kernels.)

U.S. farmers are required by law to plant regular corn in fields adjacent to Bt corn. These non-modified cornfields are intended to harbor borers. The theory behind these refuges is to slow the adaptation of the borers to Bt. It has been found that wind-borne pollen from the GM corn fertilizes regular corn in the adjacent fields, and its cobs then contain traces of Bt. This casual introduction of Bt in the normal corn decreases with distance from the GM cornfield.

Aventis patented a variety of Bt corn called *StarLink*® (Aventis CropScience GmbH, Frankfurt, Germany) in 2002. This seed was permitted to be sown in the United States with the government stipulation that the corn was not to be used for human consumption, only for animal feed. The possibility existed that some persons would be allergic to the Cry9C (Bt) protein in the kernels. According to Brown [2003], the GM Cry9C (Bt) protein turned out to be highly resistant to mammalian stomach acid and digestive enzymes; it was found to be still intact 30 minutes after swallowing, whereas other dietary proteins were completely broken down by 7 minutes.

StarLink corn did in fact end up in the human food supply—specifically, Taco Bell taco shells. This produced a scandal that prompted investigations by the U.S. FDA and EPA, as well as protests by various consumer groups. Ultimately, Taco Bell recalled all taco shells, and other companies stopped production of corn-based foods to check for the presence of StarLink Cry9C protein. In 2001, StarLink Cry9C protein was also found to contaminate Kellog's Morningstar Farms meat-free Corn Dogs, forcing their recall from the market [Brown 2003]. The unforeseen inability to promptly digest the Cry9C (Bt) protein is a good example of the *Law of Unintended Consequences.*

10.2.2.4 Genetically Modified Soybeans

GM soybeans have also had a controversial history. Monsanto developed what is known commonly as the *Roundup Ready® (Monsanto Technology LLC, St. Louis, Missouri) Soybean* in 1996. These beans are also known as the *glyphosphate-tolerant soybean line 40-3-2;* they are claimed to be safe for human and animal consumption [Monsanto 1998]. The Monsanto GM soybeans are tolerant to Monsanto's Roundup herbicide because a single modified gene has been inserted into their genome. The wild-type native EPSPS enzyme *(5-enolpyruvyl shikimate-3-phosphate synthetase)* has been replaced with a gene that makes CP4 EPSPS protein. The wild-type EPSPS is sensitive to Roundup, the CP4 EPSPS is tolerant to the glyphosphate weed killer. The CP4 EPSPS gene was derived from *Agrobacterium sp. strain CP4,* a common soil microorganism. Monsanto claims it "… is similar to other EPSPS proteins found in plants, bacteria, fungi, and microorganisms such as baker's yeast, establishing a history of safe consumption."

Because of the way that the CP4 EPSPS gene was incorporated into the wild soybean genome, several other genes are also present: *Cauliflower mosaic virus 35S promoter,* the *EPSPS chloroplast transit protein* (CTP) *sequence* from petunias, and the 3′ untranslated region of the *nopaline synthetase* gene (NOS3′). The question is, are the proteins made by these genes all degraded by cooking and digestion? [Carman 2006]. Carman points out serious flaws in the tests used to determine the safety of RR-soybeans and recommends a series of critical studies be done on the biochemistry and long-term nutritional effects of RR-soybeans. The soybean genome is about 1.5 M bases in length.

The economics of growing RR-soybeans for export also appears to be bleak. The present EU consumer boycott of GM products caused U.S. soybean sales to Europe to fall by at least $150 M in 1996. (In 1997, the United States sold about 40% of its soybeans to Europe.) Also, there is evidence from Australia that weeds, under persistent exposure to Roundup, are developing resistance to that herbicide, similar to the charlock weed in the United Kingdom (see Section 10.2.5). It does not take long for selective pressure to "evolve" resistant species of weeds and insects from the continued use of insecticides and weed killers.

We note that RR-soybeans and cotton *only* have immunity to the Monsanto weed killer *Roundup.* Hence, a farmer who buys the Monsanto RR seeds also *must* buy Monsanto Roundup to treat the fields they are grown in. In 1997, Roundup accounted for 17% of Monsanto's total annual sales [Arax & Brokaw 1997]. Another problem with Roundup is incomplete chemical degradation in the field in which it was used. Trace amounts may inhibit the growth of a normal second-year crop following the planting of RR-soybeans.

10.2.2.5 Genetically Modified Cotton

A persistent insect pest for cotton plants in the United States and in China is the cotton bollworm. Monsanto developed a GM cotton plant by inserting the gene for the *Bacillus thuringiensis* (Bt) toxin protein in its genome. Worms that try to eat Bt-cotton plants die, at least until they evolve resistance to Bt. Already there is DNA evidence that the pink bollworm *(Pectinophora gossypiella)* has developed biochemical resistance to Bt toxin *Cry1Ac.* Morin et al. [2003] showed that the *r1, r2,* and *r3* alleles of the pink bollworm's cadherin gene **BtR** confer recessively inherited resistance to Bt toxin. It appears that nature is only one step behind Monsanto, Dow Agrosciences (Indianapolis, Indiana), and Syngenta, all producers of Bt-cotton seeds.

A major benefit of growing Bt-cotton in the United States and China is there is far less need for conventional pesticide use, and hence there is far less toxic hazard for agriworkers and higher animals near the crop fields [Hossain et al. 2004]. Growth of GM Bt-cotton appears more effective in subtropical regions, where the bollworm is more endemic.

Baier [2005] argued against the use of Bt-cotton in Africa: "...the main [cotton] pests in Africa are not affected, or at least only marginally affected by Bt-cotton. Even if the worst pests could be effectively controlled, it has been observed in practice that so-called secondary pests have the attacked crops." He points out that their are other problems: In South Africa, the American variety of Bt cotton is grown that does not have hairs on its leaves like normal cotton plants. This makes the plants more attractive to the endemic locusts. Baier concluded that there is no *economic advantage* to growing Bt-cotton in Africa.

In 2005, India banned Monsanto and its Indian partners from selling three varieties of Bt-cotton in its southern state of Andhra Pradesh because it was ineffective in controlling cotton pests there. Bt-cotton seeds can be sold in other Indian states, however. (Monsanto sold 1.3 M packets of Bt-cotton seeds in 2004.)

Growing Bt-cotton is a "quick fix" that will eventually lead to insects with Bt resistance; it reduces the number of cotton species grown and reduces biodiversity. In some locations, where the major cotton pests are Bt susceptible, it does make economic, though not ecological, sense to grow Bt-cotton over the short term.

In a report from China's Xinhua News Agency [2002], Chinese scientists were not happy about Bt-cotton. The report stated:

> The plant, Bt transgenic cotton, harms natural parasitic enemies of the bollworm and seems to be encouraging other pests, according to the study by the Nanjing Institute of Environmental Sciences (NIES) under the State Environmental Protection Administration (SEPA).
>
> The report says that the diversity index of the insect community in the Bt cotton fields is lower than conventional cotton fields, while the pest dominant concentration index is higher.
>
> Scientists also verified with lab tests and field monitoring that the cotton bollworm will develop resistance to the GM cotton and concluded that Bt cotton will not resist the bollworm after eight to ten years of continuous cultivation.

Still, China continues to be second in world production (in 2004) of Bt cotton [Baier 2005], casting doubt on the sincerity of its criticism of the GM crop.

10.2.2.6 Transgenic Arabidopsis for Detection of Explosives in the Soil

In Denmark, the company *Aresa Biodetection ApS* has developed a genetically engineered plant, *Arabidopsis thaliana* (aka thale cress), for detection of explosives in the soil (e.g., buried landmines and unexploded ordnance [UXO]). Landmines and UXOs emit nitrous oxide (NO_2) at low levels as their explosives age. Danish scientists have genetically modified thale cress so that the leaves of the plant turn red *only* when the roots contact NO_2 vapors in the surrounding soil. Stress such as cold or drought will not turn the leaf color of the GM cress red, as would happen in a normal plant.

Aresa introduced the following modifications in the thale cress genome:

1. A **tt4** mutation that lacks the first enzyme required for red pigment formation (*anthocyanin*).
2. A **ga1-3** mutation that blocks the biosynthetic pathway for the formation of the growth hormone *gibberellinic acid.*
3. Inserted a plasmid containing the TFs MYB75 and MYB90. These TFs are responsible for high production of red pigments.
4. Inserted a plasmid containing the **CHS** gene that is expressed due to a promoter regulated by the presence of NO_2.

5. The **Bar** gene that confers resistance to the herbicide, *Basta,* and the firefly luciferase (**Luc**) gene were inserted as marker genes of the GM plants.

Following their controlled laboratory tests in Denmark, Aresa plans to field-test their Red Detect® GM thale cress with actual landmines in a known minefield by sowing seeds from an aircraft or by handheld seed sprayer by workers walking along known demined corridors. After allowing time for germination and maturation, the fields containing the GM cress will be oveflown and any patches of red leaves viewed and marked. Surface deminers can then target specific locations in the minefield to work on, minimizing their own risk and improving the removal rate.

To minimize the effect of the Law of Unintended Consequences, the GM plants have been modified so they will not germinate unless a growth factor has been applied to the seeds. In addition, seeds cannot be released unless another growth hormone is supplied [Report 2006, ACFNEWSOURCE 2006].

We do not know what the threshold sensitivity for the leaf color change is in terms of ppm NO_2 at the plant roots.

10.2.3 Transgenic Animals and Animal Cells

10.2.3.1 Introduction

In the preceding section, we examined GM plants and why their long-term use is controversial and can damage ecosystems. In this section, genetically modified or *transgenic animals* (TGAs) are considered. A TGA is one that carries a foreign gene that has been inserted into its genome. The foreign gene is generally assembled using *recombinant DNA* (rDNA) technology. In addition to the inserted gene that codes a protein, the rDNA usually includes other base sequences that enable it to be incorporated into the host DNA and to be expressed correctly by the cells of the host animal at the correct time and location.

10.2.3.2 Uses of TGAs and Cells

Most of the products of TGAs are directed toward the production of proteins, vaccines, enzymes, and hormones for use in therapeutic *genomic medicine.* In other cases, genes have been inserted to stimulate growth (e.g., farmed salmon) and for experimental purposes (e.g., insertion of the green fluorescent protein gene in entire zebra fish, *C. elegans,* rabbits, pigs, monkeys, and mouse motor neurons) [Zimmer 2005].

Some specific examples of TGA production are as follows:

1. TG rabbits have been bioengineered to make the human immune system stimulant *interleukin-2.*
2. The genes for sheep and pig regulatory sequences have been attached to implanted genes for human clotting factor proteins VIII and IX (to treat hemophilia), which are expressed in the milk of lactating females (milking a sow is an interesting concept).
3. Other human therapeutic chemicals in milk of TGAs include *anti-thrombin III* to prevent clotting; *collagen* to treat burns and bone fractures; *fibrinogen* to treat burns and after surgery; *erythropoietin* to treat anemia; *Alpha-1-antityrpsin* to treat emphysema and cystic fibrosis; and *lactoferrin, tissue plasminogen activator, and certain monoclonal antibodies* (to treat a particular colon cancer) and certain *human fertility hormones.*

The isolation of these genetically engineered pharmaceuticals from milk appears to be surprisingly efficient, starting with the assumption that the TGAs produce 1 g of the protein per liter of expressed milk. Also, the extraction efficiency is assumed to be about 30%. Then a pig can produce 75 g of the protein/year, a goat 100 g, a sheep 125 g, and a cow 3 kg. (This assumes that daily milk production is constant over the year, not a particularly realistic assumption.) Because the national need for blood clotting factor IX is about 2 kg/year, one cow can do the job [Arizona 2006].

An important example of a TGA raised for food, and the controversy it has created, is the *TG salmon*. The *AquaAdvantage*® (A/F Protein, Waltham, Massachusetts) salmon was developed by *Aqua Bounty Farms,* Fortune Bay, Prince Edward Island, Canada. The science behind the TG salmon was discovered by accident about 1980 by Choy Hew, PhD, then a researcher at Memorial University of Newfoundland in Canada. Hew accidentally froze a tank containing a flounder species. When the tank was thawed, the flounders were alive. It turned out that this fish species has a gene that produces a protein that serves as a natural antifreeze. (This antifreeze protein is found in many types of polar fish.) The flounder also had an *activator gene* that turned on the expression of the antifreeze protein. Not unexpectedly, this gene is normally triggered by cold. Hew and colleagues isolated and cloned the flounder activator gene and attached it to a gene from Chinook salmon that produces a seasonal surge of *growth hormone* (GH). Using transgenic techniques, they inserted the new paired gene (flounder activator + salmon GH) into fertilized salmon eggs. In the resulting TG salmon, the activator became permanently active, producing a continuous supply of GH. The continuous expression of GH causes the farmed salmon to grow to market weight (6–10 lbs) in about 18 months, rather than the 24 to 30 months for normal farmed salmon. Also, the TG salmon process their food 10 to 30% more efficiently than wild fish.

In spite of all this good news, no TG salmon are on the market today. The Canadian government has required that Aqua Bounty must ensure that no breeding TG salmon escape their pens and breed in the wild. There is a justifiable fear that they would replace natural, wild salmon in the ecosystem because of their more rapid growth and lower breeding age. Accordingly, work is under way to ensure that the TG salmon grown in ocean net pens will be sterile (triploid [**3n**] females, incapable of breeding). Thus, if they escape, they cannot breed with wild fish. (It is not known if triploid, male TG salmon can produce viable sperm.)

Research on *cell-based vaccines* for renal cell carcinoma using genetically engineered tumor cells and monocyte-derived dendritic cells has been reported by Frankenberger, Regn et al. [2005]. These authors have used gene modification of allogenic renal cell carcinoma (RCC) cell lines in vitro to create generic RCC vaccines. *Dendritic cells* (DCs) were found to be useful because of their excellent stimulatory capacity combined with their ability to process and present antigens to both naive CD4 and CD8 leukocytes. The idea of transfecting DCs with RNA to create vaccines for cancer, viral, and bacterial infection was reviewed by Cannon and Weissman [2002].

By identifying candidate molecules expressed by RCC cells using cDNA arrays and protein arrays, and identification of peptides presented by MHC molecules, it is possible to create specific anticancer vaccines to be tailored to the antigenic profile of individual tumors. Thus, patient-specific antitumor vaccines can be created with DCs, and the patient's immune system will fight the tumor. Quoting Frankenberger, Regn et al. [2005]:

> Future vaccines may well employ pools of defined antigens in order to induce multiplex immune responses that will help to protect against immune selection of antigen-loss tumor cell variants. Use of DCs loaded with synthetic peptides derived from different proteins and presented by different MHC class I molecules would be efficient and could be tailored to individual patients, but this necessitates defining the HLA molecules of the patients and does not provide epitopes for DC4 T-cell responses. … As an alternative, pools of defined [recombinant] RNAs can be used for DC loading, a method which allows antigens to be directed to both the class I and class II pathways for priming of CD4 and CD8 T-cells. This approach does not require information on specific peptide sequences or the HLA type of the patients. [Quoted with kind permission of Springer Science and Business Media.]

It is our opinion that development of the cell-based vaccine approach to fighting cancer has considerable promise, but it will be labor intensive and hence very expensive.

Current fear of the possible mutation of the H5N1 avian (bird) flu virus to a form causing a human pandemic has led to the study of faster ways to immunize both birds and humans. Traditionally, fertile developing chicken eggs have been used in vaccine production. However, H5N1 is lethal to the chicken embryos (as well as the adult hens required to lay the millions of eggs required to make vaccine the traditional way), so other cell-based vaccine sources have been explored. *Protein Sciences Corp.*, in Meriden, Connecticut, has designed a vaccine strategy that uses *transgenic caterpillar cells*. The gene responsible for producing the *hemagglutinin* coat protein in the H5N1 virus is isolated and introduced into a *baculovirus*, which is then used to infect the caterpillar cells. (*Baculoviridae* are dsDNA, polyhedral, enveloped viruses. They infect insects [fatally] and human liver cells with low toxicity. GM baculoviruses [GMBVs] have been used experimentally for gene therapy in the human liver and to control insect pests with high species specificity [Cummins 2000; D'Amico 2005].) The GMBV-infected caterpillar cells produce the hemagglutinin protein, which is harvested and used to make a vaccine that induces natural immunity to the wild H5N1 coat in birds and humans.

Cell-based vaccines using transgenic human cells are also being used to develop influenza vaccines. Recently, Sanofi-Pasteur received a $97M contact from the U.S. Department of HHS to develop flu vaccines using human cells [Medical 2005].

10.2.4 Recombinant DNA Technology

10.2.4.1 Introduction

In this section, we will examine how transgenic animals and plants are made.

The recombinant process begins with the isolation of the coding DNA gene of interest. The gene is enzymatically inserted into a *vector*, forming *recombinant DNA* (rDNA). A vector is a piece of DNA that is capable of independent growth; commonly used vectors are bacterial *plasmids* and viral phage dsDNA. The gene + vector rDNA is then cloned (amplified). The PCR is used to clone the rDNA. Once the rDNA vector is isolated in large quantities, it can be introduced into the host cells; these can be bacteria, yeast, plant cells, or animal cells (usually eggs, sperm, or zygotes).

10.2.4.2 rDNA Insertion Methods

There are two popular methods by which rDNA can be inserted into the target cell's genome. *The first method* is called *pronuclear injection*. A zygote (single-celled embryo) is microinjected with 200 to 500 copies of the vector, which incorporates itself randomly into the zygote's DNA. After a few cell divisions in vitro, the microinjected embryos are transferred to the oviducts of female animals that have been induced by hormones to act as surrogate mothers for the TG animal. In *the second method*, *gene targeting in embryonic stem cells* (ESCs) is used. Cells from the inner cell mass (ICM) of an in vitro–fertilized blastocyst are nominally pluripotent; they are collected and microinjected with the rDNA vector. A small percentage of those cells will have the vector inserted at the targeted site in the genome. The targeted ESCs are selected and injected back into the blastocyst's ICM. These modified embryos are now implanted into surrogate mothers. Because of the random nature of the rDNA incorporation, a fraction of the embryos will develop into GM animals. This method has been largely restricted to mice.

10.2.4.3 Preparation of a Plasmid Vector

Plasmids are small circles of dsDNA (a few thousand bps) found in bacterial cells. They are smaller than the bacterial chromosome and separate from it. They are useful in transgenic technology because they can easily pass from one cell to another, even when the cells are from different species, distant on the evolutionary scale. Consequently, plasmids can be used as *vectors*, permitting the reproduction of an inserted rDNA by use of the natural bacterial replicating system.

A gene to be used in recombinant DNA technology is prepared as a coding sequence of dsDNA (cDNA). To obtain this coding sequence, we start with a chromatographically purified sample of the target protein. The protein is analyzed to find the identity of the first 30 AAs at its amino end. The genetic code is used to predict the DNA Nt sequence coding this AA sequence. This DNA oligo is synthesized and used to identify the target complementary mRNA from which the entire cDNA sequence is synthesized using *reverse transcriptase*. The cDNA is cloned or amplified. The DNA produced from each clone is hybridized to total cellular mRNA. Now, purified mRNA corresponding to each cloned cDNA sequence is then produced in vitro. Having checked by cell-free protein synthesis procedures which mRNA codes for the desired protein, the appropriate cDNA is then sequenced and used to determine the entire protein's AA sequence. Lastly, the cloned cDNA is incorporated into a plasmid vector.

The plasmid is cut across both DNA strands by a *restriction enzyme* (RE), leaving loose, *sticky ends* where the cDNA can be attached. On average, restriction enzymes with four-base recognition sites will give DNA oligos about 256 bases long. Six-base recognition sites yield oligos of about 4,000 bases, and REs with eight-base recognition sites give DNA fragments about 64,000 Nts long.

Special *linking sequences* of Nts are added to the cDNA so that it will fit precisely to the cut, loose, sticky ends of the plasmid and complete the circle again. *DNA ligase enzymes* fix the cDNA + linkers into the cut plasmid. The genetically modified plasmids can now be inserted into bacteria such as *E. coli*, which will reproduce the GM plasmid, transfer it to daughter cells, and make many clones of the GM plasmid. If the cDNA in the plasmid is expressed, the bacteria will make the protein it codes, which can then be harvested. Human insulin is made this way by GM *E. coli*.

10.2.4.4 The Use of Phages as cDNA Vectors

Two types of *bacteriophage* viruses have been used extensively for cloning purposes in bacteria, λ, and M13. The λ phage has a capsid or head that encloses its dsDNA genome and also has a "tail." The M13 phage has a filamentous structure; a protein coat covers its ssDNA genome. At one end is its gene 3 protein, which is important in both adsorption and extrusion of this phage. The λ phage is a so-called *temperate phage*. Its genome is only 48.5 kb in length, encoding 46 genes. The entire λ genome has been sequenced. At the ends there are short (12 bp), single-stranded, sticky ends that enable circularization of the genome (into a plasmid) following infection. A nonessential region lies in the center of the phage's DNA. This central region controls the lysogenic properties of the phage, and much of this region can be eliminated without impairing the functions required for the lytic infection cycle [Nicholl 2002]. Once inside the bacterial host, the λ phage genome can make many copies of itself and also make copies of its structural proteins. Then, lysis of the cell releases the daughter phages.

An important drawback of using λ vectors is that the capsid places a size restriction on the cDNA that can be spliced into the λ genome. During DNA packaging, viable phage particles can be produced from DNA that is between about 38 and 51 kb in length. Thus, a wild-type phage genome can hold only about 2.5 kb of cDNA before becoming too large for viable phage production. Note that the native human insulin gene is about 1.4 kb in length and has three exons (coding regions). (Recall that the GM λ phages infect bacteria, which then make the protein coded by the inserted cDNA, as well as new vector phages. A high rate of phage infection is required to spread the vector to many bacteria in order to make lots of the desired protein.)

Packaging GM phage DNA in vitro is possible because, during the lytic cycle of normal λ phage, the phage's DNA is produced in a long sausage-like chain called a *concatemer*. N copies of the phage DNA are linked together at *cos* sites. When the wild phage particles are assembled in the bacterium, a phage-encoded endonuclease cuts the concatemer at the *cos* sites; the N individual molecules of phage DNA are enclosed in capsids, and tails attached before bacterial lysis. This entire process can be carried out in vitro; however, the components of the capsid, tail, and endonuclease must be present. Now, if a concatemer of the recombinant DNA is used, the phage reassembles with the modified DNA in its genome and can be used directly to infect *E. coli* cells, which now make more phage vectors and the desired protein, such as insulin.

The M13 phage can only infect *E. coli* that have external *F-pili*. The M13 genome is only 6,407 bps in length. When it enters the cell, it becomes dsDNA and replicates itself until there are about 100 copies in the cell. At this point in its cycle, it makes single-stranded copies of its genome and extrudes them as complete M13 phages. Oddly, the infected bacterium is not lysed. Again, a cDNA coding a desired protein can be attached to the M13 DNA and introduced into target cells that make more vector M13 phages as well as the protein.

10.2.4.5 Other Viral Vectors in Gene Therapy

Retroviruses and *adenoviruses* have been used for gene therapy in humans and other mammals [Gilbert 2003]. Recall that retroviruses have an ssRNA genome, which is made into dsDNA upon viral internalization by reverse transcriptase. The virally made DNA finds its way into the infected cell's genome where, when it is expressed, it makes a protein. It is possible to induce a cell to make a therapeutic protein by inserting a therapeutic gene into viral RNA. The viral RNA also can be modified so that the virus cannot reproduce.

Adenoviruses can be given recombinant dsDNA, with which they infect the host cell's nuclear envelope directly (after entering the cell membrane). When the host cell expresses the recombinant gene, it makes the therapeutic protein, which can be a hormone, enzyme, receptor, etc.

The difficulty in using viruses as vectors is that they have to be modified from wild types, a multistep procedure.

10.2.4.6 The Gene Gun

The bioballistic (biolistic) *gene gun* is an alternative means of introducing plasmids with recombinant DNA into both plant and animal cells. The gene gun was first developed in the early 1980s by plant scientists at Cornell University. Horticultural scientists John Sanford and Theodore Klein wanted a device to expedite plant-breeding practices by directly injecting genes into plant cells. A bioballistic particle delivery system, or gene gun, was developed toward this end. It was used to do research on GM rice plants; rice plants were developed that incorporated genes from potato plants to give them insect resistance, and genes from barley plants to make them salt and drought tolerant. The gene gun was also used to make transgenic papayas virus resistant [Voiland & McCandless 1999].

The original gene gun developed at Cornell consisted of two stainless steel chambers, 6″ × 7″ × 10″, connected to a vacuum pump. When the operator flicks a switch on the outside of the second chamber, helium is released into it at 1000 psi. The blast wave ruptures a thin disk about 2 cm in diameter. The ruptured disk releases a shock wave that travels 1 cm until it hits a second disk that is free to move. Attached to the front of that disk are 1-μm-diameter tungsten or gold particles coated with recombinant DNA (rDNA). The disk and particles travel another centimeter at about 1,300 fps, striking a screen that retains the disk but launches the DNA-coated particles toward the target cells. The high-velocity inertia-driven particles penetrate the cell membranes and release their rDNA, which diffuses into the nuclei. As one might imagine, the early version of the gene gun damaged cell membrane, killed some cells, and left tungsten or gold beads inside the cells that were inoculated with the rDNA. Cells farther from the nozzle had beads penetrate their membranes, which were able repair themselves.

It wasn't long before other bioengineers developed versions of the gene gun that were more suitable for injecting DNA into soft animal cells. Thomas et al. [2001] reported on the design of an improved helium burst device that successfully injected DNA into silkworm moth eggs and other fragile tissues. Their gun allowed both an aqueous delivery of particles and an ethanolic dry delivery. The basic Cornell gene gun was modified by replacing the original floppy macrocarrier with a rigid disk. Thomas et al. also added an improved "focusing" nozzle to better direct the particles + rDNA to the cellular target.

Many workers have made improvements on the original gene gun design. For example, DiLeo et al. [2003] used a modified biolistic device to deliver high-velocity rDNA-coated gold beads having enough kinetic energy to penetrate mouse skin and deliver rDNA to subdermal tissues. DiLeo's

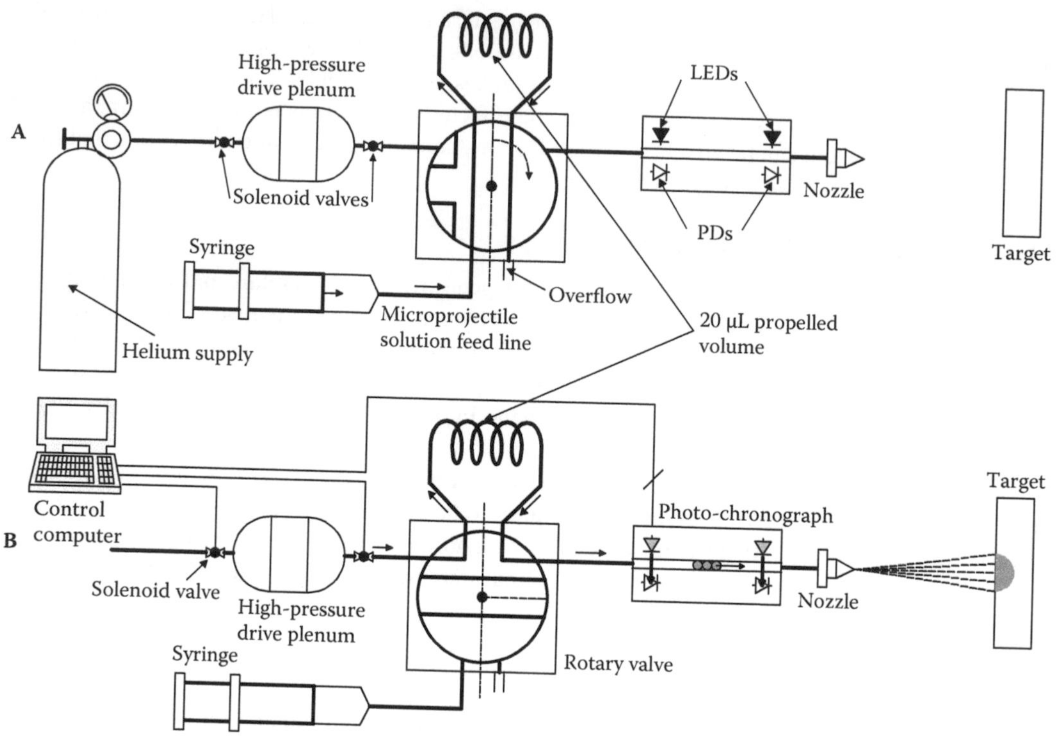

FIGURE 10.1 Schematic of the Dow gene gun. Top: The gene gun is loaded with 20 μL of microprojectiles to be injected. Bottom: High-pressure gas is discharged from a plenum tank, propelling the microprojectiles into the target cells. A photochronograph is used to measure the injection velocity.

bioballistic device was designed and built by the Dow Chemical Co., and technical details can be found in U.S. Patent #5,141,131. The system uses no high-velocity moving parts other than the genetic payload itself. Helium is loaded under high pressure into a plenum chamber, from which it is abruptly released to propel the "charge." An LED/phototransistor chronograph allows velocity measurement of the 20-μL payload up to 1,760 fps. 140 μg of rDNA were coated onto 60 mg of 3-μm-diameter gold particles. A two-way valve permitted the 20-μL charge to be loaded into a reservoir tube. The valve is then rotated so a helium pressure wave can drive the charge past the chronograph and out the nozzle into the target tissues. Figure 10.1 shows a simplified schematic of the Dow gene gun. The gene gun was capable of deep tissue delivery of its rDNA particle load. Quoting DiLeo et al. [2003]:

> The ability of [subdermal] muscles to serve as a reservoir for protein expression also has made it the target of choice for the expression of genes designed for systemic delivery. [Genetic] Diseases such as [clotting] factor IX [protein] deficiency and other such inborn errors of metabolism-type diseases are now potentially treatable using gene guns whereas before more invasive methods of gene delivery were needed.
>
> Current cancer gene therapies using gene gun methods are limited to the use of secreted products such as cytokine or secreted factors that are intended to have a systemic effect. However, the majority of antitumor gene therapies require the direct transfection of tumor cells. The direct modification of tumor cells by gene gun is now a possibility. While more work is necessary to develop gene guns further for wide-scale clinical use, the results reported here open up the possibility of applying this method to a wider range of human diseases than previously thought. [Quoted with permission.]

As presently configured, gene guns applied to humans and animal skin and tissues are necessarily painful. The gold particles are necessary to penetrate cell membranes and introduce the rDNA.

Unfortunately, they damage a certain amount of tissue on the surface where their kinetic energy is highest. Surface tissue damage should quickly heal, however. Local anesthesia is indicated for topical use on humans and animals.

10.2.4.7 Electroporation and the Effects of Nanosecond Pulsed Electric Fields on Cells

Electroporation is a means of causing cell membranes to transiently develop "pores" in their cell membranes having large enough diameters to allow DNA plasmids, oligos, and other macromolecules such as proteins in the extracellular space to enter the cell's cytoplasm. It provides a means of introducing "designer" genes to produce GMOs. When electroporation is done correctly, the pores close and the membrane "heals," reestablishing continuity. Electroporation can be applied to mammalian cells (both in vivo and in vitro) as well as bacteria.

A typical in vitro electroporation system consists of a dielectric chamber (e.g., a glass cuvette) with noble metal (Au or Pt) foil electrodes (see Figure 10.2). The bacteria to be electroporated plus a solution containing the plasmid are placed in the chamber. A high-voltage capacitor is charged to a dc voltage of about 2.5 kV. Then the capacitor is discharged through the chamber. The conductivity of the chamber depends on several factors, including the density of bacteria, the temperature, and the concentration of mobile ions such as Na^+, Cl^-, K^+, etc., in the solution. The voltage across the chamber thus has an exponential, R-C, capacitor discharge waveform starting with a peak of 2.5 kV and decaying to zero with a time constant of about 10 msec. The initial voltage gradient (electric field) in the cuvette is about 12.5 kV/cm. This peak field creates electrostatic forces that exceed the structural capability of the cell membrane's phospholipid bilayer, causing aqueous pores to form and then quickly heal as the field decays exponentially. The transient appearance of pores, and the field across the cell membrane, allows ions, plasmids, and proteins to enter the cells before the pores vanish. The time constant of the exponential capacitor discharge and its peak voltage are critical parameters for electroporation and are affected by the cell size and type and the chamber dimensions [McCord 2003; Jordan et al. 2004].

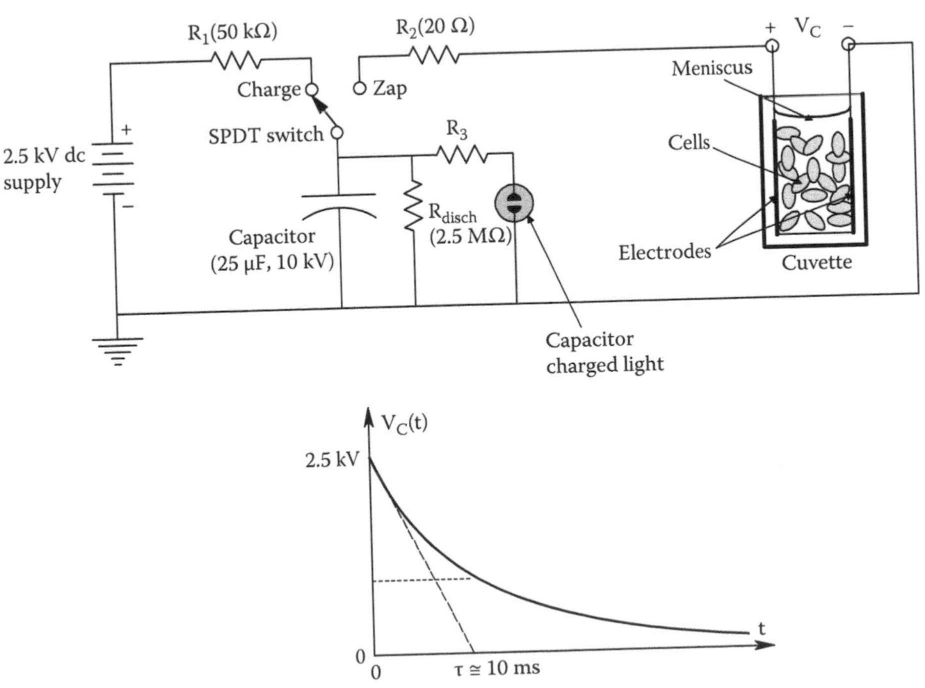

FIGURE 10.2 Circuit of an in vitro electroporation system. See the text for a description of this technique.

In the past 5 years, a new approach has been developed to alter the structure of cell membranes and internal structures in eukaryotic cells with the goal of fighting cancer cells. Nanosecond Pulsed Electric Fields (nsPEFs) have been applied to selected types of tumor cells in vivo and in vitro with mixed success [Schoenbach et al. 2006, 2004]. The target cells form a load in a modified *Blumlein transmission line pulse generator*. Figure 10.3B illustrates the schematic of a coaxial transmission line design of a Blumlein pulse generator. Note the complex equivalent circuit of the cells in vitro is shown in Figure 10.3A (one cell is shown for simplicity; there are many in series parallel). In one Blumlein system, the transmission line lengths were chosen so that the fast rectangular pulses were $\tau = 300$ ns in duration and had a peak voltage around 40 kV (see Figure 10.3C). The brief duration of the pulse prevents polarization of the plasma membrane and allows the electric field to interact at a molecular level with intracellular organelles and the nucleus. Using 300-nanosecond nsPEFs, Schoenbach's group 2006 was able to cause mouse skin melanomas to completely regress. In an earlier in vitro study, Hall, Schoenbach, and Beebe [2005] studied the effects of nsPEFs on human colon carcinoma cells; results showed direct electric field effects and biological effects involving the p53 cell protective protein.

There is some evidence that tumor regression following nsPEF treatment is due in part to induced apoptosis resulting from intracellular damage from high E-field effects. In the mouse studies, melanoma destruction may also have been the result of nsPEFs destroying the tumor's blood supply. The precise molecular biological mechanisms responsible for nsPEF-induced tumor cell death have yet to be clarified. (It is evidently not heat or rupture of the plasma membrane.)

Different pulse and field parameters can cause treated cell membranes to temporarily develop pores (electroporation) that can allow the entrance of rDNA in solution outside the cells.

10.2.4.8 Ultrasonic Poration

Another physical means of creating momentary, repairable pores in cell membranes is by ultrasonic energy. At sufficiently high energy levels, cells suspended in an ensonified liquid can have shear stresses developed in their membranes from the ultrasonic waves so that transient pores develop. The shear stresses may be from the pressure transients accompanying the formation and collapse of cavitation bubbles. Heat is also generated from the intense sound field, which can be deleterious to cells.

10.2.5 Problems with GMOs

One of the serious concerns about human consumption of GM foods is allergy to their GM proteins (e.g., Bt) and certain haptens. More that 90% of food allergies occur in response to specific proteins in milk, eggs, wheat, shellfish, tree nuts, peanuts, soybeans, and plant oils, such as cottonseed or peanut (oils are haptens). In the United States, the FDA requires that the developer of each new GMO provide scientific evidence that it does not incorporate any allergenic substance. If the developer cannot present such evidence, the law requires that they label the product to alert consumers of its possible allergic hazard [Dresbach et al. 2001].

Many high-profile NGOs, such as Greenpeace, are *totally* opposed to genetic engineering. In a 2006 Web article, Greenpeace stated:

> These genetically modified organisms (GMO) can spread through nature and interbreed with natural organisms, thereby contaminating non "GE" environments and future generations in an unforeseeable and uncontrollable way.

They go on:

> We believe: GMOs should not be released into the environment as there is not adequate scientific understanding of their impact on the environment and human health. We advocate immediate interim

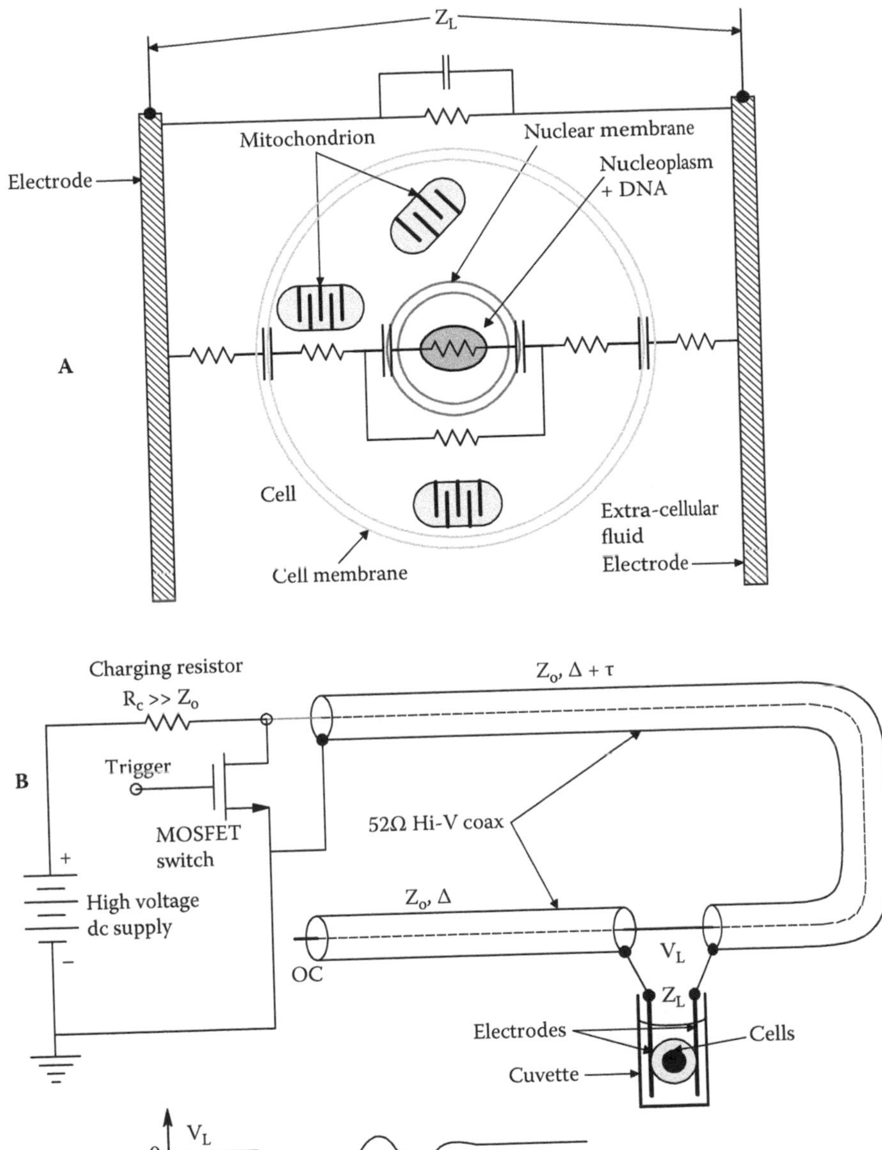

FIGURE 10.3 (See color insert following page 44.) **(A)** Equivalent circuit of a representative cell between a pair of electrodes. **(B)** A Blumlein transmission line pulse generator. It can produce very narrow, high-voltage pulses that do not cause electroporation but instead affect molecular exchanges between intracellular organelles and the nucleus.

> measures such as labeling of GE ingredients, and the segregation of genetically engineered crops and seeds from conventional ones. We also oppose all patents on plants, animals and humans, as well as patents on their genes. Life is not an industrial commodity. When we force life-forms and our world's food supply to conform to human economic models rather than their natural ones, we do so at our own peril. [Quoted with permission from Greenpeace.]

We emphasize that the Law of Unintended Consequences applies to GMOs, as discussed further in Chapter 11. *An example:* Brown [2005] reported on an unintended and unforseen consequence of growing a GM oilseed-rape (also known as canola, a brassica) crop in the United Kingdom that had enhanced herbicide resistance to permit "chemical weeding" of the crop. A distantly related plant, *charlock,* was found growing in the field that had been previously used to grow the GM rape. There was evidently cross-fertilization between the GM rape and the charlock weed that gave it strong herbicide resistance. Also from the same field, two wild turnips were found to be herbicide resistant. The *glufosinate-ammonium* herbicide used on the canola put huge selective pressure on the charlock, much the same way the profligate use of antibiotics has led to resistant strains of bacteria in hospitals. The charlock that had captured the herbicide-resistant genes from the GM canola's pollen survived and prospered. After 4 years, the resistant charlock plants had multiplied to some 103,000 plants. Brown noted that charlock seeds can remain in the ground for 20 to 30 years before germination, making the elimination of the herbicide-resistant charlock almost impossible.

Brown remarked that currently no GM crops are grown in Britain, and there are strict licensing requirements for GM crops in both the United Kingdom and the EU in general.

A GM crop that has been grown in Spain, Portugal, France, and Germany is *insect-resistant maize,* used only as animal food. Because the insect resistance is inherent in the all of plants' cells, there is a very real possibility that the insects will develop a natural resistance to the poison through natural selection (remember flies and DDT?). It is as though the cornfield were drenched in the insecticide, 24/7.

GM cotton is another crop that yields high returns/acre for the first few years, until the bollworms develop resistance to Bt, or the natural insect predators of the bollworm are killed off by the Bt. Note that the Bt insecticide is omnipresent in the cotton plants, whereas conventional insecticides are sprayed at certain intervals. Not having to handle insecticides is beneficial to farmworkers and a cost saving to the farmer. However, as we have seen, not all insect pests of cotton find Bt toxic. The issues are complex; so are biochemical and ecological systems.

Besides economic factors, Calgene took the *FlavrSavr tomato* off the market because of the stomach lesions it caused in rats in feeding tests. It is not known why this cell damage occurred in rats fed *FlavrSavr tomatoes* and not normal tomatoes. Was it associated with the gene for kanamycin resistance? Were the FS tomatoes more acidic than normal tomatoes?

Recently, small amounts of a variety of GM rice were discovered with natural rice in storage facilities in California and also on the East Coast [Elias 2006]. The GM rice was identified as an herbicide-resistant strain engineered by *Bayer CropScience AG,* a German company. A common weed plant, *red rice,* competes with normal rice and reduces crop yield. Bayer (Bayer CropScience, Research Triangle Park, North Carolina) genetically engineered normal rice to make it resistant to the *Liberty*® herbicide that normally would kill both red rice and normal crop rice, enabling "selective weeding" of the GM rice. This GM rice had never been approved for human consumption in the United States; presumably it was destined for animal food. (The GM rice problem has an eerie similarity to the case of StarLink corn [see Section 10.2.2.3].)

Quoting Elias:

> Federal officials say that the company's signature genetically engineered rice came from storage bins in Arkansas and Missouri, but they don't know where it was grown. Federal officials and company executives say the strain posed no health threat and was similar to biotech rice that had been approved [by USDA].

Subsequent investigation by the USDA showed that the GM rice contamination may have come from cross-pollination with normal rice grown in Louisiana in a field adjacent to the field where Bayer was growing its GM herbicide-resistant rice. Great pains were taken to isolate the GM crop, but evidently contamination did occur.

Nevertheless, Japan, a major importer of U.S. rice, and also the EU threatened to ban *all* U.S. rice imports. This caused U.S. rice futures to plummet and has threatened the U.S. rice export market. Rice exports are worth about $200 million annually in California, which exports mostly to Japan. Japanese and European consumers have a long-standing aversion to GM foods. Those consumers fear that not enough is known about the *long-term* effects of GM crops on the ecosystem and human health. GM food and animal imports are highly regulated, and most are already banned. Now all export rice is being tested by the USDA for GM contamination, and U.S. rice exports are no longer frozen.

The public will only accept GM foods when it can be conclusively demonstrated that there are no untoward long- or short-term effects on people consuming them, and that they have negligible impact on the environment. The first criterion can be met by extensive animal feeding trials, followed by human volunteers. The environmental impact is harder to evaluate; the whole ecosystem around the GM crop field must be studied before and after growth. Plants, animals, groundwater, and soil microbes must be examined. If GM crops, such as tomatoes, are grown in greenhouses, this makes the environmental impact simpler to evaluate. The world population is expanding, and sustainable food production will rely heavily on GM crops in the future; they must be safe.

10.3 ANIMAL REPRODUCTIVE CLONING

10.3.1 Introduction

Animal cloning is the creation of a viable genetically identical copy of an existing (or previously existing) organism using the diploid genome from one of the organism's cells (either adult or embryonic cells). The formation of human identical twins is an example of normal, natural cloning caused by separation of the cells of a diploid 2-cell embryo, which then mature into two separate trophoblasts. Identical twins occur in about 0.25% of human births [Gilbert 2003].

The first, and most widely known, synthetic mammalian clone was Dolly the sheep. Dolly was the first animal cloned from an adult diploid cell nucleus inserted into an enucleated oocyte. Dolly represented a cloning efficiency of 0.23%. That is, she was the result of 434 tries [Di Berardino 1998]. Since Dolly was cloned in 1997, many other animals have been cloned, including mice, calves, goats, Rhesus monkeys, dogs, and cats.

The United States and many European countries prohibit the cloning of humans. *Regenerative medicine* has focused on stem cell research. There is no economic incentive to clone humans; the world has enough humans by sexual reproduction.

Then why clone animals? The primary reason is economic. A sheep that gives a particularly fine grade of wool, a racehorse that has great stamina and speed, a cow that gives more milk of higher quality, a chicken with larger breast muscle, etc., are all economically valuable. Having one or more clones of such animals means there is a greater chance to reproduce their phenotype by natural breeding in further generations. Animal cloning has also been suggested for sentimental reasons; to replace a favorite dog, for example. Natural cross-breeding of clones is not as efficient as mass cloning would be, once the efficiency of the process can be improved to be cost effective.

Currently, there are three major techniques that have been used in reproductive cloning, described as follows.

10.3.2 Altered Nuclear Transfer

In altered nuclear transfer (ANT), two cells are required: an unfertilized host oocyte and a diploid donor cell. The haploid egg is enucleated. This still leaves its ribosomes and their rDNA. The donor (diploid) somatic cell is then forced into the dormant, *Gap Zero* (G0) stage. In the G0 stage, the cell is dormant; almost all gene expression is shut down. The altered donor cell's nucleus is then transferred to the enucleated egg. The egg is now stimulated to divide and form an embryo, which is implanted into the uterus of the surrogate mother.

The problem here is that the donor nucleus can have imprinted genes, especially those required for embryonic development. This is why the ANT cloning success rate is so low.

10.3.3 The Roslin Technique

In the Roslin technique (used on Dolly the sheep), a method was used to synchronize the *cell cycles* of the donor cell and the oocyte. A donor cell was taken from the udder cells of a Finn Dorset ewe and cultured in vitro. The udder cells were then grown in a sparse medium that had only enough nutrients to keep the udder cells alive. This environment caused the cell to begin shutting down all active gene expression and enter the G0 stage. The regulatory protein *cyclin D* is evidently involved in causing the G0 state [Gilbert 2003].

The egg of a black-face ewe was then enucleated and placed next to a donor cell. One to 8 hours after this juxtaposition of egg and donor, an electric pulse was used to fuse the two cell membranes together and simultaneously activate the in vitro development of an embryo. Electrical activation is not very effective because only a few electrically activated cells survive long enough to produce a viable embryo.

If the embryo survives, it is allowed to grow in a sheep's oviduct for about 6 days. It was found that embryos incubated in oviducts were much more likely to survive than those grown in vitro. The 6-day embryo is then implanted into the uterus of the surrogate mother ewe and, if all goes well, a clone of the donor ewe is born.

10.3.4 The Honolulu Technique

This was used by Wakayama et al. [1998] to clone mice. These workers initially used three types of donor cells: *sertoli cells* (epithelial cells of the male fetal seminiferous tubules), *brain cells,* and *cumulus cells* (a layer of ovarian follicular cells surrounding a mature ovum in an ovary). *Sertoli cells* and brain cells are naturally in the G0 state; the cumulus cells are almost always in either the G0 or G1 states.

Wakayama found that the cumulus cells gave the highest success rate in forming clones. His procedure was to directly implant the donor nuclei into enucleated eggs and wait an hour until the egg had "accepted" the donor nucleus. After an additional 5 hours, the cell was placed into a special culture medium to chemically shock the egg into dividing. This medium contained *cytochalasin B,* which stopped the formation of a *polar body* (a haploid second cell that normally forms during ovum meiosis before normal fertilization). The treated egg was then directly implanted into a mouse uterus.

Wakayama and Yanagimachi went on to make clones of their clones and bred the original clones naturally to demonstrate that they had full reproductive function. At the time of publication (1998), they had made about 50 clones. Wakayama et al. reported in 2000 that they had cloned mice by nuclear transfer to six generations [Wakayama et al. 2000]. They examined the cloned mice for signs of accelerated aging by a battery of physical tests and examining their *telomeres.* The physical tests demonstrated that all the cloned mice were normal compared to age-matched controls. They did observe progressive lengthening of the telomeres with successive cloned generations. They remarked that the reason for this may lie in the *telomeres* of the cumulus cell nuclei used.

In 2000, Ogura et al. reported that they had a high success rate in cloning mice by forming the zygotes through electrofusion of tail tip fibroblast cells with enucleated eggs [Ogura et al. 2000]. They claimed a high success rate for this modified cloning process (68% for fusion and nuclear formation, 40.8% for development to morulae/blastocyts, and 2.5% normal fetuses).

10.3.5 Problems in Cloning Mammals

Following the normal formation of a zygote, embryological development is orchestrated by *selector genes* that encode *transcription factor proteins.* Selector genes turn on and off the expression of other genes. Most selector genes are *homeobox* (Hox) *genes.* All animals have at least one Hox gene cluster; mice and humans have four Hox clusters. There are 39 Hox genes in humans. The Hox genes act in sequential zones of the embryo in the same order that they occur on the chromosome on which they are located.

In cloning using adult donor cells, it is possible that some Hox genes that are normally active in embryos have been inactivated by *imprinting.* Such imprinting can cause cloned embryos to die in utero or postpartum clones to die prematurely. Some cloned mice have expressed an obese phenotype [Tamashiro et al. 2002; Cibelli et al. 2002].

Note that imprinting could account for the low (0.23%) success rate in cloning Dolly the sheep.

10.3.6 Therapeutic Cloning

Therapeutic cloning has the purpose of growing human stem cells for use in regenerative medicine. It uses human donor cells for *somatic cell nuclear transfer* (SNCT). SNCT is basically the ANT technique described in Section 10.3.2.

Therapeutic cloning using SNCT/ANT differs from reproductive cloning only in that the embryo obtained in vitro is never intended to be implanted or has been altered so it will not grow if implanted. In any case, the development of the cloned embryo is not allowed to proceed past the about 100-cell stage. The sole purpose of SNCT/ANT is to harvest pluripotent stem cells from the inner cell mass of the cloned blastocyst.

When the technique is perfected, a person of either sex with a damaged pancreas, for example, could donate an epithelial cell, which could be used in the SNCT/ANT process with a donated egg. Pluripotent stem cells from the inner cell mass of the cloned blastocyst would then be transformed into pancreatic beta cells and injected into the donor patient's bone marrow, where they would live and secrete insulin. Because the MHC markers on these beta cells would be the same as the donor patient's, they would not be attacked by the patient's immune system.

10.4 STEM CELLS

10.4.1 Introduction

Mammalian stem cells are primal undifferentiated cells that retain the ability, when appropriately stimulated, to turn into specialized cell types such as neurons, blood cells, bone cells, muscle cells, skin cells, corneal cells, etc. This ability allows them to act as a natural repair system for the body's differentiated cells, hence serving as effectors in *regenerative medicine (RM).*

Stem cells have captured much attention from scientists, religious groups, legislators, politicians, and the media. One reason for the attention is their promise to be effective in RM. In one RM scenario, undifferentiated *pluripotent stem cells* would be introduced into the body and somehow caused to differentiate into specific tissues to, in theory, replace damaged organs and tissues such as the liver, pancreatic beta cells, dopamine-producing neurons in the brain, skin, hip joints, hair, spinal cord neurons, cardiac muscle cells, etc.

Another significant reason that stem cell research has attracted so much attention is the fact that a major source of human stem cells has been the inner cell mass (ICM) of a blastocyst-stage embryo, shown schematically in Figure 10.4. Removal of the inner cell mass destroys the embryo, and hence, a *potential* human life. A major source of the embryos used is the many extra zygotes formed in the process of in vitro fertilization (IVF). These by-products of the IVF process normally are either disposed of or frozen for potential further use.

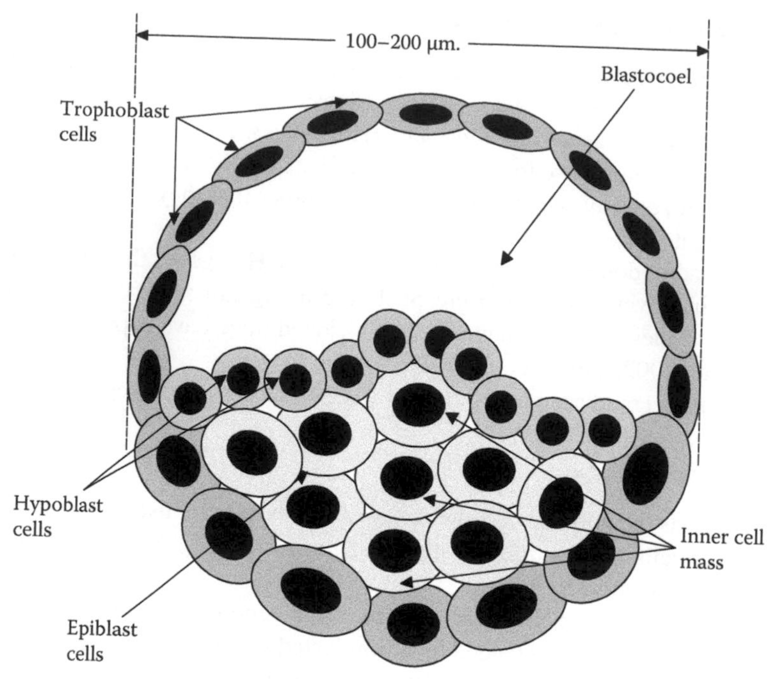

FIGURE 10.4 (See color insert.) Cross-sectional schematic of a 6-day mammalian blastocyst, with typical dimensions. Stem cells are harvested from the inner cell mass, generally destroying the blastocyst.

Certain religious groups find IVF objectionable because they believe in the sanctity of human life, which they postulate begins as soon as the one-celled, diploid zygote is formed by fusion of a sperm with an egg (see Chapter 11). IVF leads to the loss of *potential* human life in the form of the excess blastocysts grown in vitro that are never implanted or allowed to grow into embryos.

The harvesting of stem cells from the inner cell mass causes the destruction of blastocysts left over from IVF and is objected to strenuously by certain religious spokespersons. It should be noted that the blastocyst does not exhibit any of the differentiated cell masses associated with later (implanted) embryonic development; there is no heart, nerve cord, CNS, limb buds, gill slits, etc. A blastocyst is incapable of feeling pain.

10.4.2 Types of Stem Cells

Stem cells are categorized by their *potency,* which describes their ability to differentiate. True stem cells can reproduce themselves in vitro. *Totipotent stem cells* are cells that can differentiate into *all* forms of fetal cells, including germ cells and placental cells. Totipotent cells are taken from the two- or four-cell embryos. A two-cell embryo can develop into identical twins, illustrating that the cells of the two- and four-cell embryos are totipotent. Totipotent cells are not really stem cells in that they cannot make more of themselves, in vitro.

Pluripotent stem cells (PSCs) appear in the ICM of the blastocyst a few days later in embryonic development. The pluripotent stem cells are of primary interest in RM because of their theoretical potential to form any differentiated cell type in the body, except extraembryonic tissues (e.g., the placenta). There are currently three sources of pluripotent stem cells:

1. The ICM of the blastocyst embryo (this can be a blastocyst resulting from IVF, cloning, or parthenogenetic manipulation—all of which are controversial).

2. *Embryonic germ* cells, which can be isolated from the diploid precursor cells to the gonads in aborted *fetuses.*
3. *Embryonic carcinoma* (EC) cells, which can be isolated from *teratocarcinomas,* tumors that rarely occur in the gonads of a fetus. These EC cells are usually *aneuploid* [Kimball 2006a].

All these three types of PSCs can only be isolated from embryonic or fetal tissues.

Multipotent stem cells can only produce cells of a closely related family of cells. An example is the well-known adult *hematopoietic stem cells*, which are found on the outer edge of bone marrow in postpartum animals. These continually give rise to red blood cells, leukocytes, and platelets.

Unipotent stem cells by definition can produce only one body cell type when stimulated but, similar to other types of stem cells, have the property of self-renewal in undifferentiated form.

10.4.3 Sources of Stem Cells

We have already discussed the harvesting of PSCs from the ICM of blastocysts and why it is controversial. A potentially viable human embryo has to be destroyed to obtain a clone of stem cells. Fortunately, there are other sources of SCs.

Less controversial are the *multipotent stem cells* obtained from the blood in *placentas* and *umbilical cords* obtained in natural childbirth. Cord blood stem cells have been used therapeutically since 1988. Today they can be used to treat bone marrow failure disorders, hemoblobinopathies (such as sickle-cell anemia), inherited immune system disorders, inherited metabolic disorders (such as Tay Sachs disease), leukemias and lymphomas, and several other classes of diseases [Cord Blood Registry 2007]. Cord blood stem cells are administered to the patient intravenously, in coordination with other aspects of the treatment [National Marrow Donor Program 2007; Agura 2007].

Another source of stem cells that can be harvested at childbirth is *human amniotic epithelial cells* (HAECs). These are cells taken from the amniotic membrane that protects the fetus during pregnancy and from the amniotic fluid that surrounds the fetus. Given the proper cell culture factors, it has been possible to get them to display cell surface markers that suggest that they can be differentiated in vitro into cultures of all three principal germ layers: endoderm (liver and pancreas), mesoderm (cardiomyocytes), and ectoderm (neural cells). The HAECs were found to express the pluripotent stem cell–specific transcription factors *octamer-binding protein 4* (Oct) and *Nanog-4* [Miki et al. 2005]. Current (2006) research on HAECs' properties as stem cells has also been under way at the Monash Institute in Australia. Dr. Ursula Manuelpillai reported similar findings as Miki et al. In contrast to stem cells from umbilical cords, HAECs are easier to isolate and can be made to divide more easily. One amniotic membrane can yield about 60 million cells. [*The Scientist*: blog: *Epi-embryonic stem cells?* 04/01/2006]. Note that umbilical cords and amniotic membranes are normally discarded following birth. Amniotic fluid samples can be safely withdrawn without harming the fetus.

Also, researchers at the Keio University School of Medicine in Tokyo headed by Dr. S. Miyoshi have recently reported that they were able to isolate uterine epithelial (lining) cells from *human menstrual blood* that could be differentiated in culture to become functional *cardiomyocytes*. Stem cells from umbilical cord blood have already been used to treat hearts with damaged muscle; however, the density of SCs in menstrual blood is about 30 times higher than in cord blood [ABC Online 2006].

Currently, by presidential executive order, the U.S. Federal Government will only fund stem cell research on stem cells taken from *adult* human blood and organs. The so-called *adult stem cells* are largely multi- or unipotent types.

The U.S. government will continue to fund research on already-established stem cell lines (about 78 lines obtained before 2001, derived from embryos) but not fund any SC research involving human embryos. *Embryo* was defined in the 1998 Fiscal Year Appropriations Act as: "...any organism, not protected as a human subject under 45 CFR 46 as of the date of the enactment of this

Act, that is derived by fertilization, parthenogenesis, cloning, or any other means from one or more human gametes or human diploid cells" [Maienschein 2001].

In July 2006, and again in June 2007, U.S. President G.W. Bush vetoed a "stem cell bill" that would have provided U.S. government funding for stem cell research on excess IVF embryos. The bill otherwise encouraged research on nonembyronic sources of stem cells for use in regenerative medicine.

Adult stem cells (ASCs) are sparse and hard to find in adult tissues. First discovered in hematopoietic tissues, they are believed to exist in all classes of tissues that have been examined [Chepko 2005]. They serve to maintain tissue integrity, effect wound healing, possess unlimited mitotic potential, and continually replace aged and dying cells undergoing apoptosis in adult animals and plants. ASCs are long lived, are specialized for cell division, and function throughout the life of an organism.

Once an ASC is located, it must be stimulated to grow in vitro in its undifferentiated form to obtain a colony of cells large enough to be therapeutically effective when injected into the patient or reinjected into the donor. Although the U.S. government will not fund research on new lines of ESCs, many private corporations are investing in this research because of its potential medical payoff. Stem cell research is actively going on in other countries, with support from their respective governments [Yamanaka 2007]. It appears that the United States is in danger of being left behind in this medically important research.

A proposed source of PSCs is from *altered nuclear transfer* (ANT). In ANT, the diploid genome of an adult cell nucleus to be transplanted into an enucleated egg would be altered to prevent the embryo from maturing past the blastocyst stage, preventing its viability if ever implanted. (The *Cdx2* gene encoding a *homeobox transcription factor* necessary for implantation is turned off by RNA interference in the donor cell nucleus before it is transplanted to the enucleated egg.)

The motivation behind ANT was based on the thought that if the embryo could not grow to a viable fetus and if it did not involve the fusion of a sperm and egg, it was, de facto, not a human life and, therefore, when the ICM stem cells were harvested and the blastocyst destroyed, it would not be morally objectionable. However, religious critics of embryonic sources of stem cells do not see it that way. The diploid cell nucleus used is originally the product of a zygote formed by fusion of sperm and egg. However, so is a skin cell.

Another possible source of pluripotent stem cells is from a method called *oocyte assisted reprogramming* (OAR). OAR is a proposed technique in which *no* embryonic development takes place and no ICM needs to be harvested from a blastocyst, hopefully avoiding most religious criticism. In OAR, a diploid oocyte would be converted directly to a pluripotent stem cell. This will be accomplished in the following way.

It has been found that several TFs are necessary for establishing and propagating undifferentiated stem cells. Significantly, some of these TFs are only expressed in pluripotent stem cells from the ICM of the 7-day blastocyst; they are not expressed in zygotes. *Nanog* is one such *homeodomain transcription factor* [Mitsui et al. 2003]. Nanog is not expressed in oocytes or zygotes, but first becomes expressed weakly in the *morula,* then strongly in the ICM of the blastocysts. Chambers [2004] found that ESC identity is determined by the activity of cell-specific TFs such as Nanog and Oct-4, which need to be expressed at particular levels in order to function appropriately. It has been found that ESCs in which Nanog is blocked lose their pluripotency, clearly demonstrating that Nanog is a necessary factor in keeping ICM cells pluripotent. It has also been found that ESCs that constitutively express Nanog can no longer be differentiated; i.e., they are forced to remain in their undifferentiated state, which is desirable for the clonal propagation of undifferentiated ESCs [Yates & Chambers 2005].

The *altered nuclear transfer-oocyte assisted reprogramming* (ANT-OAR) process will combine the epigenetic reprogramming of a somatic (diploid) nucleus with forced expression of TFs characteristic of ESCs in order to produce a single pluripotent stem cell that can be propagated directly as a stem cell. In this process, Nanog or other similar TFs would be expressed at high levels in somatic cells prior to nuclear transfer. This procedure is intended to bias the somatic nucleus toward a pluripotent stem cell state. Such transformed nuclei would then be epigenetically reprogrammed by transplantation into enucleated oocytes. Now, when the diploid oocyte is triggered into mitosis, the product will be pluripotent stem cells with controlled genetic characteristics; the development will never pass into any form of embryonic development. A *parthenote* is not/cannot be produced. Whether this technique works remains to be seen. Because it avoids any phase of embryonic development, it may find some acceptance with the persistent religious critics of ESC research.

In 2004, Donald Landry and Howard Zucker of Columbia University presented four proposals to the President's Council on Bioethics on alternative sources of pluripotent stem cells (PSCs) [Council 2005]. We will examine their *second proposal* here, because it presents several controversial features. In this protocol, SCs will be removed from "dead" human embryos, in vitro—those nonviable embryos that are the surplus product of IVF. After IVF, a high percentage of the four- or eight-cell stage embryos undergo spontaneous "cleavage arrest," in which the embryonic cells stop dividing. The vast majority of these arrested embryos do not resume mitosis, never form blastocysts, and are incapable of implanting in a uterus. In most cases, spontaneous cleavage arrest is associated with genetic damage to the embryo's cells. It is observed that, although development has halted, some of the cells in the embryo appear to be "normal" *blastomeres*. In other words, although the embryo with arrested development is postulated to be organismically dead, some of its cells may not be. It is these cells that Zucker and Landry proposed to harvest and propagate as stem cells.

One problem with implementing ANT-OAR is that finding "normal" blastomeres in these "dead" embryos will be a process of trial and error and use a lot of resources. How can a normal blastomere best be found? What will be the criteria for "normality"? Will there be particular molecular markers on the normal cell surface that are absent on a moribund cell?

A second problem with the proposal is that a viable, normal single cell taken from an arrested four- or eight-cell embryo may be totipotent and, if cultured in vitro, could develop into a complete embryo capable of implantation. Still another problem lies with establishing criteria to determine the "death" of an embryo. Certainly, it will be more complex than just the cessation of cell division. Embryo "death" is in no way congruent with harvesting organs from a recently deceased patient. Death for an embryo where the cells are intact (have not yet lysed or undergone apoptosis) is due to unknown, complex changes in the biochemical command, communication, and control network among the cells that coordinate its development. Much work has yet to be done to understand this C^3 network and how to control it. Ding and Schultz [2004] give a list of 21 small molecules that they have found are active in regulating stem cell differentiation. How these molecules work remains to be elucidated.

It should be noted that a clone of embryonic stem cells (ESCs) can also be obtained from a *single*, undifferentiated, totipotent stem cell removed in vitro from an eight-cell *morula embryo.* Removal of one cell does not destroy the morula; the remaining seven cells can go on to form a viable blastocyst that can be implanted and mature into a healthy fetus. In fact, the one-cell removal is routinely done in certain IVF cases to test the genome of that cell for markers of genetically transmitted diseases. Recently, Klimanskaya et al. [2006] have reported in *Nature* that they have cultured pluripotent stem cells from a single cell taken from 8- to 10-cell human embryos from IVF surplus. In this technique, the ethical dilemma of killing a blastocyst embryo is avoided because the remaining cells, if implanted, evidently develop normally. These authors cultured two stable, undifferentiated hESC lines for 8 months that showed normal karyotypes and pluripotency markers, including Nanog, Oct-4, SSEA-3, SSEA-4, TRA-1-60, TRA-1-81, and alkaline phosphatase. This research may prove a less controversial source of hESCs, but religious critics already want assurance that the removal of only one cell will have no untoward effect on embryonic and fetal development, a reasonable concern.

As we have seen earlier, one source of SCs is to genetically cause a mature, diploid, differentiated (**2n**) body cell to unspecialize (or *dedifferentiate*) and, through several cell divisions, revert to a line of multi- or pluripotent, self-propagating stem cells. (This cell can be something as ubiquitous as an epithelial cell.) This is not the same as cloning because no diploid egg is made to divide into an embryo that might be implanted. Such research should find little opposition from most religious

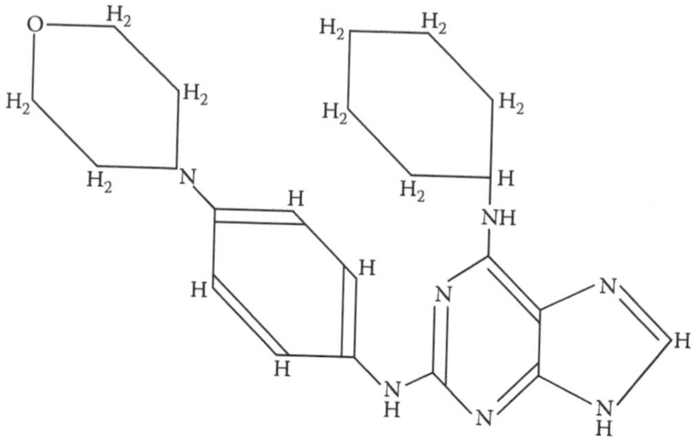

FIGURE 10.5 The *reversine* molecule. How does it work?

groups because an existing, adult, human cell line would be changed to SCs and then used for medical and research purposes for the betterment of humans.

Progress in the area of cellular dedifferentiation was reported by workers at the Scripps Research Institute in La Jolla, California. Chen et al. [2004] worked with cultures mouse C2C12 *myoblasts.* They screened a "library" of 50,000 small molecules [!] to see if any would cause the C2C12 myoblasts to lose their specialized structure and revert to a more general stem cell characteristic. Chen and coworkers found that *one* compound they called *reversine* [2-(4-morpholinoanilino)-6-cyclohexylamino-purine] was highly effective in causing the myoblasts to revert to multipotent stem cells. By treating these stem cells with selective differentiation inducers, they were able to redifferentiate the stem cells to establish new lines of fat cells (adipocytes) and bone cells (osteocytes). Figure 10.5 illustrates the *reversine* molecule. Not known, and of great interest, is the molecular biology of reversine. How does it work? Will it work on fully differentiated muscle cells? How about other adult tissue cells? Clearly, much research must be done to unlock the molecular mechanisms of the control of cellular differentiation and dedifferentiation.

Kim et al. [2004] pointed out that the C2C12 myoblasts could be redifferentiated into osteoblasts by incubating them with bone morphogenic proteins (e.g., BMP-2), and *transfection* of the C2C12 cells with a dominant-negative form of T-cell factor 4 (TCF4) converts them into adipocytes. So, perhaps the C2C12 cell line is more "potent" than Chen et al. [2004] thought.

Progress has recently been made in dedifferentiating plentiful, easily accessed adult diploid (**2n**) fibroblast (skin) cells using epigenetic manipulation of cellular genomic control mechanisms, including TFs, to cause the cells to become pluripotent stem cells (PSCs). This in vitro "reprogramming" of adult skin cells is just beginning to be understood and manipulated in mice [Wernig et al. 2007; Yamanaka 2007]. When perfected, the reprogramming of adult cells in humans will lead to life-saving regenerative medical techniques that are patient specific, without the moral dilemma of having to use a human oocyte or embryo, whatever its source and potential viability.

Jaenisch et al. at the Whitehead Institute at MIT, working with adult mouse skin (fibroblast) cells and using a retroviral vector, spliced three genes into the genomes of mouse fibroblasts. One gene was for resistance to the toxic antibiotic *neomycin.* It was placed next to genes for the TFs Oct-4 and Nanog. (The name *Nanog* comes from, *Tir na nog,* the mythological Celtic land of eternal youth. Nanog regulates early development and differentiation in ICM cells. The human Nanog protein is 52% homologous to mouse Nanog, and has 85% identity in its *homeodomain.* Both human and mouse Nanog contain Trp-rich repeats in which Trp-x-x-x is repeated 8 and 10 times, respectively. Human Nanog has 305 AAs [Nanog FLJ12581 2004], and its gene is found on chromosome 12p13.31.) Thus, if the genes for Nanog and Oct-4 are expressed in the fibroblasts, so is the gene for neomycin resistance. A neomycin bath was used to select only those fibroblasts expressing Nanog and Oct-4; all other skin cells were killed by the neomycin. About 1 out of 10^3 GM fibroblasts were selected.

Jaenisch's team next tested molecular markers on the transformed mouse skin cells versus those on pluripotent stem cells from normal mouse embryos. There were no differences.

The antepenultimate proof by Jaenisch et al. that the GM skin cells were indeed pluripotent stem cells was done by fluorescent labeling the reprogrammed cells, then injecting them into early-stage mouse embryos that were implanted, carried to term, and resulted in live mice. These mice were *chimeras* in that they consisted of the natural cells from the embryos plus the reprogrammed PSCs. Fluorescence indicated that the reprogrammed cells contributed to all tissue types in the F1 pups; blood, internal organs, hair color, etc.

The penultimate proof that the reprogrammed cells were stem cells came from breeding the F1 mice and finding lineages of the reprogrammed cells in the F2 generation, demonstrating that the GM skin cells had contributed to the F1 germ cell lines.

Jaenisch's ultimate proof that the GM skin cells were indeed reprogrammed into pluripotent stem cells involved making mouse embryos with four chromosome sets (**4n** = tetraploid). Implanted **4n** embryos can form a placenta but cannot develop into a full-term fetus. When the researchers injected the reprogrammed skin cells into the **4n** embryos and then implanted them, live, late-gestation fetuses could be recovered, created exclusively by the reprogrammed skin cells [Cameron 2007].

A slightly different approach to reprogramming mouse skin cells (taken from tail tips) to be pluripotent stem cells was used by Yamanaka et al. [Vogel 2006]. Through a process of elimination, they narrowed down a group of 24 TF genes expressed in mouse ES cells to a group of four TFs, **Oct-4, Sox-2, c-Myc, and Klf4,** that would give rise to skin cells reprogrammed to become pluripotent stem cells. A retrovirus vector was used to insert the four TF genes into the fibroblasts, in vitro. These TFs turned on many other regulatory genes used in embryonic development that are normally permanently turned off following embryonic differentiation. The GM fibroblasts were thus verifiably genetically reprogrammed into mouse PSCs [Okita, Ichisaka & Yamanaka 2007]. (Also see discussion in Chapter 11.)

It is now only a matter of time before researchers will learn the TF "gene mix" that will reprogram differentiated, adult, GM *human* cells into pluripotent stem cells. Then there will be no need to harvest pluripotent stem cells from embryos, thus destroying them. Nor will human eggs be needed. Once the problem of reprogramming adult human cells into pluripotent stem cells has a solution, researchers can concentrate on the TF signals that will cause these stem cells to redifferentiate into the desired tissue types for applications in regenerative medicine.

10.4.4 Stem Cell Applications in Regenerative Medicine

Current medical treatments using stem cells have been mostly limited to the use of adult multipotent stem cells. Adult *hematopoietic stem cells* are now routinely transplanted into bone marrow in cancer patients to treat the destructive effects of chemotherapy medicines and radiation treatments on blood cell formation. Hematopoietic SCs are also used to treat certain cancers such as leukemia and lymphoma and noncancerous aplastic anemia. Damaged corneas have been treated with adult epithelial SCs.

The future applications of ASCs and ESCs will depend on the ability of molecular biologists to unravel the complex sequence of molecular control signals that direct differentiation of SCs. Regenerative medicine using stem cells may permit the treatment of spinal cord injuries, brain injuries including dopamine-secreting neurons damaged in Parkinson's disease, neurons damaged by stroke, pancreatic beta cells, damaged heart muscle, certain cancers, liver damage, etc. By deriving SCs from adult diploid cells, the moral dilemma of tampering with early embryos, whatever their source, can be avoided. However, the easier path (molecular-biologically speaking) still appears to be to work with pluripotent stem cells (PSCs) taken from the inner cell mass of IVF embryos, a research area currently not government funded in the United States, solely on moral grounds (see Chapter 11).

Clearly, PSCs hold great promise for regenerative medicine. Their application lies in the future, however. For their effective use, developmental and molecular biologists must learn how to precisely control the differentiation and expansion of PSC clones. To regain U.S. government funding for this important branch of cell biology/medicine, they must also find less controversial sources of PSCs than IVF embryos. The research using mouse skin cells to generate PSCs shows promise in this regard.

Recently (2006), certain state legislatures (e.g., Connecticut) have voted funds to support in-state research on IVF embryonic stem cells, allowing researchers denied federal funding to continue this important research.

The failure of the U.S. government to support IVF stem cell research in the United States simply means that scientists in other countries, such as in China, South Korea, Japan, etc., may make the scientific breakthroughs and benefit most from them. The United States will end up having to buy the technology. The motivation to find and perfect stem cell–based regenerative medicine is international. All life is based on cells, and we are just beginning to understand the workings of cells and the complex genomic systems that regulate human embryonic development.

10.4.5 Cancer Stem Cells

10.4.5.1 Introduction

The generally agreed-on definition of a *cancer stem cell* (CSC) is a cell within a tumor that is able to expand clonally (self-renew), differentiate, and produce the heterogeneous lineages of cancer cells that comprise the tumor [Hermann et al. 2007].

We have seen how stem cells promise to have an enormous therapeutic potential in regenerative medicine, once cell and molecular biologists learn the details of the exogenous and endogenous molecular signal sequences that control their differentiation and proliferation. However, stem cells also have a dark side. There is an emerging body of evidence that suggests that many types of cancer are propagated by their own abnormal adult stem cells. A current flurry of research in molecular biology has been spurred by a growing body of evidence supporting the *cancer stem cell hypothesis* [Kucia et al. 2005; Wicha, Liu & Dontu 2006; Hermann et al. 2007].

Wicha et al. elaborated on the cancer stem cell hypothesis. They stated that tumors could originate in "tissue stem cells or their immediate progeny through dysregulation of the normally tightly regulated process of self-renewal [of normal cells]." They also stated that "tumors contain a cellular subcomponent that retains key stem cell properties. These properties include self-renewal, which drives tumorigenesis, and differentiation albeit aberrant, that contributes to cellular heterogeneity."

Three key questions about cancer stem cells that need to be answered concern where and how they arise:

1. Are they the result of a developmental aberration in a normal, differentiated cell?
2. Do they result from a mutation to normal adult stem cells whose role is to replace normal, differentiated cells?
3. Are they the end result of a mutation in the *niche cells* that regulate and nurture normal stem cells [Li & Neaves 2006]?

In any case, understanding the role of cancer stem cells in the growth and metastasis of a wide variety of cancers is currently a very active research topic. Cancer stem cells were first described in various human leukemias about 40 years ago [Fialkow et al. 1967]. They have also been found in a wide variety of human (and mouse) cancers, including glioblastoma (brain) [Sanai et al. 2005], breast, colon, lung, melanoma [Petrou 2006], pancreatic, prostate, and thyroid. Because the real culprits in cancer growth and metastasis appear to be "bad" stem cells, their selective elimination could prevent cancer regrowth following chemo or radiation therapies that destroy the more susceptible differentiated cancer cells [Clarke & Becker 2006]. The fraction of a cancer's cells that are classifiable as stem cells is generally very small. For example, about 1 in 10^3 cells in *malignant melanomas* are $CD20^+$ cancer stem cells [Petrou 2006]. These stem cells can easily migrate to and differentiate into melanoma cells in diverse body organs. Malignant melanoma is highly resistant to radiation and chemotherapy.

Cancer stem cells are often marked by the surface protein CD133 [Perkel 2007]. Human pancreatic cancer stem cells have the CD133 marker [Hermann et al. 2007], and it has also been shown that the CXCR4 receptor on stem cells (normal and CSCs) is a link in their proliferation and differentiation regulatory system [Kucia et al. 2005].

10.4.5.2 New Pathways for Cancer Therapy

A key specialized microenvironment for a normal adult stem cell is a group of cells surrounding it and in intimate contact with it called the *stem cell niche* [Clarke & Fuller 2006; Li and Xie 2005; Scadden 2006]. One function of the niche is to anchor stem cells. Both *N-cadherin* and other adhesion molecules, including *integrins*, play an important role in anchoring and signaling stem cells. Niche cells are thought to regulate a stem cell's state, including the regulatory pathways mediating the choice between differentiation and self-renewal [Arai et al. 2005]. Stem cells require paracrine regulatory signals from the surrounding niche cells in order to maintain their identity and self-renewal properties. Niche cells are one source of signaling molecules that activate stem cell signal transduction pathways controlling stem cell fate. They also receive feedback signals from the stem cells and their differentiation products [vetscite 2006].

A comprehensive review of known normal human stem cell niche structures and functions in hematopoietic tissue, epithelial (hair follicle) tissue, intestinal epithelial tissue, and neural (CNS) tissue is given by Li and Xie [2005]. Li and Xie also describe stem cell niches in mouse testis, *Drosophila* testis, and *C. elegans* gonads.

Li and Neaves [2006] cited the features common to *all* stem cell niches:

1. "[The niche is a] group of cells in a special location that…maintain[s] stem cells."
2. "[It] is a physical anchoring site for stem cells, [utilizing] adhesion molecules [*cadherins* and *catenins*]."
3. "[It] generates extrinsic [paracrine] factors that control stem cell number, proliferation [rate] and fate determination."
4. "[It] controls normal asymmetrical division of stem cells [at least in invertebrates]."

Li and Neaves go on to discuss the details of the control molecules cited earlier and how they act to regulate various kinds of adult stem cells. These authors proposed that another function for the stem cell niche was to prevent tumorigenesis by controlling stem cell proliferation. Conversely, they speculated that deregulation of the niche may lead to uncontrolled proliferation of stem cells, resulting in tumorigenesis and metastasis. They cited an example of this loss of feedback control in the "...mammary gland stem cell niche [leading] to abnormal expression of TFFα, resulting in the development of breast cancer."

Figure 10.6 illustrates schematically two putative scenarios of stem cell regulation: Figure 10.6A illustrates normal control of stem cells, where their proliferation is controlled by a negative feedback loop from stem cells to niche cells; Figure 10.6B shows how abnormal parametric signals might upset the normal feedback regulation of stem cell development, leading to possible oncogenesis.

The ultimate goal of understanding the origin and function of cancer stem cells is to develop cancer therapies that target cancer stem cell lines. The first step toward this goal involves mathematically

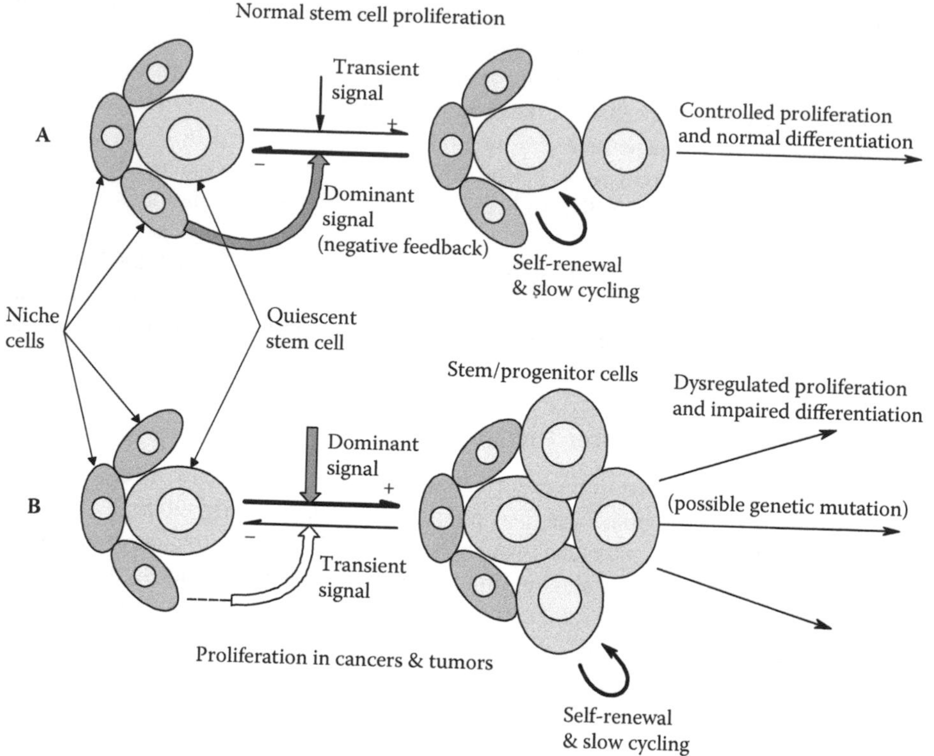

FIGURE 10.6 (See color insert.) Adult stem cells (ASCs) nurtured by niche cells. **(A)** Normal stem cell proliferation: A putative negative feedback pathway between the niche cells and the ASC downregulates its clonal expansion. A transient signal input causes limited self-renewal. Other input signals cause normal, controlled ASC differentiation. **(B)** Oncogenesis: In one scenario for tumor growth, the niche-ASC negative feedback pathway is interrupted, allowing rapid proliferation of an oncogenic stem cell. In addition, there are dysfunctional regulatory pathways that permit proliferation of differentiated cancer cells. Does the initial oncogenic mutation affect the niche cells, the ASC, or the progenitor cells?

modeling the population dynamics of normal stem cells and their differentiated product cells, coexisting cancer stem cells and their differentiated (cancer) cells, and how these cells interact.

Dingli and Michor [2006] have published an interesting paper in which they present a simple four-state nonlinear population-dynamic mathematical model for cell proliferation in a stem cell–based cancer. Their model describes the two layers of the differentiation hierarchy of the hematopoietic system in bone marrow (we have taken the liberty to change their notation for clarity).

Let n_s and n_d represent the populations of normal stem cells and differentiated cells, respectively. Also, let c_s and c_d represent the population of cancer stem cells and differentiated cancer cells, respectively. Normal stem cells are assumed to reproduce (divide) at a rate r_n cells/day and die at a rate d_o cells/day. Normal differentiated cells are produced from normal stem cells at a rate a_n cells/day, and die at a rate of d_1 cells/day. Cancer stem cells divide at a rate r_c cells/day and die at a rate d_o cells/day. Differentiated (mature) cancer cells arise from cancer stem cells at a rate a_c cells/day and die at a rate d_1 cells/day. Dingli and Michor write their system's four state equations as follows:

$$\dot{n}_s = [r_n\phi - d_o]n_s \tag{10.1}$$

$$\dot{n}_d = a_n n_s - d_1 n_d \tag{10.2}$$

$$\dot{c}_s = [r_c \quad - d_o]c_s \tag{10.3}$$

$$\dot{c}_d = a_c c_s - d_1 c_d \tag{10.4}$$

where the hyperbolic functions φ and ϕ introduce competition within the stem cell compartment. φ and ϕ are defined as

$$\phi = \frac{1}{1 + k_n(n_s + c_s)} \tag{10.5}$$

$$\varphi = \frac{1}{1 + k_c(n_s + c_s)} \tag{10.6}$$

The constants k_n and k_c are dimensionless constants "... used to simulate the crowding effect that is seen in the bone marrow microenvironment."

The authors used the following initial conditions and parameter values:

- The normal (steady state = SS) number of hematopoietic stem cells $n_s = 2 \times 10^4$ in the bone marrow compartment.
- The SS number of normal differentiated cells $n_d = 10^{12}$, both in the absence of cancer cells. $r_n = 0.005$/day, $d_o = 0.002$/day.
- Homeostasis is achieved by setting $k_n = 0.75 \times 10^{-4}$. Normal differentiated cells n_d are produced by normal stem cells n_s at a rate $a_n = 1.065 \times 10^7$/day and die at a rate $d_1 = 1$/day. Tumor stem cells can have a selective advantage compared to normal stem cells. They might have a larger growth rate ($r_c > r_n$) or might be less sensitive to environmental crowding ($k_n < k_c$).

Dingli and Michor's simulations covered a number of therapeutic scenarios. They determined that therapies directed only at mature tumor cells will fail. "Therapy that only increases apoptosis in the mature tumor cells… will ultimately fail due to the continuous amplification that occurs in the bone marrow." They reached a similar conclusion regarding therapies aimed at reducing the rate of generation of mature cells. They concluded: "*Therefore any treatment that is designed to eradicate tumor must target the tumor stem cells*" [our italic].

Dingli and Michor determined that therapies directed at tumor stem cell clones are far more promising. They stated: "A therapeutic agent designed to selectively inhibit the replication of tumor stem cells (e.g. interferon) can in principle lead to tumor eradication. In the best-case scenario, tumor cell proliferation is completely inhibited, $r_c = 0$, and the tumor stem cell pool decreases exponentially at a rate $\exp(-d_o t)$." They warned that agents such as interferon must be administered over a prolonged period of time. This is because the presence of the tumor cell clone was viewed to increase the risk of additional mutations and, therefore, more aggressive disease.

These authors concluded that the stem cell killer, when developed, should be highly specific. Also, an agent to attack the differentiated tumor cells, c_d, should be given simultaneously with the c_s killer (i.e., bimodal therapy). Note that the mathematical model given is extremely simple, yet highly instructive in its results. It can be elaborated on by including the pharmacokinetics of the therapeutic agents and the dynamics for specific known signal substances and their receptors that figure in stem cell proliferation and differentiation.

Another more complex mathematical model of cancer stem cell proliferation in the CNS was developed by Ganguly and Puri [2006]. Figure 10.7 illustrates the system architecture used in their simulations. Their model includes populations of one type of normal stem cell (SC), a population of

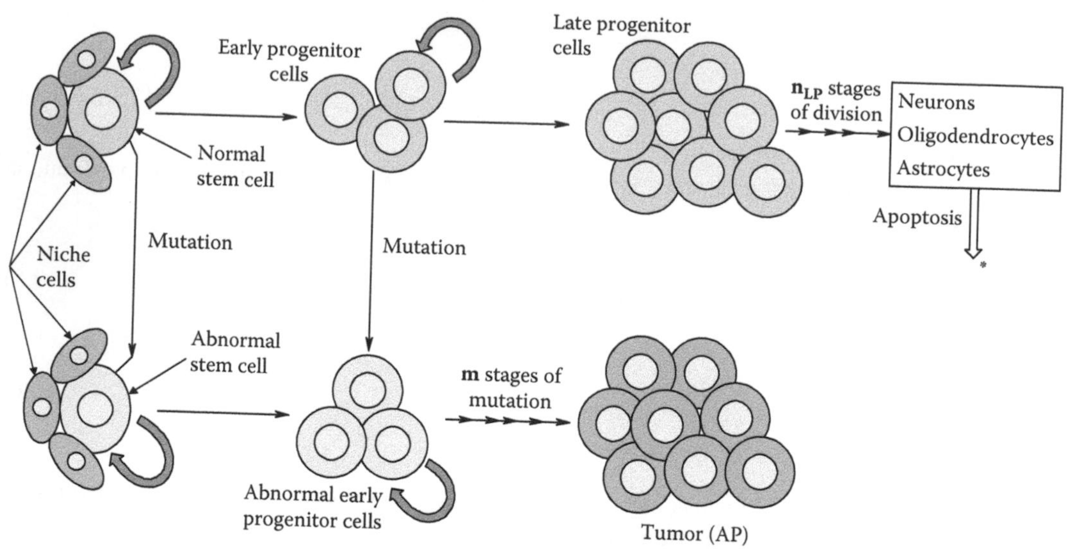

FIGURE 10.7 (See color insert.) System architecture for a complex mathematical model describing oncogenesis in the CNS from cancer stem cells proposed by Ganguly and Puri [2006]. Populations of both normal and oncogenic stem cells are shown, as well as populations of normal and abnormal early progenitor cells, late (normal) progenitor cells, and their differentiated (normal) product cells (neurons, oligodendrocytes, and astrocytes). Also, *m* populations of stages of mutation of the early abnormal progenitor cells, and finally, a population of differentiated tumor cells are shown. (The red feedback arrows represent self-renewal [growth] probabilities for the stem and progenitor cell populations.)

early progenitor (EP) cells derived from the SCs, and also a population of late progenitor (LP) cells derived from the EPs. The LPs differentiate into *neurons, oligodendrocytes*, and *astrocytes* that eventually undergo apoptosis. There is also a population of abnormal stem cells (SC_A) derived from mutations in the normal stem cells. The SC_As develop into abnormal early progenitor cells (EP_A), and thence by *m* stages of mutation form a clone of abnormal (tumor) progeny cells (AP).

The Ganguly and Puri model is too complex (over seven states) to describe in detail here. However, their conclusion is very interesting. In noncancer circumstances, the model converges to a steady population of cells. However, an "oncogenic event" in stem or early progenitor cells leads not only to the production of abnormal cells, but to a more rapid growth of those abnormal cells. Further, an injury to mature cells can increase the proliferation rates of stem cells and EP cells (including abnormal ones). This does not always lead to cancer as long as the frequency of damage is low enough to allow the restoration of a steady state between incidents.

The usefulness of dynamic models such as that of Dingli and Michor and of Ganguly and Puri is that they let oncologists explore novel therapies quantitatively, in silico, before resources are committed to animal trials and eventual clinical studies. Once the biochemical details of the regulatory signaling pathways are known, specific, theoretical immunotherapies can be developed and tested by computer, before "going wet."

Some of the known, major genetic pathways involved in cancer stem cell regulation:

- **Bmi-1** (aka BMI1): This is a common oncogene activated in lymphoma and shown to specifically regulate hematopoietic and neural stem cells. Bmi-1 is a *Polycomb group* transcriptional repressor. Bmi-1 expression is increased in human medulloblastoma, directly linking the deregulation of Bmi-1-dependent self-renewal of stem cells to tumor development [Vescovi et al. 2006].
- **Notch:** The Notch pathway has a regulatory role in the proliferation of hematopoietic, neural, and mammary stem cells. Excessive expression of Notch receptors and their ligands,

delta-like-1 and *jagged-1,* correlate with the proliferative capacity of human glioma cells. Thus, Notch appears to function as a positive effector of self-renewal in adult neural stem cells [Vescovi et al. 2006; Chiba 2006].

- **Sonic hedgehog and Wnt:** Initially described as embryonic developmental pathways, they are also implicated in adult stem cell regulation. Both *Wnt* and *Sonic hedgehog* pathways are hyperactivated in tumors and are required to sustain tumor growth, either together or singly [Zhou & Hung 2005; Reya & Clevers 2005]. It has been observed that the expression pattern of Wnt ligands in malignant melanoma correlates with its histopathological features and may provide a basis for the molecular classification for this type of cancer [Pham et al. 2003].
- **PTEN** is a tumor suppressor controlling the proliferation of neural stem cells and progenitor cells in vivo and in vitro. PTEN is often inactivated in glioblastomas, and maintenance of its expression is closely associated with a favorable prognosis in the treatment of human *glioblastoma multiforme.*

10.4.5.3 Summary

Manipulation of adult stem cell and niche cell regulatory pathways promises to lead to improved therapies for cancer. Cancer stem cells are apoptosis resistant, and current radiation and chemotherapeutic modalities can kill the more susceptible, differentiated cancer cells but have little effect on the cancer stem cells. Effective destruction of cancer stem cells may depend on the molecular biological manipulation of the niche cells surrounding them and the use of selected TFs to target the stem cells. Malignant melanoma is particularly resistant to radiation and conventional chemotherapy once it metastasizes; clearly, the ability to target its wandering cancer stem cells will be required to fight it effectively.

10.5 PARTHENOGENESIS

10.5.1 Introduction

The term *parthenogenesis* comes from the Greek words for *virgin* and *birth*, i.e., the production of viable young without the introduction of male semen. In parthenogenesis (PG), a female animal becomes pregnant and produces one or more viable offspring; it is one form of asexual reproduction. *Natural parthenogenesis* occurs in about 300 species of plants spanning 35 different families, in many invertebrates (e.g., water fleas, aphids, and bees) and in several vertebrates including lizards, salamanders, fish, and even turkeys. In the fruit fly, *Drosophila mangabeirai,* one of the *polar bodies* acts as a sperm and "fertilizes" the oocyte following the second meiotic division. The germ cells of the grasshopper, *Pycnosceleus surinamensis,* dispense with meiosis entirely, directly forming diploid ova by two mitotic divisions. All such progeny are genetically female. In the hymenopterous insects (bees, ants, and wasps), unfertilized haploid eggs develop into males, whereas fertilized eggs (from the queen), being diploid, develop into females. The haploid males are able to produce sperm by eliminating the first meiotic division, thereby forming two sperm cells through second meiosis [Gilbert 2003]. About 15 species of the whiptail lizard, *Cnemidophorus,* reproduce exclusively by PG. In this PG, the embryos (and adults) are also genetically female.

10.5.2 Artificial Parthenogenesis

Artificial parthenogenesis has been induced in vitro using normal unfertilized, haploid mammalian eggs. Diploid human, monkey, and mouse eggs have been stimulated to divide and grow. The 46(XX) eggs are subjected to a chemical or electric shock to stimulate parthenogenetic division. In early experiments, mouse eggs were placed in a culture medium that artificially activates the oocyte while suppressing the formation of the second polar body. Thus, diploid mouse eggs are formed containing the genome of the egg alone. These XX diploid cells are then induced to divide. If, at the blastocyst stage, they are implanted in surrogate mouse mothers, they go on to form diploid differentiated embryos with spinal cords, muscles, skeletons, beating heart, and organs. Halfway through the mouse's gestation (day 11 or 12 postegg stimulation), the parthenogenetic embryos deteriorate and die in utero. Clearly, something inhibits normal fetal development.

Recently, Kono et al. [2004] stunned the world of developmental biology with their report that they had successfully produced full-term mouse *parthenotes*, one of which was reared into adulthood, bred, and bore live young. Kono et al. made a total of 598 diploid oocytes by serial nuclear transfer from oocytes taken from newborn mice. That is, they added, in vitro, the *pronucleus* from another egg to each egg to make it diploid.

Kono et al. reported that "...78% of the oocytes resulted in diploid, one-cell parthenogenetic embryos that formed two second polar bodies and pronuclei." Of the 457 zygotes cultured in vitro, 91.2% reached the morula/blastocyst stage. 371 parthenogenetic morula/blastocysts were then implanted in 26 surrogate mother mice. Of these, 24 grew viable fetuses. A total of 10 live and 18 dead pups were extracted surgically after 19.5 days of gestation. Of these, only two pups showed apparently normal neonate morphology and survived. One was sacrificed for gene expression analysis; the larger pup was nursed by a foster mother and grew into adulthood, was eventually bred, and had viable pups herself. Assuming the sacrificed live pup would have lived to adulthood, too, this represents a survival rate of 0.33% pups/diploid oocyte created.

What Kono et al. did to improve the survival rate of their mouse parthenotes was to ensure expression of the **Igf2** gene, which encodes the growth-promoting factor, IGF-II, a well-known regulator of fetal growth. Monoallelic expression of the **H19** gene active in development causes the neutral genome to express Igf2 protein.

Kono et al. remarked: "The most fascinating riddle arising from these results is how appropriate expression of the *Igf2* and *H19* genes caused the modification of a wide range of genes and normal development in the $ng^{H19\Delta13}/fg^{wt}$ parthenotes." It is unclear exactly how the process works, although Kono et al. speculated that it could involve altered signaling. Their research demonstrated that "... normal development in mice is subject to a rigorous 'conflict' and differences due to imprinting of maternal and paternal genomes."

A major goal of research on mammalian PG has been to grow "normal" parthenote blastocysts in order to harvest stem cells from the inner cell mass to be used for experimental *regenerative medicine*. Kono's motivation appears to have been to understand why mature mammalian parthenote development has heretofore failed. If human partheno-genetic blastocysts are to have viable stem cells harvested from them, it is mandatory that they have normal imprinting on their genomes. (See, for example, the review on imprinting by Swales and Spears [2005].) The secret to imposing this normality appears to be controlling the known (and as yet unknown) genetic factors that sequence normal early embryonic development.

10.5.3 Human Parthenotes?

Growing a human parthenote blastocyst has been more challenging than for mice. It is also far more controversial. Even though no sperm is involved, in one form of artificial parthenogenesis two (haploid) human ova are required (to make a diploid zygote), which is objectionable to certain prolife religious groups (the fact that human parthenote blastocysts grown in vitro are unable to survive as implanted embryos or fetuses and no one intends to implant them appears irrelevant to some objectors). Two haploid eggs are required, but women are born with about 10^6 eggs. Each year following menarche, about 20,000 eggs die in the ovaries, and about 12 mature and ovulate [Kiessling 2005]. Why not use a few thousand ova (obtained by informed consent from women undergoing sterilization, or from deceased donors) for the medically and morally good purpose of parthenote stem cell research [Fangerau 2005]?

Besides growing parthenote blastocysts for their stem cells, blastocyst cells will also allow researchers to study the cellular mechanisms of certain genetic diseases, especially those involving imprinting errors [Swales & Spears 2005]. Stem cell research based on the 46(XX) cells taken from the inner cell mass of parthenote blastocysts grown in vitro can eventually be used to treat women for various diseases. To be used in applications in regenerative medicine on male humans, the stem cells need to be 46(XY) (i.e., **2n**). Such stem cells require the haploid male genome (e.g., from a 23(Y) sperm) in the creation of the blastocyst. The fusion of sperm and egg genomes to form a 46(XY) blastocyst just to harvest stem cells is morally more objectionable than is the creation of a *nonviable*, 46(XX) parthenote. 46(XX) and 46(XY) blastocysts formed in IVF (sperm + egg) are routinely used in creating human pregnancies when there is a fertility problem. That is, they can grow to a normal fetus when implanted.

Attempts have been made to grow human parthenote blastocysts by Karl Swann of the University of Wales College of Medicine, United Kingdom; by David Wininger at Wake Forest School of Medicine in North Carolina [Westphal 2003], and by Paul de Sousa at the Roslin Institute in the United Kingdom [LifeSite 2005]. In 2005, de Sousa announced that he had succeeded in growing human parthenote embryos. de Sousa and coworkers took about 300 human eggs from volunteer donors to create six parthenote blastocysts (a 2% success rate). de Sousa used electric shock to cause the 46(XX) eggs to divide. The 2% that reached blastocyst stage had about half the normal number of cells, however. In 2004, Swann et al. used a protein enzyme produced by sperm, *phospholipase C-zeta* (PLC-ζ), to trigger human diploid oocytes to divide for 4 or 5 days until they reached 50–100 cell blastocysts. The parthenote embryos appeared to undergo the same early changes as naturally fertilized eggs, producing waves of Ca^{++} across the cell membrane every 20 to 30 minutes. Swann's goal was to harvest viable stem cells from the ICM of the blastocysts.

Gynogenesis (GG) is a form of asexual reproduction related to PG. In GG, the egg is stimulated into dividing by contact with a sperm. However, the sperm's genome does not enter the egg. The diploid embryo is again female (46(XX)). The sperm is generally from a closely related sexual species that has mated with the female egg bearer. GG occurs in salamanders of the genus *Ambystoma*. It is thought that as a very rare event, actual sexual mating does occur, introducing new material into the salamander's gene pool.

Another rare, bizarre developmental anomaly in mammals is the *hydatiform mole*. (This is a uterine growth, not the mammal.) A so-called *complete hydatiform mole* arises naturally when *two* sperms enter an egg in which the female pronucleus is absent (an "empty egg"). After entering the egg, the sperm chromosomes duplicate themselves, restoring the diploid chromosome number (46(YY)). The diploid zygote now divides, but development is abnormal. Only a placenta-like mass of cells grows in the uterus, and no embryo. Ninety percent of complete hydatiform moles are 46(XX), and 10% are 46(XY).

Recently, there has been interest in in vitro, parthenogenetic zygote development as a source of stem cells. We know that parthenogenetic embryos are generally not viable in mammals. However, viable pluripotent stem cells can theoretically be harvested from the blastocyst stage of in vitro PG zygote growth. Such lines of PG stem cells have been cultured from monkey cells for over 2 years by Vrana [2003].

Thus, we see that, so far, it is difficult to grow human PG blastocysts in vitro. The research remains to determine whether the inner mass cells taken from those that do survive are useful as stem cells. Some researchers in this area had thought that perhaps growing parthenotes for their stem cells would avoid objections from prolife groups; the eggs are donated, no sperm is involved, and the embryos are not viable. However, work on parthenogenetic sources of stem cells has been criticized by certain prolife groups who feel that any use of ova and PG blastocysts (even though they cannot become viable human embryos) is objectionable on moral and religious grounds.

As we have seen in the preceding sections of this chapter, there are many possible sources for pluripotent stem cells and potential therapeutic applications for them.

10.6 CHIMERAS

10.6.1 Introduction

The original *chimera* was a figure in Greek mythology; a fire-breathing she-monster with the head of a lion, the body of a goat, and the hind quarters of a dragon or serpent. It was probably used by parents as an example of what would carry their children off if they didn't play quietly or go to sleep. Modern chimeras are animals or plants with two or more distinct genotypes of cells in their bodies. Do not confuse chimeras with *hybrids,* in which all body cells have a single genotype which is the result of the fusion of the gametes from two different species. The mule is a primary example of a hybrid, being the result of the fusion of a donkey's sperm with the egg of a horse mare. (The mare gives birth to the mule.) Modern chimeras can be divided broadly into two subclasses: *cellular chimeras* and *somatic chimeras.* All chimeras are *genetic mosaics.*

10.6.2 Cellular Chimeras

Cellular chimeras, also called *mosaics,* are created when two simultaneously developing fraternal twin blastomeres fuse to form a single blastomere, which then contains a mixture of two cell genotypes. This mosaic blastomere develops normally through morula, blastocyst, and fetal stages. The animal born has two distinct genomes; for example, its blood cells can have one genotype, and its somatic tissues the other. Cellular chimeras in mammals can also result from *fraternal twins* sharing the same placental blood supply. Thus, there is cross-mixing of *hemopoeitic stem cells in utero,* so the twins will each have a mosaic of two different genotypes of blood cells. Other cells are normal.

Note that *all* female mammals are mosaic for the genes on the X chromosome because of the inactivation (imprinting) of one or the other of the X chromosomes in all of their somatic cells. Thus, all normal females have roughly equal populations of two genetically different cell types (with respect to their X chromosomes only). In approximately half their cell types the paternal X has been inactivated, and in the other half the maternal X has been turned off. A classic example of X chromosome mosaicism can be seen in female tortoiseshell (mottled orange and black) cats. The gene encoding orange coat color is X-linked. Black color is either encoded by either a codominant allele on the X chromosome or perhaps an *autosomal gene* that is masked by the orange gene. Owing to the randomness of X chromosome inactivation in females, in black fur patches, the X chromosome bearing the orange allele has been inactivated and the X chromosome carrying the nonorange allele is active. In patches of orange fur, precisely the opposite X inactivation has occurred. Sometimes we see distinct black and orange patches as well as regions of mixed black and orange fur in female tortoiseshell cats. Note that male cats are either black or orange, never tortoiseshell.

Anyone who has received a whole blood transfusion is probably a *blood cell chimera*. They may have received multipotent hemopoeitic stem cells from the donor and, if any survived and multiplied in their bone marrow, they may have some genetically different white and red cells circulating in their blood.

An example of a *human cellular chimera* was given by Kimball [2003e]. He cited a case study reported in the *New England Journal of Medicine* [Yu et al. 2005, 16 May] in which a woman who needed a kidney transplant was typed for the HLA antigens on chromosome 6. Kimball described the results:

- "Tissue typing which is done with white blood cells, showed her to have inherited the '1' HLA region of her father (who was 1,2) and the '3' region of her mother (who was 3,4).
- She had two brothers, one who inherited 1 from their father and 3 from their mother; the other who inherited 2 from their father and 3 from their mother.
- Her husband typed 5,6.

- Of her three sons: One was 1,6 which was to be expected, but the other two were both 2,5. The 5 they got from their mother, but where did the 2 come from?
- The first thought was that she could not have been their mother, but clearly she knew better. (Paternity may sometimes be in doubt, but not maternity.)
- A clue came from typing [her] other tissues. DNA analysis of her skin cells, hair follicles, thyroid cells, bladder cells and cells scraped from inside her mouth revealed not only [HLA] 1 and 3, but also 2 and 4. It is not clear why her bone marrow was an exception- containing only 1,3 stem cells."

Kimball goes on to explain how these seemingly paradoxical results arose:

> Her mother had simultaneously ovulated two eggs: One containing a chromosome 6 with HLA 3; the other with HLA 4.
> Her father would, of course have produced equal numbers of 1-containing and 2-containing sperm.
> A 1-sperm fertilized the 3-egg.
> A 2-sperm fertilized the 4-egg.
> Soon thereafter the resulting early embryos *fused into a single embryo.*
> As this embryo developed into a fetus, both types of cells [1,3 and 2,4] participated in constructing her various organs including her oogonia (but not, apparently the blood stem cells in her bone marrow [1,3]).
> Although she was a [cellular] mosaic for the HLA (and other) genes on chromosome 6, all her cells were XX. So both her father's successful sperm had carried his X chromosome.

Kimball continues:

> However tetraparental humans have been found that were mosaic for sex chromosomes as well; that is, some of their cells were XX; the other XY. In some cases this mosaic pattern results in a hermaphrodite- a person with a mixture of male and female sex organs. [Quoted with the kind permission of John Kimball.]

Kimball points out that the woman has all four sets of transplantation antigens and thus could accept a kidney from any one of her brothers as well as her mother (her father was dead) without fear of rejection. Subsequent laboratory tests confirmed that she was unable to generate T-cells reactive against the cells of either her brothers or mother.

10.6.3 Somatic Chimeras

Synthetic *somatic chimeras* have led to some bizarre results. Once again, this is a controversial area of genetic research, particularly because chimera research has focused on growing human organs (in the sense that they are tissue compatible with humans) in large animals such as pigs and sheep and also human stem cells in rabbit-human hybrids. At the University of Nevada in Reno, researchers added human stem cells to sheep fetuses to try to create human-compatible livers that can be used in liver transplants to humans. This appears to be a "shotgun" approach to science. The sheep livers were reported to be about 80% human cells, not pure enough to avoid host immune rejection. Also, the sheep were found to have human cells in their hearts and brains. These researchers have a long way to go to control in vivo stem cell differentiation.

At the Mayo Clinic in Minnesota, researchers are examining how chimeric pigs can be used to grow "human" organs for transplant, as well as blood cells. They have developed chimeras that have a mixture of pig, human, and pig/human blood cells. They are concerned about the risk of transmitting pig viral diseases to potential human organ recipients (see Chapter 11).

Irving Weissman and coworkers at Stanford University have injected human neural stem cells into mouse brains. The result has been mouse brains with about 1% human neurons. Their goal is to make mice with 100% human neurons in order to have an animal model to better understand the development of Parkinson's, Alzheimer's, and Lou Gehrig's diseases. Weissman et al. have already bioengineered mouse/human chimeras having nearly human immune systems that were used to screen new drugs for human HIV/AIDS.

Monkey–human brain chimeras are being experimented with at the St. Kitts Biomedical Foundation in the Caribbean. Immature human brain cells are transplanted into the brains of vervet monkeys. Their goal, too, is to develop an animal model to explore treatments for Parkinson's disease. The cells are injected into the dopamine producing regions of the monkeys' brains to see whether the cells will grow and increase dopamine production [Bioethics 2005].

Note that a mouse or monkey with genetically human neurons in its brain does not in any way have a human brain. The architectures for the mouse and monkey brain are inherent in their own developmental control genes, not in the human neural stem cells used. On the other hand, chimpanzees have a 98% genomic homology with humans and have brains that give them the mental abilities and consciousness of a 4-year-old human (but without speech). Making a "humanzee" somatic chimera could produce a creature so human that it would certainly create ethical, moral, and religious dilemmas (see Chapter 11).

Somatic chimeras have been created artificially by humans for research purposes in developmental biology and medicine. Here, the four- or eight-cell blastomeres of two species are surgically fused in vitro, then implanted into the uterus of a foster mother of one of the species to undergo fetal development. A classic example of this technique is the "geep," a sheep-goat chimera. This animal had big patches of skin on its front and rear legs covered with wool (clearly the sheep contribution). The head and remainder of the body is morphologically (and genetically) goat. Of course, the cells in the geep would be genomic mosaics as well. That is, the sheep skin cells would be genomically sheep, etc.

Somatic mouse chimeras have also been created. In one example cited by Gilbert [2003], three four-cell blastomeres were fused and then implanted in a host mother. The live-born female chimera had a white, brown, and black calico coat, reflecting the solid coat-color contributions from the three blastomere donors. When this chimeric mouse was bred with a white male, she had offspring that were all white, brown, or black, reflecting her mixed genomes.

There might be some practical or economic applications for somatic domestic animal chimeras in farming, but fortunately that end does not yet seem to justify the means. Another possible use for cellular chimeras would be for nonhuman hosts to grow differentiated human stem cells, that is, hemopoeitic, neural, liver, pancreatic beta cells, etc. The key to all these goals lies in understanding the molecular biology of developmental C^3.

10.7 SUMMARY

In this chapter we have considered some of the controversial aspects of "genetic engineering," that is, genetically modified organisms (GMOs), how they are made, and their applications. The problems and potential benefits of transgenic plants including tomatoes, rice, soybeans, corn, cotton, and rape (canola) were considered. Transgenic animals were shown to have potential benefits in the production of proteins, antibodies, medicines, etc., especially in milk. The greater problem of raising GM animals for food and having them mix their genetically engineered genomes with those of wild animals was illustrated in the case of the TG salmon.

We have also reviewed the techniques used in recombinant DNA technology by which genetic engineers can insert "designer DNA" into a plant or animal's genome. These included plasmids, phages, the "gene gun," and electroporation with fast dc pulses.

The kinds and sources of stem cells were described, and their potential uses in regenerative medicine were reviewed. The role of mutant stem cells in the proliferation of cancers was discussed, and mathematical models for tumor proliferation based on cancer stem cells were described. It was argued that a successful cancer therapy must kill not only the cancerous tissues, but also the cancer stem cells giving rise to it.

Chimeras, cloning, and parthenotes and their controversial and noncontroversial sources and applications were also considered.

CHAPTER 10 HOME PROBLEMS

10.1 Describe how insects can become resistant to an insecticide during chronic application of an insecticide on a crop they eat (a classic example involves DDT). Contrast acquired insecticide resistance in GM agriculture with the evolution of "super bacteria" in hospitals (e.g., MRSA) associated with chronic exposure to antibiotics.

10.2 Why does repetitively growing a GM crop containing a systemic insecticide such as Bt lead to the development of insect pests resistant to that insecticide? Contrast the long-term effectiveness of insect protection of the GM crop with that for spraying the same non-GM crop at selected intervals in the pest's life cycle with a nonpersistent insecticide.

10.3 Describe how intracellularly injected antisense mRNA can inhibit gene expression.

10.4 A whole blood transfusion from a donor contains some of the donor's hematopoietic stem cells, which set up residence in the recipient's bone marrow. The recipient has his STR "signatures" read from both his mouth epithelial cells and his blood. They are different. Explain how this apparent paradox has happened.

10.5 Discuss putative C^3 signals between niche cells and stem cells.

10.6 Use Matlab/Simulink® or Simnon® to simulate the Dingli and Michor [2006] cancer stem cell model given in the text by Equations 10.1–10.6. Use the parameters in the text to begin with. By manipulating parameters, demonstrate Dingli and Michor's conclusion: "Therefore any treatment that is designed to eradicate tumor *must* target the tumor stem cells."

10.7 In one approach to animal cloning, the haploid nucleus of a mature egg is removed and is replaced with the diploid nucleus of another animal's somatic cell. There is limited success in achieving live birth following implantation of the cloned blastocyst (cf Section 10.3.5). Discuss possible reasons for this low rate of birth of live viable fetuses.

10.8 How many animal species to date have been cloned successfully? (Interpret "success" as a healthy animal with a normal life span that has been bred successfully.)

11 Ethical Issues in Genetic Engineering

11.1 INTRODUCTION

When there is a trade-off between costs and benefits, there will always be a debate (often heated) about how to proceed. One source of this controversy is that there is a lack of universal agreement on what is a cost and what is a benefit for a given issue. Individuals assign different weights to perceived costs and benefits. Ethical debate in the fields of molecular biology and genetics is related to diverse issues such as novelty, the unknown, religious beliefs, and nascent technical abilities. Individuals considering genetic engineering and related fields would be well-served to develop a clear ethical framework for decision making.

11.2 FRAMEWORK FOR ETHICAL DECISION MAKING

In considering ethical issues in the biosciences, it is helpful to have a theoretical framework for addressing ethics. Although it is beyond the scope of this textbook to offer a full treatment of ethical theory (entire courses are taught on this), we present summaries of several ethical models. The models discussed in the following text offer several steps to guide decision making. Note that these models do not end with the initial decision but include a step to follow up on the outcomes of one's decisions. In other words, these models indicate that ethical persons have a responsibility for assessing the results of their choices. A major difficulty in making ethical choices in genetic engineering is that we cannot always foresee all consequences. "The law of unintended consequences" encapsulates the idea that every action has unintended as well as intended outcomes, sometimes positive, sometimes harmful. For this reason, follow-up is especially important.

11.2.1 Defining Ethics

The Markkula Center for Applied Ethics at Santa Clara University has published the excellent "*Framework for Thinking Ethically*" on their Web site at http://www.scu.edu/ethics/practicing/decision/ framework.html. The authors define ethics as "standards of behavior that tell us how human beings ought to act in the many situations in which they find themselves" [Velasquez, et al. 2005].

Helpfully, the authors also identify what ethics is not:

1. *Feelings*. However, feelings can provide an important clue to the correct ethical choices.
2. *Religion*. Although most religions prescribe ethical standards, the two are not synonymous. There are many people who are not religious but who practice and are subject to ethics.
3. *Following the law*. Although most laws are based on ethical principles, corruption can introduce unethical laws into a system of government.

4. *Following culturally accepted norms.* Not all cultural norms are ethically sound (for example, slavery, which was once an accepted norm in the United States, is now recognized by those in mainstream American culture to be unethical).
5. *Science.* The authors state, "Science may provide an explanation for what humans are like. But ethics provides reasons for how humans ought to act" [Velasquez et al. 2005].

The Josephson Institute of Ethics defines ethics as "principles that define behavior as right, good and proper. Such principles do not always dictate a single 'moral' course of action, but provide a means of evaluating and deciding among competing options" [Josephson 2002].

11.2.2 Ethical Frameworks

When faced with different options, one must first recognize that there is an ethical issue at stake. It is important to gather complete and accurate information about a situation in order to make an informed decision. Once the decision has been evaluated from several perspectives and has been made and implemented, it is important to review the consequences of the decision.
Velasquez et al. [2005] list five well-known ethical perspectives or approaches:

- *Utilitarian approach:* This approach dates back to the ancient Greeks and is "the ethical doctrine that the moral worth of an action is solely determined by its contribution to overall utility" [Velasquez et al. 1989]. In other words, the Utilitarian approach requires the practitioner to ask which option will yield the greatest good for the greatest number. It implies that the needs of the many outweigh the needs of the few, but it also requires that the ethical option minimize harm. Some forms of utilitarianism, however, hold that the *end justifies the means.*
- *Rights approach:* It implies that the object of one's actions has fundamental and absolute ("inalienable") rights. The Rights approach is process oriented, concerned with respecting the dignity and rights of others, regardless of the utilitarian outcome. The United States *Bill of Rights* is an example of a well-known document written from the Rights approach.
- *Fairness or justice approach:* This approach also dates back to the ancient Greeks and seeks to impart a fair or equal outcome for all parties. The ethical action is the one that "treats people equally—or, if unequally, that treats people proportionately and fairly" [Velasquez et al. 2005]. For example, this approach would advocate for equal pay for equal work or for better pay for the most productive workers [*New Business World* 2007].
- *Common good approach:* This approach revolves around making it possible to live together as a society. The common good is understood as that which "pertains to human life shared in community," [Lovin 2003] so that "the good of the individual and the common good are inseparable" [Hollenbach 2002]. This approach is similar in concept to the Utilitarian approach but is more concerned with social order and includes a dose of the Rights approach as well.
- *Virtue approach:* This approach dates back to Plato and Aristotle [Stanford 2007]. It emphasizes virtues such as courage, honesty, and integrity. A key question is, "Would you want to become the sort of person who acts this way?" [Velasquez et al. 2005].

The Josephson Institute of Ethics offers a similar model in their booklet *Making Ethical Decisions* (available online at http://www.josephsoninstitute.org/MED/MED-intro+toc.htm) that has seven steps instead of five. The author calls this "*The Seven-Step Path to Better Decisions.*" The steps are (1) stop and think, (2) clarify goals, (3) determine facts, (4) develop options, (5) consider consequences, (6) choose, and (7) monitor and modify [Josephson 2002]. The interested reader should see their Web site for more details.

11.2.3 Other Resources

Several books have been published on the subject of ethics in the biosciences for readers wishing to delve deeper into this subject. Of note is *The Human Embryo Research Debates: Bioethics in the Vortex of Controversy* by Ronald M. Green [Green 2001]. In 1996 and 1997, Dr. Green served as director of the Office of Genome Ethics at the National Human Genome Research Institute of the National Institutes of Health and has served on the NIH Human Embryo Research Panel. He is currently the director of the Dartmouth College Ethics Institute.

11.3 GENETICALLY MODIFIED FOOD CROPS

You may have seen products in the supermarket bearing a label, "Contains no GMOs (Genetically Modified Organisms)." This labeling is a response to concerns over the known or yet-to-be discovered untoward consequences of genetically engineered agricultural crops. GMOs are most commonly found in food crops in the form of transgenic (TG) plants. A transgenic plant variety is one that has had an exogenous gene spliced into its DNA. This creates a variety of plant that could not have been achieved using selective breeding, because the gene comes from another species.

Benefits of *transgenic crops* include disease resistance; longer shelf life; new flavors; more attractive appearance; faster growth; higher yields; greater nutritional value; resistance to one specific herbicide, so that a field can be sprayed for weeds without harming the food crop; and less need for pesticides overall [CTAHR 2007].

Risks include *hidden allergens*. In 1993, for example, a gene from a Brazil nut was transferred into a soybean to improve its nutritional value. Although the gene did not code for any known allergen, nut-allergic people who had never previously had any trouble with soybeans had allergic reactions to the transgenic soy [Nordlee et al. 1996].

Transgenic varieties can trigger food allergies in one of two ways: The transferred gene or genes can code for proteins already known to trigger allergic reactions, or they can introduce allergenic proteins completely new to the food supply [Pew 2006]. Other examples of TG allergen sources are discussed in Chapter 10.

Try applying one of the ethical frameworks outlined above to this situation. What are some options that could reduce the possible harm associated with this new strain of soybean?

- Label the product to warn those with nut allergies.
- Pull the transgenic soybean off the market if it harms humans, animals, or the environment.
- Use the case to increase scientific understanding of epigenetic subtleties in transgenic engineering, preventing similar situations from arising in the future.

As with many ethical issues in genetic engineering, the TG allergen issue could also be resolved through scientific advances—in this case, finding a way to neutralizing the allergenic properties of the GM plant varieties.

11.3.1 Rogue Herbicide Resistance

A transgenic strain of canola was produced that was resistant to the pesticide *Roundup*®. Farmers in Alberta, Canada, planted the crop. They found that the canola transferred its pesticide-resistant gene to nearby weeds, creating "superweeds." When the field was sprayed with Roundup, the weeds persisted [NYT 2003].

Another problem with herbicide-resistant crops is what happens when the farmer wants to rotate a different crop in the same field—and the previous year's crop "volunteers" or persists without a new planting. The transgenic crop becomes, in this situation, a type of weed. It would be undesirable to have canola growing in a wheat field because it could not be easily separated at harvest time [Canola 2005].

Is it worth the risk to use herbicide-resistant crops? If we had the scientific knowledge to prevent them from transferring this gene to other species, would that reduce the ethical problems?

11.3.2 Antibiotic-Resistant Bacteria

Almost all transgenic organisms are modified using *plasmids*. The gene to be transferred is inserted in a plasmid that contains a gene for resistance to antibiotics. The plasmids are then inserted into bacteria through an artificial process that breaks down the bacterial cell membrane. The bacteria are treated with antibiotics so that only bacteria that "accept" the plasmids survive. This way, scientists know that all the remaining bacteria contain the desired gene. Then, the bacteria can be used to transfer the gene to the newly transgenic organism [Kimball 2006g].

The problem with this method is that the antibiotic-resistant gene is transferred along with the agriculturally desirable gene. What happens to the antibiotic-resistant gene? Does it break down during digestion, or could be transferred to the endogenous bacteria in our guts? There is some controversy about whether such "horizontal transfer" occurs, and so far the research has not been conclusive one way or the other [Nielsen et al. 1998]. One British study reported in the *Journal of Antimicrobial Chemotherapy* showed that antibiotic marker genes from transgenic plant-based chicken food were not found in the intestinal bacteria of the chickens [Chambers et al. 2002]. A small study at the University of Newcastle Upon Tyne, however, found that horizontal transfer to gut bacteria did occur in human subjects [Netherwood et al. 1999]. It should be noted that this study used subjects who had had colostomies, and that no horizontal transfer has been detected in human subjects with complete intestinal tracts. Still, it does suggest that caution should be exercised.

What are some ways that the potential harm of using antibiotic-resistant plasmids for gene transfer could be reduced? Using an ethical framework, which option or options would you choose?

- Developers of GMOs could use antibiotics not commonly used for treatment of diseases in humans. That way, if horizontal transfer did occur, it would be unlikely to compromise medical treatment for an infection.
- Developers of GMOs could use other ways of identifying which bacteria have taken up plasmids. Rather than using antibiotic resistance as a "screen," they could use markers such as green fluorescent protein (GFP).
- Researchers could seek ways to remove the antibiotic-resistant genes before plants are released to the agricultural market.

The key question in the debate over GMOs is whether it is ethical to use a new technology before we fully understand and can minimize harm from its unintended consequences.

11.4 CHIMERAS

A chimera is a transgenic animal with genes and characteristics (even entire body parts) from other species (cf Chapter 10, Section 10.6). The development and use of chimeras is a potentially bizarre field fraught with novelty and a complicated cost-benefit equation.

11.4.1 Transgenic Organ Farming

There are over 3,000 people in the United States waiting for a heart transplant, over 17,000 waiting for a liver, and over 60,000 waiting for a kidney [Mayo Clinic 2006]. In all, about 150,000 people around the world are waiting for an organ transplant, and half will die waiting [Bryan & Clare 2005].

In 2003, researchers at the University of Nevada–Reno successfully grew livers and hearts with some human cells in sheep. Clearly, the potential benefits for those on the organ transplant list could be enormous. A doctor could take bone marrow cells from a person whose liver is failing. Bone marrow cells are capable of a great deal of differentiation, much similar to stem cells. What if the DNA of these bone marrow cells were inserted into the DNA of a sheep embryo, developed into a liver that was genetically compatible with the patient [Westphal 2003]? This would reduce the chance of the body rejecting the organ. Most importantly, that patient would not have to wait for another human to die in order to receive a lifesaving liver transplant.

Another example of the use of trans-species genetic therapy is the 2006 breakthrough at the University of Minnesota in using pig islet cells to cure diabetes in monkeys. Human application, a real possibility, would solve a huge problem of supply-and-demand: Twenty million Americans have diabetes, far more than the available number of donated pancreases [Lerner 2006].

It is also easy to see how chimeras could make people uncomfortable. Growing an internal organ is one matter because we cannot see it; what if sheep grew human limbs to be used for amputees? Would we want to see sheep walking around with human arms or legs coming out of their shoulders? The instinctive discomfort that this creates in many people is referred to as the "yuck factor" [Shreeve 2005], and is an example of how feelings sometimes provide input into ethical decision making.

Other objections include those of animal-rights activists, such as PETA (People for the Ethical Treatment of Animals), who describe any animal experimentation as "inherently unethical" because of the effects on the animals [PETA 2007].

11.4.2 How Human Is Too Human?

There are other ethical issues involved with the creation of chimeras. One is "how human" another species should become, a subject touched upon in Chapter 10. What about human brains in chimpanzees or mice? If an animal with a human brain exhibited human behavior (such as speech), would we extend human rights to it?

What about reproduction? Should human–animal chimeras be allowed to reproduce? Consider the unlikely scenario in which two human–mouse chimeras mated and the result was a human fetus growing inside a mouse [Weiss 2004]. Even though the fetus could never grow to term, this would be particularly alarming to those who believe that human life begins at conception.

11.4.3 Religious Concerns

There are many religious objections to research with chimeras. Some cite the Bible's statement that animals should multiply "after their kind" [*Genesis* 1:21]. Others have concluded that the core problem is not necessarily the creation of chimeras but the way they are likely to be treated. Some (particularly the Roman Catholic Church) are concerned that it will create a class of creatures without a clear moral status [Weiss 2004].

11.4.4 Retrovirus Transfer

There could be some practical health-related dangers associated with chimeras. For example, what about the transfer of retroviruses through the mingling of DNA? Researchers at the Mayo Clinic (2004) found that human stem cells injected into pig fetuses yielded hybrid cells containing both human and porcine DNA—including porcine endogenous retroviruses (PERVs) [Vince 2004].

Transgenic organ transplants could potentially spread animal retroviruses into the human population. It is difficult to predict whether these retroviruses would attack human cells and whether our immune system would be equipped to handle them. Many researchers believe that HIV originated as *Simian Immunodeficiency Virus* (SIV), a primate retrovirus that affects apes and monkeys [Dawson 1999; Gao et al. 1999]. However, SIV did not require genetic engineering to spread to humans but managed this feat on its own.

11.5 HUMAN EMBRYONIC STEM CELLS

The debate over the use of pluripotent stem cells from human embryos (e.g., blastocysts) is the most controversial ethical area in genetic research today, and much is at stake. Stem cell studies can potentially yield great research and medical benefits. However, some people believe that the cost is too great.

11.5.1 Religious Concerns

Although religion and ethics are not identical, religious principles have a strong influence on much ethical debate. In 1869, Pope Pius IX declared that "ensoulment," hence life, begins at conception. On the other hand, Jewish and Muslim religious scholars generally feel that "human life" begins in utero at "quickening," about 40 days postconception [Maienschein 2001]. Because some religious groups (not only Roman Catholics) believe that human life begins at conception (i.e., the initial formation of a zygote), they believe that harvesting stem cells from a human embryo in vitro (i.e., blastocyst, which destroys it) is essentially taking human life. They are concerned that allowing stem cell research will encourage the "farming" of human embryos with the intent to destroy them [Weiss 2005]. They are also concerned that stem cell research will encourage women to have abortions, knowing that the aborted fetuses can be used for research and therapies [Institute for Civil Society 2006].

Wasting an existing supply. Stem cell research proponents point out that thousands of human embryos sitting in freezers at fertility clinics will never be implanted in a womb and will be flushed down the drain when they are no longer needed or no longer considered viable. If they are not used for research and medical therapies, they will be wasted [Robinson 2004].

Why embryonic stem cells? Blastocyst-derived stem cells, or embryonic stem cells (ESCs), are unique because they are "pluripotent"—they can differentiate into any type of tissue. They are also self-renewing (able to generate additional SCs). A pluripotent stem cell can become any of the more than 200 different kinds of human cells [USDHHS 2001].

Scientists are just beginning to understand the signals inside and outside of cells that trigger stem cell differentiation. These signals include gene products, chemical signals from other cells, transcription factors, and physical contact with other cells (e.g., niche cells) [NIH 2006]. An important factor in determining cell states and behaviors is the stem cell microenvironment, or niche. Jang and Shaffer [2006] describe this complex environment: "The in vivo stem cell microenvironment or niche is composed of an intricate blend of extracellular matrix proteins, soluble protein factors, immobilized protein factors, proteoglycans, small molecule signals, in some cases mineralized tissue, and numerous adjacent cell types, all of which likely vary in space and time." Jang and Shaffer caution against reductionist or single-cause thinking in determining the causes of stem cell differentiation.

A substantial amount of recent work has been dedicated to determining what makes a stem cell a stem cell. There are several genes coding for transcription factors (TFs) (proteins that control the expression of large banks of genes) that apparently determine "stemness." Three TF genes that are expressed at particularly high levels in ESCs are Nanog, Oct-4 and Sox-2 [Carlin et al. 2006]. According to Carlin et al: "The downregulation of these three transcription factors correlates with the loss of pluripotency and self-renewal, and the beginning of subsequent differentiation steps." Oct-4 and Sox-2 are active in stem cells during the early formation of an embryo, but they shut down once the blastocyst starts to differentiate into different types of tissues [Cameron 2007].

In animals, stem cells injected into a particular organ will help regenerate that organ if it is damaged or diseased. This is one reason stem cells are so potentially valuable for therapy in regenerative medicine. They have the potential to treat or cure conditions such as Parkinson's and Alzheimer's diseases, diabetes, heart disease, stroke, spinal cord injuries, and burns [RDS 2004].

11.5.1.1 Alternative Sources of Stem Cells

Recent research may lead to the technological solution for the religious and ethical dilemmas surrounding the use of ESCs. In June 2007, the journal *Nature* reported an astonishing breakthrough. Scientists at Kyoto University (Japan) were able to "reprogram" (differentiated) skin cells from a mouse to become pluripotent stem cells having all the properties of ESCs. It is a simple technique, requiring the insertion of just four genes into a skin cell, which can be accomplished by established genetic engineering techniques. Not surprisingly, two of those genes are for the Oct-4 and Sox-2 TFs mentioned earlier that are common in ESCs. The four genes used in this study affect the behavior of chromatin in determining gene expression. Similar to flipping switches, they turned off genes determining "skin cellness" [Okita et al. 2007].

The Kyoto researchers identified the four genes that were essential to stemness through a process of elimination, starting with a set of 24 genes that had been identified through previous research. The Kyoto research has been replicated by research teams at the Whitehead Institute at MIT, the University of California and Massachusetts General Hospital [Wade 2007]. At the Whitehead Institute, the mouse skin-derived stem cells (aka induced pluripotent stem cells [IPS cells]) were injected into early-stage mouse embryos. Fluorescent labels were used to track the IPS cells as the embryos grew. The tagged IPS cells contributed to all tissue types in the resulting mice [Cameron 2007].

The IPS technique has not yet been applied to humans, and there are many technical challenges to be overcome before this is possible. The benefits of overcoming those challenges would be very great. In fact, causing adult skin cells to revert to pluripotent stem cells would allow therapies that would not even be possible using ESCs. Skin cells can be harvested from an adult patient, meaning that the DNA in the resulting stem cells would be an exact match for the patient. This is important because cells from other ESC sources might be rejected by the patient's immune system, but the new cells derived from IPS cells will have histocompatibility and should not be rejected. Therefore, doctors might someday be able to take a sample of a patient's skin cells, convert them to IPS cells, thence to liver cells, for example, and then replace the patient's damaged liver cells with new liver cells that have a perfect major histocompatibility match.

The ethical implications of IPS technique are encouraging. Without destroying an embryo, we could someday realize the benefits of pluripotent stem cell therapies for humans. The U.S. Conference of Catholic Bishops has already stated that the technique "raises no serious moral problem" because it does not create, harm, or destroy human lives [Wade 2007]. Should it be possible to overcome the technological barriers, this approach could solve both medical and ethical problems.

Another study released in June 2007 may be more problematic ethically but has the advantage of being successful in human subjects. An international team in Russia and the Unites States (Maryland) has induced stem cells from unfertilized human eggs [Revazova et al. 2007]. The technique included the brief creation of blastocysts, hence, the ethical concerns. Other concerns include the quality of the stem cells (one out of six lines had chromosomal abnormalities). Still, this is the farthest we have yet advanced in developing IPS cells from another type of cell in humans.

11.5.2 Embryonic versus Adult Stem Cells

Although only embryonic stem cells (or IPS cells, discussed earlier) can differentiate into any type of cell, there are several types of stem cells contained in adult bodies that have a limited ability to differentiate, primarily within their tissue of origin (see Chapter 10). For example, an adult stem cell from the brain can differentiate into a neuron or a glial cell but not into a liver cell [NIH 2006]. Adult stem cells have been found in the brain, blood, cornea, retina, heart, fat, skin, dental pulp, bone marrow, blood vessels, skeletal muscle, and intestines [USDHHS 2001b]. Another type of stem cell considered "adult" because it does not come from embryos is the stem cell found in umbilical cord blood at birth [Schu 2005].

In addition to being less versatile, adult stem cells are scarcer in the body and harder to culture than embryonic cells. However, recent research has shown that adult bone marrow stem cells are much more versatile than previously thought:

- At Johns Hopkins University (2001), researchers found that in mice, bone marrow stem cells could become cells in intestines, lungs, and skin [Krause et al. 2001].
- A study at the University of Minnesota (2002) found that human bone marrow stem cells could become blood vessels [Schwartz et al. 2002].
- A study at the Washington University School of Medicine in St. Louis (2003) found that human bone marrow stem cells could form liver cells in a damaged liver [Wang et al. 2003].
- Also in 2003, the journal *Pediatric Transplantation* reported a study of successfully regenerating myocardial (heart) cells using human bone marrow stem cells [Orlic et al. 2003].

Current federal restrictions: In 2001, in response to ethical concerns, the United States government announced that federal funds could be used to study embryonic stem cells but only using the stem cell lines that existed as of that date [Davis et al. 2006]. Anyone in the United States wishing to work with the stem cell lines created after that date can do so only with private or state funding. As of 2003, only about 22 stem cell lines were suitable for research and *only about a dozen* were still available for research [CBS 2003].

The global picture: The United States is not the only place conducting stem cell research. There are about 155 embryonic stem cell lines worldwide. As of July 2005, the United States had 70, Sweden 33, South Korea 24, India 10, Singapore 7, the United Kingdom 3, Spain 2, and Iran 1 [Weiss 2005]. The most rapid growth in stem cell research is currently in Singapore, whose research-friendly environment has lured a large number of scientists there from other countries [Luman 2004; De Trizio & Brennan 2004].

11.6 SUMMARY

We have introduced the concept of a framework for ethical decision making and have summarized two models and several ethical approaches, with references to sources for more in-depth study.

We have examined the costs (or risks) and benefits (or potential benefits) of several areas of genetic research and engineering. This is a controversial field because the degree of cost or benefit is not always clear. In some cases, the ethical concerns could be allayed through advances in science and technology (such as developing IPS cells).

Transgenic food crops, and other GM plants, such as thale cress, offer many potential benefits yet also carry risks such as new allergens, herbicide resistance that may be transferred outside the genetically modified crop, and antibiotic resistance related to the use of antibiotic-resistant bacteria in genetic engineering.

Chimeras, or combinations of different animal species (especially human-animal combinations), offer large potential benefits such as the possibility of increasing the number of organs available for transplant. Many people find the concept of growing human "parts" in animals to be disturbing and question at what point an animal crosses the line into being human. There are also religious concerns over the creation of "unnatural" creatures. Finally, there is the possibility that chimeras could encourage the crossover of retroviruses from animals to humans.

Human embryonic stem cells potentially offer great research and medical benefits. The ethical issues include religious concerns about whether the use of embryonic stem cells necessitates the death of an embryo. Other ethical concerns include the waste of an existing supply of IVF stem cells. Until a way to induce pluripotent stem cells in humans is developed without the involvement of embryos, ethical and religious objections will continue.

Finally, the differences between embryonic and adult stem cells are discussed; adult stem cells offer a less objectionable source of stem cells from a religious viewpoint but a less useful source from a scientific viewpoint as their function apparently is more limited. However, the restrictions on embryonic stem cell research have encouraged research that has revealed that adult stem cells are more versatile than once thought, and that IPS cells have promise—necessity being the mother of invention.

CHAPTER 11 HOME PROBLEMS

11.1 Which of the five Ethical Approaches listed in Chapter 11 is PETA utilizing when they argue against organ farming in animals? On what assumption does this argument depend?

11.2 The authors refer to the "law of unintended consequences" in Chapter 10. Knowing that we cannot foresee all harm that might be done by new technologies, which step in the Josephson Institute's "Seven-Step Path to Better Decisions" is the most important for addressing this problem?

11.3 You are a scientist who has made a breakthrough in growing "human" hearts in pigs, with an extremely low rejection rate among human recipients. You know that your technology could help save the lives of thousands of people waiting for heart transplants, and every day that goes by, more people die waiting. However, your tests have indicated that the genetic engineering required to grow these hearts makes it somewhat likely that a particular *PERV* affecting the immune system will be transferred to humans. This virus could affect some of the recipients and could even be spread from them to other people—but it is only a possibility. Should you release the technology or wait until the PERV issue has been resolved, and why? Which of the five Ethical Approaches outlined in Chapter 11 is the most relevant to this decision?

11.4 You are the CEO of a biosciences company that has developed a new transgenic soybean that contains genes from a tree nut. You engaged in an ethical decision-making process and decided to label the product to warn consumers of possible allergic reactions, even though you knew this might cost you some sales. What would you do to "monitor and modify" or follow up on this decision?

11.5 As discussed in Chapter 10, Bt corn has been genetically modified to be toxic to corn borers, thus increasing yield. Weigh this benefit against the costs described in Chapter 10. If you were the head of a government agency in charge of agriculture, what policy solutions would you have recommended to make Bt corn safe for the public?

11.6 Should human embryos left over in fertility clinic freezers be used as a source of stem cells for lifesaving therapies if they would otherwise be discarded? To help reach this decision, answer the Markkula Center's five questions listed in Step 3 of the ethical framework. Do you get conflicting answers from any of the five Ethical Approaches? On what assumptions are you basing your answers?

Appendix

A.1 SIMNON® PROGRAMS USED IN CHAPTER 4

A.1.1

```
CONTINUOUS SYSTEM snell2 " 8/24/98 Cross-coupled enzymatic
oscillator.*
STATE a b p i ap bp pp ip " 8 states.
DER da db dp di dap dbp dpp dip " State derivatives.
TIME t
"
da = - k1*ra*re + k1r*rb + Ja " Ja is input rate of A.
"
db = k1*ra*re -(k1r + k2)*rb
"
dp = k2*rb - k3*rp - k4p*rp^n*rep + k4pr*rip*n
"
dip = k4p*rp^n*rep - k4pr*rip
"
re = e0 - ri - rb " Initial, constant amount of E = e0.
"
rep = ep0 - rip - rbp " Initial, constant amount of E' = ep0.
"
dap = - k1p*rap*rep + k1pr*rbp + Jap
"
dbp = k1p*rap*rep - k1pr*rbp - k2p*rbp
"
dpp = -k3p*rpp + k2p*rbp - k4*rpp^n*re + k4r*ri*n
"
di = k4*re*rpp^n - k4r*ri
"
ra = IF a > 0 THEN a ELSE 0 " Rectification of states.
rb = IF b > 0 THEN b ELSE 0
rp = IF p > 0 THEN p ELSE 0
ri = IF i > 0 THEN i ELSE 0
rap = IF ap > 0 THEN ap ELSE 0
rbp = IF bp > 0 THEN bp ELSE 0
rpp = IF pp > 0 THEN pp ELSE 0
rip = IF ip > 0 THEN ip ELSE 0
"
Ja:3 " Constants used in the model.
Jap:1
e0:5
ep0:5
k1:0.1
k1r:0.05
```

```
k1p:0.1
k1pr:0.05
k2:1
k2p:1
k3:0.3
k3p:0.3
k4:0.5
k4p:0.5
k4r:0.5
k4pr:0.5
n:2
"
b:4 " ICs. All others zero by default.
bp:4
END
```

A.1.2

```
CONTINUOUS SYSTEM CHANCE3 " 9/23/98. Mod 1/07/99
STATE b c p ea ei
DER db dc dp dea dei
TIME t
"
" STATE EQS:
"
db = J0 + k4*rc - k1*rb*rea
"
dc = k1*rb*rea - k4*rc - k2*rc
"
dp = k2*rc - k3*rp
"
dea = (k2 + k4)*rc - k1*rb*rea + k6*rp*rei - k5*rea
"
" dei = k5*rea - k6*rp*rei
"
" after substituting ea = ea0 - ei - c
"
dei = k5*ea0 - k5*rei - k5*rc - k6*rp*rei
"
" RECTIFY STATES:
"
rb = IF b > 0 THEN b ELSE 0
rc = IF c > 0 THEN c ELSE 0
rp = IF p > 0 THEN p ELSE 0
rea = IF ea > 0 THEN ea ELSE 0
rei = IF ei > 0 THEN ei ELSE 0
"
" CONSTANTS:
J0:.04
k1:1
k2:1
```

```
k3:1
k4:.5
k5:1
k6:1
e0:1
«
« ICs:
b:0.2
c:1
p:1
Ea:1
Ei:0
«
END
```

A.1.3

```
continuous system HODHUX " 6/06/99
" Run w/ Euler integration w/ dt = 0.00001.
STATE v m n h " v in mV. v is depolarization if < 0.
DER dv dm dn dh " v = Vmo - Vm. Vm is actual transmembrane V.
" Vmo = resting potential = - 70 mV.
TIME t " t in ms.
"
" HH membrane patch ODE. (1 cm^2)
dv = Jin/Cm - Jk/Cm - Jna/Cm - Jl/Cm
"
" Ionic current densities.
Jk = gko*(n^4)*(v - Vk)
Jna = gnao*(m^3)*h*(v - Vna)
Jl = glo*(v - Vl)
Jc = Cm*dv
Jnet = Jc + Jk + Jna + Jl " Microamps/cm^2
«
« Ionic conductances:
gk = gko*(n^4)          " Potassium conductance
gna = gnao*(m^3)*h      " Sodium conductance
gl = glo                " Leakage conductance
gnet = gk + gna + gl    " Net membrane conductance
negv = -v
"
dn = - n*(an + bn) + an " K+ activation parameter.
dm = - m*(am + bm) + am " Na+ activation parameter.
dh = - h*(ah + bh) + ah " Na+ inactivation parameter
"
" VOLTAGE-DEPENDENT PARAMETER FUNCTIONS:
an = 0.01*(v + 10)/(exp(0.1*v + 1) - 1)
bn = 0.125*exp(v/80)
am = 0.1*(v + 25)/(exp(0.1*v + 2.5) - 1)
bm = 4*exp(v/18)
ah = 0.07*exp(v/20)
```

```
bh = 1/(exp(0.1*v + 3) + 1)
"
Vm = -(v + 70)
Jksc = Jk/10 " Scaled current densities
Jnasc = Jna/10
Jcsc = 10*Jc " Scaled capacitor current density.
Vmsc = Vm/70 " Scaled membrane voltage.
"
" CONSTANTS:
zero:0
Vk:12 " mV
Vna:-115 " mV
VL:-10.613 " mV
"
glo:0.3 " milliS/cm^2.
gko:36 " "
gnao:120 " "
Cm:.001 " milliF/cm^2 = 1 microF/cm^2.
f:0.10 " Hz
pi:3.14159
"
" INPUTS:
"Jin1 = IF t > to THEN K*(t - to) ELSE 0 " Delayed current ramp
input.
"Jin2 = IF t > t2 THEN -K*(t - t2) ELSE 0 " Jin in microamps/cm^2.
t2:10 " End of ramp
to:1 " Jin < 0 depolarizes membrane.
K:-0.3 " Jin inward is + charges inward
Jin1 = IF t > to THEN Jino ELSE 0 " Delayed current pulse input.
Jin2 = IF t > (to + delt) THEN -Jino ELSE 0
delt:5 " Duration of current pulse, ms.
Jino:-2 " Microamps.
Jinac:1
" Jin1 = IF t > to THEN Jino ELSE 0 " Delayed current step input.
" Jin2 = Jinac*sin(2*pi*f*t) " Sinusoidal current input.
Jin = Jin1 + Jin2
"
" INITIAL CONDS. (Also final values for v = 0.)
m:0.052932
n:0.31768
h:0.59612
v:0
"
END
```

***NB: Lines preceded by quotes are comments, not part of code.**

A.2 SOME NOTATIONS AND SYMBOLS USED IN THIS TEXT

[A] Concentration of A (in moles/liter, mg/dL, etc.). **$[A]_o$:** Concentration of A outside the cell membrane. **$[A]_i$:** Concentration of A inside the cell membrane.

$$\dot{x} \equiv \frac{dx}{dt} \text{ (}x\text{-dot is the time derivative of } x.\text{)}$$

E Electric field vector. Units are volts/meter at a point, x,y,z.

E_{Na} Potential energy of a single sodium ion outside a nerve axon membrane, generally given in eV (electron-volts).

ΔG Gibbs free energy of a chemical reaction; generally given in Calories/mole.

ΔG^0 Gibbs free energy measured under "standard" conditions of temperature, pressure, and concentrations.

J_{in} Input current density injected across a unit area of nerve axon membrane.

P Electrical power in Watts.

v Transmembrane voltage change: $v \equiv [V_{m0} - V_m(t)]$ volts. Used in Hodgkin–Huxley equations and simulation.

V_{m0} Resting (steady-state) transmembrane potential across a cell membrane. Usually *about* –70 mV (inside negative with respect to outside the cell).

$V_m(t)$ Actual dynamic transmembrane voltage, measured with voltmeter + lead inside the cell. ΔV_m: See v.

Bibliography and References

ABC Online. 2006. *Uterine lining stem cells act like heart cells.* 14/03/2006. www.abc.net.au/news/newsitems/200603/s1590816.htm

Access Excellence. 1999. *Protein Synthesis.* www.accessexcellence.org//RC/VL/GG/protein_synthesis.html

ACFNEWSOURCE. 2006. *Mine-sniffing Plants.* XM Satellite Radio 4/30/06. (Accessed 5/31/07.) www.acf-newsource.org/science/mine_sniffing_plants.html

Adachi, H. et al. 2004. Two-liquid hanging-drop vapor-diffusion technique of protein crystallization. *Jap. J. Appl. Phys.* 43(1A/B): L79–L81.

Adachi, S-i. et al. 2005. Current status of 50-picosecond resolved X-ray diffraction at Photon Factory Advanced Ring (PF-AR). *J. Phys.: Conference Series.* 21: 101–105.

Adler, M. et al. 2005. Detection of femtogram amounts of biogenic amines using self-assembled DNA-protein nanostructures. *Nature Methods.* 2(2): 174–149.

Afford, S. & S. Randhawa. 2000. Demystified apoptosis. *J. Clin. Pathol: Mol. Pathol.* 53: 55–63.

Agre, P. et al. 2002. Aquaporin water channels—from atomic structure to clinical medicine. *J. Physiol.* 542: 3–16.

Agura, E. 2007. *Bone Marrow and Stem Cell Transplantation.* (Retrieved 22 April 2007.) www.plwc.org/portal/site/PLWC/menuitem.169f5d85214941ccfd748f68ee37a01d/?vgnextoid=2a6e80efecd51110VgnVCM100000ed730ad1RCRD

Airey, J.A. 2004. *Human Mesenchymal Stem Cells Form Purkinje Fibers in Fetal Sheep Heart. Circulation.* 109: 1401–1407. http://circ.ahajournals.org/cgi/content/full/109/11/1401

Alexeeva, I. et al. 2006. Absence of *Spiroplasma* or other bacterial 16S rRNA genes in brain tissue of hamsters with scrapie. *J. Clin. Microbiol.* 44(1): 91–97.

Allison, D.B., G.P. Page, T.M. Beasley & J.W. Edwards. 2006. *DNA Microarrays and Related Genomic Techniques.* CRC Press, Boca Raton, FL.

Alon, U. 2006. *An Introduction to Systems Biology.* Chapman and Hall/CRC, Boca Raton, FL.

ALSNews. 2002. *Aquaporin Structure Elucidates Water Transport.* (Accessed 5/18/07.) www.als.lbl.gov/als/science_archive/54aquaporin.html

Anderson, P.A.V., J. Roberts-Misterly & R.M. Greenberg. 2005. The evolution of voltage-gated sodium channels: Were algal toxins involved? *Harmful Algae.* 4: 95–107. (Available online at: www.sciencedirect.com)

Andressoo, J., J. Hoeijmakers & J. Mitchell. 2006. Nucleotide excision repair disorders and the balance between cancer and aging. *Cell Cycle.* 5(24): 2886–2888.

Antigenics. 2007. *Heat Shock Proteins: Basics.* (Retrieved 30 April 2007.) www.antigenics.com/products/tech/hsp/

Antoniotti, M. et al. 2002. XS-systems: *Extended S-Systems and Algebraic Differential Automata for Modeling Cellular Behavior.* fsv.dimi.uniud.it/downloads/HiPC-2002-final.pdf

Arai, F., A. Hirao & T. Suda. 2005. Regulation of hematopoietic stem cells by the niche. *TCM.* 15(2): 75–79.

Arax, M. & J. Brokaw. 1997. *No Way Around Roundup.* Foundation for National Progress. www.motherjones.com

Arizona. 2006. *Transgenic Animals (Chapter 19).* Arizona State University lecture notes for MBB 343 and BIO 343. //photoscience.la.asu.edu/photosyn/courses/BIO_343/lecture/transan.html

Armstrong, W.P. 2007. Wayne'sWord. *Transposons (Jumping Genes): A Simplified Explanation for Jumping Genes and Their Effect on the Kernel Color of Indian Corn.* http://waynesword.palomar.edu/transpos.htm

________ 1997. Wayne'sWord. *Archaebacteria: A life form on Mars?* http://waynesword.palomar.edu/ploct97.htm

Atkins, D.L. 1998. *Structure of the Plasma Membrane: Step 1 in Reviewing the Neuron.* http://sky.bsd.uchicago.edu/lcy_ref/synap/plasmalemma.html

Babini, E. & M. Pusch. 2004. A two-holed story: Structural secrets about ClC proteins become unraveled? *Physiology.* 19: 293–299.

Bae, H.C., E.H. Costa-Robles & L.E. Cassida. 1972. Microflora of soil as viewed by transmission electron microscopy. *Applied Microbiol.* 23: 637–648.

Baez, J. 2005. *Subcellular Life Forms.* //math.ucr.edu/home/baez/subcellular.html

Baier, A. 2005. *Genetically Modified Cotton and its Implications for Africa.* Pestizid Aktions-Netzwerk e.V. (PAN Germany).

Bannister, A. 2006. *The Role of Epigenetics in Cancer.* (Web paper accessed 6/14/07.) www.abcam.com/index.html?pageconfig=resource&rid=10755&pid=10628

Bárány, M. 1996. *Biochemistry of Smooth Muscle Contraction.* Academic Press, New York.

Bastian, F.O. & J.W. Foster. 2001. *Spiroplasma sp.* 16S rDNA in Creutzfeldt-Jakob disease and scrapie as shown by PCR and DNA sequence analysis. *J. Neuropathol. Exp. Neurol.* 60(6): 613–620.

Bastian, F.O., D.M. Purnell & J.G. Tully. 1984. Neuropathology of spiroplasma infection in the rat brain. *Amer. J. Pathol.* 114: 496–514.

Beard, D.A., H. Qian & J.B. Bassingthwaighte. 2004. "Stoichiometric Foundation of Large-Scale Biochemical System Analysis". In: *Modelling in Molecular Biology.* G. Ciobanu & G. Rozenburg, Eds. Springer Natural Computing Series. pp. 1–19.

Belay, E.D. et al. 2004. Chronic wasting disease and potential transmission to humans. *Emerging Infectious Dis.* 10(6): 977–984. www.cdc.gov/eid

Belfort, M. 2005. Back to basics: Function, evolution and application of homing endonucleases and inteins. *Nucl. Acids Mol. Biol.* 16: 1–10.

Belgrader, P. et al. 1999. Infectious disease: PCR detection of bacteria in seven minutes. *Science.* 284(5413): 449–450.

Bellouquid, A. & M. Delitalia. 2006. *Mathematical Modeling of Complex Biological Systems: A Kinetic Theory Approach.* Birkhäuser Boston.

Berggren, P. et al. 2000. P53 intron 7 polymorphisms in urinary bladder cancer patients and controls. *Mutagenesis.* 15(1): 57–60. www.ncbi.nlm.nig.gov/entrez/query.fcgi?cmd=Retrieve&db=PubMed&list_uids=10640531&dopt=Abstract

Berman, H.M. et al. 2000. The protein data bank. *Nucl. Acids Res.* 28: 235–242.

Bernstein, B.E. et al. 2006. A bivalent chromatin structure marks key developmental genes in embryonic stem cells. *Cell.* 125: 315–326.

Best, B. 2006. *Mechanisms of Aging.* (web paper) www.benbest.com/lifeext/aging.html

Bidwai, A. 1999. What is known about the function of introns, the nonencoding sequences in genes? In *Scientific American* online. www.sciam.com/askexpert_question.cfm?articleD=000DB4D5-63A9-IC72-9EB7809EC588F2D7

Biello, D. 2007. *Sick Genes.* (Scientific American online article at: www.sciam.com/article.cfm?id=sick=genes.)

Biocrawler. 2006. *Retrovirus.* www.biocrawler.com/encyclopedia/Retrovirus

Bioethics. 2005. *Chimeras.* www.bioethics.umn.edu/resources/topics/chimeras.shtml

Blakeslee, S. 1996. *Protein Culprit Acts to Cause Huntington's Disease.* http://query.nytimes.com/gst/fullpage.html?sec=health&res=9D01E6D61E39F936A35750C0A960958260

Blossey, P. 2006. *Computational Biology: A Statistical Mechanics Perspective.* CRC Press. Boca Raton, FL.

Boldyrev, A.A. 2005. Protection of proteins from oxidative stress. *Ann. N.Y. Acad. Sci.* 1057: 193–205.

Bonner, J.T. 1988. *The Evolution of Complexity: By Means of Natural Selection.* Princeton Univ. Press, Princeton, NJ.

Bouley, J. 2006. Giving DNA's second code first priority. *Drug Discovery News: Genomics and Proteomics.* Online at: www.drugdiscoverynews.com/print.php?newsarticle=694

Bowen, R. 2005. *Aquaporins: Water Channels.* (Accessed 5/18/07.) www.vivo.colostate.edu/hbooks/molecules/aquaporins.html

Bowen, N. & I.K. Jordan. 2002. Transposable elements and the evolution of eukaryotic complexity. In *Current Issues in Molecular Biology.* 4: 65–76. www.horizonpress.com/cimb/abstracts/v4/07.html

BrightSurf. 2006. *Mad cow protein aids creation of brain cells.* www.brightsurf.com/news/headlines/view.article.php?ArticleID=23045

BrightSurf. 2004. *Researchers discover gene mutations for Parkinson's disease.* www.brightsurf.com/news/oct_04/NIA_news_102504.php

BrightSurf. 2003. *Possible diagnosis, treatment, vaccine for mad cow, prion diseases.* www.brightsurf.com/news/june_03/EDU_news_06023_c.html

Brinkman, F.S.L, J.L. Blanchard et al. 2002. Evidence that plant-like genes in *Chlamydia* species reflect an ancestral relationship between Chlamydiaceae, Cyanobacteria and the chloroplast. *Genome Research,* July, 1159–1167.

Brown, D.R. 1999. Normal prion protein has an activity like that of superoxide dimutase. *Biochem. J.* 344: 1–5.

Brown, J.L. 2003. *StarLink.* (Article on the Penn. State Univ. *Biotechnology, Food and Agriculture* web site. Accessed 7/03/07.) http://biotech.cas.psu.edu/articles/starlink.htm

Brown, P. 2005. *GM Crops Created Super Weed, Say Scientists.* www.connectotel.com/gmfood/gu250705.txt

Bruchez, M. Jr. et al. 1998. Semiconductor nanocrystals as fluorescent biological labels. *Science.* 281: 2013–2016.

Bryan, J. & J. Clare. 2005. *Organ Farm: Frequently Asked Questions.* www.pbs.org/wgbh/pages/frontline/shows/organfarm/etc/faqs.html

Bryce, C.F.A. ca. 2002. *Alternative View on the Putative Organism,* Nanobacterium sanguineum. Web essay. www.heartfixer.com/Nanobacterium/Nanobacterium-Report.htm

Buczek, O., D. Krowarsch & J. Otlewski. 2002. Thermodynamics of single peptide bond cleavage in bovine pancreatic trypsin inhibitor (BPTI). *Protein Sci.* 11: 924–932.

Budryk, M. et al. 2000. Direct transfer of IL-12 gene into growing Renca tumors. *Acta Biochimica Polonica.* 47(2): 385–391.

Bullock, T.H. et al. 2005. Neuroscience: The neuron doctrine, redux. *Science.* 310(5749): 791–793.

BUMC (Boston University Medical Center). 2005. *Direct DNA Tests.* www.bumc.bu.edu/Dept/ContentPF.aspx?PageID=2194&DepartmentID=118

Bürglin, T. 2005. *The Homeobox Page.* (Retrieved 5 May 2007 from the Karolinska Institute Website.) http://homeobox.biosci.ki.se/

Burke, W. 2002. Genetic testing. *New Engl. J. Med.* 347(23): 1867–1875.

Button, D.K. et al. 1993. Viability and isolation of marine bacteria by dilution culture: Theory, procedures, and initial results. *Applied Environmental Microbiology* 4: 259–270.

Calbiochem. 2005. *Lectins.* (On-line catalog of lectins.) www.calbiochem.com

Calò, C.M., L. Garofano, A. Mameli, M. Pizzamiglio & G. Vona. 2003. Genetic analysis of a Sicilian population using 15 short tandem repeats. *Human Biol.* 75(2): 163–178.

Calzone, L., F. Fages & S. Soliman. 2006. BIOCHAM: An environment for modeling biological systems and formalizing experimental knowledge. *Bioinformatics Advance Access:* OUP. 3 May 2006.

Cameron, D. 2007. *Scientists create embryonic stem cells without destroying embryos.* Whitehead Institute press release 6/6/07. http://web.mit.edu/newsoffice/2007/stemcells-0606.html

Cann, A. 1999. *Retroviruses.* (Retrieved 5 May 2007 from Tulane Univ. class notes.) www.tulane.edu/~dmsander/WWW/335/Retroviruses.html

Cann, R., M. Stoneking & A. Wilson. 1987. Mitochondrial DNA and human evolution. *Nature.* 325: 31–36.

Cannon, G. & D. Weissman. 2002. RNA-based vaccines. *DNA Cell Biol.* 21(12): 953–961.

Canola. 2005. (Canola Council of Canada). *Herbicide Tolerant Volunteer Canola: A Study of Grower Management Practices.* www.canola-council.org/PDF/Herbicide_Tolerant_Fact_Sheet.pdf

Carlin, R. et al. 2006. Expression of early transcription factors Oct-4, Sox-2 and Nanog by porcine umbilical cord (PUC) matrix cells. *Reprod. Biol. Endocrinol.* 4: 8. (Published online 6 Feb. 2006. doi: 10.1186/1477-7827-4-8.)

Carman, J. (printout 3/06). *The Problem with the Safety of Roundup Ready Soybeans.* Web paper. www.psrast.org/subeqau.htm

Carmen, A.A. 2007. *Biochips as Pathways to Drug Discovery.* CRC Press, Boca Raton, FL.

Carmo-Fonseca, M. 2002. New clues to the function of the Cajal body. *EMBO Rep.* 3(8): 726–727.

Carpenter, R.H.S. 2004. Homeostasis: a plea for a unified approach. *Adv. Physiol. Education.* 28: 180–187.

Catterall, W.A. 2000. Structure and regulation of voltage-gated Ca^{2+} channels. *Annu. Rev. Cell. Dev. Biol.* 16: 521–555.

CBS. 2003. Researchers cite stem cell shortage. www.cbsnews.com/stories/2003/05/09/tech/main553079.shtml

Chalfie, M. et al. 1994. Green fluorescent protein as a marker for gene expression. *Science.* 263: 802–805.

Chambers, I. 2004. The molecular basis of pluripotency in mouse embryonic stem cells. *Cloning and Stem Cells.* 6(4): 386–391.

Chambers, P.A. et al. 2002. The fate of antibiotic resistance marker genes in transgenic plant fed material fed to chickens. *J. Antimicrobiobial Chemotherapy.* 49: 161–164. http://jac.oxfordjournals.org/cgi/content/full/49/1/161

Chan, P., T. Yuen, F. Ruf, J. Gonzalez-Maeso & S.C. Sealfon. 2005. Method for multiplex cellular detection of mRNAs using quantum dot fluorescent *in situ* hybridization. *Nucl. Acids Res.* 33(18): *e161.* 8 pp. (Published online.)

Chan, W.C.W. & S.M. Nie. 1998. Quantum dot bioconjugates for ultrasensitive nonisotopic detection. *Science.* 281: 2016–2018.

Chance, B., B. Schoener & S. Elaesser. 1964. Control of the waveform of oscillations of the reduced pyridine nucleotide in a cell-free extract. *Proc. Natl. Acad. Sci.* (USA). 52: 337–341.

Chance, B., K. Pye & J. Higgins. 1967. Waveform generation by enzymatic oscillators. *IEEE Spectrum*. 4: 79–86.

Chaplin, M. 2005. *Cellulose*. www.lsbu.ac.uk/water/hycel.html

Chayen, N.E. & J.R. Helliwell. 2002. Microgravity protein crystallization: Are we reaping the full benefit of outer space? *Ann. N.Y. Acad. Sci*. 974: 591–597.

Chen, C-Z. & H.F. Lodish. 2005. MicroRNAs as regulators of mammalian hematopoeisis. *Semin. Immunol.* 17: 155–165.

Chen, S. et al. 2004. Dedifferentiation of lineage-committed cells by a small molecule. *J. Am. Chem. Soc.* 126: 410–411.

Chen, K. & N. Rajewski. 2007. The evolution of gene regulation by transcription factors and microRNAs. *Nat. Rev. Genetics*. 8: 93–103.

Chen, T-Y. 2005. Structure and function of CLC channels. *Annu. Rev. Physiol*. 67: 809–839.

Chepko, G. 2005. Breast cancer, stem cells, and the stem cell niche. *Breast Cancer Online*. ISSN 1470–9031. 5 pp. www.bco.org

Chess, L. & H. Jiang. 2004. Resurecting CD8$^+$ suppressor T cells. *Nat. Immunol*. 5(5): 469–471.

Chiba, S. 2006. Concise review: Notch signaling in stem cell systems. *Stem Cells*. 24: 2437–2447.

Childs, G.V. 2003a. *Cilia and Flagella*. //cellbio.utmb.edu/cellbio/cilia.htm

________. 2003b. *Membrane Structure and Function*. //cellbio.utmb.edu/cellbio/membrane_intro.htm

________. 2003c. *The Mitochondrial Life cycle*. //cellbio.utmb.edu/cellbio/mitoch2.htm

________. 2003d. *Endoplasmic Reticulum: Structure and Function*. //cellbio.utmb.edu/cellbio/rer.htm

________. 2003e. *How do proteins translocate into the lumen of the rough endoplasmic reticulum?*//cellbio.utmb.edu/cellbio/rer2.htm

________. 2003f. *Golgi Complex: Structure and Function* //cellbio.utmb.edu/cellbio/golgi.htm

________. 2003g. *How are lysozomes and peroxisomes produced?* //cellbio.utmb.edu/cellbio/lysosome.htm

________. 2001. *Smooth Endoplasmic Reticulum*. //cellbio.utmb.edu/cellbio/ser.htm

Choudhuri, S. 2004. The nature of gene regulation. *Int. Arch. Biosci*. 1001–1017.

Christie, W.W. 2007. *Phosphatidylcholine and related lipids*. Scottish Crop Research Institute. www.lipidlibrary.co.uk/Lipids/pc/index.htm

Chuang, J.C. & P.A. Jones. 2007. Epigenetics and microRNAs. *Pediatric Res*. 61(5): Pt. 2. 24R–29R.

Cibelli, J.B. et al. 2002. The health profile of cloned animals. *Nat. Biotechnol*. 20: 13–14.

Cisar, J.O., D-Q Xu, J. Thompson, W. Swaim, L. Hu & D.J. Kopecko. 2000. An alternative interpretation of nanobacteria-induced biomineralization. *PNAS* 97(21): 11511–11515.

Clark, A. 2000. Post-transcriptional regulation of pro-inflammatory gene expression. *Arthritis Res*. 2: 172–174.

Clark, H.F. 1964. Suckling mouse cataract agent. *J. Infect. Dis*. 114: 476–487.

Clark, J. 2002. *Definitions of Oxidation and Reduction (Redox)*. Online at: www.chemguide.co.uk/inorganic/redox/definitions.html

Clarke, M.F. & M.W. Becker. 2006. Stem cells: The real culprits in cancer? *Scientific American*. on-line @: //sciam.com/article.cfm?chanID=sa006&articleID=000B1BED-0C0A-1498-8C0A83414B7F0000&pageNumber=6&catID=2

Clarke, M.F. & M. Fuller. 2006. Stem cells and cancer: Two faces of eve. *Cell*. 124: 1111–1115.

Clayton, A.H.A., Q.S. Hanley & P.J. VerVeer. 2004. Graphical representation and multicomponent analysis of single-frequency fluorescence lifetime imaging microscopy data. *J. Microscopy*. 213(Pt 1): 1–5.

Codoñer, F.M., J-A. Daròs, R.V. Solé & S.F. Elena. 2006. The fittest versus the flattest: Experimental confirmation of the quasispecies effect with subviral pathogens. *PloS Pathogens*. 2(12): 1187–1193.

Coffin J. et al. 1997. Overview of retroviral assembly. *Retroviruses*. Cold Spring Harbor Lab. Press. Plainview, New York. Online at: www.ncbi.nlm.nih.gov/books/bv.fcgi?rid=rv.section.2497

Cogdon, K. 2007. *Cells Hope for Diabetes*. (29 June). www.news.com.au/heraldsun/story/0,21985,21984076-662,00.html

Cohen, J. 2007. Sequencing in a flash. *Technology Review*. 110(3): 72–77.

Collins, R.D. 1968. *Illustrated Manual of Laboratory Diagnosis*. J.B. Lippincott, Philadelphia.

Columbia University Press. 2003. Retrovirus. *Columbia Electronic Encyclopedia, 6th ed*. Online at: www.answers.com/topic/retrovirus

Complement. 1998. *The Complement System*. (Pathology Lecture 10 notes). Online at: www.bio.cam.ac.uk/dept/path/teaching/ittmac/lectures/Lec10/lec10.html

Connor, S. 10 Sept. 2005. SCIENCE FESTIVAL: Embryos created by 'virgin conception'. *The (London) Independent*. Online at: www.findarticles.com/p/articles/mi_qn4158/is_20050910/ai_n15369764

Cooper, G.M. 2000. *The Cell: A Molecular Approach. I. The Origin and Evolution of Cells.* Sinauer Associates, Sunderland, MA.

Cord Blood Registry. 2007. *Diseases Treated with Umbilical Cord Blood Stem Cells.* (Retrieved 22 April 2007.) www.cordblood.com/cord blood banking with cbr/banking/diseases treated.asp

Coughlan, A. 2006. *Killer tomatoes attack human diseases. New Scientist*, 29 June. Online at: www.newscientist.com/article.ns?id=mg19125584.600&print=true

Council. 2005. *Alternative Sources of Pluripotent Stem Cells.* The President's Council on Bioethics. Online at: www.bioethics.gov

Crick, F. 1981. *Life Itself: Its Origin and Nature.* Simon and Schuster, New York.

Crick, F. 1970. Central dogma of molecular biology. *Nature.* 227: 561–563. Online at: http://profiles.nlm.nih.gov/SC/B/C/C/H/_/scbcch.pdf

Csankovszki, G., A. Nagy & R. Jaenisch. 2001. Synergism of Xist RNA, DNA methylation and histone hypoacetylation in maintaining X chromosome inactivation. *J. Cell. Biol.* 153(4): 773–783.

CTAHR. 2007. (College of Tropical Agriculture and Human Resources). *What are the Benefits of Biotechnology in Agriculture?* (Retrieved 15 April 2007 from the Univ. of Hawaii at Manoa Biotechnology and Agriculture Education Program web site.) www.ctahr.hawaii.edu/gmo/risks/benefits.asp

CTDEP. 2004. *Chronic Wasting Disease.* (On-line bulletin from the Conn. Dept. of Environmental Protection.) Online at: //dep.state.ct.us/burnatr/wildlife/fguide/cwd.htm

Cullen, M., M. Malasky, A. Harding & M. Carrington. 2003. High-density map of short tandem repeats across the human major histocompatibility complex. *Immunogenetics.* 54: 900–910.

Cullen, B.R. 2004. Derivation and function of small interfering RNAs and microRNAs. *Virus Research.* 102: 3–9.

Cummins, J. 2000. *Genetically Modified (GM) Baculovirus Vectors to Control Insect Pests and for Gene Therapy.* Online at: www.biotech-info.net/gene_therapy.html

Cutler, P. 2003. Protein arrays: The current state-of-the-art. *Proteomics.* 3: 3–18.

Cyranowski, D. 2007. Simple switch turns cells embryonic. 6/6/07. *news@nature.com.* http://news.nature.com//news/2007/070604/447618a.html

D'Amico, V. 2005. *Baculoviruses (Baculoviridae).* Online at: www.nyseas.cornell.edu/ent/biocontrol/pathogens/baculoviruses.html

Datta, A. & E.R. Dougherty. 2007. *Introduction to Genomic Signal Processing with Control.* CRC Press, Atlanta. ISBN: 978-0-8493-7198-1.

Davis, B.G. 2000. Hand in Glove (Genetics). *Chem. Ind.*. 21 Feb. 2000. 134–138.

Davis, B. et al. 2006. Ethical and public policy issues concerning stem cell research. *Developmental Biology Online.* www.devbio.com/article.php?ch=21&id=258

Dawson, D. 1999. Origin of AIDS discovered. 6th Conference on Retroviruses and Opportunistic Infections (CROI). *Perspective.* 99(1): 8. Online at: www.aegis.org/pubs/step/1999/STEP9902.html

Degterev, A. et al. 2005. Chemical inhibitor of nonapoptotic cell death with theraputic potential for ischemic brain injury. *Nat. Chem. Biol.* 1(2): 112–119.

Deininger, P.L. & M.A. Batzer. 2002. Mammalian retroelements. *Genome Res.* 12(10): 1455–1465.

DeLamarche, C. et al. 1999. Visualization of AqpZ-mediated water permeability in *Escherichia coli* by cryo-electron microscopy. *J. Bacteriol.* 181(14): 4193–4197.

de la Rosa, M., S. Rutz, H. Dorninger & A. Scheffold. 2004. Interleukin-2 is essential for $CD4^+CD25^+$ regulatory T cell function. *Eur. J. Immunology.* 34: 2480–2488.

Demidov, V.V. 2005. Rolling-circle amplification: (RCA). *Encyclopedia of Diagnostic Genomics and Proteomics.* Marcel Dekker, N.Y. 1175–1179.

DeSouza, S.J. 1999. *What is Known About the Function of Introns, the Nonencoding Sequences in Genes?* www.sciam.com/askexpert_question.cfm?articleD=000DB4D5-63A9-IC72-9EB7809EC88F2D7

De Trizio, E. & C. Brennan. 2004. The business of human embryonic stem cell research and an international analysis of relevant laws. *J. of Biolaw and Business, Reprint Series.* 7(4): www.dechert.com/library/BusinessofStemCellResearch.pdf

Di Berardino, M.A. 1998. *Cloning: Past, Present, and the Exciting Future.* Online at: www.faseb.org/opar

Dileo, J. et al. 2003. Gene transfer to subdermal tissues via a new gene gun design. *Hum. Gene Ther.* 14: 79–87. (Published on-line.)

Ding, S. & P. Schultz. 2004. A role for chemistry in stem cell biology. *Nat. Biotechnol.* 22(7): 833–840.

Dingli, D. & F. Michor. 2006. Successful therapy must eradicate cancer stem cells. *Stem Cells.* Published online 24 Aug. 2006. www.StemCells.com

Diwan, J.J. 2006. *Lipids and Membrane Structure.* Online at: www.rpi.edu/dept/bcbp/molbiochem/MBWeb/mbl/part2/lipid.htm

Diwan, J.J. 2005. *Membrane Transport.* Online at: www.rpi.edu/dept/bcbp/molbiochem/MBWeb/mbl/part2/carriers.htm

doegenomes. 2004. *How Many Genes are in the Human Genome?* Online at: www.ornl.gov/sci/techresources/Human_Genome/faq/genenumber.shtml

doegenomes. 2003. *CFTR: The Gene Associated with Cystic Fibrosis.* Online at: www.ornl.gov/sci/techresources/Human_Genome/posters/chromosome/cftr.shtml

Dolphin, A.C. 2006. A short history of voltage-gated calcium channels. *British J. Pharmacol.* 147: S56–S62.

Domingo, E. et al. 1998. Quasispecies structure and persistence of RNA viruses. *Emerging Infectious Dis.* 4(4): 10pp. www.cdc.gov/ncidod/eid/vol4no4/Domingo.htm

Drake, J. 1993. Rates of spontaneous mutation among RNA viruses. *PNAS USA.* 90(9): 4171–4175.

Drennan, B. & R.D. Beer. 2006. Evolution of repressilators using a biologically-motivated model of gene expression. *Artificial Life X: 10th Intl. Conf. on the Simulation and Synthesis of Living Systems.*

Dresbach, S.H., H. Flax, A. Sokolowski & J. Allred. 2001. The impact of genetically modified organisms on human health. *Ohio State University Extension Fact Sheet.* HYG-5058-01.

Drickamer, K. 2005. Part I: *The stucture and functions of animal lectins. Classification of animal lectins.* www.imperial.ac.uk/animallectins/ctld/lectins.html

Eastham, J.A., E. Reidel & P.T. Scardino. 2003. Variation of serum prostate-specific antigen levels: An evaluation of year-to-year fluctuations. *JAMA.* 289: 2695–2700.

Easton, D.F. et al. 2007. Genome-wide association study identifies novel breast cancer susceptibility loci. *Nature.* 447(7148): 1087–1093.

Eberharter, A. & P.B. Becker. 2004. ATP-dependent nucleosome remodeling: factors and functions. *J. Cell Science.* 117: 3707–3711. doi:10.1242/10.1242/jcs.01175Epel, E.S. et al. 2000. Stress and body shape: Stress-induced cortisol secretion is consistently greater among women with central fat. *Psychosomatic Medicine.* 62: 623–632.

Economist. 2005. Brain development may be influenced by genetic parasites. www.economist.com/science/displayStory.cfm?story_id=4078826

Edinger, A.L. & C.B. Thompson. 2004. Death by design: apoptosis, necrosis and autophagy. *Curr. Opin. Cell Biol.* 16: 663–669.

Eggink, L.L., H. Park & J. K. Hoober. 2001. The role of chlorophyll b in photosynthesis: Hypothesis. *BMC Plant Biology.* 1:2 www.biomedcentral.com/1471-2229/1/2

Ehrhartd, C., C. Kneuer & U. Bakowsky. 2004. Selectins – an emerging target for drug delivery. *Adv.Drug Delivery Rev.* 56: 527–549.

Elias, P. 2006. *Biotech Rice Puts Harvest at Risk.* AP article in *The Hartford Courant.* Section E. 21 October 2006.

Elowitz, M.B. & S. Leibler. 2000. A synthetic oscillatory network of transcriptional regulators. *Nature.* 403: 335–338.

ENCODE. 2007. Identification and analysis of functional elements in 1% of the human genome by the ENCODE pilot project. *Nature.* 447(7146): 799–816. Available at: www.nature.com/nature/journal/v447/n7146/nature05874.pdf

Engdahl, W. 2006. Monsanto Buys “Terminator” Seeds Company. *Global Research.* 26 Aug. 2006. www.globalresearch.ca/index/php?context=viewArticle&code=ENG20060827&articleId=3082

Engels, B.M. & G. Hutvagner. 2006. Principles and effects of microRNA-mediated post-transcriptional gene regulation. *Oncogene.* 25: 6163–6169.

Epigenetic Station. 2007. *Cellular Methylation Machinery.* (Retrieved 1 May 2007.) www.epigeneticstation.com/cellular-methylation-machinery/

Estévez, R. & T.J. Jentsch. 2002. CLC chloride channels: Correlating structure with function. *Curr. Opin. Struct. Biol.* 12: 531–539.

Etheridge et al. Eds. 2006. *Computational Biology.* Chapman and Hall/CRC, Atlanta.

ETV. 2004. *ETV Joint Verification Statement: PathAlert® Detection Kit.* www.epa.gov/etv/pdfs/vrvs/01_vs_invitrogen.pdf Also see: www.epa.gov/etv/pdfs/vrvs/01_vr_invitrogen.pdf

Evans, W.H. & P.E.M. Martin. 2002. Gap junctions: structure and function. *Mol. Membrane Biol.* 19: 121–136.

Evolvingcode. 2006. *Central Dogma.* www.evolvingcode.net/index.php?page=central_dogma

Fahlke, C. 2004. Molecular physiology and pathophysiology of CLC-type chloride channels. *Adv. Mol. Cell Biol.* 32: 189–217.

Fangerau, H. 2005. Can artificial parthenogenesis sidestep ethical pitfalls in human therapeutic cloning? An historical perspective. *J. Med. Ethics*. 31:733–735.

Farabee, M-J. 2001. *Control of Gene Expression*. (On-line lecture notes.) www.emc.maricopa.edu/faculty.farabee/biobk/BioBookGENCTRL.html

________. 2001. *Cellular Metabolism and Fermentation. Glycolysis, the Universal Process*. (Retrieved 12/1/2005.) www.emc.maricopa.edu/faculty/farabee/BIOBK/BioBookGlyc.html

Fatt, P. & B. Katz. 1953. The effect of inhibitory nerve impulses on a crustacean muscle fiber. *J. Physiol*. 121: 374–389.

Fedorov, A. & L. Fedorova. 2004. Introns: Mighty elements from the RNA world. *J. Mol. Evol*. 59: 718–721.

Feinberg, A.P. 2007. Phenotypic plasticity and the epigenetics of human disease. *Nature*. 447: 433–440.

Feinberg, A.P., R. Ohlsson & S. Hemikoff. 2006. The epigenetic progenitor origin of human cancer. *Nat. Rev.—Genetics*. 7: 21–33.

Ferré-D'Amare, A.R. & P.B. Rupert. 2002. Ribozymes and RNA catalysis. *Biochemical Society*. 1105–1109. (Available online.)

Fialkow, P.J., S.M. Gartler & A. Yoshida. 1967. Clonal origin of chronic myelocytic leukemia in man. *Proc. Natl. Acad. Sci*. 58: 1468–1471.

Field Museum. 2007. *Gregor Mendel: Planting the Seeds of Genetics*. www.fieldmuseum.org/mendel/story pea.asp

Fire, A. & S-Q. Xu. 1995. Rolling replication of short DNA circles. *Proc. Natl. Acad. Sci. (USA)*. 92(10): 4641–4645.

Folk, R.L. 1992. Bacteria and nannobacteria revealed in hardgrounds, calcite cements, native sulfur, sulfide minerals, and travertines. *Geological Society of America, Abstracts with Programs*, v. 24, p. 104.

________. 1993. SEM imaging of bacteria and nannobacteria in carbonate sediments and rocks. *J. Sedimentary Petrology*. 63: 990–999.

Forster, T. 1948. Intermolecular energy migration and fluorescence. *Ann. Phys*. 2: 55–75.

Foster, R.G. & L. Kreitzman. 2005. *Rhythms of Life: The Biological Clocks that Control the Daily Lives of Every Living Thing*. Yale University Press, New Haven.

Fournier, R.L. 2006. *Basic Transport Phenomena in Biomedical Engineering*. Taylor & Francis, New York.

Frankenberger, B., S. Regn et al. 2005. Cell-based vaccines for renal cell carcinoma: Genetically-engineered tumor cells and monocyte-derived dendritic cells. *World J. Urol*. 23: 166–174.

Friedman, J. & D. Xue. 2004. To live or die by the sword: The regulation of apoptosis by the proteasome. *Developmental Cell*. 6: 460–461.

Friedman, R. 2001. A Theme Among Brain Diseases. www.genomenewsnetwork.org/articles/02_01/Stuttering _genes.shtml

Fueyo, J. et al. 2003. Preclinical characterization of the antiglioma activity of a tropism-enhanced adenovirus targeted to the retinoblastoma pathway. *J. Nat. Cancer Institute*. 95(9): 652–660.

Fujimoto, Y., K. Yagita & H. Okamura. 2006. Does mPER2 protein oscillate without its coding mRNA cycling?: post-transcriptional regulation by cell clock. *Genes to Cells*. 11: 525–530.

Furshpan, E.J. & D.D. Potter. 1959. Transmission at the giant motor synapses of the crayfish. *J. Physiol*. 145: 289–325.

Gains, M.J. & A.C. LeBlanc. 2007. Prion protein and prion diseases: the good and the bad. (Review article) *Can. J. Neurol. Sci*. 34: 126–145.

Ganguly, R. & I.K. Puri. 2006. Mathematical model for the cancer stem cell hypothesis. *Cell Prolif*. 39: 3–14.

Gao, F. et al. 1999. Origin of HIV-1 in the chimpanzee *Pan troglodytes*. *Nature*. 397: 436–444.

Gao, X. et al. 2005. *In vivo* molecular and cellular imaging with quantum dots. *Curr. Opin. Biotechnol*. 16: 63–72.

Garner, A.L. et al. 2006. Implications of a simple mathematical model to cancer cell population dynamics. *Cell Prolif*. 39: 5–28.

Garner, M. 1987. Effects of DNA supercoiling on the topological properties of nucleosomes. *PNAS USA*. 84(9): 2620–2623.

Genetics. 2004. (Center for Genetics Education). *Genetics Fact Sheet 14: Genetic Imprinting*. www.genetics.com.au/pdf/factSheets/FS14.pdf

Genetics Home Reference. 2007. *Proteomics*. US National Library of Medicine: Glossary. http://ghr.nlm.nih.gov/ghr/glossary/proteomics;

Genomes. 2001. *The Human Genome*. //exobio.ucsd.edu/Space_Sciences/genomes.htm

Germenis, A.E. & A. Patrikidou. 2005. Proteomics challenging medicine. *Haema*. 8(2): 183–192.

Gibbs, W. 2003. The unseen genome: gems among the junk. *Scientific American.* 289(5): November 2003. www.mindfully.org/GE/2003/Junk-GenomeNov03.htm

Gilbert, S.F. 2003. *Developmental Biology, 7th ed.* Sinauer Associates, Sunderland, MA.

Gilon, P., M.A. Ravier, J-C Jonas & J-C Henquin. 2002. Control mechanisms of the oscillations of insulin secretion *in vitro* and *in vivo. Diabetes.* 51: Supp. 1. S144–S151.

Gingeras, T.R. 2007. Origin of phenotypes: Genes and transcripts. *Genome Res.* 17: 682–690.

Giovannoni, S.J. et al. 2005. Proteorhodopsin in the ubiquitous marine bacterium SAR11. *Nat. Lett.* 438: 82–85.

Glick, D.M. 2003. *Glossary of Biochemistry and Molecular Biology.* www.portlandpress.com/pp/books/online/glick/search.htm

Glickman, M.H. & N. Adir. 2004. The proteasome and the delicate balance between destruction and rescue. *PloS Biol.* 2(1): 0025–0027. //biology.plosjournals.org

Gogarten, J.P. et al. 2002. Inteins: structure, function and evolution. *Annu. Rev. Microbiol.* 56: 263–287.

Goldman, B. 2005. *Heat Shock Protein's Vaccine Potential: From Basic Science Breakthroughs to Feasible Personalized Medicine.* web paper. www.antigenics.com/whitepapers/hsp_potential.html

Goodsell, D.S. 2004. *The Calcium Pump.* (March 2004 Molecule of the Month). www.pdb.org/pdb/static.do?p=education_discussion/molecule_of_the_month/pdb51_3.html

Gordon, A.M., A.F. Huxley & F.J. Julian. 1966. The variation in isometric tension with sarcomere length in vertebrate muscle fibers. *J. Physiol.* 184: 17–192.

Gore, P.J.W. 1999. *Radiometric Dating.* www.gpc.edu/~pgore/geology/geo102/radio.htm

Gorvel, J-P. & C. de Chastellier. 2005. Bacteria spurned by self-absorbed cells. *Nat. Med.* 11(1): 18–19.

Gozuacik, D. & A. Kimchi. 2004. Autophagy as a cell death and tumor suppressor mechanism. *Oncogene.* 23: 2891–2906.

Grandin, H.M., B. Städtler, M. Textor & J. Vörös. 2005. Waveguide excitation fluorescence microscopy: A new tool for sensing and imaging the biointerface. Paper available on-line at: www.textorgroup.ch/pdf/publications/journals/128/Grandin_et%20al_inPress.pdf

Grant, B. 2007. The literature: Stimulating discoveries: Two groups reveal an essential messenger in store-operated calcium entry. *The Scientist.* 21(10): p. 79.

Green, R. 2001. *The Human Embryo Research Debates: Bioethics in the Vortex of Controversy.* Oxford Univ. Press, New York.

Greenpeace. 2006. *Say No to Genetic Engineering.* www.greenpeace.org/international/campaigns/genetic-engineering

Grossman, A., B. Zeiler & V. Sapirstein. 2003. Prion protein interactions with nucleic acid: Possible models for prion disease and prion function. *Neurochem. Res.* 28(6): 955–963.

Guillemin, E. 1949. *The Mathmatics of Circuit Analysis.* MIT Press, Cambridge, MA.

Guttmacher, A.E. & F.S. Collins. 2002. Genomic medicine—A primer. *New Engl. J. Med.* 347(19): 1512–1520.

Guyton, A.C. 1991. *Textbook of Medical Physiology. 8th ed.* W.B. Saunders Co., Philadelphia.

Haab, B.B., M.J. Dunham & P.O. Brown. 2001. Protein microarrays for highly parallel detection and quantification of specific proteins and antibodies in complex solutions. *Genome Biol.* 2(2): 1–13.

Hackert. 2001. *Lipid Metabolism I – Catabolism.* //courses.cm.utexas.edu/archive/Fall2001/CH339K/Hackert/Beta_Oxid/beta_oxid.html

Hajra, K.M. & J.R. Liu. 2004. Apoptosome dysfunction in human cancer. *Apoptosis.* 9: 691–704.

Halim, N. 1999. *Methylation: Gene Expression at the Right Place and Time.* In *The Scientist.* 13(24): 21. www.the-scientist.com/1999/12/6/21/1/

Hall, E.H., K.H. Schoenbach and S.J. Beebe. 2005. Nanosecond pulsed electric fields (nsPEF) induce direct electric field effects and biological effects on human colon carcinoma cells. *DNA Cell Biol.* 24(5): 283–291.

Hall, T.M.T. 2005. Structure and function of argonaute proteins. *Structure.* 13: 1403–1408.

Hall, J., M. Mattern & T. Butt. 2005. Mining the Ubiquitin Pathway. *The Scientist.* 19(23): 5 Dec. 2005. www.the-scientist.com

Hardman, J.G. & L.E. Limbird, Eds. 1996. *Goodman and Gilman's The Pharmacological Basis of Medical Practice,* 9th ed. McGraw-Hill, New York.

Harms, J.M, & F. Schlünzen. 2005. *Ribosome Structure: The path from preparation to structure determination—for ribosomes only.* www.riboworld.com/structure/protkrist-eng.html

Harris, D.F, MD. 2007. *Vaccination: Edward Jenner and vaccination.* Online essay. www.worldwideschool.org/library/books/tech/medicine/EdwardJennerAndVaccination/Chap1.html

Hasebe, M., T. Murata et al. 2003. Division of speciation mechanisms. *Annual Report 2002,* National Institute for Basic Biology (Japan). www.nibb.ac.jp/annual_report/2003/03ann404.html

Hayward, S., J. Rinehart & D. Denlinger. 2004. Desiccation and rehydration elicit distinct heat shock protein transcript responses in flesh fly pupae. *J. Exp. Biol.* 207: 963–971.

He, L. & G.J. Hannon. 2004. MicroRNAs: Small RNAs with a big role in gene regulation. *Nature Reviews—Genetics.* 5: 522–531. www.nature.com/reviews/genetics

Heegaard, E.D. & K.E. Brown. 2002. Human parvovirus B19. *Clin. Microbiol. Rev.* July. pp. 485–505.

Heiner, K.R. 2005. *Hemoglobin.* www.bio.davidson.edu/Courses/Molbio/MolStudents/spring2005/Heiner/hemoglobin.html

Held, P. 2005. An Introduction to Fluorescence Resonance Energy Transfer (FRET) Technology and its Application in Bioscience. *Application Note:* Bio-Tek Instruments. (Available online.) www.biotek.com/resources/docs/Fluorescence_Resonance_Energy_Transfer_Technology_App_Note.pdf

Held, T.K. et al. 1999. Gamma interferon augments macrophage activation by lipopolysaccharide by two distinct mechanisms, at the signal transduction level and via an autocrine mechanism involving Tumor Necrosis Factor Alpha and Interleukin-1. *Infect. Immunol.* 67(1): 206–221.

Henderson, J. 2005. Ernest Starling and 'Hormones': an historical commentary. *J. Endocrinol.* 184: 5–10.

Hermann, P.C. et al. 2007. Distinct populations of cancer stem cells determine tumor growth and metastatic activity in human pancreatic cancer. *Cell Stem Cell.* 1: 313–323.

Heslop-Harrison, J. 1963. Structure and morphogenesis of lamellar systems in grana-containing chloroplasts. *Planta.* 60(3): 243–260.

Hess, B. & A. Boiteux. 1971. Oscillatory phenomena in biochemistry. *Ann. Rev. Biochem.* 40: 237–258.

Hill, M.M, C. Adrain & S.J. Martin. 2003. Portrait of a killer: the mitochondrial apoptosome emerges from the shadows. *Mol. Interventions.* 3(1): 19–26.

Hillebrecht, J.R. et al. 2006. Structure, function, and wavelength selection in blue-absorbing proteorhodopsin. *Biochemistry.* 45: 1579–1590.

Hodgkin, A.L. & A.F. Huxley. 1952. A quantitative description of membrane current and its application to conduction and excitation in nerve. *J. Physiol.* (London), 117: 500–544.

Holden, C. & G. Vogel. 2004. A technical fix for an ethical bind? *Science.* 306: 24 Dec. 2174–2176. (Sources for human ESCs.)

Holland, J.H. 1995. *Hidden Order: How Adaptation Builds Complexity.* Helix Books, New York.

Hollenbach, D. 2002. *The Common Good and Christian Ethics.* Cambridge Univ. Press, Cambridge, UK. Accessed at: http://assets.cambridge.org/97805218/02055/sample/9780521802055.pdf

Horack, J. 1997. *Protein Crystal Growth.* http://science.nasa.gov/pcg_why.htm

Hossain, F., C.E. Pray, Y. Lu, J. Huang, C. Fan & R. Hu. 2004. Genetically modified cotton and farmers' health in *China. Int. J. Occup. Environ. Health.* 10: 296–303.

Hummer, G., F. Schotte & P.A. Anfinrud. 2004. Unveiling functional protein motions with picosecond X-ray crystallography and molecular dynamics simulations. *PNAS.* 101(43): 15330–15334.

Hunter, D.J. et al. 2007. A genome-wide association study identifies alleles in FGFR2 associated with risk of sporadic postmenopausal breast cancer. *Nat. Genet.* 39(7): 870–874. (Epub 27 May 2007.)

Hunter, J. & R. Mitchell. 1997. *Genetics and Inheritance: Gregor Mendel.* Dartmouth College Biology 4 Website. (Retrieved 1 May 2007.) www.dartmouth.edu/~cbbc/courses/bio4/bio4-1997/01-Genetics.html

Huskey, R.J. 1998. *Summary of Porphyrin and Heme Synthesis.* www.people.virginia.edu/~rjh9u/porphsyn.html

Huxley, A.F. 2000. Mechanics and models of the myosin motor. *Phil. Trans. Royal Soc. Lond. B.* 355: 433–440.

Institute for Civil Society. 2006. *Moral Issues Surrounding Stem Cells.* www. counterbalance.net/stemcell/moral1-body.html

Isenberg, A.R. & J.M. Moore. 1999. *Mitochondrial DNA Analysis at the FBI Laboratory.* www.fbi.gov/hq/lab/fsc/backissu/july1999/dnatext.htm#Step%204:%20Amplification%20by%the%20Polymerase%20Chain%20Reaction

Isis. 2000. *The function of introns-iRNAs.* (Web article.) //isis.bit.uq edu/irnas.html

Jang, J. & D. Shaffer. 2006. Microarraying the cellular microenvironment. *Mol. Syst. Biol.* 2: 39 (Published online 4 July 2006. doi: 10.1038/msb4100079.)

Jankovic, J. 2006. *Huntington's Disease.* www.bcm.edu/neurol/struct/parkinson/huntington.html

Jeffares, D.C. & A. Poole. 2006. *Were Bacteria the First Forms of Life on Earth?* www.actionbioscience.org/newfrontiers/jeffares_poole.html

JGI. 2004. *How sequencing is done.* DOE Joint Genome Institute. www.jgi.doe.gov/education/how/

Ji, L., J.D. Minna & J.A. Roth. 2005. 3p21.3 tumor suppressor cluster: prospects for translational applications. *Future Oncol.* 1(1): 79–92.

Jobling, M.A. & P. Gill. 2004. Encoded evidence: DNA in forensic analysis. *Nat. Rev.—Genetics.* 5: 739–751.

Johnson, V. et al. 2004. Multiple organ dysfunction syndrome in humans and dogs. *J. Veterinary Emergency and Critical Care.* 14(3): 158–166.

Jones, R.W., D.G. Green & R.B. Pinter. 1962. Mathematical simulation of certain receptor and effector organs. *Fed. Proc.* 21(1): 97–102.

Jordan, D.W. et al. 2004. Effect of pulsed, high-power radiofrequency radiation on electroporation of mammalian cells. *IEEE Trans. on Plasma Science.* 32(4): 1573–1578.

Josephson, M. 2002. *Making Ethical Decisions.* W. Hanson, ed. http://josephsoninstitute.org/MED/MED-intro+toc.htm

Josephson, R.K., J.G. Malamud & D.R. Stokes. 2000. Asynchronous muscle: A primer. *J. Exp. Biol.* 203: 2713–2722.

Joyce, D.E. 2002. *Phylogeny and Reconstructing Phylogenetic Trees.* //aleph0.clarku.edu/~djoyce/java/Phyltree/intro.html

Jurisica, I. & D. Wigle. 2006. *Knowledge Discovery in Proteomics.* Chapman and Hall/CRC, Boca Raton, FL.

Kaftan, D. et al. 2002. From chloroplasts to photosystems: in situ scanning force microscopy on intact thykaloid membranes. *EMBO J.* 21(22): 6146–6153.

Kajander, E.O. & N. Ciftcioglu. 1998. An alternate mechanism for pathogenic intra- and extracellular calcification and stone formation. *PNAS* 95(14): 8274–8279.

Kandel, E.R., J.H. Schwartz & T.M. Jessell. 1991. *Principles of Neural Science,* 3rd ed. Appleton and Lange, Norwalk, CT.

Kaneko, K. 2006. *Life: An Introduction to Complex Systems Biology.* Springer, New York.

Karp, G. 2007. *Cell and Molecular Biology: Concepts and Experiments,* 5th ed. Wiley/Life and Medical Sciences, New York.

Katz, B. 1966. *Nerve, Muscle, and Synapse.* McGraw-Hill, New York.

Kaufmann, B., A.A. Simpson & M.J. Rossmann. 2004. The structure of human parvovirus B19. *PNAS.* 101(32): 11628–11633.

Kemp, J.T. et al. 2005. A novel method for STR-based DNA profiling using microarrays. *J. Forensic Sci.* (Technical Note) 50(5): 1–5.

Kent, J.T. et al. 2005. A novel method for STR-based DNA profiling using microarrays. *J. Forensic Sci.* 50(5): 1–5. (Paper available online @ www.astm.org)

Kenyon. 2006. *Methanogen Overview.* //biology.Kenyon.edu/Microbial_Biorealm/archaea/methanogens/methanogens.html

Kiang, H. et al. 2005. Comparative effect of oncolytic adenoviruses with E1A or E1B-55 kDa deletions in malignant gliomas. *Neoplasia.* 7(1): 48–56.

Kiessling, A.A. 2005. Eggs alone. Human parthenotes: An ethical source for therapies? *Nature.* 434: 145. doi: 10.1038/434145a.

Kim, S., G.R. Rosania & Y-T. Chang. 2004. Dedifferentiation? What's next? *Mol. Interventions.* 4(2): 83–85.

Kimball, J. 2007a. *Genetic Variability of HIV.* (Retrieved from *Kimball's Biology Pages* 5 May 2007.) //users.rcn.com/jkimball.ma.ultranet/BiologyPages/A/AIDS.html#GeneticVariability

________. 2007b. *Photosynthesis: The Role of Light. (Chemiosmosis in Chloroplasts).* //users.rcn.com/jkimball.ma.ultranet/BiologyPages/L/LightReactions.html

________. 2007c. *Cellular Respiraition (Chemiosmosis in Mitochondria).* //users.rcn.com/jkimball.ma.ultranet/BiologyPages/C/CellularRespiration.html

________. 2007d. *The Cell Cycle.* //users.rcn.com/jkimball.ma.ultranet/BiologyPages/C/CellCycle.html

________. 2007e. *Tumor Suppressor Genes (p53).* //users.rcn.com/jkimball.ma.ultranet/BiologyPages/T/TumorSuppressorGenes.html

________. 2007f. *Antisense RNA.* //users.rcn.com/jkimball.ma.ultranet/BiologyPages/A/AntisenseRNA.html

________. 2006a. *Stem Cells.* //users.rcn.com/jkimball.ma.ultranet/BiologyPages/S/Stem_Cells.html

________. 2006b. *Transport Across Cell Membranes.* //users.rcn.com/jkimball.ma.ultranet/BiologyPages/D/Diffusion.html

________. 2006c. *The Protists.* //users.rcn.com/jkimball.ma.ultranet/BiologyPages/P/Protists.html

________. 2006d. *Telomeres.* //users.rcn.com/jkimball.ma.ultranet/BiologyPages/T/Telomeres.html

________. 2006e. *Meiosis.* //users.rcn.com/jkimball.ma.ultranet/BiologyPages/M/Meiosis.html

________. 2006f. *Antigen Receptor Diversity.* //users.rcn.com/jkimball.ma.ultranet/BiologyPages/A/AgREceptorDiversity.html

________. 2006g. *Recombinant DNA and Gene Cloning.* //users.rcn.com/jkimball.ma.ultranet/BiologyPages/R/RecombinantDNA.html

________. 2006h. *Taxonomy: Classifying Life.* //users.rcn.com/jkimball.ma.ultranet/BiologyPages/T/Taxonomy.html

________. 2005b. *Gene Translation: RNA—Protein.* //users.rcn.com/jkimball.ma.ultranet/BiologyPages/T/Translation.html

________. 2005c. *Gene Expression: Transcription.* //users.rcn.com/jkimball.ma.ultranet/BiologyPages/T/Transcription.html

________. 2005d. *Sex Chromosomes.* //users.rcn.com/jkimball.ma.ultranet/BiologyPages/S/SexChromosomes.html

________. 2005e. *DNA Repair.* //users.rcn.com/jkimball.ma.ultranet/BiologyPages/D/DNArepair.html

________. 2005f. *Mutations.* //users.rcn.com/jkimball.ma.ultranet/BiologyPages/M/Mutations.html

________. 2005g. *Ribozymes.* //users.rcn.com/jkimball.ma.ultranet/BiologyPages/R/Ribozymes.html

________. 2005h. *The Complement System.* //users.rcn.com/jkimball.ma.ultranet/BiologyPages/C/Complement.html

________. 2005i. *The Hormones of the Human.* //users.rcn.com/jkimball.ma.ultranet/BiologyPages/H/HormoneTable.html

________. 2005j. *Glycoproteins.* //users.rcn.com/jkimball.ma.ultranet/BiologyPages/G/Glycoproteins.html

________. 2005k. *Antigen Receptors.* //users.rcn.com/jkimball.ma.ultranet/BiologyPages/A/AntigenReceptors.html

________. 2005l. *Antigen Receptor Diversity.* //users.rcn.com/jkimball.ma.ultranet/BiologyPages/A/AgReceptorDiversity.html

________. 2005m. *Autophagy.* //users.rcn.com/jkimball.ma.ultranet/BiologyPages/A/Autophagy.html

________. 2005n. *The Nucleus.* //users.rcn.com/jkimball.ma.ultranet/BiologyPages/N/Nucleus.html

________. 2005o. *Lysosomes and Peroxisomes.* //users.rcn.com/jkimball.ma.ultranet/BiologyPages/L/Lysosomes.html

________. 2005p. *Synapses.* //users.rcn.com/jkimball.ma.ultranet/BiologyPages/S/Synapses.html

________. 2005q. *Ribozymes.* //users.rcn.com/jkimball.ma.ultranet/BiologyPages/R/Ribozymes.html

________. 2004a. *The Origin of Life: Abiotic Synthesis of Organic Molecules.* //users.rcn.com/jkimball.ma.ultranet/BiologyPages/A/AbioticSynthesis.html

________. 2004b. *Apoptosis.* //users.rcn.com/jkimball.ma.ultranet/BiologyPages/A/Apotosis.html

________. 2004c. *Bacteria.* //users.rcn.com/jkimball.ma.ultranet/BiologyPages/E/Eubacteria.html

________. 2004d. *Enzymes.* //users.rcn.com/jkimball.ma.ultranet/BiologyPages/E/Enzymes.html

________. 2004e. *The Cytoskeleton.* //users.rcn.com/jkimball.ma.ultranet/BiologyPages/C/Cytoskeleton.html

________. 2004f. *The Proteasome.* //users.rcn.com/jkimball.ma.ultranet/BiologyPages/P/Proteasome.html

________. 2004g. *The Golgi Apparatus.* //users.rcn.com/jkimball.ma.ultranet/BiologyPages/G/Golgi.html

________. 2003a. *Ribosomes.* //users.rcn.com/jkimball.ma.ultranet/BiologyPages/R/Ribosomes.html

________. 2003b. *Proteins.* //users.rcn.com/jkimball.ma.ultranet/BiologyPages/D/DenaturingProtein.html

________. 2003c. *Protein Kinesis.* //users.rcn.com/jkimball.ma.ultranet/BiologyPages/P/ProteinKinesis.html

________. 2003d. *Bond Energy.* //users.rcn.com/jkimball.ma.ultranet/BiologyPages/B/BondEnergy.html

________. 2003e. *Genetic Mosaics.* //users.rcn.com/jkimball.ma.ultranet/BiologyPages/M/Mosaics.html

________. 2003f. *Photosynthesis: Pathway of Carbon Fixation.* //users.rcn.com/jkimball.ma.ultranet/BiologyPages/C/CalvinCycle.html

________. 2003g. *Testing the Chemiosmosis Theory.* //users.rcn.com/jkimball.ma.ultranet/BiologyPages/C/Chemiosmosis_demo.html

________. 2003h. *Protein Kinesis: Getting Proteins to Their Destination.* //users.rcn.com/jkimball.ma.ultranet/BiologyPages/P/ProteinKinesis.html

________. 2002. *Serine Proteases.* //users.rcn.com/jkimball.ma.ultranet/BiologyPages/S/Serine_Proteases.html

King, M.W. 2006a. *Lipid Metabolism.* //web.indstate.edu/thcme/mwking/lipid-synthesis.html

________. 2006b. *Amino Acid Metabolism.* //web.indstate.edu/thcme/mwking/amino-acid-metabolism.html

________. 2006c. *Muscle Biochemistry.* //web.indstate.edu/thcme/mwking/muscle.html

________. 2005a. *RNA Synthesis.* //web.indstate.edu/thcme/mwking/rna.html

________. 2005b. *Myoglobin and Hemoglobin.* //web.indstate.edu/thcme/mwking/hemoglobin-myoglobin.html

________. 2005c. *Peptides and Proteins.* //web.indstate.edu/thcme/mwking/protein_structure.html

________. 2005d. *Biological Oxidations, Oxidative Phosphorylation.* //web.indstate.edu/thcme/mwking/oxidative-phosphorylation.html

Kirschner, M & J. Gerhart. 2005. *The Plausibility of Life.* Yale University Press, New Haven, CT.

Kiss, I.Z., M. Quigg, S-H.C Chun, H. Kori & J.L. Hudson. 2007. Characterization of synchronization in interacting groups of oscillators: Application to seizures. *Biophys. J. BioFAST.* 5 Oct. 2007. doi:10.1529//biophysj.107.113001

Kiss, T. 2001. Small nucleolar RNA-guided post-transcriptional modification of cellular RNAs. *EMBO J.* 20(14): 3617–3622.

________. 2005. *Small Nucleolar RNAs Guide rRNA Modification.* (On-line review.) www.bioinfo.org.cn/book/Great%20Experiments/great10.htm

Klimanskaya, I., Y. Chung, S. Becker, S-J Lu & R. Lanza. 2006. Human stem cell lines derived from single blastomeres. Preprint from 23 August 2006 *Nature.* www.nature.com

Knopf, G.K. & A.S. Bassi. 2007. *Smart Sensor Biotechnology.* CRC Press, Boca Raton, FL.

Kono, T. et al. 2004. Birth of parthenogenetic mice that can develop into adulthood. *Nature.* 428: 860–864.

Koonin, E. 2006. The origin of introns and their role in eukaryogenesis: A compromise solution to the introns-early versus intronslate debate. *Biology Direct.* 1: 22.

Koupparis, P. 2002. *DNA Evidence: Is it Safe to Convict?* www.scandals.org/articles/pk020929.html

Krause, D.S. et al. 2001. Multi-organ, multi-lineage engraftment by a single bone marrow-derived stem cell. *Cell.* 105(3): 369–377.

Kroemer, G. & S.J. Martin. 2005. Caspase-independent cell death. *Nat. Med.* 11(7): 725–730.

Kubista, M. et al. 2006. The real-time polymerase chain reaction. *Mol. Aspects Med.* 27: 95–125. www.elsevier.com/locate/mam

Kucia, M. et al. 2005. Trafficking of normal stem cells and metastasis of cancer stem cells involve similar mechanisms: Pivotal role of the SDF-1–CXCR4 axis. *Stem Cells.* (published online 11 May 2005; doi:10.1434/stemcells.2004-0342.) www.StemCells.com

Kumar, V., J. Heikkonen & P. Englehardt. 2005. *Cryo-EM Particle Reconstruction.* www.lce.hut.fi/research/sysbio/cryoem

Kuska, B. 1998. Beer, Bethesda and biology: How "Genomics" came into being. *JNCI.* 90(2): 93. doi:10.1093/jnci/90.2.93

Kutter, E. 1997. *Phage Therapy: Bacteriophages as Antibiotics.* www.evergreen.edu/user/T4/PhageTherapy/Phagethea.html

________. 2000. *Phage Therapy Review—March 2000 Addendum.* www.evergreen.edu/user/T4/PhageTherapy/webaddition.html

Lander, E.S. et al. 2001. Initial sequencing and analysis of the human genome. Unknown on-line paper title. *Nature.* 409(6822): 860–921.

Landrau, M. 2004. Early step in protein-folding revealed by bacterial mutant. *Microbiology: Harvard Focus. 6 Feb.* //focus.hms.harvard.edu/2004/Feb6_2004/microbiology.html

Larsson, C. et al. 2004. *In situ* genotyping individual DNA molecules by target-primed rolling-circle amplification of padlock probes. *Nature Methods.* 1(3): 227–232.

LBNL. 2006. (Lawrence Berkeley National Laboratory) *All About Breast Cancer Genes.* www.lbl.gov/Education/ELSI/Frames/cancer-genes-f.html

Lefkowitz, R.J. & S.K. Shenoy. 2005. Transduction of receptor signals by β-arrestins. *Science.* 308: 512–517.

Leicester Microbiology. 2005. *Parvoviruses.* www-micro.msb.le.ac.uk/3035/Parvoviruses.html

Leicester Microbiology. 2004. *Prion Diseases.* www-micro.msb.le.ac.uk/3035/prions.html

Lerner, M. 2006. Scientists reverse monkeys' diabetes with cells of pigs: Findings called a promising step to help humans. *Houston Chronicle.* 20 Feb. www.chron.com/disp/story.mpl/nation/3671243.html

Levine, B. 2005. Eating oneself and uninvited guests: Autophagy-related pathways in cellular defense. *Cell.* 120: 159–162.

Levine, J. 2006. *Timing is Everything.* www.pbs.org/wgbh/nova/odyssey/timing.html

Li, L. & W.B. Neaves. 2006. Normal stem cells and cancer stem cells: The niche matters. *Cancer Res.* 66(9): 4553–4557.

Li, L & T. Xie. 2005. Stem cell niche: Structure and function. *Annu. Rev. Cell Dev. Biol.* 21: 605–631.

LifeSite. 2005. First human "clones" created without use of sperm. LifeSiteNews.com. 9 September. www.lifesite.net/ldn/printerfriendly.html

Lilley, D.M. 2003. Ribozymes—a snip too far? *Nat. Struct. Biol.* 10(9): 672–673.

Lindell, D. et al. 2005. Photosynthesis genes in marine viruses yield proteins during host infection. *Nature,* on-line. www.nature.com

________. 2004. Transfer of photosynthesis genes to and from *Prochlorococcus* viruses. *PNAS,* 27 July. [PubMed Abstract]

Line, M.A. 2005. A hypothetical pathway from the RNA to the DNA world. *Origins of Life and Evolution of Biosphere.* 35: 395–400.

Little, P.F.R. 2005. Structure and function of the human genome. *Genome Research.* www.genome.org

Liu, D. et al. 1996. Rolling circle DNA synthesis: Small circular oligonucleotides as efficient templates for DNA polymerases. *J. Am. Chem. Soc.* 118(7): 1587–1594.

Liu, X-Q. 2000. Protein-splicing intein: Genetic mobility, origin, and evolution. *Ann. Rev. Genetics.* 34: 61–76.

Lodish, H. et al. 2003. *Molecular Cell Biology,* 5th ed. W.H. Freeman, New York.

Lovin, R. 2003. The common good and Christian ethics. in *Theology Today.* October 2003. Cambridge University Press.

Lowe, T.M. 2000. *The Small Nucleolar RNAs.* (online paper) http://lowelab.ucsc.edu/~lowe/thesis/node9.html

Luman, S. 2004. Singapore wants you! *Wired Magazine.* Aug. Issue 12.8. www.wired/wired/archive/12.08/singapore.html

Lynch, M. 2002. Intron evolution as a population-genetic process. *PNAS USA.* 99(9): 6118–6123.

Lyngstadaas, S.P. 2001. Synthetic hammerhead ribozymes as tools in gene expression. *Crit. Rev. Oral Biol. Med.* 12: 469–478.

MacBeath, G. & S.L. Schreiber. 2000. Printing proteins as microarrays for high-throughput function determination. *Science.* 289: 1760–1763.

Machné, R. et al. 2006. The SBML ODE Solver Library: a native API for symbolic and fast numerical analysis of reaction networks. *Bioinformatics.* 22(11): 1406–1407.

Mad Cow. 1998. *Prion Molecular Phylogeny.* www.mad-cow.org/moly_phylo.html

Madsen, M.F., S. Danø, P.G. Sørensen. 2005. On the mechanisms of glycolytic oscillations in yeast. *FEBS J.* 272: 2648–2660.

Maduke, M., C. Miller & J.A. Mindell. 2000. A decade of CLC chloride channels: Structure, mechanism, and many unsettled questions. *Ann. Rev. Biophys. Biomol. Structure.* 29: 411–438.

Mahadevan, L.C. 2004. Phosphorylation and acetylation of histone H3 at inducible genes: two controversies revisited. *Novartis Found. Symp.* 259: 102–114.

Mai, E. et al. 2002. Comparison of ELISA and AlphaScreen® Assay Technologies for Measurement of Protein Expression Levels. *AlphaScreen Application Note ASC-017.* PerkinElmer® Life Sciences. www.perkinelmer.com/lifesciences

Maienschein, J. 2001. What's in a name? Embryos, clones, and stem cells. *Amer. J. Bioethics.* 2(1): 1–8.

Malerczyk, C. et al. 2005. Ribozyme targeting of the growth factor pleiotrophin in established tumors: a gene therapy approach. *Gene Therapy.* 12: 339–346.

Malone, K.E. et al. 1998. BRCA1 mutations and breast cancer in the general population: analysis in women before age 35 years and in women before age 45 years with first-degree family history. *JAMA.* 279(12): 922–929.

Marahrens, Y., J. Loring & R. Jaenisch. 1998. Review of the role of the Xist gene in X chromosome choosing. *Cell.* 92(5): 657–664.

Maramorosch, K. 1981. Spiroplasmas: agents of animal and plant diseases. *BioScience.* 31(5): 374–380.

Margulis, L. 1981. *Symbiosis in Cell Evolution,* 1st ed. W.H. Freeman and Co., New York.

________. 1970. *Origin of Eukaryotic Cells. Evidence and Research Implications for a Theory of the Origin and Evolution of Microbial, Plant, and Animal Cells on the Precambrian Earth.* Yale University Press, New Haven.

Markwell, J. & D. Namuth. 2003. *Plant Pigments and Photosynthesis.* Peer Reviewed Web Lesson. http://croptechnology.unl.edu/printLesson.cgi?lessonID=939154129

Marmarelis, P.Z. & V.Z. Marmarelis. 1978. *Analysis of Physiological Systems.* Plenum Press, New York.

Maron, S.H. & C.F. Prutton. 1958. *Principles of Physical Chemistry.* Ch. 19. Macmillan Co., New York.

Martinez, A. et al. 2007. Proteorhodopsin photosystem gene expression enables photophosphorylation in a heterologous host. *PNAS.* 104(13): 5590–5595.

Martz, E. 2005. *Earliest Solutions for Macromolecular Crystal Structures.* (Available online.) www.umass.edu/microbio/rasmol/1st_xtls.htm

Masum, H.F., F. Oppacher & G. Carmody. Spring 2001. Genomic algorithms: Metaphors from molecular genetics. *Carleton J. Comp. Sci.* Carleton Univ., Ottawa.

Mattick, J.S. 2003. Challenging the dogma: The hidden layer of non-protein coding RNAs in complex organisms. *BioEssays.* 25: 930–939. (Published on-line at: www.interscience.wiley.com)

________. 2001. Non-coding RNAs: the architects of eukaryotic complexity. *EMBO Reports.* 2(11): 986–991.

Mayer, G. 2004. *Bacteriology—Chapter Seven: Bacteriophage.* //pathmicro.med.sc.edu/mayer/phage.htm

Mayo Clinic. 2006. *U.S. Transplant and Waiting List Statistics by Organ.* www.mayoclinic.com/health/organ-transplant/DA00098

MCB 411. 2000. Module 15. *Post-Transcriptional Events: RNA Splicing* (on-line lecture notes). www.blc.Arizona.edu/marty/411/Modules/mod15.html

McCaffrey, P. 2004. *Water Pore Structure Reveals Junction Function.* Harvard Research Briefs. (Accessed 5/21/07.) http://focus.hms.harvard.edu/2004/June4_2004/research_briefs.html

McClean, P. 1998. *Transcription: Introns, Exons, and Splicing in RNA.* (From online notes for a course in plant molecular genetics at North Dakota State University.) Retrieved 4 May 2007. www.ndsu.edu/instruct/mcclean/plsc731/transcript/transcript4.htm

McClean, P. 1997. *Recombination and Estimating the Distance between Genes.* www.ndsu.nodak.edu/instruct/mcclean/plsc431/linkage/linkage2.htm

McCord, R.P. 2003. *Molecular Tool: Electroporation.* www.bio.davidson.edu/Courses/Molbio/MolStudents/spring2003/McCord/electroporation

McCudden, C.R. et al. 2005. G-Protein signaling: back to the future. *Cell. Mol. Life Sci.* 62: 551–577.

McRee, D.E. 1993. *Practical Protein Crystallography.* Academic Press.

Medical. 2005. HHS *Awards $97 Million Contract to Develop Culture-Based Influenza Vaccine, USA.* (press release) www.medicalnewstoday.com/printerfriendlynews.php?newsid=22230

Meert, J.M. 2004. *Consistent Radiometric Dates.* //gonwandanaresearch.com/radiomet.htm

Melcher, U. 2005. *DNA Repair.* (On-line excerpt from *Molecular Genetics* by Melcher.) //opbs.okstate.edu/~melcher/MG/MGW3/MG313.html

Mellors, R.C. 1998. *Immunopathology: normal immune system.* //edcenter.med.cornell.edu/CUMC_PathNotes/Immunopathology/Immuno_01.html

Mendes, P. 1993. GEPASI: A software package for modelling the dynamics, steady states and control of biochemical and other systems. *Comput. Applicat. Biosci.* 9: 563–571.

Mendes, P. & D.B. Kell. 1998. Nonlinear optimization of biochemical pathways: Applications to metabolic engineering and parameter estimation. *Bioinformatics.* 14: 869–883.

Merck. 2007. *Osteoporosis. Merck Manual Home Edition.* (Accessed on line 3/3/08.) 7 pp. http://www.merck.com/mmhe/print/sec05/ch060/ch060a.html

Meroueh, S.O. et al. 2006. Three-dimensional structure of the bacterial cell wall peptidoglycan. *PNAS* (USA). 102(12): 4404–4409.

Miki, T. et al. 2005. Stem cell characteristics of amniotic epithelial cells. *Stem Cells.* 23: 1549–1559.

Mikkelsen, T. S. et al. 2007. Genome of the marsupial *Monodelphis domestica* reveals innovation in non-coding sequences. *Nature.* 447: 167–177. (10 May 2007)

Miller, S.L. 1953. Production of amino acids under possible primitive Earth conditions. *Science.* 117: 528–529.

Miller, S.L. & H.C. Urey. 1959. Organic compound synthesis on the primitive Earth. *Science.* 130: 245–251.

MIT. 2006. *The Central Dogma of Biology.* http://web.mit.edu/esgbio/www/dogma/dogma.html

Mitsui, K. et al. 2003. The homeoprotein Nanog is required for maintenance of pluripotency in mouse epiblast and ES cells. *Cell.* 113: 631–642.

Mohn, J. 1998. The Serial Endosymbiosis Theory of Eukaryotic Evolution. (Web essay accessed 24 May 2007.) www.geocities.com/jjmohn/endosymbiosis.htm?200725

Monsanto. 1998. Roundup Ready® soybeans: Food and feed safety. *Sight.* www.FARMSOURCE.com

Moore, J. 1972. *Heredity and Development,* 2nd ed. Oxford University Press, New York. http://scienceclassics.nas.edu/moore/moore.pdf

Moore, P.B. & T.A. Steitz. 2002. The involvement of RNA in ribosome function. *Nature.* 418: 229–236. (Available on-line at www.Nature.com/nature.)

Mootz, H.D. et al. 2003. Conditional protein splicing: A new tool to control protein structure and function *in vitro* and *in vivo. J. Am. Chem. Soc.* 125: 10561–10569.

Morin, S. et al. 2003. Three cadherin alleles associated with resistance to *Bacillus thuringiensis* in pink bollworm. *PNAS.* 100(9): 5004–5009.

Morris, W., ed. 1973. *The American Heritage Dictionary of the English Language.* Houghton-Mifflin, Boston.

MSGN. 1994a. (Mountain States Genetics Network). *Uniparental Disomy and Genomic Imprinting*. In *Genetic Drift*. 10: Winter 1994. www.mostgene.org/gd/gdvol10c.htm

MSGN. 1994. (Mountain States Genetics Network). *Triplet Repeat Disorders*. In *Genetic Drift*. 10: Winter 1994. www.mostgene.org/gd/gdvol10d.htm

Nam, J-M., S-J. Park & C.A. Mirkin. 2002. Bio-barcodes based on oligonucleotide-modified nanoparticles. *J. Amer. Chem. Soc.* 124: 3820–3821.

Nam, J-M., Stoeva, S. & C.A. Mirkin. 2004. Bio-barcode-based DNA detection with PCR-like sensitivity. *J. Amer. Chem. Soc.* 126: 5932–5933.

Nam, J-M., S. Thaxton & C.A. Mirkin. 2003. Nanoparticle-based bio-barcodes for the ultrasensitive detection of proteins. *Science*. 301: 1884–1886.

Nanog FLJ12581. 2004. *Homeobox Transcription Factor Nanog*. www.ncbi.nlm.nih.gov/entrez/query

NASA. 1996. *Fluorometer for Analysis of Photosynthesis in Phytoplankton*. NASA Technical Support Package SSC-00110. (Work done by: Fernandez, S.M., E.F. Guignon, R. Kersten and E. St. Louis at Ciencia, E. Hartford, CT.)

National Agricultural Laboratory. 2000. *Commercialized Biological-Based Products in the Food and Agricultural Industries*. (Web site updated 1/2000.) www.nal.usda.gov/bic/Misc pubs/bioprod.html

National Institutes of Health. 2007. *Opossum Genome Shows Value in "Junk" DNA. NIH Research Matters*. 21 May 2007. www.nih.gov/news/research_matters/may2007/05212007dna.htm

National Marrow Donor Program. 2007. *Cord Blood Transplants*. (Retrieved 22 April 2007.) www.marrow.org/PATIENT/Donor SelectTxProcess/ReceivingYourNewCells/index.html

Netherwood, T. et al. 1999. Gene transfer in the intestinal tract. *Applied and Environmental Microbiology*. 65(11): 5139–5141. www.pubmedcentral.nih.gov/picrender.fcgi?artid=91691&blobtype=pdf

Neuberger, M. 2005. *FAQ X-Ray Structure Determination*. www.xray-structures.ch/xray-faq.html

New Business World. 2007. *Ethics in Business: What is it?* 5 Feb 2007. http://thenewbusinessworld.blogspot.com/2007/02/ethics-in-business-what-is-it.html

Nicholl, D.S.T. 2002. *An Introduction to Genetic Engineering*, 2nd ed. Cambridge University Press.

Nielsen, B.G. 2003. Unfolding transitions in myosin give rise to the double-hyperbolic force-velocity relation in muscle. *J. Physics: Condens. Matter.* 15: S1759–S1765.

Nielsen, K.M. et al. 1998. Horizontal gene transfer from transgenic plants to terrestrial bacteria—a rare event? *FEMS Microbiol. Rev.* 22(2): 79–103.

NIH. 2006. *Stem Cell Basics*. http://stemcells.nih.gov/info/basics/basics4.asp

Nikon. 2006. *Introduction to Fluorescent Proteins*. (Online tutorial.) www.microscopyu.com/articles/livecellimaging/fpintro.html

Nilges, M. & J.P. Linge. 2002. *Definition of Bioinformatics*. (Accessed 11 May 2007.) www.pasteur.fr/recherche/unites/Binfs/definition/bioinformatics_definition.html

Nilsson, M. et al. 2006. Analyzing genes using closing and replicating circles. *TRENDS in Biotechnol*. 24(2): 83–88.

NIOSH. 1995. *The Registry of Toxic Effects of Chemical Substances: Abrin*. www.cdc.gov/niosh/rtecs/aa501bd0.html

Nobel. 1999. *Günter Blobel*. (Press release that Günter Blobel was awarded the Nobel Prize in Medicine, 11 October 1999, for the discovery that "proteins have intrinsic signals that govern their transport and localization in the cell." http://nobelprize.org/nobel_prizes/medicine/laureates/1999/press.html

Nordlee, J.A. 1996. Identification of a Brazil-nut allergen in transgenic soybeans. *New Engl. J. Med.* 334(11): 688–692. http://content.nejm.org/cgi/content/abstract/334/11/688

Northrop, R.B. 2005. *Introduction to Instrumentation and Measurements, 2nd ed.* Taylor and Francis/CRC, Boca Raton, FL.

________. 2003. *Signals and Systems in Biomedical Engineering*. CRC Press, Boca Raton, FL.

________. 2002. *Noninvasive Instrumentation and Measurements in Medical Diagnosis*. CRC Press, Boca Raton, FL.

________. 2001. *Introduction to Dynamic Modeling of Neuro-Sensory Systems*. CRC Press, Boca Raton, FL.

________. 2000. *Endogenous and Exogenous Regulation and Control of Physiological Systems*. Chapman and Hall/CRC Press, Boca Raton, FL.

Northrop, R.B. 1964. *A Study of the Neural Control of a Molluscan Smooth Muscle*. PhD Dissertation. University of Connecticut. E.G. Boettiger, adviser.

Northrop, R.B., X-Z Liu, & Q. Li. 1989. A modeling study of human immune deficiency disease. *Proc. 15th Ann. Northeast Bioengineering Conf.* Northeastern Univ., Boston, 3/27-28. pp. 173–174.

Novitski, C. 2004. *Revision of Fisher's Analysis of Mendel's Garden Pea Experiments.* In *Genetics.* 166: 1139–1140. March 2004. www.genetics.org/cgi/content/full/166/3/1139

Nunnally, B.K. 2005. *Analytical Techniques in DNA Sequencing.* CRC Press, Boca Raton, FL.

NYT. 2003. *(New York Times). Roundup Unready.* (Editorial) www.nytimes.com/2003/02/19/opinion/19WED4.html?ex=1139202000&en=0a55c309c800de6e&ei=5070

OARDC. 2003a. *Genome Sequences of Spiroplasma kunkelii.* (Accessed 5/15/07.) www.oardc.ohio-state.edu/spiroplasma/genome.htm

________. 2003b. Spiroplasmas as insect pathogens. (Accessed 5/15/07.) www.oardc.ohio-state.edu/spiroplasma/insect.htm

O'Brien, J.A. et al. 2001. Modifications to the hand-held gene gun: Improvements for *in vitro* biolistic transfection of organotype neuronal tissue. *J. Neurosci. Methods.* 112: 57–64.

Ogata, K. 1970. *Modern Control Engineering,* 1st ed., Prentice-Hall, Englewood Cliffs, NJ.

Ogawa, M. et al. 2005. Escape of intracellular *Shigella* from autophagy. *Science.* (4 Feb. 2005) 307(5710): 727–731.

Ogura, A. et al. 2000. Birth of mice after nuclear transfer by electrofusion using tail tip cells. *Mol Reproduction Dev.* 57: 55–59.

Ogburn, S. & L. Brogdon. 2004. Natural history of the grey short-tailed opossum (*Monodelphis domestica*). *Comparative Mammalian Anatomy.* (Published on the Duke University web site.) www.baa.duke.edu/companat/BAA_289L_2004/Natural_History/Opossum/opossum_natural_history.htm

O'Hagan, W.J. et al. 2002. MHz LED source for nanosecond fluorescence sensing. *Meas. Sci. Technol.* 13: 84–91.

Ohno, S. 1972. So much "junk" DNA in our genome. *Brookhaven Symp. Biol.* 23: 366–370.

Okita, K., T. Ichisaka & S. Yamanaka. 2007. Generation of germline-competent induced pluripotent stem cells. *Nature.* 448: 313–317. (Published online 6/6/07. doi: 10.1038/nature05934 (2007).]

OMIM. 2006a. *Danon Disease.* #300257. www.ncbi.nlm.nih.gov/entrez/ 8pp.

________. 2006b. *Lysosome-Associated Membrane Protein 2; LAMP-2.* #309060. www.ncbi.nlm.nih.gov/entrez/ 9pp.

________. 2006c. *Online Mendelian Inheritance in Man* (entrez). (Main site) www.ncbi.nlm.nih.gov/entrez/query.fcgi?db=OMIM.

________. 2006d. *Prion Protein; PRNP* #176640. www.ncbi.nlm.nih.gov/entrez/dispomim.cgi?id=176640

________. 2005. *Creutzfeldt-Jakob Disease.* #123400 www.ncbi.nlm.nih.gov/entrez/query.fcgi?cmd=Retrieve&db=OMIM&dopt=Detailed& 16pp.

Onken, M. 1997. Why does uracil replace thymine in RNA? *Biochemistry Posts.* www.madsci.org/posts/archives/dec97/879354206.Bc.r.html

Oparin, A.I. 1953. *The Origin of Life,* 2nd ed. Dover, New York.

Oren, M. 1999. Regulation of the p53 tumor suppressor protein. *J. Biol. Chem.* 274(51): 36031–36034.

Orgel, L.E. 1973. *The Origins of Life.* John Wiley & Sons, New York.

Orlic, D. et al. 2003. Bone marrow stem cells regenerate infarcted myocardium. *Pediatric Transplantation.* 7: 86–88. www.blackwell-synergy.com/links/doi/10.1034%Fj.1399-3046.7.s3.13.x

Otsuki, T. et al. 2004. Expression of protein gene product 9.5 (PGP 9.5)/ubiquitin-C-terminal hydrolase 1 (UCHL-1) in human myeloma cells. *British J. Haematology.* 127(3): 292–298.

Padman, R. 2004. *Cystic Fibrosis.* KidsHealth. //kidshealth.org

Palaeos. 2002. *The Atmosphere.* www.palaeos.com/Earth/Atmosphere/atmosphere.htm

Palanisamy, N. & R.S.K. Chaganti. 2001. Novel *in situ* hybridization (FISH) probe for simultaneous detection of t(9;22)(q34.1;q11.2) and associated deletions on der(9) and der(22) chromosomes. *Proc. 43rd Ann. Meeting of the Amer. Soc. of Hematology.* Dec. 7–10, 2001. Orlando, FL. Poster Session 1, Abstr. No. 622.

Paterlini, M. 2005. *'Dead' DNA feeds deep sea life. The Scientist: Daily News.* 5 October 2005. www.the-scientist.com

Patton, W.F. 2000. A thousand points of light: The application of fluorescence detection technologies to two-dimensional gel electrophoresis and proteomics. *Electrophoresis.* 21: 1123–1144.

Pawlak, M., E. Schick, M.A. Bopp, M.J. Schneider, P. Oroszian & M. Ehrat. 2002. Zeptosens' protein microarrays: A novel high performance microarray platform for low abundance protein analysis. *Proteomics.* 2: 383–393.

PDB (RCSB Protein Data Bank). 2006, 2007. *Welcome.* www.pdb.org/pbd/Welcome.do. *PDB Current Holdings Breakdown.* www.pdb.org/pdb/statistics/holdings.do

PDBTM. 27 April 2006. *Protein Data Bank of Transmembrane Proteins.* //pdbtm.enzim.hu

Penner, R. & A. Fleig. 2005. Store-operated calcium entry: A tough nut to CRAC. *Science STKE*. www.stke.org/cgi/content/full/sigtrans;2004/243/pe38

Perkel, J.M. 2007. Cancer stem cells drive metastasis. *The Scientist*. 12 September 2007. http://www.the-scientist.com/news/home/53583/

Perler, F.B. 2005. Protein splicing mechanisms and applications. *Life*. 57(7): 469–476.

Pesce, A. 2000. Prostate cancer: Prostate specific antigen. *Net Wellness*. www.netwellness.org/healthtopics/prostate/faq10.cfm

PETA (People for the Ethical Treatment of Animals). 2007. *Animal Research: Overview*. http://www.peta.org/mc/factsheet_display.asp?ID=126

Petrou, I. 2006. Examining stem cell characteristics in melanoma. *Dermatology Times*. 1 Nov 2006. www.dermatologytimes.com/dermatologytimes/articleDetail.jsp?id=383168

Pew Charitable Trusts. 2006. *Unlikely Reactions: Identifying Allergies to GM Foods*. http://pewagbiotech.org/buzz/display.php3?StoryID=12

Pham, K. et al. 2003. Wnt ligand expression in malignant melanoma: pilot study indicating correlation with histopathological features. *J. Clin. Pathol: Mol. Pathol.* 56: 280–285.

Phizicky, E. et al. 2003. Protein analysis on a proteomic scale. *Nature* (13 March 2003). 422: 208–215.

Pierce Biotechnology, Inc. 2002. *Avidin-Biotin Chemistry: A Handbook*. Available in pdf form at: www.piercenet.com

Pietrokovski, S. 2004. *Inteins-Protein Introns*. http://bioinformatics.weizmann.ac.il/~pietro/inteins/

Plath, K. et al. 2004. X chromosime inactivation. *Advanced Molecular Genetics-Biology 566*. (on-line notes). //bioweb.wku.edu/courses/biol566/L9XchromSilencing.html

Pockley, A.G. 2002. Heat shock proteins, inflammation and cardiovascular disease. *Circulation*. 105: 1012–1017.

Polavarapu, N., N. Bowen & J. McDonald. 2006. Newly identified families of human endogenous retroviruses. *J. Virol.* 80(9): 4640–4642.

Polisetty, P.K., E.O. Voit & E.P. Gatzke. 2006. Identification of metabolic system parameters using global optimization methods. *Theoretical Biology and Medical Modeling*. 3(4): 1–15.

Prasher, D.C. et al. 1992. Primary structure of the *Aequorea victoria* green fluorescent protein. *Gene*. 111: 229–233.

Preininger, A.M. & H.E. Hamm. 2004. G protein signaling: Insights from new structures. *Science's STKE*. www.stke.org/cgi/content/full/sigtrans;2004/218/re3

Prieto, J. et al. 2003. The promise of gene therapy in gastrointestinal and liver diseases. *Gut*. 52: (suppl. II). ii49–ii54. www.gutjnl.com

ProVitaMinRice Consortium. 2006. *Grand Challenges in Global Health Project*. www.goldenrice.org/Content5-GCGH/GCGH1.html

Pusch, M. 2004. Structural insights into chloride and proton-mediated gating of CLC chloride channels. *Biochemistry*. 43(5): 1135–1144.

Raftopoulou, M. 2005. Nucleosomes regulate gene expression. *Nature*. 1 Dec. 2005. Retrieved 1 May 2007. www.nature.com/milestones/geneexpression/milestones/articles/milegene16.html

Rakyan, V.K., J. Preis, H.D. Morgan & E. Whitelaw. 2001. The marks, mechanisms and memory of epigenetic states in mammals. *Biochem. J.* 356: 1–10.

Rassier, D.E., B.R. MacIntosh & W. Herzog. 1999. Length dependence of active force production in skeletal muscle. *J. Amer. Physiol.* (invited review). //www.jap.org

Raven, J.A. & J.F. Allen. 2003. Genomics and chloroplast evolution: what did cyanobacteria do for plants? *Genome Biol.* 4(209): 209.1–209.5.

RDS. 2004. *What are Stem Cells and Why are they Important?* www.rds-online.org.uk/pages/page.asp?i_ToolbarID=5&i_PageID=1498

Report. 2006. Notification Report to the EU re: the release of GM *A. thaliana*. (Accessed online 5/25/07.) http://gmoinfo.jrc.it/gmp_report.aspx?CurNot=B/DK/06/01

Revazova, E.S. et al. 2007. Patient-specific stem cell lines derived from human parthenogenic blastocysts. *Cloning and Stem Cells*. 9(3): (doi: 10.1089/clo.2007.0033)

Reya, T. & H. Clevers. 2005. Wnt signalling in stem cells and cancer. *Nature*. 434: 843–850.

Rib-X Pharmaceuticals. 2005. *Ribosomes as a drug target*. www.rib-x.com/target.html

Rich, K.A., C. Burkett & P. Webster. 2003. Cytoplasmic bacteria can be targets for autophagy. *Cell. Microbiol.* 5(7): 455–468.

Ridley, M. 1999. *Genome: The Autobiography of a Species in 23 Chapters*. Fourth Estate, London.

Rizza, A., L.J. Mandarino & E. Gerich. 1982. Cortisol-induced insulin resistance in man: Impaired suppression of glucose production and stimulation of glucose utilization due to a postreceptor defect of insulin action. *J. Clin. Endrocrin. and Metab.* 54: 131–138.

Robinson, B.A. 2004. *Human Embryos and Fertility Clinics.* Ontario Consultants on Religious Tolerance. www.religioustolerance.org/abo_inco.htm

Roche Applied Science. 1993. *Biochemical Pathways Chart,* 3rd ed., Parts 1 and 2. G. Michal, ed. (Available from: Roche Diagnostics Corp., 9115 Hague Road, Indianapolis, IN 46250-0414. 800-263-1640.)

Roche Molecular Diagnostics. 2006. *Chronology of PCR Technology.* www.roche-diagnostics.com/ba_rmd/pcr_evolution.html

Rosi, N.L. & C.A. Merkin. 2005. Nanostructures in biodiagnostics. *Chem. Rev.* 105: 1547–1562.

RSBS. 2006. *Oxygen and the Geological Record.* www.rsbs.anu.edu.au/ResearchGroups/PBE/Oxygen/O2_2_Abundence.htm

rvt. 2005. *Cyanobacteria.* (Accessed 10/10/05.) www.rvt.com/~lucas/school/cyano.html

Saccharomyces. 2005. *S. cerevisiae Pathway: glycolysis.* http://pathway.yeastgenome.org:8555/YEAST/new-image?type=PATHWAY&object=GLYC

Sage, L. 1997. *Neuroscientists Find New Trigger for Nerve Cell Death.* //record.wustl.edu/archive/1997/11-20-97/7246.html

Sage, R.F. 2002. Variation in the *kcat* of rubisco in C_3 and C_4 plants and some implications for photosynthetic performance at high and low temperature. *J. Exp. Botany.* 53(369): 609–620.

Sahni, S. ed. 2006. *Handbook of Computational Molecular Biology.* Chapman and Hall/CRC. Atlanta.

Saito, Y. et al. 2006. Specific activation of microRNA-127 with downregulation of the proto-oncogene BCL6 by chromatin-modifying drugs in human cancer cells. *Cancer Cell.* 9: 435–443.

Salk Institute. 2005. *'Jumping Genes' Contribute to the Uniqueness of Human Brains.* http://genome.wellcome.ac.uk/doc_WTD020792.html

Sanai, N., A. Alvarez-Buylla & M.S. Berger. 2005. Neural stem cells and the origin of gliomas. *N. Engl. J. Med.* 353(8): 811–822. (25 Aug. 2005)

Sander, D. 2005. *Viral Diseases and Their Etiologic Agents.* www.virology.net/Big_Virology/BVDiseaseList.html

Sandow, A. 1947. Latency relaxation and a theory of muscular mechano-chemical coupling. *Ann. N.Y. Acad. Sci.* 47: 895–929.

Sano, T., C.L. Smith & C.R. Cantor. 1992. Immuno-PCR: Very sensitive antigen detection by means of specific antibody-DNA conjugates. *Science.* 258: 120–122.

Satir, P. 1999. The cilium as a biological nanomachine. *FASEB Journal (Supplement).* 13: S235–S237.

Sato, N. 2002. Comparative analysis of the genomes of cyanobacteria and plants. *Genome Informatics* 13: 173–182.

Scadden, D.T. 2006. The stem-cell niche as an entity of action. *Nature.* 441: 1075–1079.

Schaeffer, A. et al. 2003. L-Arginine: An ultradian-regulated substrate coupled with insulin oscillations in healthy volunteers. *Diabetes Care.* 26(1): 168–171.

Science Daily. 2005. USC researchers link cellular stress to drug resistance in tumors. *Science Daily.* 4 July 2005. www.sciencedaily.com/releases/2005/07/050701062000.htm

Schier, A. & W. Gehring. 1993. Functional specificity of the homeodomain protein fushi tarazu: The role of DNA-binding specificity *in vivo. PNAS USA.* 90: 1450–1454.

Schlindwein, B. 2006. *Chloroplast DNA (ctDNA).* www.weihenstephan.de/genglos/asp/genreq.asp?nr=317

________. 2002. *A Hypermedia Glossary of Genetic Terms.* www.weihenstephan.de/~schlind/genglos.html

Schmitt, F., E. Oakley & J.P. Jost. 1997. Antibiotics induce genome-wide hypermethylation in cultured *Nicotianum tabaccum* plants. JBC Online. 272(3): 1534–1540. 17 Jan 1997. Retrieved 1 May 2007. www.jbc.org/cgi/content/full/272/3/1534

Schoenbach, K.H., R. Nuccitelli & S.J. Beebe. 2006. Zap. *IEEE Spectrum.* 43(8): 20–26.

Schoenbach, K.H. et al. 2004. Ultrashort electrical pulses open a new gateway into biological cells. *Proc. IEEE.* 92(7): 1122–1137.

Schrodinger, E. 1967. *What is Life?* Cambridge University Press, Cambridge, UK.

Schu, B. 2005. Can science bridge the stem cell divide? *Genomics and Proteomics.* April. www.genpromag.com/ShowPR.aspx?PUBCODE=018&ACCT=1800000100&ISSUE=0504&RELTYPE=PR&ORIGRELTYPE=CV&PRODCODE=00000000&PRODLETT=AQ

Schulz, K.L. 2006. *Redox Reactions.* www.esf.edu/EFB/schulz/Limnology/Redox.html

Schwartz, R.E. et al. 2001. Multipotent adult progenitor cells from bone marrow differentiate into functional hepatocyte-like cells. *J. Clinical Investigation.* 109: 1291–1302. www.jci.org/cgi/content/full/109/10/1291

Science Daily. 2007. Exploring "Junk DNA" in the Genome. Science Daily news, 16 June 2007. www.sciencedaily.com/releases/2007/06/070615091210.htm

Science. 2007. *Exploring 'Junk DNA' in the Genome*. Science Daily news, 16 June 2007. www.sciencedaily.com/releases/2007/06/070615091210.htm

Scripps. 1997. *Myoglobin Structure*. //metallo.scripps.edu/PROMISE/1MBO.html

Sears, F.W. 1949. *Optics*. Addison-Wesley Press, Cambridge, MA.

Selvin, P.R. 2000. The renaissance of fluorescence resonance energy transfer. *Nat. Struct. Biol.* 7(9): 730–734.

Sempere, L.F. et al. 2004. Expression profiling of mammalian microRNAs uncovers a subset of brain-expressed microRNAs with possible roles in murine and human neuronal differentiation. *Genome Biol.* 5(3): Article R13.

Sengbusch, P. 2003. *C3, C4 and CAM. Regulation of the Activity of Photosynthesis*. www.biologie.uni-hamburg.de/b-online/e24/24b.htm

SenseLab. 2006 printout. *Membrane Properties: Receptors*. //senselab.med.yale.edu/senselab/NeuronDB/receptors2.asp

SEP. 2006. (Science Education Partnership). *The Genetics of Human Eye Color*. www.seps.org/cvoracle/faq/eyecolor.html

Shapiro, B.E. et al. 2003. Cellerator: extending a computer algebra system to include biochemical arrows for signal transduction simulations. *Bioinformatics*. 19(5): 677–678.

Shapiro, B.E. & E.D. Mjolsness. 2001. Developmental simulation with Cellerator. In: *Proc. of the 2nd Intl. Conf. on Systems Biology (ICSB)*, Pasadena, CA.

Shing-Sheu, S., V.K. Sharma & S.P. Banerjee. 1984. Measurement of cytosolic free calcium concentration in isolated rat ventricular myocytes with quin 2. *Circulation Res.* 55: 830–834.

Shreeve, J. 2005. The other stem cell debate. *New York Times Magazine*. 10 April. www.nytimes.com/2005/04/10/magazine/10CHIMERA.html?ex=1139202000&en=b91519f186de6331&ei=5070

Sickle. 2001. *Iron Transport and Cellular Uptake*. //sickle.bwh.harvard.edu/iron_transport.html

Sickle. 1999. *Thalassemia*. //sickle.bwh.harvard.edu/thalover.html

Sidwell. 2006. *Kingdom Protista*. www.sidwell.edu/us/science/vlb5/Labs/Classification_Lab/Eukarya/Protista/

Siegal, R. 1998. *Parvovirus*. www.Stanford.edu/group/virus/parvo/parvovirus.html

Singer, E. 2006. Stem cells reborn. *Technology Review*. May/June. 58–65.

Sirohi, R.S. & M.P. Kothiyal. 1991. *Optical Components, Systems, and Measurement Techniques*. Marcel Dekker, New York.

Sizes. 2006. *Bacteria*. www.sizes.com/natural/bacteria.htm

Skretas, G. & D.W. Wood. 2005. Regulation of protein activity with small-molecule-controlled inteins. *Protein Sci.* 14: 523–532.

Slack, A. et al. 2005. The p53 regulatory gene MDM2 is a direct transcriptional target of MYCN in neuroblastoma. *PNAS (USA)*. 102(3): 731–736.

Smith, L. et al. 2006. Epigenetic regulation of the tumor suppressor gene TCF21 on 6q23-q24 in lung and head and neck cancer. *PNAS (USA)*. 103(4): 982–987.

So, P.T.C. & C.Y. Dong. 2002. Fluorescence spectrophotometry. *Encyclopedia of Life Sciences*. (Online) www.els.net

Söhl, G., S. Maxeiner & K. Willecke. 2005. Expression and functions of neuronal gap junctions. *Nat. Rev.—Neuroscience*. 6: 191–200.

Soil Association. 2005. *Flavr Savr Tomato and GM Tomato Puree: Problems With the First GM Foods. Information Sheet Version 3*. www.soilassociation.org/web/sa/saweb.nsf/home/index.html

Soukup, G.A. & R.R. Breaker. 1999. Nucleic acid molecular switches. *Tibtech*. 17: 469–476.

Southan, C. 2005. *Has the Yo-Yo Stopped? Reviewing the Evidence for a Low Basal Human Protein Number and the Implications for Proteomics and Drug Discovery*. 6th Swedish Ann. Bioinformatics Workshop, Göteborg, 11/05.

Southan, C. 2004. Has the yo-yo stopped? An assessment of human protein-coding gene number. *Proteomics*. 4: 1712–1726.

Spahn, C.M.T., E. Jan, A. Mulder, R.A. Grassucci, P. Sarnow & J. Frank. 2004. Cryo-EM visualization of a viral internal ribosome entry site bound to human ribosomes: The IRES functions as an RNA-based translation factor. *Cell*. 118: 465–475.

Spangler, R.A. & F.M. Snell. 1961. Sustained oscillations in a catalytic chemical system. *Nature*. 191(4787): 457–458.

Spusta, S.C. & M.A. Goldman. 2005. XISTential wanderings: The role of XIST RNA in X-chromosome inactivation. (on-line paper, 16 pp) www.ias.ac.in/currcsi/aug25/articles21.htm

Stacey, S.N. *et al.* 2007. Common variants on chromosomes 2q35 and 16q12 confer susceptibility to estrogen receptor-positive breast cancer. *Nat. Genet.* 39(7): 865–869. (Epub 27 May 2007.)

Stanford University. Stanford Encyclopedia of Philosophy. 2005. *Molecular Biology.* 20 pp. (Accessed 11 may 2007.) http://plato.stanford.edu/

Stanford University. Stanford Encyclopedia of Ethics. 2007. *Virtue Ethics.* http://plato.stanford.edu/entries/ethics-virtue/

Stark, L. 1968. *Neurological Control Systems.* Plenum Press, New York.

Stassen, C. 2005. *The Age of the Earth.* www.talkorigins.org/faqs/faq-age-of-earth.html

Stewart, G.R. & D.B. Young. 2004. Heat-shock proteins and the host-pathogen interaction during bacterial infection. *Current Opin. Immunol.* 16(4): 506–510.

Stresemann, C. et al. 2006. Functional diversity of DNA methyltransferase inhibitors in human cancer cell lines. *Cancer Research.* 66(5): 2794–2800.

Struhl, K. 1999. Fundamentally different logic of gene regulation eukaryotes and prokaryotes. *Cell.* 98: 1–4.

Sullivan, B.A. et al. 2001. Positive selection of a Qa-1-restricted T cell receptor with a specificity for insulin. *Immunity:* 17(1): 95–105.

Swales, A.K.E. & N. Spears. 2005. Genomic imprinting and reproduction. *Reproduction.* 130: 389–399.

SWBIC. 2005. *Protein Structure* [hot links]. (from the Southwest Biotechnology and Informatics Center) www.swbic.org/links/1.12.php

SWBIC. 2005b. *Structure Analysis* [hot links]. www.swbic.org/links/1.12.4.php

Sweeting, L.M. 1997. *Organic Redox Reactions.* //pages.towson.edu/ladon/orgrxs/reagent/redox.htm

Sydor, J.R. & S. Nock. 2003. Protein expression profiling arrays: tools for the multiplexed high-throughput analysis of proteins. *Proteome Science.* 1: 3–7. www.Proteomesci.com/content/1/1/3

Tabler, M. & M. Tsagris. 2004. Viroids: petite RNA pathogens with distinguished talents. *TRENDS Plant Sci.* 9(7): 339–348.

Takahashi, K. et al. 2002. Computational challenges in cell simulation: A software engineering approach. *IEEE Intelligent Systems.* Sept./Oct. 2002. 64–71.

Tamashiro, K.L. et al. 2002. Cloned mice have an obese phenotype not transmitted to their offspring. *Nat. Med.* 8: 262–267.

Taylor, W.R. & A. Aszódi. 2005. *Protein Geometry, Classification, Topology and Symmetry.* CRC Press, Boca Raton, FL.

Teng, X. et al. 2005. Structure-activity relationship study of novel necroptosis inhibitors. *Bioorganic Med. Chem. Lett.* (Available online at www.sciencedirect.com)

Thomas, D.C., G.A. Nardone & S.K. Randall. 1999. Amplification of padlock probes for DNA diagnostics by cascade rolling circle amplification or the polymerase chain reaction. *Arch. Pathol. Lab. Med.* 123: 1170–1176.

Thomas, J-L. et al. 2001. A helium burst biolistic device adapted to penetrate fragile insect tissues. *J. Insect Sci.* 1.9: 10pp. (Available online at: www.insectscience.org/1.9)

Thompson, R., ed. 2005. *Fluorescence Sensors and Biosensors.* CRC Press. Boca Raton, FL.

Ting, A.H., K.M. McGarvey & S.B. Baylin. 2006. The cancer epigenome-components and functional correlates. *Genes Dev.* 20: 3215–3231.

Todar, K. 2004. The diversity of metabolism in procaryotes. *Todar's Online Textbook of Bacteriology.* (Accessed 5/1/2006.) http://textbookofbacteriology.net/metabolism.html

Törnroth-Horsefield, S. et al. 2006. Structural mechanism of plant aquaporin gating. *Nature.* 439: 688–694.

Trachtenberg, S. 2004. Shaping and moving a Spiroplasma. *J. Mol. Microbiol. Biotechnol.* 7: 78–87.

Trachtenberg, S., R. Gilad & N. Geffen. 2002. The bacterial linear motor of *Spiroplasma melliferum* BC3: from single molecules to swimming cells. *Mol. Microbiol.* 47(3): 671–697.

Traut, T. 1988. Do exons code for structural or functional units in proteins? *PNAS USA.* 85(9): 2944–2948.

Trautinger, F. et al. 1996. Stress proteins in the cellular response to ultraviolet radiation. *J. Photochem Photobiol B.* 35(3): 141.

Turner, J.S. 2007. *The Tinkerer's Accomplice: How Design Emerges from Life Itself.* Harvard Univ. Press, Cambridge, MA.

Turbin, D.A. et al. 2006. MDM2 protein expression is a negative prognostic marker in breast cancer. *Modern Pathol.* 19: 69–74.

Twyman, R. 2003. *Mutation or Polymorphism?* http://genome.wellcome.ac.uk/doc_WTD020780.html

Tyers, M. & M. Mann. 2003. From genomics to proteomics. *Nature.* 422: 193–197.

UCSD. 2001. *The Genetic Code.* http://exobio.ucsd.edu/Space_Sciences/genetic_code.htm

UFL. 2006. *Toxins of Biological Origin.* www.ehs.ufl.edu/Bio/toxin.htm

University of California Museum of Paleontology. 2007. *Morphology of the Tetrapods.* (Retrieved 5 May 2007.) www.ucmp.berkeley.edu/vertebrates/tetrapods/tetramm.html

University of Idaho. 2007. *Cells, Evolution and Life.* (Course notes retrieved 5 May 2007.) www.sci.uidaho.edu/bionet/biol115/glossary/wordlist/hwords.htm

UPHAR. 2006. *Voltage-Gated Calcium Channels.* //calcium.ion.ucl.ac.uk/UPHAR.doc

USDHHS. 2001. *Stem Cells: Scientific Progress and Future Research Directions.* June. http://stemcells.nih.gov/info/scireport

USDHHS. 2001b. *Stem Cells: Scientific Progress Report and Future Research Directions.* Chapter 4: The Adult Stem Cell. http://stemcells.nih.gov/info/scireport/chapter4.asp

USGS 2004. *The Age of the Earth.* //wrgis.wr.usgs.gov/docs/parks/gtime/ageofearth.html

USGS. 2001. *Radiometric Time Scale.* //pubs.usge.gov/gip/geotime/radiometric.html

Uttamchandani, M., J. Wang & S.Q. Yao. 2006. Protein and small molecule microarrays: powerful tools for high-throughput proteonomics. *Mol. BioSyst.* 2: 58–68.

van Holde, K.E., K.I. Miller & H. Decker. 2001. Hemocyanins and invertebrate evolution. *J. Biol. Chem.* 276(19): 15563–15566. (Available on line @: http://www.jbc.org)

Vasudevan, S., Y. Tong & J.A. Steitz. 2007. Switching from repression to activation: micro RNAs can up-regulate translation. *Sciencexpress.* 29 Nov. 2007. 4 pp + figs. http://www.sciencexpress.org

Velasquez, M. et al. 2005. *A Framework for Thinking Ethically.* Markkula Center for Applied Ethics at Santa Clara University. http://www.scu.edu/ethics/practicing/decision/framework.html

Velasquez, M., C. Andre, T. Shanks & M. Meyer. 1989. Calculating consequences: The utilitarian approach to ethics. *Issues in Ethics.* 25(1): Markkula Center for Applied Ethics at Santa Clara University. Online at: www.scu.edu/ethics/practicing/decision/calculating.html

Vellai, T. & G. Vida. 1999. The origin of eukaryotes: the difference between prokaryotic and eukaryotic cells. *Proc. Royal Soc.* B. 266: 1571–1577.

Venkatachalam, K. et al. 2002. The cellular and molecular basis of store-operated calcium entry. *Nature Cell Biology.* 4: E263–E272. www.nature.com/ncb/journal/v4/n11/full/ncb1102-e263.html#B53#B53

Vera, J. & N.V. Torres. 2003. MetMAP: An integrated Matlab™ package for analysis and optimization of metabolic systems. *In Silico Biology.* 4: 0010. (Available on-line at www.bioinfo.de/isb/2003/04/0010/main.html)

Vescovi, A.L., R. Galli & B.A. Reynolds. 2006. Brain tumour stem cells. *Nat. Rev. Cancer.* 6: 425–436. www.nature.com/reviews/cancer

vetscite. 2006. Stem cells engage in dialogue with the cells that regulate their future. *Science Daily.* 21 Nov. 2006. www.vetscite.org/publish/items/003329/index.html

Vince, G. 2004. Pig-human chimeras contain cell surprise. *New Scientist.* 13 Jan. www.newscientist.com/article.ns?id=dn4558

Vincent, M., Y. Xu & H. Kong. 2004. Helicase-dependent isothermal DNA amplification. *EMBO Reports.* 5(8): 795–800.

Vogel, G. 2006. Four genes confer embryonic potential. *Science.* 313: 27.

Voiland, M. & L. McCandless. 1999. *Development of the "Gene Gun" at Cornell.* (Press release online.) www.nysaes.cornell.edu/pubs/press/1999/genegun.html

Voit, E.O. 2000. *Computational Analysis of Biochemical Systems: A Practical Guide for Biochemists and Molecular Biologists.* Cambridge University Press.

Voit E.O. 1988. "New Nonlinear Methodologies for Modeling Molecular and Cellular Systems." In: *Modelling and Control in Biomedical Systems,* C. Cobelli and L. Mariani, Eds. (IFAC Symposium Papers, Venice, Italy, 6–8 April 1988.) Pergamon Press, Oxford. pp. 217–228.

Voit, E.O & T. Radivoyevitch. 2000. Biochemical systems analysis of genome-wide expression data. *Bioinformatics.* 16(11): 1023–1037.

Voit, E.O. & M.A. Savageau. 1982. Power-law approach to modeling biological systems; III. Methods of analysis. *J. Fermentation Technol.* 60: 233–241.

von Heijne, G. 2003. *Membrane Channels.* Advanced Information on the Nobel Prize in Chemistry, 8 October 2003. Peter Agre, recipient. Available at: www.kva.se

Vrana, K.E. et al. 2003. Nonhuman primate parthenogenetic stem cells. *PNAS.* 100 Suppl.1: 11911–11916.

Wade, N. 2007. Biologists make skin cells work like stem cells. *The New York Times.* 7 June 2007. Online at: www.nytimes.com/2007/06/07/science/07cell.html?hp=&pagewanted=print

Wakayama, T. et al. 2000. Cloning of mice to six generations. *Nature.* 407: 318–319.

Wakayama, T. et al. 1998. Full-term development of mice from enucleated oocytes injected with cumulus cell nuclei. *Nature*. 394: 369–374.

Wallis, M. 2001. Episodic evolution of protein hormones in mammals. *J. Mol. Evolution*. 53(1): 10–18.

Walls, G.L. 1967. *The Vertebrate Eye and its Adaptive Radiation*. Haffner, New York.

Walter, J. et al. 2007. Light-powering *Escherichia coli* with proteorhodopsin. *PNAS*. 104(7): 2408–2412.

Walz, T. T. Hirai, K. Murata, J.B. Heymann, K. Mitsuoka et al. 1997. The three-dimensional structure of aquaporin-1. *Nature*. 387: 624–627.

Wanatabe, A. 1958. The interaction of electrical activity among neurons of lobster cardiac ganglion. *Jap. J. Physiol.* 8: 305–318.

Wang, C-W. & D.J. Klionsky. 2003. The molecular mechanism of autophagy. *Mol. Med.* 9(3/4): 65–76.

Wang, D. et al. 2002. Carbohydrate microarrays for the recognition of cross-reactive molecular markers of microbes and host cells. *Nat. Biotechnol.* 20(3): 275–281.

Wang, X. et al. 2003. Albumin-expressing hepatocyte-like cells develop in the livers of immune-deficient mice that received transplants of highly purified human hematopoietic stem cells. *Blood*. 101(10): 4201–4208. www.bloodjournal.org/cgi/content/abstract/2002-05-1338v1

Washington State University. 2007. *Definition and Scope: What is Biotechnology?* Washington State University Center for Integrated Biotechnology. www.biotechnology.wsu.edu/definition_scope/

Weaver-Missick, T. 2000. Transgenics for a better tomato. *Agricultural Research Magazine*. 48(9): 14–15

Weber-Dubrowska, B., M. Mulczyk & A. Górski. 2000. Bacteriophage therapy of bacterial infections: An update of our institute's experience. *Archivum Immunologiae et Therapiae Experimentalis*. 48: 547–551. www.evergreen.edu/user/T4/PolishUpdate.htm

Weinstock, G.M. 2007. ENCODE: More genetic empowerment. *Genome Res.* 17: 667–668.

Weiss, R. 2005. The power to divide. *National Geographic*. July. ww7.nationalgeographic.com/ngm/0507/feature1/

Weiss, R. 2004. Of mice, men and in between. *Washington Post*. 20 Nov. www.washingtonpost.com/ac2/wp-dyn/A63731-2004Nov19?language=printer.

Weller, G.R., V.L. Brandt & D.B. Roth. 2004. Doing more with less in bacterial DNA repair. *Nat. Struct. Mol. Biol.* 11(12): 1158–1159.

Wells, A.L. et al. 1999. Myosin VI is an actin-based motor that moves backwards. *Nature*. 401: 505–508.

Wernig, M. et al. 2007. *In vitro* reprogramming of fibroblasts into a pluripotent ES cell-like state. *Nature*. (in press). Published online 6/6/07. doi:10.1038/nature05944.

Westermeier, R. & R. Marouga. 2005. Protein detection methods in proteomics research. *Biosci. Rep.* 25(1/2): 19–32.

Westphal, S.P. 2003. 'Virgin birth' method promises ethical stem cells. *New Scientist*. 28 April. www.newscientist.com/article.ns?id=dn3654&print=true

________. 2003. "Humanized" organs can be grown in sheep. *New Scientist*. 17 December. www.newscientist.com/article.ns?id=dn4492

Whitmarsh, A.J. & R.J. Davis. 2000. Regulation of transcription factor function by phosphorylation. *CMLS, Cellular Mol. Life Sci.* 57: 1172–1183.

Whitworth, J.A., G.J. Mangos & J.J. Kelly. 2000. Cushing, cortisol, and cardiovascular disease. *Hypertension*. 36: 912–916.

Wicha, M.S., S. Liu & G. Dontu. 2006. Cancer stem cells: An old idea—A paradigm shift. *Cancer Res.* 66(4): 1183–1890.

Wickelgren, I. 2002. *The Gene Masters*. Times Books, Henry Holt and Co., New York.

Wilkinson, D.J. 2006. *Stochastic Modelling for Systems Biology*. Chapman and Hall/CRC. Atlanta.

Winter, K. 2005. *PrP Orthologs*. www.bio.Davidson.edu/Courses/Molbio/MolStudents/spring2005/Winter/PrionOrtho.htm

Wintzer. et al. 2004.

Wolf, S.J. 2006. *Introduction to Evolution*. (Course notes: Biology 3020. Calif. State. Univ. Stanislaus). http://arnica.csustan.edu/biol3020/meiosis/meiosis.htm

Wood, R.D., M. Mitchell & T. Lindahl. 2005. *Human DNA Repair Genes*. www.cgal.icnet.uk/DNA_Repair_Genes.html

Wood, R.D., M. Mitchell, J. Sgouros & T. Lindahl. 2001. Human DNA Repair Genes. *Science*. 291: 1284–1289.

Woods, M. 2000. *'Imprinting' Affects How Our Parents Genes Work in US*. www.post-gazette.com/healthscience/20000762heredity1.asp

Wooster, R.E. & B.L. Weber. 2003. Breast and ovarian cancer. *New Engl. J. Med.* 348: 2339–2347.

Wu, T. & R. Tanguay. 2006. Antibodies against heat shock proteins in environmental stresses and diseases: friend or foe? *Cell Stress Chaperones*. 11(1): 1–12.

WU. 2003. *mRNA Splicing*. www.neuro.wustl.edu/neuromuscular//////pathol/spliceosome.htm

Xinhua News. 2002. *Genetically Modified Cotton Damages Environment*. News release, 4 June 2002. www.china.org.cn/english/2002/Jun/33779.htm

Xu, Q. & K.S. Lam. 2003. Protein and chemical microarrays—powerful tools for proteomics. *J. Biomed. Biotechnol.* 5: 257–266.

Yagi, N. 2003. An X-ray diffraction study on early structural changes in skeletal muscle contraction. *Biophys. J.* 84: 1093–1102.

Yale University. 2005. *Yale Scientists Find MicroRNA Regulates Ras Cancer Gene*. www.sciencedaily.com/print.php?url=/releases/2005/03/050325222545.htm

Yamanaka, S. 2007. Strategies and new developments in the generation of patient-specific pluripotent stem cells. *Cell Stem Cell*. 1: 39–49.

Yang, C-R, B.E. Shapiro, E.D. Mjolsness & G.W. Hatfield. 2005. An enzyme mechanism language for the mathematical modeling of metabolic pathways. *Bioinformatics*. 21(6): 774–780.

Yang, L. et al. 2005. Event specific qualitative and quantitative polymerase chain reaction detection of genetically-modified MON863 maize based on the 5′-transgene integration sequence. *J. Agric. Food Chem.* 53: 9312–9318.

Yates, A. & I. Chambers. 2005. The homeodomain protein nanog and pluripotency in mouse embryonic stem cells. *Biochem. Soc. Trans.* 33(6): 1518–1521.

Yellen, G. 2002. The voltage-gated potassium channels and their relatives. *Nature*. 419: 35–42.

Yeo, D.S.Y. et al. 2004. Strategies for immobilization of biomolecules in a microarray. *Combinatorial Chem. High Throughput Screening*. 7: 213–231.

Yonath, A. 2001. *The Ribosome: A molecular machine with brains*. www.weizmann.ac.il/ICS/booklet/9/pdf/yonat.pdf

Yoo, C.B. & P.A. Jones. 2006. Epigenetic therapy of cancer: past, present and future. *Nat. Rev.—Drug Discovery*. 5: 37–50.

Yu, F.H., V. Yarov-Yarovoy, G.A. Guttman & W.A. Catterall. 2005. Overview of molecular relationships in the voltage-gated ion channel superfamily. *Pharmacological Rev.* 57(4): 387–395.

Yukihiro, M. et al. 2003. Nonviral gene gun mediated transfer into the beating heart. *ASAIO J.* 49(6): 641–644.

Zhang, C.C. et al. 2006. Prion protein is expressed on long-term repopulating hematopoietic stem cells and is important for their self-renewal. *PNAS Early Edition*. 30 Jan. (Published online: www.pnas.org)

Zhang, L. et al. 2002. Cloning and sequencing of quail and pigeon prion genes. *Anim. Biotechnol.* 13(1): 159–162.

Zheng, Z. 2006. *Viral RNA Splicing and Oncogenesis*. (NCI Web Site). http://ccr.ncicrf.gov/Staff/Staff.asp?profileid=5572

Zhou, B.P. & M.C. Hung. 2005. Wnt, hedgehog and snail: Sister pathways that control by GSK-3beta and beta-Trcp in the regulation of metastasis. *Cell Cycle*. 4(6): 772–776. (Epub 2005 Jun 13.)

Zheng, J., C. Zhang & R.M. Dickson. 2004. Highly fluorescent, water-soluble, size-tunable gold quantum dots. *Phys. Rev. Lett.* 93(7): 1–4.

Zhu, H. & M. Snyder. 2001. Protein arrays and microarrays. *Curr. Opinion Chem. Biol.* 5: 40–45.

Zhu, M. et al. 2001. Global analysis of protein activities using proteome chips. *Science*. 293: 2102–2105.

Zimmer, M. 2005. *Glowing Genes: A Revolution in Biotechnology*. Prometheus Books, Amherst, New York.

Zou, H., Y. Li & X. Wang. 1999. An Apaf-1.cytochrome-c multimeric complex is a functional apoptosome that activates procaspase-9. *J. Biol. Chem.* 274: 11549–11556.

Zuckerkandl, E. & G. Cavalli. 2006. Combinatorial epigenetics, "junk DNA", and the evolution of complex organisms. *Gene*. 390(1–2): 232–242.

Glossary

AA Abbreviation for *amino acid.*

AARS Enzymes Acronym for aminoacyl-tRNA synthetase enzymes.

Ab Acronym for antibody.

Adenine One of the four bases in DNA. Adenine always pairs with thymine in DNA. In RNA, adenine pairs with uracil.

Adenovirus A group of DNA-containing viruses that cause respiratory disease, including one form of the common cold. Adenoviruses can be genetically modified and used in gene therapy to treat cystic fibrosis; cancer; and, potentially, other diseases.

Allele One of the variant forms of a gene at a particular *locus*, or location on a chromosome. A single allele for each locus is inherited separately from each parent. Alleles produce variation in inherited characteristics such as eye color or blood type. There are dominant alleles and recessive alleles.

Alu elements Members of the *SINE* class of *retrotransposons.* Alu elements consist of 300 DNA bps containing a site that is recognized by the restriction enzyme, **Alu I**. There are over 1 million Alu elements in the human genome (about 10% of the total DNA).

Amino acid Any of a class of 20 small organic molecules that are assembled in long chains to form polypeptides or proteins by ribosomes. The sequence of AAs in a protein is determined by the *genetic code* of a particular *gene.* (See *Codon.*) There are also other noncoded AAs created enzymatically.

Aneuploid A defect in the spindle mechanism of a dividing cell prevents the daughter cell from receiving the correct number of chromosomes in cell division. Single chromosomes can be affected so that a daughter cell does not have a total chromosome number that is exactly 46(XX) or 46(XY). This abnormal cell is said to be aneuploid.

Animal model A laboratory animal useful for medical research because it has specific characteristics that resemble a human disease or disorder. Scientists can create animal models, usually laboratory mice, by transferring new genes into them.

ANT (Acronym for Altered Nuclear Transfer) A means of producing pluripotent stem cells (PSCs) without creating an embryo. ANT uses the technology of nuclear transfer (NT); however, the egg's cytoplasm or the nucleus of the somatic cell are altered to prevent the formation of a viable embryo. The alteration silences or deletes genes essential for the coordinated development of a maturing embryo. However, PSCs are produced. In oocyte-assisted reprogramming (OAR), the oocyte cytoplasm is chemically treated so the ANT oocyte behaves directly like a PSC. This process is called *ANT-OAR.*

Antibody A blood protein that is produced in response to, and counteracts, an antigen. Antibodies (Abs) are produced clonally by certain immune system cells in response to disease and help the body fight the particular disease. Memory B-cells store the genetic code for a particular Ab. In this way, antibodies help the body develop immunity to disease.

Antigen An antigen is any molecule that stimulates an animal's immune response. More specifically, it is a region of an extracellular molecule to which the immune system's free- or cell-surface bound antibodies can bind. The antigenic molecule (Ag) can be from an *exogenous* source, that is, from a viral coat protein, a bacterial protein, a parasite protein, a pollen glycoprotein, etc. Ags can also be *endogenous*; that is, they can arise within the body from damaged cells. In this case the Ag can be a protein, nucleic acid, glycoprotein, etc. They can also be associated with surface proteins on cancer cells. When the immune system responds abnormally to normal cell-surface molecules in the body, they are said to be *autoantigens,* and the response is called *autoimmuniity.*

Antisense DNA (asDNA) The noncoding strand in double-stranded DNA that does not code a gene into a protein. Its sequence is complimentary to that of the coding or sense strand. The sense strand codes the protein. It also refers to cloned (synthetic DNA) containing codons in the reverse direction from the normal gene being expressed. It is used to make antisense mRNA (see the following entry). Example:

5′...ATGGCCTGGACTTCA...3′ Sense strand of DNA
3′...TACCGGACCTGAAGT...5′ Antisense strand of DNA

Antisense RNA Produced from a gene sequence inserted in the opposite orientation so that the transcript is complementary to normal mRNA and can therefore bind to it and prevent translation. asRNA can be used to artificially "turn off" the expression of a gene and thus treat certain genetic diseases. Example:

5′.**CAUG**... 3′ mRNA
3′...**GUAC**... 5′ asRNA

Apoptosis Also known as *programmed cell death* (PCD). An internally, genetically regulated form of cell suicide that can be triggered by (1) external injurious agents such as heat shock, free radicals, UV irradiation, and various drugs and toxins; and (2) biological apoptosis-inducing factors including tumor necrosis factor alpha (TNF-α), which binds to the TNF receptor on the cell membrane; *lymphotoxin* (TNF-β), which also binds to the TNF receptors; and *Fas ligand* (FasL), which binds to a cell-surface receptor named Fas (aka CD95). Programmed cell death is necessary in embryonic development, and in maintenance of healthy adult tissues such as the skin, the uterine lining, and epithelial cells in the alimentary canal.

Aptamer A chemically synthesized (man-made) strand of oligonucleotide (DNA or RNA) that can adopt a highly specific, 3-D conformation. Aptamers are designed to have appropriate binding affinities and specificities toward certain target molecules.

Archaea (Archaebacteria) Microscopic, single-cell prokaryotes found living in extreme environments of temperature (heat, cold), pressure, salinity, pH (acid, alkaline), and zero pO_2. They have unique bilayer cell walls made from isoprene molecules, ether linked to L-glycerol phosphate. Archaea have a dsDNA ring genome that is unique in prokaryotes in having some introns. (This suggests they must have the means of editing out the intron information in their pmRNA, similar to eukaryote cells.) The tRNA and ribosomes of archaea are more like those of eukaryotes than bacteria. Also, the RNA polymerases of archaea resemble those of eukaryotic cells, not those of eubacteria. Archaea have been classified as a third life domain (with eukaryotes and prokaryotes).

Argonaute proteins Argonaute (AGO) proteins are an ubiquitous family of multidomain proteins expressed both in prokaryotic, archaean, and eukaryotic cells. In eukaryotes, AGOs are the catalytic components of the RNA-induced silencing complex (RISC) responsible for the gene-"silencing" mechanism known as RNA interference (RNAi). AGOs combine with small interfering RNA (siRNA) oligos and have endonuclease activity against mRNAs that are complementary to their bound siRNA oligo. AGOs probably have similar gene regulatory functions in prokaryotes [Hall 2005].

Astrocyte A type of CNS *glial cell* so named because it is star shaped. Its dense dendritic-type processes radiate from its cell body. Some processes have "end feet" that make contact with capillaries and neurons.

Attomole 1×10^{-18} mole.

Autacoids (From the Greek *autos* = "self" and *akos* = "medicinal agent" or "remedy") Locally acting signal molecules secreted by cells of the immune system and damaged body cells. Some are associated with inflammation. They are used to regulate local immune system responses, although they may have more general hormonal effects once in the circulatory system. Histamine is an autacoid. (See *cytokine*.)

Autopoiesis (From Greek *autos* = "self" and *poien* = "making") The continual internal chemical metabolism of a cell in which energy substrates are input through the cell's outer membrane (e.g., in the form of certain molecules such as sugars, O_2, H_2, CO_2, and photons). Then highly structured and regulated molecular systems within the cell make "fuel" molecules such as ATP to power the internal chemical reactions by which the cell makes more complex molecules such as proteins, glycoproteins, lipids, etc., that it uses to grow and maintain homeostasis. Autopoiesis cannot occur without the stored information in the cell's genome; it is necessary for life.

Autosomal dominant gene A pattern of Mendelian inheritance whereby an affected individual possesses one copy of a mutant allele (or gene) and one normal allele (or gene). (In contrast, recessive diseases require that the individual have two copies of the mutant allele.) Individuals with autosomal dominant diseases have a 50:50 chance of passing the mutant allele, and hence the disorder, to their children. Examples of autosomal dominant diseases include Huntington's disease, neurofibromatosis, (inherited) Creutzfeld–Jakob disease, and polycystic kidney disease.

Autosome A chromosome not involved in sex determination. The diploid human cell nucleus contains a total of 46 chromosomes, 22 pairs of autosomes, and 1 pair of sex chromosomes (XX or XY).

Avogadro's number The number of molecules in a gram-mole: $N = 6.023 \times 10^{23}$ (gm-mole)$^{-1}$.

Bacterial artificial chromosome (BAC) Large segments of DNA, 100,000 to 200,000 bases, from another species cloned into bacteria. Once the foreign DNA has been cloned into the host bacteria, many copies of it can be made.

Basal transcription factors Proteins required for the initiation of gene transcription. These include TFIID, TATA binding protein (TBP), TFIIA, and TFIIB. The transcription factor proteins TFIIE, IIF, and IIH release the *RNA Polymerase* from the TATA *box* complex so it can unwind sections of the dsDNA helix for transcription.

Base (As used in genomics) The purines *adenine and guanine* and the pyrimidines *uracil, thymine,* and *cytosine*. (Also *nitrogenous base*.)

Base pair (bp) Two bases located opposite each other on the opposing *(sense–antisense)* strands of a dsDNA or dsRNA helix molecule. They cross-link with hydrogen bonds to form "rungs of the DNA ladder." Adenine always pairs with thymine in DNA; adenine pairs with uracil in RNA; guanine always pairs with cytosine in both DNA and RNA. (See Figures 6.1 and 6.3 in the text.)

Biological clock The source of periodic behavior in eukaryotes. Periodic behavior may be *circadian* (period 24 h), *ultradian* (period > 24 h), or *infradian* (period < 24 h). Endogenous biological clocks can involve feeding, ovulation, egg laying, etc., or involve the secretion of certain hormones. Certain *zeitgebers,* such as photoperiod, lunar cycle, or tides, have been shown to entrain or synchronize circadian clocks. Biological clocks may be related to oscillations in closed-loop, biochemical systems.

Biotinylation The covalent linkage of biotin to nucleic acids or proteins. The linked biotin may be detected with avidin or streptavidin linked to an enzyme or a fluorescent molecule. Biotin forms unusually strong, stable chemical bonds with avidin.

Blastocyst A preimplantation embryo of 150 cells. It consists of a hollow ball of cells (the *trophectoderm*), a fluid-filled cavity (the *blastocoel*), and an interior cluster of cells adhering to the inner *trophectoderm* wall (the inner cell mass).

Blastomere A cell taken from an early embryo, such as the eight-cell or morula stages.

bps Base pairs (of dsDNA or dsRNA).

BRCA1, BRCA2 The first breast cancer genes to be identified. Mutated forms of these genes are believed to be responsible for about one-half the cases of inherited breast cancer, especially those that occur in younger women. Both are tumor suppressor genes; the mutations prevent the suppression.

C^3 (C cubed) In cell physiology: Abbreviation for Command, Communication, and Control (of cellular processes).

5′-Cap After gene transcription (and splicing), eukaryotic mRNA is modified by the addition of a methylated guanine Nt in a special 5′-5′ linkage to the 5′ end of the mRNA and by methylation of two 5′ terminal Nts of the mRNA. These additions are called the cap. The cap stabilizes the mRNA from enzymatic degradation and serves as a recognition site for binding to a *ribosome*.

Cajal body (CB) Spherical bodies found in association with chromatin in the nuclei of metabolically active eukaryotic cells (e.g., neurons and tumor cells). One to five CBs occur in a nucleus; they are from 0.1 to 2 μm diameter. CBs are possibly sites of assembly or regulation of the transcription machinery of the nucleus.

Carrier An individual who possesses one copy of a mutant allele that causes a genetic disease only when two copies are present. Although carriers are not affected by the disease, two carriers can produce a child who has the disease.

cDNA Protein-coding sequences of DNA (i.e., *genes*). cDNA can be obtained by the reverse transcription of a protein's mRNA. For example, the transcription and translation of human insulin cDNA allows the production of functional, human insulin molecules in nonhuman cells, such as *E. coli*. See, for example, *ORF*.

Cell The basic structural unit of life and of any multicellular organism. It is a small watery compartment filled with various molecules and a complete copy of the organism's *genome* on *DNA*. Its inner structures are protected by an external cell membrane. There are three major classes of cells: *Archaea, Prokaryotes,* and *Eukaryotes*. A cell, to be alive, must be *autopoietic*. Viruses are not cells.

Cell cycle (of early blastomeres and somatic cells) The reproduction of rapidly dividing embryonic and somatic cells is controlled by a family of *Cyclin regulatory proteins* (*Cyclin A, B, D, and E*). Cyclin B is encoded by mRNA stored in the cytoplasm of the ovum. Cyclin B is responsible for triggering mitosis. Mitosis is followed by an *interphase* stage. The interphase stage is subdivided into G1, S (protein and RNA synthesis), and G2 phases. Cells that are differentiating are usually taken out of the cell cycle and are in an extended G1 phase called G0. Cyclin D is associated with a cell moving into the G0 phase cells in the G0. Phase can be used in animal cloning. (See ANT-OAR.)

Centimorgan (cM) In genetics, a unit of recombination frequency. One cM is equal to a 1% chance that a *marker* at one genetic *locus* on a chromosome will be separated from a marker at a second locus due to a *crossing over* in a single generation. In human beings, 1 cM is equivalent, on average, to 10^6 DNA base pairs.

Central dogma The complex sequence of nuclear and ribosomal chemical events involved in the assembly of a protein beginning with the "reading" of a gene's DNA *codons* in the nucleus by pre-mRNA. pmRNA is "edited" to form mRNA, which leaves the nuclear membrane and joins ribosomes that assemble AAs into the linear peptide chain of a protein, following the genetic code (cf Table 5.4). The central dogma assumes a one-way flow of information, from gene to protein.

Centromere The constricted ("waist") region near the center of a human chromosome pair seen under the light microscope. This is the region of the chromosome where the two sister chromatids are joined to one another. The shorter arm is called the "p" arm, the longer arm is called the "q" arm. In human chromosomes, a centromere contains about 10^6 noncoding Nt pairs. Most of this DNA is repetitive short sequences (e.g., 171 base pairs) repeated over and over in tandem arrays. The *kinetochore* forms at the centromere during cell division. The fibers of the mitotic (or meiotic) *spindle* attach at the chromosomes' centromeres.

Centrosome A bundle of *microtubules* located in the cytoplasm of eukaryotic cells attached to the outside of the nucleus. Just before *mitosis,* the centrosome duplicates, and the two centrosomes move apart to the opposite ends of the nucleus. As mitosis continues, microtubules called *spindle fibers* grow out from each centrosome toward the *metaphase plate.* Some spindle fibers attach to the arms of the chromosomes, some attach to one *kinetochore* of a chromosomal *dyad*, and still others continue growing from the two centrosomes until they overlap. All three classes of spindle fibers participate in the complex mechanical events in the nucleus during mitosis, separating the chromosomes into the two daughter cells.

Chaperones and chaperonins Chaperones include chaperonins. Chaperones are complex proteins whose function is to ensure that a newly assembled polypeptide is bent, folded, and cross-linked to form the desired active 3-D protein structure. Some chaperone proteins, the *heat shock proteins,* have the role of repairing proteins whose 3-D structures are damaged by elevated temperature, chemicals, or ionizing radiation. Other chaperones are involved in the transport of new proteins across membranes (e.g., in mitochondria and the ER). Some chaperones may be involved in protein degradation.

Chaperonins are molecular machines that use chemical energy (ATP) to fold and bond new polypeptides into their mature, functional 3-D protein forms. The structure of a chaperonin is generally in the form of two stacked protein *tori* ("donuts") that form a hollow "barrel" in which the folding occurs. Each chaperonin torus is composed of seven to nine protein subunits, depending on the organism. Group I chaperonins are found in prokaryotes, as well as eukaryotic chloroplasts and mitochondria. Group II chaperonins occur in *Archaea* and the eukaryotic cytosol.

Chemiosmosis Takes place in mitochondria and chloroplasts. In mitochondria, *protons* from the *citric acid cycle* and *glycolysis* are actively pumped from the *matrix* into the *outer compartment* where they achieve a high concentration. H^+ ions (protons) then flow down a concentration and potential gradient past the enzyme, *ATP synthase* (ATPS), back into the matrix (at low H^+ concentration). The enzyme and pump molecules are embedded in the bilayer membrane separating the matrix from the outer compartment of the mitochondrion. The potential energy in the protons is used by the ATPS enzyme molecules to phosphorylate 32 ADP molecules to 32 ATPs per glucose molecule oxidized in the matrix.

Chemotaxis When a mobile single cell or complex animal follows an external chemical gradient toward its source. The chemical involved can be a hormone, a pheremone, or a cytokine, or an energy source can be followed.

Chimera A condition in which cells having two or more different genetic makeups appear in the same living, multicellular organism (plant or animal). A chimera can be produced artificially or can occur naturally by the fusion of two zygotes (or two or three 8-cell embryos) to form a single chimeric morula, which then develops normally as an embryo.

Chirality Chirality refers to how organic molecules in solution rotate the E-vector of linearly polarized light (LPL), which is passed through a glass vessel containing the solution. D-glucose rotates the E-vector of the emergent LPL light clockwise (looking toward the source) and is said to be *dextrorotary.* The AA, L-glycine, is laevorotary and rotates the emergent E-vector counterclockwise. Optical rotation φ is proportional to analyte concentration [C] times the optical path length, L, through the solution and is a function of wavelength and temperature: $\varphi = \alpha^{T},_{\lambda}$ [C] L. $\alpha^{T},_{\lambda}$ is the *specific optical rotation* of an optically active analyte in degrees/(mole meter), φ is the optical rotation in degrees, and $\alpha^{T},_{\lambda}$ is a function of temperature and wavelength.

Chromatid A single (stained) dsDNA molecule in a cell. Each of a diploid (**2n**) cell's chromosomes are composed of two identical chromatids joined at a *centromere.*

Chromatin The interphase form of nuclear DNA. A complex of two-stranded DNA and histone proteins that enables the inactive DNA to be stored compactly. Subunits of chromatin are called *nucleosomes*. In eukaryotes, the chromatin is found inside the nuclear membrane, whereas in prokaryotes it is associated with the nucleoid. Chromatin is made visible under the light microscope by staining cells with a methyl green/pyronin (MGP) stain. The methyl green colors the DNA green to blue, and the pyronine stains any nuclear RNA (such as pre-mRNA) red. Dense chromatin is called *heterochromatin. Euchromatin* is lightly stained chromatin.

Chromosome One of the threadlike strands of dsDNA carrying genes that can be visualized in the nucleus of a prophase (dividing) cell using light microscopy and a stain such as aceto-orcein, quinacrine (fluorescent), or bisbenzamide (fluorescent). Nucleated (diploid) human cells have 22 pairs of chromosomes plus two sex chromosomes. (The human female has 22 pairs of autosomes + 1 pair of X chomosomes; a human male has 22 pairs of autosomes + 1X + 1Y, so **2n** = 46.) The chromosome in a prokaryote cell (e.g., a bacterium) is a single ring of dsDNA. Other animals have different numbers of chromosomes.

cis-regulatory elements DNA sequences that regulate the expression of a gene (e.g., *promoters, enhancers,* and *silencers*) located on the same strand of DNA as the gene being controlled. Regulatory proteins (transcription factors) bind to cis-regulatory DNA elements to effect regulation. These may be located in the promoter region 5′ to the gene it regulates, in an intron within the gene, or in the 3′ region.

Cloning The process of making copies of a specific piece of DNA, usually a gene. When geneticists speak of cloning, they do not usually mean the process of making genetically identical copies of an entire organism. Animal cloning is also used to mean making a genomically identical embryo from an adult organism, then implanting that embryo into a surrogate mother in which it is grown.

Cloning vector A DNA molecule originating from a virus, phage, a plasmid, or the cell of a higher organism into which another DNA fragment (oligo) of appropriate size can be integrated (spliced) without loss of the vector's capacity for self-replication. Vectors introduce foreign DNA into host cells in which it is replicated autonomously in large quantities. Examples are *plasmids, cosmids,* and yeast artificial chromosomes (YACs). Vectors are often recombinant molecules containing DNA sequences from several sources.

Codon A sequence of three DNA bases in a gene's DNA strand that codes for one unique amino acid (AA) being incorporated into a linear peptide molecule, which goes on to form a protein. The total number of codons possible, given four different nucleic acid bases, taken three at a time, is $4^3 = 64$ (e.g., **CAT** codes for the AA histidine, but so does **CAC**).

Compartment (As used in physiology and biochemical modeling) A contiguous volume in which a substance has a uniform (or near uniform) concentration or density. An example of a compartment for glucose is the extracellular fluid volume. Blood volume is another glucose compartment. There is exchange of substances between compartments governed by diffusion and chemical kinetics.

Complementary sequence (of DNA or RNA) In the DNA or RNA helix, hydrogen bonds form across the helix between specific opposite bases, for example, between adenine and thymine, and between guanine and cytosine in DNA; and between adenine and uracil, and between guanine and cytosine in RNA. The same type of pairing can occur between the bases of single-strand *oligos* and their complements.

Complex An adjective used to denote a system consisting of many (e.g., greater than seven) causally interconnected or interwoven parts or components. A complex system generally has multiple inputs and outputs. The large number of components and a high degree of causal interconnectivity make a complex system difficult to understand and model. Biochemical systems are, in general, complex (and nonlinear). The word "complex" comes from the Latin *com-* (together) + *plectere* (to weave or braid).

Contig A chromosome map showing the locations of those regions of a chromosome where contiguous DNA segments overlap. Contig maps provide the ability to study a complete, and often large, segment of the genome by examining a series of overlapping clones which then provide an unbroken succession of information about that region.

Copy number variation (CNV) (of a gene) A diploid cell that derives its genome from male and female parents (sperm and egg) normally has two copies of a given gene, one from each parent. However, the number of copies can vary: A person can be born with one, two, three, or more copies of a gene. This is known as copy number variation (CNV). CNVs can result from insertions, deletions, and duplications of large segments of DNA. These segments are large enough to include entire genes and their control regions. Many body functions, health, and survival can be affected by CNVs. There are about 2,000 CNVs reported in the about 3.1×10^9 bps of the human genome; these include about 10% of the genes in the human genome. Most CNVs are benign, others have been linked to kidney disease, Parkinson's disease, Alzheimer's disease, and HIV susceptibility.

Cosmid An artificially constructed cloning vector containing the **cos** gene of phage lambda. Cosmids can be packaged in lambda phage particles for infection into *E. coli*. This permits cloning of larger DNA fragments (≤45 kb) than can be introduced into bacterial hosts by plasmid vectors.

CpG islands A contiguous region of DNA having a higher-than-expected concentration of the dinucleotide **CG** (cytosine + guanine). The "island" definition is made statistically, and varies. The cytosine in CpG is commonly found methylated in nature (5′-methylcytosine, or 5m**C**). Only 5m**C** is found in eukaryotes. CpG islands are often found associated with proximal regions of genes (viz., promoters and first exons). CpG dinucleotides are the site of >90% of all methylation in mammals. (The p in CpG stands for phosphate.)

Crossing (A term from Mendelian genetics) Fertilization of an organism by another organism with a different genetic constitution (i.e., different alleles).

Crossing over A term for the occurrence of new combinations of linked *characters*. The term is applied to the breaking during *meiosis* of one maternal and one paternal chromosome, the exchange of corresponding sections of DNA, and the rejoining of the chromosomes. This process can result in an exchange of *alleles* between chromosomes and gives rise to new character combinations.

C-value The amount of DNA (in picograms or kilobases) contained within a *haploid* (or 1C), eukaryote nucleus. In some cases, C-value and genome size are used interchangeably. C-values vary enormously among species. In animals they range 4,400-fold, and in terrestrial plants the range is about 1,000-fold. The smallest animal C-value is reported as 0.03 pg in a root-knot nematode; the largest is 132.83 pg in the marbled lungfish.

Using base numbers (the same as *base-pair* numbers in diploid cells), the worm *C. elegans* has a C-number of about 1.00×10^8 bases (19,427 genes). The fruit fly *Drosophila sp.* has about 1.22×10^8 bases (13,379 genes), and humans have about 3.2×10^9 bases (about 22,000 genes).

Cytogenetic map The visual appearance of a metaphase cell's chromosomes when stained and examined under a light microscope. Of particular interest are the light and dark bands exhibited on each chromosome. See *karyotype*.

Cytokine Certain proteins and peptides used for intercellular signaling, for example, by the immune system. They are secreted by a wide variety of cell types and can have effects on nearby cells with appropriate cell-surface receptors or on cells throughout the body. *Autocrine* cytokines act on the cells that secrete them. *Paracrine* cytokines act locally on surrounding cells, and *endocrine* cytokines act at a distance, carried by blood or lymph. Cytokines are generally smaller, water-soluble glycoproteins having masses between 8 and 30 kDa. *Autacoids* are cytokines released specifically by cells of the immune system.

Cytoplasm Everything outside the nuclear membrane and inside the cell membrane of a eukaryote cell. It is the cytosol plus the membranes and organelles (ribosomes, mitochondria, Golgi apparatus, etc.).

Cytosine A pyrimidine base found in both DNA and RNA. Cytosine pairs with guanine.

Cytosol The liquid phase filling the inside of a cell, excluding any internal membranes and organelles in eukaryote cells. It is the fluid in which cellular organelles and the nucleus are suspended. The cytosol contains many protein molecules, enzymes, ions, sRNAs, etc.

Deletion A particular kind of mutation: loss of a piece of DNA from a chromosome. Deletion of a gene or part of a gene can lead to a developmental abnormality, disease, or death.

Deoxyribonucleic acid (DNA) The chemical inside the nucleus of a cell that carries the genetic instructions for making living organisms.

Dicer A large-molecular-weight protein enzyme that is involved in extranuclear synthesis of miRNAs.

Diploid The number of chromosomes in normal nucleated cells derived from the *zygote,* excluding the *gametes* derived by *meiosis.* Diploid human cells have 46 chromosomes.

DMNTs DNA methyltranferases. These enzymes attach methyl groups to the *CpG dinucleotides* in DNA. Three families of DMNTs are known. (Sometimes written Dmnt.)

DNA Acronym for *deoxyribonucleic acid.* DNA carries the information for cell and organism maturation and growth, as well as information for its own reproduction. It is found in all cells and most viruses.

Dominant A gene that almost always results in a specific physical characteristic, such as a disease, even though the patient's genome possesses only one copy. With a dominant gene, the chance of passing on the gene (and therefore the disease) to children is 50–50 in each pregnancy.

Double helix The structural arrangement of eukaryotic cell DNA, which looks something like an immensely long ladder twisted into a helix, or coil. The sides of the "ladder" are formed by a backbone of sugar and phosphate molecules, and the "rungs" consist of nucleotide bases joined weakly across the middle by hydrogen bonds.

Drosha A nuclear enzyme (aka RNase III) involved in the first intranuclear step in the synthesis of miRNAs.

dsDNA, dsRNA Double-stranded DNA or RNA. Also, **ssDNA** is single-stranded DNA, etc.

Electrophoresis The process in which molecules (such as proteins, DNA, or RNA fragments) can be separated according to their size and electrical charge by applying a dc electric field to a medium containing them. In gel electrophoresis, the field forces the molecules through pores in a thin layer of a firm jelly-like substance. The gel can be made so that its pores are just the right dimensions for separating molecules within a specific range of sizes and shapes. Smaller fragments usually travel farther than large ones in a given time. Paper wetted with a solvent can also be used in electrophoresis.

Electroporation The exposure of living cells, in vivo or in vitro, to a transient pulsed dc electric field that exceeds the dielectric strength of the cell membrane and causes it to develop "pores" through which macromolecules such as hormones or DNA plasmids can enter the cell. In a correctly designed electroporation system, the membrane reseals itself in milliseconds following the pulse. The transient electric field is often generated by discharging a high-voltage capacitor (e.g., 2.5 kV) through a chamber with electrodes containing the target cells.

ELISA Acronym for Enzyme-Linked Immunoabsorbent Assay. A highly sensitive method for detection and measurement of antigens (or antibodies) in solution. In one embodiment, monoclonal Abs with an affinity to the analyte to be measured are fixed to a solid surface such as a glass slide. Next, a preparation of the same MAbs is coupled to an enzyme such as β-galactosidase that produces a colored product from a colorless substrate. The slide is immersed in a solution with the antigen (Ag) analyte in it, and the Ag binds to the fixed MAbs. Any excess Ag is washed off. Next the MAb-enzyme solution is added.

The MAb part of the conjugate molecules binds to the previously bound Ag, forming a fixed, Ab-Ag-Ab-enzyme molecular "sandwich." Any unbound conjugate is washed off, and a colorless substrate solution is added to react with the enzyme, causing color to develop. After a fixed time, the reaction is stopped, and the amount of color developed is measured by spectrophotometry. The concentration of bound Ag is a monotonic function of the color intensity.

Embryo (Definitions vary) We will define a human embryo as existing from the moment the zygote undergoes its first mitotic cell division, through its implantation in the uterus, to the 8th week postconception. After the 8th week, it becomes a fetus.

Endonuclease (aka restriction endonuclease [RE]) A protein enzyme that cleaves a single-stranded DNA molecule at a specific site (recognition sequence of four, five, or six nucleotides) in its nucleotide sequence. Many different endonucleases exist. REs are used in DNA forensic analysis.

Enhancers Molecules that promote the rate of gene expression. Enhancer molecules bind to *cis-acting* DNA regions up to 1,000s of bp away from the gene they control. Binding increases the rate of transcription of the gene by RNA. Enhancer Nt sequences can be located "upstream" or "downstream" or within the gene they control.

Enzyme A catalyst that expedites biochemical reactions. Enzymes are generally proteins, protein complexes, or RNA molecules (*ribozymes*). The chemical reactions catalyzed by enzymes must spontaneously run in the direction that they will with the enzyme present. The enzyme just increases the reaction rate. An enzyme can run a normally spontaneous reaction "backwards" by coupling it to another spontaneous one, as long as the net free energy from the sum of both reactions is negative. Most enzyme names end in *-ase* or *-yme.*

Epigenetic code Genetic information that is stored in proteins, histones, and other molecules that surround the DNA alpha helix in eukaryotic cells. Epigenetic molecules do not alter the underlying DNA sequence but cause the *silencing or activation* of certain genes. Epigenetic control of gene expression is also accomplished by the enzymatic *methylation* of cytosine (to *5-methy-cytosine*) and adenosine (to *N-6-methyladenosine*) in DNA.

Epigenetics Heritable changes in gene expression that operate outside base changes (e.g., mutations) in DNA itself. They involve DNA methylation and attachment or removal of radicals (methy-, acetyl-, phosphoryl-, etc.) to certain AAs in histone proteins, especially H3.

Epitope (Used in an immune system context) A specific structural domain on an antigen (Ag) molecule to which an *antibody* (Ab) molecule has a high chemical affinity. The paratope region of the Ab binds chemically to the epitope on an antigen. Certain B- and T-cells can also have cell-membrane-bound Abs having paratopes to an antigenic epitope.

epsp Abbreviation for excitatory postsynaptic potential in a neuron. An epsp is seen as a positive-going shift in the transmembrane potential of the subsynaptic membrane following the arrival of an excitatory presynaptic nerve spike.

ER Acronym for *endoplasmic reticulum,* a cytoplasmic structure in eukaryotic cells.

Euchromatin A less dense, more lightly stained form of chromatin in which there is a high density of genes that may be being actively transcribed. The DNA in euchromatin is loosely wrapped around the histone octamer, *nucleosome beads.* This allows pre-mRNA access to the genetic DNA. The loose structure of the euchromatin is the result of epigenetic chemical modifications to certain nucleosome AA residues.

Eukaryote A cell, or organism with cells, having a discrete nucleus surrounded with a membrane and having other well-developed subcellular organelles. Eukaryotes include all organisms *except* bacteria, blue-green algae, and viruses, including phages.

Evanescent field (EF) (Of an optical waveguide) A time-varying electromagnetic (EM) field that exists in a refractive medium of refractive index n_2 surrounding an optical waveguide core with index n_1, ($n_1 > n_2 > 1$). The intensity of the evanescent field decreases monotonically as a function of the radial distance from the surface of the core but without an accompanying phase shift. The EF is a surface wave that is coupled to the EM wave or mode propagating in the core. The EF from a flat core (planar waveguide) can be used in optical microarrays for low-noise excitation of fluorophores.

Exons The protein-coding DNA sequences of a genome.

F1 In Mendelian genetics, the first- or filial-one generation progeny after mating or genetically crossing two parents (the P-generation) with different genotypes or phenotypes.

F2 The second filial generation, obtained by mating members of the **F1** generation.

FADH2 Flavin Adenine Dinucleotide.

Femtomole 1×10^{-15} mole.

Fetus (Definitions vary) A developing human embryo implanted in the uterus, generally taken from the end of the 8th week postconception to the time of birth.

FISH Acronym for Fluorescence In Situ Hybridization. A physical mapping technique that uses fluorescent molecular tags to detect *hybridization* of *probes* with *metaphase chromosomes* and with the less condensed *somatic interphase* chromatin.

Fluorescence Certain molecules (e.g., *GFP*), when excited by short-wavelength light such as UVA, violet, or blue visible light, reradiate light at a longer wavelength in a relatively narrow spectral band. Fluorescence is widely used as a tool to tag specific molecules or organisms.

FRET Acronym for Fluorescence Resonance Energy Transfer. An analytical technique used to detect a particular molecular species. In one FRET system used to sense glucose, the intensity of the emitted, NIR fluorescence is proportional to the glucose concentration.

Gamete A mature female or male reproductive cell (egg or sperm) with a haploid set of chromosomes. Gametes are the result of the cell division process known as *meiosis*. Human gametes have 23 chromosomes (an egg has 22 + X; a sperm has 22 + Y or 22 + X chromosomes).

Gene By the central dogma, genes code proteins. A gene is an ordered sequence of nucleotides located in a particular region of a particular chromosome that encodes a particular functional product (i.e., an RNA or a protein).

Gene battery Sets of coregulated genes with common *cis*-regulatory elements that define the differentiated state of a cell.

Gene expression The complex process by which the molecular structure encoded in the DNA nucleotide bases of a gene is transcribed into either the amino acid sequence of a protein, or an RNA molecule. Understanding the control of gene expression is a current major problem in cell biology.

Gene gun An electromechanical/pneumatic, bioballistic device used to project genetic material into cells through their cell membranes for the purposes of genetic engineering or therapy. Gene guns were originally used in botanical applications. In most designs, a supersonic shock wave of helium gas accelerates about 1-μm-diameter metal particles (e.g., gold and tungsten) coated with genetic material (e.g., ribozymes and DNA plasmids) to supersonic velocity (e.g., 1,760 fps). A stopping device abruptly decelerates the gold, allowing the DNA to continue on to bombard cell surfaces at velocities that penetrate the cell membranes. A gene gun requires a direct contact with the cell surfaces. Modern gene gun designs have minimized kinetic cell damage while increasing transfection efficiency.

Genetic Code (for proteins) A degenerate code in which 64 triplets of DNA nucleotides called *codons* found in genes specify the sequence in which various AAs are added to a protein being synthesized in a ribosome. In DNA the code is as follows:

DNA Codons	Amino Acid or Operation
ATG	**Start coding** when at the 5′ beginning of a gene, else methionine
TTT, TTC	Phenylalanine

DNA Codons	Amino Acid or Operation
TTA, TTG, CTT, CTC, CTA, CTG	Leucine
ATT, ATC, ATA	Ileucine
GTT, GTC, GTA, GTG	Valine
TCT, TCC, TCA, TCG	Serine
CCT, CCC, CCA, CCG	Proline
ACT, ACC, ACA, ACG	Threonine
GCT, GCC, GCA, GCG	Alanine
TAT, TAC	Tyrosine
CAT, CAC	Histidine
CAA, CAG	Glutamine
AAT, AAC	Asparagine
AAA, AAG	Lysine
GAT, GAC	Aspartic acid
GAA, GAG	Glutamic acid
TGT, TGC	Cysteine
TGG	Tryptophan
CGT, CGC, CGA, CGG, AGA, AGG	Arginine
GGT, GGC, GGA, GGG	Glycine
TAA, TAG, TGA	**Stop coding** (at the 3′end of gene)

Genome All of the DNA contained in an organism or a cell, which includes both the chromosomes within the nucleus and the DNA in mitochondria and chloroplasts. Only a small fraction (about 1%) of the human genome codes for proteins; the balance (the introns) codes RNAs, indicates where genes start and stop along the DNA, consists of promoter sequences, or codes nothing at all.

Glial cell Also called neuroglia or simply glia, these ubiquitous cells provide support and protection for CNS and spinal cord neurons in vertebrates. They also provide nutrition, maintain homeostasis, form myelin sheaths, and modulate neuron function in the CNS. Glia are connected to each other and to neurons by gap junction synapses. There are two major categories of glial cells. *Microglia* are specialized macrophages constituting about 15% of the total cells in the CNS, capable of phagocytosis, which protect the CNS from infections. *Macroglial* cells include a number of types—*astrocytes, oligodendrocytes, ependymal cells, radial glia, schwann cells,* and *satellite cells.* There are probably 10 times as many total glial cells in the CNS than neurons.

Glycoprotein A glycoprotein is a polypeptide to which carbohydrate molecules (oligosaccharides) are attached. In *O-glycosylation,* the sugar chains can attach to the AAs hydroxylysine, hydroxyproline, serine, or threonine. When a sugar chain attaches to the amino group on asparagine's γ-carbon, it is called *N-glycosylation.* Examples of glycoproteins are antibodies, *lysosomal-associated membrane proteins* (LAMPs) and MHC molecules.

GM Acronym for *genetically modified.* For example, GM tomato.

Green fluorescent protein (GFP) GFP and its synthetic derivative molecules have revolutionized experimental genetics, genomics, and cell biology by permitting the in vivo tagging of various genes, oligos, and proteins, as well as viruses, bacteria, and even whole animals. GFP was first isolated from the jellyfish, *Aequorea victoria,* in the 1960s and cloned in 1992. The entire 27-kDa peptide structure (in the form of a spiral-walled barrel) is required for fluorescence. The basic GFP fluorophore consists of three linked AAs: Ser65 + Tyr66 + Gly67. Its fluorescent radiation wavelength is broad with a peak at 507 nm.

Guanine A purine molecule found in DNA and RNA. Guanine pairs with cytosine.

Half-life The time it takes the concentration of a molecular species in a compartment to decay by a factor of 0.5.

Haploid Refers to the number of chromosomes in an egg or sperm (**n** = 23 in humans).

Heterochromatin Generally, a genetically inactive region of chromosomes in which the histone protein structure permits tight "supercoiling" of the nucleosome–DNA complex for compact protected storage. This high density of DNA causes dark staining by methyl green. The DNA in heterochromatin regions either lacks genes or has repressed (silenced) genes. Centromeres, telomeres, and the inactivated (female) X-chromosome (Barr body) are heterochromatic. Heterochromatin is found only in the nuclei of eukaryote cells.

Heterozygous Possessing two different forms of a particular gene, one inherited from each parent.

Histones Histones are the highly conserved, major protein components of nuclear chromatin found in eukaryotes and euryarchaea. (Bacteria lack histones.) Histone proteins form the *nucleosome* "beads" around which the organism's dsDNA is wound and folded in its "storage form." They also play a role in regulating gene expression. Six major histone classes are known. Two each of four types of histone protein (H2A, H2B, H3, and H4) form the octameric *nucleosome* core. dsDNA is wrapped around each nucleosome, and the nucleosomes are strung together like beads along the linear, H1 linker histone. About 50 bps of dsDNA connect each nucleosome bead.

Histone proteins can be modified by methylation, acetylation, phosphorylation, etc., to effect such processes as gene regulation, DNA repair, mitosis, and meiosis.

Homeobox genes A DNA sequence of about 180 DNA base pairs long, found within genes that are involved with the regulation of development (morphogenesis). The sequence encodes a protein domain (a *homeodomain*) that can bind to DNA. Homeobox genes encode *transcription factors,* which in turn switch on cascades of other genes, for example, those required to make an eye. Homeodomain proteins generally act in the promoter region of their target genes as complexes with other transcription factors and/or homeodomain proteins.

Homeobox genes were first found in the fruit fly and appear ubiquitous in all animals. They have also been found in fungi, yeasts, and plants. In an analogy to computing, a homeobox gene is like a call to a subroutine. It switches on the expression of an entire subsystem, the code for which must already be present elsewhere in the genome. The development-controlling Hox genes are a subgroup of homeobox genes.

Homeodomain A highly conserved "motif" or sequence of AAs found, with minor variations in many proteins, which is involved in the sequential regulation of gene expression in embryological development. The motif consists of about a 60-AA protein alpha helix. The homeodomain binds to DNA at a **TAAT** base sequence, similar to the **TAATAAA** sequence used by RNA polymerase to locate the transcription initiation site. Homeodomain-containing proteins also contain a *third* alpha helical segment that supplies the detailed sequence specificity. The homeodomain segment aligns the protein in the dsDNA groove so that the third, recognition protein helix is brought into contact with specific sequences of the dsDNA's base pairs—a clever design.

Homeostasis (cellular) The dynamic maintenance of steady-state conditions in the internal compartment, or cytosol of a cell, by the programmed, regulated expenditure of chemical energy. Homeostasis stabilizes the interior of the cell against chemical and physical changes in its external environment. It requires the dynamic equilibria of many biochemical systems and is generally the result of the action of multiple chemical negative-feedback control systems. An example of homeostasis is the maintenance of the concentration of sodium ions in the cytosol of a cell within "normal" bounds against extracellular changes in sodium concentration. Sodium "leaks" into the cell down concentration and membrane potential gradients. Sodium homeostasis in cells is effected by sodium "pumps," specialized proteins fixed in the cell membrane that have the capability of moving sodium ions from the cell's cytosol to the extracellular fluid at the expense of metabolic energy. The sodium ion concentration in a cell determines the osmotic pressure across its cell membrane, and hence the amount of water that leaks in or out of the cell. Even water passage through a cell membrane is regulated by transmembrane *aquaporin proteins* as a part of homeostasis.

Homing Homing is defined as the enzyme-mediated transfer of a parasitic genetic element (intein or intron) to a cognate allele (another gene in the nucleus that lacks that element) (i.e., transfer will not occur if the gene already contains the parasitic element's Nts). The endonuclease proteins coded by either an intron or the ORF of the endonuclease portion of a large intein make a site-specific, double-strand break in the DNA of the target gene. Gene conversion repairs the break to generate a new, intron- or large intein-containing product.

Homoplasy Appearance of similar structure in organisms not inherited from a common ancestor or developed from a common region of differentiated cells (primordium) in an embryo.

Homozygous Possessing two identical forms of a particular gene, one inherited from each parent.

Hox genes Hox genes are a subset of *homeobox genes*. They are essential for the sequential control of embryological development in segments along the body axis; they are the "master regulators" for embryological development. Hox genes are highly conserved in all forms of life. Over millions of years of evolution, they have retained their function of assigning particular positions in the embryo. However, the structures (organs, systems) actually built depend on a different set of genes specific for a particular species. An error in the expression of a Hox gene generally leads to a nonviable embryo, a major factor in their conservation through natural selection. Exogenous retinoic acid has been shown to cause Hox gene mutations and birth defects in mammals.

Hox genes encode *transcription factors,* each with a specific, DNA-binding homeodomain protein. They act in sequential zones of the embryo in the same order that they occur on the chromosome. Humans have four Hox gene clusters (HOXA–HOXD) with a total of 39 genes. Human Hox genes (A–D) are located on chromosomes 7p15, 17q21.2, 12q13e and 2q31, respectively.

Hybridization The process of enzymatically joining two complementary strands of DNA or one each of DNA and RNA to form a double-stranded molecule. One strand is often created synthetically, labeled, and used as a *probe* to detect a specific base sequence in the target strand.

Imprinted genes Genes whose expression is determined by the parent that contributed them. They violate the usual rule of inheritance that both alleles in a heterozygote are equally expressed. The process of imprinting starts in the (haploid) gametes (egg or sperm) where the allele destined to be turned off is "marked." The marking appears to be by *methylation* of cytosine bases in the DNA in the gene's promoter region (CpG islands). Methylation of promoters appears to prevent binding of *transcription factors* to the promoter bases, and thus prevents expression of the gene.

Innate immunity In animals, a group of general biochemical processes that protect the organism from a wide variety of bacterial, viral, fungal, and parasitic pathogens. These processes include *phagocytosis* of bacteria and other invaders by immune system *leukocytes* and *macrophages;* destruction of pathogens by certain enzymes (such as lysozyme) in blood, saliva, and tears and the acid secretions of the stomach; and protection from invading organisms by the skin. In the blood, the *complement system* of about 20 proteins can be activated in various ways to destroy pathogens. Complement works in concert with *lectin proteins* that bind to certain sugar residues found on bacterial cell walls. Mammalian blood also contains *natural killer lymphocytes* that can recognize foreign cells, tumor cells, and some infected cells and kill them.

Insulator Section of DNA, 42 bps or more, located between *enhancer and promoter* or between *silencer* and *promoter* regions of adjacent genes or clusters of adjacent genes. An *insulator* prevents the expression of a gene from being influenced by the activation or repression of its neighbors. All insulators contain the Nt sequence: CCCTC. Insulators work only when bound to the *CTCF protein* (CCCTC binding factor).

Intein A segment of a posttranslational protein that is able to excise itself and rejoin the remaining portions (the *exteins*) with a peptide bond. Inteins have also been called "protein introns." Most reported inteins also contain an endonuclease domain that plays a role in intein propagation. Many genes have unrelated intein-coding segments inserted at different positions (these are not introns). The gene segments coding inteins are sometimes called *selfish genetic elements,* but it may be more accurate to call them *parasitic elements.* Inteins are found in all three domains of life: prokaryotes, archaea, and eukaryotes.

Intron The non–protein-coding DNA base sequences interrupting the otherwise contiguous codon sequences of a *genome.* Introns lying within a gene are initially coded into pre-messenger RNA (pmRNA) and then are edited out by the *spliceosme complex* to form mRNA. Other intron DNA sequences lie between the genes. Introns have been called "junk DNA." Some base sequences in some introns carry regulatory information for gene expression; others have no apparent purpose.

Isoprene A ubiquitous molecular building block for biopolymers, found in plants and animals. It is a subunit in *terpenes,* terpenoids, carotenes, coenzyme-Q, phytol, retinol (vitamin A), tocopherol (vitamin E), etc. Its formula is:

$$H_2C=C(CH_3)-CH=CH_2$$

Isoprene must be activated in the form of a pyrophosphate or a diphosphate in order to enter biosynthetic pathways. Isoprene is released by the leaves of certain tree species and is considered a natural air pollutant.

Junk DNA (Also see *Intron*) Early in the development of genomic science, it was found that a large percentage of human nuclear DNA did not code proteins. With incredible scientific hubris, this was called "junk DNA." It is now a major scientific challenge to discover what this DNA actually does code for. A large part of intron DNA is composed of *transposable elements*—parts may code for specialized, active RNAs; parts may be regulatory for gene expression; and others are actually noncoding. About 99% of human nuclear DNA is non–protein coding.

Karyotype A photomicrograph of a cell's *metaphase chromosomes* arranged in a standard format showing the number, size, and shape of each chromosome. The karyotype is used in low-resolution physical mapping to correlate gross chromosomal abnormalities with specific diseases or developmental abnormalities. (The human female karyotype number is 23 chromosome *pairs*, including XX.) See also *cytogenetic map.*

kb Abbreviation for kilobase, that is, 10^3 bases (of DNA or RNA).

Kinase (protein kinase) A specialized class of protein enzyme that adds phosphate groups ($PO_4^{\equiv}$) to other proteins (phosphorylation). The human genome contains about 500 protein kinase genes. Most kinases remove a phosphate group from ATP and covalently attach it to one of three AAs that have a free hydroxyl group (serine, threonine, and tyrosine). The human genome codes for 90 different tyrosine kinases (TKs). The phosphorylation of these target AAs alters the balance of noncovalent, intramolecular interactions that affect tertiary and quaternary protein structure, hence protein function. Phosphorylation can lead to the splitting off or addition of polypeptide subunits to the phosphorylated protein. It can thus lead to activation (or inactivation) of a protein. For example, some cytosolic TKs enter the nucleus and transfer their phosphate to *transcription factor proteins,* activating them and hence the transcription of certain genes.

Kinetochore A complex of proteins that forms at the *centromere* of each pair of chromosomes during mitosis. The kinetochore is the attachment point for the microtubules (spindle fibers) that originate in the two *centrosomes.* It appears as a distinct "waist" on a chromosome pair.

Knockout Inactivation of specific genes. Knockouts are often created in laboratory organisms (such as yeast or mice) so that scientists can study the knockout organism as a model for a particular disease.

Lamellae Layers of folded membranes (in a cell), such as in chloroplasts, where an extension of a *thykaloid* links one *granum* to another. The *Golgi apparatus* has a lamellar structure.

Lectins Nonenzymatic proteins found throughout animal and plant kingdoms, in viruses and bacteria, prokaryotes and eukaryotes, which bind to sugars and carbohydrates. Lectins are found in both soluble and cell-associated forms. They are involved in a variety of cellular processes, including cell adhesion, bacterial infection, erythrocyte agglutination, leukocyte mitosis, lymphocyte homing, cell apoptosis, and innate immunity. Major subclasses of lectin proteins include *selectins, collectins, and galectins.*

Library A collection of cloned DNA, usually from a specific organism.

Ligand A ligand is the complementary molecule that binds strongly to a cell *receptor molecule.* The bimolecular specificity generally involves specific regions on each molecule. Ligands include steroid, protein, and corticoid hormones, neurotransmitters, cytokines, serotonin, GABA, ACTH, NO, lectins, various antigen molecules, etc.

Limit cycle A limit cycle (LC) in a nonlinear system describes the amplitudes and period of self-sustained oscillations in its states. LCs can be *stable* (once started, oscillating state amplitudes converge on finite steady-state [S-S] values) or *unstable* (once started, oscillating state amplitudes die out to constant S-S values, or they can grow to a larger- or smaller-amplitude LC in the S-S). A nonlinear system with a stable limit cycle is an *oscillator.*

Limit of detection (LOD) The threshold concentration of a biochemical analyte that can be sensed with relative certainty. The LOD of the analyte is related to the signal-to-noise ratio and the precision of the measurement system. Assume an additive signal and noise model. N readings in the absence of the analyte, should ideally give a reading of zero. Noise and offset in the measurement system will give N readings with a mean of y_o and a standard deviation of σ_{yo}. In one signal detection criterion, the analyte's LOD concentration must produce an output $y_a \geq 3.3\,\sigma_{yo} + y_o$.

LINES Acronym for long interspersed elements. LINES are varieties of *retrotransposons* found in eukaryotes' genomes.

Linkage The association of genes and/or markers that lie near each other on a chromosome. Linked genes and markers tend to be inherited together.

Locus The position on a *chromosome* of a *gene* or other chromosome *marker.* The use of the term "locus" is sometimes restricted to mean regions of DNA that are *expressed.* (See *gene expression.*)

LOD score A statistical estimate of whether two loci are likely to lie near each other on a chromosome and are therefore likely to be inherited together. An LOD score of 3 or more is generally taken to indicate that the two loci are close.

LTI Acronym for linear, time-invariant (system).

Lysosome An organelle found in both animal and plant (eukaryotic) cells that contains a variety of digestive enzymes (acid hydrolases) used to digest unwanted macromolecules (proteins, lipids, carbohydrates, and nucleic acids).

Lysosomes are assembled in the *Golgi apparatus* and bud off it. The low pH (4.8) of the lysosome is maintained by proton pumps and Cl^- ion channels in its single membrane. Lysosomes have a variety of functions: In embryonic development, they digest unwanted structures (e.g., a tadpole's tail, finger webs in a 3- to 6-month-old fetus). They also digest phagocytosed bacteria and damaged mitochondria and even break down cellular organelles in autophagic cell death.

Marker An identifiable physical location on a chromosome (e.g., restriction enzyme–cutting site, gene, minisatellite, or microsatellite) whose inheritance can be monitored. Markers can be expressed regions of DNA (genes) or some segment of DNA with no known coding function (intron) but whose pattern of inheritance can be determined.

Mdm2 protein An important component in the downregulation of the p53 tumor suppressor protein/transcription factor. The Mdm2 gene is found on human chromosome 12q13. The p53 protein/TF binds to the Mdm2 gene and activates its transcription. The Mdm2 protein product binds to the p53 molecule and blocks its activity, forming a local negative feedback loop. The p53*Mdm2 complex acts as a p53-specific, E3 ubiquitin-protein ligase, leading to proteosomal degradation of p53 protein. Because Mdm2 causes p53 degradation, it is called a *proto-oncogene.*

Mediator complex About 25 proteins that modulate the activity of *RNA Polymerase II.*

Meiosis The process by which *gametes* are produced. Two consecutive cell divisions occur starting with *diploid* progenitor cells. The result is four daughter cells each with a *haploid* set of chromosomes.

Mendelian inheritance A manner in which genes and traits are passed from parents to children. Examples of Mendelian inheritance include autosomal dominant, autosomal recessive, and sex-linked genes.

Messenger RNA (mRNA) RNA template for protein synthesis. Each set of three bases, called codons, specifies a certain protein in the sequence of amino acids that comprise the protein. The sequence of a strand of mRNA is based on the sequence of a complementary strand of DNA. mRNA is exported from the nucleus to the cytoplasm after *introns* are edited out of it by the *spliceosome complex* of enzymes.

Metagenomics The study of pooled genomes recovered from environmental samples, such a sea water, rather than from clonal (monospecies) cultures or single cells.

Metaphase The phase of mitosis, or cell division, when the chromosomes align along the center of the cell. Because metaphase chromosomes are highly condensed, scientists use these chromosomes for gene mapping and identifying chromosomal aberrations.

Methylation A process where methyl groups are added to the DNA base cytosine in the *promoter region* of a gene. A methyl group is bound to the 5-position of the pyrimidine ring. Methylation can inhibit expression of a gene. Methylation is catalyzed by the enzyme *DNA methyltransferase.*

Microarray An analytical device used in proteomic and genomic research. Basically, it is a glass microscope slide to which a rectangular array of different biomolecules (DNA oligos, mRNA, proteins, or antibodies, etc.) is attached in small wells. The array is used for rapid, parallel screening for molecular reactions or molecular structures. Array readout is generally by emitted fluorescence, but arrays have been devised that have electrical readouts, photoradiographic readouts, and surface plasmon resonance outputs. Arrays exploit the chemical affinity of specific molecular species to chemically bind to one another.

Micro-RNAs (miRNAs) These are ssRNA molecules of about 22 Nts. They combine with an RNA-induced silencing complex (RISC) of proteins and are involved in gene expression downregulation (silencing) as well as upregulation. In eukaryotes, they act on mRNA to inhibit its action, or destroy it, silencing expression of a particular gene. miRNAs are cleaved from larger RNA precursors that are coded in the organism's genome by the *Dicer* enzyme complex. Humans have about 230 miRNA genes.

Microsatellites Genomic regions in which short DNA sequences or a single Nt are repeated. Often used as genetic markers to track inheritance in families. There are hundreds of thousands of microsatellites in the human genome. During DNA replication (cell division), microsatellites may pick up mutations due to the misalignment of their repetitive subunits, which results in elongation or truncation of the original sequence. This is called "instability" which can be used as a marker for deficiency in the genomic mismatch–repair proteins.

Microtubules The microtubules of the *centrosomes* are straight, hollow cylinders about 25 μm in diameter. They are of variable length but can grow to 25,000 μm in length. They are assembled by a helical wrap of *α-tubulin* + *β-tubulin* dimer molecules and are found in both animal and plant cells. Curiously, they grow by the polymerization of tubulin dimers powered by the free energy from the hydrolysis of *Guanosine Triphosphate* (GTP) molecules and shrink in length when the tubulin dimers are enzymatically released from the end of the microtubule. Thus, microtubules are linear "motors" inside the cell because they can push or pull other molecules to which they are attached. Their most notable operations are moving chromosomes during mitosis and causing cilia to wave.

Mitochondrial DNA (mtDNA) mtDNA is passed from one generation to the next solely through the maternal line of the family. It has been used for forensic identification.

Mitosis The most common form of nuclear division in cells that produces *diploid* daughter cells that are genctically identical to each other and the parent cell. Mitosis is divided into five stages: *Prophase, Prometaphase, Metaphase, Anaphase, and Telophase.* The nondividing cell is in *Interphase.* When a eukaryotic cell divides in two by mitosis, each daughter cell receives (1) a complete set of genes (for diploid cells, this means two complete genomes [**2n**]); (2) a pair of centrioles (in animal cells); (3) some of the parent cell's mitochondria (plants also receive chloroplasts); and (4) some ribosomes, portion of the endoplasmic reticulum, and other organelles.

Moiety In biochemistry, a coherent part of a molecule, especially of a protein or polypeptide. For example, a phenyl group or a carboxyl group.

Molar concentration A 1-molar solution of a molecular substance is when 1 mole of the substance is dissolved in a solvent (generally water) to make exactly one liter of solution.

Mole A gram-molecular weight of a molecular substance. That is, the weight of a molecular substance in grams numerically equal to the molecular weight of a substance. There are always an *Avogadro's number* of molecules in a mole (6.0225×10^{23}). A mole of water weighs 18.016 grams.

Molecular scissors See "restriction enzymes."

Morula The solid, 16-cell stage of an embryo, before embryonic differentiation occurs.

Mosaicism A condition in which an organism harbors two or more genetically distinct cell lines. Mosaicism results from a genetic change after formation of a *zygote*. For example, a mosaic individual can be formed by the fusion of two different zygotes into one embryo. That fused zygote is called a *chimera*.

Mutation Any heritable, permanent structural alteration in the DNA base sequence. In most cases, a mutation has little or no effect or causes a challenge to the survival of the mutated life form. In some cases, a mutation can lead to an increased chance of an organism's survival (e.g., immunity to a certain invading microorganism). There are many physical and chemical causes of mutations.

MYCN Another oncogenetic protein/TF. MYCN can directly bind to the promoter of the Mdm2 gene and stimulate Mdm2 expression. It also acts to inhibit p53-triggered apoptosis.

NAD+ Nicotine Adenine Dinucleotide.

Nanog A protein transcription factor expressed in embryonic stem cells, thought to be a critical factor in maintaining their pluripotency and self-renewal. Its gene is found on human chromosome 12p13.31.

Nanomole 1×10^{-9} mole.

Nanospheres Chemically inert spherical particles with diameters ranging from nanometers to micrometers. Gold and semiconductor nanospheres of various sizes are used as fluorescent molecular tags by affixing molecules to them, such as antibodies or nucleic acid oligos, which then react with target molecules, labeling them. Larger, magnetic nanospheres are used to physically isolate target molecules (also virions and bacteria) by having an affinity to their coatings.

Nitrogenous base (As applied to nucleic acids) A chemically basic nitrogenous ring compound. There are two purine bases—adenine and guanine—and three pyrimidine bases—cytosine, thymine, and uracil.

Nonlinear Adjective applied to a system that does not obey *superposition*. All biochemical systems are nonlinear because there are no negative concentrations.

NT (Abbreviation for Nuclear Transfer) A means of forming pluripotent stem cells and complete embryos from a single, differentiated, somatic cell. The haploid (**n**) nucleus of an egg is removed and replaced with the diploid (**2n**) nucleus removed from an adult somatic cell. Then the diploid egg is electrically and chemically stimulated to divide like a naturally fertilized egg and form an embryo that can be implanted, or from which PSCs can be harvested. NT was used to produce Dolly the sheep.

Nt Abbreviation for Nucleotide.

Nuclease An enzyme that hydrolyzes phosphodiester bonds that link adjacent nucleotides in DNA and RNA. An **exonuclease** enzyme progressively cleaves nucleotides from *free ends* of a linear nucleic acid substrate. An **endonuclease** enzyme cleaves a nucleic acid chain at *internal sites* in the nucleotide or base sequence. An endonuclease is also a *restriction enzyme.*

Nucleolus A nuclear organelle in eukaryotic cells. It is a roughly spherical structure that occurs in association with a particular point (the nucleolar organizer) on a specific chromosome, inside the nuclear envelope. When a cell divides, the nucleolus disappears during late prophase, is absent through metaphase and anaphase, and reappears during telophase. Mammalian nuclei contain between one and four nucleoli. The nucleolus is the site for rRNA gene transcription and ribosomal assembly. It is also implicated in regulation of the cell cycle, cell senescence, and aspects of molecular transport.

Nucleoside A nitrogenous base bound to a sugar (no phosphate).

Nucleosome The basic unit of *chromatin* structure; it is a DNA–histone complex composed of an octamer of *histone proteins* (two molecules each of H2A, H2B, H3, and H4) wrapped with two loops containing about 140 bps of DNA. Histone H1 draws nucleosomes together in a more compact form, forming chromosomes such as viewed under the light microscope. (See also *Histones.*)

Nucleotide (Nt) A complete subunit of ssDNA consisting of three major components: one of four *nitrogenous bases* (adenine, thymine, guanine, or cytosine), a *deoxyribose sugar molecule,* and a phosphate radical. Hundreds of thousands of different (**A**, **T**, **G**, or **C**) nucleotides are linked to form a two-stranded DNA molecule. The sequential order of the different nucleotides carries genetic information. Many Nts are joined together with *3′–5′ phosphodiester bonds.*

Oct-4 Abbreviation for *Octamer-4.* A homeodomain *transcription factor* that is critically required for the self-renewal of undifferentiated embryonic stem cells. Oct-4 is expressed in preimplantation, developing embryos. Removal of Oct-4 from an embryo promotes the cells to differentiate. Its gene is found on human chromosome 6p21.31.

Oligo (oligonucleotide) A short length of single-stranded DNA or RNA. Fluorescent-tagged synthetic oligos are often used as *probes* for detecting complementary sequences of DNA or RNA bases. Oligos bind readily to their complements.

Operator The site at one end of an *operon* where a repressor molecule binds to the DNA and thus inhibits the *transcription* process.

Operon A group of closely linked genes that appear to function as an integrated unit to affect different steps in a single biosynthetic pathway. The genes of an operon are transcribed in one mRNA molecule.

Opsonin A protein that binds to the surface of a particle and enhances its uptake by a phagocytic cell.

ORF Acronym for Open Reading Frame—any contiguous sequence of DNA or mRNA codons that can be translated into the polypeptide sequence of a protein molecule. If a gene contains one or more introns, the mRNA must be considered to find the gene's ORF. A DNA ORF is bounded on its beginning by the "start" codon (**ATG**) and at its end by one of the three "stop" codons (**TAA**, **TAG**, or **TGA**). Finding a protein's DNA ORF can be a challenge, given its polypeptide sequence; this can be like looking for a needle in a haystack—for example, one about 400 Nt ORF out of 3.1×10^9 Nts. Software such as GENIE® is generally used to build an ORF gene model. GENIE uses a generalized *hidden Markov model approach.*

p53 protein An important intracellular protein *transcription factor* that is involved in the regulation of the cell cycle, control of apoptosis, and the suppression of tumor growth. Human p53 is a 393 AA protein of 54 kDa. p53 is normally found in inactive form in a cell, bound to the protein MDM2. Various types of *cell stress*, including damage from external chemicals, DNA damage, hypoxia, viral infection, nucleotide pool depletion, etc., cause p53 activation and stabilization. Oncogenes can also trigger p53 activation mediated by the p14AF protein. Thus p53 integrates various cellular stress signals.

p53 activation affects a number of genes, leading to cell-cycle arrest in G1 by p24, or in G2, or *apoptosis* mediated by BAX, PUMA, or NOXA. The cell growth arrest allows the cell's DNA repair system to be activated. The gene for p53 is located on human chromosome 17p13.1. p53 was voted *molecule of the year* in 1993 by the journal *Science.*

Palindrome sequence (of DNA) A palindrome is a sequence of letters and/or words that reads the same backwards as it does forwards. *Able was I ere I saw elba* is a well-known English palindrome. Palindromes also occur in dsDNA on opposite strands of the same section of the DNA helix.

Example: 5′...GGAATTCC...3′
3″..CCTTAAGG...5′

This type of palindrome serves as a target for most *restriction enzymes. Enhancer* and *silencer* elements are often palindromic or inverted repeat sequences.

Parametric control Found in physiological and molecular biological systems. Regulation and control is effected by using certain internal state values to adjust the magnitude of the rate constants (gains) in the set of ODEs used to describe the system's behavior. Adjustment is generally by altering the concentration of a molecule that affects a rate constant's magnitude, or by altering the concentration of an enzyme. (In a conventional, linear control system, all gains have fixed values, and regulation is accomplished by subtracting a feedback signal from a desired input to generate an error signal that is processed to obtain the desired output.)

Paratope See *Epitope.*

Parthenogenesis Reproduction in animals without benefit of male sperm. A haploid egg is transformed to a viable diploid embryo. Parthenogenetic reproduction has been observed in certain insects and reptiles. *Imprinting* is the reason that parthenogenesis does not occur naturally in mammals. Two complete female genomes cannot produce viable young because of the imprinted genes. For example, the embryo needs the father's *Igf2* gene because the mother's copy has been imprinted and is inactive.

Parthenote The viable product of parthenogenesis.

Pascal SI unit of pressure = 1 newton/meter^2.

Patch clamp A technique in which a micropipette electrode is placed over an ion channel on a cell surface and used to measure the very small ion current passing through the channel. The channel current and the transmembrane potential can be used to calculate the channel conductance.

PBS Abbreviation for *phosphate buffered solution.*

PCR Abbreviation for Polymerase Chain Reaction. A technique for increasing the concentration of a specific DNA *oligo.* A polymerase enzyme creates in vitro multiple copies of a specific segment of DNA; billions of copies can be made. PCR is used in forensic DNA analysis and is a fast, relatively inexpensive way of copying DNA.

Peptidoglycan A polymer making up the middle of the prokaryotic bacterial cell wall. This unusual polymer, also called *murein,* is made up from two sugar subunits: *N-acetyl glucosamine* (NAG) and *N*-acetyl muramic acid (NAM). Alternating NAG and NAM are assembled in chains using β-(1,4) glycosidic bonds. The chains are cross-linked by a 4 or 5 AA peptide attached to the NAM molecules. These peptide oligos contain D-*alanine,* D-*glutamic acid*, and *mesodiaminopimelic acid.* The peptidoglycan layer is 20–80 nm thick in Gram-positive bacteria but only 7–8 nm thick in Gram-negative bacteria.

Phage (bacterophage) A virus that uniquely infects and reproduces in bacteria.

Phenotype The observable traits or characteristics of an organism (physical, biochemical, physiological); for example, hair color, blood type, weight, or the presence or absence of a disease. Phenotypic traits are not necessarily genetic.

Pheremone In higher animals, a chemical odorant released to signal sexual receptivity or mark territory.

Phosphatase A class of enzyme that removes phosphate groups from other molecules (dephosphorylation), altering their biochemical and physiological activity. In the case of p53, dephosphorylation inactivates this tumor suppressor protein.

Phosphodiester bond Each chain of ssDNA or dsDNA has its bases linked together with a covalent atomic bond formed between the 5′ phosphate and the 3′ hydroxyl groups of adjacent DNA nucleotides. An example of a 4-base, DNA *oligo* is shown in Figure 6.1 in text. Note the 5′ end of the oligo has the $PO_4^=$ group, and the 3′ end of the oligo has the -OH group on the 3′ carbon of the ribose.

Picomole 1×10^{-12} mole.

Pinocytosis A process whereby a cell (such as an amoeba) forms an invaginated pocket in its cell membrane, then pinches it off and internalizes the contents of the pocket, which contains water, dissolved ions, sugars, O_2, etc. The contents of the new vesicle can be absorbed into the cytosol and metabolized.

Plasmid Extrachromosomal elements of double-stranded DNA found in bacteria, which are distinct from the normal bacterial genome. Some plasmids carry genes that can produce advantageous phenotypes, increasing the chance for cell survival by conferring antibiotic resistance, antibiotic production, sugar fermentation for energy, degradation of aromatic compounds for energy, resistance from heavy metals, toxin production, etc. Bacterial plasmids are generally closed, circular, dsDNA molecules having from about 4 kb to 240 kb but are straight in some bacterial species such as *Borrelia.* Plasmids replicate autonomously. From 1 to about 40 copies of a given plasmid can be found in a single bacterial host.

DNA oligos from 50 to 500 bps can be artificially inserted into a host cell's plasmid ring and used as a DNA *cloning vector* in genetic engineering.

Pleiotropy The condition whereby a signal molecule (e.g., a cytokine) can have different physiological effects on different kinds of cells that have receptors specific for that signal molecule.

Pluripotent stem cells Pluripotent stem cells can differentiate and develop into any of the three major tissue types: *Endoderm* (interior gut lining), *Mesoderm* (muscle, bone, blood, organs), and *Ectoderm* (epidermal tissues and the nervous system). Unlike *totipotent* embryonic cells, pluripotent cells lack the ability to divide, specialize, and form a mature animal. Pluripotent cells further differentiate into *multipotent* cells as the embryo develops.

Pol II (aka *RNA polymerase II,* or *RNAP II*). A complex of 12 different proteins that binds to the transcription start site of a DNA gene.

Polar Body A polar body is one of three, small, nonfunctional haploid cells found in close association with a mature animal ovum. Polar bodies arise during oogenesis. As an egg matures, it goes through a two-step meiotic division process. At meiotic telophase I, one of the two daughter cells *(the first polar body)* contains very little cytoplasm, whereas the other cell *(the secondary oocyte)* contains nearly all of the premeiotic contents of the oocyte. In the second meiotic division, the larger secondary oocyte divides again, retaining nearly all its volume, producing a mature ovum with a haploid pronucleus and a secondary polar body. The small, first polar body can also divide, producing a total of three polar bodies which arc nonfunctional haploid cells. The polar bodies are eventually broken down.

Polymerase chain reaction (PCR) An in vitro technique for enzymatically replicating and "amplifying" DNA oligos. PCR was invented in 1985 by K.B. Mullis. It is widely used in medical, molecular biological, and forensic research labs for a variety of applications, including diagnosing infectious diseases and hereditary diseases, cloning genes, paternity testing, genomic research, anthropology and evolutionary biology, and amplifying evidenciary DNA so that *STR analysis* can be used to identify criminals. The PCR process involves a DNA primer, DNA polymerase enzyme, and thermal cycling. A competitive means of DNA amplification is called *rolling-circle amplification.*

Polyploidy A eukaryotic organism is polyploid if its cells contain more than two haploid sets of chromosomes. A haploid gamete contains n chromosomes. A diploid (normal) cell contains 2n, a triploid cell contains 3n, and a tetraploid cell contains 4n chromosomes. Many plants exhibit polyploidy; it is rare in animals but is found in some insects, fishes, amphibians, and reptiles. One case of a tetraploid (4n) rat in Argentina has been reported.

Primer A short oligonucleotide sequence used in a polymerase chain reaction (PCR).

Probe A synthetic DNA oligo used to detect a specific complementary DNA sequence by *hybridization*. Probes are generally labeled by a radioisotope or a fluorescent molecule.

Prokaryote A cell or organism that lacks a membrane-enclosed, discrete nucleus and other membrane-bound subcellular components, such as ribosomes and mitochondria (e.g., viruses and bacteria).

Promoter (aka promoter sequence) DNA Nt sequences lying upstream (5′) from a gene, to which the *RNA Polymerase* complex binds. (See also *TATA box.*) It contains the TATA box surrounded by a large complex of various proteins, including *Transcription Factor IID* (TFIID), which is a complex of TATA-binding protein (TBP) that recognizes and binds to the TATA box. There are 14 other protein factors that bind to TBP and each other but not to the DNA. *Transcription Factor IIB* (TFIIB) binds to both the DNA and pol II. (TFII A, D, E, F, and H also participate in gene promotion.) Promoters essentially turn a gene on or off. Transcription is initiated at the promoter.

Pronucleus The haploid nucleus of the ovum or sperm after fertilization but before they fuse to form the diploid (2n) nucleus of the zygote. Thus there is a male and female pronucleus.

Protein kinase (See *Kinase.*)

Proteome The *complete* set of all proteins coded by the genome of a species.

Proteomics The quantitative and qualitative study of the properties (chemical, structural, and physical) of the *proteome* of a living system. Proteomics considers the expression, function, and interaction of proteins in health and disease.

Proteorhodopsin (PR) PR is a transmembrane, light-transducing, 27-kD protein found in many marine bacterial species (eukaryotes and archaea). It uses photon energy to pump protons against a potential and concentration gradient to outside the plasma membrane. Different varieties of PR have broad absorption maxima at different visible wavelengths (e.g., 527 nm, 487 nm, etc.). PR works in conjunction with the ubiquitous chromophore molecule *retinene.*

Proteosome A cytoplasmic complex of proteins. A *core particle* (CP) consists of two copies each of 14 different proteins. A regulatory particle (RP) consists of 14 different proteins, six of which are *ATPases.* There are two identical RPs, one at each end of the CP. Proteosomes destroy endogenous proteins such as *transcription factors, cyclins,* proteins encoded by viruses and other intracellular parasites, proteins that are folded incorrectly because of translation errors or mutations or are damaged by other molecules in the cytosol. These proteins are marked for destruction by the proteosome by attachment of *ubiquitin* molecules.

Pyruvate kinase An enzyme involved in the *glycolysis* metabolic pathway. It catalyzes the transfer of a phosphate group from *phosphoenolpyruvate* (PEP) to ADP, making one ATP molecule + pyruvate. Mg^{++} ions are a cofactor for this reaction. (Glycolysis is described in detail in Section 4.2.2.)

Quasispecies A large population (e.g., in a compartment) of self-replicating machines that behave almost (*quasi-*) like a single type of machine (*species*). *Quasispecies models* have been used to simulate viral evolution. Here, the quasispecies is a viral, self-replicating nucleic acid (NA) (usually RNA) composed of four NA bases. New, slightly different NA sequences enter the system as the result of random errors in the replication process, which can be either correct (i.e., follow the original, "master" molecule) or erroneous. Replication rates of different molecular quasispecies can be made to differ. A trade-off exists between mutational robustness and replication rate. In a binary system, a robust, slowly replicating viral species can outcompete the faster, more error-prone species. This is called "survival of the flattest" [Codoñer et al. 2006].

All daughter virions (correct or erroneous) can be made to decay with Poisson statistics. (The probability of decay is age independent.) Neutral mutations do not lead to genetic drift of the population. Natural selection acts on the mutant distribution *as a whole,* rather than on individual variants. The quasispecies model is an equilibrium, mutation-selection process that describes a heterogeneous distribution of whole genomes ordered around a set of fittest sequences known as "master" sequences.

Receptors Protein molecules found throughout cells and their organelles and on their membranes that provide a *ligand-binding site* for endogenous and exogenous cell messenger molecules, including steroid hormones, protein and peptide hormones, cytokines, interleukins, etc. Receptors on immune system cell surface proteins and antibodies also bind to ligand regions of antigen molecules.

Recessive A type of gene that is expressed in individuals who have received two copies of the gene, one from each parent.

Regenerative medicine The use of *stem cells* to replace specialized body tissues that have been lost or functionally damaged. Such tissues include, but are not limited to, heart muscle cells, liver cells, pancreatic beta cells, CNS neurons, spinal cord neurons, bone, etc.

Regulatory region A DNA base sequence that controls gene expression.

Restriction enzymes A bacterially derived protein enzyme that recognizes a specific, short sequence of double-stranded DNA and cleaves the DNA at that site. Restriction enzymes are sometimes referred to as molecular scissors. Some restriction-enzyme-cutting sites occur relatively frequently in DNA, e.g., about every 100 bps; others occur rarely, for example, every 10^4 bps.

Retrotransposon See *Transposon.*

Retrovirus A class of virus (e.g., the human immunodeficiency virus [HIV]) in which the information for self-replication is stored in ssRNA. After the virus enters its target cell, it uses the enzyme *reverse transcriptase* to generate a DNA copy of its genome, then uses the cell's internal biochemical machinery to make copies of itself inside the cell. The cell is destroyed when the daughter virions are released (viral eclosion).

Reverse transcriptase An RNA-dependent, DNA polymerase enzyme that makes a DNA copy from an ssRNA Nt sequence. Found in retroviruses such as HIV.

Ribonuclease (RNase) A protein enzyme that hydrolyzes (breaks down) RNA.

Ribosomal frameshifting A directed change in a translational reading frame that allows the production of a single protein molecule from two or more overlapping genes. The process is programmed by the nucleotide sequence of the mRNA and is sometimes also affected by the secondary or tertiary mRNA structures. It has been described in some retroviruses, retrotransposons, bacterial insertion elements, and also in some cellular genes.

Ribosome Small, cytoplasmic cellular organelles that are the sites of protein synthesis. There are four kinds of ribosomal RNA (rRNA): 18S, 28S, 5.8S, and 5S. 18S rRNA + about 30 different protein molecules make up the small subunit of the ribosome. One each of the 28S, 5.8S, and 5S rRNA molecules + about 45 different proteins make up the large ribosomal subunit. (S refers to the sedimentation rate of a species of rRNA in an ultracentrifuge; the larger the S-number, the larger the molecule.)

Ribozyme An RNA molecule that catalyzes a chemical reaction. Many natural ribozymes can catalyze either their own cleavage or the cleavage of other RNAs. They have also been found to catalyze the aminotransferase activity of ribosomes. Examples of ribozymes are *RNase P, Group I* and *II introns,* and *hairpin ribozyme.*

RISC Acronym for RNA-induced silencing complex.

RNA Ribonucleic acid. RNA has the pyrimidine bases uracil and cytosine and the purines adenine and guanine. See Figure 5.4 in the text for the 2-D structure of tRNA. There are a number of categories of RNA that have many diverse functions, for example:

dsRNA: Double-stranded RNA
siRNA: Small interfering RNA
mRNA: Messenger RNA
miRNA: Micro-RNA
pmRNA: Premessenger RNA (operated on by spliceosome complex)
snRNA: Small nuclear RNA; constituents of the spliceosome complex
snoRNA: Small nucleolar RNA
stRNA: Small temporal RNA

RNA interference (RNAi) A form of posttranscriptional gene "silencing" in which dsRNA induces the degredation of the homologous mRNA, mimicking the effect of the reduction (or loss) of gene activity.

RNA polymerase II A complex of at least six proteins that binds to a DNA *TATA box* and which is required for gene transcription. It unwinds a region of the DNA helix.

Rolling-circle amplification (of DNA) An isothermal, in vitro method of enzymatically making hundreds to billions of copies of small, circular ssDNA probes. RCA can use a single- or double-primed approach. The double-primed RCA uses one complementary primer to a DNA minicircle (as in the single-primed approach), whereas the other is targeted to the repeated, ssDNA sequence of the primary RCA product. Thus, the double-primed RCA-process-generated product DNA with geometric kinetics.

RUBISCO Acronym *for RibUlose BIphoSphate Carboxylase Oxygenase*. Enzyme that catalyzes the first step on CO_2 fixation by all plant cells practicing photosynthesis. Possibly the most abundant protein on Earth.

Second messenger systems A multistep method of intercellular signaling where an extracellular signaling molecule (e.g., hormone, neurotransmitter, cytokine, etc.) binds with a complementary receptor protein on the cell surface. The surface receptor is part of a TMP complex with seven transmembrane domains. The TMP complex is activated by the docking, which, in turn, activates a binding site near the TMP's C-terminal for an internal *G-protein* (guanine nucleotide binding proteins). The G-protein has three subunits, $G\alpha$, $G\beta$, and $G\gamma$. When the TMP is activated, it undergoes a conformational change that allows the $G\alpha$-protein to release a bound molecule of GDP and bind a new molecule of GTP. This exchange causes the $G\alpha$ subunit to break free from the $G\beta\gamma$ dimer and the TMP. The free GTP + $G\alpha$ interacts with a nearby, membrane-bound *guanylyl cyclase* enzyme that forms free *cyclic GMP* (cGMP) from the GTP, releasing the $G\alpha$ for recycling. In this system, it is the cGMP that is the "second messenger," which has the ability to initiate internal cellular events such as gene expression. cGMP can stimulate the production of *protein kinase* ***G***, which phosphorylates target proteins in the cell, altering certain biosynthesis kinetics.

There are several second messenger systems that basically have the same architecture as the cGMP system. The TMP and G-protein composition can vary, and *cyclic adenosine monophosphate* (cAMP), *inositol 1,4,5-triphosphate* (IP_3), NO (nitric oxide), and Ca^{++} ions (responsible for muscle contraction) are other second messengers used by different cells. A second messenger signaling system can have "molecular leverage" in which one ligand binding to one cell surface receptor causes extensive changes in certain steady-state, intracellular biochemical activities.

Selfing (A term from Mendelian genetics) Self-fertilization of the blossoms of a *monoecious* plant.

Sex chromosome One of the two chromosomes that specify a eukaryotic organism's genetic sex. Humans have two kinds of sex chromosomes, one called X and the other Y. Normal females possess two X chromosomes, and normal males possess one X and one Y.

Silencer *Cis*-acting control regions of DNA that can be located up to 1,000s of bps away from the gene they control. When certain transcription factors (*repressor* molecules) bind to the silencer site, expression of the gene is inhibited. Silencers can also exist downstream of the gene they regulate or reside in introns in or near the gene. *Silencing*, then, is turning gene expression off.

SINEs Acronym for *Short Interspersed Elements*. These are Class I transposed DNA sequences of 100 to 400 bps. The most abundant SINEs are the *Alu elements*. SINES are non-protein-coding DNA sequences.

Single nucleotide polymorphism (SNP) A gene in which only one DNA base pair is different from the DNA sequence in the same gene in another similar cell from the same species. An SNP can affect the functionality of the protein coded by that gene or have no effect.

Sox-2 A transcription factor essential to the self-renewal of undifferentiated embryonic stem cells. Its gene is found on human chromosome 3q26.3-q27. Sox-2 is an acronym for "SRY-related HMG-box gene 2."

Spliceosome complex A five-molecule, protein/RNA complex that processes premessenger RNA (pmRNA) to remove noncoding Nt sequences from introns lying in a gene's Nts. The output of a gene's spliceosome complex is protein-coding mRNA. The details of how a spliceosome works have yet to be elucidated; its function is truly amazing.

SPUAMOB Acronym for Society for the Prevention of the Use of Acronyms in Molecular Biology.

ssDNA, ssRNA Single-stranded DNA or RNA.

Start codon Signals the beginning of the codons composing a gene; ATG. ATG inside a gene sequence codes for the AA methionine.

STAT Proteins The Signal Transduction and Activator of Transcription proteins regulate many aspects of cell growth, survival, and differentiation by binding selectively to certain target genes. STAT proteins are activated by the *Janus Kinase* (Jak) enzymes. Bound, they act as *transcription factors* to induce (or inhibit) gene transcription. There are seven known human STAT proteins: STAT-1, 2, 3, 4, 5a, 5b, and 6. The cytoplasmic STAT proteins regulate a number of diverse genes involved in the development and function of the immune system (maintaining immune tolerance and tumor cell surveillance), as well as more general development, differentiation and cell proliferation, survival, and apoptosis. It has been shown that inappropriate activation of certain STAT protein pathways (e.g., STAT-3, 5) is associated with the growth of a number of human malignancies, including hematologic, breast, prostate, and head and neck cancers. When constitutive STAT-3 signaling is blocked in vivo or in vitro, there is growth inhibition and apoptosis of STAT-3^+ tumor cells.

Stem cell There are both adult and embryonic stem cells. A stem cell is an undifferentiated cell that through certain physical, biochemical, and genomic signals can be caused to differentiate to a specialized cellular form and function, such as neurons, heart muscle cells, skin cells, blood cells, etc. An undifferentiated stem cell can divide indefinitely in that form until signaled to differentiate. *Totipotent* stem cells are the earliest embryonic cells that can become *pluripotent* stem cells and the placenta, and thus they can give rise to a twin embryo. A *pluripotent* stem cell can generate all the structures of the developing embryo except the placenta and surrounding tissues, as well as reprode themselves. *Committed* stem cells derive from *pluripotent* stem cells; they can differentiate into fixed categories of embryonic tissues (e.g., the *hemangioblasts* can form blood vessels and blood cells and lymphocytes; the *mesenchymal stem cells* form connective tissues, cartilage, muscle, fat, etc.). The restriction on the potency of stem cells in embryonic development is gradual, and depends on poorly understood internal and external signaling factors.

Stop codon One of three DNA codons in the genetic code that signal the termination of a particular gene's coding sequence. The stop codons are TAA, TAG, TGA.

Superposition A necessary property of *linear systems.* Linear systems also allow scaling real convolution to find the system output, and they enjoy shift invariance. These properties are summarized as follows. Note LS = linear system, x_k are inputs, y_k are outputs, $\otimes$ means real convolution, h is the LS's weighting function (impulse response), and a_k are constants.

$$\xrightarrow{a_1x_1+a_2x_2} \text{LS} \xrightarrow{y=a_1y_1+a_2y_2} \quad \text{superposition} \tag{1.7}$$

$$x_1 \rightarrow \text{LS} \rightarrow y_1 = x_1 \otimes h \quad \text{real convolution} \tag{1.5}$$

$$\xrightarrow[\text{then } a_2x_2]{\text{if}} \text{LS} \xrightarrow[y_2'=a_2(x_2\otimes h)]{y_2=x_2\otimes h} \quad \text{scaling} \tag{1.6}$$

$$\xrightarrow[+x_2(t-t_2)]{x_1(t-t_1)} \text{LS} \xrightarrow[+y_2(t-t_2)]{y=y_1(t-t_1)} \quad \text{shift invariance} \tag{1.8}$$

Svedberg onstant (S-number) Used to describe the size (mass) of protein molecules and nucleic acid oligos. It is the number of gs required by a centrifuge to precipitate purified molecules out of a standard solution. For example, we talk of the 28S component of a ribosome.

System A group of interacting, interrelated, or interdependent *elements* (also *entities, states*) forming or regarded as forming a collective entity. (There are many definitions of *system,* dependent on the type of system. The preceding definition is broad and generally acceptable.)

TaqMan® probes TaqMan probes are dual-labeled nucleotide hydrolysis probes. They utilize the activity of the *Taq polymerase* to measure the amount of a DNA target sequence in a sample. They are DNA oligos of 18–22 bps, with a reporter fluorophore attached to the 5′ end and a fluorescence quencher molecule fixed to the 3′ end. TaqMan probes are added to a PCR reaction mixture to measure its progress. During PCR, the probe anneals specifically between the forward and reverse primer to an internal region of the PCR product. The polymerase then carries out the extension of the primer and replicates the template to which the TaqMan is bound. The 5′ exonuclease activity of the polymerase cleaves the probe, releasing the reporter fluorophore away from the vicinity of the quencher molecule. Thus, the fluorescence of the free reporter molecules increases in proportion to the growth of the PCR product. This process repeats every PCR cycle and does not interfere with the accumulation of the PCR product. (TaqMan probes are a registered trademark of Roche Molecular Systems, Inc.)

TATA box Nucleotide sequence found in eukaryotic gene *promoters*. It has the 7-Nt sequence TATA**AAA** and is found about 30 bps upstream from the *start codon* (ATG) of the gene. One gene can have >1 TATA box, thus several different mRNA start sites that can be regulated independently.

Telomerase An enzyme that adds the telomere repeat sequence to the 3′ ends of DNA strands. By lengthening this strand, the enzyme *DNA polymerase* is able to complete the synthesis of the "incomplete ends" of the opposite (complementary) DNA strands. Telomerase is a ribonucleoprotein (RNA + protein). In mammals, its single strand RNA oligo is AAUCCC, which serves as a template for the DNA oligo TTAGGG that is added. The protein component of telomerase, telomere reverse transcriptase (TERT), catalyzes the synthesis of the DNA from the RNA template. Telomerase is generally found in embryonic stem cells and cancer cells. Telomerase and the maintenance of telomere length may be one key to cell longevity.

Telomeres The two end regions of a linear, eukaryotic chromosome. The DNA of human telomeres consists of as many as 2,000 repeats of the six-base DNA sequence 5′TTAGGG 3′. For example, for dsDNA:

5′...TTAGGG TTAGGG TTAGGG TTAGGG TTAGGG TTAGGG-3′
3′...AATCCC AATCCC AATCCC AATCCC AATCCC AATCCC-5′

Telomeres keep the ends of the various chromosomes in the cell from becoming attached to one another by nonhomologous end-joining.

Terpenes, terpenoids Terpenoids are chemically modified terpenes, although the terms are sometimes used interchangeably. They are hydrocarbons resulting from the combination of several isoprene units. Methyl groups may have been moved or removed and/or oxygen atoms added. From two to over eight *isoprene* units can be combined to make terpenoids. Plant terpenoids are generally aromatic and contribute to the odors of cinnamon, eucalyptus, cloves, ginger, citral, menthol, camphor, as well as to the cannabinoids found in *Cannabis sp.*

TF (or TFP) Transcription factor protein.

Thymine A pyrimidine base found only in DNA. It pairs with Adenine. (T ⇔ A).

TMP Transmembrane protein (TMP).

Totipotency The ability of a single stem cell to divide and produce all the differentiated cells in a fetus, including the placenta and various membranes. The cells formed by the initial mitotic cell divisions of a human zygote are totipotent for about four divisions (i.e., 16 cells) after fertilization, then they begin to specialize into *pluripotent* cells and then to *multipotent* cells.

Transcription The first step in gene expression (protein synthesis) in which a single-strand DNA exon is copied by premessenger RNA.

Transcription factors (TFs) TF proteins that specifically bind to *cis*-regulatory (DNA) elements and that either stimulate or inhibit gene transcription by interacting with RNA polymerase. Some transcription factors are encoded by *Hox* (selector) genes. The Signal Transducers and Activation of Transcription (STAT) proteins that regulate cell growth, differentiation, and survival are examples of TFs. STAT TFs are involved in the development and function of the immune system.

Transcriptome The set of all mRNAs ("transcripts") made in a cell, or a population of cells, or a metazoan organism. The transcriptome number is generally less than the number of genes in a cell because it reflects the number of genes being expressed at any given time. The transcriptomes of cancer and stem cells are important for the understanding of oncogenesis and cell differentiation, respectively.

Transfection Genetically engineered DNA, generally in the form of plasmids, or siRNA, is introduced into a target cell by a physical *poration* process, either by exposing the cell to a dc electric field (electroporation) or by a *gene gun*. A virus can also be used to insert DNA; this process is called viral transduction. The DNA introduced generally does not enter the cell's genome but is transiently expressed.

Translation The protein-synthetic step whereby a linear polypeptide is assembled in the cytoplasm of a cell by ribosomes from messenger RNA (mRNA) and transfer RNAs (tRNAs) carrying specified AAs. (See Section 6.2.3.)

Transmembrane potential All cells have a steady-state transmembrane potential, V_{m0}, between the inside and outside of the cell membrane. V_{m0} is on the order of tens of millivolts; the inside of the cell is generally negative with respect to the outside. V_{m0} exists because all cell membranes are semipermeable; large, trapped molecules with fixed charges are inside the membrane, whereas small ions such as Na^+, K^+, Ca^{++}, and Cl^- can move through the membrane with different effective ionic current densities through specific ion conductance channels. Also, in the process of cellular homeostasis, the cell may actively transport ions through its membrane against concentration and/or potential gradients. The transmembrane potential of neurons can reverse to tens of mV positive during a nerve action potential or spike.

Transposable element See Transposon.

Transposon A segment of DNA that can move around (or be moved around) to different positions in the genome of a single cell. In doing so, it may replicate itself and may encode its own *promoter sequence.* Transposons may cause mutations, destroy or alter the ability of genes to be expressed, and increase or decrease the amount of DNA in the genome. There are three classes of transposons: *Class I,* also called *Retrotransposons,* in which a DNA segment is first transcribed into RNA, and then the enzyme *reverse transcriptase* makes a DNA copy that is inserted into a new DNA location; *Class II, DNA transposons,* in which a DNA segment moves directly from site to site; and *Class III,* which is called *Miniature Inverted-Repeat Transposable Elements* (MITEs). MITEs are too small to encode any protein; how they are copied and moved is uncertain. Mobile retrotransposons may function epigenetically as codeterminants of the order of chromatin structure.

Trophoblast cells The outer layer of cells in an early mammalian embryo (morula) that develop into the cells that anchor the placenta of the embryo to the uterine wall (chorionic villi).

Tyrosine kinase An enzyme that can attach a phosphate group to a tyrosine residue in a targeted protein.

Ubiquitin A small protein (only 76 AAs), which is conserved throughout all the kingdoms of life (vertebrates, bacteria, yeasts, etc.). Ubiquitin is used to target certain proteins for destruction by *proteosomes*. Ubiquitin is a necessary component of all cells' C^3 structure. It is necessary for *apoptosis*, or programmed cell death.

Uracil A pyrimidine base found in RNA. It pairs with adenine.

Vector An agent, such as a virus or a small piece of DNA called a plasmid, that carries a modified or foreign gene. When used in gene therapy or genetic engineering, a vector delivers the desired gene or genes to a target cell.

VEGF Acronym for vascular endothelial growth factor proteins. An important signaling protein that stimulates endothelial cell mitogenesis and cell migration. It is involved in embryonic vasculogenesis, and its overexpression is associated angiogenesis in tumors and tumor metastasis. The variants of VEGF result from alternate splicing of mRNA from a single, eight exon VEGF gene at chromosome 6p21. All VEGF variants stimulate cellular responses by binding to cell-surface tyrosine kinase receptors (VEGFRs).

Viroid A small, nonliving, infectious pathogen (obligate parasite) composed entirely of a small, single-stranded RNA loop (plasmid) without a protein coat. They are much smaller than a typical virus and generally use plant cells to reproduce. Viroids consist of about 246 to 401 Nts, depending on the species. They contain information for self-replication inside a host cell using the cell's own RNA polymerase. Viroid infection causes the generation of specific short interfering RNAs (siRNAs).

Virus (including bacteriophages) A nonliving, self-replicating genomic system. Viruses require living eukaryotic or prokaryotic cells to reproduce. They are essentially nonliving endoparasites. The information for viral reproduction is found in their internal DNA or RNA (retroviruses). They generally have a protein coat (not a cell membrane) and have external proteins that allow them to bind to the external membranes of specific classes of host cells. Other viral proteins allow them to enter their target cell and copy their genome into the cell's DNA and initiate their reproduction, using the cell's enzyme systems and RNAs. Viruses have no metabolism outside their target cell and do not exhibit homeostasis or environmental reactivity.

XIST A non–protein-coding gene that produces an active RNA that coats the unwanted (second) X-chromosome in females. Its binding leads to nearly complete methylation of the cytosine in all the unwanted CpG dinucleotides, as well as deacetylation of histone "tails." The X-chromosome's *chromatin* compacts into an RNA-coated, inaccessible mass, "silencing" its genes.

Yeast artificial chromosome (YAC) Extremely large segments of DNA from another species spliced into DNA of yeast. YACs are used to clone up to 1 million bases of foreign DNA into a host cell, where the DNA is propagated along with the yeast cell's other chromosomes, creating GM yeast.

Yoctomole 1×10^{-24} mole. (Note that this is a fraction [0.602] of a molecule.)

Zeptomole 1×10^{-21} mole.

Zinc finger (Zif) proteins A component of Transcription Factor IIIA. Zif proteins can bind to selected sites in dsDNA. Each Zif protein has several "fingers," each containing about 30 AAs. Each finger has a pair of cysteine and a pair of histidine molecules that serve as zinc molecule ligands. Fingers typically occur as tandem repeats, with two, three, or more fingers comprising the specific DNA-binding domain of a transcription factor. The "fingers" bind in the major groove of the DNA helix.

Zygote The initial, single-celled product of the fusion of a male and female gamete (e.g., sperm and egg). A single, diploid zygote develops into an embryo through mitotic cell divisions involving the sequential and often simultaneous expression of genes from its genome.

Index

A

AA, *see* Amino acid
AARS, *see* Aminoacyl-tRNA synthetase
Abiogenesis theory, 29
ABRM, *see* Anterior byssus retractor muscle
Acetylcholine (ACh), 69, 80, 82, 217
ACh, *see* Acetylcholine
Acholeplasma sp., 53
Aconitase, 105
ACP, *see* Acyl carrier protein
ACTH, *see* Adrenocorticotropic hormone
Actin, 227
Action potentials, 77, *see also* Hodgkin–Huxley model
Acute hemorrhagic cystitis, 275
Acyl carrier protein (ACP), 112
ADCC, *see* Antibody-dependent cellular cytotoxicity
ADCs, *see* Analog-to-digital converters
Adenine nucleotide translocase exchanger, 69
Adenosine deaminase, 182, 330
Adenosine triphosphate (ATP), 14
    hydrolysis, 68
    synthase, 63, 125
S-Adenosylmethionine (SAM), 115
Adenoviruses, 339
Adenyl cyclase, 216
ADH, *see* Antidiuretic hormone
Adrenocorticotropic hormone (ACTH), 205
Adrenocorticotropin, 216
Adult stem cells (ASCs), 349, 350
Advanced glycation end product (AGE), 249
Advanced Nucleic Acid Analyzer (ANAA), 288
Aedes triseratus, 23
Afferent neurons, 77
Aflatoxins, 189
Agarose gel electrophoresis, 299
AGE, *see* Advanced glycation end product
Agglutinins, 241
AIF, *see* Apoptosis-inducing factor
Alanine, 116
Alanine transaminase (ALT), 115
Aldehydes, 251
Allosteric regulation, 203
Alpers syndrome, 252
Alpha-thalassemia/mental retardation-X-linked (ATRX) syndrome, 281
ALT, *see* Alanine transaminase
Altered nuclear transfer (ANT), 345, 350
Alzheimer's disease, 264, 267, 268
Amadori product, 250
AMH, *see* Antimullerian hormone
Amino acid (AA), 3, 14, 168
    branched-chain, 117
    catabolism, 116–118
    covalent linkage, 199
    essential, 115
    hydrophobic, 201
    metabolism, 115–116
        gluconeogenesis, 115
        purine nucleotides, 116
    recycling, 259–265
        autophagy, 261–265
        proteolysis in apoptosis and cell necrosis, 261
        proteolytic enzymes, 259
        proteosome, 259–261
    sequences, as degradation signals, 38
    sequence variability, 247
    sulfur-containing, catabolism of, 117
    transit sequences, 202
Aminoacyl-tRNA synthetase (AARS), 168, 182
Aminoguanidine, 249
Amplitude saturation, 7
ANAA, *see* Advanced Nucleic Acid Analyzer
Analog-to-digital converters (ADCs), 297
Anaphase-promoting complex (APC), 85
Aneuploidy, 89
Angiotensin, 205
Angiotensinogen, 205
Anode break excitation, 140
ANP, *see* Atrial-natriuretic hormone
ANT, *see* Altered nuclear transfer
Anterior byssus retractor muscle (ABRM), 215
Anthocyanin, 334
Antibody(ies), 6, 245–247
    constant regions, 247
    hypervariable regions, 247
    monoclonal, 316
    primary, 309
    receptors, 246
Antibody-dependent cellular cytotoxicity (ADCC), 226
Antidiuretic hormone (ADH), 67
Anti-idiotypic antibodies, 240
Antimullerian hormone (AMH), 205
Antiporter ion pump, 69
APC, *see* Anaphase-promoting complex
Apoptosis, 14, 35, 230, 261
    -inducing factor (AIF), 39
    promoter protein, 40
Apoptosome, 37, 261
AQPs, *see* Aquaporins
Aquaporins (AQPs), 66
Arabidopsis
    sp., 4, 195
    thaliana, 67, 334
Arachidonic acid, 46, 113, 217
Archaeans, 17
    biological features, 51
    cell shapes, 51
    composition, 46
    translation machinery, 51

Arginine, 260
Argonaute proteins, 186
Arrhenius relation, 99
ASBV, *see* Avocado Sunblotch Viroid
ASCs, *see* Adult stem cells
Asparagine, 260
Astrocytes, 80
Ataxia telangiectasia mutated (ATM) gene, 155
ATM gene, 155
ATP, *see* Adenosine triphosphate
Atrial-natriuretic hormone (ANP), 205
ATRX syndrome, *see* Alpha-thalassemia/mental retardation- X-linked syndrome
Autophagosomes, 61
Autophagy, 261
Avidin, 283
Avocado Sunblotch Viroid (ASBV), 23
Axoneme, 207

B

Bacillus
  subtilus, 259
  thurigensis, 332
Bacteria
  classification, 51
  genome sequencing, 51
  green sulfur bacteria, 118
  iron, 53
  methanogen, 14
  purple sulfur 52
Bacteriophages, 21, 338
Bacteriorhodopsin, 32
Barr body, 188
Basal promoter region (BPR), 193
Base excision repair (BER), 189
BBC, *see* Bio-bar code
BCAAs, *see* Branched-chain amino acids
B-cell receptor (BCR), 231
BCR, *see* B-cell receptor
Beckwith–Wiedemann syndrome, 281
BER, *see* Base excision repair
Bile salts, 112
Bio-bar code (BBC), 316, 317
Biochemical oscillators, models for, 128–136
  approximate periods, 130
  behavior, 131, 132
  biological clocks, 130, 136
  central circadian oscillator, 129
  cross-competitive inhibition, 130
  cyclical secretion of insulin, 129
  green fluorescent protein, 133
  parametric feedback, 129
  plasmid, 133
  reaction simulation, 135
  repressilator, 134, 135
  saturating amounts of repressor, 133
Biochemical Pathway Simulator (BPS), 147
Biochemical systems modeling, *see* Physical biochemistry and biochemical systems modeling
Bioinformatics, definition of, 3
Biological systems
  amplitude saturation, 7
  dynamic modeling of, 3
  dynamic nonlinearities, 7
  evolution, 6
  mathematical example, 8
  parametric feedback, 7
  static nonlinearity, 7
Biopoesis theory, 29
Biotechnology, coining of term, 4
Biotin, 283
Blastomeres, 351
Blood proteins, *see* Proteins, blood proteins
Blue-absorbing proteorhodopsin, 218
Blumlein transmission line pulse generator, 342
Bomb calorimeter, 96
Borrelia burgdorferi, 52
Botulinum toxin A, 82, 93
Bovine spongiform encephalopathy (BSE), 252, 269
Bovine virus diarrhea virus, 277
BPR, *see* Basal promoter region
BPS, *see* Biochemical Pathway Simulator
Bragg's law, 321
Branched-chain amino acids (BCAAs), 117
British thermal unit, 96
Brugia malayi, 289
BSE, *see* Bovine spongiform encephalopathy
Bursa of Fabricus, 224

C

Cadherins, 64, 355
Calcitonin, 205
Calcium ions, 216
Calmodulin, 216, 295
Calsequestrin, 211
cAMP, *see* Cyclic AMP
cAMP response element binding protein (CREB), 84
CAMs, *see* Cell adhesion molecules
Cancer
  autophagy regulation and, 264
  environmental factors, 267
  epigenetic therapy for, 278–281
  GM viruses used to fight, 275–276
  heritability, 268
  patients, analysis of point mutations in genes of, 4
  stem cell (CSC), 278, 354–359
Carbohydrate metabolism, 109–111
  gluconeogenesis, 110, 111
  insulin, 110
  phosphoenolpyruvate, 111
Carbon dating, 26–29
Carboxymyoglobin, 248
Carboxypeptidases, 259
Cardiopulmonary resuscitation (CPR), 41
Cardiovascular system (CVS), 41
Carnosine, 249
CAS, *see* Citric acid synthetase
Cascade rolling circle amplification (CRCA), 305
Caspases, 37, 259
Catenins, 63, 355
Cathepsins, 259
Caulerpa sp., 50
cDNA, *see* Coding DNA
Cell adhesion molecules (CAMs), 63

Index

Cell anatomy and physiology, basic, 45–93
cell membrane, 45–49
Archaeans, 46
categories, 45
cholesterol, 46
electroporation, 48
Eubacteria, 46
glucose receptor, 48
lipid categories, 45
transmembrane proteins, 48
cell reproduction, 84–90
cell cycle (mitosis), 84–86
chromosome shuffling, 87
genetic diversity, 86
genomic errors, 88
meiosis, 86–90
mitotic events in cell cycle, 85
oogenesis, 88
spermatogenesis, 88
chemical synapses, 80–84
G-protein-coupled receptors, 83–84
neural transmembrane chemical signal receptors, 83
summary, 84
cytosol, 50
electrical synapses (gap junctions), 78–80
eukaryotic cells, 53–56
cytosol, 56–57
endoplasmic reticulum, 59–60
Golgi apparatus, 60–61
lysosomes, 61
mitochondria, 61–63
multicelled eukaryotes, 56
nucleus and nucleolus, 57–59
peroxisomes, 61
ribosomes, 59
single-cell eukaryotes (protists), 54–56
home problems, 92
insides of cell membrane, 50
prokaryotic cells, 50–53
Archaeans, 51
Eubacteria, 51–53
summary, 90
transmembrane proteins, roles of, 63–84
antiport carrier, 69
aquaporins, 66–68
basis for DC transmembrane potential, 73–76
cell adhesion molecules, 63–64
cell signaling by changes in transmembrane potential, 76–78
chloride channels, 70
Cochlear hair cells, 70
conductance states, 71
connexon, 78, 79
difference in osmotic pressure, 65
diffusion and osmosis, 65–66
Fick's diffusion equation, 65
glial cells, 80
impermeable charged molecules, 75
internal ligands, 70
ion and molecular pumps, 68–69
Laplacian operator, 65
mammalian aquaporins, 68
membrane receptor categories, 84
molecular pumps, 67
Nernst potential, 70
neurotransmitters, 69
parietal cells, 69
propagation of nerve impulses, 77
regulated ion channels in membranes, 69–73
second messenger molecules, 70
second-messenger system, 82
semipermeable membrane, 65
squid giant neuron, 74
total osmolarity, 66
transport of substances across cell membrane, 65–78
uniport carriers, 68
voltage-gated Ca^{++} channel proteins, 72
voltage-gated cation channels, 72
voltage-gated K^+ channels, 73
voltage-gated sodium ion channels, 72
Cell death, *see* Life and death
Cell surface receptor proteins, 232
Cellular homeostasis, 83
Cellulose, 109
Central circadian oscillator, 129
Central nervous system (CNS), 15, 41, 77
CF, *see* Cystic fibrosis
cGMP, *see* Cyclic guanosine monophosphate
Chaperone proteins, 54
Chaperones, 306
Chaperonins, 170
Chemical bond energy, 95–97
average bond energies, 97
bomb calorimeter, 96
British thermal unit, 96
kilocalories, 95
net free energy, 96
potential energy, 95
Chemical bond types, 97–98
covalent bonds, 98
hydrogen bonding, 98
ionic bonds, 97
noncovalent bonds, 97
polarized covalent bonds, 98
Chemiosmosis, 32, 63, 125
Chemotaxis, 15
Chimeras, 362–364
cellular chimeras, 362–363
ethical issues, 370–371
how human is too human, 371
religious concerns, 371
retrovirus transfer, 371
transgenic organ farming, 370–371
somatic chimeras, 363–364
Chiral molecule, 197
Chloride channels, 70
Chlorophyll, 118, 119, 120
Chloroplasts, 52
Chloroplast transit protein (CTP) sequence, 333
Cholecystokinin, 205
Cholesterol, 46
ester hydrolase, 112
molecule, schematic of, 48, 49
Cholinergic neurons, 80

Cholinesterase, 259
Chromatids, 86
Chromatin, 57, 86, 159, 278
Chronic wasting disease (CWD), 252, 254, 270
Citric acid synthetase (CAS), 104
CJD, *see* Creutzfeld–Jakob disease
Clostridia sp., 53
Clostridium
  botulinum, 93
  tetani, 93
Closure temperature, 26
CNS, *see* Central nervous system
CoA, *see* Coenzyme A
Coding DNA (cDNA), 330
Codon, 160
Coenzyme A (CoA), 105
Collagen, 250, 335
Colloid osmotic pressure (COP), 246
Complement, 6
Complex systems, 5–9
  definition of, 5
  example, 6
  modeling, 9
  parametric control, 8
  properties of nonlinear systems, 7–8
Computational biology, 3
Connexins, 77
Connexon, 78
Continuous System Modeling Program (CSMP®) software, 1
COP, *see* Colloid osmotic pressure
Core particle (CP), 38
Coronavirus, 277
Corticotropin-releasing hormone (CRH), 205
Cortisol, 238, 239
Covalent bonds, 98
Covalent technology (CT), 309
CP, *see* Core particle
CPR, *see* Cardiopulmonary resuscitation
CRCA, *see* Cascade rolling circle amplification
CREB, *see* cAMP response element binding protein
Creutzfeld–Jakob disease (CJD), 252, 267, 269
CRH, *see* Corticotropin-releasing hormone
Cryoelectron microscopy, 324
CSC, *see* Cancer stem cell
CSMP® software, *see* Continuous System Modeling Program software
CT, *see* Covalent technology
CTCF zinc finger protein, 194
CTL, *see* Cytotoxic T-lymphocyte
CTP sequence, *see* Chloroplast transit protein sequence
CVS, *see* Cardiovascular system
CWD, *see* Chronic wasting disease
Cyanobacteria, 31, 52, 54
Cyclic AMP (cAMP), 83
Cyclic guanosine monophosphate (cGMP), 216
Cyclin D, 346
Cyclins, 85
Cystic fibrosis (CF), 269, 274
Cytochrome **c**, 35, 37, 261
Cytochrome oxidase, 63
Cytokines, 240
Cytosine, 290
Cytotoxic T-lymphocyte (CTL), 35, 40, 220

D

DAF, *see* Decay accelerating factor
DAG, *see* Diacyl glycerol
Danon's myopathies, 264
Dardarin, 270
DCs, *see* Dendritic cells
Death, *see* Life and death
Death receptors (DRs), 40
Decay accelerating factor (DAF), 229
Dendritic cells (DCs), 336
Deoxyribonucleic acid, *see* DNA
Desmin, 214
Deubiquitinases, 260
Dextrose, 198
DHAP, *see* Dihydroxyacetone phosphate
Diacyl glycerol (DAG), 217
Diakinesis, 87
Dicer enzymes, 185
Dihydrouridine, 181
Dihydroxyacetone phosphate (DHAP), 102, 113
Directed Panspermia (DP) theory, 29
DNA (deoxyribonucleic acid), 177–179
  alpha helix, 177
  basal promoter region, 193
  coding, 330
  counterions, 179
  damage, 35
  glycosylases, 189
  histones, 179
  hydrogen bonding, 179
  ligase, 190, 338
  limit of detection, 288
  methylation, 158, 194
  methyl transferases (DNMTs), 279
  mitochondrial, 300
  nucleoside subunit of, 177
  phosphodiester links, 177
  polymerase (DNAP), 286
  pyrimidine base, 177
  recombinant, 335, 337
  resting, 179
  structural modification of, 159
  viruses, 21
  Z-DNA, 178
DNAP, *see* DNA polymerase
DNMTs, *see* DNA methyl transferases
DOE Joint Genome Institute (JGI), 303
Double-stranded DNA (dsDNA), 45, 50
Down syndrome, 89, 269
DP theory, *see* Directed Panspermia theory
Drosophila mangabeirai, 359
DRs, *see* Death receptors
dsDNA, *see* Double-stranded DNA
Dyneins, 206, 207

E

Early progenitor (EP) cells, 358
EBV, *see* Epstein-Barr virus
EC, *see* Embryonic carcinoma
EEG, *see* Electroencephalogram
Efferent neurons, 77

EI, *see* Epigenetic inheritance
Eicosanoids, 113
Eicosapentoic acid, 46
Elastases, 259
Electroencephalogram (EEG), 42
Electroneutrality, 75, 138
Electron transport chain, 62
Electron transport phosphorylation, 53
Electroporation, 48, 341
ELISA, *see* Enzyme-linked immunoabsorbent assay
Embden–Meyerhoff Pathway, 102
Embryonic carcinoma (EC), 349
Embryonic stem cells (ESCs), 86, 337, 351, 372
ENCODE pilot project, 161
Endoplasmic reticulum (ER), 15, 54, 57
Entamoeba histolytica, 56
Enzyme(s), 203–204
  allosteric receptor domains, 203
  chaperone, 254
  coenzymes, 203
  competitive inhibition, 204
  end-product inhibition, 204
  feedback inhibition, 204
  names, 203
  parametric control, 203
  physical chemistry, 203
  precursor activation, 204
  proteolytic, 259
  regulation, 203–204
  restriction, 338
Enzyme-linked immunoabsorbent assay (ELISA), 307
EP cells, *see* Early progenitor cells
Epigenetic inheritance (EI), 216
Epigenetic processes, 194
Epinephrine, 114
EPL, *see* Expressed protein ligation
EPSP, *see* Excitatory postsynaptic potential
Epstein-Barr virus (EBV), 40
Epulopiscium fishelsoni, 50
ER, *see* Endoplasmic reticulum
Erwinia herbicola, 288
Erythropoietin, 205
ESCs, *see* Embryonic stem cells
Essential fatty acids, 113
Ester linkages, 45
Ethics, *see* Genetic engineering, ethical issues in
Eubacteria
  cell construction, 52
  cell-covering structure, 46
  classification, 51, 52
  evolution of, 51
  iron bacteria, 53
  Lyme disease pathogen, 52
  metabolism, 52
  purple sulfur bacteria, 52
Euchromatin, 58, 215
Eukaryotes, 17, 53–56
  anatomical components of, 56–63
    cytosol, 56–57
    endoplasmic reticulum, 59–60
    Golgi apparatus, 60–61
    lysosomes, 61
    mitochondria, 61–63
    nucleus and nucleolus, 57–59
    peroxisomes, 61
    ribosomes, 59
  cytoplasm of, 50
  ester linkages of, 45
  excision repair in, 189
  internal membranes of, 45
  multicelled eukaryotes, 56
  ribosomes, 50
  single-cell eukaryotes (protists), 54–56
    alveolates, 55
    brown algae, 55
    choanoflagellates, 56
    diatoms, 55
    dinoflagellates, 55
    Euglenozoa, 55
    golden algae, 55
    Mycetozoa, 55
    red algae, 55
    Sporozoa, 55
    water molds, 55
Evo-Devo, 191
Excitatory postsynaptic potential (EPSP), 82
Expressed protein ligation (EPL), 258

F

$FADH_2$, *see* Flavin adenine dinucleotide
FAs, *see* Fatty acids
FAS, *see* Fatty acid synthase
Fast journals, 9
Fatal familial insomnia (FFI), 252
Fatty acids (FAs), 114
Fatty acid synthase (FAS), 112
Feedback inhibition, 204
FFI, *see* Fatal familial insomnia
FGFR2, *see* Fibroblast growth factor receptor 2
Fibrillarin, 184
Fibroblast growth factor receptor 2 (FGFR2), 269
Ficin, 259
Fifth disease, 22
FISH, *see* Fluorescent in situ hybridization
Flavin adenine dinucleotide ($FADH_2$), 105
FlavrSavr (FS) tomato, 330, 344
FLISA, *see* Fluorophore-linked immunoabsorbent assay
Fluorescence resonance energy transfer (FRET), 288, 294
Fluorescent in situ hybridization (FISH), 293
Fluorophore-linked immunoabsorbent assay (FLISA), 309
Follicle-stimulating hormone (FSH), 205, 216
Forensic science, *see* Instrumental methods used in genomics, proteomics, and forensic science
Free radicals, 318
FRET, *see* Fluorescence resonance energy transfer
FSH, *see* Follicle-stimulating hormone
FS tomato, *see* FlavrSavr tomato
Fuligo septica, 56

G

GA, *see* Golgi apparatus
GABA, *see* Gamma aminobutyric acid
GADP, *see* Glyceraldehyde-3-phosphate

Gallionella sp., 53
Gamma aminobutyric acid (GABA), 69, 82
Gap junctions, 77, 78–80
Gastric inhibitory peptide, 205
Gastrin, 205
G-CSF, *see* Granulocyte colony-stimulating factor
GDP, *see* Guanine diphosphate
GenBank, 256
Gene(s)
 analysis of point mutations, 4
 autosomal, 362
 defects, diseases and, *see* Genomic medicine
 designer, GMO production using, 341
 Hox, 172, 347
 linked, 156
 protein-coding, 193
 regulation, 191–195
  epigenetic gene regulation, 194–195
  eukaryotes, 193–194
  prokaryotes, 192–193
 viral, 21
Gene gun, 48, 187, 339, 364
Generalized Mass Action (GMA) System, 127
Genetically modified organisms (GMOs), 4, 329, 369
 definition of, 330
 examples of, 330
 problems with, 342–345
 use of designer genes to produce, 341
Genetic engineering, 4
Genetic engineering, ethical issues in, 367–375
 chimeras, 370–371
  how human is too human, 371
  porcine endogenous retroviruses, 371
  religious concerns, 371
  retrovirus transfer, 371
  simian immunodeficiency virus, 371
  transgenic organ farming, 370–371
 framework for ethical decision making, 367–369
  common good approach, 368
  defining ethics, 367–368
  ethical frameworks, 368
  fairness or justice approach, 368
  other resources, 369
  rights approach, 358
  utilitarian approach, 368
  virtue approach, 368
 genetically modified food crops, 369–370
  antibiotic-resistant bacteria, 370
  GMOs, 369
  green fluorescent protein, 370
  hidden allergens, 369
  plasmids, 370
  rogue herbicide resistance, 369–370
  superweeds, 369
  transgenic crops, 369
 home problems, 375
 human embryonic stem cells, 372–374
  embryonic versus adult stem cells, 373–374
  federal restrictions, 374
  global picture, 374
  religious concerns, 372–373
  reprogramming of skin cells, 373
 summary, 374–375
Genetic inheritance, basis of, 155–175
 alternative splicing, 164
 ATM gene, 155
 autosomes, 158
 chaperonins, 170
 chemical gene suppression, 158
 chromatin marks, 159
 cis-regulatory elements, 162
 control of development, 170–172
 crossing over, 158
 DNA methylation, 158
 ENCODE pilot project, 161
 epigenetic inheritance, 158–159
 genes and genome, 160–170
  introns and exons, 161–164
  protein synthesis, 164–170
  transcription, 165–169
  translation, 169–170
 genetic code, 160
 genetic imprinting, 159
 genome elements, 160
 home problems, 173–175
 Hox genes, 172
 Human Genome Project, 161
 initiator tRNA, 169
 linked genes, 156
 meiosis, 157
 Mendelian inheritance, 156–158
 Mendel's Laws of Heredity, 156
 nuclear and mitochondrial DNA, 159–160
 peptide bond, 172
 phenotype, 155
 phylotypic stage, 170
 promoter regions, 159
 release factor protein, 170
 reverse transcriptase, 163, 173
 ribosomes, 169
 summary, 172–173
 traits, 157
 transposons, 156, 173
 wobble theory, 166
Genomic medicine, 267–282, 335
 acquired immunity, 276
 acute hemorrhagic cystitis, 275
 alpha-thalassemia/mental retardation-X-linked syndrome, 281
 alpha-V-integrin, 275
 antisence mRNA, 275
 Beckwith–Wiedemann syndrome, 281
 bovine virus diarrhea virus, 277
 cancer risk, 269
 cancer stem cells, 278
 conservative missense mutation, 268
 coronavirus, 277
 cystic fibrosis, 274
 direct DNA tests, 270, 271–273
 direct gene transfection, 274
 diseases tested, 271–273
 Down syndrome, 269
 epigenetic cancer chemotherapy drugs, 280
 epigenetic therapy for cancer, 278–281
  epigenetics and cancer, 278–279

epigenetic therapies for cancer, 279
other diseases with epigenetic etiologies, 279–281
examples of genetic diseases, 268–270
Factor V Leiden, 270
frame-shift mutation, 268
gene defects and disease, 273–274
gene therapy, 274–275
genetic diagnostics, 267
genetic testing, 268
Gerstmann–Straüssler–Scheinker disease, 270
GM viruses used to fight cancers, 275–276
hemagglutinin, 276, 277
hemochromomatosis risk, 268
hereditary hemochromatosis, 267
hereditary nonpolyposis colorectal cancer, 268
home problems, 281–282
insulin-like growth factor 2, 279
keratoconjunctivitis, 275
malignant glioma, 275
medical genomics, 267
memory B-cells, 276
methyltransferase enzymes, 278
multifactorial genetic disorders, 267
neurofibromatosis, 267
nonsense mutation, 268
oncogenes, 273
oncogenic epigenetic aberrations, 278
pharyngoconjunctival fever, 275
pilocarpine, 269
prion protein, 269
recombinant vaccines, 276–277
retinoblastoma protein, 275
Rett syndrome, 279
Rubenstein–Taybi syndrome, 281
sickle-cell anemia, 268
silent mutation, 268
single-gene mutations, 267
summary, 281
thyroid medullary cancer, 268
tumor suppressor genes, 273
vaccination, 276
venous thrombosis risk, 268
Genomics, applications of, 329–365
adenoviruses, 339
altered nuclear transfer, 345, 350
animal reproductive cloning, 345–347
altered nuclear transfer, 345–346
Honolulu technique, 346
problems in cloning mammals, 347
Roslin technique, 346
therapeutic cloning, 347
bacteriophage viruses, 338
blastomeres, 351
Blumlein transmission line pulse generator, 342
chimeras, 362–364
autosomal gene, 362
blood cell chimera, 362
cellular chimeras, 362–363
fraternal twins, 362
human cellular chimera, 362
somatic chimeras, 363–364
chloroplast transit protein sequence, 333
complete hydatiform mole, 361
cyclin D, 346
dendritic cells, 336
early progenitor cells, 358
economic advantage to growing Bt-cotton, 334
electroporation, 341
Gap Zero, 346
gene gun, 339, 340
genetically modified organisms, 329–345
cotton, 333–334
FlavrSavr™ tomato, 330–332
golden rice, 332
problems with GMOs, 342–345
recombinant DNA technology, 337–342
soybeans, 333
transgenic animals and animal cells, 335–337
transgenic *Arabidopsis* for detection of explosives in soil, 334–335
transgenic maize, 332–333
transgenic plants, examples of, 330–335
glioblastoma multiforme, 359
glufosinate-ammonium herbicide, 344
glyphosphate-tolerant soybean line, 333
growth hormone, 330
homeobox transcription factor, 350
homeodomain transcription factor, 350
home problems, 365
human amniotic epithelial cells, 349
imprinting, 347
in vitro fertilization, 347
Law of Unintended Consequences, 333
marker gene, 331
morula embryo, 351
natural parthenogenesis, 359
neomycin, 352
niche cells, 354, 356
octamer-binding protein 4, 349
oncogenic event, 358
parthenites, 360
parthenogenesis, 359–361
artificial parthenogenesis, 359–360
human parthenotes, 360–361
parthenote, 351
pluripotent stem cells, 347, 348
polar body, 346
Polycomb group transcriptional repressor, 358
pronuclear injection, 337
recombinant DNA technology, 330
red rice, 344
regenerative medicine, 345, 347, 360
restriction enzyme, 338
retroviruses, 339
reversine, 352
sertoli cells, 346
somatic cell nuclear transfer, 347
stem cells, 347–359
applications in regenerative medicine, 353–354
cancer stem cells, 354–359
sources, 349–353
types, 348–349
summary, 364–365
telomeres, 346
temperate phage, 338
totipotent stem cells, 348

transcription factor proteins, 347
transgenic caterpillar cells, 337
Gerstmann–Straüssler–Scheinker syndrome (GSS), 252, 254, 270
GF, *see* Growth factor
GFP, *see* Green fluorescent protein
GG, *see* Gynogenesis
GH, *see* Growth hormone
GHRH, *see* Growth hormone releasing hormone
Giardia intestinalis, 56
Gibbs free energy, 96
Glial cells, 80
Glucagon, 110, 205
Gluconeogenesis, 115
Glucose, 198
phosphatase, 110
receptor, 48
Glucosepane, 250
Glufosinate-ammonium herbicide, 344
Glutamine, 260
Glyceraldehyde-3-phosphate (GADP), 102
Glycine, 116, 243
Glycogen, 109
Glycolipids, 46
Glycolysis, 59
Glycoprotein(s), 242–243
antibodies, 242
heterodumeric, 64
immunoglobulins, 242
O-glycosylation, 242
tetrameric, 283
zona pellucida, 242
Glycosylation, 201
Glycosyl phosphatidyl inositol (GPI), 252
GMA System, *see* Generalized Mass Action System
GM-CSF, *see* Granulocyte-macrophage colony-stimulating factor
GMCSF, *see* Granulocyte-monocyte colony stimulating factor
GMOs, *see* Genetically modified organisms
GNPs, *see* Gold nanoparticles
Gold nanoparticles (GNPs), 316
Golgi apparatus (GA), 60, 202
Gonadotropin-releasing hormone, 205
Goniometer, 320
GPCRs, *see* G-protein coupled receptors
GPI, *see* Glycosyl phosphatidyl inositol
G-protein coupled receptors (GPCRs), 83
Granulocyte colony-stimulating factor (G-CSF), 330
Granulocyte-macrophage colony-stimulating factor (GM-CSF), 330
Granulocyte-monocyte colony stimulating factor (GMCSF), 221
Green fluorescent protein (GFP), 133, 292, 370
Green sulfur bacteria, 118
Growth factor (GF), 262
Growth hormone (GH), 205, 226, 330, 336
Growth hormone releasing hormone (GHRH), 205
GSS, *see* Gerstmann–Straüssler–Scheinker syndrome
GTP, *see* Guanosine triphosphate
Guanine diphosphate (GDP), 83
Guanine nucleotide phosphates, 83
Guanosine nucleotide binding protein, 217
Guanosine triphosphate (GTP), 170, 206
Gynogenesis (GG), 361

H

HAECs, *see* Human amniotic epithelial cells
Haemophilus influenzae, 3, 50
Hantavirus, 324
Hatch–Slack cycle, 124
HATs, *see* Histone acetyltransferases
HBV, *see* Hepatitis B virus
hCG, *see* Human chorionic gonadotropin
HDA, *see* Helicase-dependent, isothermal, DNA amplification
Heat shock element (HSE), 251
Heat shock factor (HSF), 251
Heat shock proteins (HSPs), 251
Helicase-dependent, isothermal, DNA amplification (HDA), 288
Helper virus, 24
Hemagglutinin, 276, 277, 337
Hematopoietic stem cells, 349
Hemocyanins, 247
Hemoglobin, 243–245
Hemophilia A, 289
Hepatitis B virus (HBV), 331
Hepatitis C, 307
Heptahelical receptors, 83
Hereditary hemochromatosis, 267
HERVs, *see* Human endogenous retroviruses
Heterochromatin, 57, 215
Heterodimeric glycoproteins, 64
Hexokinase (HK), 1902
Histidine, 117
Histone acetyltransferases (HATs), 194
HIV, *see* Human immunodeficiency virus
HK, *see* Hexokinase
HLA, *see* Human leucocyte antigen
Hodgkin–Huxley model, 150
anode break excitation, 140
auxiliary parameters, 140
current-to-frequency converter, 140
electroneutrality, 138
injected current density, 140
ion pumps, 138
monomolecular kinetics, 139
Nernst potential, 137
pacemaker neurons, 140
potential energy, 137
sodium channel deactivation, 139
transmembrane voltage change, 139
Homeobox transcription factor, 350
Homeodomain transcription factor, 350
Homeostasis, 92, 203
Homing, 258
Homologous restriction factor (HRF), 229
Homo sapiens, single-cause mentality of, 10
Hormones
adrenocorticotropic hormone, 205
antidiuretic hormone, 67
antimullerian hormone, 205
atrial-natriuretic hormone, 205

corticotropin-releasing hormone, 205
follicle-stimulating hormone, 205, 216
gonadotropin-releasing hormone, 205
growth hormone, 205, 226, 330, 336
growth hormone releasing hormone, 205
human fertility hormones, 335
luteinizing hormone, 205, 216, 226
melanocyte-stimulating hormone, 205
parathyroid hormone, 205, 216, 330
receptor proteins, 205
second messenger system, 204
thyroid-stimulating hormone, 205, 216, 307
thyrotropin-releasing hormone, 204
Hox genes, 172, 347
HPV, *see* Human papilloma virus
HRF, *see* Homologous restriction factor
HSE, *see* Heat shock element
HSF, *see* Heat shock factor
HSPs, *see* Heat shock proteins
HTUs, *see* Hypothetical taxonomic units
Human amniotic epithelial cells (HAECs), 349
Human chorionic gonadotropin (hCG), 205, 307
Human endogenous retroviruses (HERVs), 163
Human fertility hormones, 335
Human Genome Project, 161
Human immunodeficiency virus (HIV), 40, 267, 307, 331
Human leucocyte antigen (HLA), 230
Human papilloma virus (HPV), 40
Huntington's disease, 264, 268
Hydrazine, 307
Hydrogen bonding, 98, 179
Hypothetical taxonomic units (HTUs), 16

I

IAP blocker protein, 261
IBM Continuous System Modeling Program software, 1
ICF syndrome, *see* Immunodeficiency/centromeric instability/facial anomalies syndrome
ICM, *see* Inner cell mass
IF, *see* Intermediate frequency
IFNγ, *see* Interferon gamma
Ig, *see* Immunoglobulin
IGF, *see* Insulin-like growth factor
Immune system, 6
Immune system, proteins in, *see* Proteins, immune system
Immunocytokines, 219
Immunodeficiency/centromeric instability/facial anomalies (ICF) syndrome, 279–281
Immunoglobulin (Ig), 6, 224, 242, 246
Indirect optimization method (OIM), 128
Information storage system (ISS), 29
Inheritable diseases, *see* Genomic medicine
Inheritance, *see* Genetic inheritance, basis of
Inner cell mass (ICM), 347
Inositol triphosphate (ITP), 217
Instrumental methods used in genomics, proteomics, and forensic science, 283–327
biotin, avidin, and streptavidin, 283–286
avidin, 283–284
biotin, 283
biotin–avidin arrays, 285
Diels–Alder reaction, 285
immobilization in microarrays, 284–286
membrane protein array, 285
microarrays, 283
native chemical ligation, 285
nitrilotriacetic acid arrays, 285
oxime/thiazolidine formation, 285
α-oxo semicarbazone ligation, 285
polymerase chain reaction, 283
specific immobilization, 285
Staudinger ligation, 285
streptavidin, 284
tetrameric glycoprotein, 283
fluorescent molecular tags, 291–298
acceptor fluorophore, 294
donor fluorophore, 294
excitation spectrum, 291
FLIM-FRET, 295–298
fluorescence, 291–293
fluorescence lifetime, 295
fluorescence resonance energy transfer, 294–295
fluorescent in situ hybridization, 293–294
fluorescent quantum dots, 292
fluorophore, 292
GFP, 292
gold nanodots, 293
phase fluorometry, 295–298
quencher, 294
reporter fluorophores, 293
forensic applications of DNA restriction enzymes and STRs, 298–302
agarose gel electrophoresis, 299
nucleic acid hybridization, 299
restriction fragment length polymorphism analysis, 298–299
short tandem repeat markers, 299–301
validity of STR DNA evidence, 301–302
variable-length probe array, 301
home problems, 327
polymerase chain reaction, 286–291
amplicons, 286
direct DNA tests for genetic diseases, 289–290
DNA polymerase enzyme, 283
end-point PCR measurements, 287
"454" system, 290–291
helicase-dependent, isothermal, DNA amplification, 288
hemophilia A, 289
PCR process, 286–289
real-time PCR, 288
protein structure analysis, methods for, 306–326
bio-bar codes, 316–317
Bragg's law, 321
capture agents, 311
Chimera Imperacer® system, 315
competitive ELISA assays, 309
covalent technology, 309
cryoelectron microscopy, 324
electrophoretic separation of proteins, 306
ELISA tests for specific proteins and antibodies, 307–310
epitopic ambiguity, 312
free radicals, 318

genetics, 310
immuno-PCR method, 314–315
isoelectric point, 306
molecular structure determinations, 319
Patterson function, 322
planar waveguide technology, 312
polyacrylamide gel electrophoresis, 306
proteases used to disassemble polypeptides, 308
Protein Data Bank, 318
protein microarrays, 310–314
proteomics, 310
sequencing protein AAs, 307
synchrotron radiation, 322
tangential velocity, 323
time-resolved x-ray crystallography, 323
undulator magnets, 323
water vapor diffusion, 319
Waveguide Excitation Fluorescence Microscope, 313
Western blot test, 310
x-ray crystallography, structural analysis of protein "crystals" using, 317–324
rolling-circle replication of ssDNA circles, 302–306
cascade rolling circle amplification, 305
double-primed RCA, 302
ET terminator, 304
fluorescent markers, 304
JGI RCA protocol, 303–304
padlock probes in RCA, 304–306
RCA protocols, 302–303
restriction endonucleases, 302
summary, 326
Insulin, 8, 110, 129, 205, 330
Insulin-like growth factor (IGF), 2, 205, 279
Integral membrane protein, 201, 202
Integrins, 64, 355
Interferon gamma (IFNγ), 261
Interferons, 236, 330
Interleukin, 6, 232, 330
Interleukin-2, 35, 335
Interleukin-10, 35
Intermediate frequency (IF), 296
Intermediate spermatogonium, 90
Introns, 2
In vitro fertilization (IVF), 347
3-Iodotyrosine (3-IT), 69
IOM, *see* Indirect optimization method
Ion channels, 69, 75
Ionic bonds, 97
Ionizing radiation, 35
Ion pumps, 48, 73, 138
Iron bacteria, 53
Ischemia, 41
Isocitrate dehydrogenase, 105
Isoelectric focusing, 306
ISS, *see* Information storage system
3-IT, *see* 3-Iodotyrosine
ITP, *see* Inositol triphosphate
IVF, *see* In vitro fertilization

J

JGI, *see* DOE Joint Genome Institute

K

Kanamycin, 331
KEGG, *see Kyoto Encyclopedia of Genes and Genomes*
Keratin, 206
Keratoconjunctivitis, 275
Kinetochore microtubules, 86
Krebs cycle, 62
Kuru, 252
Kyoto Encyclopedia of Genes and Genomes (KEGG), 101

L

Lactic acid, 102
Lambda phage, 24
Lamins, 206
LAMP-2, *see* Lysosomal-Associated Membrane Protein 2
Laplace transform, 1
Latency relaxation phenomenon, 211, 213
Law of mass action, 98
Law of Unintended Consequences, 4, 333
LBL, *see* Mannose-binding lectin
Leakage current density, 76
Lectins, 241–242
agglutinins, 241
categories, 241
innate immune system, 242
lectin-complement pathway, 242
mannose-binding lectin, 242
ricin, 241
selectins, 242
Toll-like receptor proteins, 242
Lepidoptera, 35
Leptin, 205, 330
Leucocytes, 219
Leukotrienes (LTs), 114, 217
Levulose, 198
LH, *see* Luteinizing hormone
Life and death, 13–44
autogeneration of life, 13
cell and animal death, 34–43
apoptosis (programmed cell death), 35–40
cell necroptosis, 40–41
dormant life, 34
metazoan death, 41–43
organismic homeostasis, 34
programmed cell death, 35
true brain death, 42
true cell death, 34
cell death, 14
cell death characterization, 44
definition of life, 13–16
evolution of definitions, 13
properties of life, 14–16
domains and kingdoms (origins), 16–19
Archaean era, 18
domains of life, 19
eukaryotes, 17
example of classification, 19
nomenclature for times and eras, 18
Phanerozoic time, 18
Precambrian time, 18
prokaryotic life, 17

Proterozoic era, 18
rock dating, 17
superkingdom approach, 19
taxonomy, 16
dormant life-forms, 16
home problems, 44
nonliving, self-replicating genomic machines, 20–25
plasmids, 24
satellites and virusoids, 24–25
viroids, 23–24
viruses, 20–23
origins of life, 25–34
Archaea phyla, 32
carbon dating, 26–29
Carboniferous era, 32
daughter product, 25
Devonian era, 32
discussion, 34
first life, 29–34
half-lives of radioisotopes, 27
Oparin theory of coacervates, 30
parent isotope, 25
Permian era, 32
radioactive decay time constant, 25
RNA → DNA + Protein molecular scenario, 30
Triassic period, 32
photon energy, 14
spontaneous generation of life, 13
summary, 43–44
thermodynamic death, 14
Linearly polarized light (LPL), 197
Linear time-variable (LTV) system, 6
Linker molecule, 63
Lipid catabolism, 114–115
beta-oxidation steps, 115
energy storage, 114
proprionyl-CoA, 115
succinyl-CoA, 115
Lipid metabolism, 112–114
bile salts, 112
cellular exocytosis, 112
cerebrosides, 114
cholesterol ester hydrolase, 112
eicosanoids, 113, 114
essential fatty acids, 113
fatty acid synthesis cycle, 112
gangliosides, 114
lipid classes, 114
malonyl-CoA, 112
plasmalogens, 114
prostaglandins, 114
sphingolipids, 114
sulfatides, 114
triglycerides, 113
Listeria monocytogenes, 262
LPL, *see* Linearly polarized light
LTs, *see* Leukotrienes
LTV system, *see* Linear time-variable system
Luciferase, 290
Luteinizing hormone (LH), 205, 216, 226
Lyme disease, 307
Lymphotoxin, 35, 237
Lysine, 260
Lysosomal-Associated Membrane Protein 2 (LAMP-2), 264

M

MA, *see* Malic acid
MAC, *see* Membrane attack complex
Macrophages, 219, 220, 221
Mad cow disease, 53, 252
Magnetic nanoparticles (MNPs), 316
Maillard reaction, 250
Major histocompatibility complex (MHC), 221, 230, 242, 252, 299
MA kinetics, *see* Mass action kinetics
Malic acid (MA), 105
Malignant glioma, 275
Malonyl-CoA, 112
Maltose, 110
Mannose-binding lectin (MBL), 242
Massachusetts Institute of Technology (MIT), 2
Mass action (MA) kinetics, 127
MBCs, *see* Memory B-cells
Medical genomics, *see* Genomic medicine
Meiosis, 86, 157
Melanocyte-stimulating hormone (MSH), 205
Membrane attack complex (MAC), 228
Memory B-cells (MBCs), 224, 247, 276
Mendelian inheritance, *see* Genetic inheritance, basis of
Messenger RNA (mRNA), 164
antisense, 275
translation, 169
Metabolic maps, 101
Metazoa, 56
Methanogens, 14, 32, 46
Methionine, 115, 169
Methionine sulfoxide reductase, 251
Methyl groups, 58
Methyltransferase enzymes, 194
MHC, *see* Major histocompatibility complex
Michaelis–Menton reaction, 100
MicroRNAs (miRNAs), 164, 186
Microtubules, 86
Miller–Urey apparatus, 28
Mirabilis jalapa, 157
miRNAs, *see* MicroRNAs
Mismatch repair, 189, 190
MIT, *see* Massachusetts Institute of Technology
Mitochondria, 15
Mitochondrial DNA (mtDNA), 300
Mitochondrial PCD pathway, 37
Mitochondrial stimulation factor (MSF), 202
MNPs, *see* Magnetic nanoparticles
Model(s)
biochemical oscillators, 128–136
Hodgkin–Huxley, 136–143, 150
verification, 126
MODS, *see* Multiple organ dysfunction syndrome
Molecular biology, introduction to, 1–11
complex systems, 5–9
modeling, 9
parametric control, 8
properties of nonlinear systems, 7–8

definitions, 2–5
bioinformatics, 3–4
genetic engineering and biotechnology, 4–5
genomics, 2–3
molecular biology, 2
proteomics, 3
systems biology, 5
home problems, 10–11
reference resources, 9
scope of text, 1–2
summary, 10
Molecular oxygen, photosynthesis and, 31
Molecular transducers, 83
Monoclonal antibodies, 316
mRNA, *see* Messenger RNA
MS, *see* Multiple sclerosis
MSF, *see* Mitochondrial stimulation factor
MSH, *see* Melanocyte-stimulating hormone
mtDNA, *see* Mitochondrial DNA
Mucopolysaccharides, 60
Multifactorial genetic disorders, 267
Multiple organ dysfunction syndrome (MODS), 41
Multiple sclerosis (MS), 236
Multipotent stem cells, 349
Muscle fibers, 208–215
accessory proteins, 210
actin, 212
anterior byssus retractor muscle, 215
asynchronous insect flight muscle, 215
cardiac muscle, 214
filaments, 209, 214
gap junctions, 212
latency relaxation phenomenon, 211, 213
motor nerve, 211
motor unit, 209
myofibrils, 209
mysoin, 212
rate of doing work, 209
sarcomere, 209
skeletal muscle, 209, 214
smooth muscle, 208
striated muscle, 209
titin protein, 212
tropomyosin, 212
vertebrate muscle types, 208
Mutations, 15, 155
Mycobacterium ligase, 189
Mycobacterium tuberculosis, 262, 263
Mycoplasma
genitalium, 50
sp., 53
Myoglobin, 248
Myosin, 227
Myosin kinase, 217
Myristic acid, 46
Mytilus edulis L., 215

N

NAD+, *see* Nicotinamide adenine dinucleotide
Nanaobacterium sanguineum, 33
Nanobes, 33
Nanosecond pulsed electric fields (nsPEFs), 342
NAs, *see* Nucleic acids
Natural killer (NK) cells, 219, 220, 226
NCSs, *see* Nonlinear complex systems
N-degron, 38
Necroptosis, 14, 40
Necrosis, 14
Neomycin, 352
NER, *see* Nucleotide excision repair
Nernst equation, 74
Nernst potential, 82, 137
Neural growth factor, 35
Neural stem cells (NSCs), 252
Neurofibromatosis, 267
Neuropeptide Y, 205
Neurotransmitters, 69, 80
NHEJ, *see* Nonhomogenous end joining
Niche cells, 354, 356
Nicotinamide adenine dinucleotide (NAD+), 102, 105, 107
Nicotinic acid, 117
Nitrilotriacetic acid (NTA) arrays, 285
NK cells, *see* Natural killer cells
NLS, *see* Nuclear localization sequence
Node equation, 76
Noncovalent bonds, 97
Nondisjunction, 89
Nonhomogenous end joining (NHEJ), 189
Nonlinear complex systems (NCSs), 7, 9
Norepinephrine, 114
Notations and symbols, 380–381
NPC, *see* Nuclear pore complex
NSCs, *see* Neural stem cells
nsPEFs, *see* Nanosecond pulsed electric fields
NTA arrays, *see* Nitrilotriacetic acid arrays
Nuclear export sequence, 59
Nuclear localization sequence (NLS), 202
Nuclear pore complex (NPC), 57, 58
Nucleic acids (NAs), 177–196, 254, *see also* Instrumental methods used in genomics, proteomics, and forensic science
chirality, 177
DNA, 177–179
alpha helix, 177
counterions, 179
histones, 179
hydrogen bonding, 179
phosphodiester links, 177
pyrimidine base, 177
Z-DNA, 178
DNA repair, 188–191
aflatoxins, 189
base excision repair, 189
discussion, 190–191
homologous recombination, 190, 191
ionizing electromagnetic radiations, 188
mechanisms, 189–190
mycobacterium ligase, 189
point mutations, 189
gene regulation, 191–195
CTCF zinc finger protein, 194
epigenetic gene regulation, 194–195
eukaryotes, 193–194
Evo-Devo, 191

prokaryotes, 192–193
transcription start site, 193
upstream promoter, 193
home problems, 195–196
hybridization, 299
RNA, 179–181
anticodon loop, 180
coding strand, 181
ribosomal RNAs, 182–184
spliceosome enzymes, 180
RNAs coded by DNA, 181–188
Argonaute proteins, 186
Barr body, 188
Cajal body, 184
discussion, 188
extra-embryonic membranes, 188
messenger RNAs, 181
micro-RNAs , 186–187
peptidyl transferase, 183
primary transcript, 186
ribosomal RNAs, 182–184
ribozymes, 187
short interfering RNAs, 185–186
small nucleolar RNAs, 184–185
Svedberg constant, 183
transfer RNAs, 181–182
wobble hypothesis, 182
XIST RNA, 188
summary, 195
Nucleotide excision repair (NER), 189, 190, 191

O

OAA, *see* Oxaloacetic acid
ODEs, *see* Ordinary differential equations
Oligodendrocytes, 80
Oligomeric proteins, 200
Oncogenes, 273
Oogenesis, 88, 89
OPCR thermal cycling, 288
Open probability, 71
Opsonins, 228
Optical pulse oximeter (POX), 245
Ordinary differential equations (ODEs), 6, 98
Organic redox reactions, 106–109
alcohol dehydrogenase, 108
cytochrome molecules, 107
electron donors, 107
electron transport chain, 107
methane gas generation, 109
oxidation of water, 106
photosynthesis, 106
reducing agent, 107, 109
secondary electrochemical cells, 107
Organismic homeostasis, 34
Ornithine, 115, 117
OSA, *see* Oxalosuccinic acid
Ostrinia nubilatus, 332
Ouabain, 217
Oxaloacetic acid (OAA), 104
Oxalosuccinic acid (OSA), 105
Oxidizer molecule, 107
Oxytocin, 199

P

Pacemaker neurons, 140
Padlock probes, 304
PAF, *see* Platelet activating factor
PAGE, *see* Polyacrylamide gel electrophoresis
Pancreatic alpha cells, 110
Pancreatic delta cells, 205
Parametric control, 8
Parathyroid hormone (PTH), 205, 216, 330
Parent isotope, 25
Parietal cells, 69
Parkinson's disease, 267
Parthenites, 360
Parthenogenesis, 15, 359–361
artificial parthenogenesis, 359–360
human parthenotes, 360–361
natural parthenogenesis, 359
Parthenote, 351
Parvoviruses, 21
Pasteurization, 13
Patch-clamp recording, 71
Patterson difference algorithm, 322
Patterson function, 322
PC, *see* Phosphatidylcholine
PCD, *see* Programmed cell death
PCES, *see* Positive control element sequence
PCGs, *see* Protein-coding genes
PCR, *see* Polymerase chain reaction
PDB, *see* Protein Data Bank
PEG, *see* Polyethylene glycol
Peligibacter, 218
PEP, *see* Phosphoenolpyruvate
PEPC, *see* Phosphoenolpyruvate carboxylase/oxygenase
Peptide bond, 98, 170, 172, 199
Peptide linkages, 197
Peptidoglycan, 17, 46, 52
Perforin, 222
Peroxisomal targeting signal (PTS), 61
Persistent vegetative state (PVS), 42, 43
PERVs, *see* Porcine endogenous retroviruses
PGAM, *see* Phosphoglyceromutase
PGCs, *see* Primordial Germ Cells
PGI, *see* Phosphoglucose isomerase
PGK, *see* Phosphoglycerate kinase
PGs, *see* Prostaglandins
Phagocytic cells, 35
Phagotrophic hypothesis, 54
Pharmacogenomics, 267
Pharyngoconjunctival fever, 275
Phenylisothiocyanate (PIT), 307
Pheophytin, 125
Phosphatases, 251
Phosphatidic acid, 45
Phosphatidylcholine (PC), 45, 48, 49, 113
Phosphatidyl glycerol, 113
Phosphatidylinositol, 46
Phosphatidyl inositol biphosphate, 217
Phosphatidylserine, 35, 45, 46
Phosphoenolpyruvate (PEP), 124
Phosphoenolpyruvate carboxylase/oxygenase (PEPC), 124
Phosphoglucose isomerase (PGI), 102
Phosphoglycerate kinase (PGK), 102

Phosphoglyceromutase (PGAM), 102
Phospholipase C, 217
Phospholipids (PLs), 113
Phosphorylase, 110
Photon energy, 14
Photosynthesis, 14, 106, 118–125
  anaerobic, 118
  antenna proteins, 122
  autotrophs, 118
  Calvin cycle, 124
  carotenoids, 118
  chemical reactions and energy, 123–124
  chemiosmosis, 124, 125
  chlorophyll, 118, 119, 120
  chloroplast structure, 121–123
  Hatch–Slack cycle, 124
  light reactions, 124–125
  nuclear genome, 123
  photopigments, 121
  pigments, 118–121
  production of molecular oxygen by, 31
  reaction center, 118
  thylakoids, 118, 122
  xanthophylls, 121
Phyllobacterium mysinacearum, 33
Phylloquinones, 125
Phylogenetic trees, 116
Physical biochemistry and biochemical systems modeling, 95–154
  challenge of modeling complex biochemical systems, 126–143
    generalized approaches, 127–128
    generalized mass action system, 127
    Hodgkin–Huxley model for action potential generation, 136–143
    indirect optimization method, 128
    kinetic order, 127, 128
    models for biochemical oscillators, 128–136
    model verification, 126
    power-law formalism, 127
    rate-limiting step in reaction pathway, 126
    simplex algorithm, 128
    S-system defining equations, 127
    steady-state equations, 128
  chemical reactions, 95–101
    activation energy of reaction, 99
    allosteric regulation, 101
    Arrhenius relation, 99
    chemical bond energy, 95–97
    chemical bond types, 97–98
    chemical kinetics and mass action, 98–101
    competitive inhibition, 101
    decomposition of carbonic acid, 100
    Michaelis constant, 101
    reaction rate constant, 99
    steady state, 101
  coupled reactions, important, 101–106
    chemically feasible pathways, 101
    citric acid cycle, 104–106
    Embden–Meyerhoff pathway, 102
    fermentation, 102
    glycolysis, 102–104
    metabolic maps, 101
    shunt reactions, 101
  home-grown programs, 150
  home problems, 150–154
  law of mass action, 150
  other metabolic pathways, 106–118
    amino acid catabolism, 116–118
    amino acid metabolism, 115–116
    carbohydrate metabolism, 109–111
    chemistry of organic redox reactions, 106–109
    lipid catabolism, 114–115
    lipid metabolism, 112–114
  photosynthesis, 118–125
    Calvin cycle, 124
    chemical reactions and energy, 123–124
    chloroplast structure, 121–123
    light reactions, 124–125
    pigments, 118–121
  simulation languages, 143–150
    Biochemical Pathway Simulator, 147
    Cellerator™, 147
    free software, 146
    GEPASI, 146, 147
    GUI modeling and simulation program, 145
    Matlab, 143
    MetMAP, 148, 149
    multicellular systems, 148
    regenerative medicine, 149
    Simnon, 145, 146
    Simulink, 145
    SUNDIALS, 149
    Teranode Design Suite Biological Modeler, 146
    VCell, 147
    whole-cell simulation, 149
  summary, 150
Phytophthora ramorum, 55
Pilocarpine, 269
Pinocytosis, 15
PIT, *see* Phenylisothiocyanate
PK, *see* Pyruvate kinase
Plasmids, 24
Plastoquinone, 125
Platelet activating factor (PAF), 227
PLF, *see* Power-Law Formalism
PLs, *see* Phospholipids
Pluripotent stem cells (PSCs), 347, 348, 353
PMF, *see* Proton-motive force
pmRNA, *see* Premessenger RNA
Podovirus, 31
Polar body, 89
Polyacrylamide gel electrophoresis (PAGE), 306
Polyethylene glycol (PEG), 293
Polygalacturonase, 330
Polymerase chain reaction (PCR), 283, 286, 310
  immuno-314
  real-time, 288
Polymorphisms, 155
Porcine endogenous retroviruses (PERVs), 371
Positive control element sequence (PCES), 192
Potato Spindle Tuber Virus (PSTV), 23
Potential energy, 95
Power-Law Formalism (PLF), 127
POX, *see* Optical pulse oximeter

Index

Premessenger RNA (pmRNA), 162, 165, 180
Primary spermatocytes (PSs), 90
Primordial Germ Cells (PGCs), 89
Prochlorococcus, 31
Programmed cell death (PCD), 35, 41, 86
  embryonic development and, 40, 43
  pathway
    mitochondrial, 37
    triggering of, 39
Prokaryotes, 50–53
  Archaeans, 51
  binary fission, 51
  colonial, 51
  electron transport phosphorylation, 53
  Eubacteria, 51–53
  gene regulation in, 192
  ribosomes, 50
Prolactin, 205
Proline, 115, 117
Prostaglandins (PGs), 6, 114, 217, 232, 237
Prostate-specific antigen (PSA), 316
Protein(s), 197–266
  acyl carrier, 112
  adaptor, 261
  amino acids and peptide linkage, 197–202
  antenna, 122
  apoptosis promoter, 40
  Argonaute, 186
  blood proteins, 243–248
    allosteric protein complex, 244
    alpha globulins, 246
    carbionic anhydrase, 243
    carboxymyoglobin, 248
    ferritin proteins, 246
    gamma globulins, 247
    hemocyanins in invertebrate circulatory systems, 247–248
    hemoglobin, 243–245
    memory B-cells, 247
    Monte Carlo transcriptional system, 246
    myoglobin, 248
    oxyhemoglobin, 245
    serum proteins and antibodies, 245–247
    sickle-cell anemia, 245
    thalassemia, 245
  cell membrane receptors and ion pumps, 216–218
    calcium ions, 216
    eicosanoids, 217
    hormones, 216
    ion pumps, 217–218
    photon-driven proton pumps, 218
    rigor mortis, 218
    second messenger systems, 216–217
  cell receptor, 84
  cell surface receptor proteins, 232
  chaperone, 54, 201
  chemical disassembly, 3
  -coding genes (PCGs), 193
  connexin, 78
  crystals, 318
  CTCF zinc finger protein, 194
  cytosolic, 85
  destination, 201
  disulfide bonds, 199
  enzymes, 203–204
    molecular specificity of, 3
    physical chemistry, 203
    regulation, 203–204
  errors in protein structure, 248–256
    aldehydes, 251
    Amadori product, 250
    aminoguanidine, 249
    carbonyl group, 249
    chaperone, 254
    collagen, 250
    endogenous enzymatic repair mechanism, 251
    frame shift, 248
    glucosepane, 250
    glycation, 249
    heat shock (stress) proteins, 251–252
    imine molecule, 249
    Maillard reaction, 250
    mammalian prion diseases, 255
    methionine sulfoxide reductase, 251
    nongenetic causes, 249–251
    phosphatases, 251
    prions, 252–256
    transmissible spongiform encephalopathy, 252
  ferritin, 246
  functions, examples of, 202–248
    blood proteins, 243–248
    cell membrane receptors and ion pumps, 216–218
    enzymes, 203–204
    glycoproteins, 242–243
    hormones and intercellular command, communication, and control, 204–205
    immune system, 219–240
    lectins, 241–242
    myoglobin, 248
    structural proteins, 206–216
  globulin, 246
  glycosylation, 201
  green fluorescent protein, 133, 370
  guanosine nucleotide binding protein, 217
  heat shock proteins, 251
  heteromeric, 200
  histone proteins, 215–216
    chromosomal matrix, 215
    epigenetic inheritance, 216
    epigenetic modulation, 215
    euchromatin, 215
    heterochromatin, 215
    nucleosome, 215
  home problems, 265–266
  homomeric, 200
  IAP blocker protein, 261
  immune system, 219–240
    activating receptor, 226
    agglutination, 225
    alveolar macrophages, 220
    antigen presentation, 222, 225, 229–232
    anti-idiotypic antibodies, 240
    apoptosis, 230
    autacoids, 219, 232–240
    autoimmunity, 229
    B-cell receptor, 231

B-lymphocytes, 224
Bursa of Fabricus, 224
cell growth stimulating factors, 232
cell surface receptor proteins, 232
circumventricular organs, 240
complement, 228–229
cortisol, 238, 239
cyclooxygenase, 237
cytokines, 240
cytotoxic T-lymphocytes, 220, 222
discussion, 240
forkhead transcription factor, 223
granulocyte-monocyte colony stimulating factor, 221
helper T-cell activation, 222
human interleukin cytokines, 233–236
immunocytokines, 219
immunoglobulin, 224
inhaled proteins, 229
interferons, 236
leucocytes, 219–220
lymphotoxin, 237
macrophages, 219, 220, 221
mast cells, 226, 227
membrane attack complex, 228
memory B-cells, 224
molecular transformation mechanism, 232
monocytes, 220
natural killer cells, 219, 220, 226
opsonins, 228
plasma cells, 224
platelet activating factor, 227
platelet cells, 227
pleiotropy, 219, 240
protease release, 238
protein kinase, 236
receptor-mediated endocytosis, 224
summary of immune system cells, 219–227
suppressor T-cells, 223, 232
T-cell receptor, 222, 230
toxic shock syndrome, 237
tumor necrosis factor proteins, 237
viral infection, 220
integral membrane, 201, 202
kinases, 56
microarrays, 284, 310–314
antibody microarrays, 311–312
first large-scale protein microarrays, 311
improvements in fluorescent microarrays, 312–314
octamer-binding protein 4, 349
oligomeric, 200
optical activity, 197
peptide bonds, 98, 199
pleiotropy, 265
posttranscription regulation of gene expression, 257–259
homing, 258
inteins and protein engineering, 257–259
mechanisms, 257
super-Mendelian inheritance, 258
prion, 252, 269
protein destruction and AA recycling, 259–265
activated clotting factors, 259
activator proteases, 259
adaptor proteins, 261
apoptosome, 261
autophagy, 261–265
cholinesterase, 259
cytoplasm-to-vacuole targeting pathway, 263
Danon's syndrome, 264
deubiquitinases, 260
IAP blocker protein, 261
interferon gamma, 261
internal proteases, 259
proteolysis in apoptosis and cell necrosis, 261
proteolytic enzymes, 259
proteosome, 259–261
serine proteases, 259
subtilisin, 259
ubiquitin-proteasome pathway, 260
xenophagy, 263
protein kinase regulator, 111
quaternary structure, 200
receptors, 30, 205
recombinant hemagglutinin, 277
reducing agents, 201
release factor, 170
small nuclear ribonuclear proteins, 162
specific rotation constant, 198
splicing factor, 59
structural proteins, 206–216
A-tubule, 207
axoneme, 207
basal body, 208
centrosome, 206
cilia and flagella, 206–208
dyneins, 206, 207
flagellum, 206
histone proteins, 215–216
intermediate filaments, 206
microfilaments, 206
microtubules, 206
muscle fibers, 208–215
myosin thick filaments, 206
neurofilaments, 206
spindle fibers, 206
structures, 3
summary, 265
supersecondary structure, 199
surface receptor, 204
synthesis, RNA types involved with, 165
target, 260
TATA binding, 51
tertiary structure, 197, 199
3-D structure prediction, 3–4
titin, 212
Toll-like receptor proteins, 242
transcription factor, 35, 57, 202, 347
transmembrane, 48, 66
tumor necrosis, factor, 237
U-proteins, 184
voltage-gated Ca^{++} channel, 72
Protein Data Bank (PDB), 318
Proteolytic enzymes, 259
Proteome, 3
Proteomics, subdisciplines, 3

Proteorhodopsins (PRs), 218
Proteosome, 38, 259
Proton-motive force (PMF), 218
PRs, *see* Proteorhodopsins
PSA, *see* Prostate-specific antigen
PSCs, *see* Pluripotent stem cells
Pseudogenes, 58
Pseudouridine, 181
PSs, *see* Primary spermatocytes
PSTV, *see* Potato Spindle Tuber Virus
PTH, *see* Parathyroid hormone
PTS, *see* Peroxisomal targeting signal
PVS, *see* Persistent vegetative state
Pycnosceleus surinamensis, 359
Pyruvate kinase (PK), 102

R

RCA, *see* Rolling-circle amplification
RCR, *see* Rolling-circle replication
rDNA, *see* Recombinant DNA
RE, *see* Restriction enzyme
Receptor-mediated endocytosis, 224
Recombinant DNA (rDNA), 335, 337
Recombinant DNA technology, 337–342
  electroporation and effects of nanosecond pulsed electric fields on cells, 341–342
  gene gun, 339–341
  other viral vectors in gene therapy, 339
  plasmid vector, preparation of, 337–338
  rDNA insertion methods, 337
  ultrasonic poration, 342
  use of phages as cDNA vectors, 338–339
Recombinant hemagglutinin (rHA) protein, 277
Red algae, 118
Red tides, 55
Reducing agent, 107
Regenerative medicine (RM), 1, 5, 347, 360
Regulatory particle (RP), 38, 260
Release factor protein, 170
Renin, 205, 259
Rennin, 259
RER, *see* Rough endoplasmic reticulum
Resting transmembrane potential, 76
Restriction endonucleases, 302
Restriction enzyme (RE), 338
Restriction fragment length polymorphism (RFLP) analysis, 298
Retinoblastoma protein, 275
Retroviruses, 163, 164, 339
Rett syndrome, 279
Reverse transcriptase, 30, 163, 173, 338
RFLP analysis, *see* Restriction fragment length polymorphism
rHA protein, *see* Recombinant hemagglutinin protein
Ribonucleic acid, *see* RNA
Ribosomal RNA (rRNA), 180, 181
Ribozymes, 30, 187
Ricin, 241
Rigor mortis, 41, 218
RISC, *see* RNA Induced Silencing Complex
RM, *see* Regenerative medicine
RNA (ribonucleic acid), 179
  Induced Silencing Complex (RISC), 186
  messenger, 164
  oligonucleotides, 31
  polymerase (RNAP), 165, 181, 192
  premessenger, 162, 165, 180
  ribosomal, 180, 181
  short interfering, 185
  small nuclear, 180, 187
  transfer, 166
  viruses, 21
  XIST, 188
RNAP, *see* RNA polymerase
Rock dating, 17
Rolling-circle amplification (RCA), 302
Rolling-circle replication (RCR), 302
Rough endoplasmic reticulum (RER), 59, 201
RP, *see* Regulatory particle
rRNA, *see* Ribosomal RNA
Rubenstein–Taybi syndrome, 281
RUBISCO, *see* Ribulose biphosphate carboxylase oxygenase

S

SA, *see* Succinic acid
Saccharomyces cerevisiae, 257
SAM, *see* *S*-Adenosylmethionine
Sarcoplasmic reticulum (SR), 59, 69, 211
Satellites, 24–25
SAv, *see* Streptavidin
Scanning electron microscope (SEM), 13
Schwann cells, 80
Scrapie, 252, 253, 270
SCs, *see* Stem cells
Secondary spermatocytes (SSs), 90
Second messenger molecule, 83
Second messenger system, 216
Secretin, 205
Secretory vesicles, 61
Selectins, 242
SEM, *see* Scanning electron microscope
Semipermeable membrane (SPM), 65
SER, *see* Smooth endoplasmic reticulum
Serial endosymbiosis theory (SET), 54
Serine, 116, 117
Serine proteases, 259
Sertoli cells, 346
SET, *see* Serial endosymbiosis theory
Short interfering RNAs (siRNAs), 185
Short tandem repeat (STR)
  markers, 298
  sequences, 2–3
Shunt reactions, 101
Shuttle vesicles, 61
Sickle-cell anemia, 245, 268
Signal recognition particle (SRP), 201
Signal transduction networks (STNs), 147–148
Simian Immunodeficiency Virus (SIV), 371
Simnon® program, 6, 101, 131, 377
Single-cause mentality, 10
Single-nucleotide polymorphism (SNP), 248, 267

siRNAs, *see* Short interfering RNAs
Sister chromatids, 85
SIV, *see* Simian Immunodeficiency Virus
SM, *see* Smooth muscle
Small nuclear ribonuclear proteins (snRNP), 162
Small nuclear RNAs (snRNAs), 180, 187
Smooth endoplasmic reticulum (SER), 59, 112
Smooth muscle (SM), 214
SNCT, *see* Somatic cell nuclear transfer
SNP, *see* Single-nucleotide polymorphism
snRNAs, *see* Small nuclear RNAs
snRNP, *see* Small nuclear ribonuclear proteins
SOCE, *see* Store-operated Ca^{++} entry
Sodium dodecyl sulfate, 306
Somatic cell nuclear transfer (SNCT), 347
Somatostatin, 205
Spermatogenesis, 88
Spike generator locus, 77
Spiroplasma sp., 53
Spliceosome enzymes, 180, 187
Splicing factor proteins, 59
SPM, *see* Semipermeable membrane
SPR, *see* Surface plasmon resonance
SR, *see* Sarcoplasmic reticulum
SRP, *see* Signal recognition particle
SSM, *see* Subsynaptic membrane
SSs, *see* Secondary spermatocytes
State variables (SVs), 5, 6
Stem cells (SCs), 84, 278, 347–359
  adult, 349, 350
  applications in regenerative medicine, 353–354
  cancer stem cells, 278, 354–359
  embryonic, 86, 337, 372
  epi-embryonic, 349
  hematopoietic, 349, 362
  human embryonic, 372–374
    embryonic versus adult stem cells, 373–374
    federal restrictions, 374
    global picture, 374
    religious concerns, 372–373
    reprogramming of skin cells, 373
  multipotent, 349
  neural, 252
  niche, 355
  oncogenic event, 358
  pluripotent, 347, 348
  research, 345
  sources, 349–353
  spermatogonia, 89
  totipotent, 348
  types, 348–349
STNs, *see* Signal transduction networks
Store-operated Ca^{++} entry (SOCE), 73
STR, *see* Short tandem repeat
Streptavidin (SAv), 283, 284, 312
Streptococcus sp., 41
Streptomyces
  avidinii, 284
  sp., 52
Subsynaptic membrane (SSM), 69, 77
Subtilisin, 259
Succinic acid (SA), 105
Sulfurylase, 290
Superoxide dismutase, 249
Surface plasmon resonance (SPR), 311
Surface receptor proteins, 204
Svedberg constant, 183
SVs, *see* State variables
Synaptic boutons, 80
Synchrotron radiation, 322
Syntrophic hypothesis, 54
Systems biology, 1, 5

T

Target protein (TP), 260
Taxonomic unit (TU), 16
TBS, *see* Type B spermatogonia
TCA cycle, *see* Tricarboxylic acid cycle
T-cell receptor (TCR), 222, 230
TCR, *see* T-cell receptor
TEA, *see* Tetraethylammonium
Telomere, 86, 346
Telomere reverse transcriptase (TERT), 86
TEM, *see* Transmission electron microscopy
TERT, *see* Telomere reverse transcriptase
Tetanus toxin, 82, 93
Tetraethylammonium (TEA), 40
TF, *see* Transcription factor
TGAs, *see* Transgenic animals
TG plants, *see* Transgenic plants
Thalassemia, 245
Thermodynamic death, 14
Thermus aquaticus, 286
Thiobacillus ferrooxidans, 53
Threonine, 117, 268
Threonine dehydrogenase, 117
Threshold voltage, 72
Thrombin, 259
Thrombopoietin, 205
Thromboxanes (TXs), 114, 237
Thylakoids, 118
Thymine, 290
Thyroid-stimulating hormone (TSH), 205, 216, 307
Thyrotropin-releasing hormone (TRH), 204
Thyrozine, 69
Time-resolved x-ray crystallography (TRXC), 323
TMMR, *see* Transmembrane molecular receptor
TMPs, *see* Transmembrane proteins
TMSR, *see* Transmembrane signal receptor
TMV, *see* Tobacco mosaic virus
TNF, *see* Tumor necrosis factor
TNF-α, *see* Tumor necrosis factor alpha
Tobacco mosaic virus (TMV), 24
Toxic shock syndrome, 237
Toxoplasma gondii, 55
TP, *see* Target protein
Transcription factor (TF), 35, 57, 84, 347, 372
Transferrin, 246
Transfer RNA (tRNA), 17, 166
  aminoacyl, 170
  initiator, 169
Transgenic animals (TGAs), 335
Transgenic (TG) plants, 369
Transmembrane molecular receptor (TMMR), 83

Transmembrane proteins (TMPs), 48, 66
Transmembrane signal receptor (TMSR), 83
Transmissible spongiform encephalopathy (TSE), 53, 252, 270
Transmission electron microscopy (TEM), 13, 324
Transposons, 58, 156, 173
TRH, *see* Thyrotropin-releasing hormone
Tricarboxylic acid cycle (TCA cycle), 104
Triglycerides, 113
tRNA, *see* Transfer RNA
TRXC, *see* Time-resolved x-ray crystallography
Trypanosoma
  brucei, 55
  cruzi, 55
Tryptophan, 61, 117, 199, 249
TSE, *see* Transmissible spongiform encephalopathy
TSGs, *see* Tumor suppressor genes
TSH, *see* Thyroid-stimulating hormone
TU, *see* Taxonomic unit
Tuberculosis, 267
Tumor necrosis factor (TNF), 237
Tumor necrosis factor alpha (TNF-α), 35
Tumor suppressor genes (TSGs), 273
TXs, *see* Thromboxanes
Type B spermatogonia (TBS), 90
Tyrosine, 115, 117, 199

U

Ubiquitin, 38, 260
Ubiquitination, 38
Ubiquitin-proteasome pathway (UPP), 37, 260
Ubiquitinylation, 38
Ultramicrobacteria, 33
UP, *see* Upstream promoter
UPP, *see* Ubiquitin-proteasome pathway
U-proteins, 184
Upstream promoter (UP), 193
Urokinase, 259
USDA Agricultural Research Service, 331

V

Vaccination, 276
Variable-length probe array (VLPA), 301
Variant Creutzfeldt–Jakob disease (vCJD), 255, 269
Vasopressin, 67, 199, 216
vCJD, *see* Variant Creutzfeldt–Jakob disease
Vimentin, 214
Viral diseases, list of, 20–21
Viroids, 23–24
Virus(es), 20–23
  deadly, 22
  DNA, 21
  Epstein-Barr virus, 40
  helper, 24
  La Crosse, 23
  Podovirus, 31
  potato spindle tuber virus, 23
  RNA, 21
  satellite, 24
  tobacco mosaic virus, 24
Virusoids, 24–25
Vitamin, highly conserved, 283
VLPA, *see* Variable-length probe array
Voltage-dependent opening rate constant, 71
Voltage-gated ion channels, 217

W

Waveguide Excitation Fluorescence Microscope (WExFM), 313
Western blot test, 310
WExFM, *see* Waveguide Excitation Fluorescence Microscope
White biotechnology, 4
Wobble hypothesis, 182
Wobble rules, 182
Wobble theory, 166

X

Xanthophylls, 121
Xenophagy, 263
XIST RNA, 188
X-ray crystallography, 320

Y

Yersinia pestis, 288

Z

Z-DNA, 178
z-transform methods, 1
Zygotes, 88, 89

Milton Keynes UK
Ingram Content Group UK Ltd.
UKHW051941071024
449327UK00026B/2119

9 780367 386559